EQUATIONS OF MATHEMATICAL PHYSICS

EQUATIONS OF MATHEMATICAL PHYSICS

BY

A. N. TIKHONOV

AND

A. A. SAMARSKII

TRANSLATED BY

A. R. M. ROBSON AND P. BASU

TRANSLATION EDITED BY

DR. D. M. BRINK

Clarendon Laboratory

Oxford

DOVER PUBLICATIONS, INC.

New York

This Dover edition, first published in 1990, is an unabridged, unaltered republication of the work first published by the Pergamon Press Ltd., Oxford, England, 1963, as volume 39 in the "International Series of Monographs on Pure and Applied Mathematics." It is a translation of the original Russian *Uravneniya matematicheskoi fiziki*, published by Gostekhizdat, Moscow, 1953. It is reprinted by special arrangement with Pergamon Books Ltd., Headington Hill Hall, Oxford OX3 OBW, England.

Library of Congress Cataloging-in-Publication Data

Tikhonov, A. N. (Andreĭ Nikolaevich), 1906–
[Uravneniia matematicheskoĭ fiziki. English]
Equations of mathematical physics / by A. N. Tikhonov and A. A. Samarskii ; translated by A. R. M. Robson and P. Basu ; translation edited by D. M. Brink.
p. cm.
Translation of: Uravneniia matematicheskoĭ fiziki.
Reprint. Originally published: Oxford, England : Pergamon Press ; New York : Dist. by Macmillan Co., 1963. Originally published in series: International series of monographs on pure and applied mathematics ; v. 39.
Includes index.
ISBN-13: 978-0-486-66422-4 (pbk.)
ISBN-10: 0-486-66422-8 (pbk.)
1. Mathematical physics. 2. Differential equations. I. Samarskiĭ, A. A. (Aleksandr Andreevich) II. Title.
QA401.T512 1990
530.1′5—dc20 90-3355
CIP

www.doverpublications.com

CONTENTS

IV. EQUATIONS OF ELLIPTIC TYPE

V. WAVE PROPAGATION IN SPACE

VI. HEAT CONDUCTION IN SPACE

VII. EQUATIONS OF ELLIPTIC TYPE (CONTINUATION)

Supplement. SPECIAL FUNCTIONS

I. CYLINDRICAL FUNCTIONS

PREFACE TO THE SECOND EDITION

ERRORS and inaccuracies noticed in the first edition have been removed in the second edition. Certain sections have been subjected to minor revision. The most important changes have been made in the introductory portion of the section on special functions and in Chapter IV. A new appendix to Chapter VI, devoted to the method of finite differences for the equation of heat conduction, has been written.

The authors consider it their pleasant duty to express their gratitude to V. I. Smirnov, for the large number of valuable hints, and to A. G. Sveshnikov for help in preparing the second edition.

A. Tikhonov
A. Samarskii

PUBLISHER'S NOTE

THE READER may be interested to know that an English translation of a companion volume to the present work will be published in Autumn, 1963, entitled *A Collection of Problems on Mathematical Physics*, by B. M. Budak, A. A. Samarskii, and A. N. Tikhonov.*

It contains many problems with detailed instructions and solutions and its aim is to give students the opportunity of gaining elementary skill in solving problems in the principal classes of the equations of mathematical physics.

*[Reprinted by Dover, 1988, under the title *A Collection of Problems in Mathematical Physics* (ISBN 0-486-65806-6).]

PREFACE TO THE FIRST EDITION

THE PRESENT book was written as a textbook for students of physics.

The domain of problems of mathematical physics is closely related to the study of different physical processes, phenomena studied in hydrodynamics, the theory of elasticity, electrodynamics, etc., belong to this domain. The mathematical problems arising in these cases contain many common elements and constitute the subject-matter of mathematical physics.

The method of investigation characterizing this branch of science is mathematical in its essence. But the formulation of problems of mathematical physics, being closely connected with the study of physical problems, has its specific features.

The domain of problems relating to mathematical physics is unusually extensive. In the present book we consider problems of mathematical physics leading to partial differential equations.

We have tried to select and describe topics classified under typical physical processes; for this reason the arrangement of topics corresponds to the principal types of equations.

The study of each type of equation begins with the simplest physical problems leading to equations of the type considered. Special attention is paid to the mathematical formulation of problems, rigorous accounts of the solution of the simplest problems and physical interpretation of the results obtained.*

Problems are given in each chapter mainly for the purpose of developing technical skill. Some problems are themselves of interest to physics.

But the simplest problems forming part of the fundamental course cannot give a complete idea of the variety of problems

* During lectures in the course of mathematical physics we do not confine ourselves to an abstract explanation of the physical interpretation of the results explained. Where possible, we demonstrate the mathematical conclusions on specific physical models.

or of the role and position of mathematical physics. Because of this, appendices are given at the end of each chapter. These appendices give examples of the application of the methods described in the main text for the solution of different problems of physics and technology. A number of examples beyond the scope of the problems considered in the main text are also given. The selection of such examples could undoubtedly vary greatly.

The book contains only a part of the material constituting a course of methods of mathematical physics. The theory of integral equations and variation methods are not included in the book. The approximate methods described are not quite complete.

We considered that it was advisable to confine ourselves to the material dealt with here in order to reduce the delay in publication of the book.

The lectures delivered for over ten years at the Faculty of Physics of the Moscow State University form the basis of the book. The subject-matter of these lectures was partly reflected in the syllabuses published in 1948–9. In the present book the material of the syllabuses was enlarged and subjected to a thorough revision.

We are glad to be able to express our gratitude to our students and colleagues, A. V. Vasileva, V. B. Glasko, V. A. Il'in, A. V. Lukyanov, O. I. Panych, B. L. Rozhdestvenskii, A. G. Sveshnikov and D. N. Chetaev, without whose help we would hardly have been able to prepare the book for the press within a short period, and also to Yu. L. Rabinovich who read the manuscript and made a number of valuable observations.

A. Tikhonov
A Samarskii

CHAPTER I

CLASSIFICATION OF PARTIAL DIFFERENTIAL EQUATIONS

MANY problems of mathematical physics lead to differential equations with partial derivatives. Differential equations of the second order are the ones that occur most frequently. In the present chapter we shall consider the classification of these equations.

§ 1. Classification of partial differential equations of the second order

1. *Differential equations with two independent variables*

We give the necessary definitions

A relation between an unknown function $u(x, y)$ and its partial derivatives inclusive of those of the second order* is called a partial differential equation of the second order with two independent variables x, y:

$$F(x, y, u, u_x, u_y, u_{xx}, u_{xy}, u_{yy}) = 0.$$

An equation for a larger number of independent variables is written similarly.

The equation is called linear with respect to the highest derivatives if it has the form

$$a_{11} u_{xx} + 2 a_{12} u_{xy} + a_{22} u_{yy} + F_1(x, y, u, u_x, u_y) = 0, \qquad (1)$$

where a_{11}, a_{12}, a_{22} are functions of x and y.

*We use the following notations for the derivatives:

$$u_x = \frac{\partial u}{\partial x}, \; u_y = \frac{\partial u}{\partial y}, \; u_{xx} = \frac{\partial^2 u}{\partial x^2}, \; u_{xy} = \frac{\partial^2 u}{\partial x \, \partial y}, \; u_{yy} = \frac{\partial^2 u}{\partial y^2}.$$

If the coefficients a_{11}, a_{12}, a_{22} depend not only on x and y but, like F_1, are functions of x, y, u, u_x, u_y, then the equation is called quasi-linear.

The equation is called linear if it is linear with respect to the higher derivatives u_{xx}, u_{xy}, u_{yy}, as well as with respect to the function u and its first derivatives u_x, u_y:

$$a_{11} u_{xx} + 2 a_{12} u_{xy} + a_{22} u_{yy} + b_1 u_x + b_2 u_y + cu + f = 0\,, \quad (2)$$

where $a_{11}, a_{12}, a_{22}, b_1, b_2, c, f$ are functions of x and y only. If the coefficients of equation (2) do not depend on x and y, it is a linear equation with constant coefficients. The equation is called *homogeneous*, if $f(x, y) = 0$.

By means of a transformation of variables

$$\xi = \varphi(x, y), \quad \eta = \psi(x, y)\,,$$

which allows an inverse transformation, we obtain a new equation equivalent to the initial equation. The question naturally arises: how should ξ and η be selected so that the equation with these variables has the simplest form?

At this point we answer this question for equations linear with respect to higher derivatives of the form (1) with two independent variables x and y:

$$a_{11} u_{xx} + 2 a_{12} u_{xy} + a_{22} u_{yy} + F(x, y, u, u_x, u_y) = 0\,.$$

Transforming the derivatives to new variables, we obtain

$$\left.\begin{aligned}
u_x &= u_\xi \xi_x + u_\eta \eta_x\,,\\
u_y &= u_\xi \xi_y + u_\eta \eta_y\,,\\
u_{xx} &= u_{\xi\xi} \xi_x^2 + 2 u_{\xi\eta} \xi_x \eta_x + u_{\eta\eta} \eta_x^2 + u_\xi \xi_{xx} + u_\eta \eta_{xx},\\
u_{xy} &= u_{\xi\xi} \xi_x \xi_y + u_{\xi\eta} (\xi_x \eta_y + \xi_y \eta_x) + u_{\eta\eta} \eta_x \eta_y + u_\xi \xi_{xy} + u_\eta \eta_{xy},\\
u_{yy} &= u_{\xi\xi} \xi_y^2 + 2 u_{\xi\eta} \xi_y \eta_y + u_{\eta\eta} \eta_y^2 + u_\xi \xi_{yy} + u_\eta \eta_{yy}.
\end{aligned}\right\} \quad (3)$$

Substituting the values of the derivatives from (3) in equation (1) we have

$$\bar{a}_{11} u_{\xi\xi} + 2 \bar{a}_{12} u_{\xi\eta} + \bar{a}_{22} u_{\eta\eta} + \bar{F} = 0\,, \quad (4)$$

where

$$\bar{a}_{11} = a_{11} \xi_x^2 + 2 a_{12} \xi_x \xi_y + a_{22} \xi_y^2\,,$$

$$\bar{a}_{12} = a_{11} \xi_x \eta_x + a_{12} (\xi_x \eta_y + \eta_x \xi_y) + a_{22} \xi_y \eta_y\,,$$

$$\bar{a}_{22} = a_{11} \eta_x^2 + 2 a_{12} \eta_x \eta_y + a_{22} \eta_y^2,$$

and the function $\overline{F}$ does not depend on second derivatives. It should be noted that if the initial equation is linear, i. e.

$$F(x, y, u, u_x, u_y) = b_1 u_x + b_2 u_y + cu + f,$$

then F has the form

$$\overline{F}(\xi, \eta, u, u_\xi, u_\eta) = \beta_1 u_\xi + \beta_2 u_\eta + \gamma u + \delta,$$

i. e. the equation remains linear.*

Let us select the variables ξ and η such that the coefficient $\bar{a}_{11}$ is equal to zero. We consider an equation with partial derivatives of the first order

$$a_{11} z_x^2 + 2 a_{12} z_x z_y + a_{22} z_y^2 = 0. \tag{5}$$

Let $z = \varphi(x, y)$ be any particular solution of this equation. If $\xi = \varphi(x, y)$, the coefficient $\bar{a}_{11}$ will obviously be equal to zero. Thus the problem of selecting new independent variables mentioned above is related to the solution of equation (5)

We shall prove the following lemmas.

1. *If $z = \varphi(x, y)$ is a solution of the equation*

$$a_{11} z_x^2 + 2 a_{12} z_x z_y + a_{22} z_y^2 = 0,$$

the relation $\varphi(x, y) = C$ is the general integral of the ordinary differential equation

$$a_{11}\, dy^2 - 2 a_{12}\, dx\, dy + a_{22}\, dx^2 = 0. \tag{6}$$

2. *If $\varphi(x, y) = C$ is the general integral of the ordinary differential equation*

$$a_{11}\, dy^2 - 2 a_{12}\, dx\, dy + a_{22}\, dx^2 = 0,$$

then the function $z = \varphi(x, y)$ satisfies equation (5).

We shall prove the first lemma. Since the function $z = \varphi(x, y)$ satisfies equation (5), the equation

$$a_{11}\left(\frac{\varphi_x}{\varphi_y}\right)^2 - 2 a_{12}\left(-\frac{\varphi_x}{\varphi_y}\right) + a_{22} = 0 \tag{7}$$

* It should be noted that if the transformation of variables is linear, then $\overline{F} = F$, for the second derivatives ξ and η in the formulae (3) are equal to zero and $\overline{F}$ does not get additional terms from the transformation derivatives.

is an identity, for it is satisfied for all x, y in the domain where the solution is given. The relation $\varphi(x, y) = C$ is the general integral of the ordinary differential equation, if the function y defined from the implicit relation $\varphi(x, y) = C$ satisfies equation (6).

Let

$$y = f(x, C)$$

be this function; then

$$\frac{dy}{dx} = -\left[\frac{\varphi_x(x, y)}{\varphi_y(x, y)}\right]_{y=f(x,C)}, \tag{8}$$

where the brackets and the subscript $y = f(x, C)$ denote that in the first portion of equation (8) the variable y should be replaced by $f(x, C)$. Hence it follows that $y = f(x, C)$ satisfies equation (6), because

$$a_{11}\left(\frac{dy}{dx}\right)^2 - 2a_{12}\frac{dy}{dx} + a_{22} = \\ = \left[a_{11}\left(-\frac{\varphi_x}{\varphi_y}\right)^2 - 2a_{12}\left(-\frac{\varphi_x}{\varphi_y}\right) + a_{22}\right]_{y=f(x,C)} = 0,$$

and the expression in square brackets is equal to zero for all values of x, y not for $y = f(x, C)$ only.

We shall prove the second lemma. Let $\varphi(x, y) = C$ be the general integral of equation (6). We shall prove that

$$a_{11}\varphi_x^2 + 2a_{12}\varphi_x\varphi_y + a_{22}\varphi_y^2 = 0 \tag{7}$$

for any point (x, y). Let (x_0, y_0) be any given point. If we prove that at this point equation (7) is satisfied, then from the arbitrariness of (x_0, y_0) it will follow hence that equation (7) is an identity and the function $\varphi(x, y)$ is a solution of equation (7). Let us pass an integral curve of equation (6) through the point (x_0, y_0), putting $\varphi(x_0, y_0) = C_0$ and considering the curve $y = f(x, C_0)$. It is obvious that $y_0 = f(x_0, C_0)$. For all points of this curve we have:

$$a_{11}\left(\frac{dy}{dx}\right)^2 - 2a_{12}\frac{dy}{dx} + a_{22} = \\ = \left[a_{11}\left(-\frac{\varphi_x}{\varphi_y}\right)^2 - 2a_{12}\left(-\frac{\varphi_x}{\varphi_y}\right) + a_{22}\right]_{y=f(x,C_0)} = 0.$$

Putting $x = x_0$ in the last equation, we obtain

$$a_{11}\varphi_x^2(x_0, y_0) + 2a_{12}\varphi_x(x_0, y_0)\varphi_y(x_0, y_0) + a_{22}\varphi_y^2(x_0, y_0) = 0,$$

which was to be proved.*

Equation (6) is called the *characteristic* for equation (1), and its integrals are called *characteristic* integrals.

Putting $\xi = \varphi(x, y)$, where $\varphi(x, y) = \text{const.}$ is the general integral of equation (6), we get the coefficient of $u_{\xi\xi}$ equal to zero. If $\psi(x, y) = \text{const.}$ is another general integral of equation (6) independent of $\varphi(x, y)$, then, putting $\eta = \psi(x, y)$, we get the coefficient of $u_{\eta\eta}$ also equal to zero.

Equation (6) breaks up into two equations:

$$\frac{dy}{dx} = \frac{a_{12} + \sqrt{(a_{12}^2 - a_{11}a_{22})}}{a_{11}}, \tag{9}$$

$$\frac{dy}{dx} = \frac{a_{12} - \sqrt{(a_{12}^2 - a_{11}a_{22})}}{a_{11}}. \tag{10}$$

The sign of the expression under the radical sign determines the type of the equation

$$a_{11}u_{xx} + 2a_{12}u_{xy} + a_{22}u_{yy} + F = 0. \tag{1}$$

We shall call this equation at the point M an equation of the

hyperbolic type, if at the point M $a_{12}^2 - a_{11}a_{22} > 0$,
elliptical type, if at the point M $a_{12}^2 - a_{11}a_{22} < 0$,
parabolic type, if at the point M $a_{12}^2 - a_{11}a_{22} = 0$.**

The relation

$$\bar{a}_{12}^2 - \bar{a}_{11}\bar{a}_{22} = (a_{12}^2 - a_{11}a_{22})(\xi_x\eta_y - \xi_y\eta_x)^2,$$

can easily be verified. The invariance of the type of equation under transformation of variables follows from this. The equation can belong to different types at different points of the domain of definition.

* The relation established between equations (5) and (6) is equivalent to the well-known relation (see Stepanov, *Course of Differential Equations*, 1937, p. 287; Smirnov, *Course of Higher Mathematics*, vol. II, 1948, p. 78) between a linear equation with partial derivatives of the first order and a system of ordinary differential equations. This can be verified by resolving the left part of equation (5) into a product of two linear differential expressions.

** This terminology is taken from the theory of curves of the second order.

Let us consider the domain G, at all points of which the equation has one and the same type. Two characteristic curves pass through each point of the domain G. For equations of the hyperbolic type the characteristics are real and different, for equations of the elliptical type they are complex and different, and for equations of the parabolic type both characteristics are real and coincide.

Let us deal with each of these cases separately.

1. For an equation of the hyperbolic type $a_{12}^2 - a_{11}a_{22} > 0$ and the right-hand sides of equations (9) and (10) are real and different. The general integrals $\varphi(x, y) = C$ and $\psi(x, y) = C$ define the real families of characteristic curves. Putting

$$\xi = \varphi(x, y), \quad \eta = \psi(x, y), \tag{11}$$

we reduce equation (4), after division by the coefficient of $u_{\xi\eta}$, to the form

$$u_{\xi\eta} = \Phi(\xi, \eta, u, u_\xi, u_\eta),$$

where

$$\Phi = -\frac{\bar{F}}{2\bar{a}_{12}}.$$

This is the so-called *canonical* form of equations of the hyperbolic type.* The second canonical form is often used. Let us put

$$\xi = \alpha + \beta, \quad \eta = \alpha - \beta,$$

* In order that it should be possible to introduce new variables ξ and η by means of the functions φ and ψ, it is necessary to verify the independence of these functions. A sufficient condition for this is that the corresponding functional determinant should be non-zero. Let the functional determinant

$$\begin{vmatrix} \varphi_x & \psi_x \\ \varphi_y & \psi_y \end{vmatrix}$$

at a certain point M be made equal to zero. Then the rows are proportional, i. e.

$$\frac{\varphi_x}{\varphi_y} = \frac{\psi_x}{\psi_y},$$

This, however, is impossible because

$$\frac{\varphi_x}{\varphi_y} = -\frac{a_{12} + \sqrt{(a_{12}^2 - a_{11}a_{22})}}{a_{11}} \text{ and } \frac{\psi_x}{\psi_y} = -\frac{a_{12} - \sqrt{(a_{12}^2 - a_{11}a_{22})}}{a_{11}}$$

$$(a_{12}^2 - a_{11}a_{22} > 0)$$

(here we consider $a_{11} \neq 0$, which does not restrict the generality of the statement). Thus the independence of the functions φ and ψ is established.

i. e.

$$\alpha = \frac{\xi + \eta}{2}, \quad \beta = \frac{\xi - \eta}{2},$$

where α and β are new variables. Then

$$u_\xi = \frac{1}{2}(u_\alpha + u_\beta), \quad u_\eta = \frac{1}{2}(u_\alpha - u_\beta), \quad u_{\xi\eta} = \frac{1}{4}(u_{\alpha\alpha} - u_{\beta\beta}).$$

As a result equation (4) takes the form

$$u_{\alpha\alpha} - u_{\beta\beta} = \Phi_1, \quad (\Phi_1 = 4\Phi).$$

2. For equations of the parabolic type $a_{12}^2 - a_{11}a_{22} = 0$, equations (9) and (10) coincide and we obtain one general integral of equation (6): $\varphi(x, y) = \text{const}$. In this case let us put

$$\xi = \varphi(x, y) \text{ and } \eta = \eta(x, y),$$

where $\eta(x, y)$ is any function independent of φ. With this selection of variables the coefficient

$$\bar{a}_{11} = a_{11}\xi_x^2 + 2a_{12}\xi_x\xi_y + a_{22}\xi_y^2 = (\sqrt{a_{11}}\,\xi_x + \sqrt{a_{22}}\,\xi_y)^2 = 0,$$

since $a_{12} = \sqrt{a_{11}}\sqrt{a_{22}}$. Hence it follows that

$$\bar{a}_{12} = a_{11}\xi_x\eta_x + a_{12}(\xi_x\eta_y + \xi_y\eta_x) + a_{22}\xi_y\eta_y =$$
$$= (\sqrt{a_{11}}\,\xi_x + \sqrt{a_{22}}\,\xi_y)(\sqrt{a_{11}}\,\eta_x + \sqrt{a_{22}}\,\eta_y) = 0.$$

After dividing equation (4) by the coefficient of $u_{\eta\eta}$ we obtain the canonical form for the equation of the parabolic type

$$u_{\eta\eta} = \Phi(\xi, \eta, u, u_\xi, u_\eta), \; (\Phi = -\bar{F}/\bar{a}_{22}).$$

If u_ξ does not enter the right-hand side, this equation will be an ordinary differential equation with ξ as a parameter.

3. For an equation of the elliptical type $a_{12}^2 - a_{11}a_{22} < 0$ and the right-hand sides of equations (9) and (10) are complex. Let

$$\varphi(x, y) = C$$

be a complex integral of equation (9). Then

$$\varphi^*(x, y) = C,$$

where φ^* is the complex conjugate of φ and will be the general integral of the conjugate equation (10). Let us change to complex variables, putting

$$\xi = \varphi(x, y), \quad \eta = \varphi^*(x\ y).$$

An equation of the elliptical type is then reduced to the same form as a hyperbolic equation.

In order to avoid dealing with complex variables we introduce new variables α and β of the type

$$\alpha = \frac{\varphi + \varphi^*}{2}, \quad \beta = \frac{\varphi - \varphi^*}{2i},$$

so that

$$\xi = \alpha + i\beta, \quad \eta = \alpha - i\beta.$$

In this case

$$\begin{aligned} &a_{11}\xi_x^2 + 2a_{12}\xi_x\xi_y + a_{22}\xi_y^2 = \\ &= (a_{11}\alpha_x^2 + 2a_{12}\alpha_x\alpha_y + a_{22}\alpha_y^2) - (a_{11}\beta_x^2 + 2a_{12}\beta_x\beta_y + a_{22}\beta_y^2) + \\ &\qquad + 2i[a_{11}\alpha_x\beta_x + a_{12}(\alpha_x\beta_y + \alpha_y\beta_x) + a_{22}\alpha_y\beta_y] = 0, \end{aligned}$$

i. e.

$$\bar{a}_{11} = \bar{a}_{22} \text{ and } \bar{a}_{12} = 0.$$

Equation (4), after division by the coefficient of $u_{\alpha\alpha}$ takes the form†

$$u_{\alpha\alpha} + u_{\beta\beta} = \Phi(\alpha, \beta, u, u_\alpha, u_\beta), \ (\Phi = -\bar{F}/\bar{a}_{22}).$$

Thus, depending on the sign of the expression $a_{12}^2 - a_{11}a_{22}$ the following canonical forms of equation (1) occur:

$a_{12}^2 - a_{11}a_{22} > 0$ (hyperbolic type) $u_{xx} - u_{yy} = \Phi$ or $u_{xy} = \Phi$,
$a_{12}^2 - a_{11}a_{22} < 0$ (elliptical type) $u_{xx} + u_{yy} = \Phi$,
$a_{12}^2 - a_{11}a_{22} = 0$ (parabolic type) $u_{xx} = \Phi$.

† Such a transformation is permissible only if the coefficients of equation (1) are analytical functions. Actually if $a_{12}^2 - a_{11}a_{22} < 0$, the right-hand sides of equations (9) and (10) are complex; consequently the variable y should have complex values. The solution of such equations can be considered only when the coefficients $a_{ik}(x, y)$ are defined for complex values of y. In reducing an equation of the elliptical type to the canonical form we confine ourselves to the case of analytical coefficients.

2. *Classification of equations of the second order with several independent variables*

Let us consider a linear equation with real coefficients

$$\sum_{j=1}^{n}\sum_{i=1}^{n} a_{ij} u_{x_i x_j} + \sum_{i=1}^{n} b_i u_{x_i} + cu + f = 0 \quad (a_{ij} = a_{ji}), \qquad (12)$$

where a, b, c, f are functions of $x_1, x_2, \ldots, x_n$.

Let us introduce new independent variables ξ_k, putting

$$\xi_k = \xi_k(x_1, x_2, \ldots, x_n) \qquad (k = 1, \ldots, n).$$

Then

$$u_{x_i} = \sum_{k=1}^{n} u_{\xi_k} \alpha_{ik},$$

$$u_{x_i x_j} = \sum_{k=1}^{n}\sum_{l=1}^{n} u_{\xi_k \xi_l} \alpha_{ik} \alpha_{jl} + \sum_{k=1}^{n} u_{\xi_k} (\xi_k)_{x_i x_j},$$

where

$$\alpha_{ik} = \frac{\partial \xi_k}{\partial x_i}.$$

Substituting the expressions for the derivatives in the initial equation we obtain

$$\sum_{k=1}^{n}\sum_{l=1}^{n} \bar{a}_{kl} u_{\xi_k \xi_l} + \sum_{k=1}^{n} \bar{b}_k u_{\xi_k} + cu + f = 0,$$

where

$$\bar{a}_{kl} = \sum_{i=1}^{n}\sum_{j=1}^{n} a_{ij} \alpha_{ik} \alpha_{jl},$$

$$\bar{b}_k = \sum_{i=1}^{n} b_i \alpha_{ik} + \sum_{i=1}^{n}\sum_{j=1}^{n} a_{ij} (\xi_k)_{x_i x_j}.$$

Let us consider the quadratic form

$$\sum_{i=1}^{n}\sum_{j=1}^{n} a_{ij}^{0} y_i y_j, \qquad (13)$$

whose coefficients are equal to those of a_{ij} of the initial equation at some point $M_0(x_1^0, \ldots, x_n^0)$.

Carrying out a linear transformation of the variables y

$$y_i = \sum_{k=1}^{n} \alpha_{ik} \eta_k,$$

we obtain a new expression for the quadratic form

$$\sum_{k=1}^{n}\sum_{l=1}^{n}\bar{a}_{kl}^{0}\eta_k\eta_l ,$$

where

$$\bar{a}_{kl}^{0}=\sum_{i=1}^{x}\sum_{j=1}^{n}a_{ij}^{0}\,a_{ik}\,a_{jl}.$$

Thus the coefficients of the second derivatives in equation (12) transform like the coefficients of the quadratic form (13) under a linear transformation. As is well known, by choosing the linear transformation correctly the matrix (a_{ij}^0) of the quadratic form may be reduced to the diagonal form, in which

$$|\bar{a}_{ii}^{0}|=1, \quad \text{or} \quad 0;$$

$$\bar{a}_{ij}^{0}=0 \quad (i\neq j, i=1,2,\ldots,n).$$

The number of coefficients $\bar{a}_{ij}^0$, positive, negative and equal to zero, is the same for any linear transformation which reduces (13) to the diagonal form.

We shall call equation (12) at the point M_0 an equation of the elliptical type, if all n coefficients a_{ii}^0 are of the same sign; of the hyperbolic type (or normal hyperbolic type), if $n-1$ coefficients $\bar{a}_{ii}^0$ have the same sign and one coefficient is opposite in sign to them; of the ultrahyperbolic type, if among $\bar{a}_{ii}^0$ there are m coefficients of one sign and $n-m$ of the opposite sign $(m, n-m>0)$; of the parabolic type if some of the coefficients $\bar{a}_{ii}^0$ are equal to zero.

Selecting new independent variables ξ_i such that at the point M_0

$$a_{ik}=\frac{\partial\xi_k}{\partial t_i}=a_{ik}^{0},$$

where a_{ik}^0 are coefficients of the transformation reducing the quadratic form (13) to the canonical form (e. g. putting $\xi_k = \Sigma a_{ik}^{(0)}x_i$), we find that at the point M_0 the equation is reduced to one of the following canonical forms, depending on the type:

$$u_{x_1x_1}+u_{x_2x_2}+\ldots+u_{x_nx_n}+\Phi=0 \quad \text{(elliptical type)}$$

$$u_{x_1x_1}=\sum_{i=2}^{n}u_{x_ix_i}+\Phi \qquad \text{(hyperbolic type)}$$

$$\sum_{i=1}^{m}{}' u_{x_i x_i} = \sum_{=m+1}^{n}{}' u_{x_i x_i} + \Phi \qquad \text{(ultrahyperbolic type)}$$

$$\sum_{i=1}^{n-m}{}' (\pm u_{x_i x_i}) + \Phi = 0 \ (m > 0) \qquad \text{(parabolic type).}$$

Here we shall not dwell on a more detailed division of equations of the parabolic type into elliptically parabolic, hyperbolically parabolic equations, etc.

Thus, if equation (12) at some point M belongs to a certain type it can be reduced to the corresponding canonical form at this point.

Let us consider in greater detail the problem of whether the equation can be reduced to the canonical form in some neighbourhood of the point M, if at all points of this neighbourhood the equation belongs to one and the same type.

To reduce the equation in some region to the simplest form, the non-diagonal forms of which are equal to zero, we would have to subject the functions $\xi_i(x_1, x_2, \ldots, x_n)$ $(i = 1, 2, \ldots, n)$ to the differential relations $\bar{a}_{kl} = 0$, for $k \neq l$. The number of these conditions, equal to $\frac{n(n-1)}{2}$ exceeds n — the number of functions ξ to be determined for $n > 3$. For $n = 3$ the non-diagonal elements, generally speaking, could be made equal to zero, but the diagonal elements may be different.

Consequently, for $n \geqslant 3$ the equation cannot be reduced to the canonical form in the neighbourhood of the point M. For $n = 2$, the single non-diagonal coefficient can be made equal to zero and the condition of equality of the two diagonal coefficients satisfied, as was done in section 1.

If the coefficients of equation (12) are constant, then reducing (12) to the canonical form at one point M, we obtain an equation reduced to the canonical form in the entire domain of definition of the equation.

3. *Canonical forms of linear equations with constant coefficients*

In the case of two independent variables a linear equation of the second order with constant coefficients has the form

$$a_{11} u_{xx} + 2 a_{12} u_{xy} + a_{22} u_{yy} + b_1 u_x + b_2 u_y + cu + f(x, y) = 0. \quad (14)$$

There is a characteristic equation with constant coefficients corresponding to it, and the characteristic curves are straight lines

$$y = \frac{a_{12} + \sqrt{(a_{12}^2 - a_{11}\,a_{22})}}{a_{11}}\, x + C_1\,, \; y = \frac{a_{12} - \sqrt{(a_{12}^2 - a_{11}\,a_{22})}}{a_{11}}\, x + C_2.$$

With the help of the appropriate transformation of variables, equation (14) reduces to one of the simple forms:

$$u_{\xi\xi} + u_{\eta\eta} + b_1 u_\xi + b_2 u_\eta + cu + f = 0 \quad \text{(elliptical type)} \qquad (15)$$

$$\left.\begin{aligned} &u_{\xi\eta} + b_1 u_\xi + b_2 u_\eta + cu + f = 0 \\ \text{or}\quad &u_{\xi\xi} - u_{\eta\eta} + b_1 u_\xi + b_2 u_\eta + cu + f = 0 \end{aligned}\right\} \text{(hyperbolic type)} \qquad (16)$$

$$u_{\xi\xi} + b_1 u_\xi + b_2 u_\eta + cu + f = 0 \qquad \text{(parabolic type)} \qquad (17)$$

To simplify further we introduce a new function v in place of u:

$$u = e^{\lambda\xi + \mu\eta} \cdot v\,,$$

where λ and μ are constants as yet undetermined. Then

$$\begin{aligned} u_\xi &= e^{\lambda\xi+\mu\eta}\,(v_\xi + \lambda v)\,, \\ u_\eta &= e^{\lambda\xi+\mu\eta}\,(v_\eta + \mu v)\,, \\ u_{\xi\xi} &= e^{\lambda\xi+\mu\eta}\,(v_{\xi\xi} + 2\lambda v_\xi + \lambda^2 v)\,, \\ u_{\xi\eta} &= e^{\lambda\xi+\mu\eta}\,(v_{\xi\eta} + \lambda v_\eta + \mu v_\xi + \lambda\mu v)\,, \\ u_{\eta\eta} &= e^{\lambda\xi+\mu\eta}\,(v_{\eta\eta} + 2\mu v_\eta + \mu^2 v)\,. \end{aligned}$$

Substituting the expressions for the derivatives in equation (15) and dividing by $e^{\lambda\xi+\mu\eta}$, we obtain

$$\begin{aligned} v_{\xi\xi} + v_{\eta\eta} + (b_1 + 2\lambda)\, v_\xi + (b_2 + 2\mu)\, v_\eta &+ \\ + (\lambda^2 + \mu^2 + b_1\lambda + b_2\mu + c)\, v + f_1 &= 0\,. \end{aligned}$$

We select the parameters λ and μ such that two coefficients, those of the first derivatives, for example, become equal to zero ($\lambda = -b_1/2$; $\mu = -b_2/2$). As a result we obtain

$$v_{\xi\xi} + v_{\eta\eta} + \gamma v + f_1 = 0\,,$$

where γ is a constant depending on c, b_1 and b_2; and $f_1 = fe^{-(\lambda\xi+\mu\eta)}$. Performing similar operations for the cases of (16) and (17), we arrive at the following canonical forms for equations with constant coefficients:

$$v_{\xi\xi} + v_{\eta\eta} + \gamma v + f_1 = 0 \qquad \text{(elliptical type)} \tag{17e}$$

$$\left.\begin{aligned} v_{\xi\eta} + \gamma v + f_1 &= 0 \\ \text{or} \quad v_{\xi\xi} - v_{\eta\eta} + \gamma v + f_1 &= 0 \end{aligned}\right\} \qquad \text{(hyperbolic type)} \tag{17f}$$

$$v_{\xi\xi} + b_2 v_\eta + f_1 = 0 \qquad \text{(parabolic type)} \tag{17g}$$

As mentioned in sub-section 2, the equation with constant coefficients in the case of several independent variables

$$\sum_{i=1}^{n} \sum_{j=1}^{n} a_{ij} u_{x_i x_j} + \sum_{i=1}^{n} b_i u_{x_i} + cu + f = 0$$

is reduced, with the help of linear transformation of variables, to the canonical form for all points simultaneously in the domain of its definition. Introducing in place of u the new function v

$$u = \left\{ \exp\left(\sum_{i=1}^{n} \lambda_i x_i \right) \right\} v$$

and choosing λ_i in the necessary manner, we can further simplify the equation, giving canonical forms similar to the case $n = 2$.

Problems on Chapter I

1. Find the domains of hyperbolicity, ellipticity and parabolicity of the equation

$$u_{xx} + y u_{yy} = 0$$

and reduce them to the canonical form in the domain of hyperbolicity.

2. Reduce the following equations to the canonical form;

(a) $u_{xx} + xy u_{yy} = 0.$
(b) $y u_{xx} - x u_{yy} + u_x + y u_y = 0.$
(c) $e^{2x} u_{xx} + 2 e^{x+y} u_{xy} + e^{2y} u_{yy} = 0.$
(d) $u_{xx} + (1 + y)^2 u_{yy} = 0.$
(e) $x u_{xx} + 2 \sqrt{(xy)}\, u_{xy} + y u_{yy} - u_x = 0.$
(f) $(x - y) u_{xx} + (xy - y^2 - x + y) u_{xy} = 0.$
(g) $y^2 u_{xx} - e^{2x} u_{yy} + u_x = 0.$
(h) $\sin^2 y\, u_{xx} - e^{2x} u_{yy} + 3 u_x - 5 u = 0.$
(i) $u_{xx} + 2 u_{xy} + 4 u_{yy} + 2 u_x + 3 u_y = 0.$

3. Reduce to the canonical form and write in its simplest form the equation

$$a u_{xx} + 2 a u_{xy} + a u_{yy} + b u_x + c u_y + u = 0.$$

4. Introducing the function $v = ue^{\lambda x+\mu y}$ and choosing the parameters λ and μ suitably, simplify the following equations with constant coefficients:

(a) $u_{xx} + u_{yy} + \alpha u_x + \beta u_y + \gamma u = 0.$

(b) $u_{xx} = \dfrac{1}{a^2} u_y + \alpha u_x + \beta u_x.$

(c) $u_{xx} - \dfrac{1}{a^2} u_{yy} = \alpha u_x + \beta u_y + \gamma u.$

(d) $u_{xy} = \alpha u_x + \beta u_y,$

CHAPTER II

EQUATIONS OF THE HYPERBOLIC TYPE

EQUATIONS with partial derivatives of the second order of the hyperbolic type occur most frequently in physical problems connected with vibration processes. The simplest equation of the hyperbolic type

$$u_{xx} - u_{yy} = 0$$

is usually called the wave equation. In the present chapter as well as in the following ones we shall confine ourselves to the consideration of a class of linear equations.

§ 1. Simplest problems leading to equations of the hyperbolic type. Formulation of boundary problems

1. *Equation of small transverse vibrations of a string*

Every point of a string of length l can be specified by the value of its coordinate x. The process of vibration of a string may be described by giving the position of the points of the string at different moments of time. To define the position of the string at the moment t it is enough to give the components of the displacement vector of the point x at the moment t $\{u_1(x, t), u_2(x, t), u_3(x, t)\}$.

We consider the simplest problem of vibrations of a string, assuming that the displacements of the string lie in a plane (x, u) and that the displacement vector $\boldsymbol{u}$ is perpendicular at any moment to the axis x. The process of vibration may then be described by a function $u(x, t)$ characterizing the vertical displacement of the string. We shall consider the string as a flexible elastic thread. Mathematically this means that the tension in the string is always directed along the tangents to its instantaneous profile (Fig. 1). This condition implies that the string does not resist bending.

The tension produced in the string due to elasticity can be calculated from Hooke's law. We shall consider small vibrations

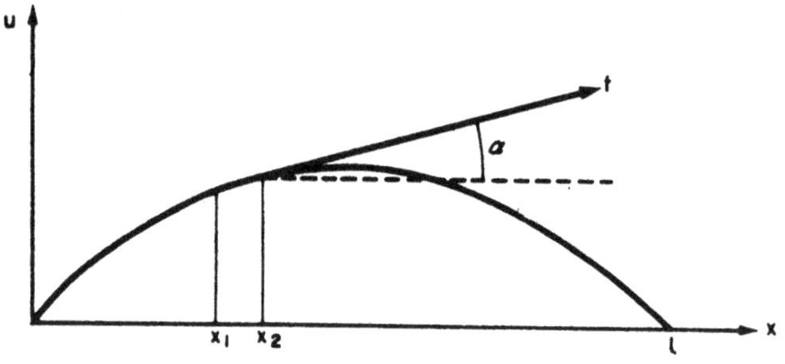

FIG. 1

of the string and neglect the square of u_x as compared with unity.

Using this condition, we calculate the elongation experienced by the segment of the string (x_1, x_2). The length of the arc of this segment is

$$S' = \int_{x_1}^{x_2} \sqrt{[1 + (u_x)^2]}\, dx \cong x_2 - x_1 = S.$$

Thus, within the limits of accuracy assumed, no elongation of the segments of the string occurs in the process of vibration. Hence, from Hooke's law it follows that the tension T at any point does not vary with time. We shall show that the tension is independent of x also, i.e.

$$T = T_0 = \text{const.}$$

The projections of the tension on the axes x and u (denoted by T_x and T_u) are

$$T_x(x) = T(x) \cos\alpha = \frac{T}{\sqrt{[1 + (u_x)^2]}} \cong T(x),$$

$$T_u(x) = T(x) \sin\alpha \cong T(x) \tan\alpha = T(x)\, u_x,$$

where α is the angle made by the tangent to the curve $u(x, t)$ with the x-axis.

Tensile forces, external forces and forces of inertia act on the segment $(x_1 x_2)$. The sum of the projections of the tensile forces on the x-axis should be equal to zero (we are considering transverse vibrations only). Since the forces of inertia and external forces are by assumption directed along the u-axis

$$T_x(x_2) - T_x(x_1) = 0 \quad \text{or} \quad T(x_1) = T(x_2). \tag{1}$$

Since x_1 and x_2 are arbitrary it follows that the tension does not depend on x, i.e. for all values of x and t

$$T \equiv T_0. \tag{2}$$

To derive the equation of transverse vibrations of a string we use Newton's second law. The component of momentum of the element (x_1, x_2) along the u-axis is

$$\int_{x_1}^{x_2} u_t(\xi, t)\, \varrho(\xi)\, d\xi,$$

where ϱ is the linear density of the string. Let us equate the change of momentum during the time interval $\Delta t = t_2 - t_1$

$$\int_{x_1}^{x_2} \varrho(\xi)\left[u_t(\xi, t_2) - u_t(\xi, t_1)\right] d\xi$$

to the impulse of the forces acting, consisting of the tension $T_0 u_x$ at the points x_1 and x_2, and external forces which we shall consider as distributed continuously with the density (load) $f(x, t)$ per unit length. As a result we obtain the equation for transverse vibrations of the element of a string in the integral form

$$\int_{x_1}^{x_2} \left[u_t(\xi, t_2) - u_t(\xi, t_1)\right] \varrho(\xi)\, d\xi =$$

$$= \int_{t_1}^{t_2} T_0\left[u_x(x_2, \tau) - u_x(x_1, \tau)\right] d\tau + \int_{x_1}^{x_2}\int_{t_1}^{t_2} f(\xi, \tau)\, d\xi\, d\tau. \tag{3}$$

To pass over to the differential equation we assume the existence and continuity of second derivatives of $u(x, t)$.† Thus,

† By assuming the double differentiability of functions, we imply that we consider only functions possessing this property. Such an assumption does not mean that functions satisfying the integral equation of vibrations and not having second derivatives do not exist. Such functions exist and are of considerable practical interest. For further details see § 2, sub-section 7.

after the mean value theorem is applied twice, formula (3) assumes the form

$$u_{tt}(\xi^*, t^*)\,\varrho(\xi^*)\,\Delta t\,\Delta x = \{T_0\,[u_{xx}(\xi^{**}, t^{**})] + f(\xi^{***}, t^{***})\}\,\Delta t\,\Delta x,$$

where

$$\xi^*, \xi^{**}, \xi^{***} \subset (x_1\, x_2), \text{ and } t^*, t^{**}, t^{***} \subset (t_1, t_2).$$

Dividing by $\Delta x\, \Delta t$ and passing to the limit when $x_2 \to x_1$,

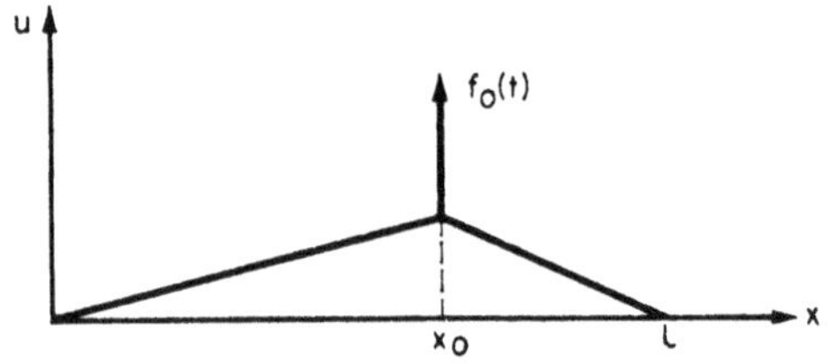

Fig. 2

$t_2 \to t_1$, we obtain the differential equation for transverse vibrations of a string

$$T_0\, u_{xx} = \varrho\, u_{tt} - f(x, t). \tag{4}$$

For the case when density is constant, $\varrho =$ const., and this equation is generally written in the form

$$u_{tt} = a^2\, u_{xx} + F(x, t) \quad \left(a = \sqrt{\frac{T_0}{\varrho}}\right), \tag{5}$$

where

$$F(x, t) = \frac{1}{\varrho} f(x, t) \tag{6}$$

is the density of the force referred to unit mass. In the absence of an external force we obtain a homogeneous equation

$$u_{tt} = a^2\, u_{xx} \text{ or } u_{xx} - u_{yy} = 0 \quad (y = at),$$

describing free vibrations of a string. This equation is the simplest example of an equation of the hyperbolic type.

If a concentrated force $f_0(t)$ (Fig. 2) is applied at the point x_0 $(x_1 < x_0 < x_2)$ equation (3) can be written as

$$\int_{x_1}^{x_2} \varrho(\xi)\,[u_t(\xi, t_2) - u_t(\xi, t_1)]\, d\xi - \int_{x_1}^{x_2}\int_{t_1}^{t_2} f(\xi, \tau)\, d\xi\, d\tau =$$
$$= \int_{t_1}^{t_2} T_0\,[u_x(x_2, \tau) - u_x(x_1, \tau)]\, d\tau + \int_{t_1}^{t_2} f_0(\tau)\, d\tau.$$

Since the velocities of points on the string are finite, the integrals on the left-hand side of this equation tend to zero as $x_1 \to x_0$ and $x_2 \to x_0$ and equation (3) assumes the form

$$\int_{t_1}^{t_2} T_0 \left[u_x (x_0 + 0, \tau) - u_x (x_0 - 0, \tau)\right] d\tau = - \int_{t_1}^{t_2} f_0 (\tau)\, d\tau . \quad (7)$$

Using the mean value theorem, dividing both sides of the equation by Δt and passing to the limit when $t_2 \to t_1$, we obtain

$$u_x (x, t) \Big|_{x_0-0}^{x_0+0} = - \frac{1}{T_0} f_0 (t).$$

Thus at the point of application of the concentrated force the first derivatives are discontinuous and the differential equation (4) loses its meaning. At this point two conditions must be fulfilled.

$$\left.\begin{aligned} u (x_0 + 0, t) &= u (x_0 - 0, t), \\ u_x (x_0 + 0, t) - u_x (x_0 - 0, t) &= - \frac{1}{T_0} f_0 (t). \end{aligned}\right\} \quad (8)$$

The first of these expresses the continuity of the string, the second determines the kink in the string at the point x_0.

2. *Equation of longitudinal vibrations of rods and strings*

Equations for longitudinal vibrations of strings, rods and springs are obtained in a similar way. Let us consider a rod of length e. Longitudinal vibrations may be described by a function $u(x, t)$ representing the displacement at the moment t of a point having the coordinate x* in the equilibrium position. In deriving

* The geometrical variable x chosen here is called Lagrange's variable. In Lagrange's variables each physical point of the rod is characterized by one and the same geometrical coordinate x during the entire process. A physical point occupying the position x at the initial moment (in the state of equilibrium) is situated at any subsequent moment t at a point with the coordinate $X = x + u(x, t)$. If we fix some geometrical point A with the coordinate X, then at different moments of time at this point there will be different physical points (with different Lagrange coordinates x). Euler's variables X, t — where X is the geometrical coordinate, are also frequently used. If $U(X, t)$ is the displacement of a point with the Eulerian coordinate X, the Lagrangian coordinate is

$$x = X - U (X, t),$$

An example of the use of Euler's coordinates is given in sub-section 6.

the wave equation we shall assume that the tension produced by the displacement follows Hooke's law.

Let us calculate the elongation of the element $(x, x + \Delta x)$ at the moment t. The coordinates of the ends of this element at time t have the values

$$x + u(x, t); \; x + \Delta x + u(x + \Delta x, t),$$

and the strain is

$$\frac{[\Delta x + u(x + \Delta x, t) - u(x, t)] - \Delta x}{\Delta x} = u_x(x + \Theta \Delta x, t)$$
$$(0 \leqslant \Theta \leqslant 1).$$

Passing to the limit when $\Delta x \to 0$, we see that the strain at the point x is determined by the function $u_x(x, t)$. By Hooke's law the tension $T(x, t)$ is

$$T(x, t) = k(x) u_x(x, t),$$

where $k(x)$ is Young's modulus at the point x.

Using Newton's law for the change in momentum of the segment (x_1, x_2), we obtain the integral wave equation

$$\int_{x_1}^{x_2} [u_t(\xi, t_2) - u_t(\xi, t_1)] \varrho(\xi) \, d\xi =$$

$$= \int_{t_1}^{t_2} [k(x_2) u_x(x_2, \tau) - k(x_1) u_x(x_1, \tau)] \, d\tau + \int_{x_1}^{x_2} \int_{t_1}^{t_2} f(\xi, \tau) \, d\xi \, d\tau, \quad (10)$$

where $f(x, t)$ is the density of the external force per unit length.

Let us assume the existence and continuity of second derivatives of the function $u(x, t)$. Applying the mean value theorem and making the limiting transition* for $\Delta x = x_2 - x_1 \to 0$ and $\Delta t = t_2 - t_1 \to 0$, we arrive at the differential equation of longitudinal vibrations of a rod**

$$[k(x) u_x]_x = \varrho u_{tt} - f(x, t). \quad (11)$$

* In what follows we shall omit details connected with limits discussed while deriving the equation of transverse vibrations of a string.

** The vibrations have to be small enough for Hooke's law to be applied. In the general case $T = k(x, u_x) u_x$ and we arrive at the quasi-linear equation

$$[k(x, u_x) u_x]_x = \varrho u_{tt} - f(x, t).$$

If the rod is homogeneous, this equation can be written as follows:

$$u_{tt} = a^2 u_{xx} + F(x, t) \qquad \left(a = \sqrt{\frac{k}{\varrho}}\right), \tag{12}$$

where

$$F(x, t) = \frac{f(x, t)}{\varrho} \tag{13}$$

3. *Vibrational energy of a string*

Let us find the energy of transverse vibrations of a string $E = K + U$, where K is the kinetic and U the potential energy. An element of the string dx moving with a velocity u_t has the kinetic energy

$$\frac{1}{2} mv^2 = \frac{1}{2} \varrho(x)\, dx\, (u_t)^2 .$$

The kinetic energy of the whole string

$$K = \frac{1}{2} \int_0^l \varrho(x) \left[u_t(x, t)\right]^2 dx . \tag{14}$$

The potential energy of transverse vibrations of a string, which has the form $u(x, t_0) = u_0(x)$ at $t = t_0$, is equal to the work which must be done in deforming the string from the equilibrium position to the position $u_0(x)$. Let the function $u(x, t)$ give the profile of the string at the instant t.

$$u(x, 0) = 0 , \quad u(x, t_0) = u_0(x) .$$

The element dx under the action of the resultant of the tensile forces

$$T \frac{\partial u}{\partial x}\bigg|_{x+dx} - T \frac{\partial u}{\partial x}\bigg|_{x} = T u_{xx}\, dx$$

moves a distance $u_t(x, t)\, dt$ in the time dt. The work done on the element during this interval is equal to the force multiplied by the distance moved. For the whole string the work done is

$$\left\{\int_0^l T_0 u_{xx} u_t\, dx\right\} dt = \left\{T_0 u_x u_t \big|_0^l - \int_0^l T_0 u_x u_{xt}\, dx\right\} dt =$$

$$= \left\{-\frac{1}{2} \frac{d}{dt} \int_0^l T_0 (u_x)^2\, dx + T_0 u_x u_t \big|_0^l \right\} dt .$$

Integrating with respect to t from 0 to t_0, we obtain:

$$-\frac{1}{2}\int_0^l T_0(u_x)^2\,dx\,\Big|_0^{t_0} + \int_0^{t_0} T_0\,u_x\,u_t\,\Big|_0^l\,dt =$$

$$= -\frac{1}{2}\int_0^l T_0\,[u_x(x, t_0)]^2\,dx + \int_0^{t_0} T_0\,u_x\,u_t\,\Big|_0^l\,dt\,.$$

It is not difficult to determine the significance of the last term in this equality. In fact, $T_0 u_x|_{x=0}$ is the tension at the end of the string $x = 0$; $u_t\,(0, t)\,dt$ is the displacement of this end, and the integral

$$\int_0^{t_0} T_0\,u_x\,u_t\,|_{x=0}\,dt \tag{15}$$

represents the work done in displacing the end $x = 0$. The term corresponding to $x = l$ has a similar significance. If the ends of the string are fastened, the work done on the ends of the string is equal to zero ($u(0, t) = 0$, hence $u_t\,(0, t) = 0$). Consequently, during the displacement of a string fastened at the ends from the equilibrium position to the position $u_0\,(x)$ the work done does not depend on the method of bringing the string to this position and is equal to

$$-\frac{1}{2}\int_0^l T_0\,[u_0'(x)]^2\,dx\,, \tag{16}$$

the potential energy of the string at the moment $t = t_0$ with the opposite sign. Thus, the total energy of the string is

$$E = \frac{1}{2}\int_0^l \left[T_0\,(u_x)^2 + \varrho\,(x)\,(u_t)^2\right]dx\,. \tag{17}$$

The expression for the potential energy of longitudinal vibrations of a rod can be obtained in an exactly similar way. It can also be obtained from the formula for the potential energy of an elastic rod

$$U = \frac{1}{2}k\left(\frac{l-l_0}{l_0}\right)^2 l_0\,.$$

Hence it follows directly that:

$$U = \frac{1}{2}\int_0^l k\,(u_x)^2\,dx\,.$$

4. *Derivation of the transmission line equation*

The passage of electrical current along a wire with distributed parameters is characterized by the current strength i and the potential v which are functions of the position of the point x and of the time t. Considering a segment of length Δx, we can write the fall in potential on the element equal to the sum of the electromotive forces

$$-v_x \Delta x = iR\,\Delta x + i_t L\,\Delta x\,, \tag{18}$$

where R and L are the resistance and the coefficient of self-induction per unit length of the wire.

The charge flowing into an element of the wire Δx in time Δt

$$\left[i\,(x, t) - i\,(x + \Delta x, t)\right] \Delta t = -i_x \Delta x \Delta t, \tag{19}$$

is equal to the sum of the charge stored on the element Δx and the amount lost because of imperfect insulation:

$$\begin{aligned} C\left[v\,(x, t + \Delta t) - v\,(x, t)\right] \Delta x + G\,\Delta x \cdot v\,\Delta t = \\ = (Cv_t + Gv)\,\Delta x\,\Delta t, \end{aligned} \tag{20}$$

where C and G are the capacitance and the leakage conductance per unit length. The losses are assumed to be proportional to the potential at the point of the wire considered.

From formulae (18), (19) and (20) we obtain the equations

$$\left.\begin{aligned} i_x + Cv_t + Gv = 0, \\ v_x + Li_t + Ri = 0, \end{aligned}\right\} \tag{21}$$

the system of telegraphic equations* or transmission line equations.

To obtain an equation for the function i, let us differentiate the first equation (21) with respect to x the second one with respect to t, multiplying it by C.

* These equations are approximate from the standpoint of the theory of the electromagnetic field, since they do not take into account electromagnetic vibrations in the medium surrounding the wire.

Carrying out the calculation assuming that the coefficients are constant we find:

$$i_{xx} + Gv_x - CLi_{tt} - CRi_t = 0.$$

Replacing v_x by its value from the second equation (21), we obtain the equation which defines the current strength in the wire

$$i_{xx} = CLi_{tt} + (CR + GL)\, i_t + GRi. \tag{22}$$

The equation for the potential has a similar appearance

$$v_{xx} = CLv_{tt} + (CR + GL)\, v_t + GRv. \tag{23}$$

Equation (22) or (23) is called the telegraphic equation. If the insulation losses can be neglected and if the resistance is very small ($G \simeq R \simeq 0$), we arrive at the well-known wave equation

$$v_{tt} = a^2\, v_{xx} \qquad \left(a = \sqrt{\frac{1}{LC}}\right). \tag{24}$$

5. *Transverse vibrations of a membrane*

A plane film which does not resist bending or shear is called a membrane. Let us consider a membrane stretched on the plane contour C. We shall study the transverse vibrations of the membrane, in which the displacement is perpendicular to the plane of the membrane.

Let ds be the element of arc of some contour taken on the surface of the membrane and passing through the point $M(x, y)$. Tensile forces $\boldsymbol{T}ds$ act on this element. The vector $\boldsymbol{T}$, due to the absence of resistance to bending and shearing, lies in the tangent plane to the instantaneous surface of the membrane and is perpendicular to the element ds. Moreover, the absence of resistance to shearing implies that the magnitude of $\boldsymbol{T}$ is independent of the direction of the element ds. Hence it follows that the magnitude of the tensile force vector $\boldsymbol{T} = \boldsymbol{T}(x, y, t)$ is not a function of x, y and t. These properties of the vector $\boldsymbol{T}$ serve as the mathematical expression of the absence of bending resistance and shearing strength.

We shall study small vibrations of the membrane, neglecting squares of the first derivatives of u_x and u_y. If follows at once from this condition that $T_h(x, y, t)$, the projection of the tensile force on the surface (x, y) is equal to the absolute value of the tensile force. In fact, for any orientation of the arc ds the angle γ' between the vector $\boldsymbol{T}$ and the plane (x, y) does not exceed the angle γ formed by the normal to the surface of the membrane at the point (x, y) with the axis z. Hence

$$\cos\gamma' \geq \cos\gamma = \frac{1}{\sqrt{(1 + u_x^2 + u_y^2)}} \cong 1, \text{ i.e. } \cos\gamma' \cong 1,$$

and

$$T_h(x, y, z, t) = T \cos\gamma' \cong T(x, y, z, t). \tag{25}$$

The vertical component of the tensile force is

$$T_u = T\frac{\partial u}{\partial n}.$$

Let us mark out on the surface of the membrane an element of area, the projection of which on the plane (x, y) is the rectangle $ABCD$ with sides parallel to the axes of coordinates (Fig. 3). The total tensile force acting on this element is

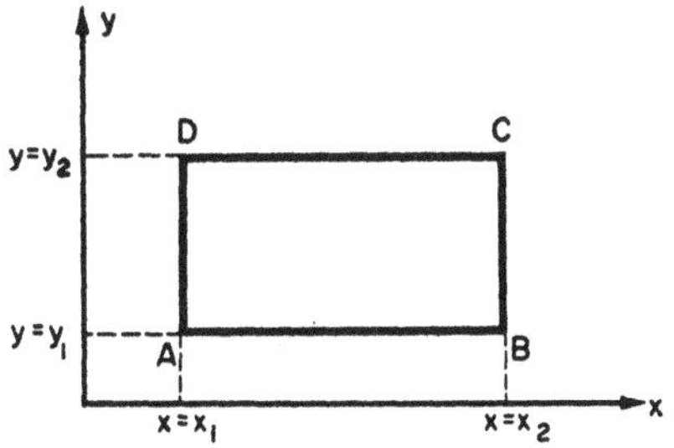

FIG. 3

$$\boldsymbol{T}^* = \oint_{ABCD} \boldsymbol{T}\, ds. \tag{26}$$

Due to the absence of displacement along the axes x, y, the projection of $\boldsymbol{T}^*$ on these axes is equal to zero.

$$T_x^* = \int_B^C T(x_2, y, t)\, dy - \int_A^D T(x_1, y, t)\, dy =$$

$$= \int_{y_1}^{y_2} \{T(x_2, y, t) - T(x_1, y, t)\}\, dy = 0.$$

Similarly

$$T_y^* = \int_{x_1}^{x_2} \{T(x, y_2, t) - T(x, y_1, t)\}\, dx = 0.$$

Using the mean value theorem and taking into account that the area $ABCD$ can be chosen arbitrarily, we obtain

$$\left.\begin{aligned} T(x, y_1, t) &= T(x, y_2, t), \\ T(x_1, y, t) &= T(x_2, y, t), \end{aligned}\right\} \tag{27}$$

i.e. the tensile force T is independent of x and y and can depend only on t.

The area of any element of the membrane at the moment of time t is, in our approximation,

$$\iint \frac{dx\,dy}{\cos\gamma} = \iint \sqrt{(1 + u_x^2 + u_y^2)}\,dx\,dy = \iint dx\,dy. \tag{28}$$

Therefore no extension occurs in the vibration process. Hence it follows, from Hooke's law, that the tensile forces are independent of time. We have thus established that the tensile force does not depend on the variables x, y and t

$$T(x, y, t) = \text{const} = T_0. \tag{29}$$

Let us derive the equation for vibrations of the membrane. Let S_1 be the projection of some segment of the membrane on the plane (x, y) and C_1 the boundary of S_1. Equating the change in momentum in the vertical direction to the impulse of the vertical components of the tensile forces and the externally acting forces with a density of $f(x, y, t)$, we get the equation for vibrations of the membrane in the integral form

$$\iint\limits_{S} [u_t(x, y, t_2) - u_t(x, y, t_1)]\,\varrho(x, y)\,dx\,dy =$$

$$= \int_{t_1}^{t_2}\int\limits_{C_1} T_0 \frac{\partial u}{\partial n}\,ds\,dt + \int_{t_1}^{t_2}\iint\limits_{S_1} f\,dx\,dy\,dt, \tag{30}$$

where $\varrho(x, y)$ is the surface density of the membrane and $f(x, y, t)$ the density of the external force.

To obtain the differential equation we assume that the function $u(x, y, t)$ has continuous second derivatives. With the help of Ostrogradskii's theorem* the contour integral can be transformed into a surface integral

$$\int\limits_{C_1} \frac{\partial u}{\partial n}\,ds = \iint\limits_{S_1} (u_{xx} + u_{yy})\,dx\,dy,$$

* See V. I. Smirnov, *A Course of Higher Mathematics*, vol. II, 1948, p. 196.

and the integral equation of vibrations reduces to the form

$$\int_{t_1}^{t_2}\iint_{S_1}\{\varrho u_{tt} - T_0(u_{xx}+u_{yy}) - f(x,y,t)\}\,dx\,dy\,dt = 0\,.$$

Making use of the mean value theorem, the arbitrary choice of S_1 and of the time interval (t_1, t_2), we conclude that the integrand is identically equal to zero. Thus we arrive at the differential equation for vibrations of the membrane

$$\varrho u_{tt} = T_0(u_{xx}+u_{yy}) + f(x,y,t)\,. \tag{31}$$

For a homogenous membrane the equation may be written in the form

$$u_{tt} = a^2(u_{xx}+u_{yy}) + F(x,y,t) \qquad \left(a^2 = \frac{T_0}{\varrho}\right), \tag{32}$$

where $F(x, y, t)$ is the force density per unit mass of the membrane.

6. *Equations of hydrodynamics and acoustics*

The functions $v_1(x, y, z, t)$, $v_2(x, y, z, t)$, $v_3(x, y, z, t)$, components of the velocity vector $\boldsymbol{v}$ at the point (x, y, z) at the moment t, are used to characterize the motion of a liquid. The density $\varrho(x, y, z, t)$, the pressure $p(x, y, z, t)$ and the density of the externally acting forces $\boldsymbol{F}(x, y, z, t)$ (if any) per unit mass, are also quantities characterizing the motion.

Let us consider some volume of the liquid T and calculate the forces acting on it. Neglecting frictional forces due to viscosity, i.e. considering an ideal liquid, we obtain an expression for the resultant of the pressures in the form of a surface integral

$$-\iint_S p\boldsymbol{n}\,dS, \tag{33}$$

where S is the surface of the volume T, $\boldsymbol{n}$ is the unit vector of the external normal. Ostrogradskii's formula gives

$$-\iint_S p\boldsymbol{n}\,dS = -\iiint_T \operatorname{grad} p\,d\tau\,. \tag{34}$$

In calculating the acceleration of any particle of the liquid the displacement of the particle has to be considered. Let

$x = x(t)$, $y = y(t)$, $z = z(t)$ be the equation of the trajectory of this particle. The time derivative of the velocity is

$$\frac{dv}{dt} = \frac{\partial v}{\partial t} + \frac{\partial v}{\partial x}\dot{x} + \frac{\partial v}{\partial y}\dot{y} + \frac{\partial v}{\partial z}\dot{z} =$$

$$= \frac{\partial v}{\partial t} + \frac{\partial v}{\partial x}v_1 + \frac{\partial v}{\partial y}v_2 + \frac{\partial v}{\partial z}v_3 = \frac{\partial v}{\partial t} + (v\nabla)\,\boldsymbol{v},$$

where

$$\nabla = \boldsymbol{i}\frac{\partial}{\partial x} + \boldsymbol{j}\frac{\partial}{\partial y} + \boldsymbol{k}\frac{\partial}{\partial z}.$$

Such a time derivative is called a *total* derivative. The equation of motion of the liquid expresses the usual relation between the acceleration of the particles and the forces acting on them.

$$\iiint\limits_T \varrho\frac{dv}{dt}\,d\tau = -\iiint\limits_T \operatorname{grad} p\,d\tau + \iiint\limits_T \varrho\boldsymbol{F}\,d\tau, \tag{35}$$

where the last integral is the resultant of external forces applied on the volume T. Hence, because the volume T is arbitrary, we obtain the equation of motion of the ideal liquid in Euler's form

$$\boldsymbol{v}_t + (\boldsymbol{v}\,\nabla)\,v = -\frac{1}{\varrho}\operatorname{grad} p + \boldsymbol{F}. \tag{36}$$

Now we derive the equation of continuity. If there are no sources or sinks inside T, the change in the amount of liquid contained inside T in unit time is equal to the flux through the boundary S

$$\frac{d}{dt}\iiint\limits_T \varrho\,dt \doteq -\iint\limits_S \varrho\boldsymbol{v}\boldsymbol{n}\,dS. \tag{37}$$

The transformation of the surface integral to a volume integral gives

$$\iiint\limits_T \left(\frac{\partial\varrho}{\partial t} + \operatorname{div}\varrho\boldsymbol{v}\right)d\tau = 0\,.$$

Since this equation is valid for an arbitrary volume T, the equation of continuity follows

$$\frac{\partial\varrho}{\partial t} + \operatorname{div}(\varrho\boldsymbol{v}) = 0$$

or

$$\frac{\partial\varrho}{\partial t} + \boldsymbol{v}\operatorname{grad}\varrho + \varrho\operatorname{div}\boldsymbol{v} = 0\,. \tag{38}$$

We should add the thermodynamic *equation of state* to equations (36) and (38). We write it in the form

$$p = f(\varrho).$$

Thus we obtain a system of five equations for five unknown functions v_x, v_y, v_z, p and ϱ. (If the equation of state had contained the temperature, it would be necessary to add the equation of heat transfer.) Thus the system of equations

$$\left.\begin{aligned} &\frac{\partial \boldsymbol{v}}{\partial t} + (\boldsymbol{v}\nabla)\,\boldsymbol{v} = \boldsymbol{F} - \frac{1}{\varrho}\operatorname{grad} p, \\ &\frac{\partial \varrho}{\partial t} + \operatorname{div}\,(\varrho \boldsymbol{v}) = 0, \\ &\qquad\qquad p = f(\varrho) \end{aligned}\right\} \tag{39}$$

represents a complete system of equations of hydrodynamics.

We apply the equations of hydrodynamics to the study of sound propagation in a gas. We make the following assumptions: (1) external forces are absent; (2) the process of sound propagation is adiabatic, therefore Poisson's adiabat serves as the equation of state

$$\frac{p}{p_0} = \left(\frac{\varrho}{\varrho_0}\right)^{\gamma} \qquad \left(\gamma = \frac{c_p}{c_v}\right),$$

where ϱ_0 and p_0 are the initial density and the initial pressure, c_p and c_v are the specific heats at constant pressure and constant volume; (3) the vibrations of the gas are small so that higher powers of the velocities, velocity gradients and the density may be neglected.

We call the quantity $s(x, y, z, t)$, equal to the relative change in density, the condensation of the gas

$$s(x, y, z, t) = \frac{\varrho - \varrho_0}{\varrho_0}, \tag{40}$$

whence

$$\varrho = \varrho_0(1 + s). \tag{41}$$

With these assumptions the equations of hydrodynamics take the form

$$\left.\begin{aligned} \boldsymbol{v}_t &= -\frac{1}{\varrho_0}\operatorname{grad}\, p, \\ \varrho_t + \varrho_0 &\operatorname{div}\, \boldsymbol{v} = 0, \\ p = p_0(1+s)^{\gamma} &\cong p_0(1 + \gamma s), \end{aligned}\right\} \tag{42}$$

since

$$\frac{1}{\varrho}\operatorname{grad} p = \frac{1}{\varrho_0}(1 - s + \ldots)\operatorname{grad} p = \frac{1}{\varrho_0}\operatorname{grad} p + \ldots ,$$

$$\operatorname{div} \varrho \boldsymbol{v} = \boldsymbol{v}\operatorname{grad}\varrho + \varrho \operatorname{div} \boldsymbol{v} = \varrho_0 \operatorname{div} \boldsymbol{v} + \ldots ,$$

where the second and higher order terms have been neglected. Introducing the notation $a^2 = \frac{\gamma p_0}{\varrho_0}$ we rewrite system (42) in the following form

$$\left.\begin{aligned} \boldsymbol{v}_t &= -a^2 \operatorname{grad} s, \\ s_t &+ \operatorname{div} \boldsymbol{v} = 0. \end{aligned}\right\} \qquad (42')$$

Applying the divergence operator to the first equation of (42) and changing the order of differentiation, we get

$$\frac{\partial}{\partial t}\operatorname{div} \boldsymbol{v} = -a^2 \operatorname{div} (\operatorname{grad} s) = -a^2 \nabla^2 s$$

where

$$\nabla^2 = \frac{\partial^2}{\partial x^2} + \frac{\partial^2}{\partial y^2} + \frac{\partial^2}{\partial z^2}$$

is the Laplace operator. Using the second equation of (42′) we get the wave equation

$$\nabla^2 s = \frac{1}{a^2} s_{tt} \qquad (43)$$

or

$$a^2 (s_{xx} + s_{yy} + s_{zz}) = s_{tt} .$$

From this and (40) we get an equation for the density

$$a^2 (\varrho_{xx} + \varrho_{yy} + \varrho_{zz}) = \varrho_{tt} .$$

We now introduce the velocity potential and show that it satisfies the same equation (43) as the condensation.

From the equation

$$\boldsymbol{v}_t = -a^2 \operatorname{grad} s$$

it follows that

$$\boldsymbol{v}(x, y, z, t) = \boldsymbol{v}(x, y, z, 0) - a^2 \operatorname{grad} \left(\int_0^t s\, dt\right), \qquad (44)$$

where $\boldsymbol{v}(x, y, z, 0)$ is the initial distribution of velocities. If initially the velocity field can be derived from a potential

$$\boldsymbol{v}|_{t=0} = -\operatorname{grad} f(x, y, z), \qquad (45)$$

then the following relation holds:

$$\boldsymbol{v} = - \text{ grad } \left[f(x,y,z) + a^2 \int_0^t s\, dt\right] = - \text{ grad } U. \qquad (46)$$

This means that there is a velocity potential $U(x, y, z, t)$ for all t. A knowledge of the velocity potential is sufficient to define the entire motion

$$\left.\begin{aligned} \boldsymbol{v} &= - \text{ grad } U, \\ s &= \frac{1}{a^2} U_t. \end{aligned}\right\} \qquad (47)$$

Substituting these values in the equation of continuity

$$s_t + \text{div } \boldsymbol{v} = 0,$$

we get the equation for the potential

$$a^2 (U_{xx} + U_{yy} + U_{zz}) = U_{tt}$$

$$U_{tt} = a^2 \Delta U. \qquad (48)$$

It is possible to obtain equations of the form (48) for the pressure o and the velocity v. Often this equation is called the equation pf acoustics.

In solving problems for two-dimensional and one-dimensional cases Laplace's operator in equation (48) has to be replaced by the operator $\frac{\partial^2}{\partial x^2} + \frac{\partial^2}{\partial y^2}$ and by $\frac{\partial^2}{\partial x^2}$ respectively.

In the case of vibrations of a gas in a confined region, certain boundary conditions must be given at the boundary. If the boundary is a solid impermeable wall, the normal component of the velocity is zero, which leads to the conditions

$$\left.\frac{\partial U}{\partial n}\right|_{\Sigma} = 0 \text{ or } \left.\frac{\partial s}{\partial n}\right|_{\Sigma} = 0. \qquad (49)$$

The constant

$$a = \sqrt{\frac{\gamma p_0}{\varrho_0}}$$

has the dimensions of velocity and, as will be shown later, it is the velocity of propagation of sound.

Let us calculate the velocity of sound in air at normal atmospheric pressure. In this case $\gamma = \frac{7}{5}$, $\varrho_0 = 0.001293$ g/cm³, $p_0 = 1.033$ kg/cm², hence

$$a = \sqrt{\frac{\gamma p_0}{\varrho_0}} = 336 \text{ m/sec.}$$

7. *Boundary and initial conditions*

For the mathematical description of a physical process it is necessary to lay down sufficient conditions to determine the process uniquely.

Differential equations have generally speaking an infinite number of solutions. Therefore in the case when the physical problem is reduced to a partial differential equation, it is necessary to supplement that equation with certain additional equations in order to specify the process uniquely.

In the case of an ordinary differential equation of the second order the solution may be defined by initial conditions, i. e. by giving the values of the function and its first derivative for the "initial" value of the argument. Other forms of additional conditions are also found, where, for example, the values of the function are given at two points. Different forms of additional conditions are also possible for a partial differential equation.

Let us consider first the simplest problem of transverse vibrations of a string fixed at the ends. In this problem, $u(x, t)$ gives the deviation of the string from the x-axis. If the ends of the string $0 \leqslant x \leqslant l$ are fixed then the "boundary condition"

$$u(0, t) = 0, \quad u(l, t) = 0. \tag{50}$$

must be fulfilled.

In addition, the "initial conditions", i. e. the shape and the velocity of the string at the initial moment t_0

$$\left.\begin{aligned} u(x, t_0) &= \varphi(x), \\ u_t(x, t_0) &= \psi(x). \end{aligned}\right\} \tag{51}$$

are given. Thus the additional conditions consist of boundary and initial conditions where $\varphi(x)$ and $\psi(x)$ are given functions of x. We shall show later that these conditions define the solution of the equation for vibrations of the string uniquely

$$u_{tt} = a^2 u_{xx}. \tag{52}$$

If the ends of the string move according to the given law, the boundary conditions assume another form

$$\left.\begin{aligned} u(0, t) &= \mu_1(t), \\ u(l, t) &= \mu_2(t), \end{aligned}\right\} \tag{50'}$$

where $\mu_1(t)$ and $\mu_2(t)$ are the given functions of time t. The problem of longitudinal vibrations of a string or a spring can be formulated in the same way.

Other types of boundary conditions are also possible. Let us consider, for example, the problem of longitudinal vibrations of a spring, one end of which is fixed (point of suspension) and the other is free. The law of motion of the free end is not given and is often an unknown function. At the point of suspension $x = 0$ the displacement

$$u(0, t) = 0;$$

at the free end $x = l$ the tension of the spring

$$T(l, t) = k \left.\frac{\partial u}{\partial x}\right|_{x=l} \tag{53}$$

is zero (there are no external forces), so that the mathematical formulation of the condition of the free end has the form

$$u_x(l, t) = 0.$$

If the end $x = 0$ moves according to a certain law $\mu(t)$ and at $x = l$ the force is given as $\bar{\nu}(t)$ then

$$u(0, t) = \mu(t), \qquad u_x(l, t) = \nu(t), \qquad \left(\nu(t) = \frac{1}{k}\bar{\nu}(t)\right).$$

The condition for elastic attachment is also typical, let us say for $x = l$

$$ku_x(l, t) = -\alpha u(l, t)$$

or

$$u_x(l, t) = -hu(l, t) \qquad \left(h = \frac{\alpha}{k}\right), \tag{54}$$

In this case the end $x = l$ may be displaced, but elastic forces at this end give rise to a tension tending to bring back the displaced end to its former position. The rigidity of attachment is characterized by the coefficient α.

If a point of elastic attachment is displaced and its deviation from the initial position is given by the function $\theta(t)$, the boundary condition assumes the form

$$u_x(l, t) = -h\left[u(l, t) - \theta(t)\right], \tag{55}$$

where

$$h = \frac{\alpha}{k}.$$

It should be noted that in the case of rigid attachment (α large) when even small displacements of the end produce large tensile forces, the latter boundary condition passes over to the first when $\mu(t) = \theta(t)$. In the case of loose attachment (α small), when large displacements of the end produce weak tensile forces, the boundary condition passes over to the condition of the free end.

We shall speak of three principal types of boundary conditions:

boundary condition of thef irst type $u(0, t) = \mu(t)$ — displacement given,

boundary condition of the second type $u_x(0, t) = \nu(t)$ — force given,

boundary condition of the third type $u_x(0, t) = h[u(0, t) - \theta(t)]$ — elastic attachment.

The boundary conditions on the second end $x = l$ are given in the same way. If the functions on the right-hand side are equal to zero, the boundary conditions are called homogeneous.

On combining the different types of boundary conditions enumerated, we get six types of the simplest boundary problems.

A more complex boundary condition exists, for example, in the case of elastic attachment not subject to Hooke's law, when the tension at the end is a non-linear function of the displacement $u(l, t)$ so that

$$u_x(l, t) = \frac{1}{k} F\left[u(l, t)\right]. \tag{56}$$

This boundary condition, unlike those considered above, is non-linear. Relations between the displacements and the tensile forces at different ends of the system are also possible. For example, in problems on the vibration of a ring, when $x=0$ and $x=l$ represent one and the same physical point, the boundary conditions assume the form

$$u(l,t)=u(0,t);\quad u_x(0,t)=u_x(l,t)\,, \tag{57}$$

i. e. they are reduced to the requirements of continuity of u and u_x. The derivatives with respect to t can also enter the boundary conditions. If the end of the spring experiences a resistance from the medium proportional to its velocity the boundary condition assumes the form

$$ku_x(l,t)=-\alpha u_t(l,t)\,. \tag{58}$$

If a load of mass m is attached to the end $x=l$ of the spring then the following condition must be fulfilled at that end

$$mu_{tt}(l,t)=-ku_x(l,t)+mg. \tag{59}$$

We shall confine ourselves below to a consideration of the three simplest types of boundary conditions, dealing principally with an example of the first type of boundary condition and noting only incidentally peculiarities associated with the second and third conditions.

Let us formulate the first boundary value problem for equation (52):

Find the function $u(x,t)$ *satisfying the equation*

$$u_{tt}=a^2u_{xx} \text{ where } 0<x<l,\ t>0$$

and the boundary and initial conditions:

$$\left.\begin{aligned} u(0,t)&=\mu_1(t)\,,\\ u(l,t)&=\mu_2(t)\,,\\ u(x,0)&=\varphi(x)\,,\\ u_t(x,0)&=\psi(x)\,.\end{aligned}\right\} \tag{60}$$

If boundary conditions of the second or third type are taken at both ends, the corresponding problems are called second or third boundary value problems. If the boundary conditions at $x = 0$ and $x = l$ are of different types, such boundary problems are called mixed without making a more detailed classification.

We now turn to a consideration of the limiting cases of this problem. The effect of boundary conditions at a point M_0 sufficiently removed from the boundary for which they are given, is felt only after a sufficiently long time.

If we are interested in the phenomenon over a small interval of time, when the effect of the boundaries is still insignificant then it is possible to consider the limiting problem *with initial conditions* for an infinite region, instead of the complete problem.

Find the solution of the equation

$$u_{tt} = a^2 u_{xx} + f(x, t) \text{ where } -\infty < x < \infty, t > 0,$$

with the initial conditions

$$\left.\begin{aligned} u(x, 0) &= \varphi(x), \\ u_t(x, 0) &= \psi(x), \end{aligned}\right\} \text{ where } -\infty < x < \infty. \tag{61}$$

This problem is often called *Cauchy's problem.*

If, however, we study the phenomenon near one boundary, and the effect of the boundary condition on the second boundary is not important during the interval of time in which we are interested, then we arrive at the formulation of the problem for a semi-infinite straight line $0 \leqslant x \leqslant \infty$, when in addition to the differential equation the following conditions are given

$$\left.\begin{aligned} u(0, t) &= \mu(t), \quad t \geq 0, \\ u(x, 0) &= \varphi(x), \\ u(x, 0) &= \psi(x) \end{aligned}\right\} 0 \leq x < \infty. \tag{62}$$

The nature of the phenomenon, for times sufficiently long after the initial time $t = 0$, is completely defined by boundary conditions, because the effect of initial conditions weakens due to friction inherent in any real system. Problems of this type are encountered frequently in cases where the system is excited by a periodic boundary condition acting over a long period. Such "steady state problems" are formulated as follows:

Find the solution of the differential equation for $0 \leqslant x \leqslant l$ and $t > -\infty$ *with the boundary conditions*

$$\left.\begin{aligned} u(0, t) &= \mu_1(t), \\ u(l, t) &= \mu_2(t). \end{aligned}\right\} \tag{63}$$

The steady state problem for the semi-finite straight line is formulated similarly.

8. *Reduction of the general problem*

In solving a complex problem it is natural to try to reduce its solution to the solution of simpler problems. We do this by representing the solution of the general boundary problem as a sum of the solutions of a number of particular boundary-value problems.

Let $u_i(x, t)$ $(i = 1, 2, \ldots, n)$ be functions satisfying the equations

$$\frac{\partial^2 u_i}{\partial t^2} = a^2 \frac{\partial^2 u_i}{\partial x^2} + f^i(x, t) \tag{64}$$

if

$$0 < x < l, \ t > 0$$

and the additional conditions

$$\left.\begin{aligned} u_i(0, t) &= \mu_1^i(t), \\ u_i(l, t) &= \mu_2^i(t); \\ u_i(x, 0) &= \varphi^i(x), \\ \frac{\partial u_i}{\partial t}(x, 0) &= \psi^i(x). \end{aligned}\right\} \tag{65}$$

It is obvious that a superposition of the solutions

$$u^{(0)}(x, t) = \sum_{i=1}^{n} u_i(x, t) \tag{66}$$

satisfies a similar equation with the right-hand side

$$f^{(0)}(x, t) = \sum_{i=1}^{n} f^i(x, t) \tag{67}$$

and additional conditions similar to (65) with

$$\left.\begin{aligned} \mu_k^{(0)}(t) &= \sum_{i=1}^{n} \mu_k^i(t) \quad (k = 1, 2), \\ \varphi^{(0)}(x) &= \sum_{i=1}^{n}{}' \varphi^i(x), \\ \psi^{(0)}(x) &= \sum_{i=1}^{n}{}' \psi^i(x). \end{aligned}\right\} \tag{68}$$

The principle of superposition applies obviously not only to this special problem but to any linear equation with linear additional conditions. We shall make use of this property repeatedly.

The solution of the general boundary problem

$$\left.\begin{aligned} & u_{tt} = a^2 u_{xx} + f(x, t) \\ & (0 < x < l, \quad t > 0); \\ & u(0, t) = \mu_1(t), \\ & u(l, t) = \mu_2(t); \\ & u(x, 0) = \varphi(x), \\ & u_t(x, 0) = \psi(x) \end{aligned}\right\} \tag{69}$$

may be represented as a sum

$$u(x, t) = u_1(x, t) + u_2(x, t) + u_3(x, t) + u_4(x, t), \tag{70}$$

where u_1, u_2, u_3, u_4 are solutions of the following particular boundary problems

$$\begin{aligned} \frac{\partial^2 u_i}{\partial t^2} &= a^2 \frac{\partial^2 u_i}{\partial x^2} \quad (i = 1, 2, 3), \\ \frac{\partial^2 u_4}{\partial t^2} &= a^2 \frac{\partial^2 u_4}{\partial x^2} + f(x, t), \end{aligned} \qquad (0 < x < l, t > 0),$$

$$\left.\begin{array}{llll} u_1(0, t) = 0, & u_2(0, t) = \mu_1(t), & u_3(0, t) = 0, & u_4(0, t) = 0, \\ u_1(l, t) = 0; & u_2(l, t) = 0; & u_3(l, t) = \mu_2(t); & u_4(l, t) = 0; \\ u_1(x, 0) = \varphi(x), & u_2(x, 0) = 0, & u_3(x, 0) = 0, & u_4(x, 0) = 0, \\ u_{1t}(x, 0) = \psi(x); & u_{2t}(x, 0) = 0; & u_{3t}(x, 0) = 0; & u_{4t}(x, 0) = 0. \end{array}\right\} \tag{71}$$

We consider this formal reduction in order to specify particular boundary-value problems constituting the fundamental stages in the solution of the general problem. A similar reduction may be carried out for limiting cases of the general boundary problem.

9. *Formulation of boundary-value problems for the case of several variables*

We have considered in detail the formulation of boundary problems for the case of one independent geometrical variable x (and time t). If the number of geometrical variables is $n > 1$ (e. g. $n = 3$), the first boundary problem is formulated in an exactly similar way:

It is required to find a function $u(M, t) = u(x, y, z, t)$ *defined within the given region* T *with the boundary* Σ, *satisfying within* T *the equation*

$$u_{tt} = a^2 \Delta u + f(x, y, z, t) \qquad (M(x, y, z) \subset T, t > 0), \qquad (72)$$

the boundary condition on Σ

$$u|_{\Sigma} = \mu(x, y, z, t) \qquad (M(x, y, z, t) \subset \Sigma, t > 0) \qquad (73)$$

($\mu(x, y, z, t)$ *is a function given on* Σ) *and the initial conditions*

$$\left.\begin{aligned} u(x, y, z, 0) &= \varphi(x, y, z), \\ u_t(x, y, z, 0) &= \psi(x, y, z) \end{aligned}\right\} \qquad (M(x, y, z) \subset T). \qquad (74)$$

The resolution of the general boundary problem into a number of simpler ones is made as in the preceding case. Limiting boundary problems for an infinite region, half-space, etc., can also be formulated.

10. *Uniqueness theorem*

In solving boundary problems:

1. It is necessary to verify that there are sufficient additional conditions to give a unique solution. This is done by proving the *uniqueness* theorem.

2. It is necessary to verify that the additional conditions do not over-define the problem, i. e. there are no incompatible conditions among them. This is done by proving the theorem of *existence*. Proof of the existence of a solution is usually closely associated with a method of finding the solution.

We shall prove the following *uniqueness theorem :*

Only one solution of the following equation can exist

$$\varrho \frac{\partial^2 u}{\partial t^2} = \frac{\partial}{\partial x}\left(k \frac{\partial u}{\partial x}\right) + F(x, t) \tag{75}$$

$$(0 < x < l, \ t > 0),$$

satisfying the initial and boundary conditions

$$\left.\begin{aligned} u(x, 0) &= \varphi(x), \\ u_t(x, 0) &= \psi(x); \\ u(0, t) &= \mu_1(t), \\ u(l, t) &= \mu_2(t). \end{aligned}\right\} \tag{76}$$

It is assumed here that the function $u(x, t)$, together with its second derivatives, is continuous in the segment $0 \leqslant x \leqslant l$ at $t \geqslant 0$ and that $\varrho(x) > 0$, $k(x) > 0$ are continuous functions.

Let us suppose that two solutions of the problem exist

$$u_1(x, t), \quad u_2(x, t),$$

and consider the difference

$$v(x, t) = u_1(x, t) - u_2(x, t).$$

The function $v(x, t)$ obviously satisfies the homogeneous equation

$$\varrho \frac{\partial^2 v}{\partial t^2} = \frac{\partial}{\partial x}\left(k \frac{\partial v}{\partial x}\right) \tag{77}$$

and the homogeneous additional conditions

$$\left.\begin{aligned} v(x, 0) &= 0, \\ v_t(x, 0) &= 0; \\ v(0, t) &= 0, \\ v(l, t) &= 0. \end{aligned}\right\} \tag{78}$$

We shall prove that the function $v(x, t)$ defined by the conditions (78) is identically equal to zero.

We consider the function

$$E(t) = \frac{1}{2} \int_0^l \left\{k (v_x)^2 + \varrho (v_t)^2\right\} dx \tag{79}$$

and show that it does not depend on t. (The function $E(t)$ is the total energy of the string at the time t.) We shall differentiate $E(t)$ with respect to t, carrying out the differentiation under the sign of the integral, which is possible because second derivatives are continuous

$$\frac{dE(t)}{dt} = \int_0^l (kv_x v_{xt} + \varrho v_t v_{tt})\, dx .$$

Integrating the first term of the right-hand side by parts we get

$$\int_0^l kv_x v_{xt}\, dx = [kv_x v_t]_0^l - \int_0^l v_t (kv_x)_x\, dx . \tag{80}$$

On substituting the values at $x = 0$ and $x = l$ the expression vanishes by the boundary conditions [from $v(0, t) = 0$ it follows that $v_t(0, t) = 0$ and similarly for $x = l$]. It follows that

$$\frac{dE(t)}{dt} = \int_0^l [\varrho v_t v_{tt} - v_t (kv_x)_x]\, dx = \int_0^l v_t [\varrho v_{tt} - (kv_x)_x]\, dx = 0 .$$

i. e. $E(t) =$ const. The initial conditions give

$$E(t) = \text{const} = E(0) = \frac{1}{2} \int_0^l [k (v_x)^2 + \varrho (v_t)^2]_{t=0}\, dx = 0 , \tag{81}$$

since

$$v(x, 0) = 0, \quad v_t(x, 0) = 0 .$$

Using formula (81) and remembering the positive sign of k and ϱ we conclude that

$$v_x(x, t) \equiv 0, \quad v_t(x, t) \equiv 0 ,$$

whence we get the identity

$$v(x, t) = \text{ const } = C_0 . \tag{82}$$

Using the initial condition we find

$$v(x, 0) = C_0 = 0 ;$$

from which it follows that

$$v(x, t) \equiv 0 . \tag{83}$$

Hence if there are two functions $u_1(x, t)$ and $u_2(x, t)$ satisfying all conditions of the theorem, then $u_1(x, t) \equiv u_2(x, t)$.

For the second boundary problem the function $v = u_1 - u_2$ satisfies the boundary conditions

$$v_x(0, t) = 0, \quad v_x(l, t) = 0, \tag{84}$$

and again the first term on the right-hand side of formula (80) vanishes. The rest of the proof of the theorem remains unchanged.

For the third boundary problem the proof requires some modification. Considering as before two solutions u_1 and u_2, we obtain equation (77) for their difference $v(x, t) = u_1 - u_2$, with the boundary conditions

$$\left.\begin{aligned} v_x(0, t) - h_1 v(0, t) = 0 \quad (h_1 \geqslant 0), \\ v_x(l, t) + h_2 v(l, t) = 0 \quad (h_2 \geqslant 0). \end{aligned}\right\} \tag{85}$$

The limit term in equation (80) becomes

$$[k v_x v_t]_0^l = -\frac{k}{2} \frac{\partial}{\partial t} [h_2 v^2(l, t) + h_1 v^2(0, t)].$$

Integrating dE/dt between the limits zero and t, we get:

$$E(t) - E(0) = \int_0^t \int_0^l v_t [\varrho v_{tt} - (k v_x)_x] \, dx \, dt -$$

$$- \frac{k}{2} \{h_2 [v^2(l, t) - v^2(l, 0)] + h_1 [v^2(0, t) - v^2(0, 0)]\}.$$

It follows from the differential equation and the initial conditions that

$$E(t) = -\frac{k}{2} [h_2 v^2(l, t) + h_1 v^2(0, t)] \leqslant 0. \tag{86}$$

Since the function under the integral in equation (79) cannot be negative $E(t) \geqslant 0$, hence

$$E(t) \equiv 0, \tag{87}$$

and consequently

$$v(x, t) \equiv 0. \tag{88}$$

Thus the uniqueness theorem is proved for the third boundary problem.

The method of proving the uniqueness theorem given here is based on properties of the total energy and is widely used in proving uniqueness theorems in various domains of mathematical physics, e. g. in the theory of electromagnetic fields, the theory of elasticity and hydrodynamics.

The proof of the uniqueness of other boundary problems will be given later when the respective problems are considered.

Problems

1. Prove that the equation of small torsional vibrations of a rod has the form

$$\Theta_{tt} = a^2 \Theta_{xx}, \quad a = \sqrt{\frac{GJ}{k}}, \tag{89}$$

where Θ is the angle of rotation of the section of the rod at the position x; G the shear modulus, and J the moment of inertia of unit length of the rod about its axis. Give a physical interpretation of the boundary conditions of the first, second and third types for this equation.

2. An absolutely flexible uniform thread is attached at one end and is in a vertical equilibrium position under the action of gravity. Derive the equation of small vibrations of the string.

Answer :

$$\frac{\partial^2 u}{\partial t^2} = a^2 \frac{\partial}{\partial x}\left[(l-x)\frac{\partial u}{\partial x}\right], \quad a^2 = g, \tag{90}$$

where $u(x, t)$ is the displacement of the point, l the length of the thread and g the acceleration due to gravity.

3. A heavy uniform thread of length l, attached by the upper end ($x = 0$) to the vertical axis, rotates about this axis with a constant angular velocity ω. Derive the equation for small vibrations of the thread about its vertical position of equilibrium.

Answer :

$$\frac{\partial^2 u}{\partial t^2} = a^2 \frac{\partial}{\partial x}\left[(l - x)\frac{\partial u}{\partial x}\right] + \omega^2 u, \text{ where } a^2 = g. \tag{91}$$

4. Derive the equation for transverse vibrations of a string in a medium of which the resistance is proportional to the first power of the velocity

$$v_{tt} = a^2 v_{xx} - h^2 v_t. \tag{92}$$

5. Derive the boundary conditions for the equation of longitudinal vibrations of an elastic rod (spring) in the case when the upper end of the rod is fixed and a load P is attached to the lower end if,

(a) the equilibrium position is taken to be the state of stress of the rod under the action of a stationary load P suspended from the lower end (statical extension);

(b) the equilibrium position is taken be the unstressed state of the rod (e. g. at the initial instant a support is removed from under the load and the rod begins to stretch).

6. Write down the equation and the boundary conditions determining the process of torsional vibrations of a rod with pulleys attached to both ends.

Answer : At $x = 0$, $x = l$ boundary conditions of the form

$$\Theta_{tt}(0, t) = a_1^2 \Theta_x(0, t), \qquad (93)$$

$$\Theta_{tt}(l, t) = -a_2^2 \Theta_x(l, t).$$

must be fulfilled.

7. At some point $x = x_0$ of the string $(0 \leqslant x \leqslant l)$ a load of mass M is suspended. Derive equations for the vibration of the system.

8. At the end $x = l$ of an elastic rod, attached elastically at the point $x = 0$, a load of mass M is suspended. Write down the equation and the boundary conditions for longitudinal vibrations of the rod, assuming that in addition an external force acts on it. Consider two cases:

(a) the force is distributed along the rod with a density $f(x, t)$;

(b) the force is concentrated at the point $x = x_0$ and is equal to $F_0(t)$

9. Consider the process of small vibrations of an ideal gas in a cylindrical tube. First derive the fundamental equations of hydrodynamics and then assuming the process to be adiabatic, derive the differentia equation for: (1) the density ϱ, (2) the pressure, p (3) the velocity potential φ, (4) the velocity v, (5) the displacement u of the particles. Give examples in which boundary conditions of the first, second and third types for these equations are realized.

10. Show the similarity between processes of mechanical, acoustic and electrical vibrations.

11. Give examples of boundary conditions of the first, second and third types for the telegraphic equations.

12. Consider the process of longitudinal vibrations of a non-homogeneous rod ($k = k_1$ at $x < x_0$, $k = k_2$ at $x > x_0$) and derive the boundary condition at the junction of the non-homogeneous portions of the rod (at $x = x_0$).

13. Give a physical interpretation of the boundary condition

$$\alpha u_x(0, t) + \beta u_t(0, t) = 0. \qquad (94)$$

14. Give an example of a mechanical model for which the equation is

$$u_{tt} = a^2 u_{xx} + b u_t + c u.$$

§ 2. Method of propagating waves

1. *D'Alembert's formula*

We begin a study of the methods of constructing solutions of boundary-value problems for equations of the hyperbolic type with a problem of an infinite string, with initial conditions

$$u_{tt} - a^2 u_{xx} = 0 . \tag{1}$$

$$\left. \begin{aligned} u(x, 0) &= \varphi(x) , \\ u_t(x, 0) &= \psi(x) . \end{aligned} \right\} \tag{2}$$

We transform this equation to the canonical form containing a mixed derivative. The equation of the characteristics

$$dx^2 - a^2 dt^2 = 0$$

breaks up into two equations

$$dx - a\,dt = 0, \quad dx + a\,dt = 0 ,$$

the integrals of which are the straight lines

$$x - at = C_1, \quad x + at = C_2 .$$

Let us introduce as usual the new variables

$$\xi = x + a', \quad \eta = x - at .$$

Calculating the derivatives

$$u_x = u_\xi + u_\eta. \qquad u_{xx} = u_{\xi\xi} + 2\,u_{\xi\eta} + u_{\eta\eta} ,$$

$$u_t = a\,(u_\xi - u_\eta), \quad u_{tt} = (u_{\xi\xi} - 2\,u_{\xi\eta} + u_{\eta\eta})\,a^2 ,$$

we see that the equation of vibrations of the string may be reduced to the form

$$u_{\xi\eta} = 0 . \tag{3}$$

Let us find the general integral of the last equation. Obviously for any solution of the equation (3)

$$u_\eta(\xi, \eta) = f^*(\eta) ,$$

where $f^*(\eta)$ is some function of the variable η only. Integrating this equation with respect to η for a fixed ξ we get

$$u(\xi, \eta) = \int f^*(\eta)\,d\eta + f_1(\xi) = f_1(\xi) + f_2(\eta) , \tag{4}$$

where f_1 and f_2 are functions of the variables ξ and η only. Conversely, for any differentiable functions f_1 and f_2 the function $u(\xi, \eta)$ determined by the formula (4) is a solution of the equation (3). Thus formula (4) is the general integral of equation (3), and the function

$$u(x, t) = f_1(x + at) + f_2(x - at) \tag{5}$$

is the general integral of equation (1).

Let us assume that a solution of the boundary-value problem exists. Then it is given by formula (5). We determine the functions f_1 and f_2 so that the initial conditions are satisfied:

$$u(x, 0) = f_1(x) + f_2(x) = \varphi(x), \tag{6}$$

$$u_t(x, 0) = af_1'(x) - af_2'(x) = \psi(x). \tag{7}$$

Integrating the second equation we get

$$f_1(x) - f_2(x) = \frac{1}{a}\int_{x_0}^{x} \psi(\alpha)\, d\alpha + C,$$

where x_0 and C are constants. From the equations

$$f_1(x) + f_2(x) = \varphi(x),$$

$$f_1(x) - f_2(x) = \frac{1}{a}\int_{x_0}^{x} \psi(\alpha)\, d\alpha + C$$

we find

$$\left.\begin{aligned} f_1(x) &= \frac{1}{2}\varphi(x) + \frac{1}{2a}\int_{x_0}^{x} \psi(\alpha)\, d\alpha + \frac{C}{2}, \\ f_2(x) &= \frac{1}{2}\varphi(x) - \frac{1}{2a}\int_{x_0}^{x} \psi(\alpha)\, d\alpha - \frac{C}{2}. \end{aligned}\right\} \tag{8}$$

Equations (8) hold for any value of the argument.*

* In formula (5) the functions of f_1 and f_2 are defined ambiguously. If a constant C_1 is substracted from f_1 and added to f_2, then u does not change. In formula (8) the constant C is not determined by φ and ψ but we can omit it without changing the value of u. On adding f_1 and f_2 the terms $\pm C/2$ cancel out.

Substituting the values found for f_1 and f_2 in (5) we get

$$u(x,t) = \frac{\varphi(x+at)+\varphi(x-at)}{2} + \frac{1}{2a}\left\{\int_{x_0}^{x+at}\psi(\alpha)\,d\alpha - \int_{x_0}^{x-at}\psi(\alpha)\,d\alpha\right\}$$

or

$$u(x,t) = \frac{\varphi(x+at)+\varphi(x-at)}{2} + \frac{1}{2a}\int_{x-at}^{x+at}\psi(\alpha)\,d\alpha. \tag{9}$$

We obtained formula (9), called *D'Alembert's* formula, by assuming the existence of a solution to the problem. This formula proves the uniqueness of the solution, for if a second solution of the problem (1)–(2) existed it would be represented by formula (9) and would coincide with the first solution.

It is easy to verify that formula (9) satisfies the equation and the initial conditions (assuming that the function φ can be differentiated twice and the function ψ once). Thus the method proves both the uniqueness and the existence of a solution of the problem.

2. *Physical interpretation*

The solution $u(x, t)$ is equal to the sum of the two functions:

$$\left.\begin{aligned} u_1(x,t) &= \frac{\varphi(x+at)+\varphi(x-at)}{2}, \\ u_2(x,t) &= \frac{1}{2a}\int_{x-at}^{x+at}\psi(\alpha)\,d\alpha. \end{aligned}\right\} \tag{10}$$

The first of these terms $u_1(x, t)$ represents the process of propagation of the initial displacement for zero initial velocity ($\psi(x) = 0$); the second term $u_2(x, t)$ corresponds to the case where vibrations are produced by the initial velocity (e. g. an impulse) with zero initial displacement. The function $u(x, t)$ can be represented geometrically as a surface in the u, x, t space (Fig. 4a). A cross-section of this surface defined by the plane $t = t_0$ and given by the formula $u = u(x, t_0)$ represents the profile of a string at the moment t_0. The section $u(x_0, t)$ represents the motion of the point x_0.

The function $u(x, t)$ defined by the formula $u = f(x - at)$ is called a propagating wave* in physics. The profiles of displacement, defined by this function at different times t, may be represented by the following method. Let us suppose that an observer moves parallel to the axis x with the velocity a (Fig. 4b).

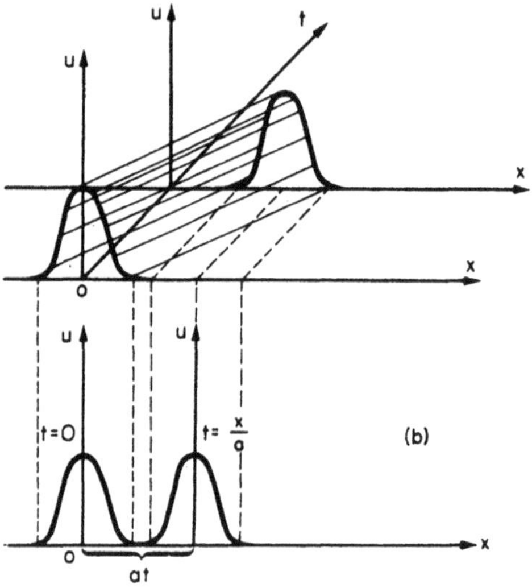

FIG. 4

If at the initial time $t_0 = 0$ the observer occupied a certain position $x = 0$, then at the time t he would have moved a distance at to the right. Let us introduce a moving system of axes moving with the observer, putting

$$x' = x - at\,,$$
$$t' = t\,.$$

Then in the new system of coordinates (x', t') the function $u = f(x - at)$ will be defined by the formula

$$u\,(x', t') = f\,(x')\,,$$

i. e. the observer will at all times see the unchanged profile $f(x')$ coinciding with the profile $f(x)$ which he saw at the initial moment.

* Another notation $u = f\left(t - \frac{x}{a}\right)$ is also frequently used in physics.

Thus the function $f(x - at)$ represents the unchanged profile $f(x')$ moving to the right the with velocity a (propagating wave).

If we consider the "phase space" (x, t) then the function $u = f(x - at)$ retains a constant value on the lines

$$x - at = \text{const}.$$

The surface $u = f(x - at)$ is a cylindrical surface with generators parallel to the line $x = at$.

Let the function $f(x)$ differ from zero only in the interval

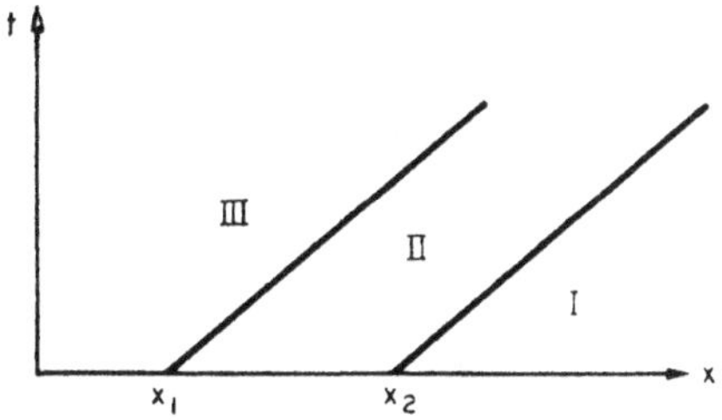

FIG. 5

(x_1, x_2). The lines $x - at = x_2$ and $x - at = x_1$ represent the forward and backward fronts of the propagating wave $f(x - at)$. These lines divide up the phase space (x, t) into the regions I, II and III.

The regions I and III consist of points (x, t) corresponding to those points of the string x, which at the given moment t lie respectively in front of and behind the wave propagating to the right (Fig. 5).

The function $f(x + at)$ represents a wave propagating to the left with velocity a. All that has been said regarding the wave propagating to the right applies to it also. The function $u_1(x_1, t)$, representing the propagation of the initial displacement $\varphi(x)$ for zero initial velocity ($\psi(x) = 0$), is given by formula (10) as a sum of two waves propagating to the right and to the left with a velocity a, the initial form of both the waves being defined by the function $\varphi(x)/2$ equal to half the initial displacement $\varphi(x)$.

As the first example we consider the propagation of an initial displacement, in the form of an isosceles triangle. This form is obtained if the string is plucked at the centre of the segment

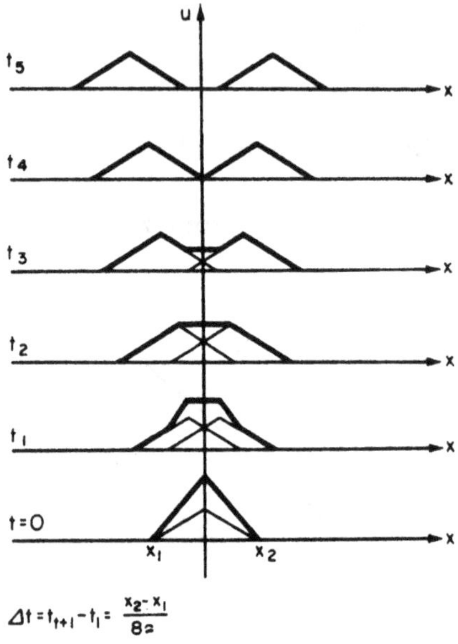

FIG. 6

(x_1, x_2) holding it at the points x_1 and x_2. On Fig. 6 we give the successive positions of this string after periods

$$\Delta t = \frac{x_2 - x_1}{8a}.$$

If the successive positions of the string after sufficiently small intervals of time are drawn, then a motion picture representing the propagation of the initial displacement may be formed.

If characteristic curves are drawn on the phase plane (x, t) through the ends of the segment $(P(x_1, 0), Q(x_2, 0))$, (Fig. 7) the plane is divided into the regions I, V, where at time t the displacement has not yet arrived, the region III which the displacements have already passed, and the regions II, IV, VI where the displacements are taking place.

As the second example we consider the case when the initial displacement is zero and the initial velocity differs from zero only in the segment (x_1, x_2) where it assumes the constant value

$$u(x, 0) = \varphi(x) = 0,$$

$$u_t(x, 0) = \varphi(x) = \begin{cases} \psi_0 & \text{where } x_1 \leqslant x \leqslant x_2, \\ 0 & \text{where } x > x_2 \text{ or } x < x_1. \end{cases}$$

In this case formula (9) becomes

$$u(x, t) = u_2(x, t) = \frac{1}{2a} \int_{x-at}^{x+at} \psi(a)\, da = [\Psi(x + at) - \Psi(x - at)].$$

The function $u(x, t)$ in this case also is given as a superposition of two waves. $\Psi(x)$ is the integral of $\psi(a)$ and represents the profile of the wave going to the left:

$$\Psi(x) = \frac{1}{2a} \int_{x_0}^{x} \psi(a)\, da.$$

It is convenient to choose $x_0 = x_1$. The auxiliary function $\Psi(x)$ is shown in Fig. 8.

$$\Psi(x) = \begin{cases} 0 & \text{where } x \leqslant x_1, \\ \frac{1}{2a}(x - x_1)\psi_0 & \text{where } x_1 \leqslant x \leqslant x_2, \\ \frac{1}{2a}(x_2 - x_1)\psi_0 & \text{where } x \geqslant x_2. \end{cases}$$

To obtain the function $u(x, t)$ we must take the difference of the waves propagating to the left and the right defined by the function $\Psi(x)$. Figure 9 shows the successive positions of these waves and their difference after a time interval $\Delta t = \frac{x_2 - x_1}{8a}$

For $t > \frac{x_2 - x_1}{2a}$ the profile of the displacement represents a trapezium extending uniformly with time. If $\psi(x)$ is not a constant, the main features of the problem remain unchanged. The phase plane (Fig. 7) shows the regions which the displacements have not yet reached (*I* and *V*), where these displacements have reached the maximum value (*III*) and where they are developing (*II*, *IV* and *VI*).

If the initial disturbance occurs in several segments the picture of propagation of vibrations may be obtained by adding the displacements corresponding to the effect of the different segments. These two examples enable us to picture the propagation of vibrations in the general case.

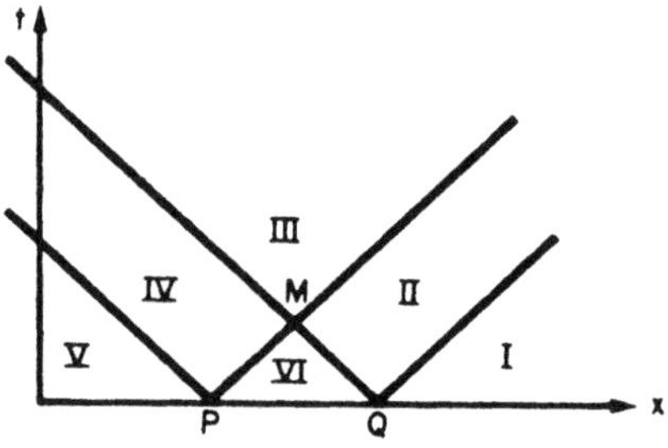

FIG. 7

As a third example we consider the problem of vibrations of a string set in motion by a concentrated impulse. By striking the string on the segment $(x, x + \Delta x)$ with some object (a hammer), we apply an impulse I to this segment equal to the change

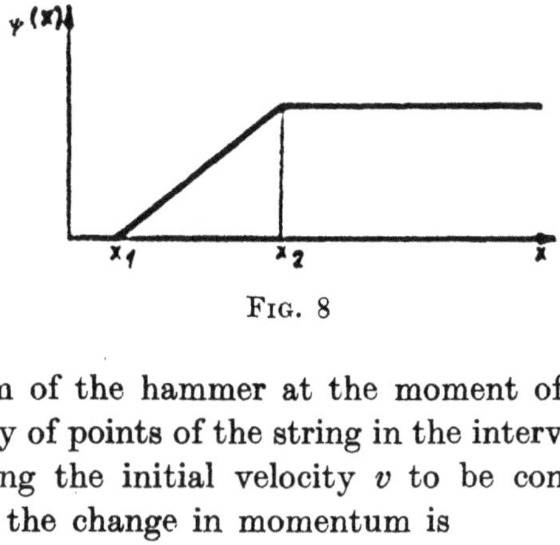

FIG. 8

in momentum of the hammer at the moment of striking. The initial velocity of points of the string in the interval Δx is equal to v. Assuming the initial velocity v to be constant on Δx, we find that the change in momentum is

$$\varrho \Delta x v = I,$$

where ϱ is the linear density of the string. Thus we must solve the problem of vibrations of a string with the initial velocity

$$\psi = \begin{cases} v = \dfrac{I}{\varrho \Delta x} & \text{inside } (x, x + \Delta x), \\ 0 & \text{outside } (x, x + \Delta x), \end{cases}$$

and zero initial displacement.

The displacement produced in this way is a trapezium whose lower side has length $2at + \Delta x$ and upper side a length $2at - \Delta x$ (for $t > \Delta x/2a$). The quantity $I/\Delta x = I_0$ can be called the impulse density. When $\Delta x \to 0$ we obtain the following picture: the displacements are equal to zero everywhere outside the interval $(x - at,\ x + at)$ and inside it are equal to $I/2a\,\varrho$. Conventionally one says that these displacements are produced by the point impulse I.

Let us consider the phase plane (x, t) and draw two characteristics

$$x - at = x_0 - at_0\,,$$

$$x + at = x_0 + at_0$$

through the point (x_0, t_0) (Fig. 10). They define the two angles a_1 and a_2, called respectively the upper and lower characteristic angles for the point $(x_0\, t_0)$.

The aetion of the point impulse at the point (x_0, t_0) produces a displacement equal to $I/2a\,\varrho$ inside the upper characteristic angle, and zero outside it.

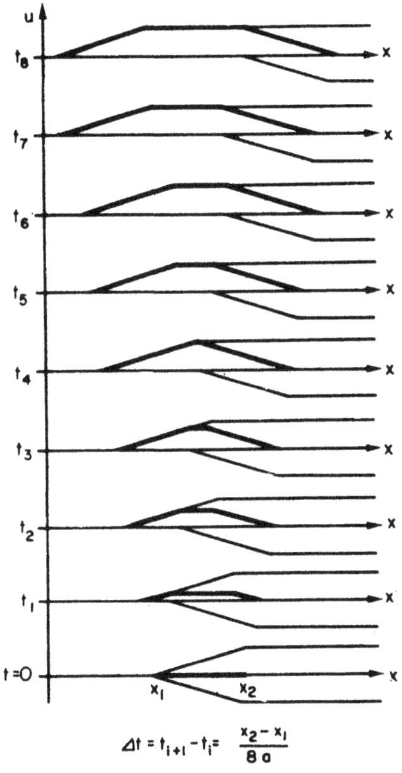

FIG. 9

Let us now deal with the question of the region in which the solution is uniquely defined by the initial conditions given in some segment PQ of the straight line $t = 0$.

Formula (9) shows that to determine the function u at some point $M(x, t)$ of the phase plane (x, t) (Fig. 7) it is sufficient to know the initial conditions in the segment PQ, where P, Q are points of the axis with the coordinates $x - at$ and $x + at$. The triangle MPQ, formed by the segments MP and MQ of the characteristics drawn from the point M and by the segment

PQ of the axis x, is called the characteristic triangle of the point M.

If the initial conditions are given only in a segment PQ,

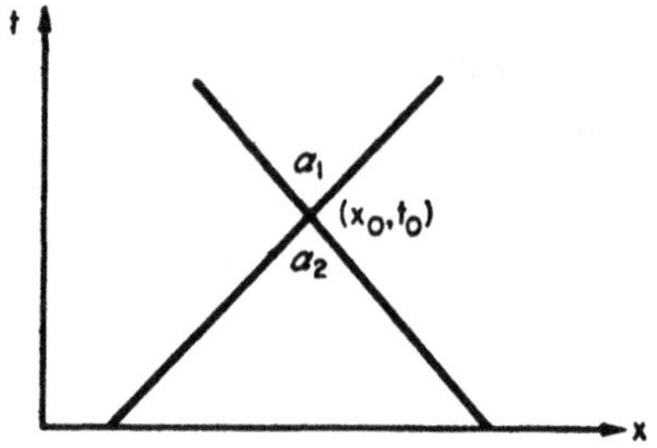

FIG. 10

but not outside it, these initial data define a unique solution inside the characteristic triangle based on the segment PQ.

3. *Continuity of solutions*

The solution of equation (1) is defined uniquely by the initial conditions (2). We shall prove that the solution changes continuously with a continuous variation of the initial conditions.

For any interval of time t_0 and any limiting error ε, a $\delta(\varepsilon, t_0)$ can be found such that two solutions of the equation (1) $u_1(x, t)$ *and* $u_2(x, t)$ *will differ by less than ε ; in the time interval t_0*

$$|u_1(x, t) - u_2(x, t| < \varepsilon \quad |(0 \leqslant t \leqslant t_0),$$

provided the initial values

$$\begin{cases} u_1(x, 0) = \varphi_1(x), \\ \dfrac{\partial u_1}{\partial t}(x, 0) = \psi_1(x), \end{cases} \quad \text{and} \quad \begin{cases} u_2(x, 0) = \varphi_2(x), \\ \dfrac{\partial u_2}{\partial t}(x, 0) = \psi_2(x) \end{cases}$$

differ by less than δ :

$$|\varphi_1(x) - \varphi_2(x)| < \delta; \quad \psi_1(x) - \psi_2(x)| < \delta. \qquad (11)$$

The proof of this theorem is simple. The functions $u_1(x, t)$ and

$u_2(x, t)$ are related to the initial values by the formula (9) so that

$$|u_1(x,t) - u_2(x,t| \leqslant \frac{|\varphi_1(x+at) - \varphi_2(x+at)|}{2} + $$
$$+ \frac{|\varphi_1(x-at) - \varphi_2(x-at)|}{2} + \int_{x-at}^{x+at} |\psi_1(\alpha) - \psi_2(\alpha)|\, d\alpha\,,$$

whence we obtain, because of the inequality (11):

$$|u_1(x,t) - u_2(x,t)| \leqslant \frac{\delta}{2} + \frac{\delta}{2} + \frac{1}{2a}\delta \cdot 2at \leqslant \delta(1+t_0)\,,$$

which proves our statement, if we put

$$\delta = \frac{\varepsilon}{1+t_0}\,.$$

Any physically defined process must be characterized by solutions depending continuously on the initial conditions. If this condition were not satisfied, two essentially different processes could correspond to almost identical initial conditions (the difference between which could lie outside the limits of accuracy of measurement). Processes of this type cannot be considered definable (physically) by such initial conditions. From the preceding theorem it follows that the process of vibration of a string is defined not only mathematically but also physically by the initial conditions.

If the solution of the boundary problem depends continuously on initial conditions, it is often said that the boundary problem has been formulated *correctly*.

We shall give an example of a problem which does not satisfy the above condition.

The function $u(x, y)$ which is a solution of Laplace's equation

$$u_{xx} + u_{yy} = 0,$$

is defined uniquely by the conditions*

$$u(x, 0) = \varphi(x), \quad u_y(x, 0) = \psi(x)\,.$$

* These conditions mathematically define the solution of Laplace's equation uniquely. Infact, giving $u_y(x, 0)$ is equivalent to giving $v_x(x, 0)$ where $v(x, y)$ is a function harmonically conjugate to $u(x, y)$. This defines uniquely the analytical function, the real part of which is the function $u(x, y)$, correct except for a constant (see Chapter IV, § 1, sub-section 5).

Let us consider the functions

$$u^{(1)}(x, y) \equiv 0 \text{ and } u^{(2)}(x, y) = \frac{1}{\lambda} \sin \lambda x \cdot \operatorname{ch} \lambda y,$$

satisfying Laplace's equation. The function $u^{(2)}(x, y)$ depends on λ as on a parameter. The initial values

$$u^{(1)}(x, 0) = 0, \quad u^{(2)}(x, 0) = \varphi(x) = \frac{1}{\lambda} \sin \lambda x,$$

$$u_y^{(1)}(x, 0) = 0, \quad u_y^{(2)}(x, 0) = \psi(x) = 0$$

differ as little as desired for sufficiently large values of λ. But the solution $u^{(2)}(x, y)$ may become as large as desired for any

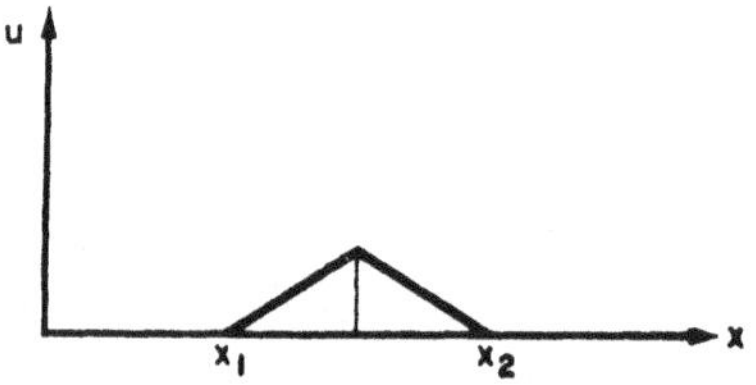

FIG. 11

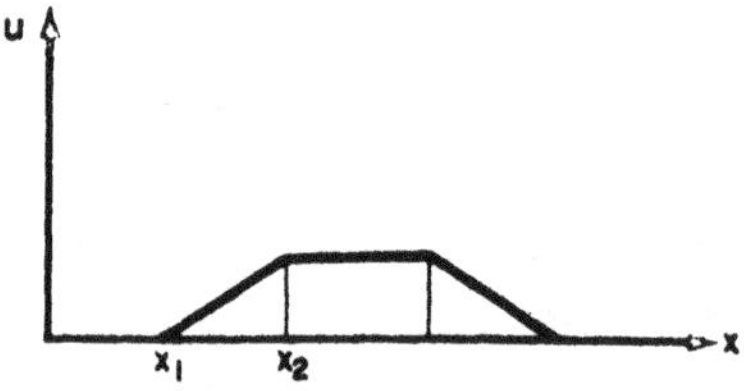

FIG. 12

fixed value of y. Hence the solution does not depend continuously on the additional conditions and the boundary-value problem has been formulated incorrectly.

We observe the following fact. It is obvious that the function $u(x, t)$ defined by the formula

$$u(x, t) = \frac{\varphi(x + at) + \varphi(x - at)}{2} + \frac{1}{2a} \int_{x-at}^{x+at} \psi(\alpha)\, d\alpha,$$

can be a solution of equation (1) only if the function $\psi(x)$ is differentiable and the function $\varphi(x)$ can be differentiated twice. It is clear from what has been said that the functions shown in Figs. 11 and 12 cannot be a solution of equation (1) because they cannot be differentiated twice everywhere. We can show that no solution of the wave equation satisfying the conditions (2) exists if the functions $\varphi(x)$ and $\psi(x)$ do not have the necessary derivatives. We know that if a solution of the wave equation exists it must be represented by formula (9). If the functions

φ, ψ are not differentiable a sufficient number of times then formula (9) defines a function which does not satisfy equation (1), i. e. no solution of the problem exists.

But if these initial conditions are slightly altered, replacing φ and ψ by differentiable functions $\varphi'(x)$ and $\psi'(x)$, then there will be a solution of equation (1) corresponding to these new initial functions. In the course of the proof of the theorem in this section we have proved that functions defined by formula (9) depend continuously on the initial functions φ and ψ (irrespective of whether these functions are differentiable or not). Thus if the functions φ and ψ are not differentiable a sufficient number of times the function defined by formula (9) is the limit of solutions of the wave equation with somewhat smoothened initial conditions.

4. *Semi-infinite region and the method of extending initial data*

Let us consider the problem of the propagation of waves in a semi-infinite region $x \geqslant 0$. This problem is of particular importance in the study of processes of reflection of waves from a boundary and is formulated as follows:

Find the solution of the wave equation

$$a^2 u_{xx} = u_{tt} \text{ where } 0 < x < \infty, t > 0,$$

satisfying the boundary condition

$$u(0, t) = \mu(t) \quad (\text{or } u_x(0, t) = \nu(t))$$

and the initial conditions

$$u(x, 0) = \varphi(x),$$
$$u_t(x, 0) = \psi(x).$$

We consider first the homogeneous boundary condition

$$u(0, t) = 0 \quad (\text{or } u_x(0, t) = 0),$$

i. e. the problem of the propagation of the initial disturbance on a string with a fixed end $x = 0$ (or free end).

Let us note the following two lemmas on the properties of the solution of the wave equation defined in an infinite region.

1. *If the initial data in the problem of the propagation of waves in an infinite region are odd functions with respect to some point x_0, then the corresponding solution at this point x_0 is zero.*

2. *If the initial data are even functions with respect to some point x_0, then the derivative with respect to x of the corresponding solution at this point is zero.*

We shall prove Lemma 1. Let us take x_0 as the origin of coordinates, $x_0 = 0$. In this case the conditions of oddness of the initial data are

$$\varphi(x) = -\varphi(-x); \quad \psi(x) = -\psi(-x).$$

The function $u(x, t)$ defined by formula (9), at $x = 0$ is

$$u(0, t) = \frac{\varphi(at) + \varphi(-at)}{2} + \frac{1}{2a}\int_{-at}^{at} \psi(\alpha)\, d\alpha = 0,$$

since the first term is equal to zero because of the oddness of $\varphi(x)$, and the second is equal to zero because the integral of an odd function between limits symmetrical with respect to the origin is always equal to zero.

Lemma 2 is proved similarly. The conditions of evenness of the initial data have the form

$$\varphi(x) = \varphi(-x); \quad \psi(x) = \psi(-x).$$

We note that the derivative of this even function is an odd function

$$\varphi'(x) = -\varphi'(-x).$$

From formula (9) it follows that

$$u_x(0, t) = \frac{\varphi'(at) + \varphi'(-at)}{2} + \frac{1}{2a}[\psi(at) - \psi(-at)] = 0,$$

since the first term is equal to zero because $\varphi'(x)$ is odd and the second term because $\psi(x)$* is even.

With the help of these two lemmas the following problems can be solved.

* These two lemmas are a consequence of the fact that if the initial conditions are even (or odd), then for $t > 0$ the function $u(x, t)$, given by D'Alembert's formula, possesses the same property. (We leave this to the reader to prove.) Geometrically it is obvious that the odd continuous function and the derivative of the even differentiable function are equal to zero for $x = 0$.

It is required to find the solution of the equation (1) satisfying the initial conditions

$$\left.\begin{aligned} u(x,0) &= \varphi(x), \\ u_t(x,0) &= \psi(x) \end{aligned}\right\} \quad 0 < x < \infty, \tag{2}$$

and the boundary condition

$$u(0,t) = 0$$

(first boundary problem).

Let us consider the functions $\Phi(x)$ and $\Psi(x)$ which are odd continuations of $\varphi(x)$ and $\psi(x)$.

$$\Phi(x) = \begin{cases} \varphi(x) & \text{where } x > 0, \\ -\varphi(-x) & \text{where } x < 0, \end{cases}$$

$$\Psi(x) = \begin{cases} \psi(x) & \text{where } x > 0, \\ -\psi(-x) & \text{where } x < 0. \end{cases}$$

The function

$$u(x,t) = \frac{\Phi(x+at) + \Phi(x-at)}{2} + \frac{1}{2a}\int\limits_{x-at}^{x+at} \Psi(\alpha)\, d\alpha$$

is determined for all x and $t > 0$. Because of Lemma 1

$$u(0,t) = 0.$$

Moreover this function satisfies the following initial conditions at $t = 0$:

$$\left.\begin{aligned} u(x,0) &= \Phi(x) = \varphi(x), \\ u_t(x,0) &= \Psi(x) = \psi(x) \end{aligned}\right\} \quad x > 0.$$

Thus, considering the function obtained $u(x,t)$ only for $x > 0$, $t > 0$ we obtain a function satisfying all conditions of the problem.

Returning to the former functions, we can write

$$u(x\ t) = \begin{cases} \dfrac{\varphi(x+at) + \varphi(x-at)}{2} + \dfrac{1}{2a}\displaystyle\int\limits_{x-at}^{x+at} \psi(\alpha)\, d\alpha & \text{where } t < \dfrac{x}{a},\ x > 0, \\[2ex] \dfrac{\varphi(x+at) - \varphi(at-x)}{2} + \dfrac{1}{2a}\displaystyle\int\limits_{at-x}^{x+at} \psi(\alpha)\, d\alpha & \text{where } t > \dfrac{x}{a},\ x > 0. \end{cases} \tag{12}$$

In the region $t < x/a$ the effect of the boundary conditions is not felt and the expression for $u(x, t)$ coincides with solution (9) for an infinite straight line.

Similarly, if at $x = 0$ we have a free end

$$u_x(0, t) = 0,$$

then, taking the even extension of the functions $\varphi(x)$ and $\psi(x)$

$$\Phi(x) = \begin{cases} \varphi(x) & \text{where } x > 0, \\ \varphi(-x) & \text{where } x < 0; \end{cases}$$

$$\Psi(x) = \begin{cases} \psi(x) & \text{where } x > 0, \\ \psi(-x) & \text{where } x < 0, \end{cases}$$

we get the solution of the wave equation

$$u(x, t) = \frac{\Phi(x + at) + \Phi(x - at)}{2} + \frac{1}{2a} \int_{x-at}^{x+at} \Psi(\alpha)\, d\alpha$$

or

$$u(x, t) = \begin{cases} \dfrac{\varphi(x + at) + \varphi(x - at)}{2} + \dfrac{1}{2a} \displaystyle\int_{x-at}^{x+at} \psi(\alpha)\, d\alpha & \text{where } t < \dfrac{x}{a}, \\ \dfrac{\varphi(x + at) + \varphi(at - x)}{2} + \\ \quad + \dfrac{1}{2a} \left\{ \displaystyle\int_0^{x+at} \psi(\alpha)\, d\alpha + \int_0^{at-x} \psi(\alpha)\, d\alpha \right\} & \text{where } t > \dfrac{x}{a}, \end{cases}$$

satisfying the initial conditions (2) in the region $x > 0$ and the boundary condition

$$u_x(0, t) = 0.$$

Later, in the solution of other problems we shall often have to use the method of extension of the initial data defined for some part of an infinite region to the whole region. For this reason we state the following results.

To solve the problem for a semi-infinite region with the boundary condition $u(0, t) = 0$ the initial data have to be extended as an odd function on the entire region.

To solve the problem for a semi-infinite region with the boundary condition $u_x(0, t) = 0$ the initial data have to be extended as an even function on the entire region.

We consider two examples. Let the initial data in a semi-

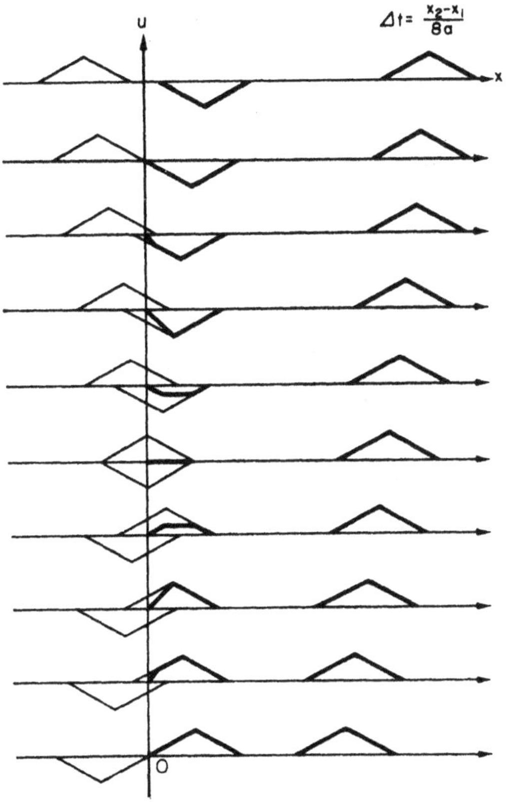

FIG. 13

infinite region, fixed at $x = 0$, differ from zero only in the interval (a, b), $0 < a < b$ in which the initial displacement given by the function $\varphi(x)$ is represented by an isosceles triangle and $\psi(x) = 0$. The solution of this problem will be obtained if the initial data are extended oddly to the infinite straight line. The process of wave propagation is shown in Fig. 13. In the beginning the process occurs as in an infinite region. The initial displacement breaks up into two waves moving in opposite directions with constant velocity and this continues till the half-wave moving left reaches the point $x = 0$ (Fig. 13). At this moment a half-wave with "inverse phase" approaches from the left side ($x < 0$) (in which similar processes occurred) towards $x = 0$. Then reflection of the half-waves from the fixed end

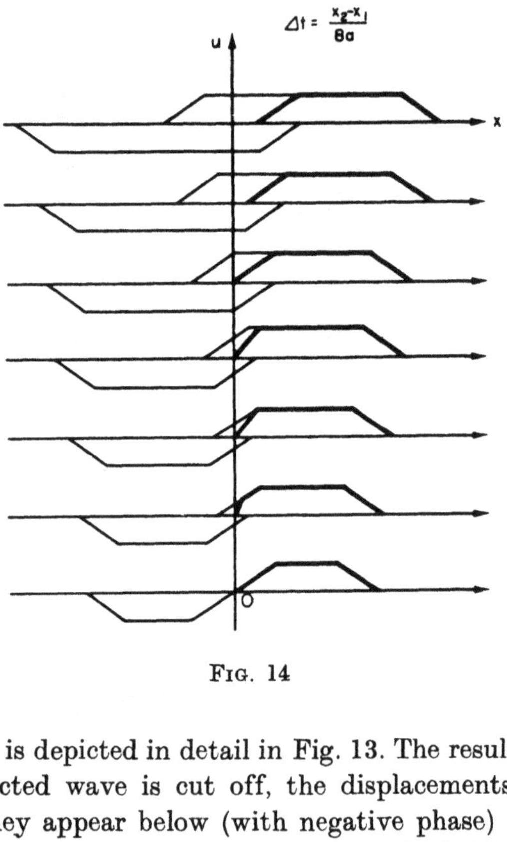

FIG. 14

occurs. This is depicted in detail in Fig. 13. The resulting profile of the reflected wave is cut off, the displacements disappear and then they appear below (with negative phase) and finally the reflected half-wave will move to the right behind the half-wave already propagating there with the same velocity. Thus during the reflection of a wave from a fixed end of a string its phase changes sign.

Let us now consider the second example. Let the initial deflection on a semi-infinite string $x \geqslant 0$, fixed at $x = 0$, be zero everywhere, and the initial velocity $\psi(x)$ be different from zero only in the interval (x_1, x_2) $(0 < x_1 < x_2)$. $\psi(x) =$ const. there.

Let us extend the initial data oddly. Displacements similar to those shown in Fig. 14 propagate from each interval (x_1, x_2) and $(-x_1, -x_2)$. As can be seen from this figure, in the initial stage in the region $x > 0$ the process occurs as on an infinite string. Reflection from the fixed end occurs next and finally a wave with a profile in the shape of an isosceles trapezium moves to the right with constant velocity.

Reflection from a free end can be studied similarly, only the initial data have to be extended evenly, so that the reflection of the wave occurs with no change of phase.

We have considered problems with uniform boundary conditions

$$u(0, t) = \mu(t) = 0$$

or

$$u_x(0, t) = \nu(t) = 0 .$$

In the general case of non-uniform boundary conditions the solution can be written in the form of a sum, each term of which satisfies only one of the additional conditions (either boundary or initial).

Now we solve a problem with zero initial conditions and the given boundary conditions

$$\bar{u}(x, 0) = 0, \quad \bar{u}_t(x, 0) = 0, \quad \bar{u}(0, t) = \mu(t), \quad t > 0 .$$

Obviously the boundary condition gives rise to a wave propagating along the string towards the right with velocity a, which suggests the analytical form of the solution

$$\bar{u}(x, t) = f(x - at) .$$

We determine the function f from the boundary condition

$$\bar{u}(0, t) = f(-at) = \mu(t) ,$$

whence

$$f(z) = \mu\left(-\frac{z}{a}\right) .$$

so that

$$\bar{u}(x, t) = \mu\left(-\frac{x - at}{a}\right) = \mu\left(t - \frac{x}{a}\right) .$$

But this function is defined only in the region $x - at \leq 0$ since $\mu(t)$ is defined only for $t \geqslant 0$. In Fig. 15 this region is shown by the shaded portion of the phase plane. To find $\bar{u}(x, t)$ for all

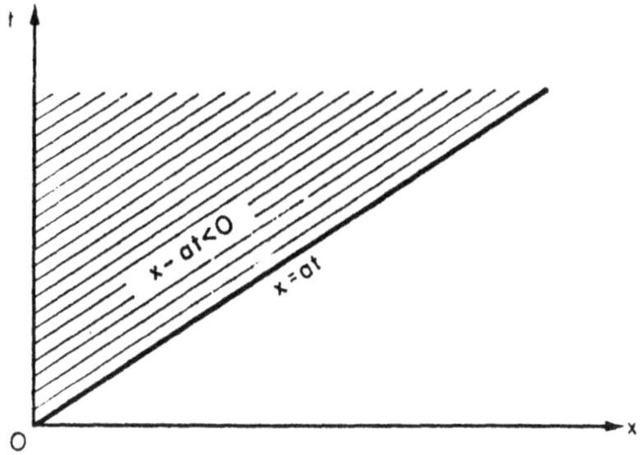

FIG. 15

values of the arguments we extend the function $\mu(t)$ to negative values of t, putting $\mu(t) = 0$ for $t < 0$. Then the function

$$\bar{u}(x, t) = \mu(t - x/a)$$

is defined for all values of the arguments and satisfies the zero initial conditions.

The sum of this function and the function (12) defined at the beginning of this section represents the solution of the first boundary problem for the homogeneous wave equation

$$u(x,t) = \begin{cases} \dfrac{\varphi(x+at) + \varphi(x-at)}{2} + \\ \quad + \dfrac{1}{2a} \displaystyle\int_{x-at}^{x+at} \psi(a)\, da \quad \text{where } t < \dfrac{x}{a}, \\ \mu\left(t - \dfrac{x}{a}\right) + \dfrac{\varphi(x+at) - \varphi(at-x)}{2} + \\ \quad + \dfrac{1}{2a} \displaystyle\int_{at-x}^{x+at} \psi(a)\, da \quad \text{where } t > \dfrac{x}{a}. \end{cases} \tag{13}$$

The solution of the second boundary problem may be formed similarly. For the third boundary problem see sub-section 7, p. 73.

We have considered here only solutions of the boundary problem for the homogeneous wave equation. For the solution of the non-homogeneous equation see sub-section 8.

5. *Problems for a finite segment*

Let us consider the boundary problems for a finite segment $(0, l)$. We shall try to find the solution of the equation

$$u_{tt} = a^2 u_{xx},$$

satisfying the boundary conditions

$$\left.\begin{aligned} u(0, t) &= \mu_1(t), \\ u(l, t) &= \mu_2(t) \end{aligned}\right\} \quad t \geq 0$$

and the initial conditions

$$\left.\begin{aligned} u(x, 0) &= \varphi(x), \\ u_t(x, 0) &= \psi(x) \end{aligned}\right\} \quad 0 \leq x \leq l.$$

Let us consider first the case of homogeneous boundary conditions

$$u(0, t) = u(l, t) = 0.$$

We shall investigate the solution of this problem by assuming the possibility of the following representation:

$$u(x, t) = \frac{\Phi(x + at) + \Phi(x - at)}{2} + \frac{1}{2a} \int_{x-at}^{x+at} \Psi(\alpha)\, d\alpha,$$

where Φ and Ψ are functions to be determined. The initial conditions

$$\left.\begin{aligned} u(x, 0) &= \Phi(x) = \varphi(x), \\ u_t(x, 0) &= \Psi(x) = \psi(x) \end{aligned}\right\} \quad 0 \leq x \leq l$$

define the values of Φ and Ψ in the interval $(0, l)$.

To satisfy the zero boundary conditions, we impose the condition of oddness with respect to the points $x = 0$, $x = l$ on the functions $\Phi(x)$ and $\Psi(x)$:

$$\Phi(x) = -\Phi(-x), \quad \Phi(x) = -\Phi(2l - x),$$

$$\Psi(x) = -\Psi(-x), \quad \Psi(x) = -\Psi(2l - x).$$

Comparing these equations we get

$$\Phi(x') = \Phi(x' + 2l) \qquad (x' = -x)$$

and similarly for $\Psi(x)$, i. e. Φ and Ψ are periodic functions with a period $2l$.

It is not difficult to see that the conditions of oddness with respect to the origin of coordinates and the conditions of period-

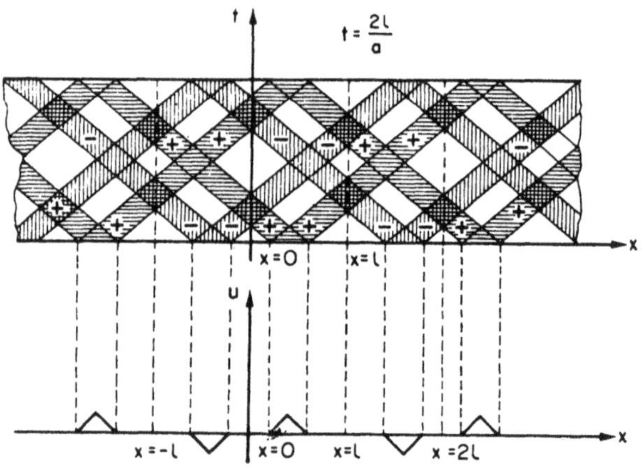

FIG. 16

icity define the extension of $\Phi(x)$ and $\Psi(x)$ in the entire region $-\infty < x < \infty$. Substituting these in formula (9) we obtain the solution of the problem.

In Fig. 16 the phase plane (x, t) and the plane (x, u), in which the initial displacement and its extension are given, are shown together. Bands within which the displacement differs from zero (see Fig. 7) are marked off by shading on the phase plane. The signs plus and minus in these bands indicate the sign (phase) of the displacement (in the form of an isosceles triangle). Using this diagram the profile of the displacement of the string at any moment t can be visualized easily. Thus at the moment $t = 2l/a$ we shall obtain displacements coinciding with the initial displacements. The function $u(x, t)$ is thus a periodic function t with the period $T = 2l/a$.

Let us now consider the problem of the propagation of the boundary condition. We shall investigate the solution of the equation

$$u_{tt} = a^2 u_{xx}$$

with the zero initial conditions

$$u(x, 0) = \varphi(x) = 0,$$
$$u_t(x, 0) = \psi(x) = 0$$

and the boundary conditions

$$\left.\begin{aligned} u(0, t) &= \mu(t), \\ u(l, t, &= 0 \end{aligned}\right\} t > 0.$$

From the results of sub-section 4 it follows that when $t < l/a$, the function

$$u(x, t) = \bar{\mu}\left(t - \frac{x}{a}\right), \quad \text{where} \quad \bar{\mu}(t) = \begin{cases} \mu(t), & t > 0, \\ 0, & t < 0. \end{cases}$$

is the solution.

But this function does not satisfy the boundary condition

$$u(l, t) = 0 \text{ when } t > l/a.$$

Let us consider the "reflected" wave going left and having the deflection $\bar{\mu}(t - l/a)$ at $x = l$. Its analytical expression is given by the formula

$$\bar{\mu}\left(t - \frac{l}{a} - \frac{l - x}{a}\right) = \mu\left(t - \frac{2l}{a} + \frac{x}{a}\right).$$

It is easy to verify that the difference of the two waves

$$\bar{\mu}\left(t - \frac{x}{a}\right) - \bar{\mu}\left(t - \frac{2l}{a} + \frac{x}{a}\right)$$

is the solution of the equation for $t < 2l/a$.

Continuing this process further we obtain the solution in the form of a series

$$u(x, t) = \sum_{n=0}^{\infty} \bar{\mu}\left(t - \frac{2nl}{a} - \frac{x}{a}\right) - \sum_{n=1}^{\infty} \bar{\mu}\left(t - \frac{2nl}{a} + \frac{x}{a}\right), \qquad (14)$$

containing only a finite number of terms different from zero. With each new reflection the argument is reduced by $2l/a$, the function $\bar{\mu}(t) = 0$ for $t < 0$. It can be proved directly that the boundary conditions are fulfilled. Let us put $x = 0$ and separate from the first sum the term with $n = 0$, $\mu(t)$. The remaining terms of the first and second sums, corresponding to identical values of n, cancel in pairs. This shows that $u(0, t) = \mu(t)$.

Replacing n, by $n - 1$ and changing the limits of summation we convert the first sum to the form

$$\sum_{n=1}^{\infty} \bar{\mu}\left(t - \frac{2\,nl}{a} + \frac{2l - x}{a}\right).$$

Now putting $x = l$, we verify directly that the terms of the first and second sums cancel.*

Formula (14) has a simple physical significance. The function

$$\bar{\mu}\left(t - \frac{x}{a}\right)$$

represents a wave excited by the boundary condition at $x = 0$, independently of the effect of the end $x = l$, as if the string were infinite $(0 < x < \infty)$. The following terms represent subsequent reflection from the fixed end $x = l$ (second sum) and from the end $x = 0$ (first sum).

Similarly, the function

$$u(x, t) = \sum_{n=0}^{\infty} \bar{\mu}\left(t - \frac{(2\,n+1)\,l}{a} + \frac{x}{a}\right) - \sum_{n=0}^{\infty} \bar{\mu}\left(t - \frac{(2\,n+1)\,l}{a} - \frac{x}{a}\right)$$

gives the solution of the homogeneous equation with the zero initial conditions $u(x, 0) = 0$, $u_t(x, 0) = 0$ and the boundary conditions $u(0, t) = 0$ and $u(l, t) = \mu(t)$. We shall not dwell on the proofs of the uniqueness of the solution and of the continuous dependence of the solution on the initial and the boundary conditions.

6. *Dispersion of waves*

We have seen that the equation

$$u_{tt} = a^2\, u_{xx}$$

* The initial conditions can also be verified directly since the arguments of all the functions are negative for $t = 0$.

has a solution in the form of a propagating wave of any shape. Let us investigate the class of equations with partial derivatives having solutions in the form of waves of any shape. We confine ourselves to a consideration of the class of linear equations of the second order with constant coefficients

$$a_{11} u_{xx} + 2 a_{12} u_{xt} + a_{22} u_{tt} + b_1 u_x + b_2 u_t + cu = 0 . \quad (15)$$

Our problem is to determine relations between the coefficients which must be satisfied if the equation has a solution of the type

$$u(x,t) = f(x - at), \quad (16)$$

where f is an arbitrary function and a a constant.

Substituting (16) in (15) we get the differential equation

$$f''(x - at)[a_{11} - 2a_{12}a + a_{22}a^2] + f'(x - at)[b_1 - b_2 a] + \\ + cf(x - at) = 0,$$

which the profile of the wave must satisfy. It has a solution for an arbitrary f only in the case when all its coefficients are equal to zero.

$$\left.\begin{aligned} a_{11} - 2a_{12}a + a_{22}a^2 &= 0, \\ b_1 - ab_2 &= 0, \\ c &= 0. \end{aligned}\right\} \quad (17)$$

If the equation has a solution in the form of a wave of arbitrary shape, we say that dispersion is absent. For absence of dispersion it is necessary and sufficient that the conditions (17) should be satisfied.

The wave velocity

$$a = \frac{a_{12} \pm \sqrt{(a_{12}^2 - a_{11}a_{22})}}{a_{22}}$$

is determined from the first relation.

For an equation of the hyperbolic type ($a_{12}^2 - a_{11}a_{22} > 0$) two velocities of wave propagation are possible. The requirement of compatability of all three relations for both values of a gives

$$b_1 = b_2 = c = 0 .$$

Hence a solution in the form of propagating waves with two possible velocities exists only for an equation of the type

$$a_{11} u_{xx} + 2 a_{12} u_{xt} + a_{22} u_{tt} = 0 . \quad (18)$$

If $a_{22} \neq 0$, equation (18) is the equation for vibrations of a string in a moving system of coordinates. Putting

$$\xi = x - \gamma t,$$
$$\eta = t,$$

we get the equation

$$(a_{11} - 2a_{12}\gamma + a_{22}\gamma^2)u_{\xi\xi} + (2a_{12} - 2a_{22}\gamma)u_{\xi\eta} + a_{22}u_{\eta\eta} = 0,$$

which coincides with the equation for vibrations of a string if $\gamma = a_{12}/a_{22}$.

In this case we have

$$u_{\eta\eta} = a^2 u_{\xi\xi},$$

where

$$a^2 = \frac{a_{12}^2 - a_{11}a_{22}}{a_{22}^2} > 0.$$

Equations of the elliptical type ($a_{12}^2 - a_{11}a_{22} < 0$) do not admit solutions in the form of waves with real velocities.

For equations of the parabolic type ($a_{12} - a_{11}a_{22} = 0$) solutions in the form of waves with real velocities are also impossible. The degenerate case is an exception, when equation (15) reduces to an ordinary differential equation.

In the parabolic case we have $a_{12} = \sqrt{a_{11}}\sqrt{a_{22}}$ and $a = \frac{a_{12}}{a_{22}} = \sqrt{\frac{a_{11}}{a_{22}}}$, $c = 0$, and also the equation $a = \frac{b_1}{b_2} = \sqrt{\frac{a_{11}}{a_{22}}}$, we can write

$$b_1 u_x + b_2 u_t = \frac{b_2}{\sqrt{a_{22}}}\left(\sqrt{[a_{11}]}\, u_x + \sqrt{[a_{22}]}\, u_t\right).$$

The original equation may be written in the form

$$\left(\sqrt{[a_{11}]}\frac{\partial}{\partial x} + \sqrt{[a_{22}]}\frac{\partial}{\partial t}\right)\left(\sqrt{[a_{11}]}\frac{\partial}{\partial x} + [\sqrt{a_{22}}]\frac{\partial}{\partial t}\right)u +$$
$$+ \frac{b_2}{\sqrt{a_{22}}}\left(\sqrt{[a_{11}]}\frac{\partial}{\partial x} + \sqrt{[a_{22}]}\frac{\partial}{\partial t}\right)u = 0$$

and it becomes an ordinary differential equation

$$\frac{d^2u}{d\xi^2} + b\frac{du}{d\xi} = 0 \qquad \left(b = \frac{b_2}{\sqrt{a_{22}}}\right),$$

if the new variables ξ and η are introduced

$$x = \sqrt{(a_{11})}\,\xi\,,$$
$$t = \sqrt{(a_{22})}\,\xi + \eta\,,$$

since

$$\frac{d}{d\xi} = \sqrt{(a_{11})}\,\frac{\partial}{\partial x} + \sqrt{(a_{22})}\,\frac{\partial}{\partial t}\,.$$

Thus the initial equation in this case has a solution in the form of an arbitrary function of the variable

$$\eta = t - \frac{\sqrt{a_{22}}}{\sqrt{a_{11}}}\,x = t - \frac{x}{a}\,.$$

Thus a propagating wave solution is possible only for the simplest wave equation in a fixed or moving system of coordinates.

In physics the concept of dispersion of waves is usually introduced omewhat differently.

Consider a harmonic wave

$$u\,(x, t) = e^{i(\omega t - kx)}, \qquad (*)$$

where ω is the frequency, $k = 2\,\pi/\lambda$ the wave number and λ the wavelength.

The velocity with which the phase of the wave

$$\alpha = \omega t - kx$$

is propagated is called the phase velocity and is obviously equal to

$$a = \frac{\omega}{k}\,.$$

If the phase velocity of the harmonic wave depends on the frequency we speak of dispersion.

An impulse or signal of arbitrary form may be formed by the superposition of harmonic waves of the type (*) (by resolution into Fourier's integral). If the phase velocity depends on the frequency, the harmonic components of the signal are displaced with respect to one another, and as a result the profile of the signal is distorted.

This means that dispersion in the sense of the definition given on p. 69 also takes place.

Conversely, if the equation has a solution in the form of waves of arbitrary shape, the phase velocity of the harmonic wave is determined from the first equation of (17) and does not depend on the frequency.

Thus, the term "dispersion of waves" in the sense used by us coincides with the condition of dependence of velocity on frequency.

Let us take the problem of finding the class of equations (15) having a solution in the form of damped waves

$$u(x, t) = \mu(t) f(x - at),$$

where $\mu(t)$ is some function of t.

Substituting this expression in equation (15) we get:

$$f'' \mu(t)(a_{11} - 2a_{12}a + a_{22}a^2) + f'[(b_1 - b_2 a)\mu + \\ + 2(a_{12} - a_{22}a)\mu'] + f(c\mu + b_2\mu' + a_{22}\mu'') = 0$$

Since f is arbitrary the coefficients of all terms are equal to zero. That is, $\mu(t)$ satisfies an ordinary differential equation with constant coefficients and must have the form

$$\mu = e^{-kt}.$$

Equating the coefficients to zero we obtain the relations for the determination of a and k:

$$\left.\begin{aligned} a_{11} - 2a_{12}a + a_{22}a^2 &= 0, \\ (b_1 - b_2 a) - 2k(a_{12} - a_{22}a) &= 0, \\ a_{22}k^2 - b_2 k + c &= 0. \end{aligned}\right\} \qquad (19)$$

Eliminating a and k from equations (19) we find conditions of compatibility. The first equation shows that solutions in the form of damped waves are possible only for an equation of the hyperbolic type. The damping decrement k is found from the second relation. Substituting next the value of k in the third equation we get a relation between the coefficients

$$4(a_{12}^2 - a_{11}a_{22})c + (a_{11}b_2^2 - 2a_{12}b_1b_2 + a_{22}b_1^2) = 0. \qquad (20)$$

When this is satisfied a solution of the equation in the form of damped waves exists.

Example. The telegraphic equation

$$u_{xx} = CLu_{tt} + (CR + LG)u_t + GRu \qquad (21)$$

does not have a solution in the form of a propagating wave, if only G or R differ from zero. Let us see if it has any solution in the form of a damped wave The velocity of damped waves is determined from the first equation of (19)

$$1 - a^2 CL = 0.$$

From the second equation we determine the damping decrement

$$k = \frac{CR + LG}{2CL}.$$

The compatibility condition (19) gives

$$4\,CLGR - (CR + LG)^2 = -\,(CR - LG)^2 = 0$$

or

$$CR = LG,$$

When this is fulfilled equation (15) has a solution in the form of a damped wave

$$u(x, t) = e^{-kt} f(x - at), \quad k = \frac{R}{L} = \frac{G}{C}, \qquad a = \sqrt{\frac{1}{CL}},$$

where f is an arbitrary function.

The absence of distortion of waves during propagation along a cable is of particular importance for long-distance telephone communication. If the signal is transmitted without distortion, then the acoustic effect to be transmitted can be reproduced without distortion by means of amplification. If dispersion takes place then the purity of transmission breaks down independently of the quality of the telephone apparatus. The question of dispersion for long-distance telegraphic communication has a similar significance.

7. *The integral wave equation*

In deducing the differential wave equation (5) in § 1 we started from the law of conservation of momentum which led us to the wave equation in the integral form (3). To pass from the integral equation to the differential we assumed that the function $u(x, t)$ has second derivatives. Any assumption restricting the class of functions considered to those with a certain property, implies a refusal to study functions not possessing this property. Thus in passing from the integral wave equation to the differential equation we exclude from consideration vibrations not satisfying the requirement of double differentiability.

We shall show that the whole theory can be developed for the class of continuous piecewise differentiable functions starting from the integral wave equation

$$\int_{x_1}^{x_2} \left[\left(\frac{\partial u}{\partial t} \right)_{t_2} - \left(\frac{\partial u}{\partial t} \right)_{t_1} \right] \varrho \, d\xi =$$

$$= \int_{t_1}^{t_2} \left[\left(k \frac{\partial u}{\partial x} \right)_{x_2} - \left(k \frac{\partial u}{\partial x} \right)_{x_1} \right] d\tau + \int_{x_1}^{x_2} \int_{t_1}^{t_2} F \, d\xi \, d\tau. \qquad (22)$$

This equation may be generalized. Let us consider the region G bounded by the piecewise smooth curve C in the plane (x, t) and show that for this region there is an integral relation

$$\int_C \left(\varrho \frac{\partial u}{\partial t} dx + k \frac{\partial u}{\partial x} dt \right) + \int_G \int F \, dx \, dt = 0. \qquad (23)$$

For a homogeneous region this formula reduces to

$$\int_C \left(\frac{\partial u}{\partial t} dx + a^2 \frac{\partial u}{\partial x} dt \right) + \int_G \int f \, dx \, dt = 0 \left(f = \frac{F}{\varrho} \right), \qquad (23')$$

If the curve C is the boundary of a rectangle with sides parallel to the coordinate axes, formula (23) concides with formula (22). If the curve C consists of pieces parallel to the axes, the region G can be represented as the sum of rectangles. Summing up the contour integrals corresponding to the different rectangles we find that terms relating to the internal boundaries cancel out, because integration is carried out in opposite directions and the remaining terms will give formula (23). Further, let the curve C contain the arcs $\overline{C}$ which are not parallel to the axes and are not lines of discontinuity of the function under the integral. We can take a network with sides parallel to the coordinate axes and consider cells of the network intersecting the region G. Let G^* denote the ensemble of these cells and C^* the boundary of the region G. Formula (23) is applicable to G^*. Passing to the limit as the dimensions of the network tend to zero, it is not difficult to verify the validity of formula (23) for the limiting curve C.

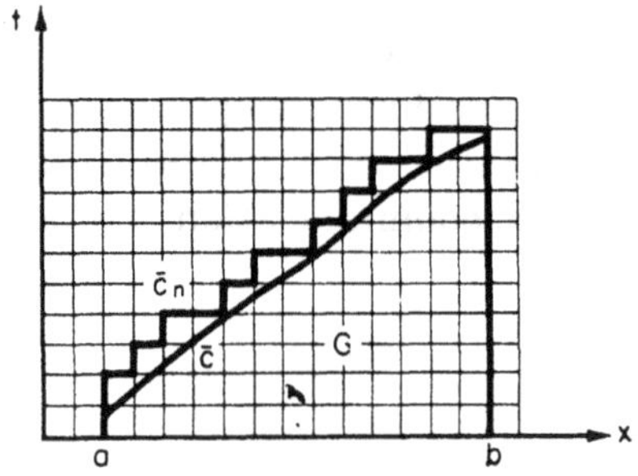

FIG. 17

The first terms of formula (23) applied to the region G^* consist of terms of the type

$$\int_{\bar{C}_n} \Phi(x, t)\, dx \text{ or } \int_{\bar{C}_n} \Phi(x, t)\, dt,$$

where $\Phi(x, t)$ is a continuous function and $\bar{C}_n$ is an arc of contour C^* approximating the arc $\bar{C}$ (Fig. 17).

Let $t = t_n(x)$ be the equation of the curve $\bar{C}_n$ and $t = t(x)$ is the equation of the curve $\bar{C}$. It is obvious that $t_n(x)$ converges uniformly to $t(x)$ and

$$\lim_{n\to\infty} \int_a^b \Phi[x, t_n(x)]\, dx = \int_a^b \Phi[x, t(x)]\, dx,$$

which proves the validity of the limiting transition.†

If the curve C contains arcs which are lines of discontinuity of the function under the integral, formula (23) retains its validity if we take for the values of the function under the integral its limiting values from the internal side of the region G. Thus the validity of the integral formula (23) is proved.

Let us consider the following problem:

Find the function $u(x,t)$ defined and piecewise smooth in the region $-\infty < x < \infty$ satisfying the equation

$$\int_C \left(\frac{\partial u}{\partial t}\, dx + a^2 \frac{\partial u}{\partial x}\, dt\right) + \int_G\int f(x, t)\, dx\, dt = 0 \tag{23'}$$

and the initial conditions

$$u(x, 0) = \varphi(x),$$
$$u_t(x, 0) = \psi(x),$$

where $\varphi(x)$ is a piecewise smooth function and $\psi(x)$ and $f(x, t)$ are piecewise continuous. Here C is an arbitrary piecewise smooth contour lying in the region $t \geqslant 0$. We shall show that this problem has a unique solution given by D'Alembert's formula.

Let us assume that the function $u(x, t)$ represents a solution of our problem. Consider the triangle ABM based on the axis $t = 0$, with its vertex at the point $M(x, t)$ and with sides which are segments of the characteristics $x - at =$ const. and $x + at =$ const. (Fig. 18) and apply formula (23′) to it. The equation $dx/dt = a$ holds along the segment AM so that

$$\frac{\partial u}{\partial t}\, dx + a^2 \frac{\partial u}{\partial x}\, dt = a\left(\frac{\partial u}{\partial t}\, dt + \frac{\partial u}{\partial x}\right) = a\, du.$$

† Since $dx = 0$ on the vertical sides of the open polygon $\bar{C}_n$, in this formula $t = t_n(x)$ is the equation of the horizontal sides of the curve $\bar{C}_n$.

Along the segment MB we have $dx/dt = -a$, so that

$$\frac{\partial u}{\partial t}\,dx + a^2\,\frac{\partial u}{\partial x}\,dt = -\,a\left(\frac{\partial u}{\partial t}\,dt + \frac{\partial u}{\partial x}\right) = -\,a\,du\,.$$

Hence the expression under the integral along the characteristics is a complete differential. Carrying out the integration along the segments BM and MA we get:

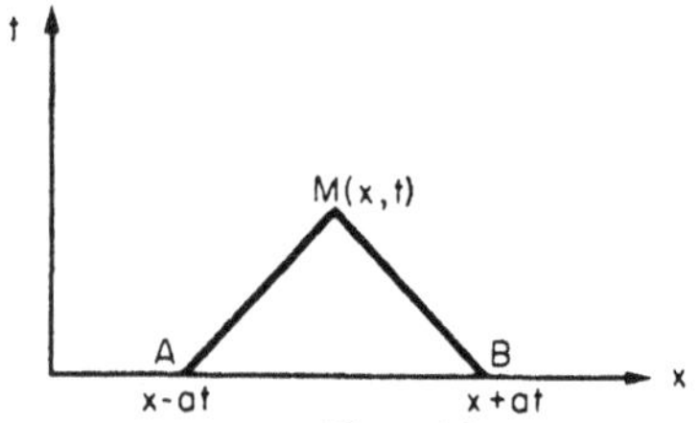

FIG. 18

$$\int\limits_B^M \left(\frac{\partial u}{\partial t}\,dx + a^2\,\frac{\partial u}{\partial x}\,dt\right) = $$

$$= -\,a\left[u\,(M) - u\,(B)\right],$$

$$\int\limits_M^A \left(\frac{\partial u}{\partial t}\,dx + a^2\,\frac{\partial u}{\partial x}\,dt\right) = a\left[u\,(A) - u\,(M)\right],$$

so that formula (23′) assumes the form:

$$u\,(M) = \frac{u\,(B) + u\,(A)}{2} + \frac{1}{2a}\int\limits_A^B \frac{\partial u}{\partial t}\,dx + \frac{1}{2a}\iint\limits_{ABM} f\,dx\,dt$$

or

$$u\,(x, t) = \frac{\varphi\,(x + at) + \varphi\,(x - at)}{2} + \frac{1}{2\,a}\int\limits_{x-at}^{x+at} \psi\,(\xi)\,d\xi +$$

$$+ \frac{1}{2\,a}\int\limits_0^t d\,\tau \int\limits_{x-a(t-\tau)}^{x+a(t-\tau)} f\,(\xi, \tau)\,d\xi\,. \qquad (24)$$

Thus if a solution of the problem exists it is uniquely defined by its initial values. In the case of a homogeneous equation ($f = 0$) this formula coincides with D'Alembert's formula.

It is not difficult to verify by direct substitution that a function of type

$$u\,(x, t) = f_1\,(x + at) + f_2\,(x - at) + \int\limits_0^t d\,\tau \int\limits_{x-a(t-\tau)}^{x+a(t-\tau)} f_3\,(\xi, \tau)\,d\,\xi\;,$$

where f_1 and f_2 are piecewise smooth functions, and f_3 is a piecewise continuous function, satisfies the equation (22) and therefore equation (23′). Solutions of the examples considered in sub-section 4 are piecewise smooth functions and come within the scope of the theory.

Next we consider the first boundary problem in a semi-infinite region. We shall look for the solution of the equation (23) at some point $M(x, t)$

for $t > x/a$ since in the region $t < x/a$ (under the characteristic curve $x = at$) the effect of the boundary conditions is not felt and the solution is given by formula (24). Let us apply formula (23′) to the quadrilateral $MAA'B$ in which MA, MB and AA' are segments of the characteristics (Fig. 19). Carrying out the integration along the characteristics MA, AA' and BM we get:

$$2au\,(M) = 2\,au\,(A) + au\,(B) - au\,(A') + \int_{A'}^{B} \frac{\partial u}{\partial t}\,dx + \iint_{MAA'B} f\,dx\,dt\,.$$

Substituting here the coordinates of the points M, A, B and A' we have

$$u\,(x,t) = u\left(0, t - \frac{x}{a}\right) + \frac{u\,(x + at, 0) - u\,(at - x, 0)}{2} +$$

$$+ \frac{1}{2\,a} \int_{at-x}^{x+at} \frac{\partial u}{\partial t}\,dx + \frac{1}{2\,a} \int_0^t d\,\tau \int_{|x-a(t-\tau)|}^{x+a(t-\tau)} f\,(\xi, \tau)\,d\,\xi\,,$$

or

$$u\,(x,t) = \mu\left(t - \frac{x}{a}\right) + \frac{\varphi\,(x + at) - \varphi\,(at - x)}{2} + \frac{1}{2\,a} \int_{at-x}^{x+at} \psi\,(\xi)\,d\,\xi +$$

$$+ \frac{1}{2\,a} \int_0^t d\,\tau \int_{|x-a(t-\tau)|}^{x+a(t-\tau)} f\,(\xi, \tau)\,d\,\xi \left(t > \frac{x}{a}\right). \qquad (25)$$

For $f = 0$ this formula coincides with formula (13) of § 2, sub-section 4. The second boundary-value problem as well as problems for a finite segment can be solved by a similar method.

While studying the first boundary-value problem we saw that the two initial conditions

$$u\,(x, 0) = \varphi\,(x), \quad u_t\,(x, 0) = \psi\,(x)$$

and the one boundary condition

$$u\,(0, t) = \mu\,(t)$$

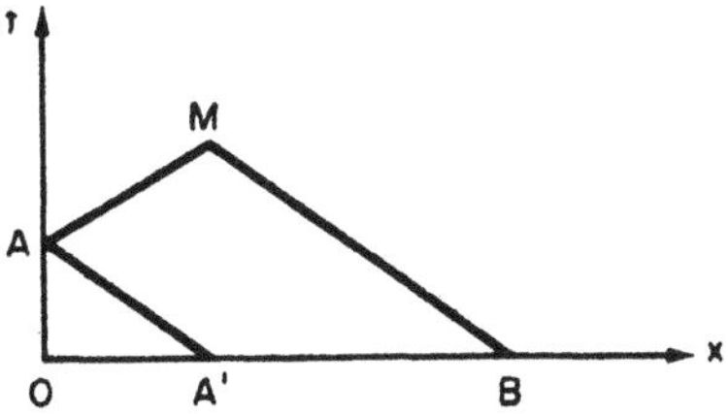

FIG. 19

are sufficient to determine the solution completely. It follows that there is a relation connecting the functions φ, ψ, μ and ν, where $\nu(t) = u_x(0, t)$. Differentiating formula (25) with respect to x and putting $x = 0$, we get

$$\nu\,(t) = \frac{1}{a}\left\{\psi\,(at) - \left[\mu'\,(t) - a\varphi'\,(at)\right]\right\}, \qquad (26)$$

where for simplicity we put $f = 0$. Using formula (26) the third boundary problem may be reduced to the first boundary problem.

8. *Propagation of discontinuities along characteristic curves*

In this section we consider discontinuities of derivatives of solutions of the equation (23). We shall show that only members of families of characteristic curves

$$x - at = \text{const}, \quad x + at = \text{const}$$

can be lines of discontinuity of the functions $u(x, t)$ satisfying equation (23).

Let some differentiable curve defined by the equation

$$x = x(t),$$

be a line of discontinuity of the derivatives of a continuous, piecewise differentiable function $u(x, t)$. Let us assume that $x(t)$ is an increasing function. Let us apply formula (23′) to the rectangle $ABCD$ (Fig. 20):

$$\int_{BA+AD} \left(\frac{\partial u}{\partial t}\, dx + a^2 \frac{\partial u}{\partial x}\, dt\right) + \int_{DC+CB} \left(\frac{\partial u}{\partial t}\, dx + a^2 \frac{\partial u}{\partial x}\, dt\right) = 0,$$

and also to the curvilinear triangles $\Delta_1 = BAD$ and $\Delta_2 = BDC$

$$\int_{BA+AD} \left(\frac{\partial u}{\partial t}\, dx + a^2 \frac{\partial u}{\partial x}\, dt\right) + \int_{DB} \left(\frac{\partial u}{\partial t}\, x' + a^2 \frac{\partial u}{\partial x}\right)_1 dt = 0\,.$$

$$\int_{DC+CB} \left(\frac{\partial u}{\partial t}\, dx + a^2 \frac{\partial u}{\partial x}\, dt\right) - \int_{DB} \left(\frac{\partial u}{\partial t}\, x' + a^2 \frac{\partial u}{\partial x}\right)_2 dt = 0\,,$$

where the brackets $(\quad)_{1,2}$ show that the limiting values have to be taken inside the triangles Δ_1 and Δ_2. Subtracting the first equation from the sum of the second and third we get:

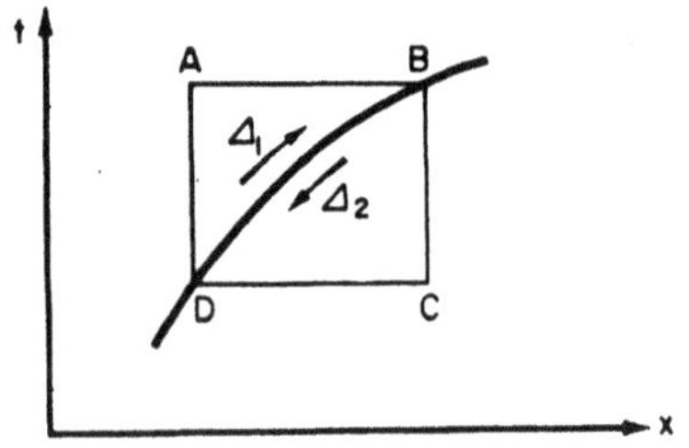

FIG. 20

$$\int_{DB} \left\{\left(\frac{\partial u}{\partial t}\, x' + a^2 \frac{\partial u}{\partial x}\right)_1 - \left(\frac{\partial u}{\partial x}\, x' + a^2 \frac{\partial u}{\partial x}\right)_2\right\} dt = 0$$

or, because the arc DB is arbitrary

$$\left[\frac{\partial u}{\partial t}\right] x' + a^2 \left[\frac{\partial u}{\partial x}\right] = 0\,, \qquad (27)$$

where the brackets denote the magnitude of discontinuity of the function

$$[f] = f_2 - f_1\,.$$

across the curve $x = x(t)$.

Let us take the derivative of $u(x, t)$ with respect to t along the line of discontinuity of the derivatives of $u(x, t)$

$$\frac{d}{dt}\, u\,[x(t), t] = \left(\frac{\partial u}{\partial x}\right)_i x' + \left(\frac{\partial u}{\partial t}\right)_i \quad (i = 1, 2)\,,$$

where we take the limiting values of the derivatives from Δ_1 and from Δ_2. The difference of the right-hand sides for $i = 1$ and $i = 2$ gives

$$\left[\frac{\partial u}{\partial t}\right] + x' \left[\frac{\partial u}{\partial x}\right] = 0 .$$

Comparing this equation with the previous one (27) and assuming that at least one of these discontinuities $\left[\frac{\partial u}{\partial t}\right], \left[\frac{\partial u}{\partial x}\right]$ is different from zero, we see that these equations are true simultaneously only if the determinant of the coefficients is equal to zero:

$$\begin{vmatrix} x' & a^2 \\ 1 & x' \end{vmatrix} = (x')^2 - a^2 = 0$$

or

$$x = \pm\, at + \text{const}.$$

Thus the lines of discontinuity of the derivatives of the solution of the wave equation are characteristic curves.

Problems

1. Draw the profile of the string for different moments of time in the following cases:

I. Infinite String

(a) the initial velocity is equal to zero ($\psi(x) = 0$) and the initial profile of the string is given in the form of Fig. 21;

(b) the initial displacement is equal to zero, and the initial velocity has a constant value $u_t(x, 0) = \psi_0$ on the portion of the string (x_1, x_2), and is equal to zero outside this portion;

(c) the initial conditions have the form

$$\varphi(x) = 0, \psi(x) = \begin{cases} 0 & \text{where } x < c, \\ \dfrac{h}{2c^2}\, x(2c - x) & \text{where } c < x < 2c, \\ 0 & \text{where } x > 2c. \end{cases}$$

II. Semi-infinite String

(d) the initial velocity is equal to zero ($\psi(x) = 0$) and the initial displacement is given in the form of a triangle shown in Fig. 21. The end of the string is assumed to be fixed;

(e) the same problem for a string with a free end $x = 0$;

(f) the initial conditions have the form

$$\varphi(x) = 0, \quad \psi(x) = \begin{cases} 0 & \text{where } 0 < x < c, \\ \psi_0 = \text{const} & \text{where } c < x < 2c, \\ 0 & \text{where } x > 2c; \end{cases}$$

the end of the string $x = 0$ is fixed;

(g) a similar problem for a string with a free end $x = 0$. The profile of the string for all the problems (a)–(g) shoud be drawn for the times

$$t_0 = 0, \quad t_k = \frac{c}{8a} k \qquad (k = 1, 2, \ldots, 8).$$

Mark the zones corresponding to the different stages of the process for the problems (a)–(g) on the phase plane (x, t).

2. Find the solution of the problem (1a) for all values of the variables x and t (formulae expressing the function $u(x, t)$ are different for different zones of the phase plane).

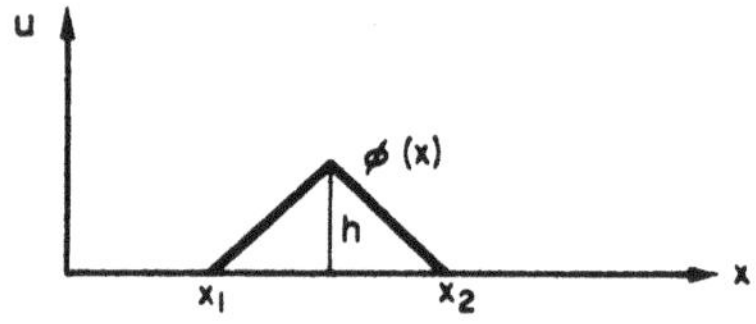

FIG. 21

3. Determine the displacement at some point x_0, t_0 using the phase plane (x, t) and the plane (x, u) in which (Fig. 21) the initial displacements ($\psi = 0$) are given, for the case of an infinite string and for a semi-infinite string with a fixed (or free) end.

4. At the mouth of a long cylindrical tube filled with gas there is a piston moving according to the law $x = f(t)$ with a velocity $v = f'(t) < a$. The initial displacement and the velocity of gas particles is zero. Find the displacement of the gas in a section with coordinate x. Consider the case of the motion of the piston with a constant velocity $c < a$. What can be said about the solution of the problem if, beginning from a certain moment, the velocity of the piston is $v > a$? (See Appendix 5.)

5. Let a wave $u(x, t) = f(x - at)$ run along an infinite string. Taking the state of the string at the moment $t = 0$ as the initial state solve the wave equation under the corresponding initial conditions.
(Compare with problem 1a.)

6. An infinite elastic rod is made up of two rods joined at $x = 0$ with the characteristics

$$k_1, \varrho_1, a_1 = \sqrt{\frac{k_1}{\varrho_1}} \text{ if } x < 0,$$

$$k_2, \varrho_2, a_2 = \sqrt{\frac{k_2}{\varrho_2}} \text{ if } x > 0.$$

(a) Let the wave

$$u(x, t) = f(t - x/a),$$

approach from the region $x < 0$. Find the coefficients of reflection and transmission of the wave at the junction ($x = 0$). Determine under what conditions the reflected wave is absent.

(b) Solve a similar problem of the initial local displacement

$$u(x, 0) = \begin{cases} 0 & \text{if } x < x_1, \\ \varphi(x) & \text{if } x_1 < x < x_2 < 0, \\ 0 & \text{if } x > x_2, \end{cases}$$

and the initial velocity is zero.

7. Let a load of mass M be suspended from some point of the string $x = x_0$ and let the wave

$$u(x, t) = f(t - x/a),$$

approach from the region $x < 0$.

Find the transmitted and the reflected wave.

8. A semi-infinite tube ($x > 0$) filled with an ideal gas has at one end ($x = 0$) a freely moving piston of mass M. At the moment $t = 0$ the

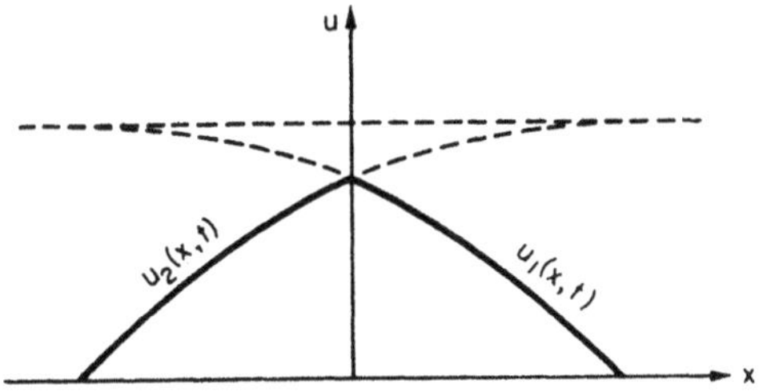

FIG. 22

piston is given an initial velocity v_0. Find the process of wave propagation in the gas if the initial displacements and the initial velocity of the gas particles is zero.

Hint. Consider the solution of the equation of vibrations in the region $x > 0$. Use the boundary condition

$$Mu_{tt}(0, t) = S\gamma p_0 u_x(0, t)$$

(p_0 is the initial pressure of the gas, S the cross-section of the tube, $\gamma = c_p/c_v$) and the initial conditions at the boundary $u(0, 0) = 0$, $u_t(0, 0) = v_0$.

Answer :

$$u(x, t) = \frac{aMv_0}{\gamma p_0 S}\left[1 - e^{\frac{\gamma p_0 S}{Ma^2}(x-at)}\right] \quad \text{where } (x - at) < 0;$$

$$u(x, t) = 0 \quad \text{where } (x - at) > 0.$$

9. An infinite string having a concentrated mass M at the point $x = 0$ is in equilibrium. At $t = 0$ the mass M is given an initial velocity v_0.

Prove that for $t > 0$ the string has the shape shown in Fig. 22 where $u_1(x, t)$ and $u_2(x, t)$ are determined by the formulae

$$u_1(x, t) = \begin{cases} \dfrac{Mav_0}{2T}\left[1 - e^{\frac{2T}{Ma^2}(x-at)}\right] & \text{when } x - at < 0 \\ 0 & \text{when } x - at > 0\,; \end{cases}$$

$$u_2(x, t) = \begin{cases} \dfrac{Mav_0}{2T}\left[1 - e^{-\frac{2T}{Ma^2}(x+at)}\right] & \text{when } x - at < 0 \\ 0 & \text{when } x - at > 0\,. \end{cases}$$

Hint. Use the condition

$$M \frac{\partial^2 u_1}{\partial t^2}(0, t) = M \frac{\partial^2 u_2}{\partial t^2}(0, t) = T \frac{\partial u_1}{\partial x}(0, t) - T \frac{\partial u_2}{\partial x}(0, t)\,.$$

10. Solve the problem of propagation of electrical waves in an infinite wire when

$$\frac{G}{C} = \frac{R}{L}$$

and for any initial conditions.

Answer :

$$v(x, t) = e^{-\frac{R}{L}t}[\varphi(x - at) + \psi(x + at)]\,,$$

$$i(x, t) = \sqrt{\frac{C}{L}} \cdot e^{-\frac{R}{L}t}[\varphi(x - at) - \psi(x + at)]\,.$$

If

$$v(x, 0) = f(x), \; i(x, 0) = \sqrt{\frac{C}{L}} \cdot F(x),$$

then

$$\varphi(x) = \frac{f(x) + F(x)}{2}\,, \quad \psi(x) = \frac{f(x) - F(x)}{2}\,.$$

11. Find the solution of the integral wave equation for a semi-infinite string for boundary conditions of the third kind.

12. A membrane is attached at the end $x = 0$ of a semi-infinite rod producing a resistance proportional to the velocity $u_t(0, t)$ to longitudinal vibrations of the rod. Find the process of vibration if the initial displacements are given and $u_t(x, 0) = \psi(x) = 0$.

§ 3. Method of separation of variables

1. *Equation of free vibrations of a string*

The *method of separation of variables*, or *Fourier's method*, is one of the most widely used methods of solving partial differential equations We shall apply this method to the problem of vibrations of a string fixed at both ends. The solution of this problem will be examined in exhaustive detail and later on in the book we shall refer to this section to avoid repetition of the proofs.

We look for the solution of the equation

$$u_{tt} = a^2 u_{xx}, \tag{1}$$

which satisfies the homogeneous boundary conditions

$$u(0, t) = 0, \quad u(l, t) = 0 \tag{2}$$

and the initial conditions

$$\left. \begin{aligned} u(x, 0) &= \varphi(x), \\ u_t(x, 0) &= \psi(x). \end{aligned} \right\} \tag{3}$$

Equation (1) is linear and homogeneous, so that the sum of solutions is also a solution of this equation. If there is a sufficiently large number of particular solutions we can try to find the required solution by summing them with certain coefficients.

The principal auxiliary problem can be stated as:

Find the solution of the equation

$$u_{tt} = a^2 u_{xx}, \tag{1}$$

not identically equal to zero, which satisfies the homogeneous boundary conditions

$$\left. \begin{aligned} u(0, t) &= 0, \\ u(l, t) &= 0 \end{aligned} \right\} \tag{2}$$

and can be represented in the form of the product

$$u(x, t) = X(x)\, T(t), \tag{4}$$

where $X(x)$ is a function only of the variable x and $T(t)$ is a function only of the variable t.

Substituting the assumed form of the solution (4) in equation (1) we get

$$X''\, T = \frac{1}{a^2} T''\, X \tag{5}$$

or, after division by XT,

$$\frac{X''(x)}{X(x)} = \frac{1}{a^2} \frac{T''(t)}{T(t)}. \tag{6}$$

In order that function (4) may be a solution of the equation (1), equations (5) and (6) must be satisfied identically, i. e. for all values of the independent variables $0 < x < l$, $t > 0$. The right-hand side of equation (6) is a function only of the variable t, and the left-hand side only of x. Fixing, for example, some value of x and varying t (or conversely) we find that the right- and left-hand sides of equation (6) remain constant

$$\frac{X''(x)}{X(x)} = \frac{1}{a^2}\frac{T''(t)}{T(t)} = -\lambda, \tag{7}$$

when their arguments are changed. Here λ is a constant which may be positive or negative but which we take with a minus sign for convenience in subsequent treatment.

From relation (7) we obtain differential equations for the functions $X(x)$ and $T(t)$

$$X''(x) + \lambda X(x) = 0, \tag{8}$$

$$T''(t) + a^2 \lambda T(t) = 0. \tag{9}$$

The boundary conditions (2) give

$$u(0, t) = X(0) T(t) = 0,$$

$$u(l, t) = X(l) T(t) = 0.$$

It follows that the function $X(x)$ must satisfy the boundary conditions

$$X(0) = X(l) = 0, \tag{10}$$

since otherwise we would have

$$T(t) \equiv 0 \quad \text{or} \quad u(x, t) \equiv 0,$$

and we are looking for a non-trivial solution. There are no additional conditions for the function $T(t)$ in the auxiliary problem.

In the course of the determination of the function $X(x)$ we have formulated the simplest eigen-value problem:

It is required to find values λ, called eigen-values for which non-trivial solutions called eigen-functions, of the following problem exist:

$$X'' + \lambda X = 0,$$

$$X(0) = X(l) = 0, \tag{11}$$

and also to find these solutions. The problem thus formulated is called *Sturm-Liouville's problem.*

We consider separately the cases when the parameter λ is negative, zero or positive.

1. When $\lambda < 0$ the problem does not have non-trivial solutions. The general solution of the equation is

$$X(x) = C_1 e^{\sqrt{-\lambda}\cdot x} + C_2 e^{-\sqrt{-\lambda}\cdot x}.$$

The boundary conditions give:

$$X(0) = C_1 + C_2 = 0;$$

$$X(l) = C_1 e^{a} + C_2 e^{-a} = 0 \qquad (a = l\sqrt{-\lambda}),$$

i. e.

$$C_1 = -C_2 \quad \text{or} \quad C_1 (e^{a} - e^{-a}) = 0.$$

But in the case considered a is real and positive so that $e^{a} - e^{-a} \neq 0$. Hence

$$C_1 = 0, \quad C_2 = 0$$

and consequently

$$X(x) \equiv 0.$$

2. Also when $\lambda = 0$ there are no non-trivial solutions. In this case the general solution of the equation is

$$X(x) = ax + b.$$

The boundary conditions give

$$X(0) = [ax + b]_{x=0} = b = 0,$$

$$X(l) = al = 0,$$

i. e. $a = 0$ and $b = 0$ and consequently

$$X(x) \equiv 0.$$

3. When $\lambda > 0$ the general solution of the equation contains imaginary powers and may therefore be written as

$$X(x) = D_1 \cos \sqrt{\lambda}\cdot x + D_2 \sin \sqrt{\lambda}\cdot x.$$

The boundary conditions give

$$X(0) = D_1 = 0,$$

$$X(l) = D_2 \sin \sqrt{\lambda}\cdot l = 0.$$

If $X(x)$ is not identically equal to zero, $D_2 \neq 0$ hence

$$\sin \sqrt{\lambda} \cdot l = 0 \tag{12}$$

or

$$\sqrt{\lambda} = \pi n / l,$$

where n is an integer. Consequently non-trivial solutions of the problem are possible only for the values

$$\lambda_n = \left(\frac{\pi n}{l}\right)^2.$$

These eigen-values have corresponding eigen-functions

$$X_n(x) = D_n \sin \frac{\pi n}{l} x.$$

Thus only for values of λ equal to

$$\lambda_n = \left(\frac{\pi n}{l}\right)^2, \tag{13}$$

are there non-trivial solutions

$$X_n(x) = \sin \frac{\pi n}{l} x, \tag{14}$$

which can be determined up to an arbitrary factor which we shall put equal to unity. To each of these values of λ_n there corresponds a solution of equation (9)

$$T_n(t) = A_n \cos \frac{\pi n}{l} a t + B_n \sin \frac{\pi n}{l} at, \tag{15}$$

where A_n and B_n are coefficients to be determined.

Returning to the problem (1)–(3), we see that the functions

$$u_n(x, t) = X_n(x) T_n(t) = \left(A_n \cos \frac{\pi n}{l} at + B_n \sin \frac{\pi n}{l} at\right) \sin \frac{\pi n}{l} x \tag{16}$$

are particular solutions of the equation (1) satisfying the boundary conditions (2) which can be represented in the form of a product (4) of two functions, one of which depends only on x, the other on t. These solutions can satisfy the initial conditions (3) of our initial problem only for particular cases of the initial functions $\varphi(x)$ and $\psi(x)$.

Now we look for a solution of the problem in the general case. Because of the linearity and homogeneity of equation (1) a sum of the particular solutions

$$u(x,t)=\sum_{n=1}^{\infty} u_n(x,t)=\sum_{n=1}^{\infty}\left(A_n\cos\frac{\pi n}{l}at+B_n\sin\frac{\pi n}{l}at\right)\sin\frac{\pi n}{l}x \tag{17}$$

also satisfies this equation and the boundary conditions (2). We shall consider this problem in greater detail a little later (see sub-section 3 of this section). The initial conditions give:

$$\left.\begin{aligned} u(x,0)=\varphi(x)&=\sum_{n=1}^{\infty} u_n(x,0)=\sum_{n=1}^{\infty} A_n\sin\frac{\pi n}{l}x,\\ u_t(x,0)=\psi(x)&=\sum_{n=1}^{\infty}\frac{\partial u_n}{\partial t}(x,0)=\sum_{n=1}^{\infty}\frac{\pi n}{l}aB_n\sin\frac{\pi n}{l}x\,.\end{aligned}\right\} \tag{18}$$

From the theory of Fourier's series it is known that the arbitrary partially continuous and partially differentiable function $f(x)$ given in the interval $0\leqslant x\leqslant l$ can be expanded in a Fourier series

$$f(x)=\sum_{n=1}^{\infty} b_n\sin\frac{\pi n}{l}x, \tag{19}$$

where*

$$b_n=\frac{2}{l}\int_0^l f(\xi)\sin\frac{\pi n}{l}\xi\,d\xi\,. \tag{20}$$

* Generally periodic functions with the period $2l$ are considered

$$F(x)=\frac{a_0}{2}+\sum_{x=1}^{\infty}\left(a_n\cos\frac{\pi n}{l}x+b_n\sin\frac{\pi n}{l}x\right),$$

$$a_n=\frac{1}{l}\int_{-l}^{+l}F(\xi)\cos\frac{\pi n}{l}\xi\,d\xi,\quad b_n=\frac{1}{l}\int_{-l}^{l}F(\xi)\sin\frac{\pi n}{l}\xi\,d\xi.$$

If the function $F(x)$ is odd then $a_n=0$, so that

$$F(x)=\sum_{n=1}^{\infty} b_n\sin\frac{\pi n}{l}x;$$

$$b_n=\frac{1}{l}\int_{-l}^{l}F(\xi)\sin\frac{\pi n}{l}\xi\,d\xi=\frac{2}{l}\int_0^l F(\xi)\sin\frac{\pi n}{l}\xi\,d\xi.$$

If the function $F(x)$ is given only in the interval $(0,l)$ we can extend it oddly and carry out the expansion in the interval from $-l$ to $+l$ which leads us to formulae (19) and (20).

If the functions $\varphi(x)$ and $\psi(x)$ satisfy the conditions of expansion into Fourier series, then

$$\varphi(x) = \sum_{n=1}^{\infty} \varphi_n \sin \frac{\pi n}{l} x, \quad \varphi_n = \frac{2}{l} \int_0^l \varphi(\xi) \sin \frac{\pi n}{l} \xi \, d\xi, \tag{21}$$

$$\psi(x) = \sum_{n=1}^{\infty} \psi_n \sin \frac{\pi n}{l} x, \quad \psi_n = \frac{2}{l} \int_0^l \psi(\xi) \sin \frac{\pi n}{l} \xi \, d\xi. \tag{22}$$

Comparison of these series with the formula (18) shows that in order to satisfy the initial conditions we must put

$$A_n = \varphi_n, \quad B_n = \frac{l}{\pi n a} \psi_n. \tag{23}$$

These equations define the function (17) completely and give the solution of the problem.

We have defined the solution in the form of an infinite series (17). If the series (17) diverges or the function defined by this series is not differentiable then naturally it cannot represent a solution of our differential equation.

In the present sub-section we confine ourselves to a formal construction of the solution. The conditions under which the series (17) converges and represents the solution will be investigated in sub-section 3.

2. *Interpretation of the solution*

Let us give an interpretation of the solution obtained. The function $u_n(x, t)$ can be written as

$$u_n(x,t) = \left(A_n \cos \frac{\pi n}{l} at + B_n \sin \frac{\pi n}{l} at\right) \sin \frac{\pi n}{l} x =$$
$$= a_n \cos \frac{\pi n}{l} a(t + \delta_n) \sin \frac{\pi n}{l} x, \tag{24}$$

where

$$a_n = \sqrt{(A_n^2 + B_n^2)}, \frac{\pi n}{l} a \delta_n = - \operatorname{arc} \tan \frac{B_n}{A_n}. \tag{25}$$

Each point x_0 of the string executes harmonic vibrations

$$u_n(x_0, t) = a_n \cos \frac{\pi n}{l} a(t + \delta_n) \sin \frac{\pi n}{l} x_0$$

with the amplitude

$$a_n \sin \frac{\pi n}{l} x_0 .$$

This type of motion is called a *standing wave*. The points $x = m(l/n)$ $(m = 1, 2, \ldots, n-1)$ at which $\sin (\pi n/l) x = 0$, remain stationary during the entire process and are called nodes of the standing wave $u_n(x, t)$. The points $x = (2m+1) l/2n$ $(m = 0, 1, \ldots, n-1)$ at which $\sin \pi (n/l) x = \pm 1$ vibrate with the maximum amplitude a_n and are called *antinodes* of the standing wave.

The profile of the standing wave at any moment is a sine wave

$$u_n(x, t) = C_n(t) \sin \frac{\pi n}{l} x ,$$

where

$$C_n(t) = a_n \cos \omega_n (t + \delta_n) \quad \left(\omega_n = \frac{\pi n}{l} a\right).$$

At the moment t when $\cos \omega_n (t + \delta_n) = \pm 1$ the displacement reaches maximum values and the velocity is zero. At the moments t when $\cos \omega_n (t + \delta_n) = 0$, the displacement is zero and the velocity is maximum. The vibration frequencies of all points of the string are identical and are equal to

$$\omega_n = \frac{\pi n}{l} a. \tag{26}$$

The frequencies ω_n are called *eigen-frequencies*. For transverse vibrations of the string $a^2 = T/\varrho$ and hence

$$\omega_n = \frac{\pi n}{l} \sqrt{\frac{T}{\varrho}} . \tag{27}$$

The energy of the n_{th} standing wave (n_{th} harmonic) for the case of transverse vibrations of the string is

$$\begin{aligned} E_n &= \frac{1}{2} \int_0^l \left[\varrho \left(\frac{\partial u_n}{\partial t}\right)^2 + T \left(\frac{\partial u_n}{\partial x}\right)^2\right] dx = \\ &= \frac{a_n^2}{2} \int_0^l \left[\varrho \omega_n^2 \sin^2 \omega_n (t + \delta_n) \sin^2 \frac{\pi n}{l} x + T \left(\frac{\pi n}{l}\right)^2 \times \right. \\ &\quad \left. \times \cos^2 \omega_n (t + \delta_n) \cos^2 \frac{\pi n}{l} x\right] dx = \\ &= \frac{a_n^2}{2} \frac{l}{2} \left[\varrho \omega_n^2 \sin^2 \omega_n (t + \delta_n) + T \left(\frac{\pi n}{l}\right)^2 \cos^2 \omega_n (t + \delta_n)\right], \end{aligned} \tag{28}$$

since

$$\int_0^l \sin^2 \frac{\pi n}{l} x\,dx = \int_0^l \cos^2 \frac{\pi n}{l} x\,dx = \frac{1}{2}.$$

Using the expressions for a_n, ω_n and also the equation $T = a^2 \varrho$ we get

$$E_n = \frac{\varrho a_n^2 \omega_n^2}{4} l = \omega_n^2 M \cdot \frac{A_n^2 + B_n^2}{4}, \tag{29}$$

where $M = l\varrho$ is the mass of the string.

We usually detect the vibrations of a string by the sound emitted. Without dwelling on the process of propagation of sound waves in air and the perception of sound it may be said that the sound of the string is a superposition of "simple tones" corresponding to the standing waves into which the vibration can be resolved. This resolution of the sound into simple tones is not a mathematical operation. The simple tones may be separated experimentally with the help of resonators.

The pitch of a tone depends on the frequency of its vibrations. The intensity is determined by its energy and hence by its amplitude. The lowest tone which can be formed by the string is determined by the lowest eigen-frequency $\omega_1 = (\pi/l)\sqrt{(T/\varrho)}$ and is called the *fundamental tone*. The remaining tones with frequencies which are multiples of ω_1 are called *overtones*. The timbre of the sound depends on the presence of overtones in addition to the fundamental tone and on the distribution of energy over the harmonics.

The lowest tone emitted by a string and its timbre depend on the method of excitation, which determines the initial conditions

$$u(x, 0) = \varphi(x); \quad u_t(x, 0) = \psi(x), \tag{3}$$

and hence the coefficients A_n and B_n. If $A_1 = B_1 = 0$, the lowest tone will correspond to the frequency ω_n, where n is the smallest for which A_n or B_n are different from zero.

Let us make the string vibrate by pulling it to one side and releasing it without initial velocity. In this case

$$u_t(x, 0) = 0, \quad u(x, 0) = \varphi(x) > 0$$

and

$$A_1 = \frac{2}{l} \int_0^l \varphi(\xi) \sin \frac{\pi}{l} \xi \, d\xi > 0,$$

since

$$\sin \frac{\pi}{l} \xi > 0 .$$

The subsequent coefficients, generally speaking, are considerably smaller than A_1, since the function $\sin (\pi n/l) \xi$ alternates in sign for $n \geqslant 2$. In particular, if $\varphi(x)$ is symmetrical with respect to the centre, $A_2 = 0$. Thus if the string is made to vibrate by pulling it to one side ($\varphi(x) > 0$) the lowest tone usually has an energy greater than the energy of the other harmonics.

The string can be made to vibrate in other ways. For example, if the initial function is odd with respect to the centre of the string then

$$A_1 = 0$$

and the lowest tone corresponds to the frequency

$$\omega = \omega_2 = \frac{2\pi}{l} \sqrt{\frac{T}{\varrho}} .$$

If the sounding string is touched exactly at the centre, its sound changes charply and it sounds an octave above its fundamental tone. This method of varying the tone is often used in playing the violin, the guitar and other string instruments and is called flageolet. This phenomenon is perfectly comprehensible from the point of view of the theory of vibration. At the moment of contact with the centre of the string we stop the standing waves having antinodes at this point and retain only the harmonics having nodes at this point. Thus only the odd harmonics remain and the lowest frequency will be

$$\omega_2 = \frac{2\pi}{l} \sqrt{\frac{T}{\varrho}} .$$

If the string is touched at a distance of one-eighth its length from an end, the pitch of the fundamental tone is raised three times, since only harmonics having nodes at the point $x = l/3$ are retained.

The formulae

$$\omega_1 = \frac{\pi}{l}\sqrt{\frac{T}{\varrho}} \text{ and } \tau_1 = \frac{2\pi}{\omega_1} = 2l\sqrt{\frac{\varrho}{T}}, \tag{30}$$

determining the frequency and correspondingly the period of the fundamental vibration, account for the following laws of vibration of strings first discovered experimentally.

1. For a string of uniform thickness and uniform tension the period of vibration of the string is proportional to its length.

2. For a given length of the string the period is inversely proportional to the square root of the tension.

3. For a given length and tension the period is proportional to the square root of the linear density of the string.

3. *Representation of an arbitrary vibration as a superposition of standing waves*

In sub-section 1 we considered the problem of free vibrations of a string fixed at the ends and proved the existence of standing wave solutions. A formal scheme of representing an arbitrary vibration as an infinite sum of standing waves was developed. In this sub-section we justify this formal solution. In the first instance we consider the generalization of the principle of superposition, well known for finite sums, to the case of an infinite series.

Let L be a linear differential operator so that $L(u)$ is equal to the sum of derivatives of the function u (ordinary or partial) with coefficients that are functions of independent variables.

Let us prove a lemma *(generalized principle of superposition).*

If the functions u_i ($i = 1, 2, \ldots, n, \ldots$) are particular solutions of a linear and homogeneous differential equation $L(u) = 0$ (ordinary or partial), then the series $u = \sum_{i=1}^{\infty} C_i u_i$ is also a solution of this equation, if calculation of the derivatives of u figuring in the equation $L(u) = 0$ can be performed by differentiation term by term.

If the derivatives of u figuring in the equation $L(u) = 0$ can be calculated by differentiating the series for u term by term, then because of linearity of the equation

$$L(u) = L\left(\sum_{i=1}^{\infty} C_i u_i\right) = \sum_{i=1}^{\infty} C_i L(u_i) = 0,$$

This proves that the function u satisfies the equation. As a sufficient condition for the possibility of differentiation term by term we shall constantly make use of the condition of uniform convergence of the series

$$\sum_{i=1}^{\infty} C_i \frac{\partial^n u_i}{\partial x^m \partial t^{n-m}}, \tag{31}$$

obtained after differentiation*.

We return to our boundary problem. First of all we should verify the continuity of the function

$$u(x, t) = \sum_{n=1}^{\infty} u_n(x, t) = \sum_{n=1}^{\infty} \left(A_n \cos \frac{\pi n}{l} at + B_n \sin \frac{\pi n}{l} at\right) \times$$
$$\times \sin \frac{\pi n}{l} x, \tag{32}$$

whence it will follow that $u(x, t)$ continuously adjoins its initial and boundary values. It is sufficient to prove the uniform convergence of the series for $u(x, t)$ since the general term of this series is a continuous function and the uniformly converging series of continuous functions defines a continuous function. Using the inequality

$$|u_n(x, t)| \leqslant |A_n| + |B_n|,$$

we deduce that the series

$$\sum_{n=1}^{\infty} (|A_n| + |B_n|) \tag{33}$$

is a majorant for the series (32). If the majorant series (33) converges then the series (32) converges uniformly, i. e. the function $u(x, t)$ is continuous.

* See V. I. Smirnov, *Course of Higher Mathematics*, vol. II, 1937.

To verify that $u_t(x, t)$ adjoins its initial values continuously, the continuity of this function has to be proved. It suffices to prove the uniform convergence of the series

$$u_t(x,t) \sim \sum_{n=1}^{\infty} \frac{\partial u_n}{\partial t} = \sum_{n=1}^{\infty} a \frac{\pi n}{l} \left(- A_n \sin \frac{\pi n}{l} at + B_n \cos \frac{\pi n}{l} at \right) \times \times \sin \frac{\pi n}{l} x \tag{34}$$

or the convergence of the majorant series

$$\frac{a\pi}{l} \sum_{n=1}^{\infty} n(|A_n| + |B_n|). \tag{35}$$

Finally to verify that the function $u(x, t)$ satisfies the equation, i. e. that the generalized principle of superposition is applicable, we have to prove the possibility of double differentiating the series for $u(x, t)$ term by term. For this it is sufficient to prove the uniform convergence of the series

$$u_{xx} \sim \sum_{n=1}^{\infty} \frac{\partial^2 u_n}{\partial x^2} = -\left(\frac{\pi}{l}\right)^2 \sum_{n=1}^{\infty} n^2 \left(A_n \cos \frac{\pi n}{l} at + B_n \sin \frac{\pi n}{l} at\right) \times \times \sin \frac{\pi n}{l} x,$$

$$u_{tt} \sim \sum_{n=1}^{\infty} \frac{\partial^2 u_n}{\partial t^2} = -\left(\frac{\pi a}{l}\right)^2 \sum_{n=1}^{\infty} n^2 \left(A_n \cos \frac{\pi n}{l} at + B_n \sin \frac{\pi n}{l} at\right) \times \times \sin \frac{\pi n}{l} x,$$

which has

$$\sum_{n=1}^{\infty} n^2 (|A_n| + |B_n|). \tag{36}$$

as a majorant series except for a coefficient of proportionality. Since

$$A_n = \varphi_n, \quad B_n = \frac{l}{\pi n a} \psi_n,$$

where

$$\varphi_n = \frac{2}{l} \int_0^l \varphi(x) \sin \frac{\pi n}{l} x\, dx, \qquad \psi_n = \frac{2}{l} \int_0^l \psi(x) \sin \frac{\pi n}{l} x\, dx,$$

our problem reduces to a proof of the convergence of the series

$$\left.\begin{aligned}&\sum_{n=1}^{\infty} n^k |\varphi_n| \quad (k = 0,\ 1,\ 2),\\ &\sum_{n=1}^{\infty} n^k |\psi_n| \quad (k = -1,\ 0,\ 1).\end{aligned}\right\} \tag{37}$$

We use the well-known* properties of the Fourier series.

If the periodic function $F(x)$ with the period $2l$ has k continuous derivatives, and the $(k+1)$th derivative is piecewise continuous then the numerical series

$$\sum_{n=1}^{\infty} n^k (|a_n| + |b_n|), \tag{38}$$

where a_n and b_n are Fourier coefficients, converges. If we have to expand the function $f(x)$ given in the interval $(0, l)$ in a series in $\sin (\pi n/l)\, x$ then the previous requirements must be fulfilled for the function $F(x)$ obtained for the odd extension of $f(x)$. In particular, for the continuity of $F(x)$ it is necessary that $f(0) = 0$, since otherwise the odd extension has a discontinuity at the point $x = 0$; similarly at the point $x = l$ one must have $f(l) = 0$ since the extended function is continuous and periodic with a period $2l$. The continuity of the first derivative at $x = 0$, $x = l$ holds automatically for the odd extension. In general, for the continuity of even derivatives of the extended function we must have

$$f^{(k)}(0) = f^{(k)}(l) = 0 \quad (k = 0, 2, 4, \ldots, 2n). \tag{39}$$

The odd derivatives are automatically continuous.

Thus, for the convergence of the series

$$\sum_{n=1}^{\infty} n^k |\varphi_n| \quad (k = 0, 1, 2)$$

we require that the initial displacement $\varphi(x)$ should satisfy the following conditions.

* See, for example, V. I. Smirnov, *A Course of Higher Mathematics* vol. II, 1937.

1° The derivatives of the function up to the second order should be continuous, the third piecewise continuous and besides,

$$\varphi(0) = \varphi(l) = 0; \quad \varphi''(0) = \varphi''(l) = 0. \tag{40}$$

For the convergence of the series

$$\sum_{n=1}^{\infty} n^k |\psi_n| \quad (k = -1, 0, 1)$$

the following requirements should be imposed on the initial velocity $\psi(x)$.

2° The function $\psi(x)$ is continuously differentiable, has a piecewise-continuous second derivative and besides

$$\psi(0) = \psi(l) = 0. \tag{41}$$

Thus we have proved that any vibration $u(x, t)$ with the initial functions $\varphi(x)$ and $\psi(x)$ satisfying the requirements 1° and 2° can be represented in the form of superposition of standing waves. The conditions 1° and 2° are sufficient conditions for the methods of proof adopted here.

A similar problem was solved in sub-section 5, § 2, by the method of propagating waves. The solution was

$$u(x, t) = \frac{\Phi(x - at) + \Phi(x + at)}{2} + \frac{1}{2a} \int_{x-at}^{x+at} \Psi(a)\, da, \tag{42}$$

where Φ and Ψ are extensions of the initial functions $\varphi(x)$ and $\psi(x)$ given in the segment $(0, l)$, which are odd with respect to 0 and l. The functions Φ and Ψ were shown to be periodic with a period $2l$ and therefore may be represented by the series

$$\Phi(x) = \sum_{n=1}^{\infty} \varphi_n \sin \frac{\pi n}{l} x, \qquad \Psi(x) = \sum_{n=1}^{\infty} \psi_n \sin \frac{\pi n}{l} x,$$

where φ_n and ψ_n are coefficients of the Fourier functions $\Phi(x)$ and $\Psi(x)$. Substituting these series in formula (42) and making use of the theorem of the sine and cosine of a sum and difference, we get

$$u(x, t) = \sum_{n=1}^{\infty} \left(\varphi_n \cos \frac{\pi n}{l} at + \frac{l}{\pi n a} \psi_n \sin \frac{\pi n}{l} at \right) \sin \frac{\pi n}{l} x, \tag{43}$$

which agrees with the expression obtained by the method of separation of variables.

Consequently formula (43) is valid with the same assumptions as formula (42) (see sub-section 1, § 3) which was obtained under the condition that the function $\Phi(x)$ is continuously differentiable twice and the function $\Psi(x)$ once.

For this to be true the functions $\varphi(x)$ and $\psi(x)$ must satisfy the following conditions

$$\begin{aligned} &\varphi(0) = \varphi(l) = 0, \quad \psi(0) = \psi(l) = 0, \\ &\varphi''(0) = \varphi''(l) = 0 . \end{aligned} \tag{44}$$

in addition to the conditions of differentiability.

Thus, conditions 1° and 2° which are sufficient for our proof of the method of separation of variables are more stringent than those sufficient to ensure the existence of a solution.

In justifying the possibility of representing the solution as a superposition of standing waves we used the method of proving the convergence of the series. The method can easily be applied to a number of other problems, although it imposes stricter restrictions than necessary on the initial functions.

4. *Inhomogeneous equations*

Let us consider the inhomogeneous wave equation

$$u_{tt} = a^2 u_{xx} + f(x, t), \quad a^2 = k/\varrho, \quad 0 < x < l \tag{45}$$

with the initial conditions

$$\left. \begin{aligned} u(x, 0) &= \varphi(x), \\ u_t(x, 0) &= \psi(x) \end{aligned} \right\} \; 0 \leq x \leq l \tag{46}$$

and the homogeneous boundary conditions

$$\left. \begin{aligned} u(0, t) &= 0, \\ u(l, t) &= 0, \end{aligned} \right\} \; t > 0 . \tag{47}$$

We shall find a solution expressed as a Fourier series in x

$$u(x, t) = \sum_{n=1}^{\infty} u_n(t) \sin \frac{\pi n}{l} x. \tag{48}$$

To find $u(x, t)$ it is necessary to determine the function $u_n(t)$. Let us represent the function $f(x, t)$ and the initial conditions as Fourier series:

$$\left.\begin{aligned}
f(x,t) &= \sum_{n=1}^{\infty} f_n(t)\sin\frac{\pi n}{l}x, & f_n(t) &= \frac{2}{l}\int_0^l f(\xi,t)\sin\frac{\pi n}{l}\xi\,d\xi;\\
\varphi(x) &= \sum_{n=1}^{\infty} \varphi_n\sin\frac{\pi n}{l}x, & \varphi_n &= \frac{2}{l}\int_0^l \varphi(\xi)\sin\frac{\pi n}{l}\xi\,d\xi;\\
\psi(x) &= \sum_{n=1}^{\infty} \psi_n\sin\frac{\pi n}{l}x, & \psi_n &= \frac{2}{l}\int_0^l \psi(\xi)\sin\frac{\pi n}{l}\xi\,d\xi.
\end{aligned}\right\} \quad (49)$$

Substituting the trial solution (48) in equation (45)

$$\sum_{n=1}^{\infty}\sin\frac{\pi n}{l}x\left\{-a^2\left(\frac{\pi n}{l}\right)^2 u_n(t)+\ddot{u}_n(t)+f_n(t)\right\}=0,$$

we see that it is satisfied if all the coefficients of the expansion are equal to zero, i. e.

$$\ddot{u}_n(t)+\left(\frac{\pi n}{l}\right)^2 a^2 u_n(t)=f_n(t). \quad (50)$$

We have obtained an ordinary differential equation with constant coefficients for $u_n(t)$. The initial conditions give

$$u(x,0)=\varphi(x)=\sum_{n=1}^{\infty}u_n(0)\sin\frac{\pi n}{l}x=\sum_{n=1}^{\infty}\varphi_n\sin\frac{\pi n}{l}x,$$

$$u_t(x,0)=\psi(x)=\sum_{n=1}^{\infty}\dot{u}_n(0)\sin\frac{\pi n}{l}x=\sum_{n=1}^{\infty}\psi_n\sin\frac{\pi n}{l}x$$

whence it follows that

$$\left.\begin{aligned}
u_n(0)&=\varphi_n,\\
\dot{u}_n(0)&=\psi_n.
\end{aligned}\right\} \quad (51)$$

These additional conditions completely determine the solution of the equation (50). The function $u_n(t)$ can be written as

$$u_n(t)=u_n^{(I)}(t)+u_n^{(II)}(t),$$

where

$$u_n^{(\mathrm{I})}(t) = \frac{l}{\pi n a} \int_0^t \sin \frac{\pi n}{l} a (t - \tau) \cdot f_n(\tau)\, d\tau \qquad (52)$$

si the solution of the non-homogeneous equation with zero initial conditions* and

$$u_n^{(\mathrm{II})}(t) = \varphi_n \cos \frac{\pi n}{l} at + \frac{l}{\pi n a} \psi_n \sin \frac{\pi n}{l} at \qquad (53)$$

is the solution of the homogeneous equation with the given initial conditions. Thus the required solution is

$$u(x, t) = \sum_{n=1}^{\infty} \frac{l}{\pi n a} \int_0^t \sin \frac{\pi n}{l} a (t - \tau) \sin \frac{\pi n}{l} x \cdot f_n(\tau)\, d\tau +$$

$$+ \sum_{n=1}^{\infty} \left(\varphi_n \cos \frac{\pi n}{l} at + \frac{l}{\pi n a} \psi_n \sin \frac{\pi n}{l} at \right) \sin \frac{\pi n}{l} x . \qquad (54)$$

The second sum represents a solution of the problem of free vibrations of a string with the given initial conditions and was investigated earlier. The first sum represents forced vibrations of a string under the action of an external force with zero initial conditions. Using expression (49) for $f_n(t)$ we find

$$u^{(\mathrm{I})}(x, t) = \int_0^t \int_0^l \left\{ \frac{2}{l} \sum_{n=1}^{\infty} \frac{l}{\pi n a} \sin \frac{\pi n}{l} a (t - \tau) \sin \frac{\pi n}{l} x \sin \frac{\pi n}{l} \xi \right\} \times$$

$$\times f(\xi, \tau)\, d\xi\, d\tau = \int_0^t \int_0^l G(x, \xi, t - \tau) f(\xi, \tau)\, d\xi\, d\tau, \qquad (55)$$

where

$$G(x, \xi, t - \tau) = \frac{2}{\pi a} \sum_{n=1}^{\infty} \frac{1}{n} \sin \frac{\pi n}{l} a (t - \tau) \sin \frac{\pi n}{l} x \sin \frac{\pi n}{l} \xi . \qquad (56)$$

We want to determine the physical significance of this solution. Let the function $f(\xi, \tau)$ differ from zero only in a small neighbourhood of the point $M_0 (\xi_0, \tau_0)$:

$$\xi_0 \leq \xi \leq \xi_0 + \Delta\xi , \quad \tau_0 \leq \tau \leq \tau_0 + \Delta\tau .$$

* See paragraph in small type at the end of this section.

The function $\varrho f(\xi, \tau)$ represents the density of the force; the force applied to the region $(\xi_0, \xi_0 + \Delta\xi)$ is

$$F(\tau) = \varrho \int_{\xi_0}^{\xi_0+\Delta\xi} f(\xi, \tau)\, d\xi\,,$$

and

$$I = \int_{\tau_0}^{\tau_0+\Delta\tau} F(\tau)\, d\tau = \varrho \int_{\tau_0}^{\tau_0+\Delta\tau} \int_{\xi_0}^{\xi_0+\Delta\xi} f(\xi, \tau)\, d\xi\, d\tau$$

is the impulse of this force during the time $\Delta\tau$. If the mean value theorem is applied to the expression

$$u(x, t) = \int_0^t \int_0^l G(x, \xi, t - \tau) f(\xi, \tau)\, d\xi\, d\tau =$$

$$= \int_{\tau_0}^{\tau_0+\Delta\tau} \int_{\xi_0}^{\xi_0+\Delta\xi} G(x, \xi, t - \tau) f(\xi, \tau)\, d\xi\, d\tau\,,$$

we have

$$u(x, t) = G(x, \bar{\xi}, t - \bar{\tau}) \int_{\tau_0}^{\tau_0+\Delta\tau} \int_{\xi_0}^{\xi_0+\Delta\xi} f(\xi, \tau)\, d\xi\, d\tau\,, \tag{57}$$

where

$$\xi_0 < \bar{\xi} < \xi_0 + \Delta\xi, \quad \tau_0 < \bar{\tau} < \tau_0 + \Delta\tau.$$

In formula (57), going to the limit when $\Delta\xi \to 0$ and $\Delta\tau \to 0$, we get

$$u(x, t) = G(x, \xi_0, t - \tau_0) \frac{I}{\varrho}\,, \tag{58}$$

which can be considered as the effect of the instantaneous impulse of magnitude I.

If the function $(I/\varrho)\, G(x, \xi, t - \tau)$ represents the effect of a concentrated impulse, then it is immediately obvious that the effect of the continuously distributed force $f(x, t)$ must be represented by the formula

$$u(x, t) = \int_0^l \int_0^t G(x, \xi, t - \tau) f(\xi, \tau)\, d\xi\, d\tau\,, \tag{59}$$

which coincides with the formula (55).

The function giving the effect of a concentrated impulse for the infinite string was considered in the previous section. It is a piecewise-constant function equal to $I/2a\varrho$ within the upper characteristic angle for the point (ξ, τ) and zero outside this angle. The function giving the effect of the concentrated impulse for the string with fixed ends $(0, l)$ may be obtained from the corresponding function for the infinite string by odd extension with respect to the points $x = 0$ and $x = l$.

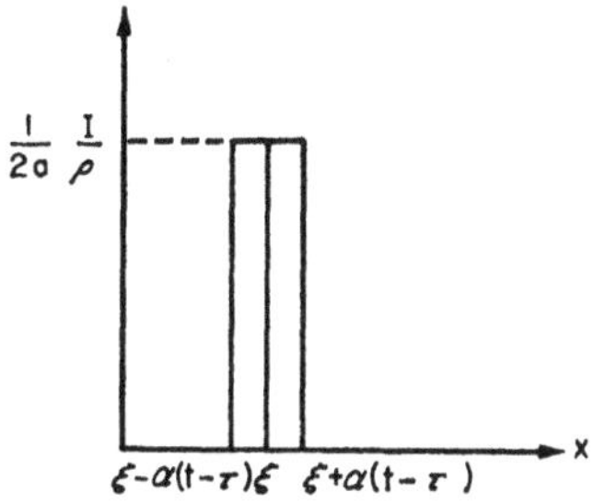

FIG. 23

Let us consider a moment t sufficiently close to τ, when the effect of reflections from the ends $x = 0$ and $x = l$ is still not felt. The impulse function is shown in the graph given in Fig. 23. We expand this function (putting $I = \varrho$) into a Fourier series in $\sin (\pi n/l) x$.

The coefficients of the Fourier series will be

$$A_n = \frac{2}{l} \int_0^t G(\alpha, \xi, t-\tau) \sin \frac{\pi n}{l} \alpha \, d\alpha = \frac{1}{al} \int_{\xi - a(t-\tau)}^{\xi + a(t-\tau)} \sin \frac{\pi n}{l} \alpha \, d\alpha =$$
$$= \frac{1}{a\pi n} \left\{ \cos \frac{\pi n}{l} [\xi - a(t-\tau)] - \cos \frac{\pi n}{l} [\xi + a(t-\tau)] \right\} =$$
$$= \frac{2}{a\pi n} \sin \frac{\pi n}{l} \xi \sin \frac{\pi n}{l} a(t-\tau).$$

Hence we obtain the formula

$$G(x, \xi, t-\tau) = \frac{2}{\pi a} \sum_{n=1}^{\infty} \frac{1}{n} \sin \frac{\pi n}{l} a(t-\tau) \sin \frac{\pi n}{l} x \cdot \sin \frac{\pi n}{l} \xi, \tag{60}$$

which coincides with formula (56) found by the method of separation of variables.

For the values $t \geqslant \tau$ when the effect of the fixed ends begins to be felt, it is cumbersome to construct the impulse function using characteristic curves; but the representation as a Fourier series is still valid.

We shall investigate the solution of the following problem without discussing the conditions of applicability of the formula obtained.

Let us consider a non-homogeneous linear equation with constant coeffcients

$$L(u) = u^{(n)} + p_1 u^{(n-1)} + \ldots + p_{n-1} u^{(1)} + p_n u = f(t) \qquad (1^*)$$

$$u^{(i)} = \left(\frac{d^i u}{dt^i}\right)$$

and the initial conditions

$$u^{(i)}(0) = 0 \quad (i = 0, 1, \ldots, n-1). \qquad (2^*)$$

Its solution is given by the formula

$$u(t) = \int_0^t U(t-\tau) f(\tau)\, d\tau, \qquad (3^*)$$

where $U(t)$ is the solution of a homogeneous equation

$$L(U) = 0$$

with the initial conditions

$$U^{(i)}(0) = 0 \ (i = 0, 1, \ldots, n-2), \ U^{(n-1)}(0) = 1. \qquad (4^*)$$

Calculating the derivatives of $u(t)$ by the differentiation of the right-hand side with respect to t, we find

$$\left.\begin{aligned}
u^{(1)}(t) &= \int_0^t U^{(1)}(t-\tau) f(\tau)\, d\tau + U(0) f(t) \quad [U(0) = 0], \\
u^{(2)}(t) &= \int_0^t U^{(2)}(t-\tau) f(\tau)\, d\tau + U^{(1)}(0) f(t) \quad [U^{(1)}(0) = 0], \\
&\cdots\cdots\cdots\cdots\cdots\cdots\cdots\cdots \\
u^{(n-1)}(t) &= \int_0^t U^{(n-1)}(t-\tau) f(\tau)\, d\tau + U^{(n-2)}(0) f(t) \quad [U^{(n-2)}(0) = 0], \\
u^{(n)}(t) &= \int_0^t U^{(n)}(t-\tau) f(\tau)\, d\tau + U^{(n-1)}(0) f(t) \quad [U^{(n-1)}(0) = 1].
\end{aligned}\right\} \qquad (5^*)$$

Substituting these derivatives in equation (1*) we get:

$$L(u) = \int_0^t L[U(t-\tau)] f(\tau)\, d\tau + f(t) = f(t),$$

i. e. the equation is satisfied. It is obvious that the initial conditions (2*) are also fulfilled.

We can give a graphic physical interpretation of the function $U(t)$ and the formula (3*). Usually the function $u(t)$ represents the displacement of some system and $f(t)$ the force acting on this system. Let our system be in a state of rest for $t < 0$, and let the force be represented by the function $f_\varepsilon(t)$ (> 0) which is different from zero only in the interval of time $0 < t < \varepsilon$. The impulse of this force will be denoted by

$$I = \int_0^t f(\tau)\, d\tau .$$

Let $u_\varepsilon(t)$ be the function corresponding to $f_\varepsilon(t)$ considering ε as a parameter and putting $I = 1$. It can be shown that when $\varepsilon \to 0$ there is a $\lim\limits_{\varepsilon\to 0} u_\varepsilon(t)$ independent of the method of selecting $f_\varepsilon(t)$ and that this limit is equal to the function $U(t)$ defined above

$$U(t) = \lim_{\varepsilon\to 0} u_\varepsilon(t),$$

if we put $U(t) \equiv 0$ for $t < 0$. Thus the function $U(t)$ can naturally be called the function of the effect of an instantaneous impulse.

Considering the formula (3*) and applying the mean value theorem we get

$$u_\varepsilon(t) = U(t - \tau_\varepsilon^*) \int_0^\varepsilon f(\tau)\, d\tau = U(t - \tau_\varepsilon^*) \quad (0 \leqslant \tau_\varepsilon^* < \varepsilon < t) .$$

Passing to the limit when $\varepsilon \to 0$, we see that there is a limit

$$\lim_{\varepsilon\to 0} u_\varepsilon(t) = \lim_{\varepsilon\to 0} U(t - \tau_\varepsilon^*) = U(t),$$

which proves our statement.

Now let us represent the solution of a non-homogeneous equation using $U(t)$ as the function of the effect of the instantaneous impulse. Dividing the interval $(0, t)$ by the points τ_i into equal parts

$$\Delta\tau = t/m,$$

we represent the function $f(t)$ in the form

$$f(t) = \sum_{i=1}^{m} f_i(t),$$

where

$$f_i(t) = \begin{cases} 0 & \text{where } t < \tau_i \text{ and } t \geqslant \tau_{i+1}, \\ f(t) & \text{where } \tau_i \leqslant t < \tau_{i+1}. \end{cases}$$

Then the function

$$u(t) = \sum_{i=1}^{m} u_i(t),$$

where $u_i(t)$ are solutions of the equation $L(u_i) = f_i$ with zero initial data'

If m is sufficiently large then the function $u_i(t)$ may be considered as the function giving the effect of the instantaneous impulse with the intensity

$$I = f_i(\tau_i)\, \Delta\tau = f(\tau_i)\, \Delta\tau,$$

so that

$$u(t) = \sum_{i=1}^{m} U(t - \tau_i) f(\tau_i) \Delta\tau \underset{\Delta\tau \to 0}{\to} \int_0^t U(t - \tau) f(\tau) d\tau,$$

i. e. we arrive at the formula

$$u(t) = \int_0^t U(t - \tau) f(\tau) d\tau,$$

which shows that the effect of the continuously acting force may be represented by the superposition of the effects of instantaneous impulses.

In the case considered above $u_n^{(1)}$ satisfies the equation (50) and the conditions $u_n(0) = \dot{u}_n(0) = 0$. For the impulse function $U(t)$ we have:

$$\ddot{U} + \left(\frac{\pi n}{l}\right)^2 a^2 U = 0, \qquad U(0) = 0, \qquad \dot{U}(0) = 1,$$

so that

$$U(t) = \frac{l}{\pi n a} \sin \frac{\pi n}{l} at.$$

Hence (3*) we obtain the formula (52)

$$u_n^{(1)}(t) = \int_0^t U(t - \tau) f_n(\tau) = \frac{l}{\pi n a} \int_0^t \sin \frac{\pi n}{l} a(t - \tau) f_n(\tau) d\tau.$$

The integral representation (3*) of the solution of the ordinary differential equation (1*), obtained above, has the same physical significance as formula (59) which gives an integral representation of the solution of the homogeneous were equation of vibrations.

5. *First general boundary problem*

Let us consider the *first general boundary problem* for the wave equation:

$$u_{tt} = a^2 u_{xx} + f(x, t) \tag{45}$$

with the additional conditions

$$\left.\begin{aligned} u(x, 0) &= \varphi(x), \\ u_t(x, 0) &= \psi(x); \end{aligned}\right\} \tag{46}$$

$$\left.\begin{aligned} u(0, t) &= \mu_1(t), \\ u(l, t) &= \mu_2(t). \end{aligned}\right\} \tag{47'}$$

Let us introduce a new unknown function $v(x, t)$ putting

$$u(x, t) = U(x, t) + v(x, t),$$

so that $v(x, t)$ represents the deviation of the function $u(x, t)$ from some known function $U(x, t)$.

This function $v(x, t)$ will be defined as the solution of the equation

$$v_{tt} = a^2 v_{xx} + \bar{f}(x, t), \quad \bar{f}(x, t) = f(x, t) - [U_{tt} - a^2 U_{xx}]$$

with the additional conditions

$$\begin{aligned} v(x, 0) &= \bar{\varphi}(x), & \bar{\varphi}(x) &= \varphi(x) - U(x, 0), \\ v_t(x, 0) &= \bar{\psi}(x); & \bar{\psi}(x) &= \psi(x) - U_t(x, 0); \\ v(0, t) &= \bar{\mu}_1(t), & \bar{\mu}_1(t) &= \mu_1(t) - U(0, t), \\ v(l, t) &= \bar{\mu}_2(t); & \bar{\mu}_2(t) &= \mu_2(t) - U(l, t). \end{aligned}$$

Let us choose the auxiliary function $U(x, t)$ in such a way that

$$\bar{\mu}_1(t) = 0 \text{ or } \bar{\mu}_2(t) = 0;$$

a simple choice would be

$$U(x, t) = \mu_1(t) + \frac{x}{l}\left[\mu_2(t) - \mu_1(t)\right].$$

The general boundary problem for the function $u(x, t)$ is thus reduced to a boundary problem for the function $v(x, t)$ with zero boundary conditions. The method of solution of this problem is described above (see sub-section 4).

6. *Boundary problems with stationary inhomogeneities*

Boundary problems with stationary inhomogeneities form a very important class of problems. Here the boundary conditions and the right-hand side of the equation do not depend on time

$$u_{tt} = a^2 u_{xx} + f_0(x), \tag{45'}$$

$$\left.\begin{aligned} u(x, 0) &= \varphi(x), \\ u_t(x, 0) &= \psi(x), \end{aligned}\right\} \tag{46}$$

$$\left.\begin{aligned} u(0, t) &= u_1, \\ u(l, t) &= u_2. \end{aligned}\right\} \tag{47'}$$

In this class it is natural to look for the solution as a sum

$$u(x, t) = \bar{u}(x) + v(x, t),$$

where $\bar{u}(x)$ is the "stationary state" defined by the conditions

$$a^2 \bar{u}''(x) + f_0(x) = 0,$$
$$\bar{u}(0) = u_1,$$
$$\bar{u}(l) = u_2,$$

and $v(x, t)$ the deviation from the stationary state. It is easy to see that the function $\bar{u}(x)$ is

$$\bar{u}(x) = u_1 + (u_2 - u_1)\frac{x}{l} +$$
$$+ \frac{x}{l}\int_0^l d\xi_1 \int_0^{\xi_1} \frac{f_0(\xi_2)}{a^2}\, d\xi_2 - \int_0^x d\xi_1 \int_0^{\xi_1} \frac{f_0(\xi_2)}{a^2}\, d\xi_2 .$$

In particular, if $f_0 =$ const. then

$$\bar{u}(x) = u_1 + (u_2 - u_1)\frac{x}{l} + \frac{f_0}{2a^2}(lx - x^2).$$

The function $v(x, t)$ obviously satisfies the homogeneous equation

$$v_{tt} = a^2 v_{xx}$$

with the homogeneous boundary conditions

$$v(0, t) = 0,$$
$$v(l, t) = 0$$

and the initial conditions

$$v(x, 0) = \bar{\varphi}(x),\ \bar{\varphi}(x) = \varphi(x) - \bar{u}(x),$$
$$v_t(x, 0) = \psi(x).$$

This v is a solution of the simplest boundary problem considered in sub-section 1 of the present section.

In deriving the wave equation for a string we did not consider the effect of the force of gravity. From what has been said above it follows that instead of considering the force of gravity (and generally forces independent of time) explicitly we can simply take the deviation from the stationary state.

Let us solve the simplest problem of this type with zero initial conditions:

$$u_{tt} = a^2 u_{xx} + f_0(x) \tag{45''}$$

$$u(x, 0) = 0, \; u_t(x, 0) = 0, \tag{46'}$$

$$u(0, t) = u_1, \; u(l, t) = u_2. \tag{47''}$$

In this case we obtain the following problem for the function $v(x, t)$

$$v_{tt} = a^2 v_{xx},$$

$$v(x, 0) = \varphi(x) = -\bar{u}(x), \;\; v(x, 0) = 0,$$

$$v(0, t) = 0, \;\; v(l, t) = 0.$$

There is no need to use the explicit analytical expression for $\bar{u}(x)$ in solving this problem. We can obtain the solution directly in terms of u_1, u_2 and $f_0(x)$.

The expression for $v(x, t)$ according to formula (17) is

$$v(x, t) = \sum_{n=1}^{m} (A_n \cos a \sqrt{\lambda_n} \cdot t + B_n \sin a \sqrt{\lambda_n} \cdot t) X_n(x),$$

where

$$X_n(x) = \sin \sqrt{\lambda_n} \cdot x \quad (\sqrt{\lambda_n} = \pi n/l)$$

is the eigen-function of the following boundary problem

$$X'' + \lambda X = 0, \tag{8}$$

$$X(0) = 0, \quad X(l) = 0. \tag{10}$$

From the initial conditions it follows that

$$B_n = 0$$

and

$$A_n = -\frac{2}{l} \int_0^l \bar{u}(x) X_n(x)\, dx.$$

The following method is very convenient for calculating this integral. Using equation (8) we find

$$X_n(x) = -\frac{1}{\lambda_n} X_n''(x).$$

Let us substitute this expression in the formula for A_n and integrate by parts twice

$$A_n = \frac{2}{l\lambda_n} \int_0^l \bar{u}(x) X_n''(x)\, dx = \frac{2}{l\lambda_n} \left\{ \bar{u} X_n'(x) \Big|_0^l - \bar{u}' X_n \Big|_0^l + \int_0^l \bar{u}'' X_n(x)\, dx \right\},$$

whence, considering the equation and the boundary conditions for $\bar{u}(x)$, we find:

$$A_n = \frac{2}{l\lambda_n}\left[u_2 X_n'(l) - u_1 X_n'(0) - \int_0^l \frac{f_0(x)}{a^2} X_n(x)\,dx\right]$$

or

$$A_n = \frac{2}{\pi n}\left[u_2(-1)^n - u_1 + \int_0^l \frac{f_0(x)}{a^2} X_n(x)\,dx\right].$$

In particular, for the homogeneous equation ($f_0(x) = 0$) we have

$$A_n = \frac{2}{l\sqrt{\lambda_n}}[u_2(-1)^n - u_1] = \frac{2}{\pi n}[u_2(-1)^n - u_1].$$

It is convenient to use this method for calculating the Fourier coefficients for boundary conditions of the second and third kind and also for the boundary-value problems for an inhomogeneous string

$$\frac{d}{dx}\left[k(x)\frac{dX}{dx}\right] + \lambda\varrho(x)X = 0,$$

if the eigen-functions and eigen-values are known.

7. *Steady state problems*

As was shown above, the problem of the vibration of a string with arbitrary boundary conditions may be reduced to the solution of an inhomogeneous equation with zero boundary conditions.

But this method often complicates the problem when the solution may be found directly.

In studying the effect of the boundary conditions it is important to find some particular solution (of a homogeneous equation) satisfying the given boundary conditions, because calculation of the effect of the initial conditions is reduced to a solution of the same equation with zero boundary conditions.

If the boundary conditions act for a sufficiently long time, then due to friction inherent in a real physical system, the effect of the initial conditions decreases. Thus we arrive naturally at a steady state problem.

To find the solution of the equation

$$u_{tt} = a^2 u_{xx} - \alpha u_t \qquad (\alpha > 0) \tag{61}$$

with the given boundary conditions

$$u(0,t) = \mu_1(t),$$
$$u(l,t) = \mu_2(t).$$

The term au_t on the right-hand side of the equation represents friction which is proportional to the velocity.

Let us consider first the problem with periodic boundary conditions:

$$u(l,t) = A \cos \omega t \quad (\text{or } u(l,t) = B \sin \omega t), \tag{62}$$

$$u(0,t) = 0. \tag{63}$$

It is more convenient later on if we write the boundary condition in the complex form

$$u(l,t) = Ae^{i\omega t}. \tag{64}$$

If

$$u(x,t) = u^{(1)}(x,t) + iu^{(2)}(x,t)$$

satisfies the equation (61) with the boundary conditions (62) and (63) then $u^{(1)}(x,t)$ and $u^{(2)}(x,t)$ — its real and imaginary parts — satisfy separately the same equation (because of its linearity), the condition (63) and the boundary conditions at $x = l$

$$u^{(1)}(l,t) = A \cos \omega t,$$
$$u^{(2)}(l,t) = A \sin \omega t.$$

Thus we look for a solution of the problem

$$\left.\begin{aligned} u_{tt} &= a^2 u_{xx} - au_t, \\ u(0,t) &= 0 \\ u(l,t) &= Ae^{i\omega t}. \end{aligned}\right\} \tag{65}$$

Putting

$$u(x,t) = X(x)\, e^{i\omega t}$$

and substituting this expression in the equation we get the following problem for the function $X(x)$:

$$X'' + k^2 X = 0 \qquad \left(k^2 = \frac{\omega^2}{a^2} - i\, a \frac{\omega}{a^2}\right), \tag{66}$$

$$X(0) = 0, \tag{67}$$

$$X(l) = A. \tag{68}$$

From equation (66) and the boundary condition (67) we find:

$$X(x) = C \sin kx.$$

The condition at $x = l$ gives

$$C = \frac{A}{\sin kl}, \tag{69}$$

so that

$$X(x) = A \frac{\sin kx}{\sin kl} = X_1(x) + iX_2(x), \tag{70}$$

where $X_1(x)$ and $X_2(x)$ are the real and imaginary parts of $X(x)$.

The required solution is

$$u(x, t) = \left[X_1(x) + iX_2(x)\right] e^{i\omega t} = u^{(1)}(x, t) + u^{(2)}(x, t),$$

where

$$u^{(1)}(x, t) = X_1(x) \cos \omega t - X_2(x) \sin \omega t,$$

$$u^{(2)}(x, t) = X_2(x) \sin \omega t + X_2(x) \cos \omega t.$$

Passing to the limit when $a = 0$ we find that

$$\bar{k} = \lim_{a \to 0} k = \frac{\omega}{a} \tag{71}$$

and correspondingly

$$\bar{u}^{(1)}(x, t) = \lim_{a \to 0} u^{(1)}(x, t) = A \frac{\sin \frac{\omega}{a} x}{\sin \frac{\omega}{a} l} \cos \omega t, \tag{72}$$

$$\bar{u}^{(2)}(x, t) = \lim_{a \to 0} u^{(2)}(x, t) = A \frac{\sin \frac{\omega}{a} x}{\sin \frac{\omega}{a} l} \sin \omega t. \tag{73}$$

The functions $u^{(1)}(x, t)$ and $u^{(2)}(x, t)$ are obviously solutions of the equation

$$u_{tt} = a^2 u_{xx}$$

satisfying the boundary conditions

$$\left.\begin{array}{ll} \bar{u}^{(1)}(0, t) = 0, & \bar{u}^{(2)}(0, t) = 0, \\ \bar{u}^{(1)}(l, t) = A \cos \omega t, & \bar{u}^{(2)}(l, t) = A \sin \omega t. \end{array}\right\} \tag{II}$$

For $a = 0$ the solution of the problem does not always exist. If the frequency of forced vibrations coincides with the eigenfrequency ω_n of the vibrations of the string with fixed ends

$$\omega = \omega_n = \frac{\pi n}{l} a,$$

then the denominator in the formulae for $u^{(1)}$ and $u^{(2)}$ becomes zero and the solution of the steady state problem does not exist.

This fact has a simple physical significance: when resonance sets in, no steady state exists. The amplitude increases without limit as t increases.

When friction is present ($\alpha \neq 0$) a steady state is possible for any ω.

It $f(t)$ is a periodic function which can be represented as a series

$$f(t) = \frac{A_0}{2} + \sum_{n=1}^{\infty} (A_n \cos \omega nt + B_n \sin \omega nt), \tag{74}$$

where $2\pi/\omega$ is the period and A_n and B_n Fourier coefficients then the solution of the problem for the case $\alpha = 0$ is

$$\bar{u}(x, t) = \frac{A_0}{2l} x + \sum_{n=1}^{\infty} (A_n \cos \omega nt + B_n \sin \omega nt) \frac{\sin \frac{\omega n}{a} x}{\sin \frac{\omega n}{a} l},$$

provided none of the frequencies ω_n coincide with the eigenfrequencies of the string with fixed ends.

If $f(t)$ is a non-periodic function then by resolving it into a Fourier's integral, it is possible to obtain a solution in the integral form by a similar method.

It should be noted that the solution of the steady state problem with $\alpha = 0$ is not determined uniquely if no additional conditions are imposed. By adding to the solution any combination of standing waves

$$\sum \left(A_n \cos \frac{\pi n}{l} at + B_n \sin \frac{\pi n}{l} at \right) \cos \frac{\pi n}{l} x,$$

where A_n and B_n are arbitrary constants we obtain a function satisfying the same equation and the same boundary conditions.

To obtain a unique solution of the problem (I) for $\alpha = 0$, we introduce the additional condition of "vanishing friction".

We say that the solution of the problem (II) satisfies the condition of "vanishing friction" if it is a solution of the problem (I) when $a \to 0$.

The problem can be solved similarly if the end $x = l$ is fixed and the boundary conditions are given for $x = 0$.

The solution of the general steady state problem

$$u(0, t) = \mu_1(t), \quad u(l, t) = \mu_2(t)$$

is determined as a sum of two terms each of which satisfies inhomogeneous boundary conditions at one end and zero boundary conditions at the other.

We shall prove the uniqueness of a bounded solution of the steady state problem for equation (61). We assume the continuity of the solution together with its derivatives up to the second order in the region $0 \leqslant x \leqslant l$, $-\infty < t < t_0$ inclusively, if the boundary conditions

$$u(0, t) = \mu_1(t), \quad u(l, t) = \mu_2(t)$$

are defined for $-\infty < t < t_0$.

Let $u_1(x, t)$ and $u_2(x, t)$ be two bounded solutions of problem (I), then

$$|u_1| < M, \quad |u_2| < M,$$

where $M > 0$ is some number.

The difference of these functions

$$v(x, t) = u_1(x, t) - u_2(x, t)$$

is bounded ($|v| < 2M$), satisfies the equation (61) and the homogeneous boundary conditions

$$v(0, t) = 0, \quad v(l, t) = 0.$$

The Fourier coefficients for the function v

$$v_n(t) = \frac{2}{l} \int_0^l v(x, t) \sin \frac{\pi n}{l} x \, dx$$

satisfy the equation

$$\ddot{v}_n + a\dot{v}_n + \omega_n^2 v_n = 0 \left(\omega_n = \frac{\pi n}{l} u\right), \tag{*}$$

since the second derivatives of the function $v(x, t)$ are continuous for $0 \leqslant x \leqslant l$.

The general solution of equation (*) is

$$v_n(t) = A_n e^{-q_n^{(1)} t} + B_n e^{-q_n^{(2)} t}, \tag{**}$$

where $q_n^{(1)}$ and $q_n^{(2)}$ are roots of the characteristic equation,

$$q_n^{(1)} = -\frac{a}{2} + \sqrt{\left(\frac{a^2}{4} - \omega_n^2\right)}, \quad q_n^{(2)} = -\frac{a}{2} - \sqrt{\left(\frac{a^2}{4} - \omega_n^2\right)} \; (a > 0). \qquad (***)$$

Only two cases are possible: (1) the roots are real and negative, (2) the roots are complex and have a negative real part. Hence it follows that any solution (**) of equation (*) is either identically equal to zero, or increases in absolute value without limit when $t \to -\infty$. In our case $|v_n| < 4M$ for and hence $v_n \equiv 0$ for all n.

Thus

$$v(x, t) \equiv 0 \text{ and } u_1(x, t) \equiv u_2(x, t). \qquad (****)$$

8. *Concentrated force*

Let us consider the problem of vibrations of a string under the action of a concentrated force applied at the point $x = x_0$. If the force is distributed over a certain region $(x_0 - \varepsilon, x_0 + \varepsilon)$ then the solution is found from formula (55). Making the limiting transition when $\varepsilon \to 0$, the solution of the given problem may be obtained.

On the other hand, while deducing the equation of vibrations we saw [see (8), sub-section 1, § 1] that at the point x_0 where the concentrated force is applied, the first derivative becomes discontinuous but the function itself remains continuous. The solution of the problem $u(x, t)$ of vibrations of a string under the action of a force concentrated at the point x_0 may be represented by two different functions:

$$\left.\begin{aligned} u(x, t) &= u_1(x, t) \text{ where } 0 \leqslant x \leqslant x_0, \\ u(x, t) &= u_2(x, t) \text{ where } x_0 \leqslant x \leqslant l. \end{aligned}\right\} \qquad (75)$$

These functions should satisfy the equation

$$u_{tt} = a^2 u_{xx}, \qquad (76)$$

the boundary and initial conditions

$$\left.\begin{aligned} u_1(0, t) &= 0, \quad u(x, 0) = \varphi(x), \\ u_2(l, t) &= 0; \quad u_t(x, 0) = \psi(x) \end{aligned}\right\} \qquad (77)$$

and the connecting conditions at the point $x = x_0$ consisting of the condition of continuity of the function $u(x, t)$:

$$u_1(x_0, t) = u_2(x_0, t), \qquad (78)$$

and of the condition relating the magnitude of the discontinuity with the force $f(t)$ acting at the point x_0:

$$\frac{\partial u}{\partial x}\Big|_{x_0-0}^{x_0+0} = \frac{\partial u_2}{\partial x}(x_0, t) - \frac{\partial u_1}{\partial x}(x_0, t) = -\frac{f(t)}{k}. \tag{79}$$

There is no need to worry about satisfying the initial conditions. If we find the particular solution of the equations (76) satisfying the boundary condition (77) and also (78) and (79), then by adding to it a solution of the homogeneous wave equation we can always satisfy the given initial conditions.

Let us consider the particular case

$$f(t) = A \cos \omega t, \quad -\infty < t < +\infty$$

and find the steady state solution satisfying the boundary conditions. We look for the solution in the form

$$u_1(x, t) = X_1(x) \cos \omega t \quad \text{where} \quad 0 \leqslant x \leqslant x_0,$$

$$u_2(x. t) = X_2(x) \cos \omega t \quad \text{where} \quad x_0 \leqslant x \leqslant l.$$

From equation (76) it follows that:

$$\left.\begin{aligned} X_1'' + \left(\frac{\omega}{a}\right)^2 X_1 = 0 \quad \text{where} \quad 0 \leqslant x \leqslant x_0, \\ X_2'' + \left(\frac{\omega}{a}\right)^2 X_2 = 0 \quad \text{where} \quad x_0 \leqslant x \leqslant l. \end{aligned}\right\} \tag{80}$$

The functions X_1 and X_2, moreover, should satisfy the boundary conditions

$$X_1(0) = 0, \quad X_2(l) = 0, \tag{81}$$

which follow from (77) and the connecting conditions

$$X_1(x_0) = X_2(x_0), \quad X_1'(x_0) - X_2'(x_0) = A/k, \tag{82}$$

which follow from (78) and (79).

From equation (80) and conditions (81) we find:

$$X_1(x) = C \sin \frac{\omega}{a} x, \quad X_2(x) = D \sin \frac{\omega}{a}(l - x);$$

the connecting conditions (82) give:

$$C \sin \frac{\omega}{a} x_0 - D \sin \frac{\omega}{a}(l - x_0) = 0,$$

$$C \frac{\omega}{a} \cos \frac{\omega}{a} x_0 + D \frac{\omega}{a} \cos \frac{\omega}{a}(l - x_0) = \frac{A}{k}.$$

Determining the coefficients C and D from these we obtain:

$$u(x,t) = \begin{cases} u_1 = \dfrac{Aa}{k\omega} \dfrac{\sin\dfrac{\omega}{a}(l-x_0)}{\sin\dfrac{\omega}{a} l} \sin\dfrac{\omega}{a} x \cos \omega t, & \text{where } 0 \leqslant x \leqslant x_0, \\[2ex] u_2 = \dfrac{Aa}{k\omega} \dfrac{\sin\dfrac{\omega}{a} x_0}{\sin\dfrac{\omega}{a} l} \sin \dfrac{\omega}{a}(l-x) \cos \omega t, & \text{where } x_0 \leqslant x \leqslant l. \end{cases}$$

A similar solution is obtained when $f(t) = A \sin \omega t$.

Thus we have the solution for the case $f(t) = A \cos \omega t$ or $f(t) = A \sin \omega t$. If $f(t)$ is a periodic function with period $2\pi/\omega$

$$f(t) = \frac{a_0}{2} + \sum_{n=1}^{\infty} (a_n \cos \omega nt + \beta_n \sin \omega nt)$$

then obviously

$$u(x,t) = \begin{cases} u_1 = \dfrac{1}{k}\left\{\dfrac{a_0 x}{2}\left(1-\dfrac{x_0}{l}\right) + \displaystyle\sum_{n=1}^{\infty} \dfrac{a \sin\dfrac{\omega n}{a}(l-x_0)}{\omega n \sin\dfrac{\omega n}{a} l} \sin\dfrac{\omega n x}{a} \times \right. \\ \qquad \times (a_n \cos \omega nt + \beta_n \sin \omega nt), \quad 0 \leqslant x \leqslant x_0; \qquad (83)^{*} \\ u_2 = \dfrac{1}{k}\left\{\dfrac{a_0 x_0}{2}\left(1-\dfrac{x}{l}\right) + \displaystyle\sum_{n=1}^{\infty} \dfrac{a \sin\dfrac{\omega n}{a} x_0}{\omega n \sin\dfrac{\omega n}{a} l} \sin\dfrac{\omega n(l-x)}{a} \times \right. \\ \qquad \times (a_n \cos \omega nt + \beta_n \sin \omega nt) \quad x_0 \leqslant x \leqslant l. \end{cases}$$

If the function $f(t)$ is not periodic then by representing it as a Fourier integral the solution may be obtained in integral form by a similar method.

* The first terms of these sums correspond to the static deflection which can be determined from the magnitude of the force

$$f(t) = \frac{a_0}{2} = \text{const},$$

by the functions

$$u = \begin{cases} u_1(x,t) = u_1(x) = \dfrac{1}{k}\dfrac{a_0}{2} x \left(1 - \dfrac{x_0}{l}\right) \text{ where } 0 \leqslant x \leqslant x_0 \\[2ex] u_2(x,t) = u_2(x) = \dfrac{1}{k}\dfrac{a_0}{2} x_0 \left(1 - \dfrac{x}{l}\right) \text{ where } x_0 \leqslant x \leqslant l. \end{cases}$$

Resonance arises if one denominator in the functions (83) is equal to zero

$$\sin \frac{\omega n l}{a} = 0, \qquad \omega n = \frac{\pi m}{l} a = \omega_m,$$

i. e. if the spectrum of frequencies of the exciting force contains one of the natural frequencies of vibration.

If the point of application of the force x_0 is one of the nodes of the standing wave corresponding to free vibration with the frequency ω_m, then

$$\sin \frac{\omega_m}{a} x_0 = 0, \quad \sin \frac{\omega_m}{a} (l - x_0) = 0 .$$

The numerators of the corresponding terms for u become zero and resonance does not occur. If the point of application of the force acting with a frequency ω_m is an antinode of the corresponding standing wave with the frequency ω_m, then

$$\sin \frac{\omega_m}{a} x_0 = 1,$$

and the resonance phenomenon is a maximum.

9. *General scheme of the method of separation of variables*

The method of separation of variables is applicable not only to the wave equation of a homogeneous string $u_{tt} = a^2 u_{xx}$, but also to the more general wave equation

$$L[u] = \frac{\partial}{\partial x}\left[k(x) \frac{\partial u}{\partial x}\right] - q(x) u = \varrho(x) \frac{\partial^2 u}{\partial t^2} . \tag{84}$$

Let us consider, for example, the first boundary problem for equation (84)

$$u(0, t) = 0, \quad u(l, t) = 0 ; \tag{85}$$

$$u(x, 0) = \varphi(x), \; u_t(x, 0) = \psi(x) . \tag{86}$$

Here k, q and ϱ depend on x and are positive ($k > 0$, $\varrho > 0$, $q \geqslant 0$).* Let us solve this problem by the method of separation of variables. To find the particular solutions we investigate the auxiliary problem:

* The case when $k(x)$ at some points becomes zero is considered separately (see Appendix).

Find a non-trivial solution of equation (84) satisfying the boundary conditions

$$u(0,t)=0, \quad u(l,t)=0$$

and which can be represented in the form of a product

$$u(x,t)=X(x)\,T(t)\,.$$

Substituting the proposed form of the solution in the equation and using the boundary conditions we get, after separation of variables:

$$\frac{d}{dx}\left[k(x)\frac{dX}{dx}\right]-qX+\lambda\varrho X==0,$$

$$T''+\lambda T=0\,.$$

We get the following eigen-value problem for the function $X(x)$:

Find the values of λ *called eigen-values for which non-trivial solutions of the problem*

$$L[X]+\lambda\varrho X=0, \tag{87}$$

$$X(0)=0, \quad X(l)=0, \tag{88}$$

*exist, and determine the corresponding eigen-functions.***

Let us formulate the fundamental properties of the eigen-functions and the eigen-values of the boundary-value problem (87) and (88).

1. There is a denumerable set of eigen-values $\lambda_1<\lambda_2<\ldots$ $\ldots<\lambda_n\ldots$ which correspond to non-trivial solutions of the problem — the eigen-functions $X_1(x)$, $X_2(x)$, $\ldots$, $X_n(x)$ . .

2. When $q\geqslant 0$ all the eigen-values of λ_n are positive.

3. The eigen-functions form an orthogonal set with the weight $\varrho(x)$ on the segment $0\leqslant x\leqslant l$:

$$\int_0^l X_m(x)\,X_n(x)\,\varrho(x)\,dx=0 \quad (m\neq n.) \tag{89}$$

** For $\varrho=\varrho_0=$ const., $k=k_0=$ const., we obtain the boundary problem of natural vibrations of a system with fixed ends:

$$X''+\mu X=0 \quad \left(\mu=\frac{\varrho_0}{k_0}\lambda\right),$$

$$X(0)=0, \quad X(l)=0,$$

investigated in § 2.

4. *(V. A. Steklov's Theorem of Expandability.)* If the arbitrary function $F(x)$ is continuously differentiable and satisfies the boundary conditions $F(0) = F(l) = 0$, it can be expanded into a uniformly and absolutely converging series of eigen-functions $\{X_n(x)\}$:

$$F(x) = \sum_{n=1}^{\infty} F_n X_n(x), \quad F_n = \frac{1}{N_n} \int_0^l F(x) X_n(x) \varrho(x)\, dx, \tag{90}$$

$$N_n = \int_0^l X_n^2(x) \varrho(x)\, dx.$$

The proof of the statements 1 and 4 is usually based on the theory of integral equations and we shall not give it here. In the present sub-section we shall deal with the proofs of the properties 2 and 3.

Before giving proofs of these properties we shall derive the formula known as Green's formula. Let $u(x)$ and $v(x)$ be arbitrary functions which can be differentiated twice in the interval $a < x < b$ and having a continuous first derivative in the segment $a \leqslant x \leqslant b$. Let us consider the expression

$$uL[v] - vL[u] = u(kv')' - v(ku')' = [k(uv' - vu')]'.$$

Integrating this equation with respect to x from a to b, we get *Green's formula*

$$\int_a^b (uL[v] - vL[u])\, dx = k(uv' - vu')\Big|_a^b. \tag{91}$$

Now we can prove property 3. Let $X_m(x)$ and $X_n(x)$ be two eigen-functions corresponding to the eigen-values λ_m and λ_n. Putting $u = X_m(x)$, $v = X_n(x)$ and remembering the boundary conditions (88) we have*:

$$\int_0^l \{X_m L[X_n] - X_n L[X_m]\}\, dx = 0 \quad (a = 0, b = l),$$

* The derivatives X_m' and X_n' are continuous everywhere in the segment $0 \leqslant x \leqslant l$, including the points $x = 0$ and $x = l$, since equation (87) gives:

$$k(x) X_m'(x) = \int_x^{x_0} (q - \lambda_m \varrho) X_m\, dx + C.$$

The existence of the derivative X_m' at $x = 0$ and $x = l$ follows from this.

whence, using equation (87) we get:

$$(\lambda_n - \lambda_m) \int_0^l X_m(x) X_n(x) \varrho(x) \, dx = 0 .$$

Thus, if $\lambda_n \neq \lambda_m$ we have the condition

$$\int_0^l X_m(x) X_n(x) \varrho(x) \, dx = 0, \tag{92}$$

expressing the orthogonality with the weight $\varrho(x)$ of the eigen-functions $X_m(x)$ and $X_n(x)$. We shall now prove that each eigen-value corresponds to only one eigen-function† (except for a constant multiplier). Any eigen-function is defined uniquely as the solution of a differential equation of the second order from the value of the function itself and its first derivative at $x = 0$ Assuming the existence of two functions $\overline{X}$ and $\overline{\overline{X}}$ which correspond to the same value of λ and become zero at $x = 0$ and taking the function

$$X^*(x) = \frac{\overline{X}'(0)}{\overline{\overline{X}}'(0)} \overline{\overline{X}}(x),$$

we see that this function satisfies the same equation (87) and the same initial conditions as the function $\overline{X}(x)$;

$$X^*(0) = \frac{\overline{X}'(0)}{\overline{\overline{X}}'(0)} \overline{\overline{X}}(0) = 0,$$

$$\frac{dX^*}{dx}(0) = \frac{\overline{X}'(0)}{\overline{\overline{X}}'(0)} \overline{\overline{X}}'(0) = \overline{X}'(0) .$$

This proves that $X^*(x) = \overline{X}(x)$ and

$$\overline{X}(x) = A \overline{\overline{X}}(x) \qquad \left(A = \frac{\overline{X}'(0)}{\overline{\overline{X}}'(0)} \right) .$$

† This property of the first boundary-value problem is based on the fact that two linearly independent solutions of a differential equation of the second order cannot become zero at one and the same point. This statement relates to the boundary-value problem with zero boundary conditions. For other boundary conditions (e. g. $X(0) = X(l)$, $X'(0) = X'(l)$) two different eigen-functions may exist, corresponding to the same eigen-value.

It should be noted that in the course of the proof we have used the condition $\overline{X}'(0) \neq 0$ which is fulfilled, since the solution of the linear equation (87) defined by the initial conditions

$$\overline{X}(0) = 0, \quad \overline{X}'(0) = 0,$$

is identically equal to zero and thus cannot be an eigen-function.

Because of the linearity and homogeneity of the equation and the boundary conditions it is obvious that if $X_n(x)$ is an eigen-function for the eigen-value λ_n, the function $A_n X_n(x)$ (A_n is an arbitrary constant) is also an eigen-function for the same λ_n. It was proved above that this completely exhausts the class of eigen-functions. Eigen-functions differing only by a factor are considered essentially the same. To remove ambiguity in the selection of the factor, the eigen-functions may be subjected to the normalization requirement

$$N_n = \int_0^l X_n^2(x)\, \varrho(x)\, dx = 1.$$

If some function $X_n(x)$ does not satisfy this requirement it can be "normalized" by multiplying by a coefficient A_n,

$$A_n \hat{X}_n(x) = X_n(x), \quad A_n = \frac{1}{\sqrt{\left[\int_0^l \hat{X}_n^2(x)\, \varrho(x)\, dx\right]}}.$$

Thus the eigen-functions $\{X_n(x)\}$ of our boundary problem (87—88) form an orthogonal and normalized system

$$\int_0^l X_m(x)\, X_n(x)\, \varrho(x)\, dx = \begin{cases} 0, & m \neq n, \\ 1, & m = n. \end{cases}$$

Let us pass on to the proof of the property 2. We shall prove that $\lambda > 0$ when $q \geqslant 0$. Let $X_n(x)$ be an eigen-function corresponding to the eigen-value λ_n, so that

$$L[X_n] = -\lambda_n \varrho(x)\, X_n(x).$$

Multiplying both parts of this equation by $X_n(x)$ and integrating with respect to x from 0 to l, we obtain:

$$\lambda_n \int_0^l X_n^2(x)\, \varrho(x)\, dx = -\int_0^l X_n(x)\, L[X_n]\, dx$$

or

$$\lambda_n = -\int_0^l X_n \frac{d}{dx}\left[k(x)\frac{dX_n}{dx}\right]dx + \int_0^l q(x) X_n^2(x)\,dx,$$

since the function $X_n(x)$ is assumed to be normalized. Integrating by parts and using the boundary conditions (88) we obtain:

$$\lambda_n = -X_n k X_n' \Big|_0^l + \int_0^l k(x)[X_n'(x)]^2 dx + \int_0^l q(x) X_n^2(x)\,dx =$$

$$= \int_0^l k(x)[X_n'(x)]^2 dx + \int_0^l q(x) X_n^2(x)\,dx \, ; \tag{93}$$

whence it follows that

$$\lambda_n > 0,$$

since $k(x) > 0$ and $q(x) \geqslant 0$.

Leaving aside the proof of the theorem of expandability, we shall deal briefly with the calculation of the coefficients of expansion.

It is easily seen that

$$F_n = \frac{\int_0^l \varrho(x) F(x) X_n(x)\,dx}{\int_0^l \varrho(x) X_n^2(x)\,dx}. \tag{94}$$

Multiplying both sides of the equation

$$F(x) = \sum_{n=1}^{\infty} F_n X_n(x)$$

by $\varrho(x) X_n(x)$, integrating with respect to x from 0 to l and remembering the orthogonality of the eigen-functions we obtain the expression given above for the coefficients F_n (Fourier coefficients).

Let us now return to the partial differential equation. The function $T(t)$ satisfies the equation

$$T'' + \lambda_n T = 0 \tag{95}$$

without any additional conditions. Because of the positive sign of λ_n (as proved) its solution has the form

$$T_n(t) = A_n \cos\sqrt{\lambda_n}\cdot t + B_n \sin\sqrt{\lambda_n}\cdot t,$$

where A_n and B_n are undetermined coefficients. Thus the auxiliary problem has an infinite number of solutions of the form

$$u_n(x, t) = T_n(t) X_n(x) = (A_n \cos \sqrt{\lambda_n} \cdot t + B_n \sin \sqrt{\lambda_n} \cdot t) X_n(x).$$

Finally we consider the problem with given initial conditions. We look for a solution in the form

$$u(x, t) = \sum_{n=1}^{\infty} (A_n \cos \sqrt{\lambda_n} \cdot t + B_n \sin \sqrt{\lambda_n} \cdot t) X_n(x). \quad (96)$$

The formal method of satisfying the initial conditions (86) is based on the theorem of expandability [4] and is carried out

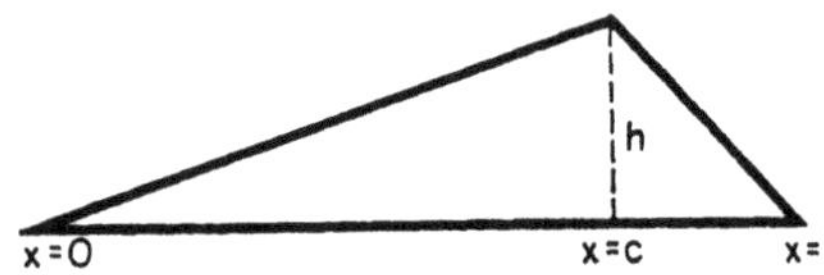

FIG. 24

exactly as in the case of the homogeneous string. From the equations

$$u(x, 0) = \varphi(x) = \sum_{n=1}^{\infty} A_n X_n(x),$$

$$u_t(x, 0) = \psi(x) = \sum_{n=1}^{\infty} B_n \sqrt{\lambda_n} \cdot X_n(x)$$

we find that

$$A_n = \varphi_n, \quad B_n = \frac{\psi_n}{\sqrt{\lambda_n}}, \quad (97)$$

where φ_n and ψ_n are the Fourier coefficients $\varphi(x)$ and $\psi(x)$ on resolving into a system of functions $\{X_n(x)\}$ which are orthogonal with the weight $\varrho(x)$ We have considered only the general scheme of the method of separation of variables and have not given conditions for its applicability.

V. A. Steklov's work on establishing this method is fundamental.*

* *Communications of the Kharkov Mathematical Society*, 2nd series, vol. 5, Nos. 1 and 2 (1896), and also *Fundamental Problems of Mathematical Physics*, vol. 1 (1922).

Problems

1. Find the function $u(x, t)$ describing the vibration of a string $(0, l)$ fixed at the ends and excited (Fig. 24) by pulling it at the point $x = c$ so that $u(c, 0) = h$ (see Appendix I).

2. A string fixed at the ends is pulled at the point $x = c$ with a force F_0. Find the subsequent vibrations of the string if at the initial moment the force ceases to act.

3. Find the function $u(x, t)$ describing the vibration of a string $(0, l)$ fixed at the ends and excited by an impulse K distributed on the segment $(c - \delta, c + \delta)$: (a) uniformly, (b) according to the law $v_0 \cos \dfrac{x - c}{2\delta} \pi$ (see Appendix I).

4. Find the function $u(x, t)$ describing the vibrations of a string $(0, l)$ fixed at the ends and excited by an impulse K applied at the point $x = c$ (see Appendix I).

5. Prove that the energy of different harmonics is additive for vibrations with the boundary conditions $u = 0$, $u_x = 0$. Consider also the case of the boundary condition $u_x + hu = 0$ (all series are to be assumed to converge uniformly). Calculate the energy of the different harmonics in the problems 1, 2, 3, 4.

6. A spring fixed at one end at the point $x = 0$, is stretched by a load of mass M suspended at the point $x = l$. Find the vibrations of the spring if at $t = 0$ the load falls off and no further forces act on the end $x = l$.

7. One end of the rod is fixed and a force F_0 acts on the other. Find the vibrations of the rod if at $t = 0$ the force ceases to act.

8. Find the vibrations of a spring, which has one end fixed and has a load of mass M suspended from the second end at the initial moment. The initial conditions are zero.

9. A homogeneous string has the ends $x = 0$ and $x = l$ fixed and mass M is attached at the point $x = c$. Find the deflection of the string $u(x, t)$ if: (a) at the initial moment the string is pulled at the point $x = c$ a distance h from the equilibrium position and released without initial velocity, (b) the initial deflection and initial velocity are zero (see Appendix III).

10. Find the vibrations of a spring with free ends under uniform initial tension (give a model of this problem).

11. Find the vibrations of a spring whose ends are attached elastically with identical coefficients of rigidity (i. e. $u_x + hu = 0$ at $x = 0$ and $x = l$) if the initial conditions are arbitrary.

Investigate the solution for small h ("loose" attachment) and large h ("rigid" attachment) and calculate the corresponding corrections to the eigen-values for a string with free and fixed ends.

12. Find the vibrations of a string $u(x, t)$ with fixed ends, if the vibra tions occur in a medium which gives a resistance proportional to the velocity. The initial conditions are arbitrary.

Answer :

$$u_{tt} = a^2 u_{xx} - 2\nu u_t, \quad \nu > 0,$$

$$u(x, t) = e^{-\nu t} \sum_{n=1}^{\infty} (a_n \cos \omega_n t + b_n \sin \omega_n t) \sin \sqrt{\lambda_n}\, x,$$

$$a_n = \frac{2}{l} \int_0^l \varphi(x) \sin \sqrt{\lambda_n}\, x dx, \qquad b_n = \nu \frac{a_n}{\omega_n} + \frac{2}{l\omega_n} \int_0^l \psi(x) \sin \sqrt{\lambda_n}\, x\, dx,$$

$$\lambda_n = (\pi n/l)^2, \quad \omega_n = \sqrt{(a^2 \lambda_n - \nu)}.$$

13. An insulated electrical wire of length L with the characteristics L, R, C and $G = 0$ is charged to some constant potential v_0. At the initial moment one end of the wire is earthed and the second remains insulated. Find the distribution of potential in the wire.

Answer :

$$v(x, t) = e^{-\frac{R}{2L}t} \sum_{n=0}^{\infty} a_n \sin \frac{2n+1}{l} \pi x \cdot \sin(\omega_n t + \varphi_n),$$

$$\omega_n = \frac{(2n+1)\pi}{2l\sqrt{(LC)}} \sqrt{\left[1 - \frac{CR^2 l^2}{L\pi^2 (2n+1)^2}\right]},$$

$$a_n = \frac{4v_0}{\pi(2n+1)\sin \varphi_n}, \qquad \tan \varphi_n = 2\omega_n \frac{L}{R}.$$

14. A string with fixed ends vibrates under the action of a harmonic force distributed with a density $f(x, t) = \Phi(x) \sin \omega t$. Find the deflection $u(x, t)$ under arbitrary initial conditions. Investigate the possibility of resonance and find the solution in the case of resonance.

15. Solve problem 14 assuming that vibrations occur in the medium with resistance proportional to the velocity. Find the steady vibrations forming the main part of the solution for $t \to \infty$.

16. A vertical elastic rod of length l is attached rigidly at its upper end to a freely falling lift which, having attained the velocity v_0, stops instantaneously. Find the vibrations of the rod, assuming that its lower end is free.

17. Solve the equation

$$u_{tt} = a^2 u_{xx} - b^2 u + A$$

with zero initial conditions and the boundary conditions

$$u(0, t) = 0, \quad u(l, t) = B,$$

where b, A and B are constants.

18. Solve the differential equation

$$u_{tt} = a^2 u_{xx} + A \sinh x$$

with zero initial conditions and the boundary conditions

$$u(0,t) = B, \quad u(l,t) = C,$$

where A, B and C are constants.

19. To a homogeneous string with rigidly attached ends $x = 0$ and $x = l$ a harmonic force is applied

$$F(t) = P_0 \sin \omega t,$$

acting from the moment $t = 0$ at the point $x = c\ (0 < c < l)$. Find the deflection of the string $u(x, t)$ assuming the initial conditions to be zero.

20. Solve the problem of vibrations of a non-homogeneous rod of length l with fixed ends, composed of two homogeneous rods, joined at the point $x = c\ (0 < c < l)$ if the initial deflection is

$$u(x,0) = \begin{cases} \dfrac{h}{c}x & \text{where } 0 \leqslant x \leqslant c, \\ \dfrac{h}{l-c}(l-x) & \text{where } c \leqslant x \leqslant l, \end{cases}$$

and the initial velocities are zero.

21. Find the steady vibrations of a spring one end of which is fixed and which has a force acting on the other end

$$F(t) = A \sin \omega_1 t + B \sin \omega_2 t.$$

22. Find the steady vibrations of an inhomogeneous rod consisting of two homogeneous rods joined at the point $x = c$ if one end of the rod is fixed and the second moves according to the law

$$u(l,t) = A \sin \omega t.$$

§ 4. Problem with data on characteristics

1. *Formulation of the problem*

We shall consider a number of problems which are developments of the first boundary-value problem for the wave equation. For simplicity we shall study the phenomena near an end, considering the other end to be removed to infinity, i. e. as an initial problem we shall take the problem for a semi-infinite straight line.

The wave equation $u_{t't'} = a^2 u_{xx}$ is symmetrical with respect to the variables x and t if we put $a^2 = 1$, i. e change the time-scale introducing the variable $t = at'$. But the additional conditions introduce asymmetry in the mathematical interpretation of x and t; in the initial conditions (at $t = 0$) two functions $u(x, 0)$ and $u_t(x, 0)$ are specified while in the boundary conditions (at $x = 0$) only one function $u(0, t)$ is given.

As pointed out in § 2, sub-section 7, there is a relation

$$u_t(0, z) + u_x(0, z) = u_t(z, 0) + u_x(z, 0) \qquad (a^2 = 1)$$

which holds for an arbitrary value of z. Hence it follows that at $x = 0$ and $t = 0$ all these functions cannot be given independently. Only three conditions are arbitrary which indicates that a symmetrical formulation of additional conditions is impossible.

The additional conditions may be given either on the lines $x = 0$, $t = 0$ (we have dealt with problems of this type before) or on some curves in the phase plane. For example, the boundary values may be given on some curve C_1 ($x = f_1(t)$); but in order that the problem may be soluble the curve C_1 must satisfy certain conditions besides being sufficiently smooth.

Let us consider the vibrations of a gas in a tube with a movable boundary (movable piston) The velocity of displacement of the boundary, moving according to the law $x = f_1(t)$, cannot be completely arbitrary. It should not exceed the velocity of sound a $\left(\frac{df_1(t)}{dt} < a\right)$. A geometrical consequence of this is that the curve C_1 ($x = f_1(t)$) should lie above a characteristic curve passing through the intersection of C_1 with the line $t = 0$ on which initial values are given (Fig. 25). If even at one point the line C_1 lies below the characteristic $x = at$, the value of the function $f_1(t)$ will be determined by the initial conditions and can not be given arbitrarily. The physical significance of this is that when the gas moves with the velocity exceeding the velocity of sound the equation of acoustics loses its validity and non-linear equations of gas dynamics* have to be used.

The initial conditions may be given not only on the axis $t = 0$, but also on some line C_2 ($t = f_2(x)$) which must satisfy the requirement $|f'_2(x)| < 1/a$. Problems of this type can be solved easily with the help of the integral wave equation (see § 2, sub-section 7).

* See Appendix IV, p. 163.

Since it is not our aim to give a complete list of all possible boundary problems, let us consider in further detail the problem of determining the solution from data on characteristics. This boundary problem is often called *Goursat's problem.* A problem with data on characteristics is of great interest from the point

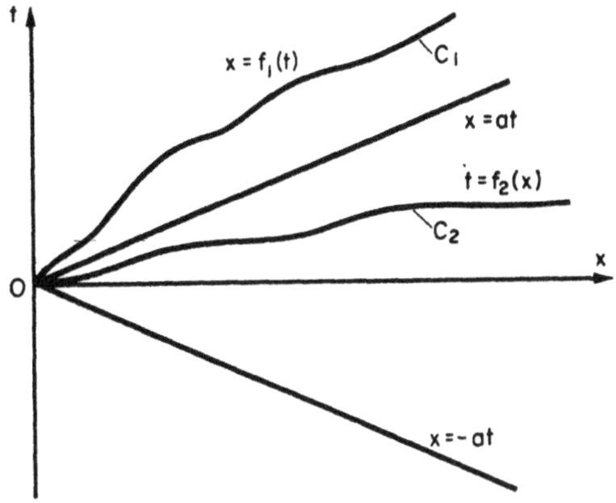

FIG. 25

of view of physical applications. It occurs, for example, in the study of processes of absorption and desorption of gases (see Appendix V), drying processes (see problem 2) and many other problems.

2. *Method of successive approximations*

Let us consider the simplest problem with data on characteristics

$$\left.\begin{aligned} u_{xy} &= f(x, y), \\ u(x, 0) &= \varphi_1(x), \\ u(0, y) &= \varphi_2(y). \end{aligned}\right\} \qquad (1)$$

The additional conditions are given for the straight lines $x = 0$ and $y = 0$ which are characteristics of equation (1). Let us assume that the functions $\varphi_1(x)$ and $\varphi_2(y)$ are differentiable and satisfy the connecting condition $\varphi_1(0) = \varphi_2(0)$. Integrating equation (1) with respect to x and y successively we get:

$$u_y(x, y) = u_y(0, y) + \int_0^x f(\xi, y)\, d\xi,$$

$$u(x, y) = u(x, 0) + u(0, y) - u(0, 0) + \int_0^y d\eta \int_0^x f(\xi, \eta)\, d\xi$$

or

$$u(x, y) = \varphi_1(x) + \varphi_2(y) - \varphi_1(0) + \int_0^y \int_0^x f(\xi, \eta)\, d\xi\, d\eta. \qquad (2)$$

Thus for the simplest equation not containing the first derivatives of u_x, u_y of the required function, the solution can be represented in an explicit analytical form (2). The uniqueness and the existence of the solution follow directly from formula (2).

We now look for a solution of the general linear equation of hyperbolic type

$$u_{xy} = a(x, y)\, u_x + b(x, y)\, u_y + c(x, y)\, u + f(x, y) \qquad (3)$$

under the additional conditions given on the characteristics $x = 0$, $y = 0$

$$u(x, 0) = \varphi_1(x),$$
$$u(0, y) = \varphi_2(y),$$

where $\varphi_1(x)$ and $\varphi_2(y)$ satisfy the conditions of differentiability and conjugation. We shall assume the coefficients a, b and c to be continuous functions of x and y.

Formula (3) shows that the functions $u(x, y)$ satisfy the integral-differential equation

$$u(x, y) = \int_0^y \int_0^x \left[a(\xi, \eta)\, u_\xi + b(\xi, \eta)\, u_\eta + c(\xi, \eta)\, u\right] d\xi\, d\eta +$$
$$+ \varphi_1(x) + \varphi_2(y) - \varphi_1(0) + \int_0^y \int_0^x f(\xi, \eta)\, d\xi\, d\eta. \qquad (4)$$

We use the method of successive approximations for its solution. As zero approximation let us take the function

$$u_0(x, y) = 0.$$

Then (4) gives the following expressions for the successive approximations:

$$\left.\begin{aligned} u_1(x, y) &= \varphi_1(x) + \varphi_2(y) - \varphi_1(0) + \int_0^y\int_0^x f(\xi, \eta)\, d\xi\, d\eta, \\ &\dots\dots\dots\dots\dots\dots\dots\dots\dots\dots \\ &\dots\dots\dots\dots\dots\dots\dots\dots\dots\dots \\ u_n(x, y) &= u_1(x, y) + \int_0^y\int_0^x \Big[a(\xi, \eta)\frac{\partial u_{n-1}}{\partial \xi} + \\ &\quad + b(\xi, \eta)\frac{\partial u_{n-1}}{\partial \eta} + c(\xi, \eta)\, u_{n-1}\Big] d\xi\, d\eta. \end{aligned}\right\} \quad (5)$$

Let us point out incidentally that

$$\left.\begin{aligned} \frac{\partial u_n}{\partial x} &= \frac{\partial u_1}{\partial x} + \int_0^y \Big[a(x, \eta)\frac{\partial u_{n-1}}{\partial x} + b(x, \eta)\frac{\partial u_{n-1}}{\partial \eta} + c(x, \eta)\, u_{n-1}\Big] d\eta, \\ \frac{\partial u_n}{\partial y} &= \frac{\partial u_1}{\partial y} + \int_0^x \Big[a(\xi, y)\frac{\partial u_{n-1}}{\partial \xi} + b(\xi, y)\frac{\partial u_{n-1}}{\partial y} + c(\xi, y)\, u_{n-1}\Big] d\xi. \end{aligned}\right\} (6)$$

We shall prove the uniform convergence of the sequences

$$\{u_n(x, y)\}, \quad \left\{\frac{\partial u_n}{\partial x}(x, y)\right\}, \quad \left\{\frac{\partial u_n}{\partial y}(x, y)\right\}.$$

To do this we consider the differences

$$z_n(x, y) = u_{n+1}(x, y) - u_n(x, y) =$$

$$= \int_0^y\int_0^x \Big[a(\xi, \eta)\frac{\partial z_{n-1}}{\partial \xi} + b(\xi, \eta)\frac{\partial z_{n-1}}{\partial \eta} + c(\xi, \eta)\, z_{n-1}(\xi, \eta)\Big] d\xi\, d\eta,$$

$$\frac{\partial z_n(x, y)}{\partial x} = \frac{\partial u_{n+1}(x, y)}{\partial x} - \frac{\partial u_n(x, y)}{\partial x} =$$

$$= \int_0^y \Big[a(x, \eta)\frac{\partial z_{n-1}}{\partial x} + b(x, \eta)\frac{\partial z_{n-1}}{\partial \eta} + c(x, \eta)\, z_{n-1}(x, \eta)\Big] d\eta,$$

$$\frac{\partial z_n(x, y)}{\partial y} = \frac{\partial u_{n+1}(x, y)}{\partial y} - \frac{\partial u_n(x, \eta)}{\partial y} =$$

$$= \int_0^x \Big[a(\xi, y)\frac{\partial z_{n-1}}{\partial \xi} + b(\xi, y)\frac{\partial z_{n-1}}{\partial y} + c(\xi, y)\, z_{n-1}(\xi, y)\Big] d\xi.$$

Let M be the upper limit of the absolute values of the coefficients $a(x, y)$, $b(x, y)$, $c(x, y)$ and H the upper limit of the absolute values of z_0 and its derivatives

$$|z_0| < H, \quad \left|\frac{\partial z_0}{\partial x}\right| < H, \quad \left|\frac{\partial z_0}{\partial y}\right| < H$$

on varying x and y within some square $(0 \leqslant x \leqslant L, 0 \leqslant y \leqslant L)$. Let us find upper bounds for the functions z_n, $\frac{\partial z_n}{\partial x}$, $\frac{\partial z_n}{\partial y}$. It is obvious that

$$|z_1| < 3HMxy < 3HM\frac{(x+y)^2}{2!},$$

$$\left|\frac{\partial z_1}{\partial x}\right| < 3HMy < 3HM(x+y),$$

$$\left|\frac{\partial z_1}{\partial y}\right| < 3HMx < 3HM(x+y).$$

Let us assume that the relations exist

$$|z_n| < 3HM^n K^{n-1}\frac{(x+y)^{n+1}}{(n+1)!},$$

$$\left|\frac{\partial z_n}{\partial x}\right| < 3HM^n K^{n-1}\frac{(x+y)^n}{n!},$$

$$\left|\frac{\partial z_n}{\partial y}\right| < 3HM^n K^{n-1}\frac{(x+y)^n}{n!},$$

where $K > 0$ is some constant number whose value we shall specify below. Using these and the formula for the $(n+1)$th approximation we get after a series of simplifications all of which increase the inequality:

$$|z_{n+1}| < 3HM^{n+1} K^{n-1}\frac{(x+y)^{n+2}}{(n+2)!}\left(\frac{x+y}{n+3}+2\right) <$$

$$< 3HM^{n+1} K^n \frac{(x+y)^{n+2}}{(n+2)!} < \frac{3H}{K^2 M}\frac{(2KLM)^{n+2}}{(n+2)!},$$

$$\left|\frac{\partial z_{n+1}}{\partial x}\right| < 3HM^{n+1} K^{n-1}\frac{(x+y)^{n+1}}{(n+1)!}\left(\frac{x+y}{n+2}+2\right) <$$

$$< 3HM^{n+1} K^n \frac{(x+y)^{n+1}}{(n+1)!} < \frac{3H}{K}\frac{(2KLM)^{n+1}}{(n+1)!},$$

$$\left|\frac{\partial z_{n+1}}{\partial y}\right| < 3HM^{n+1} K^{n-1}\frac{(x+y)^{n+1}}{(n+1)!}\left(\frac{x+y}{n+2}+2\right) <$$

$$< 3HM^{n+1} K^n \frac{(x+y)^{n+1}}{(n+1)!} < \frac{3H}{K}\frac{(2KLM)^{n+1}}{(n+1)!},$$

if we put

$$K = 2L + 2 .$$

The terms on the right-hand sides of these equations are, except for a multiplier, the general terms of the expansion of e^{2KLM}. These estimates show that the sequence of functions

$$u_n = u_0 + z_1 + \ldots + z_{n-1},$$

$$\frac{\partial u_n}{\partial x} = \frac{\partial u_0}{\partial x} + \frac{\partial z_1}{\partial x} + \ldots + \frac{\partial z_{n-1}}{\partial x} .$$

$$\frac{\partial u_n}{\partial y} = \frac{\partial u_0}{\partial y} + \frac{\partial z_1}{\partial y} + \ldots + \frac{\partial z_{n-1}}{\partial y}$$

converge uniformly to limiting functions which we denote by

$$u(x, y) = \lim_{n\to\infty} u_n(x, y),$$

$$v(x, y) = \lim_{n\to\infty} \frac{\partial u_n}{\partial x}(x, y),$$

$$w(x, y) = \lim_{n\to\infty} \frac{\partial u_n}{\partial y}(x, y).$$

Passing to the limit the formulae (5) and (6) become

$$\left.\begin{aligned} u(x, y) &= u_1(x, y) + \int_0^y\int_0^x [a(\xi, \eta) v + b(\xi, \eta) w + \\ &\qquad + c(\xi, \eta) u] \, d\xi \, d\eta, \\ v(x, y) &= \frac{\partial u_1}{\partial x}(x, y) + \int_0^y [a(x, \eta) v + b(x, \eta) w + \\ &\qquad + c(x, \eta) u] \, d\eta, \\ w(x, y) &= \frac{\partial u_1}{\partial y}(x, y) + \int_0^x [a(\xi, y) v + b(\xi, y) w + \\ &\qquad + c(\xi\ y) u] \, d\xi. \end{aligned}\right\} \qquad (7)$$

Hence the following equations hold

$$v = u_x,$$

$$w = u_y$$

and we have established that the function $u(x, y)$ satisfies the integro-differential equation

$$\begin{aligned} u(x, y) = \varphi_1(x) + \varphi_2(y) - \varphi_1(0) + \int_0^y\int_0^x f(\xi, \eta) \, d\xi \, d\eta + \\ + \int_0^y\int_0^x [a(\xi, \eta) u_\xi + b(\xi, \eta) u_\eta + c(\xi, \eta) u] \, d\xi \, d\eta, \end{aligned} \qquad (4)$$

and also the initial differential equation (3) as can be verified directly by the differentiation of (4) with respect to x and y. The function $u(x, y)$, satisfies the additional conditions also. Let us now prove the uniqueness of the solution. Assuming the existence of two solutions $u_1(x, y)$ and $u_2(x, y)$, we see that their difference

$$U(x, y) = u_1(x, y) - u_2(x, y)$$

obeys the homogeneous integro-differential equation

$$U(x, y) = \int_0^y \int_0^x (aU_x + bU_y + cU)\, d\xi\, d\eta .$$

Further, denoting by H_1 the upper bound of the absolute values

$$|U(x, y)| < H_1, \quad |U_x(x, y)| < H_1, \quad |U_y(x, y)| < H_1$$

for $0 \leqslant x \leqslant L$, $0 \leqslant y \leqslant L$ and repeating the estimate made for the functions $z_n(x, y)$ we verify that the inequality

$$|U| < 3H_1 M^{n+1} K^n \frac{(x+y)^{n+2}}{(n+2)!} < \frac{3H_1}{K^2 M} \frac{(2KLM)^{n+2}}{(n+2)!}$$

holds for any value of n. Whence it follows that

$$U(x, y) \equiv 0 \quad \text{or} \quad u_1(x, y) \equiv u_2(x, y),$$

which proves the uniqueness of the solution of the problem with data on characteristics.

If the coefficients a, b and c are constants then the equation (3) can be simplified by substituting

$$u = ve^{\lambda x + \mu y}$$

to the form

$$v_{xy} + C_1 v = f. \tag{8}$$

For $C_1 = 0$ we obtain the simplest equation (1), the solution of which is given by the formula (2).

If $C_1 \neq 0$ the solution of the problem for equation (8) can also be obtained in an explicit analytical form by a method to be described in § 5.

Problems

1. Air is blown (with a velocity v) through a tube ($x > 0$) filled with a substance containing moisture. Let $v(x, t)$ be the concentration of moisture in the absorbing substance, $u(x, t)$ the concentration of free vapour. Derive an equation for the functions $u(x, t)$ and $v(x, t)$ describing the process of drying, if: (1) the process is isothermal, and (2) the isotherm of drying has the form $u = \gamma v$, where γ is the constant of the isotherm (see also Appendix V).

2. Hot water is passed with a velocity v along a tube ($x > 0$). Let u be the temperature of water in the tube, v the temperature of the walls of the tube, u_0 the temperature of the surrounding medium. Derive the equations for u and v, neglecting the temperature distribution over the section of the tube and the walls and considering that there is a drop in temperature at the boundaries water-wall and wall-medium and heat transfer takes place according to Newton's law (see Chapter III, § 1).

§ 5. Solution of general linear equations of the hyperbolic type

1. *Conjugate differential operators*

Let us establish some auxiliary formulae required for representing solutions of boundary problems in integral form. Let

$$\mathscr{L}[u] = u_{xx} - u_{yy} + a(x, y)\, u_x + b(x, y)\, u_y + c(x, y)\, u \quad (1)$$

[$a(x, y)$, $b(x, y)$, $c(x, y)$ are differentiable functions] be a linear differential operator corresponding to a linear equation of the hyperbolic type. Multiplying $\mathscr{L}[u]$ by some function v, let us write the different terms in the form

$$\begin{aligned} vu_{xx} &= (vu_x)_x - (v_x u)_x + uv_{xx}, \\ vu_{yy} &= (vu_y)_y - (v_y u)_y + uv_{yy}, \\ vau_x &= (avu)_x - u(av)_x, \\ vbu_y &= (bvu)_y - u(bv)_y, \\ vcu &= ucv. \end{aligned}$$

Summing up the different terms we get

$$v\mathscr{L}[u] = u\mathscr{M}[v] + \frac{\partial H}{\partial x} + \frac{\partial K}{\partial y}, \quad (2)$$

where

$$\mathscr{M}[v] = v_{xx} - v_{yy} - (av)_x - (bv)_y + cv\,, \tag{3}$$

$$H = \quad vu_x - v_x u + avu = (vu)_x - (2\,v_x - av)\,u = \tag{4}$$

$$= -(vu)_x + (2\,u_x + au)\,v, \tag{4'}$$

$$K = -\,vu_y + v_y u + bvu = -\,(vu)_y + (2\,v_y + bv)\,u = \tag{5}$$

$$= (uv)_y - (2u_y - bu)\,v. \tag{5'}$$

Two differential operators are called conjugate if the difference

$$v\,\mathscr{L}[u] - u\,\mathscr{M}[v]$$

is the sum of the partial derivatives with respect of x and y of some expressions H and K.

The operators $\mathscr{L}$ and $\mathscr{M}$ considered here are obviously conjugate.

If $\mathscr{L}[u] = \mathscr{M}[u]$ then the operator $\mathscr{L}[u]$ is called self-conjugate.

The double integral of the difference $v\,\mathscr{L}[u] - u\,\mathscr{M}[v]$ over some region G bounded by a piecewise smooth contour C is

$$\int_G\!\!\int (v\,\mathscr{L}[u] - u\,\mathscr{M}[v])\,d\xi\,d\eta = \int_C (H\,d\eta - K\,d\xi)\,, \tag{6}$$

where u and v are arbitrary doubly differentiable functions (Green's formula).

2. *Integral form of solution*

We shall use formula (6) for the solution of the following problem:

Find the solution of the linear equation of the hyperbolic type

$$\mathscr{L}[u] = u_{xx} - u_{yy} + a(x,y)\,u_x + b(x,y)\,u_y + c(x,y)\,u =$$

$$= -f(x,y)\,, \tag{7}$$

satisfying the initial conditions on the curve C,

$$u|_C = \varphi(x)\,,$$

$$u_n|_C = \psi(x)$$

(u_n is the derivative along the direction of the normal to the curve).

The curve C is given by the equation

$$y = f(x),$$

where $f(x)$ is a differentiable function. Let us impose on the curve C the condition that any characteristic curve of the families $y - x =$ const. and $y + x =$ const. should intersect the curve C not more than once (for this it is necessary that $|f'(x)| < 1$). Formula (6) for the curvilinear triangle MPQ bounded by the arc PQ of the curve C and the segments of the characteristics MP and MQ (Fig. 26) gives:

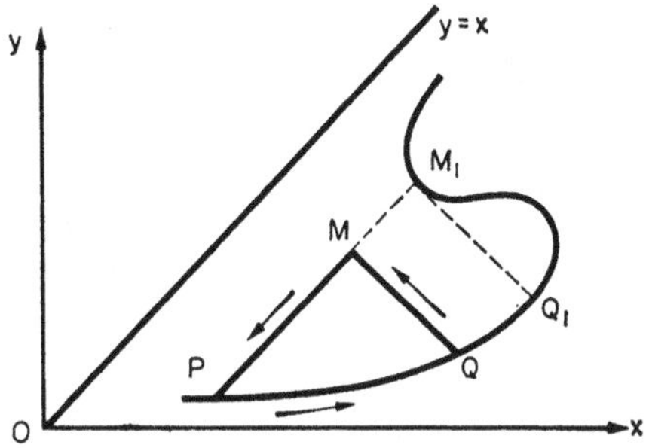

FIG. 26

$$\iint_{MPQ} (v \mathscr{L}[u] - u \mathscr{M}[v])\, d\xi\, d\eta =$$

$$= \int_Q^M (H\, d\eta - K\, d\xi) + \int_M^P (H\, d\eta - K\, d\xi) + \int_P^Q (H\, d\eta - K\, d\xi).$$

Let us transform the first two integrals taken along the characteristics MQ and MP. Taking into consideration that

$$\left.\begin{aligned} d\xi &= -d\eta = -\frac{ds}{\sqrt{2}} \text{ on } QM \\ d\xi &= d\eta \quad = -\frac{ds}{\sqrt{2}} \text{ on } MP \end{aligned}\right\} \begin{gathered} ds \text{ is an element of arc along} \\ QM \text{ and } MP \end{gathered}$$

and using the formulae (4) and (5) we get:

$$\int_Q^M (Hd\eta - Kd\xi) = -\int_Q^M d(uv) + \int_Q^M \left(2\frac{\partial v}{\partial s} - \frac{a+b}{\sqrt{2}} v\right) uds =$$

$$= -(uv)_M + (uv)_Q + \int_Q^M \left(2\frac{\partial v}{\partial s} - \frac{a+b}{\sqrt{2}} v\right) uds$$

and similarly

$$\int_M^P (Hd\eta - Kd\xi) = -(uv)_M + (uv)_P + \int_P^M \left(2\frac{\partial v}{\partial s} - \frac{b-a}{\sqrt{2}}v\right) uds \, .$$

Hence and from formula (6) it follows

$$(uv)_M = \frac{(uv)_P + (uv)_Q}{2} + \int_P^M \left(\frac{\partial v}{\partial s} - \frac{b-a}{2\sqrt{2}}v\right) uds + \int_Q^M \left(\frac{\partial v}{\partial s} - \frac{a+b}{2\sqrt{2}}v\right) uds +$$

$$+ \frac{1}{2}\int_P^Q (Hd\eta - Kd\xi) - \frac{1}{2}\iint_{MPQ} (v\mathscr{L}[u] - u\mathscr{M}[v])\, d\xi\, d\eta \, . \quad (8)$$

This formula is an identity which is true for any sufficiently smooth functions u and v.

Let u be a solution of equation (7) with the given initial conditions. The function v depends on the point M and satisfies the following requirements:

$$\mathscr{M}[v] = v_{\xi\xi} - v_{\eta\eta} - (av)_\xi - (bv)_\eta + cv = 0 \text{ inside } \triangle MPQ \quad (9)$$

and

$$\left.\begin{aligned} &\frac{\partial v}{\partial s} = \frac{b-a}{2\sqrt{2}}v \text{ on the characteristic } MP, \\ &\frac{\partial v}{\partial s} = \frac{b+a}{2\sqrt{2}}v \text{ on the characteristic } MQ, \\ &v(M) = 1 \, . \end{aligned}\right\} \quad (9a)$$

From the conditions on the characteristic curves and the last equation we find:

$$v = e^{\int_{s_0}^{s} \frac{b-a}{2\sqrt{2}} ds} \quad \text{on } MP,$$

$$v = e^{\int_{s_0}^{s} \frac{b+a}{2\sqrt{2}} ds} \quad \text{on } MQ,$$

where s_0 is the value of s at the point M. As we saw in § 4, equation (9) and the values of the function v on the characteristic curves MP and MQ completely define it in the region MPQ. The function v is often called Riemann's function.

Thus, formula (8) for the function u satisfying equation (7) assumes the following final form:

$$u(M) = \frac{(uv)_P + (uv)_Q}{2} + \frac{1}{2}\int_P^Q [v(u_\xi d\eta + u_\eta d\xi) - u(v_\xi d\eta + v_\eta d\xi) +$$

$$+ uv(a d\eta - b d\xi)] + \frac{1}{2}\int\int_{MPQ} v(M, M') f(M') \, d\sigma_{M'}$$

$$(d\sigma_{M'} = d\xi d\eta) \qquad (10)$$

This formula solves the given problem, since expressions under the sign of the integral along PQ contain functions known on the arc C. The function v was defined above and the functions

$$u|_C = \varphi(x),$$

$$u_x|_C = u_s \cos(x, s) + u_n \cos(x, n) = \frac{\varphi'(x) + \psi(x) f'(x)}{\sqrt{[1 + f'^2(x)]}},$$

$$u_y|_C = u_s \cos(y, s) + u_n \cos(y, n) = \frac{\varphi'(x) f'(x) + \psi(x)}{\sqrt{[1 + f'^2(x)]}}$$

can be calculated from the initial data.

Formula (10) shows that if the initial data are known on the arc PQ they define the function completely in PMQ if the function $f(x, y)$ is known on this region.

Formula (10), obtained by assuming the existence of a solution, determines the solution from the initial data and $f(x, y)$. This proves the uniqueness of the solution (cf. D'Alembert's formula, Chapter II, p. 45).

It may be shown that the function u defined by the formula (10) satisfies the conditions of the problem, but we shall not do this here.

3. *Physical interpretation of Riemann's function*

We want to find the physical significance of the function $v(M, M')$. To do this we look for the solution of the inhomogeneous equation

$$\mathscr{L}[u] = -2f_1 \qquad (f = 2f_1)$$

with zero initial conditions on the curve C. Using formula (10) we see that the required solution is

$$u(M) = \iint_{MPQ} v(M, M')\, f_1(M')\, d\sigma_{M'}. \tag{11}$$

Let us suppose that $f_1(M)$ is equal to zero everywhere except in the small neighbourhood S_ε of the point M_1 and satisfies the normalization condition

$$\iint_{S_\varepsilon} f_1(M')\, d\sigma_{M'} = 1. \tag{12}$$

The preceding formula in this case becomes

$$u_\varepsilon(M) = \iint_{S_\varepsilon} v(M, M')\, f_1(M')\, d\sigma_{M'} \tag{13}$$

and using the mean value theorem we get

$$u_\varepsilon(M) = v(M, M_1^*) \iint_{S_\varepsilon} f_1(M')\, d\sigma_{M'} = v(M, M_1^*),$$

where M^*_1 is some point of the region S_ε.

Contracting the ε-neighbourhood S_ε at the point M_1 ($\varepsilon \to 0$) we find:

$$u(M) = \lim_{\varepsilon \to \infty} u_\varepsilon(M) = v(M, M_1). \tag{14}$$

The function f_1, as we saw in a number of examples, is usually the force density and the variable y the time. The expression

$$\iint_{S_\varepsilon} f_1(M')\, d\sigma_{M'} = \iint_{S_\varepsilon} f_1(\xi, \eta)\, d\xi\, d\eta \tag{15}$$

is the impulse of the force. Hence, by virtue of formula (11) we conclude that $v(M, M_1)$ is the function giving the effect of a concentrated impulse applied at the point M_1.

The function $v(M, M_1) = v(x, y; \xi, \eta)$ was defined as a function of the parameters $M(x, y)$ satisfying the equation

$$\mathcal{M}_{(\xi,\eta)}[v] = 0 \tag{16}$$

in the coordinates ξ, η of the point M_1, with the additional conditions (9a).

Let us consider the function

$$u = u(M, M_1),$$

which is a function of the parameters $M_1\,(\xi, \eta)$ and satisfies the equation

$$\mathscr{L}_{(x,y)}\,[u] = 0 \tag{17}$$

in the coordinates x, y of the point M, and also the additional conditions

$$\left.\begin{aligned} &\frac{\partial u}{\partial s} = \frac{b-a}{2\sqrt{2}}\,u \text{ on the characteristic } M_1 Q_1, \\ &\frac{\partial u}{\partial s} = \frac{b+a}{2\sqrt{2}}\,u \text{ on the characteristic } M_1 P_1, \\ &u\,(M_1, M_1) = 1\,. \end{aligned}\right\} \tag{18}$$

From these conditions we find

$$u\,(M, M_1) = \left\{\begin{aligned} &e^{\int\limits_{S_0}^{S}\frac{b-a}{2\sqrt{2}}\,ds} \quad \text{on } M_1 Q_1, \\ &e^{\int\limits_{S_0}^{S}\frac{b+a}{2\sqrt{2}}\,ds} \quad \text{on } M_1 P_1, \end{aligned}\right\} \tag{19}$$

$$u\,(M_1, M_1) = 1.$$

Equation (17) and condition (18) completely define the function; u is the quadrilateral $MP_1M_1Q_1$ bounded by the segments of the characteristics MP_1, MQ_1 and MP_1, M_1Q_1 (Fig. 27).

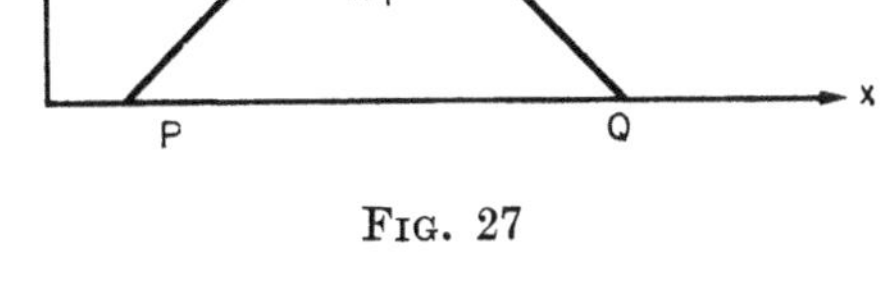

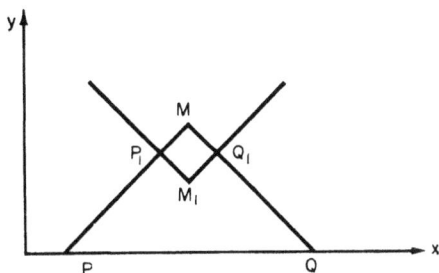

FIG. 27

Applying formula (6) to the quadrilateral $MP_1M_1Q_1$ we find:

$$\iint\limits_{MP_1M_1Q_1} \left(v\,\mathscr{L}\,[u] - u\,\mathscr{M}\,[v]\right) d\xi\, d\eta =$$

$$= \int\limits_{M}^{P_1} (H\, d\eta - K\, d\xi) + \int\limits_{Q_1}^{M} + \int\limits_{M_1}^{Q_1} + \int\limits_{P_1}^{M_1} = 0\,.$$

Using formulae (4) and (5) for K and H and the conditions (9a) on the characteristic curves for the function v, we can calculate the first two integrals of the right-hand side

$$\int_M^{P_1} (Hd\eta - Kd\xi) = -(uv)_M + (uv)_{P_1},$$

$$\int_{Q_1}^{M} (Hd\eta - Kd\xi) = -(uv)_M + (uv)_{Q_1},$$

as was done while deducing formula (10).

Similarly using equations (4'), (5') and the conditions (19) for the function $u(M, M_1)$ on the characteristic curves we find:

$$\int_{P_1}^{M_1} (Hd\eta - Kd\xi) = \int_{P_1}^{M_1} [-(vu)_\xi d\eta - (uv)_\eta d\xi] +$$

$$+ \int_{P_1}^{M_1} v[(2u_\xi d\eta + 2u_\eta d\xi) + (aud\eta - bud\xi)] = \int_{P_1}^{M_1} d(uv) +$$

$$+ \int_{P_1}^{M_1} 2\left(\frac{\partial u}{\partial s} - \frac{a+b}{2\sqrt{2}} v\right) vds = (uv)_{M_1} - (uv)_{P_1} \quad \left(d\xi = -d\eta = \frac{ds}{\sqrt{2}}\right),$$

$$\int_{M_1}^{Q_1} (Hd\eta - Kd\xi) = (uv)_{M_1} - (uv)_{Q_1} \qquad \left(d\xi = d\eta = \frac{ds}{\sqrt{2}}\right).$$

Summing up these equations we get

$$2(uv)_M = 2(uv)_{M_1}$$

or

$$u(M, M_1) = v(M, M_1), \tag{20}$$

since

$$(u)_{M_1} = (v)_M = 1$$

Thus we see that the function giving the effect of the impulse concentrated at the point M_1 may be defined as the solution of the equation

$$\mathcal{L}_{(x,y)}[v(M, M_1)] = 0$$

with the additional conditions (18).

4. *Equations with constant coefficients*

As the first example of application of formula (10) let us consider the problem of the equation of vibrations of a string with initial conditions:

$$u_{yy} = u_{xx} + f_1(x, t) \qquad (y = at,\ f_1 = f/a^2),$$

$$u(x, 0) = \varphi(x),$$

$$u_y = \psi_1(x) \qquad (\psi_1 = \psi/a).$$

In formula (10) the arc PQ is a segment of the axis $y = 0$. The operator

$$\mathscr{L} = u_{xx} - u_{yy}$$

is self-conjugate since

$$\mathscr{M} = \mathscr{L} = u_{xx} - u_{yy}.$$

As $a = 0$ and $b = 0$ the function v on the characteristic curves MP and MQ is equal to unity. It follows that

$$v(M, M') \equiv 1$$

for any point M' inside the triangle PMQ.

Next, remembering that in our case

$$d\eta = 0 \text{ on } PQ,$$

we substitute in formula (10) and get the equation:

$$u(M) = \frac{u(P) + u(Q)}{2} + \frac{1}{2}\int_P^Q u_\eta \, d\xi + \frac{1}{2}\int\!\!\int_{PMQ} f(\xi, \eta)\, d\xi\, d\eta.$$

Noting that $P = P(x - y, 0)$, $Q = Q(x + y, 0)$, where x and y are coordinates of the point $M = M(x, y)$ and using the initial conditions we have:

$$u(x, y) = \frac{u(x - y) + u(x + y)}{2} +$$

$$+ \frac{1}{2}\int_{x-y}^{x+y} \psi_1(\xi)\, d\xi + \frac{1}{2}\int_0^y \int_{x-(y-\eta)}^{x+(y-\eta)} f_1(\xi, \eta)\, d\xi\, d\eta.$$

Returning to the variables x and t we obtain the formula

$$u(x,t) = \frac{u(x-at)+u(x+at)}{2} +$$

$$+ \frac{1}{2a}\int_{x-at}^{x+at} \psi(\xi)\,d\xi + \frac{1}{2a}\int_0^t \int_{x-a(t-\tau)}^{x+a(t-\tau)} f(\xi,\tau)\,d\xi\,d\tau,$$

which we have already met in sub-section 7, § 2.

As the second example we shall consider the equation with constant coefficients

$$u_{xx} - u_{yy} + au_x + bu_y + cu = 0 \quad (a, b, c \text{ are constant numbers}) \tag{21}$$

$$u|_{y=0} = \varphi(x), \tag{22}$$

$$u_y|_{y=0} = \psi(x). \tag{23}$$

The substitution

$$U = ue^{\lambda x + \mu y} \tag{24}$$

allows reducing the equation (21) to a simpler form

$$U_{xx} - U_{yy} + c_1 U = 0, \quad c_1 = \frac{1}{4}(4c^2 - a^2 - b^2) \tag{25}$$

with the additional conditions

$$U|_{y=0} = \varphi(x)\,e^{\frac{a}{2}x} = \varphi_1(x), \tag{22'}$$

$$U_y|_{y=0} = \left(\psi(x) - \frac{b}{2}\varphi(x)\right)e^{\frac{a}{2}x} = \psi_1(x),$$

if the parameters λ and μ are chosen suitably, putting

$$\lambda = a/2, \quad \mu = -b/2. \tag{2}$$

In order to find the function $U(x, y)$ from the initial data and equation (25) we calculate Riemann's function $v(x, y;\ \xi, \eta)$.

The function v should satisfy the conditions

$$v_{xx} - v_{yy} + c_1 v = 0, \tag{27}$$

$$\left.\begin{aligned} &v = 1 \text{ on the characteristic curve } MP,\\ &v = 1 \text{ on the characteristic curve } MQ \text{ (Fig. 28).}\end{aligned}\right\} \tag{28}$$

We shall look for v in the form

$$v = v(z), \tag{29}$$

where

$$z = \sqrt{[(x-\xi)^2 - (y-\eta)^2]} \text{ or } z^2 = (x-\xi)^2 - (y-\eta)^2 . \quad (30)$$

On the characteristic curves MP and MQ the variable z

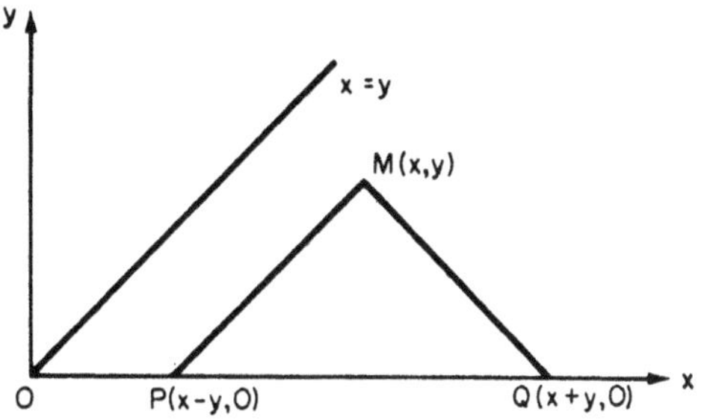

FIG. 28

vanishes so that $v(0) = 1$. Further, the left-hand side of equation (27) is transformed as follows:

$$v_{xx} - v_{yy} + c_1 v = v''(z)(z_x^2 - z_y^2) + v'(z)(z_{xx} - z_{yy}) + c_1 v = 0 .$$

Differentiating the expression for z^2 twice, with respect to x and y, we get

$$zz_x = x - \xi ,$$

$$zz_y = -(y - \eta) ,$$

$$zz_{xx} + z_x^2 = 1 ,$$

$$zz_{yy} + z_y^2 = -1 .$$

Hence from formula (30) we find

$$z_x^2 - z_y^2 = 1, \quad z_{xx} - z_{yy} = 1/z .$$

The equation for v assumes the following form:

$$v'' + \frac{1}{z} v' + c_1 v = 0$$

with the condition $v(0) = 1$. The solution of this equation is a Bessel function of zero order (see Supplement, Part I, § 1)

$$v(z) = J_0(\sqrt{c_1} \cdot z)$$

or

$$v(x, y; \xi, \eta) = J_0\left(\sqrt{\{c_1[(x-\xi)^2 - (y-\eta)^2]\}}\right) . \quad (31)$$

We shall now make use of formula (10) which in our case becomes

$$U(M) = \frac{U(P) + U(Q)}{2} + \frac{1}{2}\int_P^Q (v\,U_\eta\, d\xi - U v_\eta\, d\xi)\quad (d\eta = 0). \tag{32}$$

to find $u(x, y)$.

Let us first calculate the integral over the segment PQ ($\eta = 0$):

$$\int_P^Q (v\,U_\eta - U v_\eta)\, d\xi = \int_{x-y}^{x+y} \{J_0(\sqrt{\{c_1[(x-\xi)^2 - y^2]\}})\, U_\eta(\xi, 0) -$$

$$- \frac{U(\xi, 0)\,\sqrt{c_1}\cdot y\, J_0'\sqrt{c_1}\,(\sqrt{[(x-\xi)^2 - y^2]})}{\sqrt{\{c_1[(x-\xi)^2 - y^2]\}}}\Bigg\}\, d\xi. \tag{33}$$

Using the initial conditions (22′, 23′) we find:

$$U(x, y) = \frac{\varphi_1(x-y) + \varphi_1(x+y)}{2} + \frac{1}{2}\int_{x-y}^{x+y} J_0(\sqrt{c_1}\sqrt{[(x-\xi)^2 - y^2]}) \times$$

$$\times\, \psi_1(\xi)\, d\xi - \sqrt{c_1}\,y\,\frac{1}{2}\int_{x-y}^{x+y} \frac{J_1(\sqrt{c_1}\sqrt{[(x-\xi)^2 - y^2]})\,\varphi_1(\xi)\, d\xi}{\sqrt{[(x-\xi)^2 - y^2]}}, \tag{34}$$

whence, because of (24), (22′) and (23′),

$$u(x, y) = \frac{\varphi(x-y)\, e^{-\frac{a-b}{2}y} + \varphi(x+y)\, e^{\frac{a+b}{2}y}}{2} -$$

$$- \frac{1}{2} e^{\frac{b}{2}y} \int_{x-y}^{x+y} \Bigg\{\frac{b}{2} J_0(\sqrt{c_1}\sqrt{[(x-\xi)^2 - y^2]}) +$$

$$+ \sqrt{c_1}\cdot y\, \frac{J_1(\sqrt{c_1}\sqrt{[(x-\xi)^2 - y^2]})}{\sqrt{(x-\xi)^2 - y^2}}\Bigg\}\, e^{-\frac{a}{2}(x-\xi)}\, \varphi(\xi)\, d\xi + \tag{35}$$

$$+ \frac{1}{2} e^{\frac{b}{2}y} \int_{x-y}^{x+y} J_0(\sqrt{c_1}\sqrt{[(x-\xi)^2 - y^2]})\, e^{-\frac{a}{2}(x-\xi)}\, \psi(\xi)\, d\xi,$$

which gives the solution of the problem.

Let us consider the particular case $a = 0$, $b = 0$, i. e. the equation

$$u_{xx} - u_{yy} + cu = 0\,.$$

From formula (35) we obtain at once

$$u(x, y) = \frac{\varphi(x - y) + \varphi(x + y)}{2} +$$

$$+ \frac{1}{2} \int_{x-y}^{x+y} J_0(\sqrt{c_1}\sqrt{[(x - \xi)^2 - y^2]})\psi(\xi)\, d\xi -$$

$$- \frac{1}{2}\sqrt{c_1} \cdot y \int_{x-y}^{x+y} \frac{J_1(\sqrt{c_1}\sqrt{[(x - \xi)^2 - y^2]})}{\sqrt{[(x - \xi)^2 - y^2]}}\, \varphi(\xi)\, d\xi \,. \tag{36}$$

Putting $c_1 = 0$ and $y = at$ we arrive at D'Alembert's formula

$$u(x, t) = \frac{\varphi(x - at) + \varphi(x + at)}{2} + \frac{1}{2a} \int_{x-at}^{x+at} \overline{\psi}(\xi)\, d\xi \,, \tag{37}$$

giving the solution of the simple wave equation

$$u_{xx} - \frac{1}{a^2} u_{tt} = 0$$

under the initial conditions

$$u(x, 0) = \varphi(x) \,,$$

$$u_t(x, 0) = \overline{\psi}(x) \,,$$

$$\overline{\psi}(x) = a\psi(x) = au_y(x, 0) \,.$$

Problems on Chapter II

1. Solve problem 1 from § 4 assuming that initially the concentration of moisture is constant along the entire tube and a current of dry air is being supplied at the inlet.

2. Solve problem 2 from § 4, taking the initial temperature of the system to be equal to u_0 and the temperature at the end of the tube to be constantly maintained at $v_0 > u_0$.

3. Solve the system of telegraphic equations (see § 1 [21]):

$$i_x + Cv_t + Gv = 0 \,,$$

$$v_x + Li_t + Ri = 0$$

for an infinite line under the initial conditions

$$i(x, 0) = \varphi(x) \,,$$

$$v(x, 0) = \psi(x) \,.$$

Hint. Reduce the system of equations (§ 1 [21]) to an equation of the second order for one of the functions $i(x, t)$ or $v(x, t)$, for example

$$i_{xx} = CLi_{tt} + (CR + GL)\, i_t + GRi$$

with the initial conditions

$$i\,(x, 0) = \varphi\,(x)\,,$$

$$\left.\frac{\partial i}{\partial t}\right|_{t=0} = -\left(\frac{1}{L}v_x + \frac{R}{L}i\right)_{t=0} = -\frac{1}{L}\psi'\,(x) - \frac{R}{L}\varphi\,(x) = \psi_0\,(x)\,,$$

and then use the formula (35).

4. Investigate the solution of the telegraphic equation obtained [formula (35)] for the case of small G and R. Consider the limiting case $G \to 0$, $R \to 0$ and obtain D'Alembert's formula for the solution of the simple wave equation.

APPENDICES TO CHAPTER II

I. Vibrations of strings of musical instruments

A vibrating string excites vibrations of the air which are perceived by the ear as sound. The intensity of the sound is characterized by the energy or the amplitude of the vibrations, the tone by the period of the vibrations and the timbre by the ratio of the energies of the fundamental tone and of the overtones.* We will not dwell on the physiological processes of perception of sound and the transmission of sound through air, but shall characterize the sound of the string by its energy and period and the distribution of energy over the overtones.

Transverse vibrations of strings are usually excited in musical instruments. We distinguish three types of stringed instruments: bowed string, plucked string and struck string. In struck string instruments (e. g. piano) the vibration is excited by a blow which imparts an initial velocity without initial deflection. In plucked stringed instruments (e. g. harp, guitar) the vibrations are excited by imparting a certain initial deflection without initial velocity.

Free vibrations of a string excited by any method may be represented in the form (see Chapter II, § 3).

$$u(x,t) = \sum_{n=1}^{\infty} (a_n \cos \omega_n t + b_n \sin \omega_n t) \sin \frac{\pi n}{l} x \quad \left(\omega_n = \frac{\pi n}{l} a\right).$$

As an exercise to § 3 we set problem 1, which involves the simplest theory of excitation of plucked string instruments. The solution of this problem shows that if the initial deflection

* Rayleigh, *Theory of Sound*, vol. I, Gostekhizdat, 1940.

of the string is represented by a triangle with the height h at the point $x = c$ (Fig. 29) then,

$$a_n = \frac{2hl^2}{\pi^2 n^2 c(l-c)} \sin \frac{\pi n c}{l}, \quad b_n = 0. \tag{1}$$

The energy of the nth harmonic is

$$E_n = \frac{1}{4} \varrho l \omega_n^2 a_n^2 = Mh^2 \frac{l^2 a^2}{\pi^2 n^2 c^2 (l-c)^2} \sin^2 \frac{\pi n c}{l} \quad (M = \varrho l) \tag{2}$$

and is proportional to $1/n^2$.

In problem 4 of § 3 we consider the simplest theory of excitation of a string by striking a blow with impulse K at the point c. The solution of this problem is

$$u(x, t) = \frac{2K}{\pi a \varrho} \sum_{n=1}^{\infty} \frac{1}{n} \sin \frac{\pi n c}{l} \cdot \sin \frac{\pi n}{l} x \cdot \sin \omega_n t \quad \left(\omega_n = \frac{\pi n}{l} a\right), \tag{3}$$

$$E_n = \frac{K^2}{M} \sin^2 \frac{\pi n c}{l}. \tag{4}$$

Thus when the string is excited by a blow concentrated in a small interval of length δ, the energies of the different harmonics (for which δ is small compared to the distance between the nodes)

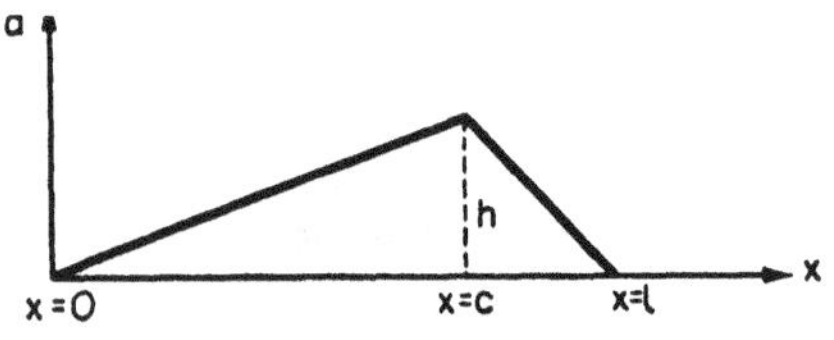

FIG. 29

will differ only slightly. The tone emitted by the string will be saturated with overtones. This conclusion is easily verified experimentally. If a stretched string is struck with the blade of a knife the string tinkles: the sound is saturated with overtones. In a piano the string is excited by a blow of a hammer wrapped in leather. Such an excitation can be represented approximately by one of the following schemes:

1. The string is excited by imparting a constant initial velocity v_0 in the interval $(c - \delta, c + \delta)$. This case corresponds to a plane, rigid hammer having a width 2δ and striking at the point c.

The process of vibration is described by the function (see problem 3, § 3)

$$u(x,t)=\frac{4v_0 l}{\pi^2 a}\sum_{n=1}^{\infty}\frac{1}{n^2}\sin\frac{\pi nc}{l}\cdot\sin\frac{\pi n}{l}\delta\cdot\sin\frac{\pi n}{l}x\cdot\sin\omega_n t,$$

and the energies of the different harmonics are

$$E_n=\frac{4Mv_0^2}{n^2\pi^2}\sin^2\frac{\pi nc}{l}\cdot\sin^2\frac{\pi n\delta}{l}.$$

2. The string is excited with an initial velocity

$$\frac{\partial v}{\partial t}(x,0)=\begin{cases} v_0\cos\frac{x-c}{\delta}\cdot\frac{\pi}{2} & (x-c)<\delta,\\ 0 & (x-c)>\delta.\end{cases}$$

This case corresponds to a rigid, convex hammer of width 2δ. Such a hammer excites the maximum initial velocity at the centre of the interval 2δ. This can be described schematically by the function given above. The vibration excited has the form (see problem 3, § 3)

$$u(x,t)=\frac{8v_0\delta}{\pi^2 a}\sum_{n=1}^{\infty}\frac{1}{n}\frac{\cos\frac{\pi n}{l}\delta\cdot\sin\frac{\pi n}{l}c}{1-\left(\frac{2\delta n}{l}\right)^2}\sin\frac{\pi n}{l}x\cdot\sin\omega_n t$$

and the energies of the harmonics are

$$E_n=\frac{16v_0^2\delta^2\varrho}{l\pi^2}\cdot\frac{1}{\left[1-\left(\frac{2\delta n}{l}\right)^2\right]^2}\cdot\cos^2\frac{\pi n\delta}{l}\cdot\sin^2\frac{\pi nc}{l}.$$

3. The hammer exciting vibrations in a string is not ideally rigid. In this case the vibrations are not determined by the initial velocity but by a force varying with time. We get an inhomogeneous equation with the force $F(x, t)$ given. For example, the force might be

$$F(x,t)=\begin{cases} F_0\cos\frac{x-c}{\delta}\cdot\frac{\pi}{2}\sin\frac{\pi t}{\tau}, & \text{if } \begin{cases}|x-c|<\delta,\\ 0\le t\le\tau,\end{cases}\\ 0 & \text{if } \begin{cases}|x-c|>\delta,\\ t>\tau.\end{cases}\end{cases}$$

The solution of this equation for $t > \tau$ is

$$u(x, t) = \frac{16 F_0 \tau \delta}{\pi^3 \varrho a} \sum_{n=1}^{\infty} \frac{1}{n} \frac{\cos \frac{\pi n \delta}{l} \cos \frac{\omega_n \tau}{2} \sin \frac{\pi n c}{l}}{\left[1 - \left(\frac{2\delta n}{l}\right)^2\right]\left[1 - \left(\frac{n a \tau}{l}\right)^2\right]} \times$$
$$\times \sin \frac{\pi n}{l} x \sin \omega_n \left(t - \frac{\tau}{2}\right).$$

These examples show that the width of the interval over which the blow is delivered and the duration of the blow have a very considerable effect on the energy of the high overtones. Moreover the presence of the factor $\sin(n\,\pi/l)c$ shows that if the centre of the blow of the hammer occurs at a node of the nth harmonic the energy of the corresponding harmonic is zero.

The presence of high overtones (beginning from the 7th) disturbs the harmonicity of the sound and causes the feeling of dissonance.* The presence of low overtones, on the other hand, produces a fulness of sound. In a piano the position of the hammer is selected close to the end of the string between the nodes of the 7th and 8th overtones in order to reduce their energy. By adjusting the width and rigidity of the hammer an attempt is made to increase the relative energy of the low (3rd and 4th) overtones. Old types of pianos with a sharp, even a tinkling sound, used narrow, rigid hammers.

II. Vibrations of rods

In courses of methods of mathematical physics fundamental importance is given to equations of the second order. But a large number of problems of the vibrations of rods, plates, etc., lead to equations of a higher order.

As an example of equations of the fourth order let us consider the problem of natural vibrations of a tuning fork, which is equivalent to the problem of vibrations of a thin rectangular

* For example, if the fundamental frequency (first harmonic) of 440 vibrations per second corresponds to "la" of the first octave, then a frequency seven times larger corresponds to "so" of the fourth octave. The interval la–so, the so-called small 7th, is dissonant and unpleasant to the ear.

rod clamped at one end. The study of the vibrations reduces to a solution of the equation

$$\frac{\partial^2 y}{\partial t^2} + a^2 \frac{\partial^4 y}{\partial x^4} = 0. \tag{1}$$

This equation occurs in many problems on the vibrations of rods, in the calculation of the stability of rotating shafts and also in the study of the vibration of ships.

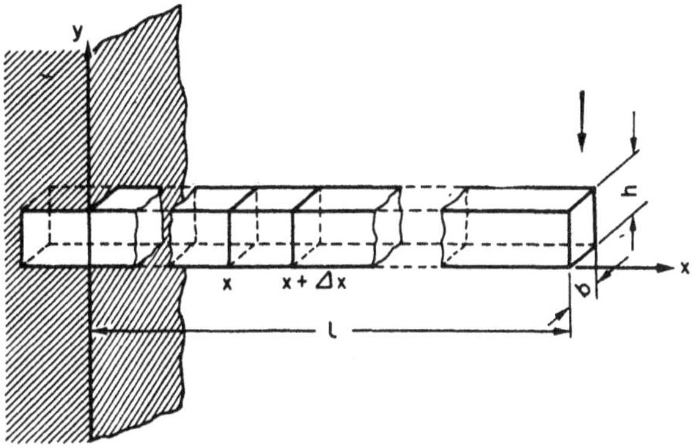

FIG. 30

We shall give an elementary derivation of equation (1). Let us consider a rectangular rod of length $l (0 \leqslant x \leqslant l)$ of height h and width b (Fig. 30). We consider an element of length dx. After bending the end sections of this element of the rod, assumed to be plane, form an angle $d\varphi$. If the deformations are small and the length of the axis of the rod does not change on bending ($dl = dx$), then

$$d\varphi = \left.\frac{\partial y}{\partial x}\right|_x - \left.\frac{\partial y}{\partial x}\right|_{x+dx} = -\frac{\partial^2 y}{\partial x^2}\, dx .$$

A layer of material at a distance η from the axis of the rod $y = 0$ changes its length by the amount $\eta d\varphi$ (Fig. 31). According to Hooke's law the tensile force acting along the layer is

$$dN = E \cdot b d\eta \cdot \frac{\eta d\varphi}{dx} = -E \cdot b \frac{\partial^2 y}{\partial x^2} \eta \, d\eta ,$$

where E is the Young's modulus of the material. The total bending moment of the forces acting in the section x is

$$M = -E\frac{\partial^2 y}{\partial x^2} b \int_{-\frac{h}{2}}^{\frac{h}{2}} \eta^2 \, d\eta = -E\frac{\partial^2 y}{\partial x^2} J, \tag{2}$$

where

$$J = b \int_{-\frac{h}{2}}^{\frac{h}{2}} \eta^2 \, d\eta = \frac{bh^3}{12}$$

is the moment of inertia of the rectangular section about a horizontal axis through the centre of mass of the section. Let us denote by $M(x)$ the moment acting on the left-hand portion of the rod at each section. At the section $x + dx$ the moment of tensile forces is equal to $-(M + dM)$.

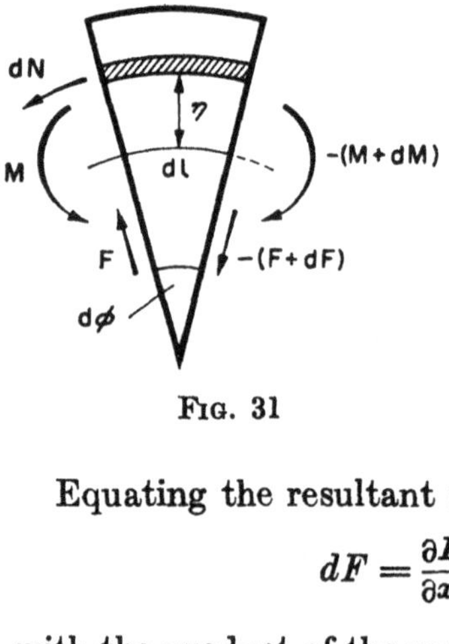

FIG. 31

The excess moment $-dM$ is balanced by the moment of the tangential forces,

$$dM = F dx.$$

Hence, from equation (2) we obtain the magnitude of the tangential force

$$F(x, t) = \frac{\partial M}{\partial x} = -EJ\frac{\partial^3 y}{\partial x^3}. \tag{3}$$

Equating the resultant force acting on the element

$$dF = \frac{\partial F}{\partial x} dx = -EJ\frac{\partial^4 y}{\partial x^4} dx$$

with the product of the mass of the element and its acceleration

$$\varrho S \frac{\partial^2 y}{\partial t^2} dx,$$

where ϱ is the density of the rod, S is the area of the cross-section (we neglect the rotating motion), we obtain the equation of transverse vibrations of the rod

$$\frac{\partial^2 y}{\partial t^2} + a^2 \frac{\partial^4 y}{\partial x^4} = 0 \qquad \left(a^2 = \frac{EJ}{\varrho S}\right). \tag{1}$$

The boundary conditions for the fixed end $x = 0$ are that the rod should be stationary and the tangent horizontal

$$y\big|_{x=0} = 0, \quad \frac{\partial y}{\partial x}\bigg|_{x=0} = 0. \tag{4}$$

At the free end the bending moment (2) and the tangential force (3) should become zero, whence it follows that

$$\frac{\partial^2 y}{\partial x^2}\bigg|_{x=l} = 0, \quad \frac{\partial^3 y}{\partial x^3}\bigg|_{x=l} = 0. \tag{5}$$

To define the motion of the rod completely, the initial conditions — initial deflection and initial velocity — must be given.

$$y\big|_{t=0} = f(x) \text{ and } \frac{\partial y}{\partial x}\bigg|_{t=0} = \varphi(x) \quad (0 \leq x \leq l). \tag{6}$$

Thus the problem is reduced to a solution of the equation (1) with the boundary conditions (4), (5) and the initial conditions (6).

We shall solve the problem by the method of separation of variables, putting

$$y = Y(x)\, T(t). \tag{7}$$

Substituting the suggested form of the solution in (1) we have:

$$\frac{T''(t)}{a^2 T(t)} = -\frac{Y^{(4)}(x)}{Y(x)} = -\lambda.$$

We obtain an eigen-value problem for the function $Y(x)$

$$Y^{(4)} - \lambda Y = 0, \tag{8}$$

$$Y\big|_{x=0} = 0, \quad \frac{dY}{dx}\bigg|_{x=0} = 0, \quad \frac{d^2Y}{dx^2}\bigg|_{x=l} = 0, \quad \frac{d^3Y}{dx^3}\bigg|_{x=l} = 0. \tag{9}$$

The general solution of equation (8) is

$$Y(x) = A\cosh\sqrt[4]{\lambda}\cdot x + B\sinh\sqrt[4]{\lambda}\cdot x + C\cos\sqrt[4]{\lambda}\cdot x + D\sin\sqrt[4]{\lambda}\cdot x.$$

From the conditions $Y(0) = 0$, $Y'(0) = 0$ we find $C = -A$, $D = -B$. Hence it follows that

$$Y(x) = A\,[\cosh\sqrt[4]{\lambda}\cdot x - \cos\sqrt[4]{\lambda}\cdot x] + [B\sinh\sqrt[4]{\lambda}\cdot x - \sin\sqrt[4]{\lambda}\cdot x].$$

The conditions $Y''(l) = 0$ and $Y'''(l) = 0$ give

$$A\,[\cosh \sqrt[4]{\lambda}\cdot l + \cos \sqrt[4]{\lambda}\cdot l] + B\,[\sinh \sqrt[4]{\lambda}\cdot l + \sin \sqrt[4]{\lambda}\cdot l] = 0\,,$$

$$A\,[\sinh \sqrt[4]{\lambda}\cdot l - \sin \sqrt[4]{\lambda}\cdot l] + B\,[\cosh \sqrt[4]{\lambda}\cdot l + \cos \sqrt[4]{\lambda}\cdot l] = 0\,.$$

This homogeneous system has non-trivial solutions A and B if the determinant of the coefficient is zero. Equating this determinant to zero we obtain the transcendental equation for the eigen-values

$$\sinh^2 \sqrt[4]{\lambda}\cdot l - \sin^2 \sqrt[4]{\lambda}\cdot l = \cosh^2 \sqrt[4]{\lambda}\cdot l + 2\cosh \sqrt[4]{\lambda}\cdot l\,(\cos \sqrt[4]{\lambda}\cdot l) + \cos^2 \sqrt[4]{\lambda}\cdot l.$$

Since $\cosh^2 x - \sinh^2 x = 1$ this equation may be written in the form

$$\cosh \mu \cdot \cos \mu = -1 \qquad (\mu = \sqrt[4]{\lambda}\cdot l). \tag{10}$$

The roots of equation (10) can be calculated numerically*

$$\mu_1 = 1\cdot 875\,,$$
$$\mu_2 = 4\cdot 694\,,$$
$$\mu_3 = 7\cdot 854\,,$$
$$\cdots\cdots$$
$$\cdots\cdots$$
$$\mu_n \approx \frac{\pi}{2}(2n-1) \qquad \text{for} \qquad n > 3.$$

The last formula gives the value of μ_n correct up to the third decimal place, beginning from $n = 3$ and correct up to the sixth decimal for $n \geqslant 7$.

Let us now find the natural frequencies of the tuning fork. The equation

$$T'' + a^2 \lambda_n T = 0$$

is satisfied by the trigonometric functions

$$T_n(t) = a_n \cos 2\pi\nu_n t + b_n \sin 2\pi\nu_n t$$

with the frequency

$$\nu_n = \frac{a\sqrt{\lambda_n}}{2\pi} = \frac{\sqrt{\lambda_n}}{2\pi}\sqrt{\frac{EJ}{\varrho S}} = \frac{\mu_n^2}{2\pi l^2}\sqrt{\frac{EJ}{\varrho S}}\,.$$

* For calculation of the roots of equation (10) see Rayleigh, *Theory of Sound*, vol. I, chap. VIII, 1940.

The frequencies ν_n of the natural vibrations are proportional to the squares of μ_n. Since

$$\frac{\mu_2^2}{\mu_1^2} = 6{\cdot}267\,, \quad \frac{\mu_3^2}{\mu_1^2} = 17{\cdot}548\,,$$

the second natural tone is higher than the fundamental tone by more than two and a half octaves, i.e. higher than the sixth harmonic of the string when the fundamental tone is the same; the third natural vibration is higher than the fundamental tone by more than four octaves. For example, if the tuning fork has a fundamental frequency of 440 vibrations per second (accepted standard for a' — "la" notes of the first octave) the next natural frequency of the tuning fork will be 2757.5 vibrations per second (between $c'''' = 2637.3$ and $f'''' = 2794.0$ — between the notes "me" and "fa" of the fourth octave of the uniformly moderated scale), the third natural frequency of 7721.1 vibrations per second is already beyond the limits of the scale of the properly musical sounds.

When vibrations of the tuning fork are excited by a blow, not only the first but also the higher harmonics are present, which explains the metallic sound at the initial moment. But in course of time the higher harmonics are damped rapidly and the tuning fork emits a pure note of the fundamental tone.

III. Vibrations of a loaded string

1. *Formulation of the problem*

Let us consider the problem of vibrations of a string $(0, l)$ fixed at the ends, at certain points of which $x = x_i$ $(i = 1, 2, \ldots$ $\ldots, n)$ are placed concentrated masses M_i.

The conditions at the point x_i may be obtained in two ways. If a concentrated force $F_i(t)$ is applied at the point x_i $(i =$ $= 1, 2, \ldots, n)$ then the following relations must be satisfied

$$u(x_i - 0, t) = u(x_i + 0, t)\,, \tag{1}$$

$$ku_x\Big|_{x_i-0}^{x_i+0} = -F_i. \tag{2}$$

In this case F_i is the force of inertia. Substituting in formula (2)

$$F_i = -M_i u_{tt}(x_i, t)$$

we get

$$M_i u_{tt}(x_i, t) = k u_x \Big|_{x_i-0}^{x_i+0}. \tag{3}$$

Another derivation of condition (3) is also possible. Let us distribute the mass M_i over the region $(x_i - \varepsilon,\ x_i + \varepsilon)$ with the constant density δ_i and use the equation of vibrations for the non-homogeneous string

$$(\varrho + \delta_i) u_{tt} = \frac{\partial}{\partial x}\left(k \frac{\partial u}{\partial x}\right), \quad x_i - \varepsilon < x < x_i + \varepsilon, \tag{4}$$

where ϱ is the density of the string. Let $u_\varepsilon(x, t)$ be the solution of this equation.

Integrating equation (4) with respect to x within the limits $x_i - \varepsilon$ and $x_i + \varepsilon$ and making the limiting transition for $\varepsilon \to 0$ we shall obtain the condition (3) for the function $u(x, t) = \lim_{\varepsilon \to 0} u_\varepsilon(x, t)$. We shall not deal with the justification for the limiting transition.

Let us formulate our problem completely:

Find the solution of the equation of vibrations

$$\varrho \frac{\partial^2 u}{\partial t^2} = \frac{\partial}{\partial x}\left(k \frac{\partial u}{\partial x}\right). \tag{5}$$

which satisfies the boundary conditions

$$\left.\begin{aligned} u(0, t) &= 0, \\ u(l, t) &= 0, \end{aligned}\right\} \tag{6}$$

the conjugation conditions (3) at the points $x = x_i$

$$\left.\begin{aligned} u(x_i - 0, t) &= u(x_i + 0, t), \\ M_i u_{tt}(x_i, t) &= k u_x \Big|_{x_i-0}^{x_i+0} \end{aligned} \quad (i = 1, 2, \ldots, n)\right\} \tag{7}$$

and the initial conditions

$$\left.\begin{aligned} u(x, 0) &= \varphi(x), \\ u_t(x, 0) &= \psi(x), \end{aligned}\right\} \tag{8}$$

where $\varphi(x)$ *and* $\psi(x)$ *are given functions.*

2. *Natural vibrations of a loaded string*

We shall deal first with the investigation of the natural frequencies and the profile of the standing waves for the loaded string. To do this we must find a solution of the given problem which can be represented in the form of a product

$$u(x,t) = X(x)\,T(t)\,. \tag{9}$$

Substituting this expression in equation (5) and using the boundary conditions we obtain after separation of variables

$$T'' + \lambda T = 0 \tag{10}$$

and

$$\left.\begin{aligned} &\frac{d}{dx}(kX') + \lambda\varrho\, X = 0\,, \\ &X(0) = 0,\ X(l) = 0\,. \end{aligned}\right\} \tag{11}$$

The conjugation conditions give:

$$X(x_i - 0) = X(x_i + 0)\,,$$

$$M_i X(x_i)\, T'' = kX'\big|_{x_i-0}^{x_i+0} T.$$

Taking equation (10) into consideration, we rewrite the latter relation in the form

$$kX'\big|_{x_i-0}^{x_i+0} = -\lambda M_i X(x_i)\,.$$

Thus for the function $X(x)$ we get the following eigen-value problem

$$\frac{d}{dx}(kX') + \lambda\varrho X = 0\,, \tag{11}$$

$$X(0) = 0, \quad X(l) = 0\,, \tag{12}$$

$$\left.\begin{aligned} X(x_i - 0) &= X(x_i + 0) \quad (i = 1, 2, \ldots, n)\,, \\ kX'(x_i + 0) - kX'(x_i - 0) &+ \lambda M_i X(x_i) = 0\,. \end{aligned}\right\} \tag{13}$$

The distinctive feature of this boundary-value problem is that the parameter λ enters not only the equation but also the additional conditions.

Let us derive the orthogonality conditions of the eigen-functions

$$X_1(x)\,,\quad X_2(x)\,, \ldots$$

As was shown in Chapter II (see § 3), the eigen-functions for the boundary problem

$$\frac{d}{dx}(kX') + \lambda\varrho X = 0\,,$$

$$X(0) = 0\,,\quad X(l) = 0$$

are orthogonal with the weight ϱ for the interval $(0, l)$:

$$\int_0^l X_m(x)\, X_n(x)\, \varrho(x)\, dx = 0 \quad (m \neq n)\,. \tag{14}$$

Distributing each mass M_i with a constant density δ_i over some interval $x_i - \varepsilon < x < x_i + \varepsilon$, where $\varepsilon > 0$ is a small number, we arrive at the problem of natural vibrations of a non-homogeneous string with the density $\varrho_\varepsilon(x)$. Let $\lambda_{\varepsilon n}$ and $\{X_{\varepsilon n}(x)\}$ be eigen-values and eigen-functions of this problem. The orthogonality condition must be fulfilled

$$\int_0^l X_{\varepsilon m}(x)\, X_{\varepsilon n}(x)\, \varrho_\varepsilon(x)\, dx = 0\,. \tag{15}$$

Separating in equation (15) the integrals over the regions $(x_i - \varepsilon, x_i + \varepsilon)$ and making the limiting transition when $\varepsilon \to 0$, we shall obtain the orthogonality relation*

$$\int_0^l X_m(x)\, X_n(x)\, \varrho(x)\, dx + \sum_{i=1}^{n} M_i\, X_m(x_i)\, X_n(x_i) = 0 \quad (m \neq n)\,, \tag{16}$$

We again leave the question of justifying this transition.

The condition of orthogonality (16) may also be obtained from the equation and the conditions (11)–(13). Let $X_m(x)$ and $X_n(x)$ be the eigen-functions of the problem (1) corresponding to the eigen-values λ_m and λ_n satisfying the equations

$$\frac{d}{dx}\left(k\frac{dX_m}{dx}\right) + \lambda_m \varrho X_m = 0\,,$$

$$\frac{d}{dx}\left(k\frac{dX_n}{dx}\right) + \lambda_n \varrho X_n = 0\,.$$

Let us multiply the first equation by $X_n(x)$, the second by $X_m(x)$ and subtract the second equation from the first. Integrating

* R. Courant and D. Hilbert, *Methods of Mathematical Physics;* vol. I, chap. VI.

the equation obtained successively over the regions $(0, x_1)$; (x_1, x_2); ...; (x_k, l) and adding we have:

$$(\lambda_m - \lambda_n) \int_0^l X_m(x) X_n(x) \varrho(x)\, dx - \\ - \sum_{i=0}^{k} \int_{x_i}^{x_i+1} \frac{d}{dx}[X_m kX'_n - X_n kX'_m]\, dx = 0\,, \quad (17)$$

putting $x_0 = 0$, $x_{k+1} = l$. Carrying out the integration in each term of the sum and combining terms corresponding to the substitutions $x = x_i - 0$ and $x = x_i + 0$ we obtain a sum of terms of the type

$$A_i = (X_m kX'_n - X_n kX'_m)_{x=x_i-0} - (X_m kX'_n - X_n kX'_m)_{x=x_i+0}.$$

The substitutions for $x = 0$ and $x = l$ in this case give zero because of the boundary conditions.

To calculate A_i we make use of the conjugation conditions

$$\left.\begin{aligned} X_j(x_i - 0) &= X_j(x_i + 0)\,, \\ kX'_j(x_i + 0) - kX'_j(x_i - 0) &= -M_i \lambda_j X_j(x_i) \end{aligned}\right\} (j = m, n). \quad (13')$$

Rewriting A_i in the form

$$A_i = X_m(x_i)\left[kX'_n(x_i - 0) - kX'_n(x_i + 0)\right] - \\ - X_n(x_i)\left[kX'_m(x_i - 0) - kX'_m(x_i + 0)\right]$$

and using formula (13) we find

$$A_i = X_m(x_i) M_i \lambda_n X_n(x_i) - X_n(x_i) M_i \lambda_m X_m(x_i) = \\ = M_i X_m(x_i) X_n(x_i)(\lambda_n - \lambda_m).$$

Now equation (17) may be written as

$$(\lambda_m - \lambda_n)\left\{ \int_0^l X_m(x) X_n(x) \varrho(x)\, dx + \sum_{i=1}^{k} M_i X_m(x_i) X_n(x_i)\right\} = 0.$$

If $\lambda_m \neq \lambda_n$, the condition of orthogonality (16) follows immediately from this.

The norm of the eigen-functions $X_n(x)$ is determined from the formula

$$N_n = \int_0^l X_n^2(x) \varrho(x)\, dx + \sum_{i=1}^{k} M_i X_n^2(x_i). \quad (18)$$

We shall not give proofs of the existence of an infinite number of eigen-values and eigen-functions, the positive sign of the eigen-values, the theorems of expandability. This boundary problem like problems of the usual type considered by us in § 3 of Chapter II is reduced to a certain integral equation which in this case is an integral equation in Stieltje's integrals.

It is obvious that in the expansion of some function $f(x)$ into a series

$$f(x) = \sum_{n=1}^{\infty} f_n X_n(x)$$

the coefficients of expansion will be defined by the formula

$$f_n = \frac{\int_0^l f(x) X_n(x) \varrho(x)\, dx + \sum_{i=1}^{k}{}' M_i f(x_i) X_n(x_i)}{N_n}. \tag{19}$$

The problem with initial conditions given in sub-section 1, is solved by the usual method of separation of variables. The problem of the vibration of a rod (or beam) with concentrated masses is extensively applied in physics and technology. Poisson himself solved the problem of longitudinal motion of a load suspended by an elastic thread. A. N. Krylov showed* that the theory of the indicator of a steam engine, the torsional vibrations of a shaft with a flywheel at the end, different types of "vibrating" valves, etc., can be reduced to this problem. A study of the torsional vibrations of a thread with a mass suspended from the end (e.g. a small mirror) is important for the theory of many measuring instruments.

A problem of this type has assumed particular importance in connection with the study of the stability of vibrations of the wings of an aeroplane. To solve this problem it is necessary to calculate the characteristic frequencies of the wing (beam of variable section), loaded with masses (motors). This problem also occurs in the calculation of the natural frequencies of antennae with concentrated capacities and self-inductances (see in this connection the appendix on the analogy between mechanical and electromagnetic vibrations).

* A. N. Krylov, *Some Differential Equations of Mathematical Physics*, chap. VII, Publ. Acad. of Sciences, U.S.S.R., 1932.

We shall not deal here with the approximate methods of finding eigen-values and functions which are similar to the approximate methods of finding the corresponding quantities for a non-homogeneous string.

3. *String with load at the end*

The problem of vibrations of a homogeneous string, one end of which ($x = 0$) is fixed and to the other end ($x = l$) of which is attached a load of mass M, is of considerable practical interest.

In this case the condition at $x = l$ assumes the form

$$Mu_{tt} = -ku_x(l, t)$$

and we obtain the following equation for the amplitude of the standing waves

$$X_n'' + \lambda_n X_n = 0$$

with the boundary conditions

$$X_n(0) = 0, \quad X_n'(l) = \frac{M}{\varrho} \lambda_n X_n(l).$$

Hence we find that

$$X_n(x) = \frac{\sin \sqrt{(\lambda_n)}\, x}{\sin \sqrt{(\lambda_n)}\, l},$$

where λ_n is determined from the equation

$$\cot(\sqrt{\lambda_n} \cdot l) = \frac{M}{\varrho} \sqrt{\lambda_n}. \tag{20}$$

The condition of orthogonality of the function $\{X_n(x)\}$ becomes

$$\int_0^l X_n(x) X_m(x) \varrho(x)\, dx + MX_n(l) X_m(l) = 0.$$

Let us calculate the norm

$$N_n = \int_0^l X_n^2(x) \varrho\, dx + MX_n^2(l).$$

Using equation (20) we get

$$N_n = \frac{l\varrho}{2} + \frac{M}{2} + \frac{M^2}{2\varrho} \lambda_n l.$$

The problem with initial data is solved by the usual method.

4. *Corrections for eigen-values*

Let us calculate the corrections for eigen-frequencies in the case of large and small loads M. We shall consider the case when the load is suspended from the end of a string. Two limiting cases are possible.

1. $M = 0$. The end $x = l$ is free. The eigen-values are determined from the formula

$$\sqrt{\lambda_n^{(1)}} = \frac{2n+1}{2} \frac{\pi}{l}.$$

2. $M = \infty$. The end $x = l$ is firmly attached: $u(l, t) = 0$. The eigen-values are determined from the formula

$$\sqrt{\lambda_n^{(2)}} = \frac{\pi n}{l}.$$

We shall be interested in the case of small $M (M \to 0)$ and large $M (M \to \infty)$.

1. M small. Let us find the correction to the eigen-value $\lambda_n^{(1)}$ putting

$$\sqrt{\lambda_n} = \sqrt{\lambda_n^{(1)}} + \varepsilon M, \tag{21}$$

where ε is some number. Substituting (21) in equation (20) and neglecting M^2 and higher powers of M we get

$$\lambda_n = \lambda_n^{(1)} \left(1 - \frac{2M}{\varrho l}\right), \tag{22}$$

i.e. the eigen-values of the loaded string when $M \to 0$ increase, approaching the eigen-frequencies of a string with a free end.

2. M is large. Choosing $1/M$ as the small parameter we put

$$\sqrt{\lambda_n} = \sqrt{{}_n^{(2)}} + \varepsilon \frac{1}{M}.$$

Equation (20) gives

$$\varepsilon = \frac{\varrho}{\sqrt{(\lambda_n^{(2)})}\, l}.$$

Here we have neglected terms containing $1/M^2$ and higher powers of $1/M$.

Thus

$$\sqrt{\lambda_n} = \sqrt{\lambda_n^{(2)}} + \frac{1}{\sqrt{(\lambda_n^{(2)})}\, l} \frac{\varrho}{M}, \quad \lambda_n = \lambda_n^{(2)} + \frac{2\varrho}{Ml}, \tag{23}$$

i. e. on increasing the load the eigen-frequencies decrease, approaching the eigen-frequencies of a string with fixed ends.

IV. Equations of gas dynamics and theory of shock waves

1. *Law of conservation of energy*

The equations of acoustics (see § 1) were obtained on the assumption that the velocities of motion in the gas and the changes in pressure were small which allowed linearizing the equations of hydrodynamics.

In problems arising in the study of the flight of rockets and high-speed aeroplanes, in the theory of ballistics, explosion waves, etc., one has to deal with hydrodynamic processes characterized by high velocities and pressure gradients. In this case the linear approximation of accustics is unsuitable and it is necessary to use the non-linear equations of hydrodynamics. Since this type of motion is found in practice in gases, it is usual to refer to hydrodynamics of high velocities as *gas dynamics*.

The equations of gas dynamics in the case of one dimensional motion of a gas (in the direction of the axis x) are

$$\frac{\partial \varrho}{\partial t} + \frac{\partial}{\partial x}(\varrho\, v) = 0 \quad \text{(equation of continuity)} \tag{1}$$

$$\varrho \frac{\partial v}{\partial t} + \varrho\, v \frac{\partial v}{\partial x} = -\frac{\partial p}{\partial x} \quad \text{(equation of motion)} \tag{2}$$

$$p = f(\varrho, T) \quad \text{(equation of state)} \tag{3}$$

Thus the equations of gas dynamics are equations of motion of an ideal compressible liquid in the absence of external forces. Viscosity is neglected.

Let us derive the law of conservation of energy. The energy of unit volume is

$$\frac{\varrho\, v^2}{2} + \varrho\varepsilon, \tag{4}$$

where the first term is the kinetic energy, the second the internal energy. Here ε obviously denotes the internal energy of unit mass.

For an ideal gas $\varepsilon = c_v T$, here c_v is the specific heat at constant volume, T the temperature. Let us calculate the change in energy in unit time

$$\frac{\partial}{\partial t}\left(\frac{\varrho\, v^2}{2} + \varrho\varepsilon\right) = \frac{\partial}{\partial t}\left(\frac{\varrho\, v^2}{2}\right) + \frac{\partial}{\partial t}(\varrho\varepsilon). \tag{5}$$

Differentiating the first term and using equations (1) and (2) we get

$$\frac{\partial}{\partial t}\left(\frac{\varrho v^2}{2}\right) = \frac{v^2}{2}\frac{\partial \varrho}{\partial t} + \varrho v \frac{\partial v}{\partial t} = -\frac{v^2}{2}\frac{\partial}{\partial x}(\varrho v) - \varrho v \frac{\partial}{\partial x}\left(\frac{v^2}{2}\right) - v\frac{\partial p}{\partial x}. \tag{6}$$

To transform the derivative $\frac{\partial}{\partial t}(\varrho\varepsilon)$ we go back to the first law of thermodynamics expressing the law of conservation of energy

$$dQ = d\varepsilon + p\,d\tau, \tag{7}$$

where dQ is the amount of heat received (or given off) by the system from outside, $p\,d\tau$ is the work done in changing the volume by the amount $d\tau (\tau = 1/\varrho$ is the specific volume).

If the process is adiabatic (there is no heat exchange with the medium) then

$$dQ = 0$$

and

$$d\varepsilon = -pd\frac{1}{\varrho} = \frac{p}{\varrho^2}\,d\varrho. \tag{8}$$

Using this equation we have

$$d(\varrho\varepsilon) = \varepsilon\,d\varrho + \varrho\,d\varepsilon = \varepsilon\,d\varrho + \frac{p}{\varrho}\,d\varrho = w\,d\varrho, \tag{9}$$

$$\frac{\partial}{\partial t}(\varrho\varepsilon) = w\frac{\partial \varrho}{\partial t}, \tag{10}$$

where

$$w = \varepsilon + p/\varrho \tag{11}$$

is the heat function or the heat content of unit mass.

The derivative $\frac{\partial \omega}{\partial x}$ satisfies the equation

$$\varrho v \frac{\partial w}{\partial x} = v\frac{\partial p}{\partial x}. \tag{12}$$

because of (9) and (11).

Taking into account equations (2), (5), (6), (10), (12) we obtain the law of conservation of energy in the differential form

$$\frac{\partial}{\partial t}\left(\frac{\varrho v^2}{2} + \varrho\varepsilon\right) = -\frac{\partial}{\partial x}\left[\varrho v\left(\frac{v^2}{2} + w\right)\right]. \tag{13}$$

To determine the physical significance of this equation we integrate it over a certain region (x_1, x_2)

$$\frac{\partial}{\partial t}\int_{x_1}^{x_2}\left(\frac{\varrho v^2}{2}+\varrho\varepsilon\right)dx = -\varrho v\left(\frac{v^2}{2}+w\right)\Big|_{x_1}^{x_2}.$$

The left-hand side represents the change in energy in unit time in the interval (x_1, x_2) — the right the energy flux from the region considered in unit time.

If the effect of thermal conductivity cannot be neglected the equation of conservation of energy becomes

$$\frac{\partial}{\partial t}\left(\frac{\varrho v^2}{2}+\varrho\varepsilon\right) = -\frac{\partial}{\partial x}\left[\varrho v\left(\frac{v^2}{2}+w\right)-\varkappa\frac{\partial T}{\partial x}\right], \qquad (14)$$

where $\varkappa$ is the coefficient of thermal conductivity.

2. *Shock waves. Conditions of dynamic compatibility*

With high velocities, motions are possible in which discontinuities in the distribution of hydrodynamic quantities (pressure, velocity, density, etc.) occur on some surfaces moving in space. These discontinuities are usually called *shock waves*.

On the surface of discontinuity (the front of the shock wave) the conditions of continuity of flow of matter, energy and momentum (Hugoniot's conditions) must be fulfilled. Let us derive these conditions.

We convert equation (2) to a form more convenient for our purposes. Multiplying (1) by v and adding to (2) we obtain

$$\frac{\partial}{\partial t}(\varrho v) = -\frac{\partial}{\partial x}(p+\varrho v^2). \qquad (2')$$

Let us now rewrite the equations of continuity and conservation of momentum and energy in the form

$$\frac{\partial \varrho}{\partial t} = -\frac{\partial}{\partial x}(\varrho v), \qquad (1')$$

$$\frac{\partial(\varrho v)}{\partial t} = -\frac{\partial}{\partial x}(p+\varrho v^2), \qquad (2')$$

$$\frac{\partial}{\partial t}\left(\frac{\varrho v^2}{2}+\varrho\varepsilon\right) = -\frac{\partial}{\partial x}\left[\varrho v\left(\frac{v^2}{2}+w\right)\right]. \qquad (13)$$

We consider the line $x = a(t)$, the "path" of the discontinuity on the plane (x, t). If AC is some arc of the line of discontinuity $x = a(t)$, where A and C are points with the coordinates x_1, t_1, and $x_2 = x_1 + \Delta x$; $t_2 = t_1 + \Delta t$ respectively, we construct the rectangle $ABCD$ with sides parallel to the coordinate axes.

Writing the law of conservation of matter in the integral form

$$\int_{x_1}^{x_2} [(\varrho)_{t_2} - (\varrho)_{t_1}]\, dx = - \int_{t_1}^{t_2} [(\varrho v)_{x_2} - (\varrho v)_{x_1}]\, dt\,, \tag{15}$$

we have on the left the change in mass in the interval (x_1, x_2) in the time interval (t_1, t_2), and on the right the amount of matter flowing out from the interval (x_1, x_2) in the time (t_1, t_2). If the functions ϱ and $\varrho\, v$ are continuous and differentiable everywhere inside $ABCD$, equation (15) is equivalent to equation (1′). In our case this is not so.

Using the mean value theorem for each term separately we get

$$[(\varrho)_{\substack{t=t_2\\x=x^*}} - (\varrho)_{\substack{t=t_1\\x=x^{**}}}] \frac{\Delta x}{\Delta t} = - (\varrho v)_{\substack{x=x_2\\t=t^*}} + (\varrho v)_{\substack{x=x_1\\t=t^{**}}}\,,$$

where x^*, x^{**}, t^*, t^{**} are the mean values of the arguments of x and t.

Passing to the limit when $\Delta x \to 0$ $(x_2 \to x_1)$ and $\Delta t \to 0$ $(t_2 \to t_1)$ and denoting by the index 1 values of the functions above the curve $x = a(t)$ (behind the shock wave front) and by the index 2 values of the functions below this curve (before the front) we get

$$(\varrho_2 - \varrho_1)\, U = - (\varrho v)_1 + (\varrho v)_2\,, \tag{16}$$

where

$$U = \frac{da}{dt} = \lim_{\Delta t \to 0} \frac{\Delta x}{\Delta t}$$

is the velocity of the shock wave.

In a coordinate system moving with the shock wave where

$$u_1 = U - v_1\,, \quad u_2 = U - v_2$$

denote the velocities of particles before and behind the front of the shock wave respectively. Equation (16) obtained above may be rewritten as

$$\varrho_1 u_1 = \varrho_2 u_2\,. \tag{16′}$$

This equation expressed the continuity of the flow of matter through the front of the shock wave.

Writing in integral form the law of conservation of momentum we have

$$\int_{x_1}^{x_2} [(\varrho v)_{t_2} - (\varrho v)_{t_1}]\, dx = - \int_{t_1}^{t_2} [(p + \varrho v^2)_{x_2} - (p + \varrho v^2)_{x_1}]\, dt\,,$$

where on the right we have the sum of the impulse of the forces (pressures) acting and the flux of momentum. Passing to the limit when $\Delta x \to 0$ and $\Delta t \to 0$ we obtain the law of conservation of the flux of momentum on the front

$$U\,[(\varrho v)_2 - (\varrho v)_1] = - (p + \varrho v^2)_1 + (p + \varrho v^2)_2$$

or

$$p_1 + \varrho_1 u_1^2 = p_2 + \varrho_2 u_2^2. \tag{17}$$

The equation of conservation of energy on the front is obtained similarly

$$\left(\frac{\varrho\, v^2}{2} + \varrho\varepsilon\right)_2 U - \left(\frac{\varrho\, v^2}{2} + \varrho\varepsilon\right)_1 U = - \varrho_1 v_1 \left(\frac{v^2}{2} + w\right)_1 + + \varrho_2 v_2 \left(\frac{v^2}{2} + w\right)_2,$$

which after simple transformations assumes the form

$$\varrho_1 u_1 \left(w_1 + \frac{u_1^2}{2}\right) = \varrho_2 u_2 \left(w^2 + \frac{u_2^2}{2}\right)$$

or, because of condition (16)

$$w_1 + \frac{u_1^2}{2} = w_2 + \frac{u_2^2}{2}\,. \tag{18}$$

Thus on the shock wave front the following equations (conditions of dynamic compatibility or Hugoniot's conditions) must be fulfilled.

$$\varrho_1 u_1 = \varrho_2 u_2\,, \tag{16'}$$

$$p_1 + \varrho_1 u_1^2 = p_2 + \varrho_2 u_2^2\,, \tag{17}$$

$$w_1 + \frac{u_1^2}{2} = w_2 + \frac{u_2^2}{2}\,. \tag{18}$$

From the first two equations (16) and (17) we express u_1 and u_2 in terms of of p and ϱ:

$$u_1^2 = \frac{\varrho_2}{\varrho_1} \cdot \frac{p_1 - p_2}{\varrho_1 - \varrho_2}; \quad u_2^2 = \frac{\varrho_1}{\varrho_2} \cdot \frac{p_1 - p_2}{\varrho_1 - \varrho_2}.$$

whence

$$u_1^2 - u_2^2 = -\frac{\varrho_1 + \varrho_2}{\varrho_1 \varrho_2}(p_1 - p_2).$$

Next, substituting this expression in equation (18) we find the relation between the values of the energy on both sides of the front

$$w_1 - w_2 = \frac{1}{2\varrho_1 \varrho_2}(\varrho_1 + \varrho_2)(p_1 - p_2)$$

and

$$\varepsilon_1 - \varepsilon_2 = \frac{1}{2\varrho_1 \varrho_2}(\varrho_1 - \varrho_2)(p_1 + p_2).$$

Let us consider an ideal gas for which

$$p = R\varrho T; \ \varepsilon = c_v T; \ w = c_p T = \frac{c_p}{c_p - c_v} RT = \frac{\gamma}{\gamma - 1} \cdot \frac{p}{\varrho},$$

i. e.

$$w = \frac{\gamma}{\gamma - 1} \cdot \frac{p}{\varrho}. \tag{19}$$

Using formula (19), after simple transformations we arrive at the equation of Hugoniot's adiabatic

$$\frac{\varrho_2}{\varrho_1} = \frac{(\gamma + 1) p_2 + (\gamma - 1) p_1}{(\gamma - 1) p_2 + (\gamma + 1) p_1} \tag{20}$$

or

$$\frac{p_2}{p_1} = \frac{(\gamma + 1) \varrho_2 - (\gamma - 1) \varrho_1}{(\gamma + 1) \varrho_1 - (\gamma - 1) \varrho_1}. \tag{21}$$

From this formula any one of the quantities p_1, ϱ_1. p_2, ϱ_2 may be determined if the three others are known.

The shock wave always moves with respect to the gas from regions of higher pressure to those of lower pressure: $p_2 > p_1$. Hence it follows that the density of the gas behind the front is greater than the density before the front.

Formula (20) expresses the relation between p_2 and ϱ_2, when p_1 and ϱ_1 are given. The function for given p_1 and ϱ_1 is a monotonic increasing function tending to a finite limit when $(p_2/p_1) \to \infty$ (shock wave of large amplitude)

$$\frac{\varrho_2}{\varrho_1} = \frac{\gamma + 1}{\gamma - 1}. \tag{22}$$

This formula gives the maximum discontinuity in density (thickening) which may exist on the front of the shock wave. For a diatomic gas $\gamma = 7/5$ and the maximum thickening is 6

$$\frac{\varrho_2}{\varrho_1} = 6.$$

Using equations 16′, 17 and 20 and putting $p_1 = 0$ we find:

$$u_1 = \sqrt{\left(\frac{\gamma + 1}{2} \cdot \frac{p_2}{\varrho_1}\right)}; \quad u_2 = \sqrt{\left[\frac{(\gamma - 1)^2}{2(\gamma + 1)} \cdot \frac{p_2}{\varrho_1}\right]}.$$

If the shock wave moves through a stationary gas $(v_1 = 0)$ the velocity of propagation of the shock wave is

$$U = \sqrt{\left(\frac{\gamma + 1}{2} \cdot \frac{p_2}{\varrho_1}\right)},$$

i.e. it increases in proportion to the square root of p_2.

Let us now consider the simplest problem of the theory of shock waves which admits of an analytical solution. In a cylindrical tube $a > 0$ infinite on one side and closed with a piston on the other $(x = 0)$ there is a gas at rest with the constant density ϱ_1 and at the constant pressure p_1. At the initial moment $t = 0$ the piston starts moving with the constant velocity v in the positive direction of the axis x. A shock wave, which coincides with the piston at the initial moment and then moves away from it with the velocity $U > v$, is formed in front of the piston. Between the piston and the front of the shock wave a region 2 develops in which the gas moves with the velocity of the piston. Before the front (region 1) the gas is in an unexcited state: $\varrho = p_1$, $\mathrm{p} = p_1$ $(v = 0)$.

Using conditions (16), (17) and (18) on the front it is easy to determine the velocity of the front as well as the magnitude of the discontinuity, the density and the pressure.

Let us introduce the non-dimensional quantities

$$\omega = \frac{\varrho_1}{\varrho_2}; \quad \tilde{U} = \frac{U}{c_1}; \quad \tilde{v} = \frac{v}{c_1}; \quad \tilde{p} = \frac{\gamma p_2}{\varrho_1 c_1^2}; \tag{23}$$

where $c_1 = \sqrt{(\gamma P_1/\varrho_1)}$ is the velocity of sound before the front (in the unexcited region 1). Then the conservation equations can be written in the form

$$\omega\tilde{U} = \tilde{U} - \tilde{v} \quad \text{or} \quad \tilde{U} = \frac{\tilde{v}}{1-\omega}, \tag{24}$$

$$\tilde{p} = 1 + \gamma\tilde{U}\tilde{v} \quad \text{or} \quad \tilde{p} = 1 + \gamma\frac{\tilde{v}^2}{1-\omega},$$

$$\tilde{p}\omega = 1 + (\gamma - 1\left(\tilde{U}\tilde{v} - \frac{1}{2}\tilde{v}^2\right). \tag{26}$$

Eliminating $\tilde{p}$ and $\tilde{U}$ from these we obtain the quadratic equation for the determination of ω:

$$2\omega^2 - \omega[4 + (\gamma + 1)\tilde{v}^2] + [2 + (\gamma - 1)\tilde{v}^2] = 0. \tag{27}$$

Since $\omega < 1$; ($\varrho_2 > \varrho_1$) as discussed above, we select the smaller root

$$\omega^2 = 1 + \frac{(\gamma+1)}{4}\tilde{v}^2 - \tilde{v}\sqrt{\left[1 + \frac{(\gamma+1)^2}{16}\tilde{v}^2\right]}. \tag{28}$$

From equations (24) and (28) we find

$$\tilde{U} = \frac{(\gamma+1)}{4}\tilde{v} + \sqrt{\left[1 + \frac{(\gamma+1)^2}{16}\tilde{v}^2\right]}, \tag{29}$$

$$\tilde{p} = 1 + \frac{\gamma(\gamma+1)}{4}\tilde{v}^2 + \gamma\tilde{v}\sqrt{\left[1 + \frac{(\gamma+1)^2}{16}\tilde{v}^2\right]}. \tag{30}$$

Returning to the original quantities we find that

$$\varrho_2 = \varrho_1 = \frac{1 + \frac{\gamma+1}{4}\cdot\frac{v^2}{c_1^2} + \frac{v}{c_1}\sqrt{\left[1 + \frac{(\gamma+1)^2}{16c_1^2}\cdot v^2\right]}}{1 + \frac{(\gamma-1)v^2}{2c_1^2}}, \tag{31}$$

$$U = \frac{\gamma+1}{4}v + c_1\sqrt{\left[1 + \frac{(\gamma+1)^2}{16c_1^2}\right]v^2}, \tag{32}$$

$$p_2 = p_1\cdot\left\{1 + \frac{\gamma(\gamma+1)}{4}\frac{v^2}{c_1^2} + \frac{\gamma c}{c_1}\sqrt{\left[1 + \frac{(\gamma+1)^2}{16c_1^2}v^2\right]}\right\}. \tag{33}$$

Since the velocity of the shock wave is constant, the position of the front at the moment t is

$$x = a(t) = \left\{\frac{(\gamma+1)}{4}v + c_1\sqrt{\left[1 + \frac{(\gamma+1)^2}{16c_1^2}v^2\right]}\right\}t. \tag{34}$$

In the limiting case $v/c_1 \gg 1$ (shock wave of high intensity), from formulae (31)–(33) we find the limiting relations

$$\varrho_2 = \varrho_1 \frac{\gamma+1}{\gamma-1}; \qquad U = \frac{\gamma+1}{2} v; \qquad p_2 = p_1 \cdot \frac{\gamma(\gamma+1)}{2} \cdot \frac{v^2}{c_1^2},$$

obtained earlier.

If $v/c_1 \ll 1$ (wave of low intensity) the terms v^2/c_1^2 may be neglected:

$$\varrho_2 = \varrho_1 \left(1 + \frac{v}{c_1}\right),$$

$$U = c_1 + \frac{(\gamma+1)}{4} v,$$

$$p_2 = p_1 \left(1 + \frac{\gamma v}{c_1}\right).$$

3. *Weak discontinuities*

The motion of a shock wave on whose front the quantities p, ϱ, v have discontinuities was discussed above. This kind of discontinuity is called *strong*.

Motions in which the first derivatives of the quantities p, ϱ, v suffer discontinuities, while these quantities themselves remain continuous are also possible. Such discontinuities are called *weak*.

In § 2, sub-section 8, the motion of discontinuities of this type was considered and it was established that these discontinuities propagate along the characteristic curves. We started from the equation of acoustics in that case. But a similar result is true for non-linear problems of gas dynamics also.

It can be shown that a surface of weak discontinuity propagates with respect to the gas with a velocity equal to the local velocity of the sound.

Let us consider a small neighbourhood of the weak discontinuity and take the mean values of hydrodynamical quantities in this neighbourhood A weak discontinuity, may be considered as a small disturbance on the background of mean values which satisfies the equation of acoustics and must propagate with the local velocity of sound.

As an example we shall consider the escape of gas into vacuum (wave rarefaction). Let the gas filling the halfspace $x > 0$ be at rest at the initial moment $t = 0$ and have constant values of the density ϱ and the pressure p_0 in the entire region $x > 0$. At $t = 0$ the external pressure applied to the plane $x = 0$ is taken off and the gas begins to move. A weak discontinuíty occurs (wave of rerefaction) propagating with the velocity of sound c_0 in the positive direction of the axis x. On the forward front of the gas $x = x_1(t)$, at $t = 0$ there is a discontinuity of density and pressure. But this discontinuity disappears immediately after the beginning of motion.
From the conditions of continuity of the fluxes of matter and momentum at $x = x_1(t)$

$$0 = \varrho_1^- (v_1 - v_1^-) = \varrho_1^+ (v_1 - v_1^+),$$

$$p_1^- + \varrho_1^- (v_1 - v_1^-)^2 = p_1^+ + \varrho_1^+ (v_1 - v_1^+)^2,$$

where $\varrho_1^-, p_1^-, v_1^-$ are limiting values from the left at the point x_1 (t and $\varrho_1^+, p_1^+, v_1^+$ are limiting values from the right at the same point we obtain

$$p_1^+ = 0 \text{ and } \varrho_1^+ = 0,$$

from the equations

$$\varrho_1^- = p_1^- = v_1^- = 0.$$

For an adiabatic process the equation of state for an ideal gas has the form

$$p = p_0 \left(\frac{\varrho}{\varrho_0}\right)^{\gamma}. \tag{35}$$

We shall look for the solution of the problem in the form

$$\varrho = \varrho(\xi); \quad p = p(\xi); \quad v = v(\xi),$$

where

$$\xi = \frac{x}{t}.$$

Calculating the derivatives

$$\frac{\partial f}{\partial t} = -\frac{1}{t} \xi \frac{df}{d\xi}; \qquad \frac{\partial f}{\partial x} = \frac{1}{t} \frac{df}{d\xi},$$

where $f = \varrho, v$ or p and substituting the results in equations (1) and (2) we get

$$\left.\begin{aligned}(v-\xi)\frac{d\varrho}{d\xi} &= -\varrho\frac{dv}{d\xi},\\ (v-\xi)\varrho\frac{dv}{d\xi} &= \frac{dp}{d\xi}.\end{aligned}\right\} \tag{36}$$

Let us multiply the first equation by $(v-\xi)$ and add it to the second

$$(v-\xi)^2\frac{d\varrho}{d\xi} = \frac{dp}{d\xi}$$

or

$$\frac{dp}{d\varrho} = (v-\xi)^2.$$

Whence we have

$$v - \xi = \pm\sqrt{\frac{dp}{d\varrho}} = \pm c,$$

where c is the local velocity of sound in the adiabatic process.

Since we are considering the motion of a weak discontinuity in the positive direction of the axis x, in the preceding formula the negative sign has to be chosen, i. e.

$$v - \xi = -c. \tag{37}$$

Substituting this solution in equation (36) we obtain

$$\frac{dv}{d\varrho} = \frac{c}{\varrho} \tag{38}$$

or, what is the same

$$\frac{dv}{dp} = \frac{1}{\varrho c}.$$

Using the equation of state (35) we find

$$c^2 = \gamma\frac{p}{\varrho}$$

and after integration of equation (38)

$$v = \frac{2}{\gamma-1}\cdot c_0\left[\left(\frac{\varrho}{\varrho_0}\right)^{\frac{\gamma-1}{2}} - 1\right]. \tag{39}$$

From the latter formula ϱ can be expressed in terms of v:

$$\varrho = \varrho_0\left(1 + \frac{\gamma-1}{2}\cdot\frac{v}{c_0}\right)^{\frac{2}{\gamma-1}}. \tag{40}$$

Here

$$c_0 = \sqrt{\frac{\gamma p_0}{\varrho_0}}$$

denotes the velocity of sound at $v = 0$ (in a gas at rest). Formula (39) can also be rewritten as

$$v = \frac{2}{\gamma - 1}(c - c_0). \tag{41}$$

Substituting expression (40) for ϱ in the equation of state (35) we find

$$p = p_0\left(1 + \frac{\gamma - 1}{2} \cdot \frac{v}{c_0}\right)^{\frac{2\gamma}{\gamma-1}}. \tag{42}$$

From equations (41) and (37) we obtain the formula

$$v = \frac{2}{\gamma + 1}\left(\frac{x}{t} - c_0\right), \tag{43}$$

which defines the dependence of v on x and t. Then substituting expression (43) for v in formulae (40) and (42) we get the dependence of ϱ and p on x and t explicitly. All the quantities are dependent on x/t. If the distances are measured in units proportional to t, the pattern of motion does not change. Such motion is called *self-model* motion.

Let us find the velocity of the forward front v_1 (t).

Putting $p = 0$ in equation (42) we have

$$v_1 = -\frac{2}{\gamma - 1}c_0. \tag{44}$$

It follows that the velocity of escape of the gas into the vacuum is finite. For diatomic gases $\gamma = 7/5$ and

$$v_1 = -5\,c_0.$$

Expression (44) for the velocity of the left front $x = x_1(t)$ may also be obtained from the equation of conservation of matter

$$\int_{x_1}^{x_2} \varrho\, dx = \varrho_0\, x_2 = \varrho_0\, c_0\, t. \tag{45}$$

Introducing the variable

$$\xi = \frac{x}{t},$$

we get

$$\int_{v_1}^{c_0} \varrho \, d\xi = \varrho_0 c_0 .$$

Next, substituting the expression for ϱ from (40) and putting

$$1 + \frac{\gamma - 1}{\gamma + 1} \cdot \frac{\xi - c_0}{c_0} = \lambda ,$$

we have

$$\int_{\lambda_1}^{\lambda_2} \lambda^{\frac{2}{\gamma-1}} \, d\lambda = \frac{\gamma - 1}{\gamma + 1} , \tag{46}$$

where

$$\lambda_1 = 1 + \frac{\gamma - 1}{\gamma + 1} \cdot \frac{v_1 - c_0}{c_0} ; \quad \lambda_2 = 1 .$$

After calculating the integral (46) we shall get

$$\lambda_2^{\frac{\gamma+1}{\gamma-1}} - \lambda_1^{\frac{\gamma+1}{\gamma-1}} = 1 ,$$

i. e.

$$\lambda_1 = 0 ,$$

whence it follows that

$$v_1 = - \frac{2c_0}{\gamma - 1} .$$

Thus the problem of escape of a gas into a vacuum is solved.

We have confined ourselves to simple problems of gas dynamics. For a more detailed discussion of the problems touched upon here we refer the reader to specialized literature.*

V. Dynamics of the absorption of gases

1. *Equations describing the process of absorption of a gas*

Let us consider the problem of absorption of a gas.** A mixture of gas and air passes through a tube (the axis of which we take as the x-axis) filled with an absorbent. Let $a(x, t)$ denote

*See N. E. Kochin, I. A. Kibel' and N. V. Roze, *Theoretical Hydromechanics*, part II, chap. I, Gostekhizdat, 1948; L. Landau and E. Lifshitz, *Mechanics of Continuous Media*, chap. VII, Gostekhizdat, 1944; Ya. B. Zeldovich, *Theory of Shock Waves and Introduction to Gas Dynamics*, Publ. Acad. of Sciences, 1946; L. I. Sedov, Propagation of strong explosion waves, *Prikladnaya Matematika i Mekhanika (Applied Mathematics and Mechanics)*, **10**, No. 2, 1946.

** A. N. Tikhonov, A. A. Zhukhovitskii and Ya. L. Zabezhiuskii, Absorption of gas from a current of air by a layer of granular material, *Zhur. Fiz. Khim. (Journal of Physical Chemistry)*, 20, No. 10, 1946.

the quantity of gas absorbed by unit volume of the absorbent and $u(x, t)$ the concentration of gas in the pores of the absorbent in the layer x.

We write the equation of conservation of matter, assuming that the velocity ν is sufficiently large so that the process of diffusion does not play an essential role in the transfer of the gas. Let us consider the absorbing layer from x_1 to x_2 in the time interval from t_1 to t_2. Evidently, for the layer we can write the equation of conservation of matter as

$$[\nu u|_{x_1} - \nu u|_{x_2}]\, S\, \Delta t = [(a+u)|_{t_2} - (a+u)|_{t_1}]\, S\, \Delta x, \qquad (1)$$

which after division by $\Delta x\, \Delta t$ and passing to the limit when $\Delta x \to 0$ and $\Delta t \to 0$ assumes the form

$$-\nu \frac{\partial u}{\partial x} = \frac{\partial}{\partial t}(a+u). \qquad (2)$$

The left-hand side of the equation* represents the amount of gas flowing into the section Δx in time Δt, and the right-hand side the change in the amount of gas contained in that section in the same time. To this conservation equation we must add the equation of the kinetics of absorption

$$\frac{\partial a}{\partial t} = \beta (u - y), \qquad (3)$$

where β is the so-called kinetic coefficient, y the concentration of the gas in "equilibrium" with the quantity a of gas absorbed.

The quantities a and y are connected by the equation

$$a = f(y), \qquad (4)$$

which is characteristic of the absorbent.

The curve $a = f(y)$ is called the *absorption* isotherm. If

$$f(y) = \frac{y}{u_0 + py},$$

then the isotherm is called *Langmuir's* isotherm. The simplest form of the function f corresponds to the so-called *Henry's*

* For the system of equations (2′) and (6) one initial condition is sufficient, since the axis $t = 0$ becomes the characteristic curve. For further details regarding this see the remark on p. 178.

isotherm which is valid in the region of small concentrations

$$a = \frac{1}{\gamma} y, \tag{5}$$

where $1/\gamma$ is Henry's coefficient.

In this case we have following problem: find the functions $u(x, t)$ and $a(x, t)$ from the equations (neglecting $\frac{\partial u}{\partial t}$ in (2))

$$-\nu \frac{\partial u}{\partial x} = \frac{\partial a}{\partial t}, \tag{2'}$$

$$\frac{\partial a}{\partial t} = \beta(u - \gamma a), \tag{6}$$

$$a(x, 0) = 0, \tag{7}$$

$$u(0, t) = u_0 \tag{8}$$

Let us eliminate the function $a(x, t)$ by differentiating the first equation with respect to t

$$-\nu u_{xt} = \beta u_t - \beta\gamma a_t = \beta u_t + \beta\nu\gamma u_x$$

or

$$u_{xt} + \frac{\beta}{\nu} u_t + \beta\gamma u_x = 0.$$

We determine the initial condition for u by putting $t = 0$ in the first equation

$$-\nu u_x(x, 0) = \beta u(x, 0), \qquad u(0, 0) = u_0,$$

whence we find

$$u(x, 0) = u_0 e^{-\frac{\beta}{\nu} x}.$$

The problem of finding the function $u(x, t)$ reduces to the integration of the equation

$$u_{xt} + \frac{\beta}{\nu} u_t + \beta\gamma u_x = 0 \tag{9}$$

with the additional conditions

$$u(x, 0) = u_0 e^{-\frac{\beta}{\nu} x}, \tag{10}$$

$$u(0, t) = u_0. \tag{8}$$

The characteristic curves of this equation are the lines

$$x = \text{const}, \quad t = \text{const}.$$

Thus the additional conditions in this problem are given on the characteristic curves. The problem is formulated similarly for the function $a(x, t)$

$$a_{xt} + \frac{\beta}{\nu} a_t + \beta\gamma a_x = 0, \tag{11}$$

$$a(x, 0) = 0, \tag{7}$$

$$a(0, t) = \frac{u_0}{\gamma}(1 - e^{-\beta\gamma t}). \tag{12}$$

It should be noted that similar equations occur in connection with a number of other problems (for example the process of drying with a current of air, heating a tube with a current of water, etc.).*

* While passing over to equation (2′) we neglected the term u_t. But we can show that we arrive at the same equation if we introduce the variables:

$$\tau = t - \frac{x}{\nu}; \quad t = \tau + \frac{\xi}{\nu},$$

$$\xi = x, \qquad x = \xi, \qquad \text{(Fig. 32)}$$

i.e. time at the point x is reckoned from $t_0 = x/\nu$, the moment of arrival of the stream of the gas-air mixture at this point. We have

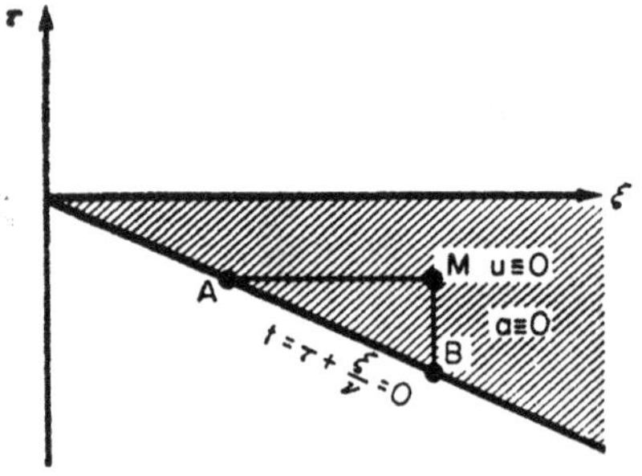

FIG. 32

$$\frac{\partial u}{\partial x} = \frac{\partial u}{\partial \xi} = \frac{1}{\nu}\frac{\partial u}{\partial \tau},$$

$$\frac{\partial}{\partial t} = \frac{\partial}{\partial \tau}$$

and equation (2) assumes the form

$$-\nu\frac{\partial u}{\partial \xi} = \frac{\partial a}{\partial \tau}, \tag{2″}$$

$$\frac{\partial a}{\partial \tau} = \beta(u - \gamma a). \tag{6}$$

The initial conditions (7) and the equations (2) and (6) give

$$\left.\begin{aligned} u(x, 0) &= 0, \\ u_t(x, 0) &= 0. \end{aligned}\right\} \tag{7′}$$

In the region between the straight line $t = 0$ and the axis ξ we obtain the problem of determination of the function u from the initial conditions (7′) (Cauchy's problem). It is obvious that in this region the function $u(x, t) \equiv 0$ (and also $a \equiv 0$). From equations (2′) and (6) it can be seen that at $\tau = 0$ the function $u(x, t)$ has a discontinuity, while the function $a(x, t)$ remains continuous. Thus, at $\tau = 0$ the function u is determined from equation (2′) when $a(x, 0) = 0$. Determining, as was done on p. 177 (see formulae (10) and (12)), the values of $u(x, 0)$ and $a(0, t)$ we obtain for the functions $u(x, t)$ and $a(x, t)$ problems with data on characteristics.

The solution of equation (9) may be obtained in an explicit form by the method described in § 5 and is given by the formula

$$u(x_1, t_1) = u_0\, e^{-x_1}\left[e^{-t_1} I_0(2\sqrt{[x_1 t_1]}) + \frac{1}{x_1}\int_0^{x_1 t_1} e^{-\frac{\tau}{x_1}} I_0(2\sqrt{\tau})\, d\tau\right], \tag{13}$$

where $x_1 = \beta\, x/v$, $t_1 = \beta\, t/\gamma$ is a non-dimensional variable, I_0 is a Bessel function of the first kind of zero order from an imaginary argument.

Using asymptotic formulae for the functions I_0 we can obtain an asymptotic representation for the solution for large values of arguments.

2. *Asymptotic solution*

We have studied above the process of absorption of a gas subject to Henry's absorption isotherm, which connects the amount of absorbed matter a with the equilibrium concentration by a linear relation

$$a = \frac{1}{\gamma}\, y\,.$$

Let us consider an absorption isotherm of the general form

$$a = f(y)\,.$$

If the non-dimensional variables

$$x_1 = \frac{x\beta}{v}, \quad t_1 = \frac{t\beta}{\gamma}, \quad \bar{u} = \frac{u}{u_0}, \quad z = \frac{y}{u_0}, \quad v = \frac{a}{u_0\gamma},$$

are introduced, then the system (2, 6, 7, 8) assumes the form

$$\left.\begin{aligned} \frac{\partial \bar{u}}{\partial x_1} &= -\frac{\partial v}{\partial t_1}, \\ \frac{\partial v}{\partial t_1} &= (\bar{u} - z), \end{aligned}\right\} \tag{14}$$

$$v = f_1(z) = \frac{1}{\mu_0\gamma} f(z u_0) \tag{15}$$

with the additional conditions

$$\bar{u}(0, t) = 1\,, \tag{16}$$

$$v(x, 0) = 0\,. \tag{17}$$

We are interested in the asymptotic behaviour of functions representing the solution of the system (14).

We shall assume the following regarding the function $f_1(z)$:

1. $f_1(z)$ is an increasing function and $f_1(0) = 0$.

2. $f_1(z)$ has a continuous derivative for all values of z, $0 \leqslant z \leqslant 1$.

3. A ray going from the origin of coordinates to the point $(1, f_1(1))$ lies below the curve $f_1(z)$ in the interval $0 \leqslant z \leqslant 1$ (Fig. 33). This always happens for convex isotherms.

Introducing the notation for the reciprocal function

$$z = f_1^{-1}(v) = F(v),$$

we shall look for the asymptotic solution of the problem in the form of a propagating wave

$$\begin{aligned} \tilde{u} &= \psi(\xi), \\ \tilde{v} &= \varphi(\xi), \end{aligned} \quad \xi = x - \sigma t, \tag{18}$$

where σ is the rate of propagation of the wave which is to be determined.

This means that at large distances (when $x \to \infty$) or after a large interval of time (when $t \to \infty$)

$$v(x, t) = \tilde{v} = \varphi(x - \sigma t); \quad \bar{u}(x, t) = \tilde{u} = \psi(x - \sigma t).$$

The concentrations $\bar{u}$ and v must satisfy the equilibrium condition

$$v = f_1(\bar{u}) \quad \text{or} \quad \bar{u} = F(v).$$

when $x = \infty$ or $t = \infty$.

Then from condition (16) it follows that:

$$\bar{u}\Big|_{\substack{x=0 \\ t=\infty}} = \psi(-\infty) = 1; \quad \varphi(-\infty) = v\Big|_{\substack{x=0 \\ t=\infty}} = f_1(1). \tag{19}$$

From condition (17) it follows that:

$$v\Big|_{\substack{x=\infty \\ t=0}} = \varphi(+\infty) = 0; \quad \psi(+\infty) = \bar{u}\Big|_{\substack{x=\infty \\ t=0}} = F(0) = 0. \tag{20}$$

Conditions (19) signify that when $t \to \infty$ ($\xi \to -\infty$) saturation should be established everywhere.

Substituting the proposed form of the solution in equation (14) we get

$$\psi' = \sigma\varphi' = 0 , \tag{21}$$

$$-\sigma\varphi' = \psi - F(\varphi) . \tag{22}$$

From (21) and (20) we conclude that

$$\psi(\xi) - \sigma\varphi(\xi) = 0 . \tag{23}$$

Then from equations (19) it follows that

$$\sigma = \frac{\psi(\xi)}{\varphi(\xi)}\bigg|_{\xi=-\infty} = \frac{1}{f_1(1)} \tag{24}$$

or in the original variables

$$\sigma = \gamma \frac{u_0}{a_0} . \tag{24'}$$

From (22) and (23) we find

$$-\sigma \frac{d\varphi}{\sigma\varphi - F(\varphi)} = d\xi . \tag{25}$$

After integration we have

$$\omega(\varphi) = \xi - \xi_0 , \tag{26}$$

where $\omega(\varphi)$ is any integral on the left-hand side and ξ_0 the constant of integration. Hence the required function $\varphi(\xi)$ is determined except for an unknown constant ξ_0:

$$\varphi = \omega^{-1}(\xi - \xi_0) , \tag{27}$$

$$\psi = \sigma\omega^{-1}(\xi - \xi_0) . \tag{28}$$

Let us see if the function ω^{-1} can be determined and if the functions φ and ψ satisfy the conditions imposed when $\xi \to \infty$ and $\xi \to -\infty$. We shall show that the derivative

$$\frac{d\omega}{d\varphi} = -\sigma \frac{1}{\sigma\varphi - f_1^{-1}(\varphi)} < 0 , \tag{29}$$

i. e.

$$\xi - \xi_0 = \omega(\varphi)$$

is a monotonic decreasing function φ. The denominator in (29) is

$$\sigma\varphi - f_1^{-1}(\varphi) = \frac{1}{f_1(1)}\varphi - f_1^{-1}(\varphi) .$$

The first term is the abscissa of the point belonging to the ordinate φ, lying on the line from the origin of coordinates to the

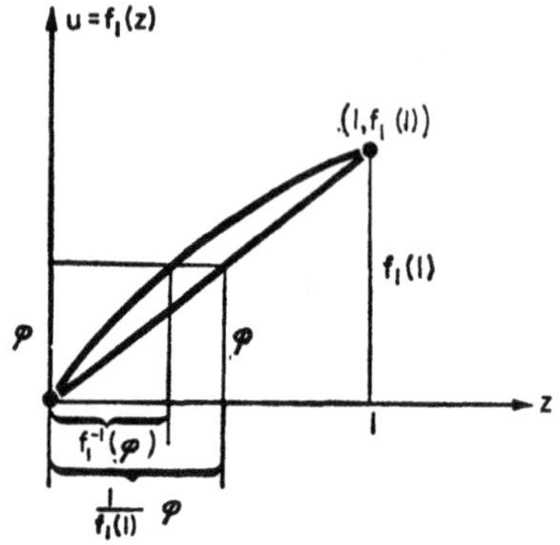

FIG. 33

point $(1, f_1(1))$ (Fig. 33). Since we have imposed the condition that the curve $\varphi = f_1(z)$ lies above this line,

$$f_1^{-1}(\varphi) < \frac{1}{f_1(1)}\varphi \qquad (0 \leqslant \varphi \leqslant f_1(1))$$

and consequently,

$$\sigma\varphi - f_1^{-1}(\varphi) > 0.$$

Moreover

$$\sigma\varphi - f_1^{-1}(\varphi) = 0 \text{ when } \varphi = 0 \text{ and when } \varphi = f_1(1).$$

Hence it follows that

$$\xi - \xi_0 = \omega(\varphi) = \infty \qquad \text{when} \quad \varphi = 0,$$
$$\xi - \xi_0 = \omega(\varphi) = -\infty \qquad \text{when} \quad \varphi = f_1(1),$$

For the reciprocal function we have

$$\varphi = \omega^{-1}(\xi - \xi_0) = f_1(1) \quad \text{when} \quad \xi = -\infty,$$
$$\varphi = \omega^{-1}(\xi - \xi_0) = 0 \qquad \text{when} \quad \xi = \infty.$$

Further, because of equation (29) we have

$$\psi = \sigma\varphi = \frac{1}{f_1(1)}\varphi = 1 \quad \text{when} \quad \xi = -\infty,$$
$$\psi = \sigma\varphi = \frac{1}{f_1(1)}\varphi = 0 \quad \text{when} \quad \xi = \infty.$$

Thus all the conditions (19) and (20) are satisfied and it is proved that the system of equations has a solution in the form of a propagating wave containing an undetermined constant ξ_0.

To determine ξ_0 we integrate the first equation with respect to t_1 from 0 to t_0 and with respect to x from 0 to x_0:

$$\Big[\int_0^{t_0} \bar{u}(x_0, \tau)\, d\tau - \int_0^{t_0} \bar{u}(0, \tau)\, d\tau\Big] + \\ + \Big[\int_0^{x_0} v(x, t_0)\, dx - \int_0^{x_0} v(x, 0)\, dx\Big] = 0. \quad (30)$$

This equation expresses the law of conservation of matter. Passing to the limit when $x_0 \to \infty$ and using the initial conditions for $\bar{u}$ and v:

$$\int_0^{\infty} v(x, t_0)\, dx = \int_0^{t_0} \bar{u}(0, \tau)\, d\tau = t_0.$$

Let us assume that for large values of t the solution of our problem approaches the functions $\tilde{u}$ and $\tilde{v}$ found above in the form of propagating waves.

If we determine ξ_0 from the condition

$$\int_0^{\infty} v(x, t_0)\, dx - t_0 \to 0 \quad (t_0 \to \infty), \quad (31)$$

then this will be the value of ξ_0 which corresponds to the functions $\tilde{u}(x, t)$ and $\tilde{v}(x, t)$.

Let us transform our integral

$$\int_0^{\infty} \tilde{v}(x, t_0)\, dx = \int_0^{\infty} \varphi(x - \sigma t_0)\, dx = \int_0^{\infty} \omega^{-1}(x - \sigma t_0 - \xi_0)\, dx = \\ = \int_{-\sigma t_0 - \xi_0}^{\infty} \omega^{-1}(\xi)\, d\xi = \int_{\zeta_1}^{\infty} \omega^{-1}(\zeta)\, d\zeta \qquad \left(\begin{array}{l}\zeta = x - \sigma t_0 - \xi_0, \\ \zeta_1 = -\sigma t_0 - \xi_0\end{array}\right).$$

Let us denote by φ^* the value of $\omega^{-1}(\zeta_1)$ when $\zeta = 0$

$$\omega^{-1}(0) = \varphi^*.$$

It is easy to see that if $\varphi = \omega^{-1}(\zeta)$ is the reciprocal function for $\zeta = \omega(\varphi)$ then (Fig. 34)

$$\int_{\zeta_1}^{\infty} \omega^{-1}(\zeta)\, d\zeta = \int_{\zeta_1}^{0} \omega^{-1}(\zeta)\, d\zeta + \int_0^{\infty} \omega^{-1}(\zeta)\, d\zeta = \\ = \Big[-\zeta_1 \omega^{-1}(\zeta_1) + \int_{\varphi^*}^{\omega^{-1}(\zeta_1)} \omega(\varphi)\, d\varphi + \int_0^{\varphi^*} \omega(\varphi)\, d\varphi\Big]. \quad (32)$$

It follows that instead of the limiting equation (31) we can write

$$\int_{-\sigma t_0-\xi_0} \omega^{-1}(\zeta)\,d\zeta - t_0 =$$

$$= \{(\sigma t_0 + \xi_0)\varphi(-\sigma t_0 - \xi_0) + \int_0^{\varphi(-\sigma t_0 - \xi_0)} \omega(\varphi)\,d\varphi\} - t_0 \to 0 \quad (t_0 \to \infty). \tag{32'}$$

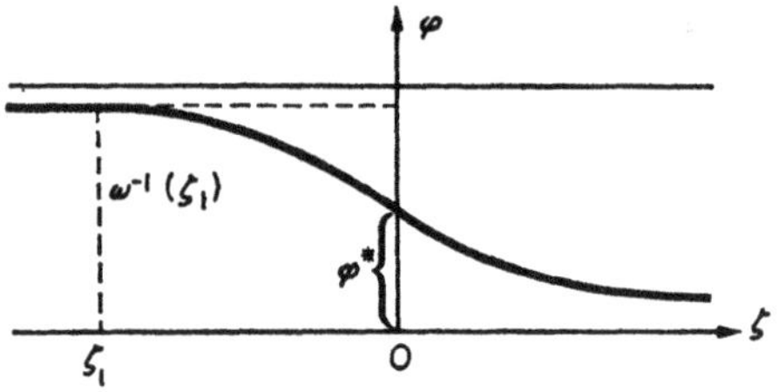

FIG. 34

Let us pass to the limit when $t_0 \to \infty$. Then

$$\sigma\varphi(-\sigma t_0 - \xi_0) \to \sigma\varphi(-\infty) = \sigma f_1(1) = 1. \tag{32''}$$

To calculate the limit of the expression

$$\sigma t_0 \varphi(-\sigma t_0 - \xi_0) - t_0,$$

let us use equation (25). Expanding $f^{-1}(\varphi) = F(\varphi)$ into a series near the point $\varphi_0 = f_1(1)$ we get

$$\sigma\varphi - F(\varphi) = \sigma(\varphi - \varphi_0) + 1 - F(\varphi) =$$
$$= \sigma(\varphi - \varphi_0) - [F(\varphi) - F(\varphi_0)] =$$
$$= [\sigma - F'(\varphi_0)](\varphi - \varphi_0) + \ldots,$$

whence

$$-\sigma \frac{d\varphi}{[\sigma - F'(\varphi_0)](\varphi - \varphi_0) + \ldots} = d\xi, \tag{33}$$

where the dots denote terms of a higher order than $(\varphi - \varphi_0)$.

From requirement 3 for the function f_1 it follows that

$$F'(\varphi_0) > \sigma = \frac{1}{f_1(1)}.$$

From equation (33) we find the form of φ when $\xi \to -\infty$

$$\varphi = Ae^{k\xi} + \varphi_0, \tag{34}$$

where A and $k > 0$ are constant.

From (34) it follows that

$$\lim_{t_0\to\infty} t_0 [\sigma\varphi(-\sigma t_0 - \xi_0) - 1] = \lim_{t_0\to\infty} t_0 A\sigma e^{-k(\sigma t_0+\xi_0)} = 0 . \quad (32''')$$

Making the limiting transition when $t_0 \to \infty$ in formula (32′) and taking (32″) and (32‴) into consideration we get:

$$\xi_0 = -\frac{1}{f_1(1)} \int_0^{f_1(1)} \omega(\varphi)\, d\varphi . \quad (35)$$

The profiles of the waves $\{\tilde{u}, \tilde{v}\}$ are thus completely determined.

Langmuir's isotherm is a case of particular interest. Let us find the asymptotic solution for the process of absorption of a gas subject to Langmuir's isotherm.

Equation (25) assumes the form

$$-\sigma \frac{d\varphi}{\sigma\varphi - \dfrac{\varphi}{1-p\varphi}} = d\xi , \quad (36)$$

where $\sigma = 1/f_1^{(1)} = 1 + p$ is the wave velocity. From (36) we find:

$$\xi - \xi_0 = \omega(\varphi) ,$$

where

$$\omega(\varphi) = \sigma \int \frac{(1-p\varphi)\, d\varphi}{\varphi - \sigma\varphi(1-p\varphi)} + A = \frac{\sigma}{\sigma - 1}\left[\frac{1}{\sigma} \ln(\sigma - 1 - p\sigma\varphi) - \ln\varphi\right] + A.$$

It is obvious that when φ varies from 0 to $f_1^{)(1}$, $\omega(\varphi)$ varies from $-\infty$ to $+\infty$. We choose A so that

$$\varphi^* = \frac{1}{2} f_1(1) ,$$

i.e. such that

$$\omega(\varphi^*) = 0 \text{ where } \varphi^* = \frac{1}{2} f_1(1) = \frac{1}{2}\,\frac{1}{1+p} .$$

Under this condition

$$A = -\frac{\sigma}{\sigma-1}\left[\frac{1}{\sigma} \ln\left(\frac{1}{2}p\right) - \ln\left(\frac{1}{2}\,\frac{1}{1+p}\right)\right]$$

or

$$\omega(\varphi) = \frac{\sigma}{\sigma - 1}\left[\frac{1}{\sigma} \ln 2(1 - \sigma\varphi) - \ln 2(1+p)\varphi\right] .$$

The value of ξ_0 is defined by the formula

$$\xi_0 = \frac{1}{f_1(1)} \int_0^{f_1(1)} \omega(\varphi)\, d\varphi = -(\ln 2 - 1)$$

and does not depend on $p = u_0/y$, i. e. on the given concentration.

The required asymptotic solution has the form:

$$\begin{aligned} \tilde{v}(x, t) &= \omega^{-1}(x - \sigma t - \xi_0), \\ \tilde{u}(x, t) &= \sigma\omega^{-1}(x - \sigma t - \xi_0), \end{aligned} \tag{37}$$

where $\omega^{-1}(\xi)$ in the reciprocal function for $\omega(\varphi)$.

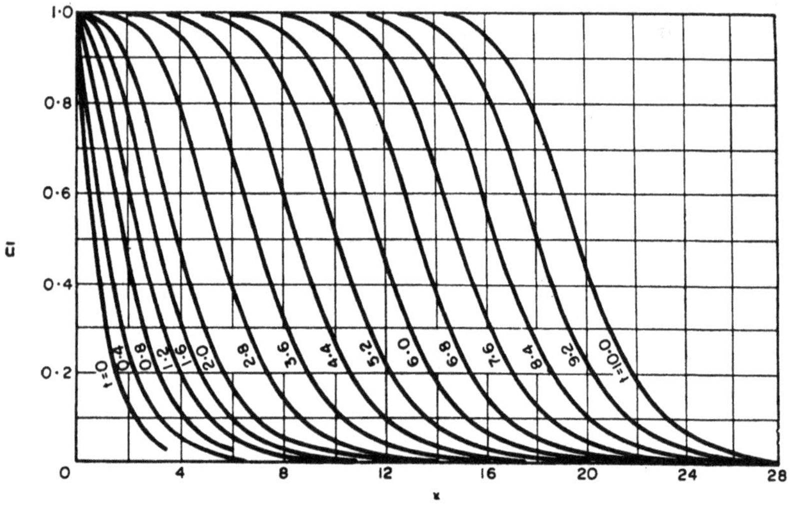

FIG. 35

Figure 35 gives the results of the numerical integration of equations (14) for Langmuir's isotherm by the method of finite differences. These graphs are given for the values $0 < t \leqslant t_1 = 10$. When $t = t_1$ the results of numerical integration coincide with the asymptotic solution within 1 per cent. For the values $t > t_1$ asymptotic formulae may be used.

VI. Physical analogies

While considering phenomena in different fields of physics we often discover certain common features. As a result in the mathematical formulation of the problems we obtain the same equations describing different physical phenomena. As the simplest example we take the equation

$$a\frac{d^2 x}{dt^2} + bx = 0\,,$$

describing oscillatory processes of the simplest systems: mathematical pendulum, oscillation of a load under the action of the elastic force of the spring, electrical oscillations in a simple circuit with inductance and capacity, etc. The similarity in the equations for different physical processes allows the drawing of inferences regarding the properties of unfamiliar phenomena from the study of a more familiar phenomenon. Thus the study of some acoustic phenomena may be facilitated by a preliminary consideration of electrical networks.

The propagation of electrical vibrations in systems with distributed constants is described by the telegraphic equations

$$\left.\begin{aligned} -\frac{\partial I}{\partial x} &= C\frac{\partial V}{\partial t} + GV\,, \\ -\frac{\partial V}{\partial x} &= L\frac{\partial I}{\partial t} + RI\,, \end{aligned}\right\} \tag{1}$$

where C, G, L, R are the capacity, the leakage, the inductance and the resistance of the system. If the resistance and leakage of the current can be neglected, then the usual wave equations are obtained for the potential and the current strength

$$\frac{\partial^2 V}{\partial x^2} - LC\frac{\partial^2 V}{\partial t^2} = 0\,,$$

$$\frac{\partial^2 I}{\partial x^2} - LC\frac{\partial^2 I}{\partial t^2} = 0\,,$$

and equations (1) assume the form

$$\left.\begin{aligned} -\frac{\partial I}{\partial x} &= C\frac{\partial V}{\partial t}\,, \\ -\frac{\partial V}{\partial x} &= L\frac{\partial I}{\partial t}\,. \end{aligned}\right\} \tag{2}$$

While solving the problem of propagation of sound we arrive at the equations

$$\left.\begin{aligned} -\frac{\partial p}{\partial x} &= \varrho\frac{\partial v}{\partial t}\,, \\ -\frac{\partial v}{\partial x} &= \frac{1}{\tau}\,\frac{\partial p}{\partial t}\,, \end{aligned}\right\} \tag{3}$$

where v is the velocity of the vibrating particles, ϱ the density, p the pressure and $\tau = p_0 \gamma$ the coefficient of elasticity of air.

The similarity of equations (2) and (3) allows us to establish a correspondence between acoustic and electrical quantities. Pressure corresponds to the potential difference, and the velocity of particles to the current. The density, which determines the inertial properties of the gas, corresponds to the inductance of the electrical circuit, and $1/\tau$, i.e. the reciprocal of the coefficient of elasticity, corresponds to the capacity of the electrical circuit. This correspondence can also be established from expressions for the kinetic and the potential energy for electrical and acoustic systems.

Returning to equations (1) we can introduce acoustical analogues of resistance and leakage. Acoustical resistance has to be taken into account in those cases where considerable friction of the gas against the walls of the vessel is present during the motion of the gas. In analogy with electrical resistance which is defined as the ratio of the voltage to the current, it is possible to introduce acoustic resistance defined by the ratio of the pressure to the current in the medium. The latter is proportional to the velocity of gas particles.

$$R_A = \frac{p}{uv} .$$

In those cases where the motion of a gas in a porous medium is considered, it is necessary to introduce a quantity similar to leakage in electrical circuits. This quantity denoted by P is called the porosity.

A mechanical analogy of the telegraphic equation is the equation of longitudinal vibrations of a rod, which like equations (2) may be written in the form

$$-\frac{\partial v}{\partial x} = \frac{1}{k}\frac{\partial T}{\partial t},$$

$$-\frac{\partial T}{\partial x} = \varrho \frac{\partial v}{\partial t},$$

where T is the tension of the rod, v the velocity of the vibrating points, ϱ the density and k the coefficient of elasticity of the rod.

Comparing this equation with equation (2) we can establish the analogy between mechanical and electrical quantities. Thus, by establishing the correspondence between electric potential and the tension, the current and the velocity of the particles we shall find that the reciprocal of the coefficient of elasticity corresponds to the capacity and the density to the inductance.

Thus a consideration of similar dynamical problems leads to the establishment of correspondence between a number of electrical, acoustical and mechanical quantities.

This correspondence may be illustrated by the following table.*

	Electrical system	Acoustical system	Mechanical system
Variables	Potential V	Pressure p	Tension (force) I
	Current I	Velocity of particles v	Rate of displacement $\dot{x}$
	Charge e	Displacement u	Displacement x
Parameters	Inductance L	Inertia (density) ϱ	Mass density ϱ_m
	Capacity C	Acoustic capacity $C_A = 1/\tau$	Flexibility $C_M = 1/k$
	Resistance R	Acoustic resistance R_A	Mechanical resistance R_M

The arguments developed above enable us to obtain some information on the nature of the phenomena in a number of acoustical problems before solving the problem itself.

Thus the problem of the movement of air in a porous medium for simple harmonic waves leads to the equations**

$$-i\omega\varrho_m u + ru = -\operatorname{grad} p,$$

$$\nabla^2 p + i\frac{\gamma P\omega}{\varrho c^2}(r - i\omega\varrho_m)p = 0,$$

where u is the mass velocity of air through the medium, p the pressure, ϱ the density, ϱ_m the effective density of air which may be greater than ϱ, since particles of the substance may vibrate together with the air, P the porosity, c and ω the velocity

* See, for example, G. Olson, *Dynamical Analogies*, Foreign Languages Publishing House, 1947.

** See V. V. Furduev, *Electro-acoustics*, Gostekhizdat, 1948.

and the frequency of sound, r the resistance to the current, characterizing the fall of pressure in the material. Putting $r = R_A$; $\varrho_m = L_A$; $\gamma P/\varrho c^2 = C_A$ we get our equations in the form

$$L_A \frac{\partial u}{\partial t} + R_A u = - \text{ grad } p,$$

$$C_A L_A \frac{\partial^2 p}{\partial t^2} + C_A R_A \frac{\partial p}{\partial t} = \nabla^2 p.$$

These equations are quite similar to the equations of propagation of electrical oscillations in wires. Therefore in analogy with the characteristic impedance of the wire

$$Z = \sqrt{\frac{R + i \omega L}{G + i \omega C}}$$

we can at once write the expression for the resistance, called the characteristic impedance of the porous material

$$Z = c\sqrt{\varrho} \sqrt{\frac{\varrho_m - i \frac{r}{\omega}}{\gamma P}},$$

taking $G = 0$. The expression for the characteristic impedance indicates damping of the waves propagating in a porous material.

The analogy established between electrical and acoustic phenomena enables replacing the study of a number of acoustical problems by a consideration of the equivalent electrical circuits.

The method of analogy has lately found wide application in analogue devices in which an equivalent electrical circuit is constructed for solving an equation corresponding to some physical process.

CHAPTER III

EQUATIONS OF THE PARABOLIC TYPE

SECOND order partial differential equations of the parabolic type are most often encountered in a study of heat conduction and diffusion. The simplest equation of parabolic type

$$u_{xx} - u_y = 0$$

is usually called *the equation of heat conduction.*

§ 1. Physical problems leading to equations of parabolic type Formulation of boundary-value problems

1. *A linear problem of heat conduction*

Let us examine a homogeneous rod of length l, thermally insulated along the surface and sufficiently thin so that at any

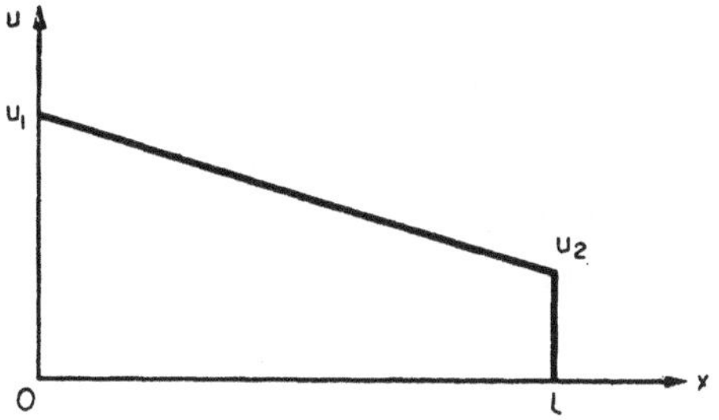

FIG. 36

instant it is possible to assume that the temperature is constant at all points of a cross-section. If the ends of the rod are maintained at the constant temperatures u_1 and u_2 then a linear distribution of temperature is established along the rod (Fig. 36).

$$u(x) = u_1 + \frac{u_2 - u_1}{l} x. \quad (1)$$

At the same time heat flows from the warmer to the cooler end of the rod. The amount of heat flowing through a cross-sectional area S in unit time, is given by the experimental formula

$$Q = -k \frac{u_2 - u_1}{l} S = -k \frac{\partial u}{\partial x} S, \tag{2}$$

where k is the coefficient of thermal conductivity, depending on the material of the rod.

The value of the heat flow is considered positive, if the heat flows in the direction of increasing x.

The process of propagation of heat in the rod may be described by a function $u(x, t)$ representing the temperature of a section x at a time t. Let us find the equation that the function $u(x, t)$ must satisfy. In order to do this, we formulate the physical laws which define the processes.

1. *Fourier's Law*. If the temperature of a body is non-uniform, heat currents arise in it, directed from points of higher temperature to points of lower temperature.

The amount of heat flowing through section x in the time interval $(t, t + dt)$ is equal to

$$dQ = qS\,dt, \tag{3}$$

where

$$q = -k(x) \frac{\partial u}{\partial x} \tag{4}$$

he density of heat flow, equals the amount of heat flowing hrough an area of 1 cm² in unit time. This law represents generalization of formula (2). It can also be given in the integral form

$$Q = -S \int_{t_1}^{t_2} k \frac{\partial u}{\partial x}(x, t)\,dt, \tag{5}$$

where Q is the amount of heat flowing through section x in the time interval (t_1, t_2). If the rod is heterogeneous, then k is a function of x.

2. The amount of heat which must be given to a homogeneous body in order to increase its temperature by Δu is equal to

$$Q = cm\,\Delta u = c\varrho V \Delta u, \tag{6}$$

where c is the specific heat, m is the mass of the body, ϱ is its density, V is the volume.

If the change of temperature is a function of x, or if the rod is heterogeneous, then

$$Q = \int_{x_1}^{x_2} c\varrho S \Delta u(x)\, dx. \tag{7}$$

3. Inside the rod heat may be produced or absorbed (e.g. by the passage of current, as the result of a chemical reaction, etc.). The generation of heat may be characterized by the density of the heat sources $F(x, t)$ at the point x at time t.* As a result in a length of rod $(x, x + dx)$ in a time interval $(t, t + dt)$, there is an amount of heat generated

$$dQ = SF(x, t)\, dx dt \tag{8}$$

or in integral form

$$Q = S \int_{t_1}^{t_2} \int_{x_1}^{x_2} F(x, t)\, dx dt, \tag{9}$$

where Q is the amount of heat generated in a length of rod (x_1, x_2) for a time interval (t_1, t_2).

The equation of heat conduction is obtained by calculation of the balance of heat in a given length (x_1, x_2) for a given time interval (t_1, t_2). Applying the principle of conservation of energy and making use of formulae (5), (7) and (9) we can write down the equality

$$\int_{t_1}^{t_2} \left[k \frac{\partial u}{\partial x}(x, \tau)\Big|_{x=x_2} - k \frac{\partial u}{\partial x}(x, \tau)\Big|_{x=x_1} \right] d\tau + \int_{x_1}^{x_2} \int_{t_1}^{t_2} F(\xi, \tau)\, d\xi\, d\tau =$$

$$= \int_{x_1}^{x_2} c\varrho\, [u(\xi, t_2) - u(\xi, t_1)]\, d\xi, \tag{10}$$

which is the equation of heat conduction in integral form.

* If, for example, the heat is generated as the result of the passage of an electric current of strength I through the rod, whose resistance per unit length equals R, then $F = 0 \cdot 24 I^2 R$.

In order to obtain the equation of heat conduction in differential form, let us assume that the function $u(x, t)$ has continuous derivatives u_{xx} and u_t.*

Making use of the mean value theorem, we obtain the equality

$$\left[k \frac{\partial u}{\partial x}(x, \tau)\big|_{x=x_2} - k \frac{\partial u}{\partial x}(x, \tau)\big|_{x=x_1}\right]_{\tau=t_3} \Delta t + F(x_4, t_4)\,\Delta x\,\Delta t =$$

$$= \{c\varrho\,[u(\xi, t_2) - u(\xi, t_1)]\}_{\xi=x_3} \Delta x, \tag{11}$$

which, with the help of the mean value theorem, can be transformed to the form

$$\frac{\partial}{\partial x}\left[k \frac{\partial u}{\partial x}(x, t)\right]_{\substack{x=x_5\\ t=t_3}} \Delta t\,\Delta x + F(x_4, t_4)\,\Delta x\,\Delta t =$$

$$= \left[c\varrho \frac{\partial u}{\partial t}(x, t)\right]_{\substack{x=x_3\\ t=t_5}} \Delta x\,\Delta t, \tag{12}$$

where, t_3, t_4, t_5 and x_3, x_4, x_5 are intermediate points of the interval (t_1, t_2) and (x_1, x_2).

Hence, after dividing throughout by the product $\Delta x\,\Delta t$, we find

$$\frac{\partial}{\partial x}\left(k \frac{\partial u}{\partial x}\right)\Big|_{\substack{x=x_5\\ t=t_3}} + F(x, t)\Big|_{\substack{x=x_4\\ t=t_4}} = c\varrho \frac{\partial u}{\partial t}\Big|_{\substack{t=t_5\\ x=x_3}}. \tag{13}$$

All these considerations refer to the arbitrary intervals (x_1, x_2) and (t_1, t_2). Proceeding to the limit as $x_1, x_2 \to x$ and $t_1, t_2 \to t$ we obtain the equation

$$\frac{\partial}{\partial x}\left(k \frac{\partial u}{\partial x}\right) + F(x, t) = c\varrho \frac{\partial u}{\partial t}, \tag{14}$$

called *the equation of heat-conduction.*

Let us examine some special cases.

1. If the rod is homogeneous, k, c, ϱ are constant, and the equation is usually written in the form

$$u_t = a^2 u_{xx} + f(x, t),$$

$$a^2 = k/c\varrho, \quad f(x, t) = \frac{F(x, t)}{c\varrho},$$

* By differentiating the function $u(x, t)$ we, generally speaking, may lose a series of possible solutions which satisfy the integral equation but not the differential equation. Nevertheless, in the case of the equation of heat conduction, by differentiating the solution we do not lose solutions because it can be proved that if a function satisfies equation (10) then it certainly must be differentiable.

where a^2 is a constant, called *the coefficient of heat conduction.* If the sources are absent, i. e. $F(x, t) = 0$, then the equation of heat conduction takes the simple form

$$u_t = a^2 u_{xx} . \tag{14'}$$

2. The density of the heat sources may depend on the temperature. In the case of a heat exchange with the surrounding medium, obeying *Newton's law,* the amount of heat which is lost by the rod, per unit length and time is equal to

$$F_0 = h(u - \theta),$$

where $\theta(x, t)$ is the temperature of the surrounding medium, h is *the coefficient of heat exchange*. Thus, the density of the heat sources,* at the point x at time t, equals

$$F = F_1(x, t) - h(u - \theta),$$

where $F_1(x, t)$ is the density of other sources of heat.

If the rod is homogeneous, then the equation of heat conduction with lateral heat exchange has the following form

$$u_t = a^2 u_{xx} - \alpha u + f(x, t), \tag{15}$$

where $\alpha = \frac{h}{c\varrho}$; $f(x, t) = \alpha\theta(x, t) + \frac{F_1(x, t)}{c\varrho}$ is a known function.

3. The coefficients k and c, as a rule, are slowly varying functions of temperature. Therefore the assumption made above about the constancy of these coefficients is possible only on condition that small temperature changes are considered. The study of conduction processes for large changes of temperatures leads to the non-linear equation of heat conduction, which for heterogeneous media is

$$\frac{\partial}{\partial x}\left(k(u, x)\frac{\partial u}{\partial x}\right) + F(x, t) = C(u, x)\,\varrho(u, x)\frac{\partial u}{\partial t}$$

(see Appendix III).

* Since in our approximation the distribution of temperature is not considered over a cross-section, the effect of the surface sources is equivalent to the effect of the volume sources of heat.

2. *Equation of diffusion*

If a region is filled non-uniformly by a gas, then diffusion takes place from points of higher concentration to points of lower concentration. This phenomenon also takes place in solutions, if the concentration of the solute is not constant throughout the volume.

Let us examine the process of diffusion in a hollow tube, or in a tube filled with a porous medium, assuming that at any moment of time the concentration of the gas (solution) in a section of the tube is the same. Then the process of diffusion may be described by a function $u(x, t)$ which represents the *concentration* at section x at time t.

According to *Nernst's* law the mass of gas flowing through section x in the time interval $(t, t + dt)$ is equal to

$$dQ = -D \frac{\partial u}{\partial x}(x, t) S dt, \tag{16}$$

where D is the coefficient of diffusion, S is the cross-sectional area of the tube.

By definition of the concentration, the quantity of gas in a volume V equals

$$Q = uV;$$

hence we find that the change in the mass of gas in a section (x_1, x_2) of the tube due to a change in concentration of Δu equals

$$\Delta Q = \int_{x_1}^{x_2} c(x) \Delta u \cdot S dx,$$

where $c(x)$ is the *coefficient of porosity.**

Let us form the equation for the conservation of the mass of gas in a section (x_1, x_2) in a time interval (t_1, t_2)

$$S \int_{t_1}^{t_2} \left[D(x_2) \frac{\partial u}{\partial x}(x_2, \tau) - D(x_1) \frac{\partial u}{\partial x}(x_1, \tau) \right] d\tau =$$

$$= S \int_{x_1}^{x_2} c(\xi) [u(\xi, t_2) - u(\xi, t_1)] d\xi .$$

* The coefficient of porosity is the ratio of the volume of the pores to the total volume V_0 equal in our case to Sdx.

Hence, as in 1, we obtain the equation

$$\frac{\partial}{\partial x}\left(D\frac{\partial u}{\partial x}\right) = c\frac{\partial u}{\partial t}, \tag{17}$$

which is *the equation of diffusion.* It is completely analogous to the equation of heat conduction. In deducing this equation we assumed that there were no sources of matter in the tube and that diffusion across the surface of the tube was absent. Taking account of these effects leads to equations similar to equations (14) and (15).

If the coefficient of diffusion is constant, then the diffusion equation takes the form

$$u_t = a^2 u_{xx},$$

where

$$a^2 = \frac{D}{c}.$$

If the coefficient of porosity $c = 1$, and the coefficient of diffusion is constant, the diffusion equation has the form

$$u_t = Du_{xx}.$$

3. *Conduction of heat in space*

The process of conduction of heat in space may be characterized by the temperature $u(x, y, z, t)$ which is a function of x, y, z and t.

If the temperature is not constant, heat currents are produced, directed from points of higher temperature to points of lower temperature. The amount of heat flowing through an area $d\sigma$ at the point (x, y, z) in a time interval $(t, t + dt)$ is defined according to Fourier's law by the formula

$$dQ = -k\frac{\partial u}{\partial \boldsymbol{n}}(x, y, z, t)\, d\sigma\, dt,$$

where k is the coefficient of thermal conductivity of the body,* n is the normal to an element of surface $d\sigma$ in the direction of heat flow. As is well known

$$\frac{\partial u}{\partial n} = \frac{\partial u}{\partial x}\cos(\widehat{n, x}) + \frac{\partial u}{\partial y}\cos(\widehat{n, y}) + \frac{\partial u}{\partial z}\cos(\widehat{n, z}) = \operatorname{grad} u \cdot \boldsymbol{n},$$

* We are examining a homogeneous isotropic body.

so it is possible to write

$$dQ = -k \operatorname{grad} u \cdot \boldsymbol{n}\, d\sigma\, dt .$$

Hence it follows that the flow of heat carried away per unit time and unit area is equal to

$$q_n = \boldsymbol{q} \cdot \boldsymbol{n} , \tag{18}$$

where $\boldsymbol{q} = -k \operatorname{grad} u$ is a vector for the density of heat flow.

For the amount of heat flowing out of a given surface S in a time interval (t_1, t_2) we have:

$$Q_1 = - \int_{t_1}^{t_2} \iint_S k \frac{\partial u}{\partial n}\, dt d\sigma$$

or

$$Q_1 = - \int_{t_1}^{t_2} \iint_S k \operatorname{grad} u \cdot \boldsymbol{n}\, d\sigma dt = \int_{t_1}^{t_2} \iint_S \boldsymbol{q} \cdot \boldsymbol{n}\, d\sigma dt , \tag{19}$$

where $\boldsymbol{n}$ is the outer normal. The amount of heat necessary to change the temperature of points of the body by $\Delta u(x, y, z) = u(x, y, z, t_2) - u(x, y, z, t_1)$ equals

$$Q_2 = \iiint_V c\varrho\, [u(\xi, \eta, \zeta, t_2) - u(\xi, \eta, \zeta, t_1)]\, dV . \tag{20}$$

Finally, denoting the density of the heat sources by $F(x, y, z, t)$ we obtain an expression for the amount of heat liberated by a volume V in a time interval (t_1, t_2)

$$Q_3 = \int_{t_1}^{t_2} \iiint_V F(\xi, \eta, \zeta, t)\, dV dt . \tag{21}$$

Now we construct the equation for the balance of heat in a given elementary volume V, the surface of which we denote by S. It is obvious that

$$Q_2 = Q_3 - Q_1 . \tag{22}$$

Assuming twofold differentiation of the function by the variables x, y, z and once with respect to t, and also the continuity of these derivatives in the region being examined, let us transform the relations (19), (20) and (21). Using Ostrogradskii's relation

$$\int_S \boldsymbol{q} \cdot \boldsymbol{n}\, d\sigma = \int_V \operatorname{div} \boldsymbol{q}\, dV^*$$

* We assume the normal in this formula, as also in (19), to be external.

and using the mean value theorem for a function of several variables, we obtain

$$Q_1 = \operatorname{div} \boldsymbol{q}(x_1, y_1, z_1, t_3)\, V\, \Delta t,$$

$$Q_2 = c\varrho \left[u(x_2, y_2, z_2, t_2) - u(x_2, y_2\, z_2, t_1)\right] V =$$

$$= c\varrho \frac{\partial u}{\partial t}(x_2, y_2, z_2, t_4)\, V\, \Delta t,$$

$$Q_3 = F(x_3, y_3, z_3\, t_5)\, V\, \Delta t.$$

After dividing by $V\,\Delta t$ the conservation equation takes the form

$$c\,\varrho \frac{\partial u}{\partial t}(x_2, y_2 \cdot z_2, t_4) =$$

$$= -\operatorname{div} \boldsymbol{q}(x_1, y_1, z_1, t_3) + F(x_3, y_3, z_3, t_5), \tag{23}$$

where all values lie inside the region being examined, viz. at some mean point of the volume V for a mean value of time in the interval (t_1, t_2).

This equality holds for an arbitrary volume V inside the body. Contracting the volume to a point with coordinates x, y, z and proceeding to the limit as $t_1, t_2 \to t$, we obtain*

$$c\,\varrho \frac{\partial u}{\partial t}(x, y, z, t) = -\operatorname{div} \boldsymbol{q}(x, y, z, t) + F(x, y, z, t), \tag{24}$$

where because of the continuity of the derivatives all values are taken at the one point.

Substituting q from formula (18) we obtain the equation of heat conduction

$$c\,\varrho\, u_t = \operatorname{div}(k \operatorname{grad} u) + F$$

or

$$c\,\varrho\, u_t = \frac{\partial}{\partial x}\left(k \frac{\partial u}{\partial x}\right) + \frac{\partial}{\partial y}\left(k \frac{\partial u}{\partial y}\right) + \frac{\partial}{\partial z}\left(k \frac{\partial u}{\partial z}\right) + F. \tag{25}$$

If the medium is homogeneous the equation is usually written in the form

$$u_t = a^2 (u_{xx} + u_{yy} + u_{zz}) + F/c\,\varrho, \tag{26}$$

where $a^2 = k/c\varrho$ is the coefficient of heat conduction, or

$$u_t = a^2\, \nabla^2 u + f \qquad (f = F/c\,\varrho,) \tag{26'}$$

where $\nabla^2 = \frac{\partial^2}{\partial x^2} + \frac{\partial^2}{\partial y^2} + \frac{\partial^2}{\partial z^2}$ is the Laplace operator.

* We assume $F(x, y, z, t)$ is continuous in the region under consideration.

4. *The formulation of boundary-value problems*

To find a unique solution of the equation of heat conduction it is necessary to add initial and boundary conditions.

The initial condition, in contrast to the equation of hyperbolic type, consists only of the given value of the function $u(x, t)$ at the initial time t_0.

The boundary conditions may differ in their dependence on the thermal condition at the boundaries. Let us examine three basic types of boundary conditions.

1. The temperature is given at the ends of the rod

$$u(0, t) = \mu(t), \tag{27}$$

where $\mu(t)$ is a function given for a certain interval $t_0 \leqslant t \leqslant T$, where T is the period of time for which the process is investigated.

2. At the boundary a derivative value is given

$$\frac{\partial u}{\partial x}(l, t) = \nu(t). \tag{28}$$

We get this condition if the value of the heat flux $Q(l, t)$ flowing across the end section of the rod is given

$$Q(l, t) = -k\frac{\partial u}{\partial x}(l, t),$$

whence $\partial u/\partial x\,(l, t) = \nu(t)$, if $\nu(t)$ is a known function, represented for a given flux $Q(l, t)$ by the formula

$$\nu(t) = -\frac{Q(l, t)}{k}.$$

3. At the boundary a linear relation between the derivative and the function is given

$$\frac{\partial u}{\partial x}(l, t) = -\lambda\left[u(l, t) - \theta(t)\right].$$

This boundary condition corresponds to the heat exchange according to Newton's law between the surface of the body and the surrounding medium, whose temperature θ is known. Making use of the two expressions for the heat flux flowing through the end $x = l$,

$$Q = h(u - \theta)$$

and

$$Q = -k\frac{\partial u}{\partial x},$$

we obtain the mathematical formulation of the third boundary condition in the form

$$\frac{\partial u}{\partial x}(l, t) = -\lambda\left[u(l, t) - \theta(t)\right], \tag{29}$$

where $\lambda = h/k$ is the coefficient of heat exchange, and $\theta(t)$ is a certain given function. For the end $x = 0$ of the rod $(0, l)$ the third boundary condition has the form

$$\frac{\partial u}{\partial x}(0, t) = +\lambda\left[u(0, t) - \theta(t)\right]. \tag{29'}$$

The boundary conditions at $x = 0$ and $x = l$ may be of different types so that the number of different boundary-value problems is large.

If the medium is heterogeneous and the coefficients of the equation are discontinuous functions, then the interval $(0, l)$ in which the solution of the problem is sought is divided by the points of discontinuity of the coefficients into several parts within which the function u satisfies the equation of heat conduction, and the conditions of connection at the boundaries. In the simplest case these conditions are deduced from the continuity of the temperature and the continuity of the heat flow

$$u(x_i - 0, t) = u(x_i + 0, t),$$

$$k(x_i - 0)\frac{\partial u(x_i - 0, t)}{\partial x} = k(x_i + 0)\frac{\partial u}{\partial x}(x_i + 0, t),$$

where x_i are points of discontinuity of the coefficients.

In addition to the problems described here limiting cases of them are frequently encountered. Let us examine the process of heat conduction in a very long rod. In the course of a small interval of time the effect of the conditions at the boundary is very slight in the central part of the rod and the temperature is determined basically by the initial distribution of temperature. In this case the precise length of the rod does not have any significance because a change in length of the rod does not substantially influence the temperature of the part of interest to us; in problems of this type one usually considers that the rod has infinite length. Thus *the problem (Cauchy's problem) of the distribution of temperature along an infinite straight rod with initial conditions* may be stated:

Find the solution of the equation of heat conduction in the region $-\infty < x < \infty$ *and* $t \geqslant t_0$ *which satisfies the condition*

$$u(x, t_0) = \varphi(x) \quad (-\infty < x < +\infty),$$

where $\varphi(x)$ is a given function.

Similarly, if the section of the rod, whose temperature interests us, is located near one end and is far from the other, then in this case the temperature is determined essentially by the thermal condition of the near end and by the initial conditions. In problems of a similar type one usually assumes that the rod is semi-infinite and the coordinate varies within the limits $0 \leqslant x \leqslant \infty$. Let us illustrate by an example the formulation *of the chief boundary-value problem for a semi-infinite rod :*

Find the solution of the equation of heat conduction in the region $0 < x < \infty$ *and* $t_0 \leqslant t$, *which satisfies the condition*

$$\left.\begin{aligned} u(x, t_0) &= \varphi(x) \quad (0 < x < \infty), \\ u(0, t) &= \mu(t) \quad (t \geqslant t_0), \end{aligned}\right\} \tag{30}$$

where $\varphi(x)$ and $\mu(t)$ are given functions.

The problems quoted above are limiting cases of the main boundary-value problems. Other limiting cases are possible which neglect the initial conditions. The effect of the initial conditions is diminished in the course of time, and after a long time the temperature of the rod is determined by the boundary conditions. In this case one may usefully assume that the experiment continues for an infinitely long time and the system reaches a steady state independent of the initial conditions.

Thus we arrive at *a steady state boundary-value problem*, when we have the solution of the equation of heat conduction for $0 \leqslant x \leqslant l$, and $-\infty < t$, which satisfies the conditions

$$\begin{aligned} u(0, t) &= \mu_1(t), \\ u(l, t) &= \mu_2(t). \end{aligned} \tag{31}$$

Depending on the nature of the boundary condition there are other possible forms of steady state problems.

One very important steady state problem for a semi-infinite rod ($l = \infty$) is the following: solve the equation of heat conduction for $0 < x < \infty$, $t > -\infty$, with the boundary condition

$$u(0, t) = \mu(t), \tag{27}$$

where $\mu(t)$ is a given function.

Most often steady state problems occur in a system with periodic boundary conditions.

$$\mu(t) = A \cos \omega t. \tag{32}$$

(see Appendix I to Chapter III).

It is natural to assume that after the lapse of a large interval of time the temperature of the rod for practical purposes also varies according to a periodic law with the same frequency. We never obtain an exactly periodic solution because the effect of the initial conditions, although they diminish in the course of time, do not reduce to zero; to allow for this effect in view of the errors of observation would not be meaningful. When considering a periodic solution we neglect the effect of the initial data.

The statement of boundary-value problems outlined above refers not only to the equation with constant coefficients. By the words *the equation of heat conduction* we understand any of the equations of the preceding paragraphs.

Besides the linear boundary-value problems outlined above, there are also problems with non-linear boundary conditions, e. g. of the form

$$k \frac{\partial u}{\partial x}(0, t) = \sigma [u^4(0, t) - \theta^4(0, t)]. \tag{33}$$

This boundary condition corresponds to the radiation according to the Stefan-Boltzmann law from the end $x = 0$ in a medium of temperature $\theta(t)$.

We will examine in detail the first boundary-value problem for a restricted region.

We shall describe *the solution of the first boundary-value problem* by a function $u(x, t)$ which possesses the following properties:

(1) $u(x, t)$ is finite and continuous in a closed region

$$0 \leqslant x \leqslant l, \ t_0 \leqslant t \leqslant T;$$

(2) $u(x, t)$ satisfies the equation of heat conduction in an open region

$$0 < x < l, \ t_0 < t;$$

(3) $u(x, t)$ satisfies the initial and boundary conditions, viz.

$$u(x, t_0) = \varphi(x), \ u(0, t) = \mu_1(t), \ u(l, t) = \mu_2(t),$$

where $\varphi(x)$, $\mu_1(t)$, $\mu_2(t)$ are continuous functions satisfying the conditions of connection at the boundaries

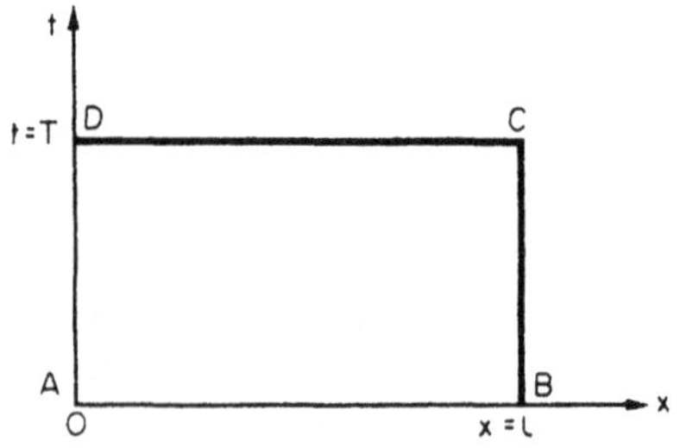

FIG. 37

$$\varphi(0) = \mu_1(t_0) \quad [= u(0, t_0)]$$

and

$$\varphi(l) = \mu_2(t_0) \quad [= u(l, t_0)],$$

necessary for the continuity of $u(x, t)$ in a closed region.

Let us examine the plane of phase points (x, t) (Fig. 37). In our problem we seek a function $u(x, t)$, defined inside the rectangle $ABCD$. This region is defined by the statement of the problem because the process of the conduction of heat is investigated in a rod $0 \leqslant x \leqslant l$, for a time interval $t_0 \leqslant t \leqslant T$, in the course of which the temperature of the end is known to us. Let $t_0 = 0$; we assume that $u(x, t)$ satisfies the equation only for $0 < x < l$, $0 < t \leqslant T$, but not at $t = 0$ (side AB) nor at $x = 0$, $x = l$ (sides AD and BC), where the values of this function are directly given by the initial and boundary conditions. If we required that the equation be satisfied, for example, at $t = 0$, then as well we would require that a derivative $\varphi'' = u_{xx}(x, 0)$ exist. This requirement would confine the range of physical phenomena under consideration, excluding from investigation those functions for which it is not fulfilled. Condition (3), without the assumption of the continuity of $u(x, t)$ in the range

$0 \leqslant x \leqslant l$, $0 \leqslant t \leqslant T$ (i. e. in the closed rectangle $ABCD$) or some other condition replacing this assumption, loses meaning.* In fact let us examine the function $v(x, t)$ defined in the following form

$$v(x,t) = C \qquad (0 < x < l,\ 0 < t \leqslant T),$$

$$v(x,0) = \varphi(x) \qquad (0 \leqslant x \leqslant l),$$

$$\left.\begin{aligned} v(0,t) &= \mu_1(t), \\ v(l,t) &= \mu_2(t) \end{aligned}\right\} \quad (0 \leqslant t \leqslant T),$$

where C is an arbitrary constant. The function $v(x, t)$ obviously satisfies condition (2) and also the boundary conditions. Nevertheless this function does not describe the distribution of temperature in a rod with initial temperature $\varphi(x) \neq C$ and boundary temperatures $\mu_1(t) \neq C$ and $\mu_2(t) \neq C$, because it is discontinuous at $t = 0$, $x = 0$ $x = l$.

The continuity of the function $u(x, t)$ in the region $0 < x < l$, $0 < t < T$ follows from the fact that the function satisfies the differential equation. Thus, the requirement of the continuity of $u(x, t)$ at $0 \leqslant x \leqslant l$, $0 \leqslant t \leqslant T$, in essence concerns only those points where the boundary and initial values are given. In the following we shall imply by the words *the solution of the equation, which satisfies boundary conditions* a function which satisfies conditions (1), (2), (3).

Similarly other boundary-value problems may be set, among them problems on an infinite rod and steady state problems.

All those problems mentioned above are relevant to examples with several independent geometrical variables. In these problems the initial temperature at $t = t_0$ and the boundary conditions on the surface of the body are given. Also it is possible to examine problems for an infinite region.

In connection with each problem the following points arise:**

(1) the uniqueness of the solution of the given problem,

(2) the existence of a solution,

* Later boundary-value problems will be set with discontinuous boundary and initial conditions. For these problems the boundary conditions will be specified precisely.

** Compare with Chapter II, § 2.

(3) the continuous dependence of the solution on additional conditions

If the given problem has several solutions then the words *the solution of the problem* do not have a definite meaning. Therefore before speaking about the solution of a problem, it is necessary to prove its uniqueness. For most purposes condition (2) is essential because the proof of the existence of a solution usually gives a method of calculating the solution.

As was noted earlier (see Chapter II, § 2, sub-section 3) a process is termed physically definite if for a small change of the initial and boundary conditions the solution is only changed a little. It may be proved that the process of conduction of heat is defined physically by the initial and boundary conditions.

5. *The maximum value principle*

We shall examine the equation with constant coefficients

$$v_t = a^2 v_{xx} + \beta v_x + \gamma v . \tag{34}$$

As we saw, this equation on substituting

$$v = e^{\mu x + \lambda t} \cdot u \text{ where } \mu = -\frac{\beta}{2a^2}, \quad \lambda = \gamma - \frac{\beta^2}{4a^2}$$

reduces to the form

$$u_t = a^2 u_{xx} . \tag{35}$$

Let us prove the following property of the solutions of this equation which we will call *the maximum value principle.*

If the function $u(x, t)$, finite and continuous in the closed region $0 \leqslant t \leqslant T$ and $0 \leqslant x \leqslant l$, satisfies the equation of heat conduction

$$u_t = a^2 u_{xx} \tag{35}$$

at points of the range $0 < x < l$, $0 < t \leqslant T$, then the maximum and minimum values of the function $u(x, t)$ occur either at the initial time or at points of the boundary $x = 0$ or $x = l$.

The physical significance of this theorem is obvious; if the temperature at the boundary or at the initial time does not exceed a certain value M, then inside the body it is not possible to produce a temperature greater than M.

The proof of the theorem is indicated by the converse. Let us denote by M the maximum value of $u(x, t)$ at $t = 0$ $(0 \leqslant x \leqslant l)$ or at $x = 0$ or at $x = l$ $(0 \leqslant t \leqslant T)$* and let us assume that at some point (x_0, t_0) $(0 < x_0 < l,\ 0 < t_0 \leqslant T)$ the function $u(x, t)$ reaches its maximum value equal to

$$u(x_0, t_0) = M + \varepsilon .$$

Let us compare the terms of equation (35) at the point (x_0, t_0). Since at the point (x_0, t_0) the function attains its maximum value then necessarily

$$\frac{\partial u}{\partial x}(x_0, t_0) = 0 \quad \text{and} \quad \frac{\partial^2 u}{\partial x^2}(x_0, t_0) \leqslant 0^{**}). \tag{36}$$

Further, since $u(x_0, t)$ reaches a maximum value at $t = t_0$ then***

$$\frac{\partial u}{\partial t}(x_0, t_0) \geq 0. \tag{37}$$

Comparing the terms of the right- and left-hand sides of equation (35) we see that they are different. Nevertheless this argument still does not prove the theorem, because the right- and left-hand sides may both be equal to zero, which would not by itself involve an inconsistency. We quoted this argument in order to distinguish more clearly the basic concept of the proof. To

* If the continuity of $u(x, t)$ in the closed interval $0 \leqslant x \leqslant l,\ 0 \leqslant t \leqslant T$ were not assumed, then the function $u(x, t)$ might not reach its maximum at any point, and furthermore the discussion would not be applicable. By virtue of the theorem that any continuous function reaches its maximum value in a closed region, we may be certain that: (1) the function $u(x, t)$ reaches a maximum value on boundary or at $t = 0$ which we denote by M; (2) if $u(x, t)$ were at one point greater than M, then there would exist a point (x_0, t_0) at which $u(x, t)$ would reach a maximum value exceeding M; $u(x_0, t_0) = M + \varepsilon (\varepsilon > 0)$ where $0 < x_0 < l,\ 0 < t_0 \leqslant T$.

** Sufficient conditions for the function $f(x)$ to have a minimum at the point x_0 in the interval $(0, l)$ are the following $\left.\frac{\partial f}{\partial x}\right|_{x=x_0} = 0$, $\left.\frac{\partial^2 f}{\partial x^2}\right|_{x=x_0} > 0$. Thus if at the point $x_0, f(x)$ has a maximum value then (1) $f'(x_0) = 0$ and (2) it is impossible that $f''(x_0) > 0$, i.e. $f''(x_0) \leqslant 0$.

*** It is clear that if $t_0 < T$ then $\frac{\partial u}{\partial t} = 0$, but if $t = T$ then $\frac{\partial u}{\partial t} \geqslant 0$.

complete the proof, let us find a point (x_1, t_1) at which $\frac{\partial^2 u}{\partial x^2} \leqslant 0$ and $\frac{\partial u}{\partial t} > 0$. In order to do this we examine the auxiliary function

$$v(x, t) = u(x, t) + k(t_0 - t), \tag{38}$$

where k is some positive number. It is obvious that

$$v(x_0, t_0) = u(x_0, t_0) = M + \varepsilon$$

and

$$k(t_0 - t) \leq kT.$$

Let us choose $k > 0$ so that kT is less than $\varepsilon/2$, i. e. $k < \varepsilon/2T$; then the maximum value of $v(x, t)$ at $t = 0$ or at $x = 0$, $x = l$ will not exceed $M + \varepsilon/2$, i. e.

$$v(x, t) \leqslant M + \varepsilon/2 \text{ (where } t = 0 \text{ or } x = 0, x = l), \tag{39}$$

since for these arguments the first term of formula (38) does not exceed M, and the second does not exceed $\varepsilon/2$.

By virtue of the continuity of the function $v(x, t)$ it must reach its maximum value at some point (x_1, t_1). It is obvious that

$$v(x_1, t_1) \geqslant v(x_0, t_0) = M + \varepsilon.$$

Therefore $t_1 > 0$ and $0 < x < l$, since at $t = 0$ or at $x = 0$, $x = l$ the inequality (39) holds. Comparing the symbols of the right- and left-hand sides of equation (35) at the point (x_1, t_1) we obtain

$$v_{xx}(x_1, t_1) = u_{xx}(x_1, t_1) \leqslant 0$$

and

$$v_t(x_1, t_1) = u_t(x_1, t_1) - k \geqslant 0 \text{ or } u_t(x_1, t_1) \geqslant k > 0.$$

Thus, at the point (x_1, t_1) equation (35) cannot be satisfied since the values of the various terms on the right and left side are continuous. In this manner, the first part of the theorem is proved.

Similarly the second part of the theorem may be proved for a minimum value. However this does not require a separate proof because the function $u_1 = -u$ has a maximum value where u has a minimum.

Let us derive some conclusions from the maximum value principle. First of all, we prove a uniqueness theorem for the first boundary value problem.

6. *Uniqueness theorem*

If the two functions $u_1(x, t)$ and $u_2(x, t)$, defined and continuous in the region $0 \leqslant x \leqslant l$, $0 \leqslant t \leqslant T$, satisfy the equation of heat conduction

$$u_t = a^2 u_{xx} + f(x, t) \text{ where } 0 < x < l, t > 0), \tag{35}$$

with the identical initial and boundary conditions

$$u_1(x, 0) = u_2(x, 0) = \varphi(x),$$
$$u_1(0, t) = u_2(0, t) = \mu_1(t),$$
$$u_1(l, t) = u_2(l, t) = \mu_2(t),$$

*then $u_1(x, t) = u_2(x, t)$**

In order to prove this theorem, we examine the function

$$v(x, t) = u_2(x, t) - u_1(x, t).$$

Since the functions $u_1(x, t)$ and $u_2(x, t)$ are continuous at

$$0 \leqslant x \leqslant l,$$
$$0 \leqslant t \leqslant T,$$

the function $v(x, t)$, equal to their difference, is continuous in this same region. As the difference of the two solutions of the equation of heat conduction in the region $0 < x < l$, $t > 0$, the function $v(x, t)$ is also a solution of the equation of heat conduction in this region.

Thus, let us apply the maximum value principle to this function, viz. it reaches its maximum and minimum value either at $t = 0$ or $x = 0$ or $x = l$. However, we have

$$v(x, 0) = 0, \qquad v(0, t) = 0, \qquad v(l, t) = 0.$$

Therefore

$$v(x, t) \equiv 0,$$

* In sub-section 3, § 2, this theorem will be strengthened and the requirement of continuity at $t = 0$ removed.

i. e.

$$u_1(x, t) \equiv u_2(x, t).$$

Hence it follows that the solution of the first boundary-value problem is unique.

Let us prove some further consequences of the maximum value principle.

1. *If two solutions of the equation of heat conduction* $u_1(x, t)$ *and* $u_2(x, t)$ *satisfy the conditions*

$$u_1(x, 0) \leqslant u_2(x, 0),$$

$$u_1(0, t) \leqslant u_2(0, t), \quad u_1(l, t) \leqslant u_2(l, t),$$

then

$$u_1(x, t) \leqslant u_2(x, t)$$

for all values $\quad 0 \leqslant x \leqslant l, \ 0 \leqslant t \leqslant T.$

In fact the difference $v(x, t) = u_2(x, t) - u_1(x, t)$ satisfies the conditions under which the maximum value principle was established and moreover

$$v(x, 0) \geqslant 0, \quad v(0, t) \geqslant 0, \quad v(l, t) \geqslant 0.$$

Therefore

$$v(x, t) \geqslant 0 \text{ where } 0 < x < l, \ 0 < t \leqslant T,$$

because otherwise the function $v(x, t)$ would have a negative minimum value in the range $0 < x < l$, $0 < t \leqslant T$.

2. *If three solutions of the equation of heat conduction*

$$u(x, t), \quad \underline{u}(x, t), \quad \bar{u}(x, t)$$

satisfy the conditions

$$\underline{u}(x, t) \leqslant u(x, t) \leqslant \bar{u}(x, t) \text{ where } t = 0, \ x = 0 \text{ and } x = l,$$

then the same inequalities are fulfilled, i.e. for all x, t *at* $0 \leqslant x \leqslant l$, $0 \leqslant t \leqslant T$.

This statement is the application of result 1 to the functions

$$u(x, t), \ \bar{u}(x, t) \text{ and } u(x, t), \ \underline{u}(x, t).$$

3. *If two solutions of the equation of heat conduction* $u_1(x, t)$ *and* $u_2(x, t)$ *satisfy the inequality*

$$|u_1(x, t) - u_2(x, t)| \leqslant \varepsilon \text{ where } t = 0, \ x = 0, \ x = l,$$

then

$$|u_1(x,t) - u_2(x,t)| \leqslant \varepsilon$$

is identical, i.e. it holds for all x, t

$$0 \leqslant x \leqslant l, \ 0 \leqslant t \leqslant T.$$

This statement follows from result 2 if it is applied to the following solutions of the equation of heat conduction

$$\underline{u}(x,t) = -\varepsilon,$$
$$u(x,t) = u_1(x,t) - u_2(x,t),$$
$$\bar{u}(x,t) = \varepsilon.$$

The question of a continuous dependence of the solution of the first boundary-value problem on the initial and boundary values is answered by the result 3. If in a certain physical problem, instead of the solution of the equation of heat conduction fitting the initial and boundary conditions

$$u(x,0) = \varphi(x), \quad u(0,t) = \mu_1(t), \quad u(l,t) = \mu_2(t),$$

we substitute a solution $u(x,t)$ corresponding to other initial and boundary values which are defined by the functions $\varphi^*(x)$, $\mu_1{}^*(t)$, $\mu_2{}^*(t)$, not distinguishable within the limits of a given degree of accuracy ε from the functions $\varphi(x)$, $\mu_1(t)$, $\mu_2(t)$:

$$|\varphi(x) - \varphi^*(x)| \leqslant \varepsilon, \quad |\mu_1(t) - \mu_1^*(t)| \leqslant \varepsilon,$$
$$|\mu_2(t) - \mu_2^*(t)| \leqslant \varepsilon,$$

then the function $u_1(x,t)$ will be indistinguishable from the function $u(x,t)$ within the same limits of accuracy

$$|u(x,t) - u_1(x,t)| \leqslant \varepsilon.$$

The above theorems for the first boundary-value problem for a finite region in two- or three-dimensional space may be proved by repetition of the arguments used above.

Similar theorems hold for problems in an infinite region and in steady state problems. In these cases, however, the proofs require modification and some additional conditions have to be imposed in order to determine a unique solution.

7. *Uniqueness theorem for an infinite region*

In the solutions of problems for an infinite one-dimensional region there is a further restriction on a solution, i.e. there exists an M such that $|u(x, t)| < M$ for all $-\infty < x < \infty$ and $t \geqslant 0$.

If $u_1(x, t)$ and $u_2(x, t)$ are continuous, bounded functions within the range of the variables (x, t) and which satisfy the equation of heat conduction

$$u_t = a^2 u_{xx} \quad (-\infty < x < \infty, \ t > 0) \tag{35}$$

and the condition

$$u_1(x, 0) = u_2(x, 0) \quad (-\infty < x < \infty),$$

then

$$u_1(x, t) \equiv u_2(x, t) \quad (-\infty < x < \infty, \ t \geqslant 0).$$

Let us examine, as is usual, the function

$$v(x, t) = u_1(x, t) - u_2(x, t).$$

The function $v(x, t)$ is continuous, satisfies the equation of heat conduction, is finite in the entire region

$$|v(x, t)| \leqslant |u_1(x, t)| + |u_2(x, t)| < 2M \quad (-\infty < x < \infty, \ t \geqslant 0)$$

and satisfies the condition

$$v(x, 0) = 0.$$

The principle of maximum value, which we used in the proof of the uniqueness of the problem for a section, is inapplicable, because in an infinite region the function $v(x, t)$ may nowhere attain a maximum value. In order to use this principle, let us examine the region

$$|x| \leqslant L,$$

where L is a large number, which subsequently we shall increase indefinitely, and the function

$$V(x, t) = \frac{4M}{L^2}\left(\frac{x^2}{2} + a^2 t\right). \tag{40}$$

The function $V(x, t)$ is continuous, satisfies the equation of heat conduction, and moreover possesses the following properties:

$$V(x, 0) \geqslant |v(x, 0)| = 0,$$
$$V(\pm L, t) \geqslant 2M \geqslant v(\pm L, t).$$

Using the principle of maximum value in the region $|x| \leqslant L$ we have

$$-\frac{4M}{L^2}\left(\frac{x^2}{2}+a^2 t\right) \leqslant v(x,t) \leqslant \frac{4M}{L^2}\left(\frac{x^2}{2}+a^2 t\right). \tag{41}$$

Let us fix certain values of (x, t) and let L increase indefinitely. Proceeding to the limit as $L \to \infty$, we obtain

$$v(x,t) \equiv 0,$$

which proves the theorem.

§ 2. Method of the separation of variables

1. *A homogeneous boundary-value problem*

Let us turn to the solution of the first boundary-value problem of the equation of heat conduction

$$u_t = a^2 u_{xx} + f(x,t) \tag{1}$$

with initial condition

$$u(x,0) = \varphi(x) \tag{2}$$

and boundary conditions

$$\left.\begin{aligned} u(0,t) &= \mu_1(t), \\ u(l,t) &= \mu_2(t). \end{aligned}\right\} \tag{3}$$

We begin the study of the general first boundary-value problem with the solution of the following simple problem (I):

Find the solution of the homogeneous equation

$$u_t = a^2 u_{xx}, \tag{4}$$

which satisfies the initial condition

$$u(x,0) = \varphi(x) \tag{2}$$

and the zero boundary conditions

$$u(0,t) = 0, \quad u(l,t) = 0. \tag{5}$$

For the solution of this problem let us examine, as is customary in the method of the separation of variables, the subsidiary problem:

Find the solution of the equation

$$u_t = a^2 u_{xx},$$

not identically equal to zero, which satisfies the homogeneous boundary conditions

$$u(0, t) = 0, \quad u(l, t) = 0 \tag{5}$$

and which has the form

$$u(x, t) = X(x)\, T(t), \tag{6}$$

where $X(x)$ is a function only of the variable x, and $T(t)$ is a function only of the variable t.

Substituting the assumed form of the solution (6) in equation (4) and dividing both sides of the equality by a^2XT, we obtain:

$$\frac{1}{a^2}\frac{T'}{T} = \frac{X''}{X} = -\lambda, \tag{7}$$

where λ is a constant, since the left-hand side of the equality depends only on t and the right-side only on x.

Hence it follows that

$$X'' + \lambda X = 0, \tag{8}$$

$$T' + a^2 \lambda T = 0. \tag{8'}$$

The boundary conditions (5) give:

$$X(0) = 0, \quad X(l) = 0. \tag{9}$$

Thus, we obtain an eigen-value problem for the function $X(x)$

$$X'' + \lambda X = 0, \quad X(0) = 0, \quad X(l) = 0, \tag{10}$$

examined in the solution of the wave equation in Chapter II (see § 3, sub-section 1). It was shown that only for certain values of the parameter λ equal to

$$\lambda_n = \left(\frac{\pi n}{l}\right)^2 \quad (n = 1, 2, 3, \ldots), \tag{11}$$

do non-trivial solutions of equation (8) exist. These are

$$X_n(x) = \sin\frac{\pi n}{l} x. \tag{12}$$

The solution of equation (8′) corresponding to these values λ_n is

$$T_n(t) = C_n e^{-a^2\lambda_n t}, \tag{13}$$

where C_n are coefficients not determined for the present.

Returning to the fundamental subsidiary problem we see that the functions

$$u_n(x, t) = X_n(x) T_n(t) = C_n e^{-a^2\lambda_n t} \sin \frac{\pi n}{l} x, \tag{14}$$

are particular solutions of equation (4), satisfying zero boundary conditions.

We return now to the solution of problem (*I*). First we construct the formal series

$$u(x, t) = \sum_{n=1}^{\infty} C_n e^{-\left(\frac{\pi n}{l}\right)^2 a^2 t} \sin \frac{\pi n}{l} x. \tag{15}$$

The function $u(x, t)$ satisfies the boundary conditions because all terms of the series satisfy them. Since we require that the initial conditions be fulfilled, we obtain:

$$\varphi(x) = u(x, 0) = \sum_{n=1}^{\infty} C_n \sin \frac{\pi n}{l} x, \tag{16}$$

i.e. C_n are the Fourier coefficients of the function $\varphi(x)$ analysed in the interval $(0; l)$:

$$C_n = \varphi_n = \frac{2}{l}\int_0^l \varphi(\xi) \sin \frac{\pi n}{l} \xi \cdot d\xi. \tag{17}$$

We will show that the series (15) with coefficients C_n, defined by formula (17), satisfies all the conditions of problem (I). In order to do this, it is necessary to prove that the function $u(x, t)$, defined by series (15) is differentiable, satisfies the equation in the region $0 < x < l$, $t > 0$ and is continuous at points of the boundary of this region (at $t = 0$, $x = 0$, $x = l$).

Since equation (4) is linear, then by virtue of the principle of superposition the series, consisting of partial solutions, will also be a solution if it converges, and if it is possible to differentiate it twice with respect to x and once with respect to t (see lemma of Chapter II, § 3, sub-section 3). Let us show that at $t \geqslant \bar{t} > 0$ ($\bar{t}$ is any auxiliary number) the series for the derivatives

$$\sum_{n=1}^{\infty} \frac{\partial u_n}{\partial t} \quad \text{and} \quad \sum_{n=1}^{\infty} \frac{\partial^2 u_n}{\partial x^2}$$

converge uniformly. Actually

$$\left|\frac{\partial u_n}{\partial t}\right| = \left| - C_n \left(\frac{\pi}{l}\right)^2 a^2 n^2 e^{-\left(\frac{\pi n}{l}\right)^2 a^2 t} \sin \frac{\pi n}{l} x \right| <$$

$$< |C_n| \left(\frac{\pi}{l}\right)^2 \cdot a^2 n^2 e^{-\left(\frac{\pi n}{l}\right)^2 a^2 t} .$$

Furthermore, the function $\varphi(x)$ must satisfy additional requirements. Let us assume in the first place that $\varphi(x)$ is bounded, $|\varphi(x)| < M$; then

$$|C_n| = \left|\frac{2}{l}\right| \left| \int_0^l \varphi(\xi) \sin \frac{\pi n}{l} \xi \, d\xi \right| < 2M ,$$

whence it follows that

$$\left|\frac{\partial u_n}{\partial t}\right| < 2M \left(\frac{\pi}{l}\right)^2 a^2 n^2 e^{-\left(\frac{\pi n}{l}\right)^2 a^2 \bar{t}} \quad \text{where } t \geqslant \bar{t}$$

and similarly

$$\left|\frac{\partial^2 u_n}{\partial x^2}\right| < 2M \left(\frac{\pi}{l}\right)^2 n^2 e^{-\left(\frac{\pi n}{l}\right)^2 a^2 \bar{t}} \quad \text{where } t \geqslant \bar{t} .$$

In general

$$\left|\frac{\partial^{k+l} u_n}{\partial t^k \partial x^l}\right| < 2M \left(\frac{\pi}{l}\right)^{2k+l} \cdot n^{2k+l} \cdot a^{2k} \cdot e^{-\left(\frac{\pi n}{l}\right)^2 a^2 \bar{t}} \text{ where } t \geqslant \bar{t}.$$

Let us examine the convergence of the major series $\{a_n\}$, where

$$a_n = N n^q e^{-\left(\frac{\pi n}{l}\right)^2 a^2 \bar{t}} . \tag{15'}$$

According to D'Alembert's criterion, this series converges because

$$\lim_{n\to\infty} \left|\frac{a_{n+1}}{a_n}\right| = \lim_{n\to\infty} \frac{(n+1)^q}{n^q} \frac{e^{-\left(\frac{\pi}{l}\right)^2 a^2 (n^2 + 2n + 1)\bar{t}}}{e^{-\left(\frac{\pi}{l}\right)^2 a^2 n^2 \bar{t}}} =$$

$$= \lim_{n\to\infty} \left(1 + \frac{1}{n}\right)^q e^{-\left(\frac{\pi}{l}\right)^2 a^2 (2n+1)\bar{t}} = 0.$$

Hence it is possible to differentiate the series (15) any number of times in the region $t \geqslant \bar{t} > 0$. Further, applying the principle of superposition, we deduce that the function defined by this series satisfies equation (4). By virtue of the arbitrary nature of $\bar{t}$, this holds for all $t > 0$. In the same way it can be proved that for $t > 0$ the series (15) describes a continuous function.*

If the function $\varphi(x)$ is continuous and has a nearly-continuous derivative and satisfies the conditions $\varphi(0) = 0$ and $\varphi(l) = 0$, then the series

$$u(x, t) = \sum_{n=1}^{\infty} C_n e^{-\left(\frac{\pi n}{l}\right)^2 a^2 t} \sin \frac{\pi n}{l} x \tag{15}$$

defines a continuous function at $t \geqslant 0$.

Actually, from the inequality

$$|u_n(x, t)| < |C_n| \text{ where } t \geqslant 0,\ 0 \leqslant x \leqslant l)$$

the uniform convergence of series (15) at $t \geqslant 0$, $0 \leqslant x \leqslant l$ at once follows, if it is remembered that for a continuous function $\varphi(x)$ with a piecewise continuous derivative the series of moduli of Fourier coefficients converges if $\varphi(0) = \varphi(l) = 0$.**

Thus the problem of finding a solution of the main boundary-value problem for a homogeneous equation with zero boundary conditions and with a continuous piecewise differentiable initial condition is completely solved.

2. *The source function*

Let us transform the solution (15) substituting the integrals for C_n

$$u(x, t) = \sum_{n=1}^{\infty} C_n e^{-\left(\frac{\pi n}{l}\right)^2 a^2 t} \sin \frac{\pi n}{l} x =$$

$$= \sum_{n=1}^{\infty} \left[\frac{2}{l} \int_0^l \varphi(\xi) \sin \frac{\pi n}{l} \xi \, d\xi\right] \cdot e^{-\left(\frac{\pi n}{l}\right)^2 a^2 t} \sin \frac{\pi n}{l} x =$$

$$= \int_0^l \left[\frac{2}{l} \sum_{n=1}^{\infty} e^{-\left(\frac{\pi n}{l}\right)^2 a^2 t} \sin \frac{\pi n}{l} x \cdot \sin \frac{\pi n}{l} \xi\right] \varphi(\xi) \, d\xi.$$

* In the proof that series (15) satisfies the equation $u_t = a^2 u_{xx}$ at $t > 0$ only the restrictions on the Fourier coefficients C_n are used, which in particular would hold for any bounded $\varphi(x)$.

** See Chapter II, § 3, sub-section 3.

The rearrangement of the order of summation and integration is always valid at $t > 0$, because the series in brackets converges uniformly in ξ for $t > 0$.***

Let us define

$$G(x, \xi, t) = \frac{2}{l} \sum_{n=1}^{\infty} e^{-\left(\frac{\pi n}{l}\right)^2 a^2 t} \sin \frac{\pi n}{l} x \cdot \sin \frac{\pi n}{l} \xi . \tag{18}$$

Using the function $G(x, \xi, t)$, it is possible to represent the solution $u(x, t)$ as

$$u(x, t) = \int_0^l G(x, \xi, t) \varphi(\xi) d\xi . \tag{19}$$

The function $G(x, \xi, t)$ is called *an instantaneous point source function*, or in more detail the function of the temperature effect of an instantaneous point source of heat. (It is often called Green's function.)

We shall show that the source function $G(x, \xi, t)$, considered as a function of x, represents the distribution of temperature in the rod $0 \leqslant x \leqslant l$ at the time t if the temperature at the initial time $t = 0$ equals zero and if at this time a certain amount of heat is instantaneously liberated at the point $x = \xi$ (the exact amount will be calculated later) and the ends of the rod are maintained at zero temperature.

The expression "the amount of heat Q liberated at the point ξ" means, as usual, that we are concerned with the heat liberated in "a small" interval around the point ξ under consideration. The change of temperature $\varphi_\varepsilon(\xi)$ produced by the liberation of heat around the point is equal to zero outside the interval $(\xi - \varepsilon, \xi + \varepsilon)$ in which the heat is liberated, and inside this interval $\varphi_\varepsilon(\xi)$ is assumed to be a positive, continuous and differentiable function for which

$$c\varrho \int_{\xi-\varepsilon}^{\xi+\varepsilon} \varphi_\varepsilon(\xi) d\xi = Q , \tag{20}$$

The left-hand side of this equation represents the amount of heat producing a change of temperature $\varphi_\varepsilon(\xi)$. The

*** The series Σa_n, where a_n is defined by formula (15′) for $q = 0$ is the maior of the series in brackets.

process of distribution of temperature is defined by formula (19):

$$u_\varepsilon(x, t) = \int_0^l G(x, \xi, t)\, \varphi_\varepsilon(\xi)\, d\xi . \tag{21}$$

Let us perform now a limiting transition for $\varepsilon \to 0$. Taking into consideration the continuity of G for $t > 0$, the equality (20) and assuming the theorem of mean value for the fixed values x, t we have:

$$u_\varepsilon(x, t) = \int_{\xi-\varepsilon}^{\xi+\varepsilon} G(x, \xi, t)\, \varphi_\varepsilon(\xi)\, d\xi =$$

$$= G(x, \xi^*, t) \int_{\xi-\varepsilon}^{\xi+\varepsilon} \varphi_\varepsilon(\xi)\, d\xi = G(x, \xi^*, t) \frac{Q}{c\varrho} , \tag{21'}$$

where ξ^* is a certain mean point of the interval $(\xi - \varepsilon, \xi + \varepsilon)$. Because of the continuity of the function $G(x, \xi, t)$ at ξ with $t > 0$ we obtain:

$$\lim_{\varepsilon \to 0} u_\varepsilon(x, t) = \frac{Q}{c\varrho} G(x, \xi, t) =$$

$$= \frac{Q}{c\varrho} \cdot \frac{2}{l} \sum_{n=1}^{\infty} e^{-\left(\frac{\pi n}{l}\right)^2 a^2 t} \sin \frac{\pi n}{l} x \cdot \sin \frac{\pi n}{l} \xi . \tag{22}$$

Hence it follows that $G(x, \xi, t)$ represents the temperature effect of an instantaneous point source of output $Q = c\varrho$, located at time $t = 0$ at the point ξ of the interval $(0, l)$.

Let us note the following property of the function $G(x, \xi, t)$: *the function* $G(x, \xi, t) \geqslant 0$ *for any* $x, \xi, t > 0$. We examine the original function $\varphi_\varepsilon(x)$ which possesses the properties described above, and the solution (21′) corresponding to it. From the positive nature of the initial and boundary conditions and from the maximum value principle it follows that

$$u_\varepsilon(x, t) \geqslant 0$$

for all $0 \leqslant x \leqslant t$ and $t > 0$. Hence, using formula (21′) we have:

$$u_\varepsilon(x, t) = G(x, \xi^*, t) \frac{Q}{c\varrho} \geqslant 0 \text{ (where } t > 0) . \tag{21''}$$

Proceeding to the limit as $\varepsilon \to 0$, from (21′) we obtain the inequality

$$G(x, \xi, t) \geqslant 0 \text{ where } 0 \leqslant x, \xi \leqslant l \text{ and } t > 0,$$

which it was required to prove.

This result has a simple physical meaning. But to determine it directly from formula (19) would be difficult, because $G(x, \xi, t)$ is represented by an infinite series.

3. *Boundary-value problems with discontinuous initial conditions*

The theorem stated above refers to the solutions of the equation of heat conduction which are continuous in the closed region $0 \leqslant x \leqslant l$, $0 \leqslant t \leqslant T$. These conditions of continuity are very stringent. Let us examine the simple problem of the uniform cooling of a heated rod with the ends at zero temperature. Additional conditions have the form

$$u(x, 0) = u_0, \quad u(0, t) = u(l, t) = 0.$$

If $u_0 \neq 0$ then the solution of this problem must be discontinuous at the points $(0, 0)$ and $(l, 0)$. This example shows that the conditions outlined above of the continuity of the initial value and the conditions of the junction of it with the boundary values exclude significant useful cases from consideration. Nevertheless formula (19) also gives the solution of the boundary-value problem in this case.

Thus, if we wish to make use of the results of the theory, beyond the apparent limits of its applicability, then we must generalize the theory stated above in order to cover the main problems. In applications one often makes use of formulae of limited applicability without specifying the limitations. A consistent basis for all formulae would be very cumbersome and often it would distract the attention of the investigator from the quantitative and qualitative aspects of the physical nature of the process.

Nevertheless we consider it necessary, at least for the simplest examples, to give the basis of a mathematical system adequate for the solution of the main problems.

Let us examine the boundary-value problems with partly-continuous initial values, not assuming that the initial function is connected with boundary conditions. This class of additional conditions is sufficiently common for practical needs and sufficiently simple for the statement of the theory. Our object is to prove that the same formula (19) gives a solution of this problem. We carry out the investigation in several stages, first proving a preliminary theorem.

The solution of the equation of heat conduction

$$u_t = a^2 u_{xx} \, (0 < x < l, \ t > 0), \tag{4}$$

continuous in the closed interval $(0 \leqslant x \leqslant l, \ 0 \leqslant t \leqslant T)$ *and satisfying the conditions*

$$u(0, t) = u(l, t) = 0, \tag{5}$$

$$u(x, 0) = \varphi(x), \tag{2}$$

where $\varphi(x)$ *is an arbitrary continuous function reducing to zero at* $x = 0$, $x = l$, *is uniquely defined and is represented by the formula*

$$u(x, t) = \int_0^l G(x, \xi, t) \, \varphi(\xi) \, d\xi. \tag{19}$$

This theorem was proved above with the additional assumption that $\varphi(\xi)$ has a partly-continuous differential coefficient.

Let us remove this condition. We examine the sequence of functions with partly-continuous differential coefficients $\varphi_n(\xi)$ $(\varphi_n(0) = \varphi_n(l) = 0)$, converging uniformly to $\varphi(x)$. (By the nature of $\varphi_n(x)$ it is possible to select, for example, functions which represent broken lines, coinciding with $\varphi(x)$ at the points $l \cdot k/n, k = 0, 1, 2, \ldots, n$.) The functions $u_n(x, t)$ defined by formula (19) from $\varphi_n(x)$ satisfy the condition of partly-continuous differentiability. These functions converge uniformly and describe in the limit the continuous function $u(x, t)$. Actually, for any ε it is possible to find $n(\varepsilon)$ such that

$$|\varphi_{n_1}(x) - \varphi_{n_2}(x)| < \varepsilon \quad (0 \leqslant x \leqslant l), \quad \text{if} \quad n_1, n_2 \geqslant n(\varepsilon),$$

because these functions by definition converge uniformly. Hence by the maximum value principle it follows also that

$$|u_{n_1}(x, t) - u_{n_2} x, t)| < \varepsilon\,(0 \leqslant x \leqslant l,\ 0 \leqslant t \leqslant T), \text{ if } n_1, n_2 \geqslant n(\varepsilon),$$

and proves the uniform convergence of the series of functions $u_n(x, t)$ to a certain continuous function $u(x, t)$.

If, having fixed the point (x, t), we proceed to the limit under the integral sign, then we find that the function

$$u(x, t) = \lim_{n\to\infty} u_n(x, t) =$$

$$= \lim_{n\to\infty} \int_0^l G(x, \xi, t)\,\varphi_n(\xi)\,d\xi = \int_0^l G(x, \xi, t)\,\varphi(\xi)\,d\xi$$

is continuous in the closed interval $0 \leqslant x \leqslant l$, $0 \leqslant t \leqslant T$ and satisfies conditions (2). By virtue of the first footnote on p. 217, it is easily seen that it also satisfies equation (4). The proof of the theorem is completed.

Formula (19) gives a unique continuous solution of the problem under consideration.

Let us turn to the proof of *the uniqueness theorem for the case of a partly continuous initial function* $\varphi(x)$, with zero boundary conditions. We prove that *the function, continuous in the region* $t > 0$, *satisfying the equation of heat conduction,*

$$u_t = a^2 u_{xx} \tag{4}$$

in the region $0 < x < l$, $t > 0$ *with zero boundary conditions*

$$u(0, t) = u(l, t) = 0 \tag{5}$$

and initial condition

$$u(x, 0) = \varphi(x), \tag{2}$$

is defined identically if

(1) *it is continuous at the points of continuity of the function* $\varphi(x)$,

(2) *it is bounded in the closed region* $0 \leqslant x \leqslant l$, $0 \leqslant \bar{t} \leqslant \bar{t}_0$, *where* $\bar{t}_0$ is *an arbitrary positive number.*

Let us assume that such a function exists. It is obvious from the preceding theorem that it may be described in the range $t > \bar{t}$ by the formula

$$u(x, t) = \int_l^0 G(x, \xi, t - \bar{t})\,\varphi_{\bar{t}}(\xi)\,d\xi \quad (t > \bar{t} > 0) \tag{19'}$$

for any $\bar{t} > 0$ where

$$0 < \bar{t} \leqslant t, \quad \varphi_{\bar{t}}(x) = u(x, \bar{t}).$$

Let us perform a limiting transition in this formula as $\bar{t} \to 0$, keeping x and t constant. We show that* the transition is possible under the integral sign and therefore the function $u(x, t)$ may be represented in the form of the integral

$$u(x, t) = \int_0^l G(x, \xi, t)\, \varphi(\xi)\, d\xi \qquad [\varphi(\xi) = u(\xi, 0)], \tag{19}$$

which is well defined.

Let $x_1, x_2, \ldots, x_n$ be points of discontinuity of the function $\varphi(x)$. Assuming $x_0 = 0$ and $x_{n+1} = l$ (Fig. 38) and assuming the closed interval $I_k(x_k + \delta \leqslant x \leqslant x_{k+1} - \delta)$ $(k = 0, 1, \ldots, n)$, where δ is a certain fixed sufficiently small number it is easily seen that the function under the integral sign of (19′) converges

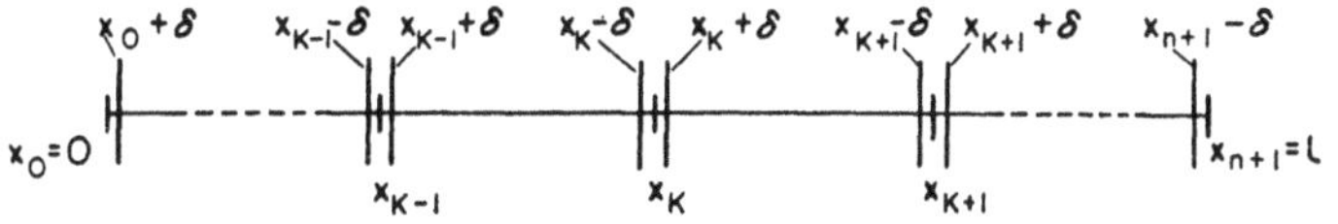

FIG. 38

uniformly to the function under the integral sign of (19) in every interval $I_k (k = 0, 1, 2, \ldots, n)$. In the intervals $\bar{I}_k(x_k - \delta \leqslant x \leqslant x_k + \delta)$ $(k = 1, 2, 3, \ldots n)$, $\bar{I}_0(x_0 \leqslant x \leqslant x_0 + \delta)$, and $\bar{I}_{n+1}(x_{n+1} - \delta \leqslant x \leqslant x_{n+1})$ the expressions under the integral sign in (19) and (19′) are bounded by a certain number N for any $\bar{t}(0 \leqslant \bar{t} \leqslant t_0)$ because of an assumed boundedness of the function $u(x, t)$ and because of the continuity of $G(x, \xi, t)$ at $0 \leqslant \xi \leqslant l$, $t > 0$. Dividing the difference of the integrals (19) and (19′) into $2n + 3$ integrals, given by $I_k (k = 0, 1, \ldots, n)$ and $\bar{I}_k (k = 0, 1, \ldots, n + 1)$, we see that this difference may be made smaller than a given number ε, if

$$\delta \leqslant \frac{\varepsilon}{2n + 3} \, \frac{1}{4N},$$

* The theorem proved below is a particular case of Lebesgue's theorem concerning the possibility of transition to a limit under the integral sign, if a series of functions $F_n(x)$ converges almost everywhere to a limiting integrable function $F(x)$ and if this series is bounded by the integrable function. This proof is quoted in order to avoid introducing many concepts. If use is made of these ideas then by complete analogy it is possible to prove the theorem that the solution of the equation of heat conduction $u(x, t)$, which satisfies the zero boundary conditions, is defined
(1) if $u(x, t) \leqslant F(x)$ where $F(x)$ is a certain integrable function and
(2) if almost everywhere

$$\lim_{t \to 0} u(x, t) = \Phi(x)$$

where $\varphi(x)$ is a given simple integrable function.

so that

$$\left| \int\limits_{\bar{I}_k} [G(x, \xi, t-\bar{t})\, \varphi_{\bar{t}}(\xi) - G(x, \xi, t)\, \varphi(\xi)]\, d\xi \right| \leqslant \frac{\varepsilon}{2n+3},$$

and if t is chosen so small that

$$| G(x, \xi, t-\bar{t})\, \varphi_{\bar{t}}(\xi) - G(x, \xi, t)\, \varphi(\xi) | <$$

$$< \frac{1}{l}\, \frac{\varepsilon}{2n+3} \text{ where } t \leqslant \bar{t} \text{ on } I_k \; (k = 0, 1, \ldots, n),$$

so that

$$\int\limits_{I_k} [G(x, \xi, t-\bar{t})\, \varphi_{\bar{t}}(\xi) - G(x, \xi, t)\, \varphi(\xi)]\, d\xi <$$

$$< \frac{\varepsilon}{2n+3} \text{ where } t \leqslant \bar{t} \; (k = 0, 1, \ldots, n).$$

Hence the inequality follows

$$\left| \int\limits_0^l [G(x, \xi, t-\bar{t})\, \varphi_{\bar{t}}(\xi) - G(x, \xi, t)\, \varphi(\xi)]\, d\xi \right| < \varepsilon \text{ where } t \leqslant \bar{t},$$

proving the validity of the transition to a limit under the integral sign. Thus, if a function $u(x, t)$ exists, satisfying the conditions of the theorem, then it is represented by the form (19), proving the uniqueness of such a function.

Let us prove now that *formula (19) represents a solution of equation (4) satisfying conditions (2) for any partly-continuous function $\varphi(x)$ and is continuous at all points of the continuity of $\varphi(x)$.*

We prove this theorem by two methods. We prove that it is true if $\varphi(x)$ is a linear function

$$\varphi(x) = cx.$$

Let us examine the series of auxiliary functions (Fig. 39)

$$\varphi_n(x) = \begin{cases} cx \text{ where } 0 \leqslant x \leqslant l\left(1 - \frac{1}{n}\right), \\ a(l - x) \text{ where } l\left(1 - \frac{1}{n}\right) \leqslant x \leqslant l, \end{cases}$$

where a is determined from the condition of continuity of the function $\varphi_n(x)$ at the point $x = l(1 - 1/n)$

$$cl\frac{n-1}{n} = a\frac{l}{n}, \text{ i.e. } a = (n-1)c.$$

The functions $u_n(x, t)$, determined with the help of formula (19) for $\varphi_n(x)$, are continuous solutions of the equation of heat conduction with zero boundary conditions and initial conditions

$$u_n(x, 0) = \varphi_n(x).$$

Because

$$\varphi_n(x) \leqslant \varphi_{n+1}(x) \quad (0 \leqslant x \leqslant l),$$

then by the maximum value principle

$$u_n(x, t) \leqslant u_{n+1}(x, t).$$

The function $U_0(x) = cx$ is a continuous solution of the equation of heat conduction.
By virtue of the maximum value principle

$$u_n(x, t) \leqslant U_0(x),$$

because this inequality holds at $x = 0$, $x = l$ and $t = 0$. Thus $u_n(x, t)$ is a monotonic increasing series bounded above by the

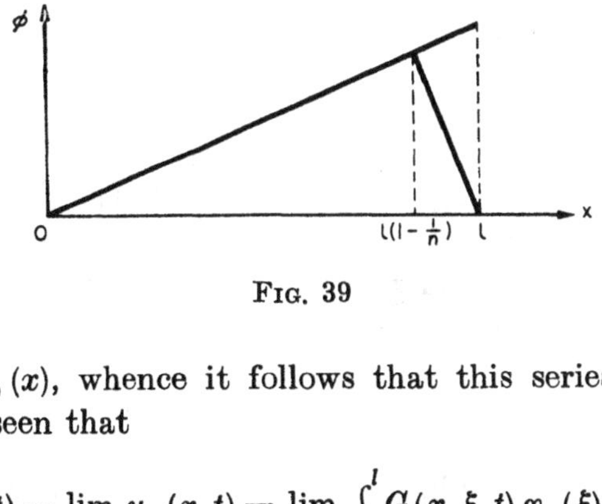

FIG. 39

function $U_0(x)$, whence it follows that this series converges. It is easily seen that

$$u(x, t) = \lim_{n\to\infty} u_n(x, t) = \lim_{n\to\infty} \int_0^l G(x, \xi, t)\, \varphi_n(\xi)\, d\xi =$$

$$= \int_0^l G(x, \xi, t)\, \varphi(\xi)\, d\xi \leqslant U_0(x),$$

since the transition to a limit under the integral sign is valid. By virtue of the first footnote on p. 217 this function satisfies the equation and the zero boundary conditions at $t > 0$. Let us show that this function is continuous at $t = 0$ for $0 \leqslant x \leqslant l$.

Let $x_0 < l$. Let us choose n so that $x_0 < l\left(1 - \frac{1}{n}\right)$. In this case $\varphi_n(x_0) = U_0(x_0)$. In view of this

$$u_n(x, t) \leqslant u(x, t) \leqslant U_0(x)$$

and

$$\lim_{\substack{x \to x_0 \\ t \to 0}} u_n(x, t) = \lim_{x \to x_0} U_0(x) = \varphi(x_0),$$

we conclude that there exists a limit

$$\lim_{\substack{x \to x_0 \\ t \to 0}} u(x, t) = \varphi(x_0),$$

not depending on the manner in which $x \to x_0$ and $t \to 0$. Hence the continuity of $u(x, t)$ at the point $(x_0, 0)$ follows. This function is bounded because it does not exceed $U_0(x)$. Thus for $\varphi(x) = cx$, the theorem is proved.

By the replacement of x with $l - x$ we see that the theorem is true for

$$\varphi(x) = b(l - x). \tag{2''}$$

Hence it follows that it is true for any function of the type

$$\varphi(x) = B + Ax,$$

because a similar function may be obtained by the addition of (2′) and (2″). Further, it follows also that the theorem is correct for any continuous function without the assumption that $\varphi(0) = \varphi(l) = 0$. Actually, it is possible to represent any function $\varphi(x)$ of such a type in the form

$$\varphi(x) = \left[\varphi(0) + \frac{x}{l}(\varphi(l) - \varphi(0))\right] + \psi(x),$$

where the component in the square brackets is a linear function, and $\psi(x)$ is a continuous function, reducing to zero at the ends of the section: $\psi(0) = \psi(l) = 0$. Since we are already satisfied that for each component the theorem is applicable, hence it follows that the theorem is correct for $\varphi(x)$.

Let us turn now to a proof of the theorem for an arbitrary partly-continuous function $\varphi(x)$. Formula (19) in this case defines a solution satisfying the equation and the zero boundary conditions.

Let the point x_0 be any point of the continuity of the function $\varphi(x)$. Let us show that for any ε it is possible to find $\delta(\varepsilon)$ such that $|u(x, t) - \varphi(x_0)| < \varepsilon$, if $|x - x_0| < \delta(\varepsilon)$ and $t < \delta(\varepsilon)$. By virtue of the continuity of the function $\varphi(x)$ at the point x_0 there exists an $\eta(\varepsilon)$ such that

$$|\varphi(x) - \varphi(x_0)| \leqslant \frac{\varepsilon}{2} \text{ where } |x - x_0| < \eta(\varepsilon),$$

whence

$$\varphi(x_0) - \frac{\varepsilon}{2} \leqslant \varphi(x) \leqslant \varphi(x_0) + \frac{\varepsilon}{2} \text{ where } |x - x_0| < \eta(\varepsilon). \quad (23)$$

Let us construct subsidiary continuous differentiable functions $\overline{\varphi}(x)$ and $\underline{\varphi}(x)$:

$$\left.\begin{aligned} \overline{\varphi}(x) &= \varphi(x_0) + \frac{\varepsilon}{2} \text{ where } |x - x_0| < \eta(\varepsilon), \\ \overline{\varphi}(x) &\geqslant \varphi(x) \text{ where } |x - x_0| > \eta(\varepsilon), \end{aligned}\right. \quad \text{(a)}$$

$$\left.\begin{aligned} \underline{\varphi}(x) &= \varphi(x_0) - \frac{\varepsilon}{2} \text{ where } |x - x_0| < \eta(\varepsilon), \\ \underline{\varphi}(x) &\leqslant \varphi(x) \text{ where } |x - x_0| > \eta(\varepsilon). \end{aligned}\right. \quad \text{(b)}$$

In the interval $|x - x_0| < \eta(\varepsilon)$ the functions φ and φ satisfy only the conditions (a) and (b) and in other respects are arbitrary. Because of the inequalities (23)

$$\underline{\varphi}(x) \leqslant \varphi(x) \leqslant \overline{\varphi}(x). \quad (24)$$

Let us examine the functions

$$\bar{u}(x, t) = \int_0^l G(x, \xi, t)\, \overline{\varphi}(\xi)\, d\xi,$$

$$\underline{u}(x, t) = \int_0^l G(x, \xi, t)\, \underline{\varphi}(\xi)\, d\xi.$$

Because of the continuity of the functions $\overline{\varphi}(x)$ and $\underline{\varphi}(x)$ the functions $\bar{u}(x, t)$ and $\underline{u}(x, t)$ are continuous at the point x_0, i.e. a $\delta(\varepsilon)$ can be found such that

$$\left.\begin{aligned} |\bar{u}(x, t) - \overline{\varphi}(x)| &\leqslant \frac{\varepsilon}{2}, \\ |\underline{u}(x, t) - \underline{\varphi}(x)| &\leqslant \frac{\varepsilon}{2} \end{aligned}\right\} \text{ where } |x - x_0| < \delta(\varepsilon), \quad t < \delta(\varepsilon),$$

whence

$$\left.\begin{aligned}\bar{u}(x,t) &\leqslant \bar{\varphi}(x) + \frac{\varepsilon}{2} = \varphi(x_0) + \varepsilon, \\ \underline{u}(x,t) &\geqslant \underline{\varphi}(x) - \frac{\varepsilon}{2} = \varphi(x_0) - \varepsilon,\end{aligned}\right\} \text{ where } |x - x_0| < \delta(\varepsilon), \; t < \delta(\varepsilon).$$

Because function $G(x, \xi, t)$ is positive from formula (24) it follows that

$$\underline{u}(x,t) \leqslant u(x,t) \leqslant \bar{u}(x,t). \tag{25}$$

Hence we obtain the inequality

$$\varphi(x_0) - \varepsilon \leqslant u(x,t) \leqslant \varphi(x_0) + \varepsilon \text{ where } |x - x_0| < \delta(\varepsilon), t < \delta(\varepsilon)$$

or

$$|u(x,t) - \varphi(x_0)| < \varepsilon \text{ where } |x - x_0| < \delta(\varepsilon), \; t < \delta(\varepsilon),$$

which it was required to prove. The boundedness of the function $|u(x, t)|$ follows from (25) and from the boundedness of the functions $\bar{u}(x, t)$ and $\underline{u}(x, t)$. Hence the theorem is proved.

4. *The inhomogeneous equation of heat conduction*

Let us examine the inhomogeneous equation of heat conduction

$$u_t = a^2 u_{xx} + f(x,t) \tag{1}$$

with initial condition

$$u(x, 0) = 0 \tag{26}$$

and boundary conditions

$$\begin{aligned} u(0,t) &= 0, \\ u(l,t) &= 0. \end{aligned} \tag{5}$$

We shall search for a solution $u(x, t)$ as a Fourier series in $\sin(\pi n/l)\, x$

$$u(x,t) = \sum_{n=1}^{\infty} u_n(t) \sin \frac{\pi n}{l} x. \tag{27}$$

In order to find the function $u(x, t)$ it is necessary to find the function $u_n(t)$. Let us represent the function $f(x, t)$ as a series

$$f(x,t) = \sum_{n=1}^{\infty} f_n(t) \sin \frac{\pi n}{l} x,$$

where

$$f_n(t) = \frac{2}{l} \int_0^l f(\xi, t) \sin \frac{\pi n}{l} \xi \, d\xi . \tag{28}$$

Substituting the assumed form of the solution in the original equation (1) we have:

$$\sum_{n=1}^{\infty} \sin \frac{\pi n}{l} x \left\{ \left(\frac{\pi n}{l} \right)^2 a^2 u_n(t) + \dot{u}_n(t) - f_n(t) \right\} = 0 .$$

This equation will be satisfied if all the coefficients in the bracket equal zero, i. e.

$$\dot{u}_n(t) = - a^2 \left(\frac{\pi n}{l} \right)^2 u_n(t) + f_n(t) . \tag{29}$$

Making use of the initial condition for $u(x, t)$

$$u(x, 0) = \sum_{n=1}^{\infty} u_n(0) \sin \frac{\pi n}{l} x = 0 ,$$

we derive the initial condition for $u_n(t)$:

$$u_n(0) = 0 . \tag{30}$$

Solving the ordinary differential equation (29) with the zero boundary condition (30),* we find:

$$u_n(t) = \int_0^t e^{-\left(\frac{\pi n}{l}\right)^2 a^2 (t-\tau)} f_n(\tau) \, d\tau . \tag{31}$$

Substituting expression (31) for $u_n(t)$ in formula (27) we obtain the solution of the original problem in the form

$$u(x, t) = \sum_{n=1}^{\infty} \left[\int_0^t e^{-\left(\frac{\pi n}{l}\right)^2 a^2 (t-\tau)} f_n(\tau) \, d\tau \right] \sin \frac{\pi n}{l} x . \tag{32}$$

Using expression (28) for $f_n(\tau)$ let us transform the solution (32):

$$u(x, t) = \int_0^t \int_0^l \left\{ \frac{2}{l} \sum_{n=1}^{\infty} e^{-\left(\frac{\pi n}{l}\right)^2 a^2 (t-\tau)} \sin \frac{\pi n}{l} x \cdot \sin \frac{\pi n}{l} \xi \right\} f(\xi, \tau) \, d\xi \, d\tau =$$

$$= \int_0^t \int_0^l G(x, \xi, t - \tau) f(\xi, \tau) \, d\xi \, d\tau , \tag{33}$$

* See the footnote on p. 99.

where

$$G(x, \xi, t-\tau) = \frac{2}{l} \sum_{n=1}^{\infty} e^{-\left(\frac{\pi n}{l}\right)^2 a^2(t-\tau)} \sin \frac{\pi n}{l} x \cdot \sin \frac{\pi n}{l} \xi \qquad (34)$$

coincides with the source function defined by formula (18).

Let us consider the physical significance of this solution

$$u(x, t) = \int_0^t \int_0^l G(x, \xi, t-\tau) f(\xi, \tau) \, d\xi \, d\tau . \qquad (33)$$

We assume that the function $f(\xi, \tau)$ differs from zero in a sufficiently small neighbourhood round the point $M_0 (\xi_0, \tau_0)$

$$\xi_0 \leqslant \xi \leqslant \xi_0 + \Delta\xi, \quad \tau_0 \leqslant \tau \leqslant \tau_0 + \Delta\tau_0 .$$

The function $F(\xi, \tau) = c \varrho f(\xi, \tau)$ represents the density of the heat sources. The total amount of heat, liberated in the section $(0, l)$ equals

$$Q = \int_{\tau_0}^{\tau_0+\Delta\tau} \int_{\xi_0}^{\xi_0+\Delta\xi} c \varrho f(\xi, \tau) \, d\xi \, d\tau . \qquad (35)$$

Let us apply the mean value theorem to the expression

$$u(x, t) = \int_0^t \int_0^l G(x, \xi, t-\tau) f(\xi, \tau) \, d\xi \, d\tau =$$

$$= \int_{\tau_0}^{\tau_0+\Delta\tau} \int_{\xi_0}^{\xi_0+\Delta\xi} G(x, \xi, t-\tau) f(\xi, \tau) \, d\xi \, d\tau =$$

$$= G(x, \bar{\xi}, t-\bar{\tau}) \cdot \frac{Q}{c\varrho} = \bar{u}(x, t) ,$$

where

$$\xi_0 < \bar{\xi} < \xi_0 + \Delta\xi, \quad \tau_0 < \bar{\tau} < \tau_0 + \Delta\tau .$$

Proceeding to the limit as $\Delta\xi \to 0$ and $\Delta\tau \to 0$, we obtain the function

$$u(x, t) = \lim_{\substack{\Delta\xi \to 0 \\ \Delta\tau \to 0}} \bar{u}(x, t) = \frac{Q}{c\varrho} G(x, \xi_0, t-\tau_0) \qquad (36)$$

which may be interpreted as a function giving the effect of an instantaneous source of heat located at the point ξ_0 at time τ_0.

If the function $Q/c\varrho \, G(x, \xi, t-\tau)$, which represents the effect of a single instantaneous centred source, is known, then it is possible to express the effect of the sources distributed continuously with density $F(x, t) = c\varrho f(x, t)$, by formula (33).

Thus, the temperature effect of the heat sources acting in the region $(\xi_0, \xi_0 + \Delta\xi)$ $(\tau_0, \tau_0 + \Delta\tau)$, is given by the expression

$$G(x, \xi, t - \tau) f(\xi, \tau) \Delta\xi \Delta\tau \quad \left(\frac{Q}{c\varrho} = f(\xi, \tau) \Delta\xi \Delta\tau\right).$$

If the sources are distributed continuously, then, summing the effects of the sources acting over the entire region $0 \leqslant \xi \leqslant l$, $0 \leqslant \tau \leqslant t$, we obtain after a limiting transition as $\Delta\xi \to 0$ and $\Delta\tau \to 0$

$$u(x, t) = \int_0^t \int_0^l G(x, \xi, t - \tau) f(\xi, \tau)\, d\xi\, d\tau.$$

Thus, having understood the physical significance of the source function $G(x, \xi, t)$, it is possible at once to write down expression (33) for the function which gives the solution of the inhomogeneous equation.

Knowing the formal solution of the problem it is possible to investigate the conditions of applicability of this formula. We will not carry out this investigation.

We have considered the non-homogeneous equation with zero initial conditions. If the initial condition differs from zero, then the solution of the homogeneous equation with the given initial condition $u(x, 0) = \varphi(x)$, found in sub-section 1 should be added to this solution.

5. *The first general boundary-value problem*

Let us examine the first general boundary-value problem for the equation of heat conduction:

Find the solution of the equation

$$u_t = a^2 u_{xx} + f(x, t) \tag{1}$$

with additional conditions

$$u(x, 0) = \varphi(x), \tag{2}$$

$$\left.\begin{aligned} u(0, t) &= \mu_1(t), \\ u(l, t) &= \mu_2(t). \end{aligned}\right\} \tag{3}$$

Let us introduce a new unknown function $v(x, t)$

$$u(x, t) = U(x, t) + v(x, t), \tag{37}$$

representing the deviation from a certain known function $U(x, t)$.

This function $v(x, t)$ will be defined as the solution of the equation

$$v_t - a^2 v_{xx} = \bar{f}(x, t),$$

$$\bar{f}(x, t) = f(x, t) - [U_t - a^2 U_{xx}]$$

with additional conditions

$$v(x, 0) = \bar{\varphi}(x), \quad \bar{\varphi}(x) = \varphi(x) - U(x, 0),$$

$$v(0, t) = \bar{\mu}_1(t), \quad \bar{\mu}_1(t) = \mu_1(t) - U(0, t),$$

$$v(l, t) = \bar{\mu}_2(t), \quad \bar{\mu}_2(t) = \mu_2(t) - U(l, t).$$

Let us choose the auxiliary function $U(x, t)$ so that

$$\bar{\mu}_1(t) = 0 \text{ and } \bar{\mu}_2(t) = 0,$$

for example*

$$U(x, t) = \mu_1(t) + \frac{x}{l}\left[\mu_2(t) - \mu_1(t)\right].$$

Thus, the determination of the function $u(x, t)$, which gives the solution of the general boundary-value problem, reduces to a determination of the solution $v(x, t)$ of the boundary-value problem with zero boundary conditions. The method of finding the function $v(x, t)$ is given in sub-section 4.

This formal system for solving problems with inhomogeneities in the equation and boundary conditions is not always the most suitable. Difficulties arise in determining the auxiliary function $v(x, t)$.

In particular, for problems with fixed inhomogeneities it is best to find a time independent solution and the deviation from this solution.**

* See Chapter II, § 3, sub-section 5.
** See Chapter II, § 3, sub-section 6.

Let us consider, for instance, the problem of a finite rod $(0, l)$, whose ends are maintained at the constant temperatures u_0 and u_1:

$$u_t = a^2 u_{xx},$$
$$u(x, 0) = \varphi(x),$$
$$u(0, t) = u_0,$$
$$u(l, t) = u_1.$$

We shall look for a solution in the form of a sum

$$u(x, t) = \bar{u}(x) + v(x, t),$$

where $\bar{u}(x)$ is time independent, and $v(x, t)$ is the deviation from this temperature.

The functions $\bar{u}(x)$ and $v(x, t)$ satisfy the conditions

$$\begin{aligned} \bar{u}'' &= 0, & v_t &= a^2 v_{xx}; \\ \bar{u}(0) &= u_0, & v(x, 0) &= \varphi(x) - \bar{u}(x) = \varphi_1(x); \\ \bar{u}(l) &= u_1, & v(0, t) &= 0. \\ & & v(l, t) &= 0. \end{aligned}$$

Hence we find:

$$\bar{u}(x) = u_0 + \frac{x}{l}(u_1 - u_0).$$

The function $v(x, t)$, satisfying an initial condition and homogeneous boundary conditions, may easily be found by the method of separation of variables.

Problems

1. Derive an equation for the process of heating a homogeneous thin wire with a constant electric current, if there is heat exchange between its surface and the surrounding medium.

Answer : $u_t = a^2 u_{xx} - hu + q$, where h and q are certain constants.

2. Derive the equation of diffusion in a medium, moving uniformly in the direction of the x-axis with speed w. Examine the case of a single independent variable.

Answer : $u_t = Du_{xx} - wu_x$ (D the coefficient of diffusion).

3. Starting from Maxwell's equation, assuming $E_x = E_z = 0$, $H_z = 0$ and neglecting displacement currents, prove that in a homogeneous conducting medium the component of the electromagnetic field E_y satisfies the equation

$$\frac{\partial^2 E_y}{\partial z^2} = \frac{4\pi\sigma}{c^2} \frac{\partial E_y}{\partial t},$$

where σ is the conductivity of the medium, c the speed of light. Derive the equation for H_x.

4. Give a physical interpretation of the following boundary conditions in problems of heat conduction and diffusion:

$$\text{a) } u(0,t)=0, \quad \text{b) } u_x(0,t)=0, \quad \text{c) } u_x(0,t)-hu(0,t)=0, \quad (h>0).$$
$$u_x(l,t)+hu(l,t)=0$$

5. Solve the problem of the cooling of a homogeneous rod with its ends at zero temperature, assuming the absence of heat exchange at the surface.

Answer :

$$u(x,t)= \\ =\frac{4U_0}{\pi}\sum_{n=1}^{\infty}\frac{e^{-\frac{a^2(2k-1)^2\pi^2}{l^2}}}{(2k-1)}\sin\frac{(2k-1)\pi}{l}x \quad [u(x,0)=U_0].$$

6. The initial temperature of a rod is $u(x,0)=u_0=\text{const.}$ for $0<x<l$. The temperature of the ends is kept constant $u(0,t)=u_1$; $u(l,t)=u_2$ for $0<t<\infty$. Find the temperature of the rod if heat exchange at the surface is absent. Find the steady state temperature.

7. Solve problem 6 for the following boundary conditions: one end is kept at a constant temperature, the other end is thermally insulated.

8. Solve problem 1 for the heating of a thin homogeneous wire with a constant electric current, if the initial temperature, boundary temperature, and also the temperature of the surrounding medium equal zero.

9. A cylinder of length l is filled with air at the pressure and temperature of the surrounding medium, and has one end open at $t=0$. A gas diffuses into the cylinder from the surrounding atmosphere, where its concentration equals u_0. Find the quantity of gas diffusing into the cylinder in a time t, if the initial concentration of the gas in the cylinder equals zero.

10. Solve problem 9 on the assumption that the left end of the cylinder is enclosed by a semi-permeable membrane.

11. Solve the problem of the cooling of a homogeneous rod thermally insulated along the surface, if its initial temperature $u(x,0)=\varphi(x)$, and a heat exchange occurs at the ends with a medium of zero temperature. Examine the special case $\varphi(x)=u_0$.

12. Solve problem 11 assuming that the temperature of the surrounding medium equals U_0.

13. Solve problem 11 assuming that a heat exchange occurs at the surface with a medium whose temperature:

(a) equals zero,

(b) is constant and equal to u_1.

14. Find the steady state temperature of a rod, neglecting heat exchange at the surface and assuming that one end is thermally insulated and that a heat flow which changes harmonically with time, is supplied to the other end.

15. Solve problem 14 assuming that one end of the rod has zero temperature, and the temperature of the other end changes harmonically with time.

16. The rod $(0, l)$ consists of two homogeneous parts of equal cross-section, touching at the point $x = x_0$ and having characteristics a_1, k_1 and a_2, k_2 respectively. Find the steady state temperature in such a rod, if one end of the rod $(x = 0)$ is maintained at zero temperature and the temperature of the other end varies sinusoidally with time.

17. The left end of the composite rod of problem 16 is maintained at a temperature equal to zero, and the right end at a temperature $u(l, t) = u_1$, the initial temperature of the rod being equal to zero. Find the temperature $u(x, t)$.

18. Find the temperature $u(x, t)$ of a rod, whose initial temperature equals zero, and whose boundary conditions have the form

$$u(0, t) = Ae^{-\alpha t}, u(l, t) = B,$$

where A, B and $\alpha > 0$ are constants.

§ 3. Problems in an infinite region

1. *The source function for an infinite region*

In the preceding section we obtained an expression for the instantaneous point source function in a finite segment $(0, l)$ in the following form.*

$$G_l(x, \xi, t) = \frac{2}{l} \sum_{n=1}^{\infty} e^{-\left(\frac{\pi n}{l}\right)^2 a^2 t} \sin \frac{\pi n}{l} x \cdot \sin \frac{\pi n}{l} \xi. \tag{1}$$

If the point source corresponds to an amount of heat Q, then the process of distribution of temperature is characterized by a function

$$u(x, t) = \frac{Q}{c\varrho} G_l(x, \xi, t). \tag{2}$$

From what has been said about propagation of heat in an infinite region, it is clear that the function $G(x, \xi, t)$ should be the limit

* We insert a suffix l in the function $G_l(x, \xi, t)$ in order to distinguish it from the source function $G(x, \xi, t)$ for an infinite region, with which we shall be concerned in this section.

of the function (1) for a finite segment when both ends recede to infinity. To calculate this limit let us transform formula (1) so that the ends of the segment have coordinates $(-l/2,\ l/2)$. This is achieved by introducing new coordinates x' and ξ'

$$x' = x - \frac{l}{2},\quad \xi' = \xi - \frac{l}{2}.$$

The source function of an instantaneous point source of magnitude $Q = c\varrho$, located at the point ξ' of the segment $(-l/2,\ l/2)$ has the form

$$G_l(x', \xi', t) = \frac{2}{l}\sum_{n=1}^{\infty} e^{-\left(\frac{\pi n}{l}\right)^2 a^2 t} \sin\frac{\pi n}{l}\left(x' + \frac{l}{2}\right)\cdot\sin\frac{\pi n}{l}\left(\xi' + \frac{l}{2}\right). \tag{1'}$$

Let us transform the products of the sines. If n is even, i. e. $n = 2m$ then

$$\sin\frac{2\pi m}{l}\left(x' + \frac{l}{2}\right)\cdot\sin\frac{2\pi m}{l}\left(\xi' + \frac{l}{2}\right) = \sin\frac{2\pi m}{l}x' \sin\frac{2\pi m}{l}\xi'.$$

If n is odd, i.e. $n = 2m + 1$, then

$$\sin\frac{(2m+1)\pi}{l}\left(x' + \frac{l}{2}\right)\cdot\sin\frac{(2m+1)\pi}{l}\left(\xi' + \frac{1}{2}\right) = \\ = \cos\frac{(2m+1)\pi}{l}x'\cdot\cos\frac{(2m+1\pi}{l}\xi'.$$

Thus,

$$G_l(x', \xi', t) = \frac{2}{l}\sum_{n=0}^{\infty}{}'' e^{-\left(\frac{\pi n}{l}\right)^2 a^2 t} \sin\frac{\pi n}{l}x' \sin\frac{\pi n}{l}\xi' + \\ + \frac{2}{l}\sum_{n=1}^{\infty}{}' e^{-\left(\frac{\pi n}{l}\right)^2 a^2 t} \cos\frac{\pi n}{l}x' \cos\frac{\pi n}{l}\xi', \tag{1''}$$

where the strokes at the side of the sum $''$ (or $'$) indicate that the summation is carried out only over even (odd) values of n.

We shall find the limit of the first sum as $l \to \infty$. It may be written in the form

$$\frac{2}{l}\sum_{n=0}^{\infty}{}'' e^{-\lambda_n^2 a^2 t}\sin\lambda_n x'\cdot\sin\lambda_n\xi' = \frac{1}{\pi}\sum_{n=0}^{\infty}{}'' f_1(\lambda_n)\,\Delta\lambda, \tag{3}$$

where

$$f_1(\lambda) = e^{-\lambda^2 a^2 t}\sin\lambda x'\cdot\sin\lambda\xi',\ \Delta\lambda = 2\pi/l \text{ and } \lambda_n = \pi n/l.$$

The sum (3) tends to the integral of the function $f_1(\lambda)$ over the interval $0 \leqslant \lambda < \infty$. When $l \to \infty$, then $\Delta\lambda \to 0$. Performing a limiting transition to the integral, we have:†

$$\lim_{\Delta\lambda\to 0} \frac{1}{\pi} \sum_{n=0}^{\infty}{}'' f_1(\lambda_n)\,\Delta l = \frac{1}{\pi}\int_0^{\infty} f_1(\lambda)\,d\lambda = \frac{1}{\pi}\int_0^{\infty} e^{-\lambda^2 a^2 t} \sin\lambda x' \cdot \sin\lambda\xi'\,d\lambda. \tag{4}$$

Similarly, the second sum may be written as

$$\frac{2}{l}\sum_{n=0}^{\infty}{}' e^{-\lambda_n^2 a^2 t} \cos\lambda_n x' \cos\lambda_n \xi' = \frac{1}{\pi}\sum_{n=0}^{\infty}{}' f_2(\lambda_n)\,\Delta\lambda\,, \tag{5}$$

where

$$f_2(\lambda) = e^{-\lambda^2 a^2 t}\cos\lambda x' \cos\lambda\xi',\ \Delta\lambda = 2\pi/l \text{ and } \lambda_n = \pi n/l\,.$$

Going to the limit we get

$$\lim_{\Delta\lambda\to 0} \frac{1}{\pi}\sum_{n=1}^{\infty}{}' f_2(\lambda_n)\,\Delta\lambda = \frac{1}{\pi}\int_0^{\infty} f_2(\lambda)\,d\lambda = \frac{1}{\pi}\int_0^{\infty} e^{-\lambda^2 a^2 t}\cos\lambda x' \cos\lambda\xi'\,d\lambda. \tag{6}$$

† In formula (4) we obtained improper integrals on the right side as the limits of the integral sums, taken over the entire infinite segment $(0, \infty)$. For proof of this transition it is necessary to show that it does not contradict the usual definition of an improper integral

$$\int_0^{\infty} f(\lambda)\,d\lambda = \lim_{L\to\infty}\int_0^{L} f(\lambda)\,d\lambda\,.$$

We will show that if the continuous function $f(\lambda)$, given for $0 \leqslant \lambda < \infty$, is such that for some division of the axis $(0, \infty)$ into the equal segments $(\alpha \leqslant \Delta\lambda_i \leqslant \beta)$ the integral sum

$$\sum_{i=1}^{\infty} f(\lambda_i^*)(\lambda_i - \lambda_{i-1}) \quad (\lambda_{i-1} \leqslant \lambda_i^* \leqslant \lambda_i)$$

converges for any choice of λ_i^*, then the improper integral exists

$$\int_0^{\infty} f(\lambda)\,d\lambda\,.$$

Let us prove that the following sequence has a limit

$$\lim_{\mu_k\to\infty}\int_0^{\mu_k} f(\lambda)\,d\lambda\,,$$

Combining these results, we have:

$$G(x, \xi, t) = \lim_{l\to\infty} G_l(x, \xi, t) =$$

$$= \frac{1}{\pi}\int_0^\infty e^{-\lambda^2a^2t} \sin\lambda x \sin\lambda\xi \, d\lambda + \frac{1}{\pi}\int_0^\infty e^{-\lambda^2a^2t} \cos\lambda x \cos\lambda\xi \, d\lambda =$$

$$= \frac{1}{\pi}\int_0^\infty e^{-\lambda^2a^2t} \cos\lambda(x - \xi) \, d\lambda .$$

Thus

$$G(x, \xi, t) = \frac{1}{\pi}\int_0^\infty e^{-\lambda^2a^2t} \cos\lambda(x - \xi) \, d\lambda. \tag{7}$$

Let us calculate the auxiliary integral

$$I = \int_0^\infty e^{-\lambda^2 a} \cos\lambda\beta \, d\lambda \quad (a > 0), \tag{8}$$

---

which does not depend on the way μ_k tends to infinity. First we examine the case $\mu_k = \lambda_k$. The difference

$$(*) \qquad \int_0^{\lambda_{k+h}} f(\lambda)\, d\lambda - \int_0^{\lambda_k} f(\lambda)\, d\lambda = \int_{\lambda_k}^{\lambda_{k+h}} f(\lambda)\, d\lambda \leq \sum_{i=k}^{k+h-1} f(\bar{\lambda}_i)(\lambda_{i+1} - \lambda_i);$$

similarly

$$(**) \qquad \int_{\lambda_k}^{\lambda_{k+h}} f(\lambda)\, d\lambda \geqslant \sum_{i=k}^{k+h-1}{}' f(\underline{\lambda_i})(\lambda_{i+1} - \lambda_i),$$

where $f(\bar{\lambda}_i)$ and $f(\underline{\lambda_i})$ are the greatest and least values of $f(\lambda)$ in the interval $(\lambda_i, \lambda_i + 1)$. From the convergence of the integral sum for the chosen sequence λ_k it follows that for any ε there exists $N(\varepsilon)$ such that for $k > N(\varepsilon)$

$$\left|\sum_{i=k}^{k+h}{}' f(\lambda_i)(\lambda_i - \lambda_{i-1})\right| < \varepsilon .$$

Hence from the relations (*) and (**) it follows that

$$\left|\int_{\lambda_k}^{\lambda_{k+h}} f(\lambda)\, d\lambda\right| < \varepsilon \text{ when } k > N(\varepsilon) .$$

Because any integral sum converges it follows that $f(\lambda)$ tends to zero as λ tends to infinity. Hence

$$\left|\int_{\mu_k}^{\mu_{k+l}} f(\lambda)\, d\lambda - \int_{\lambda_k}^{\lambda_{k+h}} f(\lambda)\, d\lambda\right| < \varepsilon \text{ when } n > N(\varepsilon) .$$

which depends on the two parameters a and β. We fix the value of a and denote this integral by $I(\beta)$. Then we calculate the derivative

$$\frac{dI}{d\beta} = -\int_0^\infty e^{-\lambda^2 a} \lambda \sin \lambda\beta \, d\lambda$$

by differentiation under the integral sign. Integrating by parts, we obtain:

$$\frac{dI}{d\beta} = \sin \lambda\beta \frac{1}{2a} e^{-\lambda^2 a} \Big|_0^\infty - \frac{\beta}{2a} \int_0^\infty e^{-2\lambda^2 a} \cos \lambda\beta \, d\lambda = -\frac{\beta}{2a} I(\beta).$$

Thus for $I(\beta)$ we get a differential equation which can be solved by separation of variables

$$\frac{I'}{I} = -\frac{\beta}{2a},$$

---

where λ_k and λ_{k+h} are points of the given sequence near to the points μ_n and μ_{n+l}.

This proves the existence of the improper integral

$$\int_0^\infty f(\lambda)\, d\lambda.$$

Now we prove that the integral sum converges that to improper integral

$$\lim_{\Delta\lambda_i \to 0} \sum_{i=0}^{\infty}{}' f(\lambda_i^*)\, \Delta\lambda_i = \int_0^\infty f(\lambda) d\lambda.$$

We consider the difference

$$\left|\int_0^\infty f(\lambda)\, d\lambda - \sum_{i=0}^{\infty} f(\lambda_i^*) \Delta\lambda \right| \leqslant$$

$$\leqslant \left|\int_0^{\lambda_k} f(\lambda)\, d\lambda - \sum_0^{\lambda_k}{}' f(\lambda_i^*)\, \Delta\lambda_i \right| + \left|\int_{\lambda_k}^\infty f(\lambda)\, d\lambda\right| + \left|\sum_{\lambda_k}^\infty f(\lambda_i^*)\, \Delta\lambda_i\right|;$$

by a suitable choice of $\Delta\lambda_i$ and λ_k each sum on the right side may be made less than $\varepsilon/3$, which proves our statement.

In our case

$$f(\lambda) = e^{-\lambda^2 a^2 t} \cos \lambda x \cos \lambda\xi \text{ or } f(\lambda) = e^{-\lambda^2 a^2 t} \sin \lambda x \sin \lambda\xi.$$

The integral sum, formed by division into the equal parts $\lambda_i - \lambda_{i-1} = \Delta\lambda$, satisfies the inequalities

$$\left|\sum_{i=0}^{\infty} f(\lambda_i^*)\, \Delta\lambda\right| \leqslant \sum_{i=0}^{\infty} \bar{f}(\lambda_{i-1})\, \Delta\lambda = \sum_{i=0}^{\infty}{}' e^{-\lambda_{i-1}^2 a^2 t}\, \Delta\lambda < \Delta\lambda \sum_{i=0}^{\infty}{}' e^{-(i-1)^2(\Delta\lambda)^2 a^2 t}.$$

On the basis of D'Alembert's principle it is easy to see that the sum converges. Taking the ratios of successive terms we get,

$$\frac{e^{-i^2 \Delta\lambda^2 a^2 t}}{e^{-(i-1)^2 \Delta\lambda^2 a^2 t}} = e^{-(2i-1)(\Delta\lambda)^2 a^2 t}.$$

whence

$$I(\beta) = Ce^{-\frac{\beta^2}{4a}}.$$

When $\beta = 0$ we find the value of the constant

$$C = I(0) = \int_0^\infty e^{-\lambda^2 a}\, d\lambda = \frac{1}{\sqrt{a}} \int_0^\infty e^{-z^2}\, dz = \frac{1}{\sqrt{a}} \cdot \frac{\sqrt{\pi}}{2},$$

since

$$\int_0^\infty e^{-z^2}\, dz = \frac{\sqrt{\pi}}{2}.$$

Hence

$$I(\beta) = \int_0^\infty e^{-\lambda^2 a} \cos \lambda\beta\, d\lambda = \frac{1}{2} \frac{\sqrt{\pi}}{\sqrt{a}} e^{-\frac{\beta^2}{4a}}. \tag{9}$$

Applying formula (9) to the integral (7) we get an expression *the source function for an infinite region*

$$G(x, \xi, t) = \frac{1}{2\sqrt{(\pi a^2 t)}} e^{-\frac{(x-\xi)^2}{4a^2 t}}. \tag{10}$$

This function is often called *the fundamental solution* of the equation of heat conduction.

By direct examination one can satisfy oneself that the function

$$G(x, \xi, t - t_0) = \frac{Q}{c\varrho\, 2\sqrt{[\pi a^2 (t - t_0)]}} e^{-\frac{(x-\xi)^2}{4a^2(t-t_0)}} \tag{10'}$$

represents the temperature at the point x at the time t, if an amount of heat $Q = c\varrho$ is liberated at the initial time $t = t_0$ at the point ξ.

(1) The function $G(x, \xi, t - t_0)$ satisfies the equation of heat conduction in the variables (x, t),* as may be verified by direct differentiation.

* In fact

$$G_x = -\frac{1}{2\sqrt{\pi}} \cdot \frac{x - \xi}{2[a^2(t - t_0)]^{3/2}} e^{-\frac{(x-\xi)^2}{4a^2(t-t_0)}},$$

$$G_{xx} = \frac{1}{2\sqrt{\pi}} \left[-\frac{1}{2} \frac{1}{[a^2(t - t_0)]^{3/2}} + \frac{(x - \xi)^2}{4[a^2(t - t_0)]^{5/2}} \right] e^{-\frac{(x-\xi)^2}{4a^2(t-t_0)}},$$

$$G_t = \frac{1}{2\sqrt{\pi}} \left[-\frac{a^2}{2[a^2(t - t_0)]^{3/2}} + \frac{a^2(x - \xi)^2}{4[a^2(t - t_0)]^{5/2}} \right] e^{-\frac{(x-\xi)^2}{4a^2(t-t_0)}},$$

i. e.

$$G_t = a^2 G_{xx}.$$

(2) The total amount of heat found on the x-axis for $t > t_0$ equals

$$c\varrho \int_{-\infty}^{\infty} G(x, \xi, t - t_0)\, dx = \frac{Q}{\sqrt{\pi}} \int_{-\infty}^{\infty} e^{-\frac{(x-\xi)^2}{4a^2(t-t_0)}} \frac{dx}{2\sqrt{[a^2(t - t_0)]}} =$$

$$= \frac{Q}{\sqrt{\pi}} \int_{-\infty}^{\infty} e^{-\alpha^2}\, d\alpha = Q = c\varrho\,,$$

since

$$\int_{-\infty}^{\infty} e^{-\alpha^2}\, d\alpha = \sqrt{\pi}$$

$$\left(\alpha = \frac{x - \xi}{2\sqrt{[a^2(t - t_0)]}}, \quad d\alpha = \frac{dx}{2\sqrt{[a^2(t - t_0)]}}\right).$$

Thus, the total amount of heat is constant. The function $G(x, \xi, t - t_0)$ depends on the time only through the argument $\theta = a^2(t - t_0)$ so that it is possible to write it as

$$G = \frac{1}{2\sqrt{\pi}} \frac{1}{\sqrt{\theta}} e^{-\frac{(x-\xi)^2}{4\theta}}. \tag{10''}$$

In Fig. 40 the graph of G is given as a function of x for different values of θ. Almost the entire area bounded by this curve lies in the interval

$$(\xi - \varepsilon,\ \xi + \varepsilon),$$

where ε is an arbitrary small number, provided $\theta = a^2(t - t_0)$ is sufficiently small. The magnitude of this area, multiplied by $c\,\varrho$, equals the amount of heat injected at the initial time. Thus, for small values of $t - t_0 > 0$, almost all the heat is concentrated in a small region around the point ξ. It follows that at time t_0 all the heat is injected at the point ξ.

Considering the temperature at the fixed point $x = \xi + h$ as a function of time. When $h = 0$, i.e. at $x = \xi$, we obtain:

$$G_{x=\xi} = \frac{1}{2\sqrt{\pi}} \frac{1}{\sqrt{\theta}}.$$

Thus the temperature at this point, where heat is liberated, becomes infinitely high for small θ.

If $x \neq \xi$, i.e. $h \neq 0$, then the function G is a product of two factors

$$G_{x\neq\xi} = \left[\frac{1}{2\sqrt{\pi}} \frac{1}{\sqrt{\theta}}\right] e^{-\frac{h^2}{4\theta}}$$

The second factor is less than unity: for large θ it ≈ 1, for small θ it ≈ 0. Hence it follows that $G_{x\neq\xi} = G_{x=\xi}$ for large θ; $G_{x\neq\xi} \ll G_{x=\xi}$ for small θ. The smaller h is, i.e. the nearer x to ξ,

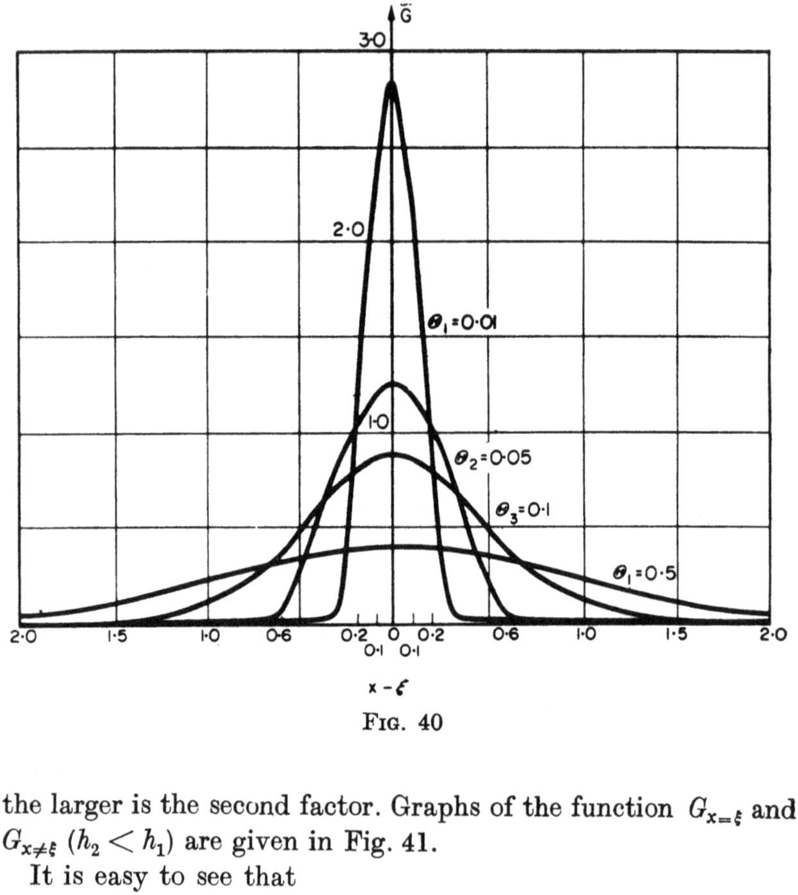

FIG. 40

the larger is the second factor. Graphs of the function $G_{x=\xi}$ and $G_{x\neq\xi}$ $(h_2 < h_1)$ are given in Fig. 41.

It is easy to see that

$$\lim_{\theta\to 0} G_{x\neq\xi} = 0.$$

Expanding near the limit, we find:

$$\lim_{\theta\to 0}\left[\frac{1}{2\sqrt{\pi}}\frac{1}{\sqrt{\theta}}\right]e^{-\frac{h^2}{4\theta}} = \frac{1}{2\sqrt{\pi}}\lim_{\theta\to 0}\frac{-\frac{1}{2}\theta^{-\frac{3}{2}}}{\frac{h^2}{4\theta^2}e^{\frac{h^2}{4\theta}}} = 0\,.$$

Formula (10′) shows that at any point x the temperature, produced by an instantaneous point source acting at the initial time $t = 0$ differs from zero for any time $t > 0$ no matter how small.

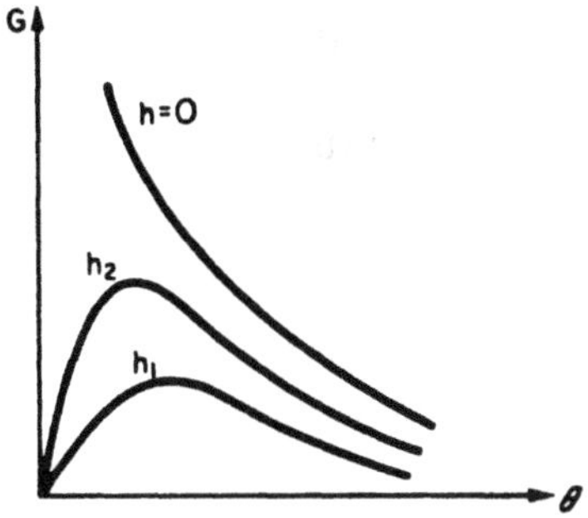

FIG. 41

This means that heat is propagated with infinite speed contradicting the molecular-kinetic theory of heat. Such a contradiction arises because in deriving the equation of heat conduction we make use of phenomenological ideas about the propagation of heat, not allowing for the inertia in the motion of the molecules.

2. *The propagation of heat in an infinite region*

The problem of the conduction of heat in this case is set in the following way:

Find the bounded function $u(x, t)$ $(-\infty < x < \infty,\ t \geqslant 0)$ *satisfying the equation of heat conduction*

$$u_t = a^2 u_{xx} \ (-\infty < x < \infty, t > 0) \tag{11}$$

and the initial condition

$$u(x, 0) = \varphi(x). \tag{12}$$

We shall, as before, restrict the initial condition by requiring continuity* of the function $u(x, t)$ at $t = 0$.

It is easy to find an analytical representation of the solution. Let an auxiliary function $\bar{\varphi}(x)$ be everywhere equal to zero, except in a small interval $(\xi_0 - \delta,\ \xi_0 + \delta)$ in which it coincides with $\varphi(x)$. In order to change the initial temperature from zero

* As we saw in sub-section 7, § 1, the solution of the equation of heat conduction is well defined by its initial conditions if it is bounded. Therefore in the formulation of the theory we introduce this restriction.

to $\varphi(x)$ it is necessary to inject an amount of heat into the interval $(\xi_0 - \delta, \xi_0 + \delta)$

$$Q = c\varrho \int_{\xi_0-\delta}^{\xi_0+\delta} \varphi(x)\,dx \approx c\varrho\,\varphi(\bar{\xi})\,\Delta\xi \qquad (\Delta\xi = 2\delta).$$

The temperature at the point x at time $t > 0$, will equal

$$\frac{Q}{c\varrho} G(x, \xi, t) = G(x, \xi, t)\,\varphi(\bar{\xi})\,\Delta\xi, \tag{13}$$

where $\bar{\xi}$ is an intermediate value in the interval $(\xi_0 - \delta, \xi_0 + \delta)$.

Dividing the entire line into small intervals and using the principle of superposition, we may represent $u(x, t)$ as a sum of components of type (13). This sum tends as $\delta \to 0$ to the integral

$$u(x, t) = \int_{-\infty}^{\infty} G(x, t, \xi)\,\varphi(\xi)\,d\xi \tag{14}$$

or

$$u(x, t) = \frac{1}{2\sqrt{\pi}} \int_{-\infty}^{\infty} \frac{1}{\sqrt{(a^2 t)}}\, e^{-\frac{(x-\xi)^2}{4a^2 t}}\,\varphi(\xi)\,d\xi. \tag{14'}$$

This function must represent the solution of the problem.

The arguments outlined above are not, of course, proofs. Let us examine the conditions of applicability of this formula.

We show that the formula

$$u(x, t) = \frac{1}{2\sqrt{\pi}} \int_{-\infty}^{\infty} \frac{1}{\sqrt{(a^2 t)}}\, e^{-\frac{(x-\xi)^2}{4a^2 t}}\,\varphi(\xi)\,d\xi,$$

called *Poisson's integral*, represents for $t > 0$ a bounded solution of the equation of heat conduction, tending to $\varphi(x)$ for $t = 0$ at all points of the continuity of this function, provided $\varphi(x)$ is bounded, $|\varphi(\xi)| < M$.

Let us prove a preliminary lemma *(the generalized principle of superposition)*.

If the function $U(x, t, \alpha)$ in the variables (x, t) satisfies the linear differential equation

$$L(U) = 0$$

for any fixed value of the parameter α, then the integral

$$u(x, t) = \int U(x, t, \alpha)\,\varphi(\alpha)\,d\alpha$$

is also a solution of the same equation $L(u) = 0$, *if the derivatives appearing in the linear differential operator* $L(U)$ *can be calculated by means of differentiation under the integral sign.*

The linear differential operator $L(U)$ represents the sum of the derivatives of the function U with certain coefficients depending on x and t. Differentiation of the function u, by hypothesis, can occur under the integral sign. The coefficients also can be brought under the integral sign. Hence it follows that

$$L(u) = \int L\big(U(x, t, a)\big)\,\varphi(a)\,da) = 0\,.$$

i.e. that the function $u(x, t)$ satisfies the equation $L(u) = 0$.

We recall the sufficient conditions for differentiation under the integral sign.

The function

$$F(x) = \int_a^b f(x, a)\,da$$

with finite limits a and b is differentiable under the integral sign if $\frac{\partial f}{\partial x}(x, a)$ is a continuous function of the variables x and a in the region of variation (see V. I. Smirnov, *Kurs vysshei matematiki (Course of Higher Mathematics)*, vol. II, chap. III, § 3, 1950).

It is easily seen also that the function

$$F_1(x) = \int_a^b f(x, a)\,\varphi(a)\,da$$

with finite limits a and b is differentiable under the integral sign for these conditions on the function $f(x, a)$ and an arbitrary, bounded (and everywhere integrable) function $\varphi(a)$. If the limits of integration are infinite then the uniform convergence of the integral, obtained as a result of differentiation of the function under the integral by the parameter, is also necessary (see above reference).

These remarks hold also for a multiple integral.

The principle of superposition holds for the linear equations $L(u) = 0$. Thus the function

$$u(x, t) = \sum_{i=1}^{n} C_i\,u_i(x, t)\,,$$

which is a sum of a finite number of solutions is also a solution of the equation. If we have a solution $u(x, t, a)$ depending on a parameter a then the integral sum

$$\sum u(x, t, a_n)\, C_n \qquad (C_n = \varphi(a_n)\Delta a) \tag{15}$$

is a solution. The lemma proved above gives the conditions for which the limit of the sum (15), i.e. the integral

$$u(x, t) = \int U(x, t, a)\, \varphi(a)\, da,$$

is a solution of the equation $L(u) = 0$. From this point of view it is natural to call the lemma, the *generalized principle of superposition.*

Let us investigate the integral (14′). We show, firstly, that if the function $\varphi(x)$ is bounded, $|\varphi(x)| < M$, then the integral (14′) converges and represents a bounded function. In fact

$$|u(x, t)| < M \frac{1}{2\sqrt{\pi}} \int_{-\infty}^{\infty} \frac{1}{\sqrt{(a^2 t)}} e^{-\frac{(x-\xi)^2}{4a^2 t}}\, d\xi =$$

$$= M \frac{1}{\sqrt{\pi}} \int_{-\infty}^{\infty} e^{-\alpha^2}\, d\alpha = M \qquad \left(\alpha = \frac{\xi - x}{2\sqrt{(a^2 t)}}\right),$$

since

$$\int_{-\infty}^{\infty} e^{-\alpha^2}\, d\alpha = \sqrt{\pi}.$$

We show further that the integral (14′) satisfies the equation of heat conduction for $t > 0$. It is sufficient to prove that the derivatives of this integral for $t > 0$ can be calculated by means of differentiation under the integral sign.

In the case of finite limits of integration this is valid, because all the derivatives of the function

$$\frac{1}{2\sqrt{(\pi a^2 t)}} e^{-\frac{(x-\xi)^2}{4a^2 t}}$$

for $t > 0$ are continuous. In order to differentiate under an integral sign with infinite limits it is sufficient that the integral, which is obtained after differentiation under the integral sign be uniformly convergent. Let us carry out this investigation using as example the first derivative with respect to x.

In order to prove the differentiability of the function (14) with respect to x and also of the equality

$$\frac{\partial u}{\partial x} = \int_{-\infty}^{\infty} \frac{\partial}{\partial x}(G(x, \xi, t))\, \varphi(\xi)\, d\xi$$

it is sufficient to prove the uniform convergence of the integral on the right-hand side; moreover, for differentiation at the point (x_0, t_0) it is sufficient to prove the uniform convergence of the integral in some region of the variables containing the values (x_0, t_0), for example in the region

$$t_1 \leqslant t_0 \leqslant t_2, \; |x| \leqslant \bar{x}.$$

A sufficient condition for the uniform convergence of the integral (similar to the condition for the uniform convergence of a sum) is the existence of a positive function $F(\xi)$, not dependent on the parameters (x, t), which dominates the function $|G_x(x, \xi, t)\, \varphi(\xi)|$:

$$\left| \frac{\partial}{\partial x} G(x, \xi, t)\, \varphi(\xi) \right| \leqslant F(\xi), \; \xi > \bar{x}, \; \xi < -\bar{x}, \qquad (15')$$

for which the following integrals converge:

$$\int_{x_1}^{\infty} F(\xi)\, d\xi < \infty, \quad \int_{-\infty}^{x_1} F(\xi)\, d\xi < \infty.$$

The number x_1 is some number greater than $\bar{x}$.

Let us determine the absolute value of the expression under the integral sign in the formula for $\frac{\partial u}{\partial x}$:

$$\left| \frac{\partial G}{\partial x}(x, \xi, t) \right| \left| \varphi(\xi) \right| = \frac{1}{2\sqrt{\pi}} \frac{|\xi - x|}{2 [a^2 t]^{3/2}} e^{-\frac{(x-\xi)^2}{4a^2 t}} |\varphi(\xi)| \leqslant$$

$$\leqslant \frac{M}{2\sqrt{\pi}} \frac{|\xi| + \bar{x}}{2 [a^2 t_1]^{3/2}} e^{-\frac{(\xi - \bar{x})^2}{4a^2 t_2}} = F(\xi) \text{ where } \xi < \bar{x} \qquad (16)$$

for any $|\bar{x}| \leqslant x$ and $t_1 \leqslant t \leqslant t_2$. It is easy to deduce the convergence of the integral (15″) from the function $F(\xi)$. The integral

$$\int_{x_1}^{\infty} F(\xi)\, d\xi = \int_{x_1}^{\infty} \frac{1}{2\sqrt{\pi}} \frac{|\xi| + \bar{x}}{2[a^2 t_1]^{3/2}} e^{-\frac{(\xi - \bar{x}^2)}{4a^2 t_2}} d\xi =$$

$$= \int_{x_1 - \bar{x}}^{\infty} \frac{1}{2\sqrt{\pi}} \frac{\xi_1 + 2\bar{x}}{2[a^2 t_1]^{3/2}} e^{-\frac{\xi_1^2}{4a^2 t_2}} d\xi$$

$$(\xi_1 = |\xi| - \bar{x})$$

converges, because the factor under the integral is of the type $(a\,\xi + b)\, e^{-c\xi^2}$. Hence we deduce that

$$\frac{\partial u}{\partial x} = \int_{-\infty}^{\infty} \frac{\partial G}{\partial x}(x, \xi, t)\, \varphi(\xi)\, d\xi\,. \tag{17}$$

The theorem may be proved for higher derivatives in the some way. Thus we have proved that the function (14′) satisfies the equation of heat conduction.

Finally we prove the result

$$u(x, t) \to \varphi(x_0) \text{ when } t \to 0 \text{ and } x \to x_0$$

at all points of the continuity of the function $\varphi(x)$.

Let $\varphi(x)$ be continuous at a certain point x_0. We must prove that

$$\lim_{\substack{t \to 0 \\ x \to x_0}} u(x, t) = \varphi(x_0)\,,$$

i.e. given $\varepsilon > 0$, it is possible to find $\delta(\varepsilon)$ such that

$$|u(x, t) - \varphi(x_0)| < \varepsilon\,,$$

if

$$|x - x_0| < \delta(\varepsilon) \text{ and } |t| < \delta(\varepsilon)\,.$$

Because the function $\varphi(x)$ is continuous at the point x_0, there exists $\eta(\varepsilon)$ such that

$$|\varphi(x) - \varphi(x_0)| < \frac{\varepsilon}{6}\,, \tag{18}$$

when

$$|x - x_0| < \eta\,.$$

Dividing the range of integration, let us represent $u(x, t)$ as a sum of three components:

$$u(x,t) = \frac{1}{2\sqrt{\pi}} \int_{-\infty}^{x_1} \frac{1}{\sqrt{(a^2 t)}} e^{-\frac{(x-\xi)^2}{4a^2 t}} \varphi(\xi)\, d\xi + \frac{1}{2\sqrt{\pi}} \int_{x_1}^{x_2} \dots d\xi +$$

$$+ \frac{1}{2\sqrt{\pi}} \int_{x_2}^{\infty} \dots d\xi = u_1(x,t) + u_2(x,t) + u_3(x,t), \tag{19}$$

where

$$x_1 = x_0 - \eta \text{ and } x_2 = x_0 + \eta. \tag{20}$$

The main component of this sum u_2 may be written as

$$u_2(x,t) = \frac{\varphi(x_0)}{2\sqrt{\pi}} \int_{x_1}^{x_2} \frac{1}{\sqrt{(a^2 t)}} e^{-\frac{(x-\xi)^2}{4a^2 t}} d\xi +$$

$$+ \frac{1}{2\sqrt{\pi}} \int_{x_1}^{x_2} \frac{1}{\sqrt{(a^2 t)}} e^{-\frac{(x-\xi)^2}{4a^2 t}} [\varphi(\xi) - \varphi(x_0)]\, d\xi = I_1 + I_2.$$

The integral I_1 can be calculated explicitly

$$I_1 = \frac{\varphi(x_0)}{2\sqrt{\pi}} \int_{x_1}^{x_2} \frac{e^{-\frac{(x-\xi)^2}{4a^2 t}}}{\sqrt{(a^2 t)}} d\xi = \frac{\varphi(x_0)}{\sqrt{\pi}} \int_{\frac{x_1 - x}{2\sqrt{(a^2 t)}}}^{\frac{x_2 - x}{2\sqrt{(a^2 t)}}} e^{-\alpha^2} d\alpha,$$

where

$$\alpha = \frac{\xi - x}{2\sqrt{(a^2 t)}}, \quad d\alpha = \frac{d\xi}{2\sqrt{(a^2 t)}}. \tag{21}$$

Because $|x - x_0| < \eta$, the upper limit is positive, and the lower negative, and for $t \to 0$ the upper limit tends to $+\infty$ and the lower to $-\infty$. Hence it follows that

$$\lim_{\substack{t \to 0 \\ x \to x_0}} I_1 = \varphi(x_0).$$

Thus, it is possible to find δ, as that

$$|I_1 - \varphi(x_0)| < \frac{\varepsilon}{6}. \tag{22}$$

when

$$|x - x_0| < \delta_1 \quad \text{and} \quad |t| < \delta_1.$$

We show that the remaining integrals: I_2, u_3 and u_2 are small. Let us investigate the integral I_2:

$$|I_2| \leqslant \frac{1}{2\sqrt{\pi}} \int_{x_1}^{x_2} \frac{1}{\sqrt{(a^2 t)}} e^{-\frac{(x-\xi)^2}{4a^2 t}} |\varphi(\xi) - \varphi(x_0)| \, d\xi.$$

From the equalities (20) we see that for

$$x_1 < \xi < x_2$$

the following inequality holds

$$|\xi - x_0| < \eta.$$

Making use of inequality (18), and also the fact that

$$\frac{1}{\sqrt{\pi}} \int_{x'}^{x''} e^{-\alpha^2} \, d\alpha < \frac{1}{\sqrt{\pi}} \int_{-\infty}^{\infty} e^{-\alpha^2} \, d\alpha = 1,$$

whatever x' and x'' may be, we obtain:

$$|I_2| \leqslant \frac{\varepsilon}{6} \cdot \frac{1}{2\sqrt{\pi}} \int_{x_1}^{x_2} \frac{1}{\sqrt{(a^2 t)}} e^{-\frac{(x-\xi)^2}{4a^2 t}} \, d\xi =$$

$$= \frac{\varepsilon}{6} \frac{1}{\sqrt{\pi}} \int_{\frac{x_1 - x}{2\sqrt{(a^2 t)}}}^{\frac{x_2 - x}{2\sqrt{(a^2 t)}}} e^{-\alpha^2} \, d\alpha < \frac{\varepsilon}{6}, \qquad (23)$$

where the new variable α is defined by formula (21). We also find that

$$|u_3(x, t)| = \frac{1}{2\sqrt{\pi}} \left| \int_{x_2}^{\infty} \frac{1}{\sqrt{(a^2 t)}} e^{-\frac{(x-\xi)^2}{4a^2 t}} \varphi(\xi) \, d\xi \right| <$$

$$< \frac{M}{\sqrt{\pi}} \int_{\frac{x_2 - x}{2\sqrt{(a^2 t)}}}^{\infty} e^{-\alpha^2} \, d\alpha \to 0 \quad \text{when} \quad \begin{matrix} x \to x_0 \\ t \to 0 \end{matrix} \qquad (24)$$

and similarly

$$|u_1(x, t)| = \frac{1}{2\sqrt{\pi}} \left| \int_{-\infty}^{x_1} \frac{1}{\sqrt{(a^2 t)}} e^{-\frac{(x-\xi)^2}{4a^2 t}} \varphi(\xi) \, d\xi \right| <$$

$$< \frac{M}{\sqrt{\pi}} \int_{-\infty}^{\frac{x_1 - x}{2\sqrt{(a^2 t)}}} e^{-\alpha^2} \, d\alpha \to 0 \quad \text{when} \quad \begin{matrix} x \to x_0 \\ t \to 0, \end{matrix} \qquad (25)$$

because if $x \to x_0$, then $x_2 - x > 0$, and $x_1 - x < 0$, and if $t \to 0$, then in the final terms of (24) and (25) the lower limit and upper limit respectively tend to $+\infty$ and $-\infty$. Hence it is possible to assign δ_2 so that

$$|u_3(x,t)| < \varepsilon/3 \quad \text{and} \quad |u_1(x,t)| < \varepsilon/3, \tag{26}$$

when

$$|x - x_0| < \delta_2 \quad \text{and} \quad |t| < \delta_2.$$

Making use of the results established above we obtain:

$$|u(x,t) - \varphi(x_0)| \leqslant |u_1 + [I_1 - \varphi(x_0)] + I_2 + u_3| \leqslant |u_1| + \\ + |I_1 - \varphi(x_0)| + |I_2| + |u_3| < \frac{\varepsilon}{3} + \frac{\varepsilon}{6} + \frac{\varepsilon}{6} + \frac{\varepsilon}{3} = \varepsilon \tag{27}$$

when

$$|x - x_0| < \delta \quad \text{and} \quad |t| < \delta,$$

where δ is the smallest of the numbers δ_1 and δ_2.

Thus, we have shown that the function

$$u(x,t) = \frac{1}{2\sqrt{\pi}} \int_{-\infty}^{\infty} \frac{1}{\sqrt{(a^2 t)}} e^{-\frac{(x-\xi)^2}{4a^2 t}} \varphi(\xi)\, d\xi \tag{14'}$$

is bounded and satisfies the equation of heat conduction and the initial condition.

If the initial value is given not at $t = 0$ but at $t = t_0$, then the expression for $u(x, t)$ takes the form

$$u(x,t) = \frac{1}{2\sqrt{\pi}} \int_{-\infty}^{\infty} \frac{1}{\sqrt{[a^2(t - t_0)]}} e^{-\frac{(x-\xi)^2}{4a^2(t-t_0)}} \varphi(\xi)\, d\xi. \tag{14''}$$

The uniqueness of the solution obtained for the continuous function $\varphi(x)$ follows from the theorem proved in § 2, sub-section 3. If the initial function $\varphi(x)$ has a finite number of points of discontinuity, then the integral (14'') represents a bounded solution of equation (1), continuous everywhere except at the points of discontinuity of the function $\varphi(x)$.*

Let us examine the following problem as an example.

* Making use of the method stated in sub-section 3, § 2, one can be satisfied that the function $u(x, t)$ is well defined by the given conditions.

Find the solution of the equation of heat conduction, if the initial temperature (at $t = t_0 = 0$) has constant but different values for $x > 0$ and $x < 0$, namely

$$u(x,0) = \varphi(x) = \begin{cases} T_1 & \text{where} \quad x > 0, \\ T_2 & \text{where} \quad x < 0. \end{cases}$$

Making use of formula (14′) we obtain the solution of the problem

$$u(x,t) = \frac{1}{2\sqrt{\pi}} \int_{-\infty}^{\infty} \frac{1}{\sqrt{(a^2 t}} e^{-\frac{(x-\xi)^2}{4a^2 t}} \varphi(\xi)\, d\xi =$$

$$= \frac{T_2}{\sqrt{\pi}} \int_{-\infty}^{0} e^{-\frac{(x-\xi)^2}{4a^2 t}} \frac{d\xi}{2\sqrt{(a^2 t)}} + \frac{T_1}{\sqrt{\pi}} \int_{0}^{\infty} e^{-\frac{(x-\xi)^2}{4a^2 t}} \frac{d\xi}{2\sqrt{(a^2 t)}} =$$

$$= \frac{T_2}{\sqrt{\pi}} \int_{-\infty}^{-\frac{x}{2\sqrt{(a^2 t)}}} e^{-\alpha^2}\, d\alpha + \frac{T_1}{\sqrt{\pi}} \int_{-\frac{x}{2\sqrt{(a^2 t)}}}^{\infty} e^{-\alpha^2}\, d\alpha =$$

$$= \frac{T_1 + T_2}{2} + \frac{T_1 - T_2}{\sqrt{\pi}} \int_{0}^{\frac{x}{2\sqrt{(a^2 t)}}} e^{-\alpha^2}\, d\alpha, \tag{28}$$

because

$$\frac{1}{\sqrt{\pi}} \int_{-\infty}^{-z} e^{-\alpha^2}\, d\alpha = \frac{1}{\sqrt{\pi}} \int_{-\infty}^{0} e^{-\alpha^2}\, d\alpha - \frac{1}{\sqrt{\pi}} \int_{0}^{z} e^{-\alpha^2}\, d\alpha = \frac{1}{2} - \frac{1}{\sqrt{\pi}} \int_{0}^{z} e^{-\alpha^2}\, d\alpha$$

and

$$\frac{1}{\sqrt{\pi}} \int_{-z}^{\infty} e^{-\alpha^2}\, d\alpha = \frac{1}{2} + \frac{1}{\sqrt{\pi}} \int_{-z}^{0} e^{-\alpha^2}\, d\alpha = \frac{1}{2} +$$

$$+ \frac{1}{\sqrt{\pi}} \int_{0}^{z} e^{-\alpha^2}\, d\alpha \qquad \left(z = \frac{x}{2\sqrt{(a^2 t)}}\right).$$

In particular, if

$$T_2 = 0, \qquad T_1 = 1,$$

then

$$u(x,t) = \frac{1}{2}\left(1 + \frac{2}{\sqrt{\pi}} \int_{0}^{z} e^{-\alpha^2}\, d\alpha\right) \qquad \left(z = \frac{x}{2\sqrt{(a^2 t)}}\right).$$

The profile of the temperature at a time t is given by the curve

$$f(z) = \frac{1}{2} + \frac{1}{\sqrt{\pi}} \int_0^z e^{-\alpha^2} d\alpha ,$$

where z represents the coordinate of the point of interest measured in units of $2\sqrt{(a^2 t)}$. The construction of this curve does not present difficulty because the integral

$$\Phi(z) = \frac{2}{\sqrt{\pi}} \int_0^z e^{-\alpha^2} d\alpha ,$$

usually called *the error integral,* often occurs in the theory of probability and there exist detailed tables for it.

Formula (28) for arbitrary T_1 and T_2 becomes

$$u(x, t) = \frac{T_1 + T_2}{2} + \frac{T_1 - T_2}{2} \Phi\left(\frac{x}{2\sqrt{(a^2 t)}}\right). \tag{29}$$

Hence the temperature is constant all the time at the point $x = 0$ and equals half the sum of the initial values from the right and left, since $\Phi(0) = 0$.

The solution of the inhomogeneous equation

$$u_t = a^2 u_{xx} + f(x, t) \qquad (-\infty < x < \infty , \ t > 0)$$

with zero initial conditions

$$u(x, 0) = 0 ,$$

obviously must be represented by the formula

$$u(x, t) = \int_0^t \int_{-\infty}^{\infty} G(x, \xi, t - \tau) f(\xi, \tau) \, d\xi \, d\tau ,$$

as follows from the physical significance of the function $G(x, \xi, t)$ (see sub-section 4, § 2). We will not discuss this formula, and the conditions which must be imposed on the function $f(x, t)$.

3. *Boundary-value problems for a semi-infinite straight line*

As we have already noted in § 1, sub-section 4, in those cases where the distribution of temperature near one end of the rod is of interest to us and the effect of the other end is negligible, it is assumed that this end is at infinity. This leads to the problem of heat conduction

$$u_t = a^2 u_{xx}$$

for a semi-infinite region $x > 0$ for values of $t > 0$, satisfying the initial condition

$$u(x, 0) = \varphi(x) \qquad (x > 0)$$

and the boundary condition which, depending on the character of the boundary region, takes one of the following forms:

$$u(0, t) = \mu(t) \quad \text{(the first boundary-value problem)}$$

$$\frac{\partial u}{\partial x}(0, t) = \nu(t) \quad \text{(the second boundary-value problem)}$$

or

$$\frac{\partial u}{\partial x}(0, t) = \lambda[\mu(0, t) - \theta(t)] \quad \text{(the third boundary-value problem).}$$

We give a detailed investigation only of the *first boundary-value problem*

$$u(x, 0) = \varphi(x), \quad u(0, t) = \mu(t). \tag{30}$$

In order that there should be a unique solution, it is necessary to impose certain conditions at infinity. We require as an additional condition that the function $u(x, t)$ be everywhere bounded

$$|u(x, t)| < M \text{ where } 0 < x < \infty \text{ and } t \geqslant 0,$$

where M is a certain constant. Hence the initial function $\varphi(x)$ must also satisfy the condition

$$|\varphi(x)| < M.$$

The solution of the problem can be represented as a sum

$$u(x, t) = u_1(x, t) + u_2(x, t),$$

where $u_1(x, t)$ represents the effect of the initial conditions, and $u_2(x, t)$ the effect of the boundary condition. These functions can be defined as the solutions of equation (11), satisfying the conditions

$$u_1(x\ 0) = \varphi(x), \quad u_1(0, t) = 0 \tag{30'}$$

and

$$u_2(x, 0) = 0, \qquad u_2(0, t) = \mu(t). \tag{30''}$$

It is obvious that the sum of these functions will satisfy conditions (30). Let us prove two lemmas concerning the function $u(x, t)$ defined by Poisson's integral

$$u(x,t) = \frac{1}{2\sqrt{\pi}} \int_{-\infty}^{\infty} \frac{1}{\sqrt{(a^2 t)}} e^{-\frac{(x-\xi)^2}{4a^2 t}} \psi(\xi)\, d\xi . \tag{31}$$

1. *If the function $\psi(x)$ is an odd function, i.e.*

$$\psi(x) = -\psi(-x),$$

then the function (31)

$$u(x,t) = \frac{1}{2\sqrt{\pi}} \int_{-\infty}^{\infty} \frac{1}{\sqrt{(a^2 t)}} e^{-\frac{(x-\xi)^2}{4a^2 t}} \psi(\xi)\, d\xi$$

reduces to zero at $x = 0$,

$$u(0,t) = 0 .$$

It is assumed that the integral, defining the function $u(x, t)$, converges, which is true if $\psi(x)$ is bounded. The function under the integral sign in the integral

$$u(0,t) = \frac{1}{2\sqrt{\pi}} \int_{-\infty}^{\infty} \frac{1}{\sqrt{(a^2 t)}} e^{-\frac{\xi^2}{4a^2 t}} \psi(\xi)\, d\xi$$

is odd with respect to ξ, since it is the product of an odd function and an even one. The integral of the odd function with limits symmetrical with respect to the origin of coordinates equals zero; hence

$$u(0,t) = 0 ,$$

which proves the lemma.

2. *If the function $\psi(x)$ is an even function i.e.*

$$\psi(x) = \psi(-x),$$

then the derivative of the function $u(x, t)$ from formula (31) equals zero at $x = 0$

$$\frac{\partial u}{\partial x}(0,t) = 0$$

for all $t > 0$.

In fact

$$\left.\frac{\partial u}{\partial x}\right|_{x=0} = -\frac{1}{2\sqrt{\pi}} \int_{-\infty}^{+\infty} \frac{(x-\xi)}{2\,(a^2 t)^{3/2}}\, e^{-\frac{(x-\xi)^2}{4a^2 t}}\, \psi(\xi)\, d\xi \Big|_{x=0} = 0\,,$$

since at $x = 0$ the function under the integral sign is odd, if $\psi(\xi)$ is even.

The function $u\,(x, t)$ may now be found. We introduce an auxiliary function $U(x, t)$ defined in the infinite region $-\infty < < x < \infty$ and satisfying the equation, and also the conditions

$$U\,(0,\ t) = 0\,,$$
$$U\,(x, 0) = \varphi\,(x) \quad \text{where} \quad x > 0.$$

Making use of the first lemma, it is possible to determine this function from an initial function $\Psi(x)$ corresponding to $\varphi(x)$ for $x > 0$ and an odd function for $x < 0$ i.e.

$$\Psi\,(x) = \begin{cases} \varphi\,(x) & \text{where} \quad x > 0\,, \\ -\varphi\,(-x) & \text{where} \quad x < 0\,, \end{cases}$$

so that

$$U\,(x, t) = \frac{1}{2\sqrt{\pi}} \int_{-\infty}^{\infty} \frac{1}{\sqrt{(a^2 t)}}\, e^{-\frac{(x-\xi)^2}{4a^2 t}}\, \Psi\,(\xi)\, d\xi\,.$$

Considering the value of the function $U(x, t)$ only in the region of interest to us, $x \geqslant 0$ we obtain:

$$u\,(x, t) = U\,(x, t) \quad \text{where} \quad x \geqslant 0.$$

Making use of the definition of $\Psi(x)$ we have

$$U\,(x, t) = \frac{1}{2\sqrt{\pi}} \int_{-\infty}^{0} \frac{1}{\sqrt{(a^2 t)}}\, e^{-\frac{(x-\xi)^2}{4a^2 t}}\, \Psi\,(\xi)\, d\xi\, +$$
$$+ \frac{1}{2\sqrt{\pi}} \int_{0}^{\infty} \frac{1}{\sqrt{(a^2 t)}}\, e^{-\frac{(x-\xi)^2}{4a^2 t}}\, \Psi\,(\xi)\, d\xi =$$
$$= -\frac{1}{2\sqrt{\pi}} \int_{0}^{\infty} \frac{1}{\sqrt{(a^2 t)}}\, e^{-\frac{(x-\xi)^2}{4a^2 t}}\, \varphi\,(\xi)\, d\xi\, +$$
$$+ \frac{1}{2\sqrt{\pi}} \int_{0}^{\infty} \frac{1}{\sqrt{(a^2 t)}}\, e^{-\frac{(x-\xi)^2}{4a^2 t}}\, \varphi\,(\xi)\, d\xi\,,$$

on substituting $\xi' = -\xi$ in the first integral and using the equality

$$\Psi(\xi) = -\varphi(-\xi) = -\varphi(\xi').$$

Adding the two integrals together, we obtain the function

$$u_1(x, t) = \frac{1}{2\sqrt{\pi}} \int_0^\infty \frac{1}{\sqrt{(a^2 t)}} \{e^{-\frac{(x-\xi)^2}{4a^2t}} - e^{-\frac{(x+\xi)^2}{4a^2t}}\} \varphi(\xi)\, d\xi \quad (32)$$

in a form not containing the auxiliary function. We observe that at $x = 0$ the expression in brackets reduces to zero and $u_1(0, t) = 0$.

Making use of lemma 2 it is easy to show that the solution of the equation of heat conduction with a homogenous boundary condition of the second kind $\frac{\partial \bar{u}_1}{\partial x}(0, t) = 0$ and the initial condition $\bar{u}_1(x, 0) = \varphi(x)$ is represented by

$$\bar{u}_1(x, t) = \frac{1}{2\sqrt{\pi}} \int_0^\infty \frac{1}{\sqrt{(a^2 t)}} \{e^{-\frac{(x-\xi)^2}{4a^2t}} + e^{-\frac{(x+\xi)^2}{4a^2t}}\} \varphi(\xi)\, d\xi . \quad (32')$$

Let us apply the formula obtained to the problem of the cooling of a uniformly heated rod, whose end is kept at zero temperature. We have to solve the equation of heat conduction, with the conditions

$$v_1(x, t_0) = T, \quad v_1(0, t) = 0 .$$

Because the initial condition is given not at $t = 0$ but at $t = t_0$, in place of formula (32) we have;

$$v_1(x, t) = \frac{T}{2\sqrt{\pi}} \int_0^\infty \{e^{-\frac{(x-\xi)^2}{4a^2(t-t_0)}} - e^{-\frac{(x+\xi)^2}{4a^2(t-t_0)}}\} \frac{d\xi}{\sqrt{[a^2(t-t_0)]}} . \quad (33)$$

Dividing the integral into two components and introducing the variables

$$\alpha = \frac{\xi - x}{2\sqrt{[a^2(t-t_0)]}}, \quad \alpha_1 = \frac{\xi + x}{2\sqrt{[a^2(t-t_0)]}},$$

we obtain:

$$v_1(x,t) = \frac{T}{\sqrt{\pi}}\left[\int\limits_{-\frac{x}{2\sqrt{[a^2(t-t_0)]}}}^{\infty} e^{-a^2}\,da - \int\limits_{\frac{x}{2\sqrt{[a^2(t-t_0)]}}}^{\infty} e^{-a_1^2}\,da_1\right] =$$

$$= \frac{T}{\sqrt{\pi}} \int\limits_{-\frac{x}{2\sqrt{[a^2(t-t_0)]}}}^{\frac{x}{2\sqrt{[a^2(t-t_0)]}}} e^{-a^2}\,da = T\frac{2}{\sqrt{\pi}} \int\limits_{0}^{\frac{x}{2\sqrt{[a^2(t-t_0)]}}} e^{-a^2}\,da$$

or

$$v_1(x,t) = T\Phi\left(\frac{x}{2\sqrt{[a^2(t-t_0)]}}\right), \tag{33'}$$

where

$$\Phi(z) = \frac{2}{\sqrt{\pi}}\int\limits_0^z e^{-a^2}\,da$$

is the error integral.

Let us now determine the function $u_2(x,t)$ which represents the second part of the solution of the first boundary-value problem.

Let

$$\mu(t) = \mu_0 = \text{const.}$$

The function

$$\bar{v}(x,t) = \mu_0\Phi\left(\frac{x}{2\sqrt{[a^2(t-t_0)]}}\right) \tag{34}$$

is a solution of the equation of heat conduction, satisfying the conditions

$$\bar{v}(x,t_0) = \mu_0, \quad \bar{v}(0,t) = 0.$$

Hence it follows that the function

$$v(x,t) = \mu_0 - \bar{v}(x,t) = \mu_0\left[1 - \Phi\left(\frac{x}{2\sqrt{[a^2(t-t_0)]}}\right)\right] \tag{35}$$

is the required solution because it satisfies the required conditions

$$v(x,t_0) = 0 \qquad (x > 0) \text{ and } v(0,t) = \mu_0 \qquad (t > t_0).$$

We can represent $v(x,t)$ in the form

$$v(x,t) = \mu_0 U(x,t),$$

where

$$U(x, t) = 1 - \Phi\left(\frac{x}{2\sqrt{[a^2(t-t_0)]}}\right) = \frac{2}{\sqrt{\pi}} \int\limits_{\frac{x}{2\sqrt{[a^2(t-t_0)]}}}^{\infty} e^{-a^2} da \qquad (36)$$

is a solution of the same problem as $v(x, t)$, but with $\mu_0 = 1$.

By definition the function $U(x, t)$ has meaning only for $t \geqslant t_0$. Let us extend the definition of this function assuming

$$U(x, t) \equiv 0 \text{ where } t < t_0 .$$

It is obvious that this definition gives the correct value of the function $U(x, t)$ at $t = 0$ and a function thus defined will satisfy the equation of heat conduction for all t at $x > 0$. The boundary value of this function (at $x = 0$) is a step function, equal to zero for $t < t_0$ and equal to unity for $t > t_0$. The function $U(x, t)$ is an auxiliary link in the determination of the function $u_2(x, t)$.

Let us examine a second auxiliary problem. Find the solution of the equation of heat conduction for the following initial and boundary conditions:

$$v(x, t_0) = 0 , \quad v(0, t) = \mu(t) = \begin{cases} \mu_0 \text{ when } t_0 < t < t_1 , \\ 0 \text{ when } t > t_1 . \end{cases}$$

By direct examination one can see that

$$v(x, t) = \mu_0 [U(x, t - t_0) - U(x, t - t_1)] .$$

In general if a bounded function $\mu(t)$ is given in the form of a step function

$$\mu(t) = \begin{cases} \mu_0 & \text{when } t_0 < t \leqslant t_1 , \\ \mu_1 & \text{when } t_1 < t \leqslant t_2 , \\ \dots & \dots \\ \dots & \dots \\ \mu_{n-1} & \text{when } t_{n-1} < t \leqslant t_n , \end{cases}$$

then, by complete analogy, we obtain the solution of the boundary-value problem II in the following form:

$$u(x, t) = \sum_{i=0}^{n-2} \mu_i [U(x, t - t_i) - U(x, t - t_{i+1})] + \\ + \mu_{n-1} U(x, t - t_{n-1}) . \qquad (37)$$

Making use of the mean value theorem, we obtain:

$$u(x,t)=\sum_{t=0}^{n-2}\mu_i\,\frac{\partial U(x,t-\tau)}{\partial t}\bigg|_{\tau_i}\Delta\tau+\mu_{n-1}U(x,t-t_{n-1}) \quad (38)$$

$$\text{when}\quad t_i\leq\tau_i\leq t_{i+1}.$$

We can now find a solution $u(x, t)$ of the equation of heat conduction with a zero initial condition and boundary condition

$$u(0,t)=\mu(t)\quad(t>0),$$

where $\mu(t)$ is an arbitrary partly-continuous function. An approximate solution of this problem is easily obtained in the form (37), if the function $\mu(t)$ is replaced by a step function. Passing to a limit we obtain that the limit of the sum (38) equals

$$\int_0^t\frac{\partial U}{\partial t}(x,t-\tau)\,\mu(\tau)\,dt,$$

since for $x>0$

$$\lim_{t-t_{n-1}\to 0}\mu_{n-1}U(x,t-t_{n-1})=0.$$

Thus the solution $u_2(x, t)$ of the second problem must equal

$$u_2(x,t)=\int_0^t\frac{\partial U}{\partial t}(x,t-\tau)\,\mu(\tau)\,d\tau. \quad (39)$$

We shall not discuss the validity of the limiting transition and the conditions of applicability of this formula in relation to the function $\mu(\tau)$.

One is easily satisfied that

$$\frac{\partial U}{\partial t}(x,t)=\frac{\partial}{\partial t}\left(\frac{2}{\sqrt{\pi}}\int_{\frac{x}{2\sqrt{(a^2t)}}}^{\infty}e^{-\alpha^2}\,d\alpha\right)=\frac{1}{2\sqrt{\pi}}\,\frac{a^2x}{[a^2t]^{3/2}}\,e^{-\frac{x^2}{4a^2t}}=$$

$$=-2a^2\frac{\partial G}{\partial x}(x,0,t)=2a^2\frac{\partial G}{\partial\xi}\bigg|_{\xi=0}\quad\left(G=\frac{1}{2\sqrt{\pi}}\,\frac{1}{\sqrt{(a^2t)}}\,e^{-\frac{(x-\xi)^2}{4a^2t}}\right).$$

Thus the solution being sought in the case of the arbitrary function $\mu(t)$ may be described in the form

$$u_2(x,t)=\frac{a^2}{2\sqrt{\pi}}\int_{t_0}^{t}\frac{x}{[a^2(t-\tau)]^{3/2}}\,e^{-\left[\frac{x^2}{4a^2(t-\tau)}\right]}\mu(\tau)\,d\tau$$

or

$$u_2(x, t) = 2a^2 \int_{t_0}^{t} \frac{\partial G}{\partial \xi}(x, 0, t - \tau)\, \mu(\tau)\, d\tau^1). \tag{40}$$

We note that in the process of obtaining formula (40) we nowhere made use of the special properties of the equation of heat conduction, except its linearity, nor did we make use of the analytical form of the function $U(x, t)$, but only the fact that is satisfies the boundary and initial conditions

$$U(0, t) = 1 \quad \text{when} \quad t > 0,$$
$$U(x, 0) = 0 \quad \text{when} \quad x > 0$$

or

$$U(0, t) = \begin{cases} 1 & \text{when} \quad t > 0, \\ 0 & \text{when} \quad t < 0. \end{cases}$$

It is obvious that if we are dealing with the solution of any linear differential equation with a boundary condition

$$u(0, t) = \mu(t) \quad (t > 0),$$

with zero initial conditions and zero additional conditions, if such occur (for instance at $x = l$), then the solution of this problem may be represented in the form

$$u(x, t) = \int_0^t \frac{\partial U}{\partial t}(x, t - \tau)\, \mu(\tau)\, d\tau, \tag{41}$$

where $U(x, t)$ is the solution of a similar boundary-value problem with

$$U(0, t) = 1.$$

The principle formulated here, called *Duhammel's principle*, shows that the fixed boundary value presents the fundamental difficulty in solving boundary-value problems. If the boundary-value problem with a fixed boundary value is solved, then the solution of the boundary-value problem with a variable boundary

[1] This description of the solution of the first boundary-value problem with zero initial conditions is given here for the sake of comparison with the solution of the same problem obtained in Chapter V, § 4, by another method.

condition is represented by formula (41). One uses this principle in solving a great many boundary-value problems. One finds a solution only for a fixed boundary condition, and remembers that the solution of the boundary-value problem with variable $\mu(t)$ is given by formula (41).

The sum of the functions

$$u_1(x, t) + u_2(x, t)$$

gives a solution of the first boundary-value problem for a semi-infinite region for a homogeneous equation.

Making use of formula (29) sub-section 2, § 3, and the principle of odd extension of a function it is easily shown that the solution of the inhomogeneous equation

$$u_t = a^2 u_{xx} + f(x, t) \qquad (0 < x < \infty, \quad t > 0)$$

with a zero initial and zero boundary condition $[u(0, t) = 0]$ is given by the formula

$$u_3(x, t) =$$

$$= \frac{1}{2\sqrt{\pi}} \int_0^\infty \int_0^t \frac{1}{\sqrt{[a^2(t-\tau)]}} \left\{ e^{-\frac{(x-\xi)^2}{4a^2(t-\tau)}} - e^{-\frac{(x+\xi)^2}{4a^2(t-\tau)}} \right\} f(\xi, \tau)\, d\xi\, d\tau . \qquad (42)$$

The sum

$$u_1(x, t) + u_2(x, t) + u_3(x, t) = u(x, t) \qquad (43)$$

gives the general solution of the first boundary-value problem

$$u_t = a^2 u_{xx} + f(x, t),$$
$$u(0, t) = \mu(t), \quad u(x, 0) = \varphi(x).$$

§ 4. Steady state problems

If the process of heat conduction is investigated long after the initial time, then the effect of the initial conditions in practice is not observable. In this case the problem becomes one of finding the solution of the equation of heat conduction which satisfies the boundary conditions of one of three types, given for all $t > -\infty$. If the rod is finite then boundary conditions are given at both ends. For a semi-infinite rod only one boundary condition is given.

Let us examine the first boundary-value problem for a semi-infinite rod:

Find the bounded solution of the equation of heat conduction in the region $x > 0$, *satisfying the condition*

$$u(0, t) = \mu(t), \tag{1}$$

where $\mu(t)$ *is a given function. Let it be assumed that the functions* $u(x, t)$ *and* $\mu(t)$ *are bounded everywhere, i.e.*

$$|u(x, t)| < M,$$
$$|\mu(t)| < M.$$

As will be shown below (see the small type), the function $u(x, t)$ is well defined. The most common boundary condition is

$$\mu(t) = A \cos \omega t. \tag{2}$$

This problem was investigated by Fourier and was applied to a determination of the temperature oscillations of the earth.*

Let us write the boundary condition in the form

$$\mu(t) = Ae^{i\omega t}. \tag{2'}$$

From the linearity of the equation of heat conduction it follows that the real and imaginary part of any complex solution of the equation of heat conduction each satisfies the same equation.

If a solution of the equation of heat conduction, satisfying relation (2') is found, then its real part satisfies relation (2) and the imaginary part the relation

$$u(0, t) = \mu_1(t) = A \sin \omega t.$$

Thus let us consider the problem

$$\left.\begin{aligned} u_t &= a^2 u_{xx}, \\ u(0, t) &= Ae^{i\omega t}. \end{aligned}\right\} \tag{3}$$

Its solution will be sought in the form

$$u(x, t) = Ae^{\alpha x + \beta t}, \tag{4}$$

where α and β are undefined constants.

* See Appendix I.

Putting expression (4) in equation (3) and the boundary condition, we find:

$$\alpha^2 = \frac{1}{a^2}\beta ,$$

$$\beta = i\omega ,$$

hence

$$\alpha = \pm \sqrt{\left(\frac{\beta}{a^2}\right)} = \pm \sqrt{\left(\frac{\omega}{a^2}\right)}\sqrt{i} = \pm \sqrt{\left(\frac{\omega}{a^2}\right)}\frac{(1+i)}{\sqrt{2}} =$$

$$= \pm \left[\sqrt{\left(\frac{\omega}{2a^2}\right)} + i\sqrt{\left(\frac{\omega}{2a^2}\right)}\right].$$

For $u(x, t)$ we have:

$$u(x, t) = Ae^{\pm\sqrt{\left(\frac{\omega}{2a^2}\right)}x + i\left(\pm\sqrt{\left[\frac{\omega}{2a^2}\right]}x + \omega t\right)}. \tag{5}$$

The real part of this solution

$$u(x, t) = Ae^{\pm\sqrt{\left(\frac{\omega}{2a^2}\right)}x}\cos\left(\pm\sqrt{\left[\frac{\omega}{2a^2}\right]}x + \omega t\right) \tag{6}$$

satisfies the equation of heat conduction and the boundary relation (2). Formula (6) depending on the choice of sign defines not one, but two functions. But only the function corresponding to the minus sign is bounded. Thus, we obtain the solution of the problem

$$u(x, t) = Ae^{-\sqrt{\left(\frac{\omega}{2a^2}\right)}x}\cos\left(-\sqrt{\left[\frac{\omega}{2a^2}\right]}x + \omega t\right). \tag{7}$$

The steady state problem for a finite segment can be solved similarly:

$$\left.\begin{aligned} u_t &= a^2 u_{xx} , \\ u(0, t) &= A\cos\omega t , \\ u(l, t) &= 0 . \end{aligned}\right\} \tag{8}$$

Rewriting the boundary condition as

$$\hat{u}(0, t) = Ae^{-i\omega t}, \quad \hat{u}(l, t) = 0 ,$$

we look for a solution in the form

$$\hat{u}(x, t) = X(x)\, e^{-i\omega t}. \tag{9}$$

Putting this expression in equation (8) we obtain an equation for the function $X(x)$

$$X'' + \frac{i\omega}{a^2} X = 0 \text{ where } X'' + \gamma^2 X = 0 ,$$

$$\gamma = \sqrt{\frac{i\omega}{a^2}} = \sqrt{\left(\frac{\omega}{2a^2}\right)} (1 + i) \tag{10}$$

with additional conditions

$$X(0) = A , \; X(l) = 0 . \tag{11}$$

Solving it we get the function $X(x)$

$$X(x) = A \frac{\sin \gamma (l - x)}{\sin \gamma l} = X_1(x) + iX_2(x) , \tag{12}$$

where X_1 and X_2 are its real and imaginary parts. For the function $\hat{u}(x, t)$ we obtain the expression

$$\hat{u}(x, t) = A \frac{\sin \gamma (l - x)}{\sin \gamma l} e^{-i\omega t}. \tag{13}$$

Separating the real part we find the solution of the steady state problem

$$u(x, t) = X_1(x) \cos \omega t + X_2(x) \sin \omega t . \tag{14}$$

We will not give an explicit expression for X_1 and X_2, although this can easily be done.

If the function $u(t)$ is a superposition of harmonics of different frequency, then the solution may be obtained as a superposed solution, corresponding to the separate harmonics.

Let us prove the uniqueness of the steady state problem for a semi-infinite region. We shall proceed from the formula

$$u(x, t) = \frac{a^2}{2\sqrt{\pi}} \int_{t_0}^{t} \frac{x}{[a^2(t-\tau)]^{3/2}} e^{-\frac{x^2}{4a^2(t-\tau)}} u(0, \tau)\, d\tau +$$

$$+ \frac{1}{2\sqrt{\pi}} \int_0^\infty \frac{1}{\sqrt{[a^2(t-t_0)]}} \{e^{-\frac{(x-\xi)^2}{4a^2(t-t_0)}} - e^{-\frac{(x+\xi)^2}{4a^2(t-t_0)}}\} u(\xi, t_0)\, d\xi = I_1 + I_2 \tag{15}$$

$$(t \geqslant t_0) ,$$

which represents any bounded solution of the equation of heat conduction in terms of its initial value $u(x, t_0)$ and boundary value $u(0, t) = \mu(t)$ in the region $x \geqslant 0, t \geqslant t_0$.

Let us prove that

$$\lim_{t_0 \to -\infty} I_2(x, t_0) = 0, \tag{16}$$

provided

$$|u(x, t)| < M$$

for any t. Actually,

$$|I_2| < \frac{M}{\sqrt{\pi}} \left\{ \int_{-\frac{x}{2\sqrt{[a^2(t-t_0)]}}}^{\infty} e^{-\alpha_1^2} d\alpha_1 - \int_{\frac{x}{2\sqrt{[a^2(t-t_0)]}}}^{\infty} e^{-\alpha_2^2} d\alpha_2 \right\} =$$

$$= \frac{M}{\sqrt{\pi}} 2 \int_0^{\frac{x}{2\sqrt{[a^2(t-t_0)]}}} e^{-\alpha^2} d\alpha,$$

where

$$\alpha_1 = \frac{\xi - x}{2\sqrt{[a^2(t-t_0)]}} \text{ and } \alpha_2 = \frac{\xi + x}{2\sqrt{[a^2(t-t_0)]}}.$$

The result (16) follows, if x and t are fixed and $t_0 \to -\infty$. If in formula (15) x and t are fixed and we let $t_0 \to -\infty$ then $u(x, t)$ will equal the limit of the first component T_1 and we obtain the formula (17) proving that two

$$u(x, t) = \frac{a^2}{2\sqrt{\pi}} \int_{-\infty}^{t} \frac{x}{[a^2(t-\tau)]^{3/2}} e^{-\frac{x^2}{4a^2(t-\tau)}} \mu(\tau)\, d\tau, \tag{17}$$

different solutions of our problem are not possible. It can also be proved that for any bounded partly-continuous function $\mu(t)$ formula (17) represents the solution of the required problem.

The steady state problem can be investigated for a finite segment $(0 \leqslant x \leqslant l)$ similarly. This problem without the conditions of boundedness has many solutions, because the function

$$u_n(x, t) = Ce^{-\left(\frac{\pi n}{l}\right)^2 a^2 t} \sin \frac{\pi n}{l} x$$

for any n is a solution with zero boundary values. But such a solution becomes infinite for $t \to -\infty$, and it is not difficult to prove the uniqueness of the bounded solution.

Problems on Chapter III

1. Find the source function for:

(a) a semi-infinite rod with boundary conditions of first and second type and with no heat exchange at the surface;

(b) an infinite rod with heat exchange at the surface;

(c) a semi-infinite rod with heat exchange at the surface and with boundary conditions of the first two types.

2. Find the source function for a semi-infinite rod thermally insulated along the surface for the third boundary-value problem [the boundary condition is of the form $\frac{\partial u}{\partial x} - hu(0, t) = f(t)$].

Answer :

$$G(x, \xi, t-\tau) = \frac{1}{2\sqrt{[\pi a^2 (t-\tau)]}} \left\{ e^{-\frac{(x-\xi)^2}{4a^2(t-\tau)}} + \right.$$

$$\left. + e^{-\frac{(x+\xi)^2}{4a^2(t-\tau)}} - 2h \int_0^\infty e^{-hz-\frac{(a+\xi+x)^2}{4a^2(t-\tau)}} da \right\}.$$

3. Solve the equation of heat conduction for cases (a), (b), (c) in problem 1, if:

(1) at the point $x = \xi_0$ there is a source of heat $Q = Q(t)$, in particular $Q = Q_0 = \text{const.}$;

(2) there is an initial distribution of temperature $u(x, 0) = \varphi(x)$, in particular

$$\varphi(x) = \begin{cases} u_0 \text{ for } 0 < x < l, \\ 0 \text{ outside } (0, l); \end{cases}$$

(3) the heat sources are distributed with density $f(x, t)$ over the entire rod, and the initial temperature equals zero; examine, in particular, the case $f = q_0 = \text{const.}$ (stationary source).

4. A semi-infinite rod thermally insulated along the surface was heated uniformly to a temperature

$$u(x, 0) = u_0 = \text{const.} \quad (x > 0).$$

The end of the rod, starting at time $t = 0$, is maintained at zero temperature

$$u(0, t) = 0 \quad (t > 0).$$

Find the temperature of the rod $u(x, t)$ and, making use of the tables of the error integral

$$\Phi(z) = \frac{2}{\sqrt{\pi}} \int_0^z e^{-a^2} da,$$

construct a graph in x over the interval $0 \leqslant x \leqslant l$ of the function $u(x, t)$ for $t = l^2/16a^2$, $t = l^2/2a^2$, $t = l^2/a^2$.

Hint. It is useful to introduce the dimensionless variables

$$x' = \frac{x}{l}, \quad \theta = \frac{a^2 t}{l^2}, \quad v = \frac{u}{u^0}.$$

5. The end of a semi-infinite cylinder opens at the initial moment $t = 0$ to the atmosphere where the concentration of a certain gas equals u_0.

Find the concentration of the gas in the cylinder $u(x, t)$ for $t > 0$, $x > 0$, if the initial concentration $u(x, t) = 0$. Making use of the tables of the error integral, find the time at which the concentration of the gas reaches 95 per cent of the external concentration in a layer, which is at a distance l from the end of the cylinder. Find the law of motion of the front of constant concentration.

6. A heat current $ku_x(0, t) = q(t)$ is supplied to the end of a semi-infinite rod, whose initial temperature was equal to zero. Find the temperature $u(x, t)$ of the rod, if:

(a) the rod is thermally insulated at the surface;

(b) heat exchange occurs at the surface of the rod (according to Newton's law) with a medium of zero temperature.

Examine the special case $q = q_0 = \text{const}$.

7. The end of a semi-infinite rod is maintained at the constant temperature u_0; heat exchange occurs at the surface of the rod with a medium whose steady temperature equals u_1. The initial temperature of the rod equals zero. Find $u(x, t)$ the temperature of the rod.

8. Solve problems 6a, 6b, assuming that $u(x, 0) = u_0 = \text{const}$.

9. Find the steady temperature along a semi-infinite rod thermally insulated along the surface, at the end of which

(a) there is an assigned temperature $u(0, t) = A \cos \omega t$;

(b) there is an assigned heat current $Q(t) = B \sin \omega t$;

(c) a heat exchange occurs according to Newton's law with a medium whose temperature varies according to the law $v(t) = C \sin \omega t$.

10. Making use of the method of images, construct the source function for a finite rod thermally insulated along the surface with boundary conditions of first and second types.

11. An infinite rod consists of two homogeneous rods, touching at the point $x = 0$ and possessing the characteristics a_1, k_1 and a_2, k_2 respectively. The initial temperature

$$u(x, 0) = \varphi(x) = \begin{cases} T_1 & \text{where } x < 0, \\ T_2 & \text{where } x > 0. \end{cases}$$

Find the temperature $u(x, t)$ of the rod for the case where the surface is thermally insulated.

APPENDICES TO CHAPTER III

I. Temperature waves

The problem of propagation of temperature waves in the earth was one of the first examples of the application of the mathematical theory of heat conduction, developed by Fourier, to the study of natural phenomena.

The temperature at the surface of the earth, as is well known, is characterized by daily and yearly periodicity. Let us turn to the problem of propagation of periodic temperature oscillations in the earth, which we will consider as a homogeneous region $0 \leqslant x \leqslant \infty$. This is a typical steady state problem, since after repeated changes of the temperature at the surface, the effect of the initial temperature will be less than the effect of other factors, which we have already neglected (e.g. the inhomogeneity of the earth). Thus we have the following problem:*

Find the bounded solution of the equation of heat conduction

$$\frac{\partial u}{\partial t} = a^2 \frac{\partial^2 u}{\partial x^2} \quad (0 \leqslant x < \infty , \quad -\infty < t), \tag{1}$$

satisfying the condition

$$u(0, t) = A \cos \omega t . \tag{2}$$

This problem was examined in Chapter III. Its solution has the form (see Chapter III, § 4, (7))

$$u(x, t) = Ae^{-\sqrt{\left(\frac{\omega}{2a^2}\right)}\, x} \cos \left(\sqrt{\left[\frac{\omega}{2a^2}\right]}\, x - \omega t\right) . \tag{3}$$

On the basis of this solution we can give the following properties of temperature waves in the earth. If the temperature of the surface varies periodically for a long time, then a temperature oscillation of the same period is also established in the earth whereupon:

* Kh. S. Carslaw, *Theory of Heat Conduction*, chap. III, Gostekhizdat 1947.

1. The amplitude of the oscillation decreases exponentially with depth $A(x) = Ae^{-\sqrt{(\omega/2a^2)}x}$, (Fourier's first law).

2. Temperature oscillations in the earth are out of phase with those at the surface. The time lag δ of the maximum (minimum) temperature in the earth compared with that at the surface is proportional to the depth

$$\delta = \sqrt{\left(\frac{1}{2\omega a^2}\right)}x$$

(Fourier's second law).

3. The depth of penetration of the temperature waves into the earth depends on the period of oscillation at the surface. The relative change of temperature amplitude equals

$$\frac{A(x)}{A} = e^{-\sqrt{\left(\frac{\omega}{2a^2}\right)}x}.$$

This formula proves that the smaller the period, the less the depth of penetration of the waves. For temperature oscillations of periods T_1 and T_2 the depths x_1 and x_2, at which the same relative change of temperature occurs, are connected by the relation

$$x_2 = \sqrt{\left(\frac{T_2}{T_1}\right)}x_1$$

(Fourier's third law). Thus, for example, comparison of daily and yearly oscillations, for which $T_2 = 365T_1$, shows that

$$x_2 = \sqrt{365}\,x_1 = 19\cdot 1\,x_1,$$

i.e. that the depth of penetration of yearly oscillations of the same amplitude at the surface would be 19.1 times greater than the depth of penetration of the daily oscillations.

As an example let us quote the results of observations on yearly temperature oscillations at the station of Gosh in Priamur*:

Depth (m)	Amplitude (°C)
1	11·5
2	6·8
3	4·2
4	2·6

* M. I. Sumgin, S. P. Kachurin, N. I. Tolstikhin, V. F. Tumel', *General Solid State*, chap. V, Publ. Acad. of Sciences, U. S. S. R., 1940.

These results show that the amplitude of yearly oscillations at a depth of 4 m is only 13.3 per cent of the value at the surface which is equal to 19.5°.

On the basis of these results it is possible to find the coefficient of thermal conductivity of the earth

$$\ln \frac{A(x)}{A} = -\sqrt{\left(\frac{\omega}{2a^2}\right)}\, x\,, \quad a^2 = \frac{\omega x^2}{2 \ln^2 \frac{A(x)}{A}}\,,$$

from which we find that the coefficient of thermal conductivity of the earth equals

$$a^2 = 4 \cdot 10^{-3} \, \frac{\text{cm}^2}{\text{sec}}\,.$$

The time lag of the maximum temperature at a depth of 4 m is 4 months.

However, one must realize that the theory developed here refers to the conduction of heat in dry earth or rock. The presence of moisture complicates the temperature phenomena. For example on freezing a liberation of latent heat occurs, not accounted for by this theory.

Thermal conductivity is one of the characteristics of a body, important for a study of its physical properties and also for various technical calculations. A study of temperature waves in rods provides one method for determining thermal conductivity.

A periodic temperature $u(t)$ is maintained at the end of a sufficiently long rod. We expand this function as a Fourier series

$$\mu(t) = \frac{a_0}{2} + \sum_{n=1}^{\infty} \left(a_n \cos \frac{2\pi n}{T} t + b_n \sin \frac{2\pi n}{T} t\right) =$$

$$= \frac{a_0}{2} + \sum_{n=1}^{\infty} A_n \cos \left[\frac{2\pi n}{T} (t - \delta_n^0)\right],$$

$$A_n = \sqrt{(a_n^2 + b_n^2)}\,,$$

$$\delta_n^0 = \frac{T}{2\pi n}\left(\pi + \arctan \frac{b_n}{a_n}\right).$$

where T is the period, and considering the temperature waves corresponding to each component, we obtain the temperature $u(x, t)$ for any x as a periodic function of time. Its nth harmonic equals

$$u_n(x, t) = a_n(x) \cos \frac{2\pi n}{T} t + b_n(x) \sin \frac{2\pi n}{T} t =$$

$$= A_n e^{-\sqrt{\left(\frac{\pi n}{Ta^2}\right)} x} \cos\left[\sqrt{\left(\frac{\pi n}{Ta^2}\right)} x - \frac{2\pi n}{T} t + \delta_n^0\right]$$

or

$$\frac{\sqrt{[a_n^2(x_1) + b_n^2(x_1)]}}{\sqrt{[a_n^2(x_2) + b_n^2(x_2)]}} = e^{-\sqrt{\left(\frac{\pi n}{Ta^2}\right)}(x_1 - x_2)}$$

This formula shows that if we measure the temperature at any two points x_1 and x_2 over a complete period and determine the coefficients $a_n(x_1)$, $b_n(x_1)$, $a_n(x_2)$, $b_n(x_2)$ by harmonic analysis, it is possible to determine the coefficient of thermal conductivity, a^2 of the rod.

Periodic oscillations of temperature in the rod can be produced, for example, in the following way. Let us place one end of the rod in an electric oven, and let us switch on and switch off a current for equal intervals of time. As a result of such a periodic heating periodic oscillations of temperature are established in the rod after a certain time; measuring the temperatures $u(x_1, t)$ and $u(x_2, t)$ at any two points x_1 and x_2 over a complete period by means of a thermocouple and subjecting u_1 and u_2 to the analysis described above, it is possible to determine a^2 the coefficient of thermal conductivity of the material from which the rod is made. In order to apply the theory, the rod must be thermally insulated along the surface and also a temperature control must be applied at the other end of the rod in order to be able to use the theory of temperature waves in a semi-infinite rod. In order to use the theory of temperature waves in a semi-infinite rod one must be satisfied that the temperature at the free end of the rod is constant. This is controlled by means of an additional thermocouple.

II. The effect of radioactive decay on the temperature of the earth's crust

All that we know about the internal thermal state of the earth is deduced from a few facts obtained from observations at its surface. The essential information about the thermal state of the earth's crust consists of the following. Daily and yearly oscillations of temperature occur in a relatively thin surface layer (of the order 10–20 m for yearly oscillations). Below this layer the temperature changes very slowly with time.

Observations in mines and holes made in the upper 2–3 km of the earth's crust, show that the temperature increases with depth by an average of 3°C per 100 m.

The first calculations made at the end of the last century, aimed to give a theoretical explanation of the observed geo thermic gradient but encountered insurmountable difficulties.* These calculations aimed to give a description of the cooling of the earth. The initial temperature must have been of the order $T_0 = 1200\,°C$ (the fusion temperature of rock), and the surface temperature of the order 0°C which must here been approximatly the same for the entire period of existence of life on the earth. The simplest quantitative theory of cooling of the earth leads to the equation of heat conduction

$$\frac{\partial u}{\partial t} = a^2 \frac{\partial^2 u}{\partial z^2}$$

in the semi-space $0 < z < \infty$ for these initial and boundary conditions:

$$u(z, 0) = T_0,$$

$$u(0, t) = 0.$$

The solution of this problem was considered in § 3 of the present chapter and was given by the formula

$$u(z, t) = T_0 \frac{2}{\sqrt{\pi}} \int_0^{\frac{z}{2\sqrt{(a^2 t)}}} e^{-\alpha^2} d\alpha .$$

The gradient of this function for $z = 0$ equals

$$\left.\frac{\partial u}{\partial z}\right|_{z=0} = \left.\frac{T_0}{\sqrt{\pi}\sqrt{(a^2 t)}} e^{-\frac{z^2}{4a^2 t}}\right|_{z=0} = \frac{T_0}{\sqrt{\pi}\sqrt{(a^2 t)}} .$$

Substituting here the known values of the geothermic gradient $\gamma = 3 \times 10^{-4}$ °C/cm, $T_0 = 1200$°C, and also the value of $a^2 = 0.006$ cm²/sec, corresponding to the experimentally determined

* Kh. S. Carslaw, *The Theory of Heat Conduction*, chap. III, Gostekhizdat, 1947.

mean coefficient of thermal conductivity of granite and basalt, we obtain a value for the time of cooling $t = 0.85 \times 10^{15}$ sec $=$ $= 27{,}000{,}000$ years. Such an estimate of the age of the earth is in complete disagreement with geological observations. The approximate nature of the theory under consideration (neglecting the curvature of the earth, the variability of the coefficient of thermal conductivity, the uncertainty of the value of T_0), cannot, of course, change the order of magnitude of the value found for the age of the earth, which according to recent data is estimated to be approximately 2×10^9 years.

The physical picture of the thermal state of the earth was subjected to substantial revision after the discovery of the phenomenon of *radioactive disintegration*. Radioactive elements, scattered in the earth's crust, produce heat during disintegration, so that the equation of heat conduction must have the form

$$\frac{\partial u}{\partial t} = a^2 \frac{\partial^2 u}{\partial z^2} + f \quad \left(f = \frac{A}{c\varrho}\right),$$

where A is the volume density of the heat sources. On the basis of many measurements on the radioactivity of rocks and the heat produced by it the accepted value is

$$A = 1.3 \times 10^{-12} \text{ cal/cm}^3 \text{ sec.}$$

This value includes the heat produced by uranium, thorium and potassium together with their products of disintegration.

Let us assume that the density of radioactive sources inside the earth's sphere is constant and equal to the value A, found for the upper layers of the earth's crust. In this case the total amount of heat produced in the earth per unit time will equal

$$Q = \frac{4}{3} \pi R^3 A\,.$$

We make a second assumption that the temperature of the earth is falling in spite of the radioactive heat produced. In this case the flow of heat across unit surface

$$q = k \left. \frac{\partial u}{\partial z} \right|_{z=0} \geqslant \frac{Q}{4\pi R^2},$$

where k and $\left. \dfrac{\partial u}{\partial z} \right|_{z=0}$ are the coefficient of heat conduction and the geothermal gradient at the surface of the earth.

Hence we obtain a value for $\frac{\partial u}{\partial z}$ at $z = 0$

$$\left.\frac{\partial u}{\partial z}\right|_{z=0} \geqslant \frac{AR}{3k} \cong 6{\cdot}3 \times 10^{-2}\ {}^\circ\mathrm{C/cm},$$

where $R = 6.3 \times 10^3$ km is the radius of the earth and $k = 0.004$ is the mean value of the coefficient of heat conduction of sedimentary rocks.

Thus the geothermal gradient, calculated on the assumption that the distribution of radioactive elements is constant and that the temperature of the earth is not rising, exceeds the observed value

$$\gamma = 3 \times 10^{-4}\ {}^\circ\mathrm{C/cm}$$

by two orders of magnitude.

Thus we reject the hypothesis of a uniform distribution of radioactive elements and assume that the radioactive elements are distributed in a layer of thickness H at the surface of the earth. Neglecting the curvature of the earth, we obtain an equation for a steady temperature

$$\frac{\partial^2 u}{\partial z^2} = \begin{cases} -\frac{A}{k} & \text{when } 0 \leqslant z \leqslant H, \\ 0 & \text{when } z > H \end{cases}$$

with boundary conditions

$$u(0) = 0,$$

$$\left.\frac{\partial u}{\partial z}\right|_{z \to \infty} = 0.$$

It is obvious that the solution of this problem is

$$u(z) = \begin{cases} \frac{A}{k}\left(Hz - \frac{z^2}{2}\right), & 0 \leqslant z \leqslant H, \\ \frac{A}{k}\frac{H^2}{2}, & z \geqslant H, \end{cases}$$

since this function is continuous together with the first derivative at $z = H$ and satisfies the boundary conditions of the problem.

Assigning a value of the gradient of this function at $z = 0$,

$$\left.\frac{\partial u}{\partial z}\right|_{z=0} = \frac{AH}{k},$$

and comparing it with the observed value

$$\gamma = 3 \times 10^{-4}\ ^\circ\text{C/cm}$$

we find that

$$H = \frac{\gamma k}{A} \cong 10^6\ cm = 10\ \text{km}\,.$$

Let us estimate the effect of the hypothesis of a steady temperature on the value of the geothermic gradient. We consider the solution of the equation of heat conduction

$$\frac{\partial w}{\partial t} = a^2 \frac{\partial^2 w}{\partial z^2} + f,$$

$$f = \begin{cases} \dfrac{A}{c\varrho}, & 0 \leqslant z \leqslant H\,, \\ 0, & z > H \end{cases}$$

with zero initial and boundary conditions

$$w(z, 0) = 0\,,$$

$$w(0, t) = 0\,.$$

The solution of this problem is represented, as we saw in § 3, by the integral

$$w(z, t) = \int_0^\infty \int_0^t G(z, \zeta; t - \tau) f(\zeta)\, d\tau\, d\zeta\,,$$

where G is the source function for a semi-infinite region

$$G(z, \zeta; t - \tau) = \frac{1}{2\sqrt{\pi}\sqrt{[a^2(t-\tau)]}} \left\{ e^{-\frac{(z-\zeta)^2}{4a^2(t-\tau)}} - e^{-\frac{(z+\zeta)^2}{4a^2(t-\tau)}} \right\}.$$

Let us calculate the value of the gradient at $z = 0$ substituting the value of the function f:

$$\left.\frac{\partial w}{\partial z}\right|_{z=0} = \frac{A}{c\varrho 2\sqrt{\pi}} \int_0^H \int_0^t \frac{\zeta}{\sqrt{[a^2(t-\tau)]^3}} e^{-\frac{\zeta^2}{4a^2(t-\tau)}}\, d\zeta\, d\tau =$$

$$= \frac{A}{c\varrho\sqrt{\pi}} \int_0^t \frac{1}{\sqrt{[a^2(t-\tau)]}} \int_0^{\frac{H^2}{4a^2(t-\tau)}} e^{-\alpha}\, d\alpha\, d\tau =$$

$$= \frac{A}{c\varrho\sqrt{\pi}} \int_0^t \frac{1}{\sqrt{(a^2\theta)}} \left[1 - e^{-\frac{H^2}{4a^2\theta}}\right] d\theta\,, \text{ where } \theta = t - \tau\,.$$

Thus

$$\left.\frac{\partial w}{\partial z}\right|_{z=0} = \frac{A}{c\varrho\sqrt{\pi}}\left\{\frac{2\sqrt{t}}{a} - \frac{H}{a^2}\int\limits_{\sigma_0}^{\infty} e^{-\sigma^2}\frac{d\sigma}{\sigma^2}\right\},$$

where

$$\sigma = \frac{H}{2\sqrt{(a^2\theta)}},\quad \sigma_0 = \frac{H}{2\sqrt{(a^2 t)}},\quad \frac{d\sigma}{\sigma^2} = -\frac{a^2}{H}\frac{d\theta}{\sqrt{(a^2\theta)}}.$$

Let us calculate the integral

$$\int\limits_{\sigma_0}^{\infty} e^{-\sigma^2}\frac{d\sigma}{\sigma^2} = -\left.\frac{e^{-\sigma^2}}{\sigma}\right|_{\sigma_0}^{\infty} - 2\int\limits_{\sigma_0}^{\infty} e^{-\sigma^2}d\sigma = \frac{e^{-\sigma_0^2}}{\sigma_0} - 2\int\limits_{\sigma_0}^{\infty} e^{-\sigma^2}d\sigma,$$

from which

$$\left.\frac{\partial w}{\partial z}\right|_{z=0} = \frac{A}{c\varrho a^2}\left\{\frac{2a\sqrt{t}}{\sqrt{\pi}}\left[1 - e^{-\frac{H^2}{4a^2t}}\right] + H\frac{2}{\sqrt{\pi}}\int\limits_{\frac{H}{2\sqrt{(a^2t)}}}^{\infty} e^{-\sigma^2}d\sigma\right\}. \qquad (1)$$

We note that

$$\lim_{t\to\infty}\left.\frac{\partial w}{\partial z}\right|_{z=0} = \frac{A}{k}H,$$

since $c\varrho a^2 = k$, the limit of the first component in the square brackets equals zero, and the limit of the second component equals H.

Let us calculate the difference between $\frac{\partial w}{\partial z}$ and its limiting value for

$$t = 2\times 10^9\ \text{years} = 6\times 10^{16}\ \text{sec}.$$

The value σ_0 is small:

$$\sigma_0 = \frac{H}{2\sqrt{(a^2 t)}} = \frac{10^6}{2\sqrt{6\times 10^{-3}\times 6\times 10^{16})}} = \frac{1}{2\times 19} \cong 0{\cdot}025.$$

Expanding the function, appearing in formula (1) in series we obtain

$$\frac{A}{k}H - \left.\frac{\partial w}{\partial z}\right|_{z=0} = \frac{A}{k}H\left\{\frac{1}{\sqrt{\pi}\sigma_0}[\sigma_0^2 + \ldots] + \frac{2}{\sqrt{\pi}}\cdot\sigma_0\right\} \cong \frac{A}{k}H\cdot 0{\cdot}04,$$

i. e. $\left.\frac{\partial w}{\partial z}\right|_{z=0}$ differs from its limiting value by 4 per cent.

It is easy to calculate the function $w(z, t)$ for $z > 0$ and to show that for $z \geqslant H$, $w(z, t)$ still does not attain its limiting value for t equal to the age of the earth* (although, as we have agreed, the gradient at the surface is practically equal to its limiting value).

The above are of course only estimates; however, taking into consideration the very steady rate of radioactive disintegration, unaffected by temperatures and pressures available to us, we arrive at the conclusion that the concentration of radioactive elements must decrease with depth. A physical theory explaining the law of decrease of concentration of radioactive elements with depth, has not yet been given.

III. The method of similarity in the theory of heat conduction

The method of similarity is very useful for the solution of a class of problems on heat conduction. At an example let us consider two problems.

1. *The source function for an infinite region*

The equation of heat conduction remains invariant under a transformation of the variables

$$\left.\begin{aligned} x' &= kx, \\ t' &= k^2 t, \end{aligned}\right\} \tag{1}$$

i.e. if the scale of length changes by k, then the time scale should change by k^2.

We will look firstly for a solution of the equation of heat conduction

$$u_t = a^2 u_{xx} \tag{2}$$

with initial conditions

$$u(x, 0) = \begin{cases} u_0 & \text{when } x > 0, \\ 0 & \text{when } x < 0. \end{cases} \tag{3}$$

* A. N. Tikhonov: *Concerning the Effect of Radioactive Disintegration on the Temperature of the Earth's Crust*, Publ. Acad. of Sciences S. S. S. R., Branch of Maths. and Nat. Science, 1937, pp. 431–59.

By the change of scale indicated above the initial condition (3) also remains invariant, therefore the function $u(x, t)$ satisfies the equality

$$u(x, t) = u(kx, k^2 t) \tag{4}$$

for any values of x, t, and k.

Assuming

$$k = \frac{1}{2\sqrt{t}}, \tag{5}$$

we obtain:

$$u(x, t) = u\left(\frac{x}{2\sqrt{t}}, \frac{1}{4}\right) = u_0 f\left(\frac{x}{2\sqrt{t}}\right). \tag{6}$$

Thus, u depends only on the argument

$$z = \frac{x}{2\sqrt{t}}. \tag{7}$$

Calculating the derivatives of u from formula (6)

$$\frac{\partial^2 u}{\partial x^2} = u_0 \frac{d^2 f}{dz^2} \cdot \frac{1}{4t}, \qquad \frac{\partial u}{\partial t} = -\frac{x \cdot u_0}{4\, t^{3/2}} \frac{df}{dz} = -u_0 \cdot \frac{z}{2t} \frac{df}{dz},$$

substituting in the equation of heat conduction (2) and dividing by the factor $u_0/4t$, we obtain:

$$a^2 \frac{d^2 f}{dz^2} = -2z \frac{df}{dz} \tag{8}$$

with additional conditions

$$f(-\infty) = 0, \quad f(\infty) = 1, \tag{9}$$

corresponding to the initial condition for function u.

Integrating equation (8) we have:

$$a^2 \frac{f''}{f'} = -2z, \quad f' = C e^{-\frac{z^2}{a^2}}, \tag{9a}$$

$$f = C \int_{-\infty}^{z} e^{-\frac{\xi^2}{a^2}} d\xi = C_1 \int_{-\infty}^{\frac{z}{a}} e^{-\zeta^2} d\zeta.$$

Here the lower limit was chosen so that the first condition (9) would be fulfilled. In order to satisfy the second condition (9), one must assume

$$C_1 = \frac{1}{\sqrt{\pi}}.$$

Thus

$$u(x,t) = \frac{u_0}{\sqrt{\pi}} \int_{-\infty}^{\frac{x}{2\sqrt{(a^2t)}}} e^{-\xi^2}\, d\xi = \frac{u_0}{2}\left[1 + \Phi\left(\frac{x}{2\sqrt{(a^2 t)}}\right)\right], \tag{10}$$

where

$$\Phi(z) = \frac{2}{\sqrt{\pi}} \int_0^z e^{-\xi^2}\, d\xi \tag{10a}$$

(the error integral). If the initial value has the form

$$u(x,0) = \begin{cases} u_0 & \text{when } x > \bar{x}, \\ 0 & \text{when } x < \bar{x}, \end{cases} \tag{11}$$

then

$$u(x,t) = \frac{u_0}{2}\left[1 + \Phi\left(\frac{x-\bar{x}}{2\sqrt{(a^2 t)}}\right)\right]. \tag{12}$$

Let us turn now to the solution of the second auxiliary problem, where the initial values are

$$u(x,0) = \begin{cases} 0 & \text{when} \quad x_2 < x \\ u_0 & \text{when} \quad x_1 < x < x_2, \\ 0 & \text{when} \quad x < x_1. \end{cases} \tag{13}$$

In this case

$$u(x,t) = \frac{u_0}{2}\left[\Phi\left(\frac{x-x_1}{2\sqrt{(a^2 t)}}\right) - \Phi\left(\frac{x-x_2}{2\sqrt{(a^2 t)}}\right)\right].$$

The initial temperature u_0 corresponds to an amount of heat

$$Q = c\varrho\,(x_2 - x_1)\, u_0.$$

If

$$Q = c\varrho, \tag{13c}$$

then

$$u(x,t) = -\frac{1}{x_2 - x_1} \cdot \frac{1}{2}\left[\Phi\left(\frac{x-x_2}{2\sqrt{(a^2 t)}}\right) - \Phi\left(\frac{x-x_1}{2\sqrt{(a^2 t)}}\right)\right]. \tag{14}$$

The source function for an instantaneous point source should be the limit of the function $u(x,t)$ as $x_2 - x_1 \to 0$.

A limiting transition in formula (14) gives:

$$u(x,t) = -\frac{\partial}{\partial \xi}\left[\frac{1}{2}\Phi\left(\frac{x-\xi}{2\sqrt{(a^2 t)}}\right)\right]_{\xi = x_1}, \tag{15}$$

since on the right side of formula (14) there is a difference whose limit is the derivative in (15).

Differentiating, we find:

$$u(x, t) = \frac{1}{2\sqrt{\pi}} \frac{1}{\sqrt{(a^2 t)}} e^{-\frac{(x-x_1)^2}{4a^2 t}}, \tag{16}$$

i.e. $u(x, t) = G(x, x_1, t)$ the source function of an instantaneous point source.

2. *Boundary-value problems for the non-linear equation of heat conduction*

Let us consider the equation

$$\frac{\partial}{\partial x}\left[k(u)\frac{\partial u}{\partial x}\right] = c\varrho \frac{\partial u}{\partial t}. \tag{17}$$

We want to find the solution of this equation, satisfying the boundary condition

$$u(0, t) = u_1 \tag{18}$$

and the initial condition

$$u(x, 0) = u_2. \tag{19}$$

In this case the transformation (1) also does not alter equation (17) and the additional conditions (18) and (19). Hence it follows that

$$u(x, t) = f\left(\frac{x}{2\sqrt{t}}\right) = f(z) \quad \left(z = \frac{x}{2\sqrt{t}}\right). \tag{20}$$

Using this expression, we obtain the equation for f

$$\frac{d}{dz}\left[k(f)\frac{df}{dz}\right] = -2c\varrho z \frac{df}{dz} \tag{21}$$

with additional conditions

$$f(0) = u_1, \ f(\infty) = u_2. \tag{22}$$

In those cases where the function f cannot be determined analytically, it may be found by means of numerical integration.

On very general assumptions regarding the functions k and $c\varrho$ equation (21) has a unique solution satisfying conditions (22). However, we will not give a proof of this.

Let us consider as an example equation (17) in which $k(u) = k_0 + k_1(u)$ is a linear function, and $c\varrho$ is a constant. Altering the time scale and the scale of u, we obtain the equation

$$\frac{\partial}{\partial x}\left[(1 + au)\frac{\partial u}{\partial x}\right] = \frac{\partial u}{\partial t} \tag{23}$$

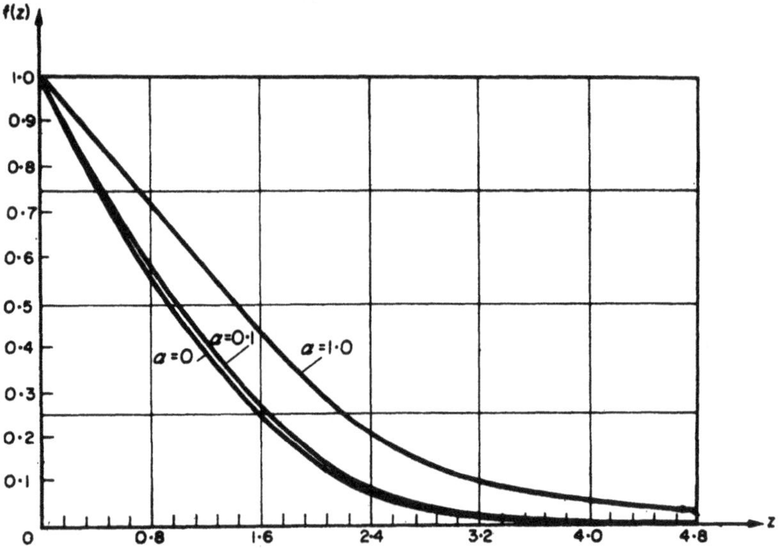

FIG. 42

with initial and boundary conditions

$$u(x, 0) = 0\,, \quad u(0, t) = 1\,. \tag{24}$$

Assuming

$$u(x, t) = f(z)\,, \quad z = \frac{x}{2\sqrt{t}}\,,$$

we obtain equations for f

$$\frac{d}{dz}\left[(1 + af)\frac{df}{dz}\right] = -\,2z\frac{df}{dz}\,, \tag{25}$$

$$f(0) = 1\,, \quad f(\infty) = 0\,. \tag{26}$$

Figure 42 shows results of the numerical integration of equation (25) for different values of a.

IV. A Problem of freezing

It is possible to produce a change in the physical state of a body by a change of temperature, for example by changing the temperature through the point of fusion we get a transition from a liquid phase to a solid phase. As the phase transition is taking place the temperature remains constant, and latent heat of fusion is liberated. Let us find the additional conditions which must be fulfilled at the surface of separation of the solid and liquid phases.*

We consider a plane problem in which the plane $x = \xi(t)$ is the surface of separation. After a time $t, t + \Delta t$, the boundary $x = \xi$ is displaced from the point $\xi = x_1$ to the point $\xi = x_2 = x_1 + \Delta\xi$. Also the mass $\varrho\Delta\xi$ solidifies (or melts if $\Delta\xi < 0$) and a corresponding amount of heat $\lambda\varrho\Delta\xi$ is liberated.

In order to achieve a heat balance, this amount of heat must be equivalent to the difference in the amounts of heat, passing through the boundaries $\xi = x_1$ and $\xi = x_2$, i.e. must fulfill the condition

$$\left[k_1 \frac{\partial u_1}{\partial x}\bigg|_{x_1} - k_2 \frac{\partial u_2}{\partial x}\bigg|_{x_2}\right] \Delta t = \lambda\varrho\Delta\xi ,$$

where k_1 and k_2 are the coefficients of heat conduction of the first and second phases, and λ is the latent heat of fusion.

Passing to a limit as $\Delta t \to 0$, we obtain a relation at the boundary of separation in the following form:

$$k_1 \frac{\partial u_1}{\partial x}\bigg|_{x=\xi} - k_2 \frac{\partial u_2}{\partial x}\bigg|_{x=\xi} = \lambda\varrho \frac{d\xi}{dt} . \tag{1}$$

This relation holds for the process of solidification (where $\Delta\xi > 0$ and $d\xi/dt > 0$) and also for the process of fusion (where $\Delta\xi < 0$ and $d\xi/dt < 0$); the direction of the process is determined by the sign of the left-hand side.

Let us consider the freezing of water, where the temperature of the phase transition equals zero. We shall consider a mass of water $x \geqslant 0$ bounded on one side by the plane $x = 0$. At the initial time $t = 0$ the water has a constant temperature $c > 0$.

* F. Frank and R. Mises, *Differential and Integral Equations of Mathematical Physics*, chap. III, Gostekhizdat, 1937.

If a constant temperature $c_1 < 0$ is maintained at the surface $x = 0$, then the boundary of freezing $x = \xi$ will move into the liquid.

The problem of the distribution of temperature inside the frozen water and of the speed of propagation of the boundary of freezing reduces to the solution of the equations

$$\left.\begin{aligned} \frac{\partial u_1}{\partial t} &= a_1^2 \frac{\partial^2 u_1}{\partial x^2} \quad \text{when} \quad 0 < x < \xi, \\ \frac{\partial u_2}{\partial t} &= a_2^2 \frac{\partial^2 u_2}{\partial x_2} \quad \text{when} \quad \xi < x < \infty \end{aligned}\right\} \tag{2}$$

with additional conditions

$$\left.\begin{aligned} u_1 &= c_1 \text{ when } x = 0, \\ u_2 &= c \text{ when } t = 0 \end{aligned}\right\} \tag{3}$$

and conditions at the solid-liquid boundary

$$u_1 = u_2 = 0 \text{ when } x = \xi, \tag{4}$$

$$k_1 \frac{\partial u_1}{\partial x}\bigg|_{x=\xi} - k_2 \frac{\partial u_2}{\partial x}\bigg|_{x=\xi} = \lambda \varrho \frac{d\xi}{dt}, \tag{1}$$

where k_1, a_1^2 and k_2, a_2^2 are the coefficients of heat conduction and thermal conductivity of the solid and liquid phases respectively.

We shall look for a solution of the problem in the form

$$u_1 = A_1 + B_1 \Phi\left(\frac{x}{2a_1 \sqrt{t}}\right),$$

$$u_2 = A_2 + B_2 \Phi\left(\frac{x}{2a_2 \sqrt{t}}\right),$$

where A_1, B_1, A_2, and B_2, are constants for the present undefined and Φ is the error integral

$$\Phi(x) = \frac{2}{\sqrt{\pi}} \int_0^x e^{-\xi^2} d\xi .$$

Satisfying conditions (3) and (4), we obtain:

$$A_1 = c_1, \quad A_2 + B_2 = c$$

from condition (3) and

$$A_1 + B_1 \Phi\left(\frac{\xi}{2a_1\sqrt{t}}\right) = 0\,,$$

$$A_2 + B_2 \Phi\left(\frac{\xi}{2a_2\sqrt{t}}\right) = 0$$

from condition (4). The latter conditions must hold for any values of t. This is possible only if

$$\xi = \alpha\sqrt{t}\,, \tag{5}$$

where α is some constant. Relation (5) gives the law of motion of the boundary of freezing.

For the constants A_1, B_1, A_2, B_2 and α the expressions

$$\left.\begin{aligned} A_1 &= c_1\,, & B_1 &= -\frac{c_1}{\Phi\left(\frac{\alpha}{2a_1}\right)}\,, \\ A_2 &= -\frac{c\Phi\left(\frac{\alpha}{2a_2}\right)}{1-\Phi\left(\frac{\alpha}{2a_2}\right)}\,, & B_2 &= \frac{c}{1-\Phi\left(\frac{\alpha}{2a_2}\right)}\,. \end{aligned}\right\} \tag{6}$$

are obtained.

In order to find the constant α, it is necessary to make use of the relation (1)

$$\frac{k_1 c_1 e^{-\frac{\alpha^2}{4a_1^2}}}{a_1 \Phi\left(\frac{\alpha}{2a_1}\right)} + \frac{k_2 c e^{-\frac{\alpha^2}{4a_2^2}}}{a_2\left[1-\Phi\left(\frac{\alpha}{2a_2}\right)\right]} = -\lambda\varrho\alpha\frac{\sqrt{\pi}}{2}\,. \tag{7}$$

The solution of this transcendental equation gives a value for α. There must be at least one solution for $c_1 < 0$, $c > 0$ because a change in α from 0 to ∞ the left-hand side of the equation changes from $-\infty$ to $+\infty$*, and the right-hand side from 0 to $-\infty$. If c is equal to the temperature of fusion ($c = 0$) then expressions (6) and (7) defining the coefficients take a simpler form:

$$A_2 = B_2 = 0\,,$$

$$A_1 = c_1\,, \quad B_1 = -\frac{c_1}{\Phi\left(\frac{\alpha}{2a_1}\right)} \tag{6'}$$

* For the asymptotic description of the function $1 - \Phi(z)$ as $z \to 0$ see p. 745.

and

$$\frac{k_1 c_1 e^{-\frac{a^2}{4a_1^2}}}{a_1 \Phi\left(\frac{a}{2a_1}\right)} = -\lambda \varrho a \frac{\sqrt{\pi}}{2}. \tag{7'}$$

Putting $a/2a_2 = \beta$. we can rewrite equation (7′) in a form such that:

$$\frac{1}{\sqrt{\pi}} \frac{e^{-\beta^2}}{\Phi(\beta)} = -D\beta,$$

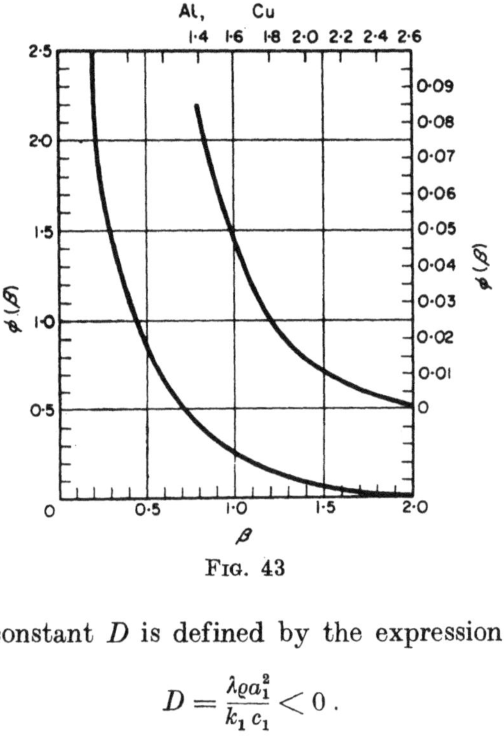

Fig. 43

where the constant D is defined by the expression

$$D = \frac{\lambda \varrho a_1^2}{k_1 c_1} < 0.$$

Using the graph of the function $\varphi(\beta') = \frac{e^{-\beta^2}}{\sqrt{[\pi]}\,\Phi(\beta)}$, given in Fig. 43, it is easy to determine the value of a graphically.

The solution of the problem on freezing may also be obtained by means of the method similarly, given in Appendix III to this chapter. The problem on freezing is in a sense a limiting case of a non-linear boundary-value problem, examined in Appendix III. In fact, the coefficients of heat conduction and specific heat in the problem on freezing are partly-continuous functions, and moreover, at $u = 0$, the specific heat has an infinitely high value. It is possible to obtain this case as a limiting one for $\varepsilon \to 0$ in which the latent heat is not produced instantaneously, but over a certain interval $-\varepsilon, +\varepsilon$, in which the condition

$$\int_{-\varepsilon}^{\varepsilon} c(u)\,du = \lambda .$$

must be fulfilled.

However it is possible to solve this problem directly, using the method of similarity. It is easy to verify that all the conditions of the problem remain invariant if the length scale is increased k times, and the time scale k^2 times. This means that the solution of the problem depends on the argument $x/\sqrt{t}$, i.e. that

$$u(x, t) = f\left(\frac{x}{\sqrt{t}}\right).$$

Hence, in particular, it follows that the motion of a zero isotherm will be described by the equation

$$\xi = a\sqrt{t},$$

where a is the value of the argument, for which

$$f(a) = 0.$$

To find the function f we have the equations

$$a_1^2 \frac{d^2 f_1}{dz^2} = -2z \frac{df_1}{dz} \quad \text{when} \quad 0 < z < a,$$

$$a_2^2 \frac{d^2 f_2}{dz^2} = -2z \frac{df_2}{dz} \quad \text{when} \quad a < z < \infty;$$

$$f_1(0) = c_1;\ f_2(\infty) = c;\ f_1(a) = f_2(a) = 0;$$

$$k_1 f_1'(a) - k_2 f_2'(a) = \lambda\varrho \frac{a}{2}.$$

Therefore the function $f(z)$ has the following form:

$$f(z) = \begin{cases} f_1(z) = A_1 + B_1 \Phi\left(\dfrac{z}{2a_1}\right), & \text{if } 0 < z < a, \\ f_2(z) = A_2 + B_2 \Phi\left(\dfrac{z}{2a_2}\right), & \text{if } a < z < \infty. \end{cases} \quad (7h)$$

To determine the constants A_1, B_1, A_2, B_2 we must utilize conditions (3) and (4) from which formula (6) results. Condition

(7) determines a. Thus the analytical part of the solution is the same by both methods.

The considerations stated here show that it is possible also to solve the problem of freezing in those cases where the latent heat is produced not at a fixed temperature but over some range of temperatures. Similarly it is possible to solve the problem if it has not one, but several critical temperatures which occur at the phase changes in the process of transition from one crystalline structure to another, for example in the recrystallization of steel.

V. The Einstein—Kolmogorov equation

Microscopic particles, existing in free suspension in a medium, perform a random motion called the Brownian movement. Let us define the probability of finding a particle which starts from point M_0 at time t_0 in a small region ΔV around the point M at time t by the function

$$W(M, t; M_0, t_0) \cdot \Delta V. \tag{1}$$

The probability is considered here in the sense that if over some small interval $t_0 + \Delta t$ a sufficiently large number of particles N (whose mutual interaction is negligibly small) start from a point M_0, then the concentration of these particles as $\Delta t \to 0$ at the point M at time t will equal $W(M, t; M_0, t_0)$.

We meet a similar phenomenon in the diffusion of a gas in any medium (for example, air). The function $W(M, t; M_0, t_0)$ represents the source function.

It is obvious that

$$\int W(M, t; M_0, t_0)\, dV_M = 1 \qquad (t > t_0) \tag{2}$$

and that if the initial concentration of the particles at a certain moment of time t_0 equals $\varphi(M)$, then the concentration $u(M, t)$ of these particles at a time $t > t_0$ will equal

$$u(M, t) = \int W(M, t; P, t_0)\, \varphi(P)\, dV_P, \tag{3}$$

where the integral is taken over all space.

From the latter relation it follows that*

$$W(M, t; M_0, t_0) = \int W(M, t; P, \theta)\, W(P, \theta; M_0, t_0)\, dV_P \quad (t_0 < \theta < t), \tag{4}$$

for any value $t_0 < \theta < t$. This latter equation is called *the Einstein–Kolmogorov equation.*

We show that if certain conditions are imposed on the function $W(M, t; M_0, t_0)$ the solution of the Einstein–Kolmogorov equation satisfies some partial differential equation of the parabolic type. We consider the case where the position of the point M is denoted by the single coordinate x. Let us assume that the function $W(x, t; x_0, t_0)$ satisfies the following conditions:

[1°] $$\lim_{\tau \to 0} \frac{\overline{x - \xi}}{\tau} = \lim \frac{1}{\tau} \int (x - \xi)\, W(x, t + \tau;\ \xi, t)\, d\xi = A(x, t). \tag{5}$$

If in time τ the particle passes from the position ξ to the position x then $\overline{(x - \xi)/\tau}$ is the average speed of the particle. Thus, the first condition implies the existence of a finite average speed for the motion of the particles.

[2°]

$$\lim_{\tau \to 0} \frac{\overline{(x - \xi)^2}}{\tau} = \lim \frac{1}{\tau} \int (x - \xi)^2\, W(x, t + \tau, \xi, t)\, d\xi = 2\, B(x, t). \tag{6}$$

The value $(x - \xi)^2$ does not depend on the direction of displacement of the point x relative to the point ξ. The mean value of the square of the difference after a time

$$\overline{(x - \xi)^2} = \int (x - \xi)^2\, W(x, t + \tau; \xi, t)\, d\xi \tag{6a}$$

is usually taken as a measure of the irregular motion over this interval of time. Condition 2° implies the assumption of a linear dependence of the mean square on the time for small τ,

[3°] $$\lim_{\tau \to 0} \frac{\overline{|x - \xi|^3}}{\tau} = \lim \frac{1}{\tau} \int |x - \xi|^3 \cdot M(x, t + \tau; \xi, t)\, d\xi = 0. \tag{7}$$

The function $W(x, t + \tau;\ \xi, t)$ is the source function of a point source and for small values of τ it must rapidly decrease when $|x - \xi| \to \infty$, and increase when $|x - \xi|$ is small.

* M. A. Leontovich, *Statistical Physics*, chap. VI, Gostekhizdat, 1944; A. N. Kolmogorov, *Analytical methods of the theory of probability. Advances in Mathematical* Sciences, no. V, 1938.

In order to obtain the Einstein–Kolmogorov differential equation let us multiply both sides of equation (4) by an arbitrary function $\psi(x)$ which reduces to zero along with its derivative at the boundaries of the region of integration, and let us integrate over this region:

$$\int W(x, t+\tau; x_0, t_0)\,\psi(x)\,dx = $$
$$= \int W(\xi, t; x_0, t_0)\,d\xi \int W(x, t+\tau; \xi, t)\,\psi(x)\,dx. \qquad (7a)$$

Now we expand the function on the right-hand side in a Taylor's series in $(x-\xi)$:

$$\psi(x) = \psi(\xi) + \psi'(\xi)(x-\xi) + \frac{\psi''(\xi)}{2}(x-\xi)^2 + \frac{\psi'''(\xi^*)}{3!}(x-\xi)^3, \qquad (7b)$$

where ξ^* is an intermediate value lying between x and ξ. Dividing by τ, after simple rearrangement we have:

$$\int \psi(x)\,\frac{W(x, t+\tau; x_0, t_0) - W(x, t; x_0, t_0)}{\tau}\,dx =$$
$$= \int W(\xi, t; x_0, t_0)\left[\psi'(\xi)\,\overline{\frac{x-\xi}{\tau}} + \psi''(\xi)\,\overline{\frac{(x-\xi)^2}{2\tau}}\right]d\xi + \qquad (7c)$$
$$+ \frac{1}{3!\,\tau}\int\int \psi'''(\xi^*)(x-\xi)^3\,W(\xi, t; x_0, t_0)\,W(x, t+\tau; \xi, t)\,d\xi\,dx.$$

Assuming that $\psi'''(x)$ is bounded

$$|\psi'''(x)| < A \qquad (7d)$$

and remembering that

$$\int W(\xi, t; x_0, t_0)\,d\xi = 1, \qquad (7e)$$

we obtain:

$$\left|\frac{1}{\tau}\int\int \psi'''(\xi^*)(x-\xi)^3\,W(\xi, t; x_0, t_0)\,W(x, t+\tau; \xi, t)\,d\xi\,dx\right| \leq$$
$$\leq \frac{A}{\tau}\int |x-\xi|^3\,W(x, t+\tau; \xi, t)\,dx = \frac{A\,\overline{|x-\xi|^3}}{\tau}. \qquad (7f)$$

From condition 3° it follows that this expression tends to zero as $\tau \to 0$. Therefore, passing to a limit as $\tau \to 0$ and using conditions 1°, 2° we obtain:

$$\int \psi(x)\,\frac{\partial W(x, t; x_0, t_0)}{\partial t}\,dx =$$
$$= \int W(\xi, t; x_0, t_0)\,[\psi'(\xi)\,A(\xi, t) + \psi''(\xi)\,B(\xi, t)]\,d\xi. \qquad (7g)$$

We integrate by parts on the right side, and use the fact that the function ψ reduces to zero together with its derivative at the boundary of the region of integration, and obtain as a result:

$$\int \psi(x)\left[\frac{\partial W}{\partial t} + \frac{\partial (AW)}{\partial x} - \frac{\partial^2 (BW)}{\partial x^2}\right] dx = 0\,. \tag{7h}$$

Because this relation must hold for an arbitrary function $\psi(x)$ we get *the Einstein–Kolmogorov differential equation* for the probability function $W(x, t;\ x_0, t_0)$

$$\frac{\partial W}{\partial t} = -\frac{\partial (AW)}{\partial x} + \frac{\partial^2 (BW)}{\partial x^2}\,. \tag{8}$$

This is an equation of parabolic type, similar to the equation of heat conduction, and may be written in the form

$$W_t = \frac{\partial}{\partial x}(BW_x) + \alpha W_x + \beta W\,, \tag{9}$$

where

$$\alpha = -A + B_x\,, \tag{9a}$$
$$\beta = -A_x + B_{xx} = \alpha_x\,.$$

From equation (9) it is apparent that the value B has a physical meaning as the coefficient of diffusion. If the process under consideration is homogeneous in space and time, i.e. if the function W depends only on the difference $\xi = x - x_0$ and $\theta = t - t_0$, then the coefficients A and B do not depend on x and t, and are constants. Equation (8) in this case is an equation with constant coefficients

$$\frac{\partial W}{\partial t} = -A\frac{\partial W}{\partial x} + B\frac{\partial^2 W}{\partial x^2}\,. \tag{10}$$

If the function W depends only on $|x - \xi|$, i.e. if the probabilities of displacement to the right and left are equal, then obviously A must equal zero. Analytically this follows from formula (5) because the function under the integral sign is odd.

In this case equation (8) is the simple equation of heat conduction

$$\frac{\partial W}{\partial t} = B\frac{\partial^2 W}{\partial x^2}\,. \tag{11}$$

VI. The δ-function

1. *Definition of the δ-function*

As well as continuously distributed physical quantities (mass charge, heat sources, mechanical impulse and so on) one often has to deal with point quantities (point mass, point charge, point source of heat, concentrated impulse, etc.). It should not be forgotten that these concepts are "limiting forms", although they are often used in physics as independent concepts, omitting the corresponding limiting transition.

Having in view the physical aspect of the problem, let us consider the potential at the point M (see Chapter IV, § 5) of a unit of mass, concentrated inside some volume T surrounding the point M_0. We take any sequence of functions $\{\varrho_n\}$ ($\varrho_n > 0$), each of which equals zero outside a sphere $S_{\varepsilon_n}^{M_0}$ of radius ε_n with centre at the point M_0, where $\varepsilon_n \to 0$ as $n \to \infty$, and for which, beginning with a certain n,

$$\iiint_T \varrho_n(P)\, d\tau_P = \iiint_{S_{\varepsilon_n}^{M_0}} \varrho_n(P)\, d\tau_P = 1. \tag{1}$$

Considering the series of functions

$$u_n = \iiint_T \frac{\varrho_n}{r}\, d\tau, \tag{1a}$$

which are potentials of a mass distributed with density ϱ_n, and performing a limiting transition as $n \to \infty$, we obtain:

$$\lim_{n\to\infty} u_n = \frac{1}{r_{M_0 M}}. \tag{2}$$

This result obviously does not depend on the choice of sequence $\{\varrho_n\}$. Although the sequence $\{u_n\}$ converges to $1/r$, nevertheless the sequence $\{\varrho_n\}$ does not have a limit. "The limiting form", corresponding to the sequence $\{\varrho_n\}$ is called the function $\delta(M, M_0)$.

A fundamental relationship defining the δ-function is the following formal operator relation:

$$\iiint_T \delta(M_0, M) f(M)\, d\tau_M = \begin{cases} f(M_0), & \text{if } M_0 \subset T, \\ 0, & \text{if } M_0 \not\subset T, \end{cases} \tag{3}$$

where $f(M)$ is an arbitrary continuous function of the point M. Remembering that as $n \to \infty$ the functions ϱ_n tend uniformly to zero in any region not containing the point M_0, and increase indefinitely in the vicinity $S_{\varepsilon_n}^{M_0}$ of the point M_0, one sometimes defines the δ-function formally by means of the relations

$$\left.\begin{aligned} \delta(M, M_0) &= 0 \quad \text{when} \quad M \neq M_0, \\ \delta(M, M_0) &= \infty \quad \text{when} \quad M = M_0 \end{aligned}\right\} \tag{4}$$

and

$$\int\int_T\int \delta(M, M_0)\, d\tau_M = \begin{cases} 1 & \text{when} \quad M_0 \subset T, \\ 0 & \text{when} \quad M_0 \not\subset T. \end{cases} \tag{5}$$

Relation (5) is an obvious result of formula (3) for $f = 1$.

By examination of sequences of functions in different problems one is led to consider various definitions of convergence.

We say that the sequence of functions

$$\{u_n(x)\} = u_1(x), \quad u_2(x), \ldots, \; u_n(x), \ldots \tag{6}$$

converges uniformly in the interval (a, b), if for any $\varepsilon > 0$ it is possible to assign N such that for $n, m > N$, for any x in (a, b) the condition

$$|u_n(x) - u_m(x)| < \varepsilon \text{ where } n, m > N. \tag{6a}$$

is fulfilled.

We say that the sequence (6) converges in the mean to a limit in the interval (a, b) if for any $\varepsilon > 0$ it is possible to assign N such that for $n, m > N$

$$\int_a^b |u_n(x) - u_m(x)|^2 \, dx < \varepsilon. \tag{6b}$$

We say that the sequence (6) converges weakly in the interval (a, b) if for any continuous function f there exists a limit

$$\lim_{n \to \infty} \int_a^b f(x)\, u_n(x)\, dx. \tag{6c}$$

In the consideration of converging sequences one usually introduces *the limiting elements* of the sequences. In the case of uniform convergence the limiting element belongs to the same class of functions as the members of the sequence. This does not always hold for convergence to a mean and weak convergence.

If the limiting element does not belong to the class of functions under consideration, then limiting elements may be introduced extending the original class. Moreover extension is understood as the combination of the original and limiting elements. For example, extensions occur in the theory of real numbers, where irrational numbers are introduced as the limiting elements, defined by a class of equivalent sequences of rational numbers.

Speaking of the limiting elements in the case of slow convergence, we say that the two sequences $\{u_n\}$ and $\{v_n\}$ have the same limiting element if these sequences are equivalent, i.e. if the sequence $\{u_n - v_n\}$ converges weakly to zero:

$$\lim_{n\to\infty} \int_a^b f(x)\,[u_n(x) - v_n(x)]\,dx = 0. \tag{6d}$$

We shall define the sequence of positive function $\{\delta_n\}$ to be a normalized local sequence at the point x_0, if the function δ_n equals zero outside the interval $(x_0 - \varepsilon_n,\ x_0 + \varepsilon_n)$, where

$$\varepsilon_n \to 0 \text{ as } n \to \infty, \tag{6e}$$

and

$$\int_a^b \delta_n(x)\,dx = 1. \tag{6f}$$

It is obvious that the sequence $\{\delta_n\}$ converges weakly. The limiting element of the sequence $\{\delta_n\}$ is usually called the δ-function of the point x_0.

If the limiting element u of a weakly convergent sequence $\{u_n\}$ comes from the class of functions u_n, then the integral of the product of some function $f(x)$ and the element u is equal to the limit

$$\lim_{n\to\infty} \int_a^b f(x)\,u_n(x)\,dx = \int_a^b f(x)\,u\,dx. \tag{6g}$$

It is obvious that for the δ-function of the point x_0 the equality

$$\int_a^b f(x)\,\delta(x_0, x)\,dx = f(x_0) \tag{6h}$$

holds. This relation is often mistaken for the definition of the δ-function.

2. *Expansion of the δ-function in a Fourier series*

The δ-function can be defined as a limiting form of another series, equivalent in the sense of weak convergence to the series $\delta_n(x)$ consisting of local normalized functions of the point x_0.

Let us consider the series of functions

$$\bar{\delta}_n(x_0, x) = \frac{1}{2l} + \frac{1}{l}\sum_{m=1}^{n}\left(\cos\frac{m\pi}{l}x_0\cdot\cos\frac{m\pi}{l}x + \sin\frac{m\pi}{l}x_0\sin\frac{m\pi}{l}x\right) =$$

$$= \frac{1}{2l} + \frac{1}{l}\sum_{m=1}^{n}\cos\frac{m\pi}{l}(x - x_0) \tag{7}$$

or in complex form

$$\bar{\delta}_n(x, x_0) = \frac{1}{2l}\sum_{-n}^{n} e^{im\frac{\pi}{l}(x-x_0)}, \tag{7'}$$

defined in the interval $(-l, l)$.

It is obvious that for any function $g(x)$, expandable as a Fourier series in $(-l, l)$, the following limit relation holds

$$\lim_{n\to\infty}\int_{-l}^{l}\bar{\delta}_n(x_0, x)\, g(x)\, dx = g(x_0)\,. \tag{8}$$

This shows that in the class of continuous functions $\{g(x)\}$, expandable in a Fourier series, the series $\delta_n(x_0, x)$ defined above is equivalent in the sense of weak convergence to the series $\bar{\delta}_n(x_0, x)$, i.e. that

$$\delta(x_0, x) = \frac{1}{2l} + \frac{1}{l}\sum_{m=1}^{\infty}\cos\frac{m\pi}{l}(x_0 - x)\,, \tag{9}$$

if this equality is considered from the point of view of weak convergence.

From the same point of view the equality

$$\delta(x_0, x) = \sum_{n=1}^{\infty}\varphi_n(x)\,\varphi_n(x_0)\,, \tag{10}$$

holds, where $\{\varphi_n(x)\}$ is a complete orthogonal and normalized set of functions, given in some interval (a, b), and also the equality

$$\delta(x_0, x) = \frac{1}{2\pi}\int_{-\infty}^{\infty} e^{ik(x_0-x)}\, dk = \frac{1}{\pi}\int_{0}^{\infty}\cos k(x_0 - x)\, dk\,. \tag{11}$$

We show that in the calculation of the integrals containing the δ-function, one can use series (9), and interchange the order of integration and summation.

Let us consider the function $g(x)$, expandable in a Fourier series, and the integral

$$\int_{-l}^{l} g(x)\,\delta(x_0, x)\,dx. \tag{11a}$$

Substituting in place of $\delta(x_0, x)$ the expression in formula (9), let us perform an integration of the series, term by term. As a result we obtain:

$$g(x) = \frac{\bar{g}_0}{2} + \sum_{m=1}^{\infty} \left(\bar{g}_m \cos\frac{\pi m}{l} x + \bar{\bar{g}}_m \sin\frac{\pi m}{l} x\right). \tag{11'}$$

where

$$\left.\begin{aligned} \bar{g}_0 &= \frac{1}{l}\int_{-l}^{l} g(x_0)\,dx_0, \\ \bar{g}_m &= \frac{1}{l}\int_{-l}^{l} g(x_0)\cos\frac{\pi m}{l} x_0\,dx_0, \\ \bar{\bar{g}}_m &= \frac{1}{l}\int_{-l}^{l} g(x_0)\sin\frac{\pi m}{l} x_0\,dx_0. \end{aligned}\right\} \tag{12}$$

Comparison of formula (11) with the equality

$$\int_{-l}^{l} \delta(x, x_0)\,g(x)\,dx = g(x_0) \qquad (-l < x_0 < l) \tag{12a}$$

shows that the integration of the series for the δ-function leads to the correct result.

Thus, in the class of continuous functions, expandable in a Fourier series, the series of partial sums

$$\frac{1}{2l}\sum_{n=-k}^{k} e^{i\frac{\pi n}{l}(x-x')} \tag{12b}$$

is equivalent to the normalized local sequence $\{\delta_n\}$.

Other representations of the δ-fun tion are also based on the use of certain functional sequences, equivalent in the sense of weak convergence to the sequence $\{\delta_n\}$.

3. *The application of the δ-function to the construction of the source function*

Let us consider the following problem:

$$u_t = a^2 u_{xx}, \tag{13}$$

$$u(x, 0) = \varphi(x), \tag{14}$$

$$u(0, t) = u(l, t) = 0. \tag{15}$$

Corresponding to the function $\varphi(x)$ there is a unique solution of the problem

$$u(x, t) = \mathscr{L}[\varphi(x)]. \tag{15a}$$

Let us assume that the operator $\mathscr{L}$ can be represented in integral form

$$u(x, t) = \mathscr{L}[\varphi(x)] = \int_0^l G(x, \xi, t)\, \varphi(\xi)\, d\xi, \tag{16}$$

where $G(x, \xi, t)$ is the kernel of the operator $\mathscr{L}$.

In order to find the kernel $G(x, \xi, t)$, let us assume:

$$\varphi(x) = \delta(x - x_0). \tag{14'}$$

Replacing $\varphi(x)$ in formula (16) by the δ-function, we obtain:

$$u(x, t) = G(x, x_0, t), \tag{17}$$

i.e. $G(x. x_0, t)$ is the solution of problem (13) with the initial condition (14′) here.

Let us represent the δ-function in the form of a Fourier series

$$\delta(x - x_0) = \sum_{n=1}^{\infty} \frac{2}{l} \sin\frac{n\pi}{l} x \sin\frac{n\pi}{l} x_0. \tag{17a}$$

and try to find the kernel G in the form of the sum

$$G(x, x_0, t) = \sum_{n=1}^{\infty} A_n(t) \sin\frac{n\pi}{l} x, \tag{18}$$

every component of which must satisfy the equation of heat conduction. It follows that

$$A_n(t) = B_n e^{-a^2 \left(\frac{n\pi}{l}\right)^2 t}. \tag{18a}$$

From the initial condition we at once obtain:

$$B_n = \frac{2}{l} \sin \frac{n\pi}{l} x_0. \tag{18b}$$

Thus we obtain the kernel G

$$G(x, x_0, t) = \frac{2}{l} \sum_{n=1}^{\infty} e^{-\left(\frac{\pi n}{l}\right)^2 a^2 t} \sin \frac{n\pi}{l} x \sin \frac{n\pi}{l} x_0, \tag{19}$$

in a form agreeing with the one derived in § 3. The solution of problems (13)–(15) is given by formula (16), where $G(x, x_0, t)$ is a function defined by formula (19).

By a similar method it is possible to find an expression for the source function in an infinite region. Function G in this case will be defined by the conditions

$$u_t - a^2 u_{xx} = 0 \qquad (-\infty < x < \infty), \tag{20}$$

$$u(x, 0) = \varphi(x) = \delta(x - x_0). \tag{21}$$

We use the expansion of the δ-function as a Fourier integral

$$\delta(x - x_0) = \frac{1}{\pi} \int_0^{\infty} \cos \lambda (x - x_0)\, d\lambda. \tag{21a}$$

and look for $G(x, x_0, t)$ in the form

$$G(x, x_0, t) = \frac{1}{\pi} \int_0^{\infty} A_\lambda(t) \cos \lambda (x - x_0)\, d\lambda. \tag{22}$$

From equation (20) we find:

$$A_\lambda(t) = A_\lambda^{(0)} e^{-a^2\lambda^2 t}. \tag{23}$$

Assuming $t = 0$ and comparing formulae (23) and (21), we obtain

$$A_\lambda^{(0)} = 1. \tag{23a}$$

Thus,

$$G(x, x_0, t) = \frac{1}{\pi} \int_0^{\infty} e^{-a^2\lambda^2 t} \cos \lambda (x - x_0)\, d\lambda. \tag{23b}$$

This integral was calculated in § 3 of Chapter III, giving:

$$G(x, x_0, t) = \frac{1}{2\sqrt{(\pi a^2 t)}} e^{-\frac{(x-x_0)^2}{4a^2 t}}. \tag{23c}$$

Hence it follows that the solution of the problem of heat conduction in an infinite region can be expressed by the formula

$$u(x, t) = \int_{-\infty}^{\infty} G(x, \xi, t)\, \varphi(\xi)\, d\xi\,. \tag{24}$$

A discussion of the limits of applicability of formulae, obtained by the method of the δ-function, requires a separate investigation.

As a further example let us consider the inhomogenous equation

$$u_t = a^2 u_{xx} + \frac{F(x, t)}{c\varrho}, \tag{25}$$

where $F(x, t)$ is the density of heat sources. If at the point $x = \xi$ at time $t = t_0$ an instantaneous source of heat of magnitude Q_0 is established, then

$$F(x, t) = Q_0\, \delta(x - \xi)\, \delta(t - t_0) \tag{26}$$

We look for the solution of the inhomogenous equation

$$u_t = a^2 u_{xx} + \frac{Q_0}{c\varrho}\, \delta(x - \xi)\, \delta(t - t_0) \quad (t_0 > 0) \tag{27}$$

with zero initial condition (27a

$$u(x, 0) = 0\,.$$

Using the integral presentation

$$\delta(x - \xi) = \frac{1}{\pi} \int_0^{\infty} \cos \lambda (x - \xi)\, d\lambda\,, \tag{27b}$$

we seek a function $u(x, t)$ in the form

$$u(x, t) = \frac{1}{\pi} \int_0^{\infty} u_\lambda(t) \cos \lambda (x - \xi)\, d\lambda\,. \tag{27c}$$

Substituting these expressions in equation (27), we obtain an equation for $u_\lambda(t)$:

$$\dot{u}_\lambda(t) + a^2 \lambda^2 u_\lambda(t) = \frac{Q_0}{c\varrho}\, \delta(t - t_0) \tag{27d}$$

with initial condition

$$u_\lambda(0) = 0\,. \tag{27e}$$

As is well known, the solution of the inhomogeneous equation

$$\dot{u} + a^2 u = f(t), \qquad u(0) = 0 \tag{27f}$$

has the form

$$u(t) = \int_0^\infty e^{-a^2(t-\tau)} f(\tau)\, d\tau . \tag{28}$$

In our case

$$u_\lambda(t) = \frac{Q_0}{c\varrho} \int_0^t e^{-a^2\lambda^2(t-\tau)} \delta(\tau - t_0)\, d\tau = \begin{cases} 0 \text{ when } t < t_0 \\ \dfrac{Q_0}{c\varrho} e^{-a^2\lambda^2(t-t_0)} \text{ when } t > t_0 . \end{cases} \tag{29}$$

Thus,

$$u(x, t) = \frac{Q_0}{c\varrho} \frac{1}{\pi} \int_0^\infty e^{-a^2\lambda^2(t-t_0)} \cos\lambda (x-\xi)\, d\lambda = \frac{Q_0}{c\varrho} G(x-\xi,\ t-t_0),$$

where (29a)

$$G(x, \xi, t - t_0) = \frac{1}{2\sqrt{[\pi a^2 (t - t_0)]}} e^{-\frac{(x-\xi)^2}{4a^2(t-t_0)}} \tag{29b}$$

is the source function for an instantaneous point source.

A similar method of constructing the source function is often used in theoretical physics.*

* See the similar account of the theory of the δ-function and numerous examples of its application in the book by D. D. Ivanenko and A. A. Sokolov, *Classical Field Theory*, chap. I, Gostekhizdat, 1949.

CHAPTER IV

EQUATIONS OF ELLIPTIC TYPE

In a study of a variety of steady state problems, (oscillations, heat conduction, diffusion and others) one often arrives at equations of elliptic type. The most common equation of this type is *Laplace's equation*

$$\nabla^2 u = 0 .$$

The function u is said to be harmonic in the domain T, if it is continuous in this region together with its derivatives up to second order and if it satisfies Laplace's equation.

§ 1. Problems reducible to Laplace's equation

1. *Steady heat flow. Statement of boundary-value problems*

Let us consider a steady heat flow. In Chapter III it was shown that the temperature of a non-steady heat field satisfies the differential equation of heat conduction

$$u_t = a^2 \nabla^2 u \qquad \left(a^2 = \frac{k}{c\varrho}\right).$$

If the process is steady, then a time independent distribution of temperature $u(x, y, z)$ is established, which satisfies Laplace's equation

$$\nabla^2 u = 0 . \tag{1}$$

In the presence of a source of heat we obtain the equation

$$\nabla^2 u = -f , \; f = F/k , \tag{2}$$

where F is the density of the heat sources, and k is the coefficient of thermal conductivity. The inhomogeneous equation of Laplace is often called *Poisson's equation*.

Let us consider some volume T, bounded by the surface Σ. The problem of a steady distribution of temperature $u(x, y, z)$ inside the volume T is formulated in the following way:

Find the function $u(x, y, z)$ satisfying the equation

$$\nabla^2 u = -f(x, y, z) \tag{2}$$

inside T and the boundary condition, which may have one of the following forms :

I. $u = f_1$ at Σ *(first boundary-value problem)*,

II. $\frac{\partial u}{\partial n} = f_2$ at Σ *(second boundary-value problem)*,

III. $\frac{\partial u}{\partial n} + h(u - f_3) = 0$ at Σ *(third boundary-value problem)*,

*where f_1, f_2, f_3, h are given functions, $\frac{\partial u}{\partial n}$ is the derivative with respect to the outer normal to the surface Σ.**

The physical significance of these boundary conditions is obvious (see Chapter III, § 1). The first boundary-value problem for Laplace's equation is often called *Dirichlet's problem*, and the second problem *Neumann's problem*. If there exists a solution in the region T_0, external to the surface Σ, then the corresponding problem is called the *external boundary-value problem*.

2. *Potential flow in a Fluid. Potential of a steady current and an electrostatic field*

As a second example let us consider the potential of a fluid flow without sources. A steady flow of incompressible fluid (density $\varrho =$ const.) occurs inside a given volume T with boundary Σ, with a characteristic velocity $\boldsymbol{v}(x, y, z)$. If the fluid flow is irrotational, then the velocity $\boldsymbol{v}$ is a vector derivable from a potential, i. e.

$$\boldsymbol{v} = -\operatorname{grad} \varphi, \tag{3}$$

* It is obvious that a steady temperature distribution can be established only when the total heat flow across the boundary of the region equals zero. It follows that the function f_2 should satisfy the additional condition

$$\iint_\Sigma f_2 \, d\sigma = 0 .$$

where φ is a scalar function, called the *velocity potential*. If sources are absent, then

$$\operatorname{div} \boldsymbol{v} = 0 . \tag{4}$$

Substituting here expression (3) for v, we obtain:

$$\operatorname{div} \operatorname{grad} \varphi = 0$$

or

$$\nabla^2 \varphi = 0 , \tag{5}$$

i.e. the velocity potential satisfies Laplace's equation.

Let a steady current of volume density $\boldsymbol{j}(x, y, z)$ exist in a homogeneous conducting medium. If there are no volume sources of current in the medium, then

$$\operatorname{div} \boldsymbol{j} = 0 , \tag{6}$$

The electric field E is related to the current density according to Ohm's law

$$\boldsymbol{E} = \boldsymbol{j}/\lambda , \tag{7}$$

where λ is the conductivity of the medium. Since the process is steady, the electric field is irrotational,* i.e. there exists a scalar function $\varphi(x, y, z)$ such that

$$\boldsymbol{E} = - \operatorname{grad} \varphi . \tag{8}$$

Hence on the basis of formulae (6) and (7) we deduce that

$$\Delta^2 \varphi = 0 , \tag{9}$$

i.e. the potential of the electric field of a steady current satisfies Laplace's equation.

Let us consider the electric field of a stationary charge. From the steady nature of the process, it follows that

$$\operatorname{curl} \boldsymbol{E} = 0 , \tag{10}$$

i.e. the field is a potential field and

$$\boldsymbol{E} = - \operatorname{grad} \varphi . \tag{8}$$

* From Maxwell's second equation $\mu/e\, \dot{H} = $ —curl E it follows that curl rot $E = 0$.

Let $\varrho(x, y, z)$ be the volume density of the charges, in a medium with a dielectric constant $\varepsilon = 1$. Proceeding from the fundamental law of electrodynamics

$$\iint_S E_n \, dS = 4\pi \sum e_i = 4\pi \iiint_T \varrho \, d\tau , \tag{11}$$

where T is some volume, S is its boundary surface, $\sum e_i$ is the sum of all the charges inside T, and using Ostrogradskii's theorem

$$\iint_S E_n \, dS = \iiint_T \operatorname{div} \boldsymbol{E} \, d\tau , \tag{12}$$

we obtain:

$$\operatorname{div} \boldsymbol{E} = 4\pi\varrho .$$

Substituting expression (8) for $\boldsymbol{E}$, we have:

$$\nabla^2 \varphi = - 4\pi\varrho , \tag{13}$$

i. e. the electrostatic potential φ satisfies Poisson's equation. If there are no volume charges ($\varrho = 0$) then the potential φ should satisfy Laplace's equation

$$\nabla^2 \varphi = 0 .$$

The main boundary-value problems for the processes under consideration are the three types outlined above. We will not give here any other boundary-value problems. Some of these problems will be considered in the appendices.

3. *Laplace's equation in a curvilinear system of coordinates*

Let us investigate the expression for the Laplacian operator in an orthogonal curvilinear system of coordinates. In place of the cartesian coordinates x, y, z let us introduce the curvilinear coordinates q_1, q_2, q_3 by means of the relations

$$q_1 = f_1(x, y, z) , \quad q_2 = f_2(x, y, z) , \quad q_3 = f_3(x, y, z) , \tag{14}$$

solving for x, y, z it is possible to write

$$x = \varphi_1(q_1, q_2, q_3) , \quad y = \varphi_2(q_1, q_2, q_3) , \quad z = \varphi_3(q_1, q_2, q_3) . \tag{15}$$

Assuming $q_1 = C_1$, $q_2 = C_2$, $q_3 = C_3$ where C_1, C_2, C_3 are constants, we obtain three sets of coordinate surfaces:

$$f_1(x, y, z) = C_1 , \quad f_2(x, y, z) = C_2 , \quad f_3(x, y, z) = C_3 . \tag{16}$$

Let us consider an element of volume in the new coordinates, bounded by three pairs of coordinate surfaces (Fig. 44). Along the edge AB $q_2 = \text{const.}$, $q_3 = \text{const.}$; along AD $q_1 = \text{const.}$, $q_2 = \text{const.}$; along AC $q_1 = \text{const.}$, $q_3 = \text{const.}$ The direction-cosines of the tangent to AB, AD, and AC are proportional to

$$\frac{\partial \varphi_1}{\partial q_1}, \quad \frac{\partial \varphi_2}{\partial q_1}, \quad \frac{\partial \varphi_3}{\partial q_1};$$

$$\frac{\partial \varphi_1}{\partial q_2}, \quad \frac{\partial \varphi_2}{\partial q_2}, \quad \frac{\partial \varphi_3}{\partial q_2};$$

$$\frac{\partial \varphi_1}{\partial q_3}, \quad \frac{\partial \varphi_2}{\partial q_3}, \quad \frac{\partial \varphi_3}{\partial q_3}.$$

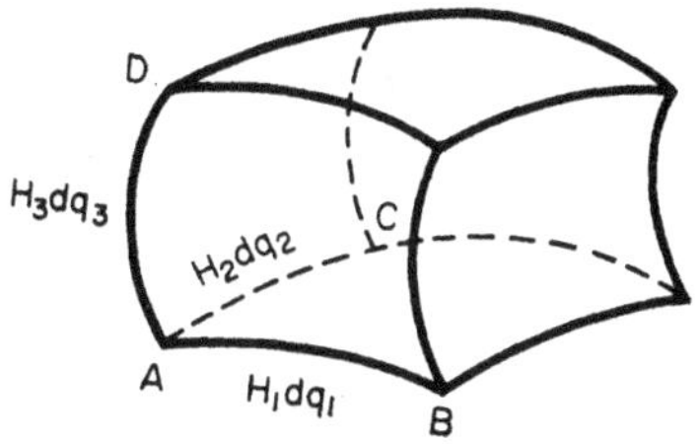

FIG. 44

The orthogonality of the curvilinear system leads to relations

$$\frac{\partial \varphi_1}{\partial q_i}\frac{\partial \varphi_1}{\partial q_k} + \frac{\partial \varphi_2}{\partial q_i}\frac{\partial \varphi_2}{\partial q_k} + \frac{\partial \varphi_3}{\partial q_i}\frac{\partial \varphi_3}{\partial q_k} = 0 \qquad (i \neq k). \tag{17}$$

Let us calculate an element of length in the new coordinates

$$ds^2 = dx^2 + dy^2 + dz^2 = \left(\frac{\partial \varphi_1}{\partial q_1} dq_1 + \frac{\partial \varphi_1}{\partial q_2} dq_2 + \frac{\partial \varphi_1}{\partial q_3} dq_3\right)^2 +$$

$$+ \left(\frac{\partial \varphi_2}{\partial q_1} dq_1 + \frac{\partial \varphi_2}{\partial q_2} dq_2 + \frac{\partial \varphi_2}{\partial q_3} dq_3\right)^2 +$$

$$+ \left(\frac{\partial \varphi_3}{\partial q_1} dq_1 + \frac{\partial \varphi_3}{\partial q_2} dq_2 + \frac{\partial \varphi_3}{\partial q_3} dq_3\right)^2. \tag{18}$$

Removing the brackets and using the orthogonality relation (17), we obtain:

$$ds^2 = H_1^2\, dq_1^2 + H_2^2\, dq_2^2 + H_3^2\, dq_3^2, \tag{19}$$

where

$$\left.\begin{aligned} H_1^2 &= \left(\frac{\partial \varphi_1}{\partial q_1}\right)^2 + \left(\frac{\partial \varphi_2}{\partial q_1}\right)^2 + \left(\frac{\partial \varphi_3}{\partial q_1}\right)^2, \\ H_2^2 &= \left(\frac{\partial \varphi_1}{\partial q_2}\right)^2 + \left(\frac{\partial \varphi_2}{\partial q_2}\right)^2 + \left(\frac{\partial \varphi_3}{\partial q_2}\right)^2, \\ H_3^2 &= \left(\frac{\partial \varphi_1}{\partial q_3}\right)^2 + \left(\frac{\partial \varphi_2}{\partial q_3}\right)^2 + \left(\frac{\partial \varphi_3}{\partial q_3}\right)^2. \end{aligned}\right\} \tag{20}$$

Along each edge of an elementary volume only one coordinate varies, therefore according to formula (19) we shall have for the length of these edges

$$ds_1 = H_1 \, dq_1, \quad ds_2 = H_2 \, dq_2, \quad ds_3 = H_3 \, dq_3, \tag{21}$$

so that an element of volume equals

$$dv = ds_1 \, ds_2 \, ds_3 = H_1 \, H_2 \, H_3 \, dq_1 \, dq_2 \, dq_3. \tag{22}$$

Let us consider now some vector field $\boldsymbol{A}(x, y, z)$. Let us calculate $\operatorname{div} \boldsymbol{A}$, defined by the familiar relation of vector analysis

$$\operatorname{div} \boldsymbol{A} = \lim_{v_M \to 0} \frac{\iint_S A_n \, dS}{v_M}, \tag{23}$$

where S is a surface, bounding some volume v_M, containing the point M under consideration. Let us apply this formula to an element of volume dv, described in Fig. 44.

Making use of the mean value theorem, it is possible to express the change in flux of the vector $\boldsymbol{A}$ between opposite faces, for example from the right to the left face, in the form

$$Q_1 = A_1 \, ds_2 \, ds_3 \,|_{q_1+dq_1} - A_1 \, ds_2 \, ds_3 \,|_{q_1}.$$

Formula (21) gives:

$$Q_1 = [H_2 \, H_3 \, A_1 \,|_{q_1+dq_1} - H_2 \, H_3 \, A_1 \,|_{q_1}] \, dq_2 \, dq_3 =$$

$$= \frac{\partial}{\partial q_1} (H_2 \, H_3 \, A_1) \, dq_1 \, dq_2 \, dq_3. \tag{24}$$

Similarly the flux changes across the two other pairs of opposite faces can be calculated

$$Q_2 = \frac{\partial}{\partial q_2} (H_3 \, H_1 \, A_2) \, dq_1 \, dq_2 \, dq_3. \tag{25}$$

$$Q_3 = \frac{\partial}{\partial q_3} (H_1 \, H_2 \, A_3) \, dq_1 \, dq_2 \, dq_3. \tag{26}$$

Substituting in formula (23)

$$\iint_S A_n \, dS = Q_1 + Q_2 + Q_3$$

and using formula (22), we obtain an expression for the divergence in orthogonal curvilinear coordinates

$$\operatorname{div} \boldsymbol{A} = \frac{1}{H_1 H_2 H_3}\left[\frac{\partial}{\partial q_1}(H_2 H_3 A_1) + \right.$$

$$\left. + \frac{\partial}{\partial q_2}(H_3 H_1 A_2) + \frac{\partial}{\partial q_3}(H_1 H_2 A_3)\right]. \tag{27}$$

Let us assume that the field $\boldsymbol{A}$ is derivable from a potential, i.e.

$$\boldsymbol{A} = \operatorname{grad} u. \tag{28}$$

Then

$$A_1 = \frac{\partial u}{\partial s_1} = \frac{1}{H_1}\frac{\partial u}{\partial q_1}; \quad A_2 = \frac{1}{H_2}\frac{\partial u}{\partial q_2}; \quad A_3 = \frac{1}{H_3}\frac{\partial u}{\partial q_3}. \tag{29}$$

Substituting (29) in (27) for A_1, A_2 and A_3, we obtain an expression for the Laplacian operator

$$\nabla^2 u = \operatorname{div}\operatorname{grad}\ u = \frac{1}{H_1 H_2 H_3}\left[\frac{\partial}{\partial q_1}\left(\frac{H_2 H_3}{H_1}\frac{\partial u}{\partial q_1}\right) + \right.$$

$$\left. + \frac{\partial}{\partial q_2}\left(\frac{H_3 H_1}{H_2}\frac{\partial u}{\partial q_2}\right) + \frac{\partial}{\partial q_3}\left(\frac{H_1 H_2}{H_3}\frac{\partial u}{\partial q_3}\right)\right]. \tag{30}$$

Thus Laplace's equation $\Delta^2 u = 0$ in orthogonal curvilinear coordinates q_1, q_2, q_3 is written in the following way:

$$\nabla^2 u = \frac{1}{H_1 H_2 H_3}\left\{\frac{\partial}{\partial q_1}\left(\frac{H_2 H_3}{H_1}\frac{\partial u}{\partial q_1}\right) + \right.$$

$$\left. + \frac{\partial}{\partial q_2}\left(\frac{H_3 H_1}{H_2}\frac{\partial u}{\partial q_2}\right) + \frac{\partial}{\partial q_3}\left(\frac{H_1 H_2}{H_3}\frac{\partial u}{\partial q_3}\right)\right\} = 0. \tag{31}$$

Let us consider two special cases.

1. *Spherical coordinates.* In this case $q_1 = r$, $q_2 = \theta$, $q_3 = \varphi$ and the transformations (15) take the form

$$x = r\sin\theta\cos\varphi, \quad y = r\sin\theta\sin\varphi, \quad z = r\cos\theta.$$

We calculate ds^2.

$$ds^2 = (\sin\theta\cos\varphi\, dr + r\cos\theta\cos\varphi\, d\theta -$$

$$- r\sin\theta\sin\varphi\, d\varphi)^2 + (\sin\theta\sin\varphi\, dr + r\cos\theta\sin\varphi\, d\theta +$$

$$r\sin\theta\cos\varphi\, d\varphi)^2 + (\cos\theta\, dr - r\sin\theta\, d\theta)^2;$$

after removal of the brackets and simplification we find:

$$ds^2 = dr^2 + r^2\, d\theta^2 + r^2 \sin^2\theta\, d\varphi^2,$$

i.e.

$$H_1 = 1, \quad H_2 = r, \quad H_3 = r\sin\theta.$$

Substituting the values H_1, H_2, H_3 in formula (31), we obtain Laplace's equation in spherical coordinates

$$\frac{1}{r^2\sin\theta}\left[\frac{\partial}{\partial r}\left(r^2\sin\theta\frac{\partial u}{\partial r}\right) + \right.$$

$$\left. + \frac{\partial}{\partial\theta}\left(\sin\theta\frac{\partial u}{\partial\theta}\right) + \frac{\partial u}{\partial\varphi}\left(\frac{1}{\sin\theta}\frac{\partial u}{\partial\varphi}\right)\right] = 0$$

or finally

$$\frac{1}{r^2}\frac{\partial}{\partial r}\left(r^2\frac{\partial u}{\partial r}\right) + \frac{1}{r^2\sin\theta}\frac{\partial}{\partial\theta}\left(\sin\theta\frac{\partial u}{\partial\theta}\right) + \frac{1}{r^2\sin^2\theta}\frac{\partial^2 u}{\partial\varphi^2} = 0. \qquad (32)$$

2. *Cylindrical coordinates.* In this case $q_1 = \varrho$, $q_2 = \varphi$, $q_3 = z$:

$$x = \varrho\cos\varphi, \quad y = \varrho\sin\varphi, \quad z = z,$$

so that

$$H_1 = 1, \qquad H_2 = \varrho, \qquad H_3 = 1.$$

Laplace's equation in cylindrical coordinates takes the form

$$\frac{1}{\varrho}\frac{\partial}{\partial\varrho}\left(\varrho\frac{\partial u}{\partial\varrho}\right) + \frac{1}{\varrho^2}\frac{\partial^2 u}{\partial\varphi^2} + \frac{\partial^2 u}{\partial z^2} = 0. \qquad (33)$$

If the unknown function u does not depend on z, then equation (33) simplifies to

$$\frac{1}{\varrho}\frac{\partial}{\partial\varrho}\left(\varrho\frac{\partial u}{\partial\varrho}\right) + \frac{1}{\varrho^2}\frac{\partial^2 u}{\partial\varphi^2} = 0. \qquad (34)$$

4. *Some particular solutions of Laplace's equation*

The solutions of Laplace's equation, possessing spherical or cylindrical symmetry, i.e. depending only on one variable r or ϱ, are of great interest.

The solution of Laplace's equation $u = U(r)$, possessing spherical symmetry, will be determined from the ordinary differential equation

$$\frac{d}{dr}\left(r^2\frac{dU}{dr}\right) = 0.$$

Integrating this equation, we find:

$$U = \frac{C_1}{r} + C_2,$$

where C_1 and C_2 are arbitrary constants. Assuming, for example, $C_1 = 1$, $C_2 = 0$, we obtain the function

$$U_0 = 1/r,$$

which is often called the *fundamental solution of Laplace's equation in* three dimensions.

Similarly, assuming

$$u = U(\varrho)$$

and using equation (33) or (34) we find a solution, having cylindrical or circular symmetry (in the case of two independent variables) in the form

$$U(\varrho) = C_1 \ln \varrho + C_2.$$

Choosing $C_1 = -1$ and $C_2 = 0$ we shall have:

$$U_0 = \ln \frac{1}{\varrho}. \tag{36}$$

The function $U_0(\varrho)$ is often called the *fundamental solution of Laplace's equation in a plane* (for two independent variables).

The function $U_0 = 1/r$ satisfies the equation $\Delta u = 0$ everywhere except at the point $r = 0$ where Δu becomes infinite. Except for a factor of proportionality it is the field of a point charge e, located at the origin of coordinates; the potential of this field equals

$$u = \frac{e}{r}.$$

Similarly, the function $\ln \frac{1}{\varrho}$ satisfies Laplace's equation everywhere except at the point $\varrho = 0$ where it becomes infinite and except for a factor it agrees with the field of a line charge (see in detail § 5, sub-section 2), the potential of which equals

$$u = 2 e_1 \ln \frac{1}{\varrho},$$

where e_1 is the charge density per unit length. These functions are of great importance in the theory of harmonic functions.

5. *Harmonic functions and analytic functions of the complex variable*

A very common method of solution of Laplace's equation in two-dimensional cases uses functions of a complex variable. Let

$$w = f(z) = u(x, y) + iv(x, y)$$

be some function of a complex variable $z = x + iy$, where u and v are functions of the variables x and y. The greatest interest lies in the so-called analytic functions for which the derivative exists

$$\frac{dw}{dz} = \lim_{\Delta z \to 0} \frac{\Delta w}{\Delta z} = \lim_{\Delta z \to 0} \frac{f(z + \Delta z) - f(z)}{dz}.$$

The increment $\Delta z = \Delta x + i \Delta y$, obviously, may tend to zero in several ways. For each way that Δz tends to zero, generally speaking, a value of its limit may be obtained. But if the function $w = f(z)$ is analytic, then the limit $\lim\limits_{\Delta z \to 0} \Delta f / \Delta z = f'(z)$ does not depend on the choice of path.

The necessary and sufficient conditions for a function to be analytic are the so called *Cauchy–Riemann relations*.

$$\left.\begin{aligned} u_x &= v_y, \\ u_y &= -v_x. \end{aligned}\right\} \tag{37}$$

The necessity of these relations may be proved in the following way.

Let $w = u + iv = f(z)$ be an analytic function. Calculating the derivatives

$$w_x = u_x + iv_x = \frac{\partial w(z)}{\partial z} z_x = \frac{dw}{dz},$$

$$w_y = u_y + iv_y = \frac{\partial w(z)}{\partial z} z_y = i \frac{dw}{dz}$$

and using the equality of the values dw/dz, determined from these two relations, we obtain:

$$u_x + iv_x = v_y - iu_y = \frac{dw}{dz},$$

from which follow the Cauchy–Riemann relations. We will not give a proof of the sufficiency of these conditions.

In the theory of functions of a complex variable it is proved that a function, analytic in some region G of the plane $z = x + iy$, has in this region derivatives of all orders and may be expanded in a power series. In particular, the functions $u(x, y)$ and $v(x, y)$ have continuous derivatives of second order with respect to x and y.

Differentiating the first equality of (37) with respect to x, and the second with respect to y, we obtain:

$$u_{xx} + u_{yy} = 0 \text{ or } \nabla_2^2 u = 0 .$$

Similarly, changing the order of differentiation we find:

$$v_{xx} + v_{yy} = 0 \text{ or } \nabla_2^2 v = 0 .$$

Thus the real and imaginary part of the analytic function, satisfies Laplace's equation. Generally it is said that u and v, satisfying the Cauchy–Riemann relations, are conjugate harmonic functions.

Let us consider the transformation

$$\left.\begin{array}{ll} x = x(u, v); & u = u(x, y), \\ y = y(u, v), & v = v(x, y), \end{array}\right\} \tag{38}$$

from some region G of the plane (x, y) to a region G' of the plane (u, v), so that a given point of the region G' has a corresponding point in the region G and, conversely, a given point of the region G has a corresponding point in the region G'.

Let

$$U = U(x, y)$$

be a twofold continuously differentiable function, given inside the region G.

Let us show how the Laplacian operator of the function $U = U[x(u, v), y(u, v)] = \widetilde{U}(u, v)$ transforms.

We calculate the derivatives of the function

$$U_x = \widetilde{U}_u u_x + \widetilde{U}_v v_x, \quad U_y = \widetilde{U}_u u_y + \widetilde{U}_v v_y,$$

$$U_{xx} = \widetilde{U}_{uu} u_x^2 + \widetilde{U}_{vv} v_x^2 + 2\widetilde{U}_{uv} u_x v_x + \widetilde{U}_u u_{xx} + \widetilde{U}_v v_{xx},$$

$$U_{yy} = \widetilde{U}_{uu} u_y^2 + \widetilde{U}_{xv} v_y^2 + 2\widetilde{U}_{uv} u_y v_y + \widetilde{U}_u u_{yy} + \widetilde{U}_v v_{yy}$$

from which we obtain:

$$U_{xx} + U_{yy} = \widetilde{U}_{uu}(u_x^2 + u_y^2) + \widetilde{U}_{vv}(v_x^2 + v_y^2) +$$
$$+ 2\widetilde{U}_{uv}(u_x v_x + u_y v_y) + \widetilde{U}_u(u_{xx} + u_{yy}) + \widetilde{U}_v(v_{xx} + v_{yy}) .$$

If u and v are conjugate harmonic functions, then the transformation (38) is equivalent to the transformation realized by the analytic function

$$w = f(z) = u + iv \qquad (z = x + iy) . \tag{40}$$

In this case because of the Cauchy–Riemann relations for the functions u and v the relations

$$u_x^2 + u_y^2 = u_x^2 + v_x^2 = v_y^2 + v_x^2 = |f'(z)|^2$$
$$u_x v_x + u_y v_y = 0 .$$

are fulfilled.

Formula (39) takes the form

$$U_{xx} + U_{yy} = (U_{uu} + U_{vv})\,|f'(z)|^2 \tag{41}$$

or

$$\nabla^2_{uv} \widetilde{U} = \frac{1}{|f'(z)|^2} \nabla^2_{x,y} U. \tag{41'}$$

It follows that the function $U(x, y)$ harmonic in the region G transforms into the function $\widetilde{U} = U(u, v)$, harmonic in the region G', provided $|f'(z)|^2 \neq 0$.

6. *The inversion transformation*

In a study of harmonic functions one often uses the inversion transformation. A transformation is said to be an inversion in a sphere of radius a, if to any point M there is a corresponding point M', lying on the same line through the origin of coordinates as the point M, whose radius-vector r' is connected to the radius-vector of the point M by the relation

$$r' r = a^2 \text{ where } r' = a^2/r . \tag{42}$$

Furthermore we shall assume $a = 1$. It is always possible to satisfy this condition by a change of the scale of length.

Let us show that a harmonic function of two independent variables $u(\varrho, \varphi)$ by an inversion transformation becomes the harmonic function

$$v(\varrho', \varphi) = u(\varrho, \varphi), \quad \text{where } \varrho = 1/\varrho'. \tag{43}$$

In fact, the function $u(\varrho, \varphi)$ and similarly the function $v\left(\frac{1}{\varrho}, \varphi\right)$ as functions of the variables ϱ and φ satisfy the equations

$$\varrho^2 \nabla^2_{\varrho, \varphi} u = \varrho \frac{\partial}{\partial \varrho}\left(\varrho \frac{\partial u}{\partial \varrho}\right) + \frac{\partial^2 u}{\partial \varphi^2} = 0$$

and

$$\varrho^2 \nabla^2_{\varrho, \varphi} v = \varrho \frac{\partial}{\partial \varrho}\left(\varrho \frac{\partial v}{\partial \varrho}\right) + \frac{\partial^2 v}{\partial \varphi^2} = 0 .$$

Changing to the variables ϱ' and φ, we obtain:

$$\varrho \frac{\partial v}{\partial \varrho} = \varrho \frac{\partial v}{\partial \varrho'} \cdot \frac{\partial \varrho'}{\partial \varrho} = -\varrho' \frac{\partial v}{\partial \varrho'},$$

from which it follows that $v(\varrho', \varphi)$ satisfies the equation $\Delta_{\varrho,\varphi}\, v = 0$ since

$$\varrho'^2 \nabla^2_{\varrho', \varphi} v = \varrho' \frac{\partial}{\partial \varrho'}\left(\varrho' \frac{\partial v}{\partial \varrho'}\right) + \frac{\partial^2 v}{\partial \varphi^2} = 0 .$$

Changing to the case of three independent variables, let us show that the function

$$v(r', \theta, \varphi) = r u(r, \theta, \varphi), \quad \text{where } r = 1/r' \tag{44}$$

satisfies Laplace's equation $\Delta_{r'\theta,\varphi,}\, v = 0$, if $u(r, \theta, \varphi)$ is an harmonic function in its variables $\Delta_{r,\theta,\varphi}\, u = 0$.

Transformation (44) is often called Kelvin's transformation.

It is easy to satisfy oneself by direct differentiation, that the first component in the Laplacian operator reduces to the form

$$\frac{1}{r^2}\frac{\partial}{\partial r}\left(r^3 \frac{\partial u}{\partial r}\right) = \frac{\partial^2 u}{\partial r^2} + \frac{2}{r}\frac{\partial u}{\partial r} = \frac{1}{r}\frac{\partial^2 (ru)}{\partial r^2}, \tag{45}$$

so that

$$r\Delta_{r,\theta,\varphi} u = \frac{\partial^2 (ru)}{\partial r^2} + \frac{1}{r}\left[\frac{1}{\sin\theta}\frac{\partial}{\partial \theta}\left(\sin\theta \frac{\partial u}{\partial \theta}\right) + \frac{1}{\sin^2\theta}\frac{\partial^2 u}{\partial \varphi^2}\right] = 0$$

or

$$\frac{\partial^2 v}{\partial r^2} + \frac{1}{r^2}\left[\frac{1}{\sin\theta}\frac{\partial}{\partial \theta}\left(\sin\theta \frac{\partial v}{\partial \theta}\right) + \frac{1}{\sin^2\theta}\frac{\partial^2 v}{\partial \varphi^2}\right] = 0 .$$

Noting that

$$\frac{\partial v}{\partial r} = \frac{\partial v}{\partial r'} \cdot \frac{\partial r'}{\partial r} = -r'^2 \frac{\partial v}{\partial r'},$$

we find that v satisfies the equation $\Delta_{r',\theta,\varphi}\, v = 0$, since

$$r'^2 \frac{\partial}{\partial r'}\left(r'^2 \frac{\partial v}{\partial r'}\right) + r'^2 \left[\frac{1}{\sin\theta}\frac{\partial}{\partial \theta}\left(\sin\theta \frac{\partial v}{\partial \theta}\right) + \frac{1}{\sin^2\theta}\frac{\partial^2 v}{\partial \varphi^2}\right] = 0,$$

or

$$r'^4 \nabla^2_{r',\theta,\varphi}\, v = 0.$$

§ 2. General properties of harmonic functions

In the present section we give an integral representation of harmonic functions which is fundamental to an investigation of their general properties. One of the most important results of this integral relation is the maximum value principle, frequently used both in the proof of the uniqueness theorem and in the solution of boundary-value problems. We also give a mathematical statement of internal and external boundary-value problems for Laplace's equation, and prove the uniqueness and stability of the solutions.

1. *Green's formula. Integral representation of a solution*

In a study of equations of elliptic type we will often use Green's theorems, which are a direct consequence of Ostrogradskii's relation.

Ostrogradskii's relation in the simplest case has the form

$$\iiint_T \frac{\partial R}{\partial z}\, dx\, dy\, dz = \iint_\Sigma R \cos\gamma \, d\sigma, \tag{1}$$

where T is some volume, bounded by a sufficiently smooth surface Σ, $R(x, y, z)$ is an arbitrary function, continuous on and inside Σ and possesing continuous derivatives inside T, γ is the angle between the direction of the z-axis and the outer normal to Σ. One may easily demonstrate the validity of this relation, by performing an integration with respect to z.

One usually writes Ostrogradskii's relation in the form

$$\iiint\limits_T \left(\frac{\partial P}{\partial x} + \frac{\partial Q}{\partial y} + \frac{\partial R}{\partial z}\right) d\tau = \iint\limits_\Sigma \{P \cos \alpha + Q \cos \beta + R \cos \gamma\}\, d\sigma, \tag{2}$$

where $d\tau = dxdydz$ is an element of volume, $\alpha = (\widehat{nx})$, $\beta = (\widehat{ny})$, $\gamma = (\widehat{nz})$ are the angles of the outer normal $\boldsymbol{n}$ to the surface Σ with the coordinate axes, P, Q, R are arbitrary differentiable functions.*

If P, Q, R are considered as components of some vector $\boldsymbol{A} = P\boldsymbol{i} + Q\boldsymbol{j} + R\boldsymbol{k}$, then Ostrogradskii's relation may be written in the following way:

$$\iiint\limits_T \operatorname{div} \boldsymbol{A}\, d\tau = \iint\limits_\Sigma A_n\, d\sigma, \tag{2'}$$

where

$$\operatorname{div} \boldsymbol{A} = \frac{\partial P}{\partial x} + \frac{\partial Q}{\partial y} + \frac{\partial R}{\partial z} \tag{2'a}$$

and

$$A_n = P \cos \alpha + Q \cos \beta + R \cos \gamma \tag{2'b}$$

is the component of the vector $\boldsymbol{A}$ along the outer normal.

Let us proceed now to the derivation of Green's formulae.

Let $u = u(x, y, z)$ and $v = v(x, y, z)$ be functions, continuous together with their first derivatives inside $T + \Sigma$ and having continuous second derivatives in T.

Putting

$$P = u\frac{\partial v}{\partial x}, \quad Q = u\frac{\partial v}{\partial y}, \quad R = u\frac{\partial v}{\partial z} \tag{2'c}$$

and using Ostrogradskii's relation, we arrive at Green's *first relation*

$$\iiint\limits_T u \nabla^2 v\, d\tau = \iint\limits_\Sigma u \frac{\partial v}{\partial n}\, d\sigma -$$

$$- \iiint\limits_T \left(\frac{\partial u}{\partial x}\frac{\partial v}{\partial x} + \frac{\partial u}{\partial y}\frac{\partial v}{\partial y} + \frac{\partial u}{\partial z}\frac{\partial v}{\partial z}\right) d\tau, \tag{3}$$

* Later we shall assume that Ostrogradskii's relations are applicable to those regions which we consider.

where $\nabla^2 = \frac{\partial^2}{\partial x^2} + \frac{\partial^2}{\partial y^2} + \frac{\partial^2}{\partial z^2}$ is the Laplacian operator, $\frac{\partial}{\partial n} = \cos\alpha \frac{\partial}{\partial x} + \cos\beta \frac{\partial}{\partial y} + \cos\gamma \frac{\partial}{\partial z}$ is the derivative with respect to the direction of the outer normal.

If one considers the relation

$$\operatorname{grad} u \operatorname{grad} v = \nabla u \cdot \nabla v = \frac{\partial u}{\partial x}\frac{\partial v}{\partial x} + \frac{\partial u}{\partial y}\frac{\partial v}{\partial y} + \frac{\partial u}{\partial z}\frac{\partial v}{\partial z}, \tag{3a}$$

then it is possible to express Green's theorem in the form

$$\iiint_T u \nabla^2 v \, d\tau = -\iiint_T \nabla u \nabla v \, d\tau + \iint_\Sigma u \frac{\partial v}{\partial n} d\sigma. \tag{3'}$$

Interchanging the functions u and v, we have:

$$\iiint_T v \nabla^2 u \, d\tau = -\iiint_T \nabla v \nabla u \, d\tau + \iint_\Sigma v \frac{\partial u}{\partial n} d\sigma. \tag{4}$$

Subtracting equation (4) from (3′), we obtain *Green's second theorem*

$$\iiint_T (u \nabla^2 v - v \nabla^2 u) \, d\tau = \iint_\Sigma \left(u \frac{\partial v}{\partial n} - v \frac{\partial u}{\partial n}\right) d\sigma. \tag{5}$$

The region T may be bounded by several surfaces. Green's formulae are applicable where the surface integrals are taken over all surfaces bounding the region T.

As we saw (§ 1, sub-section 4), the function $U_0(M) = 1/r$, where $r = \sqrt{[(x-x_0)^2 + (y-y_0)^2 + (z-z_0)^2]}$ is the distance between the points $M(x, y, z)$ and $M_0(x_0, y_0, z_0)$, satisfies Laplace's equation.

Let $u(M)$ be an harmonic function, continuous together with its first derivatives in the region $T + \Sigma$ and having second derivatives in T. Let us consider the function $v = 1/r_{MM_0}$, where M_0 is some interior point of the region T. Because this function has a discontinuity at the point $M_0(x_0, y_0, z_0)$ inside T, it is impossible to apply Green's second theorem directly to the functions u and v in the region T. However, the function $v = 1/r_{MM_0}$ is bounded in the region $T - K_\varepsilon$ by the boundary $\Sigma + \Sigma_\varepsilon$, where K_ε is a sphere of radius ε with centre at the point M_0 and with surface Σ_ε (Fig. 45).

Applying Green's second formula (5) to the functions u and $v = 1/r$ in the region $T - K_\varepsilon$, we obtain:

$$\iiint\limits_{T-K_\varepsilon} \left(u \nabla^2 \frac{1}{r} - \frac{1}{r} \nabla^2 u\right) d\tau = \iint\limits_{\Sigma} \left(u \frac{\partial}{\partial n}\left(\frac{1}{r}\right) - \frac{1}{r}\frac{\partial u}{\partial n}\right) d\sigma +$$

$$+ \iint\limits_{\Sigma_\varepsilon} u \frac{\partial}{\partial n}\left(\frac{1}{r}\right) d\sigma - \iint\limits_{\Sigma_\varepsilon} \frac{1}{r}\frac{\partial u}{\partial n} d\sigma. \qquad (6)$$

On the right-hand side of this relation only the last two integrals

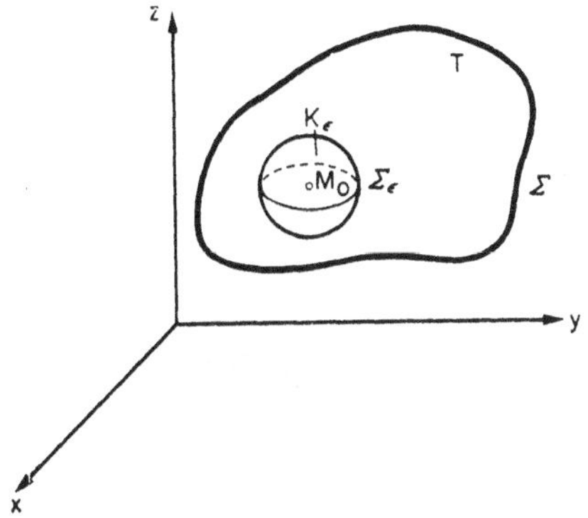

FIG. 45

depend on ε. Calculating the derivative with respect to the outer normal to the region $T - K_\varepsilon$ on Σ_ε, we find that

$$\frac{\partial}{\partial n}\left(\frac{1}{r}\right)\bigg|_{\Sigma_\varepsilon} = -\frac{\partial}{\partial r}\left(\frac{1}{r}\right)\bigg|_{r=\varepsilon} = \frac{1}{\varepsilon^2},$$

from which

$$\iint\limits_{\Sigma_\varepsilon} u \frac{\partial}{\partial n}\left(\frac{1}{r}\right) d\sigma = \frac{1}{\varepsilon^2}\iint\limits_{\Sigma_\varepsilon} u\, d\sigma = \frac{1}{\varepsilon^2} 4\pi\varepsilon^2 u^* = 4\pi u^*, \qquad (7)$$

where u^* is the average value of the function $u(M)$ on the surface Σ_ε. Let us transform the third integral

$$\iint\limits_{\Sigma_\varepsilon} \frac{1}{r}\frac{\partial u}{\partial n} d\sigma = \frac{1}{\varepsilon}\iint\limits_{\Sigma_\varepsilon} \frac{\partial u}{\partial n} d\sigma = \frac{1}{\varepsilon} 4\pi\varepsilon^2 \left(\frac{\partial u}{\partial n}\right)^* = 4\pi\varepsilon\left(\frac{\partial u}{\partial n}\right)^*, \qquad (8)$$

where $\left(\frac{\partial u}{\partial n}\right)^*$ is the mean value of the normal derivative $\frac{\partial u}{\partial n}$

on the sphere Σ_ε. Substituting expressions (7) and (8) in (6) and taking account of the fact that $\nabla^2(1/r) = 0$ in $T - K_\varepsilon$, we have:

$$\iiint\limits_{T-K_\varepsilon} \left(-\frac{1}{r}\right) \nabla^2 u \, d\tau =$$

$$= \iint\limits_{\Sigma} \left[u \frac{\partial}{\partial n}\left(\frac{1}{r}\right) - \frac{1}{r}\frac{\partial u}{\partial n}\right] d\sigma + 4\pi u^* - 4\pi\varepsilon \left(\frac{\partial u}{\partial n}\right)^*. \tag{9}$$

We now let the radius ε tend to zero, and obtain:

(1) $\lim\limits_{\varepsilon \to 0} u^* = u(M_0)$, since $u(M)$ is a continuous function, and u^* is its mean value in a sphere of radius ε with centre at the point M_0;

(2) $\lim\limits_{\varepsilon \to 0} 4\pi\varepsilon \left(\frac{\partial u}{\partial n}\right)^* = 0$, since from the continuity of the first derivatives of the function $u(M)$ inside T the same is true of the normal derivative

$$\frac{\partial u}{\partial n} = \frac{\partial u}{\partial x}\cos\alpha + \frac{\partial u}{\partial y}\cos\beta + \frac{\partial u}{\partial z}\cos\gamma$$

in the neighbourhood of the point M_0;

(3) according to the definition of a definite integral

$$\lim_{\varepsilon\to 0} \iiint\limits_{T-K_\varepsilon} \left(-\frac{1}{r}\nabla^2 u\right) d\tau = \iiint\limits_{T} \left(-\frac{1}{r}\nabla^2 u\right) d\tau .$$

Collecting these results together, we arrive at the *fundamental integral formula* of the theory of harmonic functions

$$4\pi u(M_0) = -\iint\limits_{\Sigma} \left[u(P)\frac{\partial}{\partial n}\left(\frac{1}{r_{M_0P}}\right) - \frac{1}{r_{M_0P}}\frac{\partial u}{\partial n}\right] d\sigma_P - \iiint\limits_{T} \frac{\nabla^2 u}{r} d\tau. \tag{10}$$

In the application to the harmonic function $u(M)$ ($\nabla^2 u = 0$) formula (10) gives:

$$u(M) = \frac{1}{4\pi}\iint\limits_{\Sigma} \left[\frac{1}{r_{MP}}\frac{\partial u}{\partial n} - u(P)\frac{\partial}{\partial n}\left(\frac{1}{r_{MP}}\right)\right] d\sigma_P . \tag{11}$$

Thus, the value of a harmonic function at any interior point of a region is expressed by the value of this function and its normal derivative on the surface of the region. Note that we have assumed the continuity of u and its first derivative on the boundary. We note immediately that each of the integrals

$$\iint_{\Sigma} \mu(P) \frac{1}{r_{MP}} d\sigma_P \text{ and } \iint_{\Sigma} \frac{\partial}{\partial n_P}\left(\frac{1}{r_{MP}}\right) \nu(P)\, d\sigma_P, \tag{12}$$

where μ and ν are continuous functions, is a harmonic function. In fact, since the functions under the integral sign and all their derivatives are continuous inside the surface Σ, then derivatives of the functions (12) of any order may be calculated by differentiating under the integral sign. Since, moreover, the functions

$$\frac{1}{r_{MP}} \text{ and } \frac{\partial}{\partial n_P}\left(\frac{1}{r_{MP}}\right) = \frac{\partial}{\partial \xi}\left(\frac{1}{r}\right) \cos \alpha_P + \frac{\partial}{\partial \eta}\left(\frac{1}{r}\right) \cos \beta_P + \frac{\partial}{\partial \zeta}\left(\frac{1}{r}\right) \cos \gamma_P$$

satisfy Laplace's equation with respect to the variables $M(x, y, z)$ and because of the generalized principle of superposition (see the lemma on p. 244), the functions (12) also satisfy Laplace's equation with respect to the variables x, y, z.

An important result follows: any harmonic function is differentiable any number of times* inside the region of harmonic variation. We note also that a harmonic function is analytic (it is expandable in a power series) at any point M_0 of the region T. It is possible to show this by considerations, based on the same integral representation (11).

Similar formulae hold for harmonic functions of two independent variables. Let S be some region of a plane (x, y), bounded by the contour C, and n the direction of the normal to this contour, external with respect to the region S. Assuming $v = \ln \frac{1}{r_{M_0P}}$, where $r_{M_0P} = \sqrt{[(x - x_0)^2 + (y - y_0)^2]}$ is the distance from the point P on S to the point M_0, lying inside S and using considerations similar to those which have been used above, we obtain in place of formula (10):

$$2\pi u(M_0) = -\int_C \left[u \frac{\partial}{\partial n}\left(\ln \frac{1}{r_{M_0P}}\right) - \ln \frac{1}{r_{M_0P}} \frac{\partial u}{\partial n}\right] ds_P - \\ - \iint_S \ln \frac{1}{r} \nabla^2 u\, dS, \tag{12'}$$

* If a function u, harmonic inside T, is not continuous together with its first derivative on the surface Σ, then the theorem almost maintains its effect in that it is possible to surround the point M with a region, lying together with its boundary inside T.

where M_0 is any fixed point within the region S.

If $u(M)$ is a harmonic function, then it follows that

$$u(M_0) = \frac{1}{2\pi} \int_C \left[\ln\frac{1}{r} \frac{\partial u}{\partial n} - u \frac{\partial}{\partial n}\left(\ln\frac{1}{r}\right)\right] ds. \qquad (12'')$$

2. *Some fundamental properties of harmonic functions*

Let us determine some very important properties of harmonic functions:

1. *If v is a function, harmonic in the region T, bounded by the surface Σ, then*

$$\int\int_S \frac{\partial v}{\partial n} d\sigma = 0, \qquad (13)$$

where S is any closed surface, lying entirely within the region T. In fact, substituting in Green's first formula (3) some harmonic function $v(\nabla^2 v = 0)$ and the function $u \equiv 1$, we obtain at once formula (13). From formula (13) it follows that the second boundary-value problem $\left(\nabla^2 u = 0, \text{in } T, \frac{\partial u}{\partial n} = f\Big|_\Sigma\right)$ may have a solution only on condition

$$\int\int_\Sigma f d\sigma = 0.$$

This property of harmonic functions can be interpreted as a condition of the absence of sources inside the region T.

2. *If the function $u(M)$ is harmonic in some region T, and M_0 is any point, lying inside T, then there is a relation*

$$u(M_0) = \frac{1}{4\pi a^2} \int\int_{\Sigma_a} u\, d\sigma, \qquad (14)$$

where Σ_a is a sphere of radius a with centre at the point M_0, lying entirely within the region T (the mean value theorem).

This theorem proves that the value of a harmonic function at some point is equal to the mean value of this function on any sphere Σ_a with centre at M_0, if the sphere Σ_a does not go outside the region of harmonic variation of the function $u(M)$

Let us apply formula (11) to a sphere K_a with centre at the point M_0 and with surface Σ_a:

$$4\pi u(M_0) = -\iint\limits_{\Sigma_a} \left[u \frac{\partial}{\partial n}\left(\frac{1}{r}\right) - \frac{1}{r}\frac{\partial u}{\partial n}\right] d\sigma.$$

Taking into consideration that

$$\frac{1}{r} = \frac{1}{a} \text{ on } \Sigma_a \text{ and } \iint\limits_{\Sigma_a} \frac{\partial u}{\partial n} d\sigma = 0,$$

and also

$$\frac{\partial}{\partial n}\left(\frac{1}{r}\right)\bigg|_{\Sigma_a} = \frac{\partial}{\partial r}\left(\frac{1}{r}\right)\bigg|_{r=a} = -\frac{1}{a^2}$$

(the direction of the outer normal to Σ_a coincides with the direction of the radius), we obtain the required result:*

$$u(M_0) = \frac{1}{4\pi a^2}\iint\limits_{\Sigma_a} u\, d\sigma,$$

In the case of two independent variables there is a similar theorem:

$$u(M_0) = \frac{1}{2\pi a}\int\limits_{C_a} uds, \tag{15}$$

where C_a is a circle of radius a with centre at point M_0.

3. *If the function $u(M)$, defined and continuous in the closed region $T + \Sigma$, satisfies the equation $\nabla^2 u = 0$ inside T, then maximum and minimum values of the function $u(M)$ are attained on the surface Σ (the principle of maximum value).*

Let us assume that the function $u(M)$ attains a maximum value at some interior point M_0 of the region T, so that $u_0 = u(M_0) \geqslant u(M)$, where M is any point of the region T. Let us surround the point M with a sphere Σ'_ϱ of radius ϱ, lying entirely

* In the proof of this theorem we made use of equality (13), assuming the existence of derivatives on the surface of the sphere. If the function $u(M)$, continuous in the closed region $T + \Sigma$, satisfies the equation $\Delta u = 0$ only for interior points of T, then the preceeding conclusion for the sphere Σ_{a_0}, which touches the surface Σ will be unfounded. However the theorem is valid for any $a < a_0$, and, passing to a limit as $a \to a_0$, we obtain:

$$u(M_0) = \frac{1}{4\pi a_0^2}\iint\limits_{\Sigma_{a_0}} u(M)\, d\sigma.$$

within the region T. Since, by hypothesis, $u(M_0)$ is the maximum value of the function $u(M)$ in $T + \Sigma$, then $u|_\Sigma \leqslant u(M_0)$. Using the mean value formula (14) and substituting $u(M)$ everywhere under the integral by $u(M_0)$, we obtain:

$$u(M_0) = \frac{1}{4\pi\varrho^2} \iint\limits_{\Sigma_\varrho} u(M)\, d\sigma_M \leqslant \frac{1}{4\pi\varrho^2} \iint\limits_{\Sigma_\varrho} u(M_0)\, d\sigma = u(M_0). \tag{16}$$

If it is assumed that $u(M) < u(M_0)$ at even one point M of the sphere Σ_ϱ, then obviously in place of the sign $\leqslant$ we will

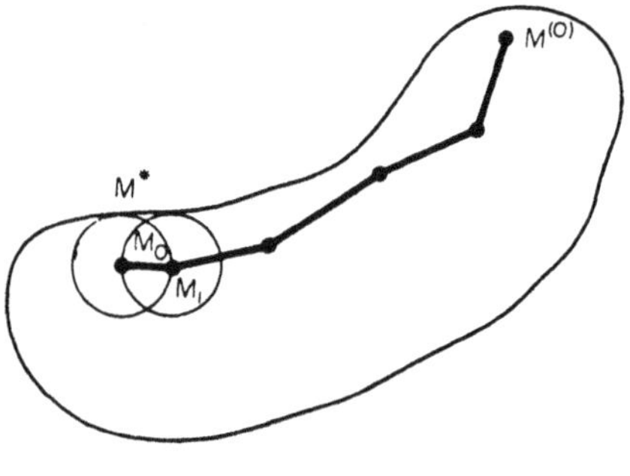

FIG. 46

have the sign $<$, which gives a contradiction. Thus, over the entire surface Σ_ϱ, $u(M) \equiv u(M_0)$.

If ϱ_0^m is the minimum distance from the point M_0 to the surface Σ, then $u(M) \equiv u(M_0)$ for all points, lying inside $\Sigma_{\varrho_0^m}$. It follows from the continuity of u that at the points M^*, which belong to the common part of $\Sigma_{\varrho_0^m}$ and Σ, $u(M^*) = u(M_0)$. This proves the theorem, since we have shown that the maximum value of $u(M_0)$ is reached at points of the boundary M^*.

One is easily satisfied that if the region T is connected and if a maximum value is reached even at only one interior point M_0, then $u(M) \equiv u(M_0)$ over the entire region. Let $M^{(0)}$ be any other point of the region T. We join the point $M^{(0)}$ to the point M_0 by a broken line L (Fig. 46), whose length we denote by l. Let M_1 be the point at which the line L leaves $\Sigma_{\varrho_0^m}$. At this point $u(M_1) = u(M_0)$. Let us describe about this point a sphere $\Sigma_{\varrho_1^m}$ of radius ϱ_1^m, touching Σ, and let M_2 be the point at which

L leaves $\Sigma_{\varrho_1^m}$; at this point $u(M_2) = u(M_0)$. Carrying this procedure further, we find that after not more than $p = l/\varrho^{(m)}$ spheres where $\varrho^{(m)}$ is the minimum distance from L to Σ, one of these spheres encloses the point $M^{(0)}$, from which it follows that $u(M^{(0)}) = u(M_0)$. Because of the arbitrary nature of $M^{(0)}$ and the continuity of $u(M)$ in the closed region $T + \Sigma$, we deduce that $u(M) \equiv u(M_0)$, everywhere including the points of the boundary. Thus, only harmonic functions which are constant can reach their maximum value at interior points of the region.

An identical theorem may be proved for a minimum value.

Conclusion 1.

If the functions u and U are continuous in the region $T + \Sigma$, harmonic in T and if

$$u \leqslant U \text{ on } \Sigma,$$

then

$$u \leqslant U \text{ everywhere inside } T.$$

In fact the function $U - u$ is continuous in $T + \Sigma$, harmonic in T and

$$U - u \geqslant 0 \text{ on } \Sigma.$$

Because of the principle of maximum value

$$U - u \geqslant 0 \text{ everywhere inside } T,$$

from which our statement follows.

Conclusion 2.

If the functions u and U are continuous in the region $T + \Sigma$ harmonic in T and if

$$|u| \leqslant U \text{ on } \Sigma,$$

then

$$|u| \leqslant U \text{ everywhere inside } T.$$

From the conditions of the theorem it follows that the three harmonic functions $-U$, u and U satisfy the relations

$$-U \leqslant u \leqslant U \text{ on } \Sigma.$$

Applying *conclusion 1* twice, we derive that

$$-U \leqslant u \leqslant U \text{ everywhere inside } T$$

or

$$|u| \leqslant U \text{ inside } T.$$

3. *Uniqueness and stability of the first boundary-value problem*

Let a region T be given, bounded by the surface Σ. The first boundary-value problem in region T for Laplace's equation consists of the following.

It is required to find a function u such that

(a) *within the region T it satisfies the equation* $\nabla^2 u = 0$;

(b) *it is defined and continuous in the closed region* $T + \Sigma$ *including the boundary ;*

(c) *it takes an assigned value on the boundary* Σ.

In condition (a) the harmonic nature of the function inside region T is assumed. The harmonic requirement on the boundary is unnecessary, because it entails additional restrictions for the boundary values.

The condition of continuity in the closed region (or any condition replacing it) is necessary for uniqueness. If one neglects this condition, then it is possible to consider any function, *equal to a constant C inside T and the given function f on the boundary*, as the solution of the problem, since it satisfies conditions (a), (c).

Let us prove the uniqueness theorem:

The first internal boundary-value problem for Laplace's equation has a unique solution.

Let us assume that there exist two different functions u_1 and u_2 which are the solutions of the problem, i.e. there exist functions, continuous in the closed region $T + \Sigma$, satisfying inside the region Laplace's equation and having the same value f on the surface Σ.

The difference between these functions $u = u_1 - u_2$ possesses the following properties:

(1) $\nabla^2 u = 0$ inside the region T;

(2) u is continuous in the closed region $T + \Sigma$;

(3) $u \mid_\Sigma = 0$.

The function $u(M)$ is continuous and harmonic in the region T and equals zero on the boundary. As is well known, any continuous function reaches its maximum value in a closed region. We must show that $u \equiv 0$. If the function $u \not\equiv 0$, and at one point $u > 0$, then it reaches a positive maximum value inside the region T, which is impossible. In precisely the same way is proved that the function u cannot assume a negative value inside T. Hence it follows that

$$u \equiv 0 .$$

Let us proceed to a proof of the continuous dependence of the solution of the first boundary-value problem on the boundary conditions. Let us recall that a problem is said to be physically defined if a small change of the additional conditions, determining the solution of the problem (in our case of the boundary conditions), produce a corresponding small change in the solution.

Let u_1 and u_2 be continuous functions in $T + \Sigma$ and harmonic inside T, for which the modulus of the difference of the boundary values does not exceed $\varepsilon > 0$. We show that

$$|u_1 - u_2| \leqslant \varepsilon$$

everywhere in the region under consideration.

This statement follows immediately from conslusion 2, p. 323, because of the fact that $U \equiv \varepsilon$ is a harmonic function.

Thus, we have proved the continuous dependence of the solution on the boundary conditions and the uniqueness of the first internal problem.

4. *Problems with discontinuous boundary conditions*

Often first boundary-value problems arise with discontinuous boundary conditions. No function, continuous in a closed region, can be the solution of this problem. Therefore we have to give a statement of the first boundary-value problem applicable to this case.

Suppose we are given a partly-continuous function $f(P)$ along the curve C, bounding the region S. It is required to find the function $u(M)$: (1) harmonic inside region S; (2) continuously adjoining the boundary values at points of continuity of the latter; (3) bounded in the closed region $S + C$.

We note that the additional requirement of boundedness refers to the neighbourhood of points of discontinuity of the function $f(P)$.

Let us prove the following theorem:

The solution of the first boundary-value problem with a partly continuous boundary condition is unique.

Let u_1 and u_2 be two solutions of the given problem. The difference

$$v = u_1 - u_2$$

(1) is a harmonic function inside S;

(2) continuously adjoins a zero value on the boundary, with the exception of points of discontinuity of $f(P)$, at which it may have a discontinuity;

(3) is bounded in $S + C$; $|v| < A$.

Let us construct the harmonic function

$$U(M) = \varepsilon \sum_{i=1}^{n} \ln \frac{D}{r_i},$$

where ε is an arbitrary positive number, D is the diameter of the region, r_i is the distance from the point M under consideration to the ith point of discontinuity P_i. The function $U(M)$ is positive, because all the components are greater than zero.

Let us construct at any point of discontinuity P_i a circle K_i of radius δ, choosing δ so that every component

$$\varepsilon \ln \frac{D}{r_i}$$

on the corresponding circumference C_i exceeds A, i. e. so that $\varepsilon \ln \frac{D}{\delta} \geqslant A$. The function v is continuous in the closed region $S - \sum_{i=1}^{n} K_i$.

In this region because of the maximum value principle U dominates the function v:

$$|v(M)| \leqslant U(M).$$

Having fixed an arbitrary point M in the region S and letting $\varepsilon \to 0$ we obtain:

$$\lim_{\varepsilon \to 0} U(M) = 0;$$

hence
$$v(M) = 0,$$
since v does not depend on ε, or
$$u_1 \equiv u_2,$$
which it was required to prove.

5. *Isolated singular points*

Let us consider the singular points of a harmonic function. Let P be an isolated singular point, lying inside the region of harmonic variation of the function u. Two cases seem possible:

(1) the harmonic function is bounded in the neighbourhood of the point P;

(2) the harmonic function is not bounded in the neighbourhood of the point P.

We have already met singular points of the second type (for instance, $\ln 1/r$). The following theorem proves that the first type of singular point does not exist.

If a bounded function $u(M)$ is harmonic inside the region S, with the exception of the point P, then it is possible to define a value $u(P)$ such that the function $u(M)$ is harmonic everywhere inside S.

Let us consider a circle K_a of radius a with centre at the point P, lying entirely within S, and let us consider a harmonic function v inside it coinciding with the function u on the circumference C_a of the circle K_a.*

We form the difference
$$w = u - v,$$
which

(1) is harmonic everywhere inside K_a, except the point P, at which w is not given,

(2) adjoins continuously the zero boundary conditions on C_a.

(3) is bounded in the closed region $K_a + C_a$ $(|v| < A)$.

As in the proof of the previous theorem (sub-section 4) let us form a positive harmonic function
$$U(M) = \varepsilon \ln \frac{a}{r}.$$

* The existence of such a function will be proved in § 3, where its definition is not based on this theorem.

Here ε is an arbitrary positive quantity, a is the radius of the circle K_a, r is the distance from the point M under consideration to the point of discontinuity P.

Let us construct a circle K_δ with centre at the point P, choosing its radius δ so that on its circumference the value U exceeds A, and let us consider the region $K_a - K_\delta$. The function w is continuous in the closed region $\delta \leqslant r \leqslant a$ and on the boundary of this region the inequality $|w| \leqslant U$ holds. Because of the maximum value principle the positive function U dominates the function w

$$|w| \leqslant U(M) \text{ where } \delta \leqslant r \leqslant a .$$

Fixing an arbitrary point M of the region K_a, not coinciding with P, and performing a limiting transition as $\varepsilon \to 0$, we obtain:

$$\lim_{\varepsilon \to 0} U(M) = 0 ,$$

hence,

$$w = 0 .$$

Thus, the function u coincides with function v everywhere in the region S, with the exception of the point P. Assuming $u(P) = v(P)$ we obtain the function $u \equiv v$, harmonic everywhere inside the region S. Hence the theorem is proved.

Similarly the theorem may be proved for the case of three dimensions, where the function $U(M) = 2\left(\frac{1}{r} - \frac{1}{a}\right)$ may be taken as the majorant function.

In the proof of the theorem in this section we assumed that the function u was bounded in the neighbourhood of the point P. But these same considerations hold if it is assumed that the function u in the neighbourhood of the point P satisfies the inequality

$$|u(M)| < \varepsilon(r) \log \frac{1}{r_{PM}} , \tag{17}$$

where $\varepsilon(r)$ is an arbitrary function, tending to zero as $r \to 0$, i.e. in the neighbourhood of the point P the function $u(M)$ grows more slowly than $\log \frac{1}{r_{PM}}$.

Thus, if the function $u(M)$ is a harmonic function inside the region S, with the exception of the point P, in the neighbourhood of which it grows more slowly than $\log \frac{1}{r_{MP}}$ as $M \to P$, then this function is bounded in the neighbourhood of the point P, and it is possible to define a value $u(P)$ such that the function u is harmonic over the entire region S.

Similarly in the case of three independent variables : if a harmonic function $u(M)$ in the neighbourhood of an isolated singular point P grows more slowly than $1/r$,

$$|u(M)| < \varepsilon(r) \frac{1}{r_{MP}} \quad \begin{pmatrix} \varepsilon(r) \to 0 \\ r \to 0 \end{pmatrix}, \tag{18}$$

then it is bounded in the neighbourhood of this point, and it is possible to define a value $u(P)$ such that the function $u(M)$ will be harmonic at the same point P.

6. *Regularity of a harmonic function at infinity*

An harmonic function in three variables $u(x, y, z)$ is said to be regular at infinity if

$$|u| < \frac{A}{r} \text{ and } \left|\frac{\partial u}{\partial x}\right| < \frac{A}{r^2}, \quad \left|\frac{\partial u}{\partial y}\right| < \frac{A}{r^2}, \quad \left|\frac{\partial u}{\partial z}\right| < \frac{A}{r^2} \tag{19}$$

for $r \geqslant r_0$.

Let us prove that if the function u is harmonic outside some closed surface Σ and tends uniformly to zero at infinity, then it is regular at infinity.

The condition of uniform convergence to zero at infinity implies that there exists a function $\varepsilon^*(r)$ such that

$$|u(M)| < \varepsilon^*(r) \quad \begin{pmatrix} \varepsilon^*(r) \to 0 \\ r \to \infty \end{pmatrix}, \tag{20}$$

where r is the radius-vector of the point M.

Performing a Kelvin transformation

$$v(r', \theta, \varphi) = r u(r, \theta, \varphi), \text{ where } r' = \frac{1}{r},$$

we have that the function v is harmonic everywhere inside the surface Σ', into which the surface Σ transforms with the exception of the origin of coordinates where it may have an isolated singular point.

From (20) it follows that in the neighbourhood of the origin, the inequality

$$|v(r', \theta, \varphi)| \leqslant \varepsilon^*\left(\frac{1}{r'}\right)\frac{1}{r'} = \varepsilon(r')\frac{1}{r'},$$

holds for function v, where

$$\varepsilon(r') = \varepsilon^*\left(\frac{1}{r'}\right) \to 0 \text{ when } r' \to 0 .$$

On the basis of the last theorem of sub-section 5 the function $v(r', \theta, \varphi)$ is bounded and harmonic for $r' \leqslant r_0'$:

$$|v(r', \theta, \varphi)| \leqslant A \text{ where } r' \leqslant r_0', \tag{21}$$

from which it follows that

$$|u(r, \theta, \varphi)| = \frac{|v(r', \theta, \varphi|}{r} \leqslant \frac{A}{r} \text{ where } r \geqslant r_\varepsilon = \frac{1}{r_0'} .$$

Because of the harmonic nature of function v at $r' = 0$ it is possible to write:

$$\frac{\partial u(x, y, z)}{\partial x} = \frac{\partial}{\partial x}\left(\frac{1}{r} \cdot v(x', y', z')\right) =$$

$$= -\frac{x}{r^3} \cdot v + \frac{1}{r}\left[\frac{\partial v}{\partial x'} \cdot \frac{\partial x'}{\partial x} + \frac{\partial v}{\partial y'} \cdot \frac{\partial y'}{\partial x} + \frac{\partial v}{\partial z'} \cdot \frac{\partial z'}{\partial x}\right], \tag{22}$$

where

$$x' = \frac{x}{r} r', \quad y' = \frac{y}{r} r', \quad z' = \frac{z}{r} r'.$$

Hence, calculating the derivatives $\frac{\partial x'}{\partial x}$, $\frac{\partial y'}{\partial x}$, $\frac{\partial z'}{\partial x}$ and taking into consideration the bounded nature of the first derivatives of function v in the neighbourhood of the point $r' = 0$, we obtain:

$$\left|\frac{\partial u}{\partial x}\right| \leqslant \frac{A}{r^2} \text{ when } r \to \infty .$$

Similar results hold for the derivatives $\frac{\partial u}{\partial y}$ and $\frac{\partial u}{\partial z}$.

7. *External boundary-value problems. Uniqueness of the solution for two- and three-dimensional problems*

Miscellaneous external boundary-value problems exist for two and three independent variables.

Let us consider a three-dimensional case. Let T be a region extending to infinity, bounded by some surface Σ. The *first external boundary-value problem* consists of the following:

It is required to find the function $u(x, y, z)$, satisfying the conditions :

(1) $\nabla^2 u = 0$ *in the infinite region* T ;

(2) u *is everywhere continuous, including the surface* Σ ;

(3) $u \,|_{\Sigma} = f(x, y, z)$, *where* f *is a function, given on the surface* Σ ;

(4) $u(M)$ *converges uniformly to* 0 *at infinity* : $u(M) \to 0$ *as* $M \to \infty$.

The final condition is essential for a unique solution, as may easily be shown by a simple example. Let it be required to solve the external first boundary-value problem for a sphere S_R of radius R with constant boundary condition

$$u\,|_{S_R} = \text{const} = f_0.$$

Omitting condition (4) we see that the functions $u_1 = f_0$ and $u_2 = f_0 \dfrac{R}{r}$ may serve as the solutions of the problem, and also any function

$$u = \alpha u_1 + \beta u_2, \text{ where } \alpha + \beta = 1.$$

Let us prove that

the external first boundary-value problem for harmonic functions with three independent variables has a unique solution. Assuming the existence of two solutions u_1 and u_2 satisfying conditions (1) — (4), we see that their difference $u = u_1 - u_2$ represents a solution of the problem with zero boundary conditions. Since condition (4) is also fulfilled for function u, then for any arbitrary $\varepsilon > 0$ it is possible to assign an R^* such that

$$|u(M)| < \varepsilon \text{ when } r \geqslant R^*.$$

If the point $\overline{M}$ lies inside region T' (Fig. 47) confined between surface Σ and the sphere S_{Σ} $(r \geqslant R^*)$, then $u(\overline{M}) < \varepsilon$, which follows from the maximum value principle, applied to the region T'. Because of the arbitrary nature of ε we deduce that $u \equiv 0$ in region T', and hence over the entire region T, which proves the uniqueness of the solution of the external first boundary-value problem in space.

The *first external problem in a plane* is stated in the following way:

it is required to find the function u *satisfying the conditions :*

(1) $\nabla^2 u = 0$ *in the infinite region* Σ *under consideration, bounded by the contour* C *;*

(2) *the function* u *is everywhere continuous, including* C *;*

(3) $u/c = f(x, y)$, *where* f *is a function, given on* C *;*

(4) $u(M)$ *is bounded at infinity, i. e. there exists a number* N *such that* $|u(M)| \leqslant N$.

The requirement that the solution tends to zero at infinity appears to be sufficient in order to show that

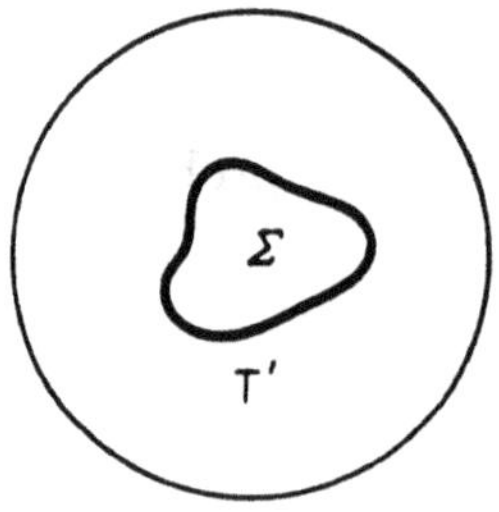

FIG. 47

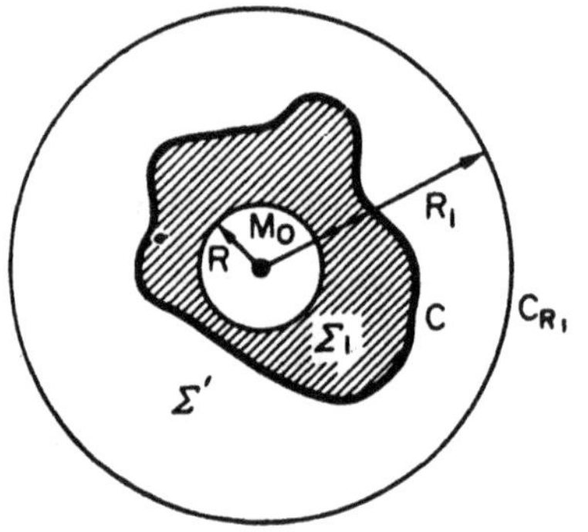

FIG. 48

two different solutions are not possible, but it is too stringent, since it usually allows no solution.

Let us prove that

the external first boundary-value problem for functions of two variables has a unique solution.

Assuming the existence of two different solutions u_1 and u_2 and considering their difference $u = u_1 - u_2$, which is a solution of the boundary-value problem with zero boundary conditions, we have by virtue of condition (4)

$$|u| \leqslant N = N_1 + N_2,$$

where N_1 and N_2 are such that $|u_1| \leqslant N_1$, $|u_2| \leqslant N_2$. We denote by Σ_1 the region complementary to the region Σ. Let us consider a point M_0 inside Σ_1 and a circle of radius R with centre at the point M_0, lying inside Σ_1 (Fig. 48). The harmonic function

$\ln \frac{1}{r_{MM_0}}$ does not have a singularity in the region Σ; the function $\ln \frac{r_{MM_0}}{R}$ is positive in the entire region Σ including C.

Let C_{R_1} be a circle of radius R_1 with centre at M_0 containing entirely the contour C, and Σ' be the region bounded by curves C and C_{R_1}. The function U_{R_1} defined by the equality

$$u_{R_1} = N \frac{\ln r/R}{\ln R_1/R}, \tag{23}$$

is a harmonic function equal to N along a circle of radius R_1, positive along C; by the principle of maximum value it follows that u_{R_1} dominate the modulus of the function $u(M)$ in the region Σ':

$$|u(M)| < u_{R_1}(M).$$

Let us fix the point M and let us increase R_1 indefinitely. It is obvious that $u_{R_1}(M) \to 0$ as $R_1 \to \infty$. Hence it follows that

$$u(M) = 0.$$

Thus, because of the arbitrary nature of M, the uniqueness of the solution of the problem is proved. The result may also be proved using the transformation of the inverse radius-vectors, transforming the region external to the contour C into a region internal to the contour C' into which the contour C changes.

Moreover the point at infinity changes into an isolated singular point, in the neighbourhood of which the function v is bounded. By the theorem of sub-section 5 the harmonic nature of function v at the origin follows, and in the same way the uniqueness of the solution.

From the above considerations it follows that a harmonic function of two variables $u(M)$ bounded at infinity, tends to a definite limit as M tends to infinity.

The difference in the statement of the first boundary-value external problem for two and three variables may be illustrated by the following physical example. Let a sphere of radius R be given, on the surface of which the temperature u_0 is maintained constant, and it is required to determine the steady distribution of temperature in the space, external to the sphere. The function

$u = u_0 (R/r)$ is a solution of this problem, reducing to zero at infinity.

Let us consider now a two-dimensional problem with a constant boundary value on a circle of radius R

$$u|_\Sigma = f_0 = \text{const.}$$

In this case $u \equiv f_0$ is the unique bounded solution of the problem and there exists no solution reducing to zero at infinity.

For a three-dimensional and plane infinite region the principle of maximum value holds. This can be shown by means of considerations similar to those used in the proof of the uniqueness theorem. Hence, in its turn, the continuous dependence of the solution on the boundary conditions may be proved.

8. *The second boundary-value problem. Condition of regularity at infinity. Uniqueness theorem*

We denote the solution of the second boundary-value problem by the function u, continuous in the region $T + \Sigma$ and satisfying the following relation on the surface Σ

$$\left.\frac{\partial u}{\partial n}\right|_\Sigma = f(M)\,.$$

Let us prove that the solution of the second internal boundary-value problem is determined except for an arbitrary constant.

We prove the result making the additional assumption that the function u has continuous first derivatives in the region $T + \Sigma$*

Let u_1 and u_2 be two continuously differentiable functions in $T + \Sigma$, satisfying the equation $\nabla^2 u = 0$ in T and the relation $\left.\frac{\partial u}{\partial n}\right|_\Sigma = f(xyz)$ on Σ. The function $u = u_1 - u_2$: satisfies the boundary condition:

$$\left.\frac{\partial u}{\partial n}\right|_\Sigma = 0\,.$$

* The assumption concerning the continuity of the first derivatives in $T + \Sigma$ is made to simplify the proof. The proof of the uniqueness on the most general assumptions was given by M. V. Keldyshem and M. A. Lavrent'ev (*Dokl. Akad. Nauk SSSR*, vol. XVI, 1937); see also Smirnov, *Course in Higher Mathematics*, vol. IV, 2nd edition.

Assuming $v = u$ in Green's first relation (3) and using the relations $\nabla^2 u = 0$ and $\left.\frac{\partial u}{\partial n}\right|_{\Sigma} = 0$, we obtain:

$$\iiint \left[\left(\frac{\partial u}{\partial x}\right)^2 + \left(\frac{\partial u}{\partial y}\right)^2 + \left(\frac{\partial u}{\partial z}\right)^2\right] d\tau = 0\,.$$

It follows from the continuity of the function u and its first derivatives that

$$\frac{\partial u}{\partial x} = \frac{\partial u}{\partial y} = \frac{\partial u}{\partial z} \equiv 0\,,$$

i.e.

$$u \equiv \text{const},$$

which was requiried to be proved.

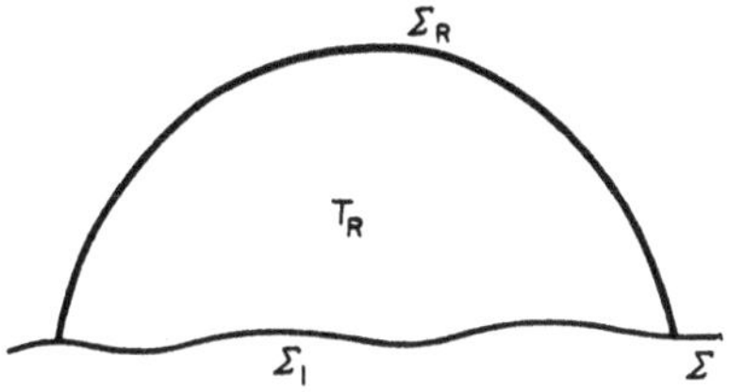

FIG. 49

The method of proof outlined here is applicable, in the case of an infinite region, for functions satisfying the requirement of regularity at infinity.

Let us prove that

in the case of infinite regions Green's formula (3) *is applicable for functions, regular at infinity.*

Let us consider a region T, extending to infinity. Let Σ be the boundary of this region. We draw a sphere of radius R and denote by T_R the part of our region, lying inside this sphere. The region T_R is bounded by part of the spherical surface Σ_R, and by the surface Σ_1, which is part of the surface Σ (Fig. 49). Applying Green's formula in region T_R to the two functions u and v, regular at infinity, we obtain:

$$\iiint_{T_R} u \nabla^2 v\, d\tau = -\iiint_{T_R} \left[\frac{\partial u}{\partial x}\frac{\partial v}{\partial x} + \frac{\partial u}{\partial y}\frac{\partial v}{\partial y} + \frac{\partial u}{\partial z}\frac{\partial v}{\partial z}\right] d\tau + \\ + \iint_{\Sigma_1} u\frac{\partial v}{\partial n}\, d\sigma + \iint_{\Sigma_R} u\frac{\partial v}{\partial n}\, d\sigma\,. \qquad (23')$$

Let us calculate the integral over Σ_R, using the property of regularity of the functions u and v:

$$\left|\iint_{\Sigma_R} u\frac{\partial v}{\partial n}\,d\sigma\right| = \left|\iint_{\Sigma_R} u\,(v_x\cos\alpha + v_y\cos\beta + v_z\cos\gamma)\,d\sigma\right| \leqslant$$

$$\leqslant\left|\iint_{\Sigma_R}\frac{A}{R}\cdot\frac{3A}{R^2}\,d\sigma\right| \leqslant \frac{3A^2}{R^3}\,4\pi R^2 = \frac{12\pi A^2}{R}.$$

Hence it follows that

$$\lim_{R\to\infty}\iint_{\Sigma_R} u\frac{\partial v}{\partial n}\,d\sigma = 0.$$

The integral over T_R on the right tends to the integral over the entire region T as $R \to \infty$. This integral exists, since the expression under the integral sign tends to infinity as $1/R^4$ because of the regularity of u and v. If the limit of the integral over Σ_1 exists and is equal to the integral over Σ, then the limit of the righthand side in (23′) exists; therefore there is a limit

$$\lim_{R\to\infty}\iiint_{T_R} u\,\nabla^2 v\,d\tau = \iiint_T u\,\nabla^2 v\,d\tau.$$

As a result we arrive at the relation

$$\iiint_T u\,\nabla^2 v\,d\tau = -\iiint_T\left[\frac{\partial u}{\partial x}\frac{\partial v}{\partial x} + \frac{\partial u}{\partial y}\frac{\partial v}{\partial y} + \frac{\partial u}{\partial z}\frac{\partial v}{\partial z}\right]d\tau +$$

$$+\iint_{\Sigma} u\frac{\partial v}{\partial n}\,d\sigma. \tag{24}$$

Thus the applicability of the first, and therefore the second, theorem of Green for infinite regions is established for functions regular at infinity.

Let us prove now that

the second boundary-value problem for an infinite region has a unique solution, regular at infinity.

Assuming $u = v = u_1 - u_2$ in relation (24) and using the fact that $\nabla^2 u = 0$ and $\left.\frac{\partial u}{\partial n}\right|_{\Sigma} = 0$, we obtain:

$$\iiint_T (u_x^2 + u_y^2 + u_z^2)\,d\tau = 0.$$

Hence because of the continuity of the derivatives of function u it follows that

$$u_x = 0, \quad u_y = 0 \quad u_z = 0 \text{ and } u \equiv \text{const.}$$

Since $u = 0$ at infinity, then

$$u \equiv 0,$$

i.e.

$$u_1 \equiv u_2,$$

which was requiring to be proved.

Naturally the question arises: is it possible to prove the uniqueness of the first boundary-value problem by the same method?

Let u_1 and u_2 be different solutions of the first boundary-value problem (internal). We apply formula (3) to the functions $u = u_1 - u_2$ and $v = u$ in region T, bounded by the surface Σ:

$$\iiint_T u \Delta u \, d\tau = - \iiint_T (u_x^2 + u_y^2 + u_z^2) \, d\tau + \iint_\Sigma u \frac{\partial u}{\partial n} \, d\sigma.$$

Hence, taking into consideration the relations

$$\nabla^2 u = 0, \quad u|_\Sigma = 0,$$

we obtain

$$\iiint_T (u_x^2 + u_y^2 + u_z^2) \, d\tau = 0$$

and, consequently

$$u_x = u_y = u_z = 0 \text{ and } u = \text{const.}$$

On the surface Σ the function u equals zero, therefore we can prove that

$$u \equiv 0 \text{ and } u_1 \equiv u_2.$$

But this proof is invalid, since in the process of the proof we assumed the existence of derivatives of the unknown function on the surface Σ which is not provided for in the statement of the problem. Proof of the uniqueness, based on the principle of maximum value, is free from this shortcoming.

§ 3. Solution of boundary-value problems for the simplest regions by the method of separation of variables

The solution of boundary-value problems for the simplest regions can be given by the method of separation of variables. For the solution of subsidiary equations, encountered in addition, special classes of functions are often necessary. In this section we confine ourselves to problems, soluble in terms of trigonometric functions. When studying special functions we consider the solution of other problems.

1. *The first boundary-value problem for a circle*

Let us solve the first boundary-value problem for a circle: *Find the function u satisfying the equation :*

$$\nabla^2 u = 0 \text{ inside the circle} \tag{1}$$

and the boundary condition

$$u = f \text{ on the boundary of the circle} \tag{2}$$

where f is a given function.

We assume firstly that function f is continuous and differentiable; later we relax the condition of differentiability and even continuity of function f (see sub-section 4, § 2). Along with the internal boundary-value problem we shall also consider the external boundary-value problem.

Let us introduce polar coordinates (ϱ, φ) with origin at the centre of the circle. Equation (1) in polar coordinates has the form

$$\nabla^2 u = \frac{1}{\varrho}\frac{\partial}{\partial \varrho}\left(\varrho \frac{\partial u}{\partial \varrho}\right) + \frac{1}{\varrho^2}\frac{\partial^2 u}{\partial \varphi^2} = 0 \tag{3}$$

(see (34), § 1). We solve the problem by the method of separation of variables, i.e. we search for a particular solution of equation (1) of the form

$$u(\varrho, \varphi) = R(\varrho)\,\Phi(\varphi)\,.$$

Substituting the assumed form of the solution in equation (3), we obtain

$$\frac{\frac{d}{d\varrho}\left(\varrho \frac{dR}{d\varrho}\right)}{\frac{R}{\varrho}} = -\frac{\Phi''}{\Phi} = \lambda ,$$

where $\lambda = \text{const}$. Hence we obtain the two equations:

$$\Phi'' + \lambda\Phi = 0 , \tag{4}$$

$$\varrho \frac{d}{d\varrho}\left(\varrho \frac{dR}{d\varrho}\right) - \lambda R = 0 . \tag{5}$$

The first of these equation gives:

$$\Phi (\varphi) = A \cos \sqrt{(\lambda\varphi)} + B \sin \sqrt{(\lambda\varphi)} .$$

We note that for a change of the angle φ by 2π the single valued function $u(\varrho, \varphi)$ must return to its original value

$$u (\varrho, \varphi + 2\pi) = u (\varrho, \varphi) .$$

Hence it follows that

$$\Phi (\varphi + 2\pi) = \Phi (\varphi) ,$$

i. e. $\Phi (\varphi)$ is a periodic function of the angle φ with period 2π. This is possible only if $\sqrt{\lambda} = n$, where n is an integer and

$$\Phi_n (\varphi) = A_n \cos n\varphi + B_n \sin n\varphi .$$

We shall seek the function $R(\varrho)$ in the form

$$R (\varrho) = \varrho^{\mu}.$$

Substituting in equation (5) and dividing by ϱ^{μ}, we find:

$$n^2 = \mu^2 \text{ or } \mu = \pm n \ (n > 0) .$$

Hence,

$$R (\varrho) = C\varrho^n + D\varrho^{-n},$$

where C and D are constants.

For the solution of the internal problem it is necessary to assume $R = C \varrho^n (\mu = + n)$, since, if $D \neq 0$ function $u = R(\varrho) \Phi(\varphi)$ tends to infinity at $f = 0$ and is not a harmonic function inside the circle. For the solution of the external problem, on the other hand, it is necessary to take $R = D \varrho^{-n} (\mu = —n)$, since the solution of the external problem must be bounded at infinity.

Thus, we have found particular solutions of our problem:*

$$u_n(\varrho, \varphi) = \varrho^n (A_n \cos n\varphi + B_n \sin n\varphi) \text{ where } \varrho \leqslant a,$$

$$u_n(\varrho, \varphi) = \frac{1}{\varrho^n} (A_n \cos n\varphi + B_n \sin n\varphi) \text{ where } \varrho \geqslant a.$$

The sums of these solutions

$$u(\varrho, \varphi) = \sum_{n=0}^{\infty} \varrho^n (A_n \cos n\varphi + B_n \sin n\varphi) \quad \text{in the internal region,}$$

$$u(\varrho, \varphi) = \sum_{n=0}^{\infty} \frac{1}{\varrho^n} (A_n \cos n\varphi + B_n \sin n\varphi) \quad \text{in the external region}$$

for sufficiently rapid convergence, will also be harmonic functions.

In order to determine the coefficients A_n and B_n we use the boundary relation

$$u(a, \varphi) = \sum_{n=0}^{\infty} a^n (A_n \cos n\varphi + B_n \sin n\varphi) = f. \tag{6}$$

Assuming that f is given as a function of the angle φ, we expand it in a Fourier series

$$f(\varphi) = \frac{\alpha_0}{2} + \sum_{n=1}^{\infty} (\alpha_n \cos n\varphi + \beta_n \sin n\varphi), \tag{7}$$

where

$$\alpha_0 = \frac{1}{\pi} \int_{-\pi}^{\pi} f(\psi)\, d\psi, \quad \alpha_n = \frac{1}{\pi} \int_{-\pi}^{\pi} f(\psi) \cos n\psi \, d\psi,$$

$$\beta_n = \frac{1}{\pi} \int_{-\pi}^{\pi} f(\psi) \sin n\psi \, d\psi.$$

* The expression for the Laplacian operator in a polar system of coordinates (3) loses meaning at $\varrho = 0$. We show that $\nabla^2 u_n = 0$ also for $\varrho = 0$. For proof of this we can no longer use polar coordinates.

Let us change to a cartesian system of coordinates; the particular solutions

$$\varrho^n \cos n\varphi \text{ and } \varrho^n \sin n\varphi,$$

being the real and imaginary parts of the functions

$$\varrho^n e^{in\varphi} = (\varrho e^{i\varphi})^n = (x + iy)^n,$$

are polynomials in x and y. It is obvious that a polynomial, satisfying the equation $\nabla^2 u = 0$ for $\varrho > 0$ satisfies this equation also for $\varrho = 0$ because of the continuity of the second derivatives.

Comparing series (6) and (7), we obtain:

$$A_0 = \frac{\alpha_0}{2}, \quad A_n = \frac{\alpha_n}{a^n}, \quad B_n = \frac{\beta_n}{a^n} \quad \text{in the internal region,}$$

$$A_0 = \frac{\alpha_0}{2}, \quad A_n = \alpha_n a^n, \quad B_n = a^n \beta^n \quad \text{in the external region.}$$

Thus, we have obtained a formal solution of the first internal problem for a circle in the form of the series

$$u(\varrho, \varphi) = \frac{\alpha_0}{2} + \sum_{n=1}^{\infty} \left(\frac{\varrho}{a}\right)^n (\alpha_n \cos n\varphi + \beta_n \sin n\varphi), \tag{8}$$

and a solution of the external problem in the form

$$u(\varrho, \varphi) = \frac{\alpha_0}{2} + \sum_{n=1}^{\infty} \left(\frac{a}{\varrho}\right)^n (\alpha_n \cos n\varphi + \beta_n \sin n\varphi). \tag{9}$$

In order to satisfy oneself that the functions obtained are in fact the solutions, one must be satisfied about the applicability of the principle of superposition, for which it is necessary to prove the differentiability of these functions term by term. Both series may be represented by the one relation

$$u(\varrho, \varphi) = \sum_{n=1}^{\infty} t^n \, (\alpha_n \cos n\varphi + \beta_n \sin n\varphi) + \frac{\alpha_0}{2},$$

where

$$t = \begin{cases} \frac{\varrho}{a} \leqslant 1 \text{ where } \varrho \leqslant a \text{ (internal problem)} \\ \frac{a}{\varrho} \leqslant 1 \text{ where } \varrho \geqslant a \text{ (internal problem)} \end{cases}$$

α_n, β_n are Fourier coefficients of the function $f(\varphi)$.

Let us prove that series (8), (9) may be differentiated any number of times for $t < 1$. Let

$$u_n = t^n (\alpha_n \cos n\varphi + \beta_n \sin n\varphi).$$

We calculate the kth derivative of function u_n with respect to φ

$$\frac{\partial^k u_n}{\partial \varphi^k} = t^n n^k \left[\alpha_n \cos\left(n\varphi + k\frac{\pi}{2}\right) + \beta_n \sin\left(n\varphi + k\frac{\pi}{2}\right)\right].$$

Hence we obtain the result

$$\left|\frac{\partial^k u_n}{\partial \varphi^k}\right| \leqslant t^n n^k 2M,$$

where the maximum of the modulus of the Fourier coefficients α_n and β_n is denoted by M

$$\left.\begin{aligned} |\alpha_n| &< M, \\ |\beta_n| &< M. \end{aligned}\right\} \tag{10}$$

Let us fix some value $\varrho_0 < a$ (for the internal problem) or $\varrho_1 = a^2/\varrho_0 > a$ (for the external problem), and put $t_0 = \frac{\varrho_0}{a} < 1$. Considering the series

$$\left|\sum_{n=1}^{\infty} t^n n^k (|\alpha_n| + |\beta_n|)\right| \leqslant 2M \sum_{n=1}^{\infty} t_0^n n^k \qquad (t \leqslant t_0),$$

we see that it converges uniformly for $t \leqslant t_0 < 1$ for any k. Therefore series (8) and (9) may be differentiated with respect to φ at any point inside (outside) the circle any number of times. Similarly it can be proved that series (8) and (9) may also be differentiated with respect to the variable ϱ inside (outside) a circle of radius $\varrho_0 < a$ as many times as desired.

Because of the arbitrary nature of ϱ_0 we deduce that series (8) and (9) are differentiable for any internal (external) point of the circle, and the principle of superposition is applicable.

Thus we have proved that functions (8) and (9) satisfy the equation $\nabla^2 u = 0$.*

In this proof we used only one property of the function $f(\varphi)$, that its Fourier coefficients are bounded (formula (10)). This holds for any bounded function (even for any absolutely integrable function). Thus, series (8) and (9), determine functions satisfying the equation

$$\nabla^2 u = 0 \text{ where } t < 1.$$

for any bounded function.

* This equation is also satisfied for $\varrho = 0$; actually, expressing the derivatives in cartesian coordinates by derivatives in polar coordinates, it is easily seen that functions (8) and (9) for $t \leqslant t_0$ may be differentiated with respect to x and y any number of times. Because of the footnote on p. 340 it follows that

$$\nabla^2 u = 0, \text{ where } \varrho = 0$$

Let us turn now to a proof of the continuity of u in a closed region ($t \leqslant 1$). It is obvious that without more detailed information concerning the properties of the function $f(\varphi)$ it is impossible to do this.

From the assumption of the continuity and differentiability of the function $f(\varphi)$ it follows that its expansion in a Fourier series exists and the following series converges

$$\sum_{n=1}^{\infty} (|\alpha_n| + |\beta_n|) < \infty. \tag{11}$$

In addition, we have:

$$|t^n \alpha_n \cos n\varphi| \leqslant |\alpha_n|,$$

$$|t^n \beta_n \sin n\varphi| \leqslant |\alpha_n|.$$

Therefore series (8) and (9) converge uniformly for $t \leqslant 1$ and, consequently, represent continuous functions on the boundary of the circle. From formula (11) we see that function (9), obtained for the external problem, is bounded at infinity.

Thus it is established that series (8) and (9) satisfy all the conditions of the problems.

2. *Poisson's integral*

Let us now reduce (8) and (9) to a simpler form. For definiteness let us consider an internal problem, and write down the result for the external one by analogy.

Substituting the expressions for the Fourier coefficients in (8) and changing the order of summation and integration, we have:

$$u(\varrho, \varphi) = \frac{1}{\pi} \int_{-\pi}^{\pi} f(\psi) \left\{ \frac{1}{2} + \sum_{n=1}^{\infty} \left(\frac{\varrho}{a}\right)^n (\cos n\psi \cos n\varphi + \right.$$

$$\left. + \sin n\psi \sin n\varphi) \right\} d\psi =$$

$$= \frac{1}{\pi} \int_{-\pi}^{\pi} f(\psi) \left\{ \frac{1}{2} + \sum_{n=1}^{\infty} \left(\frac{\varrho}{a}\right)^n \cos n(\varphi - \psi) \right\} d\psi. \tag{12}$$

Let us use the following identity:

$$\frac{1}{2}+\sum_{n=1}^{\infty} t^n \cos n(\varphi-\psi)=\frac{1}{2}+\frac{1}{2}\sum_{n=1}^{\infty} t^n\left[e^{in(\varphi-\psi)}+e^{-in(\varphi-\psi)}\right]=$$

$$=\frac{1}{2}\left\{1+\sum_{n=1}^{\infty}\left[(te^{i(\varphi-\psi)})^n+(te^{-i(\varphi-\psi)})^n\right]\right\}=$$

$$=\frac{1}{2}\left[1+\frac{te^{i(\varphi-\psi)}}{1-te^{i(\varphi-\psi)}}+\frac{te^{-i(\varphi-\psi)}}{1-te^{-i(\varphi-\psi)}}\right]=$$

$$=\frac{1}{2}\,\frac{1-t^2}{1-2t\cos(\varphi-\psi)+t^2}\qquad\left(t=\frac{\varrho}{a}<1\right).$$

Substituting the results obtained in (12), we obtain:

$$u(\varrho,\varphi)=\frac{1}{2\pi}\int_{-\pi}^{\pi} f(\psi)\,\frac{1-\dfrac{\varrho^2}{a^2}}{\dfrac{\varrho^2}{a^2}-2\dfrac{\varrho}{a}\cos(\varphi-\psi)+1}\,d\psi$$

or

$$u(\varrho,\varphi)=\frac{1}{2\pi}\int_{-\pi}^{\pi} f(\psi)\,\frac{a^2-\varrho^2}{\varrho^2-2a\varrho\cos(\varphi-\psi)+a^2}\,d\psi, \qquad (13)$$

The relation obtained, giving the solution of the first boundary-value problem inside a circle, is called *Poisson's integral*, and the expression under the integral sign

$$K(\varrho,\varphi,a,\psi)=\frac{a^2-\varrho^2}{\varrho^2-2a\varrho\cos(\varphi-\psi)+a^2}$$

Poisson's kernel. We note that $K(\varrho,\varphi,a,\psi)>0$ for $\varrho<a$, since $2a\varrho<a^2+\varrho^2$, if $\varrho\neq a$.

Poisson's integral is investigated on the assumption that $\varrho<a$; for $\varrho=a$, (8) has no meaning. However,

$$\lim_{\substack{\varrho\to a\\ \varphi\to\varphi_0}} u(\varrho,\varphi)=f(\varphi_0),$$

since the series from which Poisson's integral is derived is a continuous function in a closed region.

The function defined by the relation

$$u(\varrho, \varphi) = \begin{cases} \dfrac{1}{2\pi}\displaystyle\int_{-\pi}^{\pi} f(\psi) \frac{a^2 - \varrho^2}{\varrho^2 - 2a\varrho\cos(\varphi - \psi) + a^2} d\psi & \text{when } \varrho < a, \\ f(\varphi) & \text{when } \varrho = a, \end{cases} \qquad (13')$$

satisfies the equation $\nabla^2 u = 0$ for $\varrho < a$, and is continuous in the closed region, $\varrho = a$.

The solution of the external boundary-value problem obviously has the form

$$u(\varrho, \varphi) = \begin{cases} \dfrac{1}{2\pi}\displaystyle\int_{-\pi}^{\pi} f(\psi) \frac{\varrho^2 - a^2}{\varrho^2 - 2a\varrho\cos(\varphi - \psi) + a^2} d\psi & \text{when } \varrho > a, \\ f(\psi) & \text{when } \varrho = a. \end{cases} \qquad (14)$$

In deriving these results we assumed that the function $f(\varphi)$ is continuous and differentiable, and making use of this we proved that the solution of the problem may be represented in the form of an infinite series. Later by means of identities we arrived at Poisson's integral.

Let us prove now that Poisson's integral gives a solution of the first boundary-value problem in the case where the function $f(\varphi)$ is only continuous.

Poisson's integral represents a solution of Laplace's equation for $\varrho < a$ $(t < 1)$ for an arbitrary bounded function. In fact, for $\varrho < a$ $(t < 1)$ Poisson's integral is identical to series (8) and because of the observation made on page 342, satisfies equation $\Delta u = 0$ for an arbitrary bounded function $f(\varphi)$.

Thus, it remains to us to prove that function u continuously adjoins its boundary values. Let us choose some sequence of continuous differentiable functions

$$f_1(\varphi), f_2(\varphi), \ldots, f_k(\varphi), \ldots,$$

converging uniformly to the function $f(\varphi)$.*

$$\lim_{k \to \infty} f_k(\varphi) = f(\varphi).$$

* We will not discuss this point. It is possible to choose such a sequence by many methods.

The sequence of boundary functions will correspond to the sequence of harmonic functions $u_k(\varrho, \varphi)$, defined by formula (13) or (8). The uniform convergence of the sequence $\{f_k(\varphi)\}$ indicates that for any $\varepsilon > 0$ it is possible to assign $k_0(\varepsilon) > 0$ such that

$$|f_k(\varphi) - f_{k+l}(\varphi)| < \varepsilon \text{ when } k > k_0(\varepsilon),\ l > 0.$$

Because of the maximum value principle we have for the functions $u_k(r, \varphi)$, representing the solution of the first boundary value problem:

$$|u_k(\varrho, \varphi) - u_{k+l}(\varrho, \varphi)| < \varepsilon$$
$$\text{where } \varrho \leqslant \varrho_0, \text{ if } k > k_0(\varepsilon),\ \varepsilon > 0.$$

Thus, the sequence $\{u_k\}$ converges uniformly to some function $u = \lim_{k \to \infty} u_k$. The limit function $u(\varrho, \varphi)$ is continuous in the closed region, since all the functions u_k, represented by integrals

$$u_k(\varrho, \varphi) = \frac{1}{2\pi} \int_{-\pi}^{\pi} \frac{a^2 - \varrho^2}{\varrho^2 - 2a\varrho \cos(\varphi - \psi) + a^2} f_k(\psi)\, d\psi,$$

are continuous in the closed region. We have that

$$u(\varrho, \varphi) = \lim_{k \to \infty} u_k(\varrho, \varphi) =$$

$$= \begin{cases} \dfrac{1}{2\pi} \displaystyle\int_{-\pi}^{\pi} \frac{a^2 - \varrho^2}{a^2 - 2a\varrho(\varphi - \psi) + \varrho^2} f(\psi)\, d\psi & \text{when } \varrho < a \\ f(\varphi) & \text{when } \varrho = a \end{cases}$$

since the sequence $\{f_k\}$ converges uniformly to f and therefore a limiting transition under the integral sign is valid.

Thus the function

$$u(\varrho, \varphi) = \frac{1}{2\pi} \int_{-\pi}^{\pi} \frac{a^2 - \varrho^2}{\varrho^2 - 2a\varrho \cos(\varphi - \psi) + a^2} f(\psi)\, d\psi$$

for an arbitrary continuous function $f(\varphi)$ is the solution of Laplace's equation, continuously adjoining the given values on the boundary of the circle.

3. *Cases of discontinuous boundary values*

Let us prove that (13′) and (14) give a solution of the boundary-value problem for an arbitrary partly-continuous function $f(\varphi)$, i.e. that this solution is bounded over the whole region and continuously adjoins the boundary values at the points of continuity of the function $f(\varphi)$. Let φ_0 be any point of continuity of the function $f(\varphi)$. It is necessary to show that for any ε, δ can be found such that

$$|u(\varrho, \varphi) - f(\varphi_0)| < \varepsilon,$$

if

$$|\varrho - a| < \delta(\varepsilon) \text{ and } |\varphi - \varphi_0| < \delta(\varepsilon).$$

Because of the continuity of the function $f(\varphi)$ there exists $\delta_0(\varepsilon)$ such that

$$|f(\varphi) - f(\varphi_0)| < \frac{\varepsilon}{2}, \text{ if } |\varphi - \varphi_0| < \delta_0(\varepsilon).$$

Let us consider auxiliary continuous and differentiable functions $\overline{f(\varphi)}$ and $\underline{f(\varphi)}$, satisfying the following relations

$$\overline{f(\varphi)} = f(\varphi_0) + \varepsilon/2 \quad \text{where } |\varphi - \varphi_0| < \delta_0(\varepsilon),$$
$$\overline{f(\varphi)} \geqslant f(\varphi) \quad \text{where } |\varphi - \varphi_0| > \delta_0(\varepsilon)$$

and

$$\underline{f(\varphi)} = f(\varphi_0) - \varepsilon/2 \quad \text{where } |\varphi - \varphi_0| < \delta_0(\varepsilon),$$
$$\underline{f(\varphi)} \leqslant f(\varphi) \quad \text{where } |\varphi - \varphi_0| > \delta_0(\varepsilon),$$

and in other respect arbitrary. If by means of (13) we assign for $\bar{f}$ and $\underline{f}$ the functions $\bar{u}(\varrho, \varphi)$ and $\underline{u}(\varrho, \varphi)$ then these will be harmonic functions, continuously adjoining $\bar{f}(\varphi)$ and $\underline{f}(\varphi)$.

Because of the positive nature of the Poisson integral we have

$$\underline{u}(\varrho, \varphi) \leqslant u(\varrho, \varphi) \leqslant \bar{u}(\varrho, \varphi),$$

since

$$\underline{f}(\varphi) \leqslant f(\varphi) \leqslant \bar{f}(\varphi).$$

From the continuity of the functions $\bar{u}(\varrho, \varphi)$ and $\underline{u}(\varrho, \varphi)$ on the boundary at $\varphi = \varphi_0$ it follows that there exists $\delta_1(\varepsilon)$ such that

$$|\bar{u}(\varrho, \varphi) - \bar{f}(\varphi_0)| \leqslant \varepsilon/2$$

for

$$|\varrho - a| < \delta_1(\varepsilon), \quad |\varphi - \varphi_0| < \delta_1(\varepsilon)$$

and

$$|\underline{u}(\varrho, \varphi) - \underline{f}(\varphi_0)| \leqslant \varepsilon/2$$

for

$$|\varrho - a| < \delta_1(\varepsilon), \quad |\varphi - \varphi_0| < \delta_1(\varepsilon).$$

From these inequalities we find:

$$\left.\begin{array}{l} \bar{u}(\varrho, \varphi) \leqslant \bar{f}(\varphi_0) + \frac{\varepsilon}{2} = f(\varphi_0) + \varepsilon, \\ f(\varphi_0) - \varepsilon = \underline{f}(\varphi_0) - \frac{\varepsilon}{2} \leqslant \underline{u}(\varrho, \varphi) \end{array}\right\} \text{ when } \begin{array}{l} |\varrho - a| < \delta(\varepsilon), \\ |\varphi - \varphi_0| < \delta(\varepsilon), \end{array}$$

where $\delta = \min(\delta_0, \delta_1)$.

Comparing the inequalities obtained, we find that

$$f(\varphi_0) - \varepsilon \leqslant \underline{u}(\varrho, \varphi) \leqslant u(\varrho, \varphi) \leqslant \bar{u}(\varrho, \varphi) \leqslant f(\varphi_0) + \varepsilon$$

or

$$|u(\varrho, \varphi) - f(\varphi_0)| < \varepsilon \quad \text{when} \quad \begin{array}{l} |a - \varrho| < \delta(\varepsilon), \\ |\varphi - \varphi_0| < \delta(\varepsilon), \end{array}$$

which proves the continuity of $u(\varrho, \varphi)$ at the point (a, φ_0)

The boundedness of $u(\varrho, \varphi)$ follows from the positive nature of the Poisson series

$$u(\varrho, \varphi) < M \frac{1}{\pi} \int_0^{2\pi} \frac{a^2 - \varrho^2}{a^2 + \varrho^2 - 2a\varrho \cos(\varphi - \psi)} d\psi = M,$$

if $f(\varphi) < M$. The value of the integral

$$\frac{1}{\pi} \int_0^{2\pi} \frac{(a^2 - \varrho^2)\, d\psi}{\varrho_2 - 2a\varrho \cos(\varphi - \psi) + a^2} = 1,$$

because in an earlier proof it was shown that the left-hand part represents a harmonic function, continuously adjoining the boundary values $f \equiv 1$, and such a function is identically equal to 1. Similarly $u(\varrho, \varphi) > M_1$, if $f > M_1$, which proves the modulus of the function $u(\varrho, \varphi)$ is bounded.

§ 4. Source function

The method of source functions (Green's Functions) gives a convenient analytical representation of boundary-value problems.

In this section we give the definition and fundamental properties of the source function for Laplace's equation and also find the source function for a range of the simplest regions (circle, sphere, semispace).

1. *Source function for the equation* $\nabla^2 u = 0$ *and its fundamental properties*

It was shown in § 2, sub-section 1 that any function u, continuous together with its first derivatives in the closed region T, bounded by a sufficiently smooth surface Σ, and having second derivatives inside T, has the integral representation

$$u(M_0) = \frac{1}{4\pi}\iint\limits_{\Sigma}\left[\frac{1}{r_{PM_0}}\frac{\partial u}{\partial n} - u(P)\frac{\partial}{\partial n}\left(\frac{1}{r_{PM_0}}\right)\right] d\sigma_P - \\ - \frac{1}{4\pi}\iiint\limits_{T}\frac{\nabla^2 u}{r_{MM_0}}\, d\tau_M .$$

If the function $u(M)$ is harmonic, then the volume integral equals zero; if $u(M)$ satisfies Poisson's equation then the volume integral is a known function.

Let $v(M)$ be some harmonic function, with no singularity. Green's second theorem

$$\iiint\limits_{T}(u\nabla^2 v - v\nabla^2 u)\, d\tau = \iint\limits_{\Sigma}\left(u\frac{\partial v}{\partial n} - v\frac{\partial u}{\partial n}\right) d\sigma$$

gives:

$$0 = \iint\limits_{\Sigma}\left(v\frac{\partial u}{\partial n} - u\frac{\partial v}{\partial n}\right) d\sigma - \iiint\limits_{T} v\nabla^2 u\, d\tau . \qquad (2)$$

Adding (2) and (1) we obtain:

$$u(M_0) = \iint\limits_{\Sigma}\left[G\frac{\partial u}{\partial n} - u\frac{\partial G}{\partial n}\right] d\sigma - \iiint\limits_{T}\nabla^2 u\cdot G\, d\tau , \qquad (3)$$

where

$$G(M, M_0) = \frac{1}{4\pi r_{MM_0}} + v \qquad (3')$$

is a function of the two points: $M_0(x, y, z)$ and $M(\xi, \eta, \zeta)$. The point M is fixed and therefore x, y, z act as parameters.

Inside region T function G satisfies the equation

$$\nabla^2 G = 0$$

everywhere, except at the point $M = M_0$ where it has a singularity of the form $\frac{1}{4\pi r}$. Let us choose function v so that

$$G|_\Sigma = 0,$$

i.e.

$$v|_\Sigma = -\frac{1}{4\pi r}.$$

This Function G, we will call the source *function of the first boundary-value problem for the equation* $\Delta u = 0$. The source function gives an explicit solution of the first boundary-value problem for the equation $\nabla^2 u = 0$. In fact, (3) gives:

$$u(M_0) = -\iint_\Sigma u \frac{\partial G}{\partial n} d\sigma = -\iint_\Sigma f \frac{\partial G}{\partial n} d\sigma \quad (f = u|_\Sigma). \quad (4)$$

One should have in view the fact that (4) is obtained from Green's theorem, which assumes the functions u and G satisfy certain conditions on the surface Σ. In (4) the expression $\frac{\partial G}{\partial n}$ appears, whose existence on the surface Σ does not follow directly from the definition of G.

In obtaining relation (4) we assumed the existence of a harmonic function u, taking the value f on the surface Σ.Thus, even for those regions, for which a source function exists, satisfying the conditions of Green's theorem, relation (4) gives explicity only those solutions u of the first boundary-value problem which satisfy the same conditions. (proving the uniqueness of this class of solution of the first boundary-value problem).

Detailed study, carried out by A. M. Lyapunov, showed that for an extensive range of surfaces, called Lyapunov surfaces (see § 5) relation (4) represents the solution of the first boundary-value problem for very general conditions.

Let us look again at the definition of G. The function G is determined from v which is a solution of the first boundary-value problem for the equation $\nabla^2 v = 0$ with boundary values

$v|_\Sigma = -1/4\pi r$. One might form the impression that we have made a circular argument. In order to find u, the solution of the first boundary-value problem, it is necessary first to find v the solution of the same problem. In fact there is no circular argument, since knowledge of the source function allows us to solve the first boundary-value problem with arbitrary boundary values ($u|_\Sigma = f$), whereas to find G it is sufficient to solve the first boundary-value problem with special boundary values ($v|_\Sigma = -1/4\pi r$), which is considerably simpler, as we will see from a set of examples.

On an electrostatic interpretation the source function

$$G(M, M_0) = \frac{1}{4\pi r} + v$$

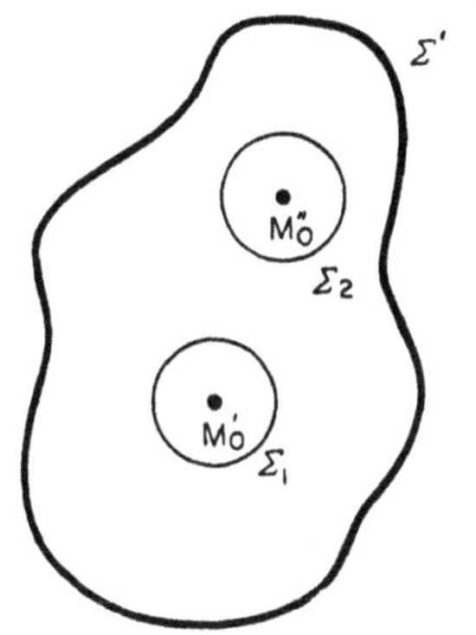

FIG. 50

represents the potential at point M of a point charge,* situated at point M_0 inside an earthed conducting surface Σ. The first component $1/4\pi r$ is, obviously, the potential of a point charge in free space, and the second component v denoted the potential field of charges, induced on the conducting surface Σ. Thus the formation of the source function reduces to a determination of the induced field.

* We recall that the potential of a point charge of value e in a medium of dielectric constant ε is defined by the relation $v = e/\varepsilon r$, if the c. g. s. system of units is used. Thus the source function corresponds to a charge of value $\varepsilon/4\pi$ absolute electrostatic units.

If a rationalized system of units is used, in which Coulomb's law has the form $f = ee'/4\pi\varepsilon r$, then in this system of units in vacuum ($\varepsilon = 1$) the function $G(M, M_0)$ corresponds to unit charge.

On a thermal interpretation the steady temperature of a point source of heat of intensity q is given by the relation

$$q/4\pi kr$$

where k is the coefficient of heat conduction. Thus, function $G(MM_0)$ is the temperature at point M, if the temperature of the surface of the body equals zero, and there is a heat source of intensity $q = k$ at the point M_0.

If the dimension of length is chosen so that $k = 1$, then function G corresponds to a source of intensity equal to unity.

We will derive some properties of the source function, assuming that the regions under consideration are such that source functions exists for them, possessing normal derivatives on the surface Σ and satisfying the conditions of applicability of Green's theorem.

1. The source function is everywhere positive inside T. In fact, function G reduces to zero on the boundary of the region Σ and is positive on the surface of a sufficiently small sphere surrounding the pole. It follows, from the maximum value principle that it is positive everywhere. We note also that $\left.\frac{dG}{dn}\right|_{\Sigma} \leqslant 0$, which immediately follows from the above and the condition $G\,|_{\Sigma} = 0$.

2. The source function is symmetrical with respect to its arguments $M_0\,(x, y, z)$ and $M(\xi, \eta, \zeta)$:

$$G\,(M, M_0) = G\,(M_0, M).$$

Let M_0' and M_0'' be fixed points of the region T. Let us consider the spheres Σ_1, and Σ_2 of radius ε with centres at the points M_0' and M_0'' (Fig. 50). Assuming

$$u = G\,(M, M_0')\,, \quad v = G\,(M, M_0'')$$

and applying Green's theorem

$$\iiint\limits_{T_\varepsilon} (u \,\Delta\, v - v \,\Delta\, u)\, d\tau = \iint\limits_{\Sigma_1+\Sigma_2+\Sigma} \left(u \frac{\partial v}{\partial n} - v \frac{\partial u}{\partial n}\right) d\sigma \tag{5}$$

to the region T_ε, obtained by excluding the spheres Σ_1 and Σ_2 from the main region we have:

$$\iint\limits_{\Sigma_1} \left[G\,(M, M_0') \frac{\partial G(M, M_0'')}{\partial n} - G\,(M, M_0'') \frac{\partial G\,(M, M_0')}{\partial n}\right] d\sigma_M +$$

$$+ \iint\limits_{\Sigma_2} \left[G\,(M, M_0') \frac{\partial G(M, M_0'')}{\partial n} - G\,(M, M_0'') \frac{\partial G(M, M_0')}{\partial n}\right] d\sigma_M = 0,$$

because the left side of equation (5) equals zero, since $\nabla^2 G = 0$. The integral over the surface Σ equals zero because of the boundary conditions.

Proceeding next to a limit as $\varepsilon \to 0$ and using the singularity of a source function, we obtain:*

$$G(M_0', M_0'') = G(M_0'', M_0')$$

or

$$G(M, M_0) = G(M_0, M).$$

This symmetry of the source function is the mathematical expression of the *principle of reciprocity* in physics: the source situated at point M_0 produces at point M an effect such as a source situated at point M produces at M_0. The principle of reciprocity is of a very general nature and occurs in different physical fields (electromagnetic, elastic, etc.).

The source function $G(M, M_0)$ for the case of two dimensions obviously will be determined by the relations:

1. $\nabla^2 G = 0$ everywhere in the region S under consideration, except at the point $M = M_0$.

2. At the point $M = M_0$ the function G has a singularity of the form

$$\frac{1}{2\pi} \ln \frac{1}{r_{MM_0}}.$$

3. $G|_C = 0$, where C is the boundary of the region S. The source function in this case has the form

$$G(M, M_0) = \frac{1}{2\pi} \ln \frac{1}{r_{MM_0}} + v(M, M_0),$$

where v is everywhere a continuous harmonic function, satisfying on the boundary the relation

$$v|_C = -\frac{1}{2\pi} \ln \frac{1}{r_{MM_0}}.$$

The solution of the first boundary-value problem for $\nabla^2 u = 0$ is given by the relation

$$u(M_0) = -\int_C f \frac{\partial G}{\partial n} ds \quad (f = u|_C).$$

* This theorem is proved by Lyapunov for a class of surfaces called Lyapunov surfaces.

2. *Method of electrostatic images and the source function for a sphere*

The most common method of obtaining the source function is the *method of electrostatic images*. This idea is as follows. In the construction of the source function

$$G(M, M_0) = \frac{1}{4\pi r_{MM_0}} + v$$

the induced field v is represented as a field of charges, situated outside the surface Σ and selected in such a way that the relation

$$v|_\Sigma = -1/4\pi r.$$

is fulfilled.

These charges are called *electrostatic images* of a unit charge, situated at point M_0 and producing in the absence of surface Σ a potential $1/4\pi r$. In many cases the choice of such charges is straightforward. We give some examples of the derivation of the source function by the method of electrostatic images.

The continuity of the first derivatives of the functions G on surface Σ is obvious for the source functions obtained in these examples.

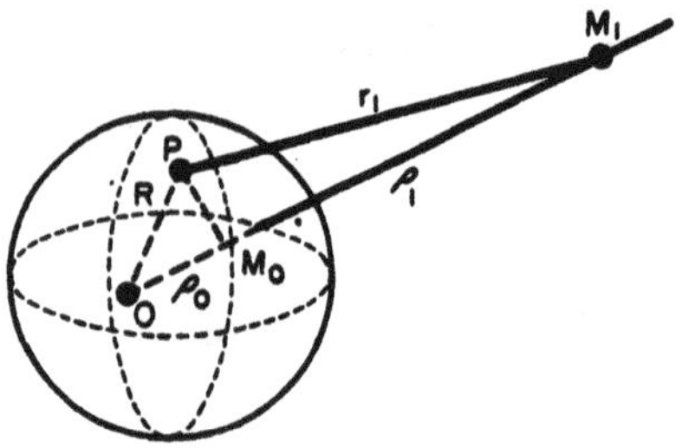

FIG. 51

As a first example let us consider the source function for a sphere.

We are given a sphere of radius R with centre at the point O and want to find its source function.

We place a unit charge at the point M_0 and mark off a section on the OM_1 on the radius passing through M_0 such that

$$\varrho_0 \varrho_1 = R^2, \tag{6}$$

where $\varrho_0 = OM_0$ and $\varrho_1 = OM_1$ (Fig. 51).

Transformation (6) is an inversion and the point M_1 is called the *conjugate* of M_0. This transformation is reciprocal, and it is possible to consider the point M_0 as the conjugate of M_1.

Let us prove that for all points P, situated on the sphere, the distances to M_0 and M_1 are proportional. In order to do this let us consider the triangles OPM_0 and OPM_1 (see Fig. 51); they are similar because the angle at O is common, and the sides adjacent to it are proportional:

$$\frac{\varrho_0}{R} = \frac{R}{\varrho_1} \quad \text{or} \quad \frac{OM_0}{R} = \frac{R}{OM_1}.$$

From the similarity of the triangles it follows:

$$\frac{r_0}{r_1} = \frac{\varrho_0}{R} = \frac{R}{\varrho_1}, \tag{7}$$

where $r_0 = |\overrightarrow{M_0 P}|$, $r_1 = |\overrightarrow{M_1 P}|$. From the ratios (7) we obtain:

$$r_0 = \frac{\varrho_0}{R} r_1$$

for all points of the sphere. Therefore the harmonic function $v = -R/\varrho_0 r_1$ takes the same value as the function $1/r_0$ on the surface. It represents the potential of a charge of value $-R/\varrho_0$, situated at point M_1.

Thus, the function

$$G(P, M_0) = \frac{1}{4\pi}\left(\frac{1}{r_0} - \frac{R}{\varrho_0}\frac{1}{r_1}\right) \tag{8}$$

is the source function of a sphere, since this is a harmonic function having a singularity $(1/4\pi)\,(1/r_0)$ at M_0 and reducing to zero on the sphere.

The solution of the first boundary-value problem is given by relation (4).

Let us calculate the derivative

$$\frac{\partial G}{\partial n} = \frac{1}{4\pi}\left[\frac{\partial}{\partial n}\left(\frac{1}{r_0}\right) - \frac{R}{\varrho_0}\frac{\partial}{\partial n}\left(\frac{1}{r_1}\right)\right], \tag{9}$$

where n is the outer normal, $\boldsymbol{r}_1 = \overrightarrow{M_1 M}$ (M, generally speaking, does not lie on the sphere).

The derivatives of $1/r_0$ and $1/r_1$in the direction $\boldsymbol{n}$ equal

$$\left.\begin{aligned} \frac{\partial}{\partial n}\left(\frac{1}{r_0}\right) &= \frac{\partial}{\partial r_0}\left(\frac{1}{r_0}\right)\frac{\partial r_0}{\partial n} = -\frac{1}{r_0^2}\cos(\widehat{\boldsymbol{r}_0, \boldsymbol{n}}), \\ \frac{\partial}{\partial n}\left(\frac{1}{r_1}\right) &= \frac{\partial}{\partial r_1}\left(\frac{1}{r_1}\right)\frac{\partial r_1}{\partial n} = -\frac{1}{r_1^2}\cos(\widehat{\boldsymbol{r}_1, \boldsymbol{n}}), \end{aligned}\right\} \tag{10}$$

since

$$\frac{\partial r_0}{\partial n} = \cos(\widehat{\boldsymbol{r}_0, \boldsymbol{n}}), \quad \frac{\partial r_1}{\partial n} = \cos(\widehat{\boldsymbol{r}_1, \boldsymbol{n}}). \tag{11}$$

It is easy to find the values $\cos(\widehat{\boldsymbol{r}_0, \boldsymbol{n}})$ and $\cos(\widehat{\boldsymbol{r}_1, \boldsymbol{n}})$:

$$\cos(\widehat{\boldsymbol{r}_0, \boldsymbol{n}}) = \frac{R^2 + r_0^2 - \varrho_0^2}{2Rr_0}, \tag{11'}$$

$$\cos(\widehat{\boldsymbol{r}_1, \boldsymbol{n}}) = \frac{R^2 + r_1^2 - \varrho_1^2}{2Rr_1}. \tag{11''}$$

Making use of (7) we have:

$$\cos(\widehat{\boldsymbol{r}_1, \boldsymbol{n}})\,|_\Sigma = \frac{R^2 + \dfrac{R^2}{\varrho_0^2} r_0^2 - \dfrac{R^4}{\varrho_0^2}}{2R\dfrac{R}{\varrho_0} r} = \frac{\varrho_0^2 + r_0^2 - R^2}{2\varrho_0 r_0},$$

since $\varrho_1 = R^2/\varrho_0$, by definition of point M_1, and $r_1 = (R/\varrho_0) r_0$ on the sphere Σ. Using formulae (10) and also expressions (9), (11′), (11″), we find:

$$\frac{\partial G}{\partial n}\bigg|_\Sigma = \frac{1}{4\pi}\left[-\frac{1}{r_0^2}\,\frac{R^2 + r_0^2 - \varrho_0^2}{2Rr_0} + \frac{\varrho_0^2}{R^2 r_0^2}\,\frac{R}{\varrho_0}\,\frac{\varrho_0^2 + r_0^2 - R^2}{2\varrho_0 r_0}\right] =$$

$$= -\frac{1}{4\pi R}\,\frac{R^2 - \varrho_0^2}{r_0^3}.$$

Thus function $u(M_0)$ given by formula (4) equals

$$u(M_0) = \frac{1}{4\pi R}\iint_\Sigma f(P)\,\frac{R^2 - \varrho_0^2}{r_0^3}\,d\sigma_P. \tag{12}$$

Let us introduce a spherical system of coordinates with origin at the centre of the sphere. Let (R, θ, φ) be the coordinates of point P, and $(\varrho_0, \theta_0, \varphi_0)$ the coordinates of point M_0; γ the angle between the radius-vectors $\overrightarrow{OP}$ and $\overrightarrow{OM}_0$. Then (12) may be written in the form

$$u(\varrho_0, \theta_0, \varphi_0) = \frac{R}{4\pi}\int_0^{2\pi}\int_0^{\pi} f(\theta, \varphi)\,\frac{R^2 - \varrho_0^2}{(R^2 - 2R\varrho_0\cos\gamma + \varrho_0^2)^{3/2}}\,\sin\theta\,d\theta\,d\varphi \tag{12'}$$

where

$$\cos\gamma = \cos\theta\cos\theta_0 + \sin\theta\sin\theta_0\cos(\varphi - \varphi_0)^*. \tag{13}$$

This formula is called *Poisson's integral for a sphere.*

In the same way the source function may be constructed, for a region external to the sphere,

$$G(M, M_1) = \frac{1}{4\pi}\left(\frac{1}{r_1} - \frac{R}{\varrho_1}\frac{1}{r_0}\right), \tag{14}$$

where $r_1 = MM_1$ is the distance from the fixed point M_1 lying outside the sphere, $r_0 = MM_0$ is the distance from point M_0, the conjugate of point M_1, ϱ_1 is the distance of M_1 from the origin of coordinates, and R is the radius of the sphere.

Taking into account the difference of the directions of the normal for internal and external problems, we obtain:

$$u(\varrho_1, \theta_1, \varphi_1) = \frac{R}{4\pi}\int_0^{2\pi}\int_0^{\pi} \frac{\varrho_1^2 - R^2}{[R^2 - 2\varrho_1 R\cos\gamma + \varrho_1^2]^{3/2}} f(\theta, \varphi)\sin\theta\, d\theta\, d\varphi,$$

where $\cos\gamma$ is given by (13) (it is only necessary to replace suffix 0 by 1).

3. *Source function for a circle*

The source function for a circle may be obtained by a method similar to that for a sphere. In this case the function should have the form

$$G = \frac{1}{2\pi}\ln\frac{1}{r} + v. \tag{15}$$

Repeating the arguments of the preceding section from formula (6) to (8) we find function G

$$G(P, M_0) = \frac{1}{2\pi}\left[\ln\frac{1}{r_0} - \ln\frac{R}{\varrho_0}\frac{1}{r_1}\right], \tag{16}$$

* In fact, the direction cosines of the vectors $\overrightarrow{OP}$ and $\overrightarrow{OM_0}$ are equal respectively to $(\sin\theta\cos\varphi,\ \sin\theta\sin\varphi,\ \cos\theta)$ and $(\sin\theta_0\cos\varphi_0,\ \sin\theta_0\sin\varphi_0,\ \cos\theta_0)$, from which

$$\cos\gamma = \cos\theta\cdot\cos\theta_0 + \sin\theta\sin\theta_0(\cos\varphi\cos\varphi_0 + \sin\varphi\sin\varphi_0) =$$
$$= \cos\theta\cos\theta_0 + \sin\theta\sin\theta_0\cos(\varphi - \varphi_0).$$

where $\varrho_0 = OM_0$, $r_0 = M_0P$, $r_1 = M_1P$, $R = OP$ is the radius of the circle (Fig. 52). It is easy to see that a harmonic function defined in this way reduces to zero on the boundary

$$G|_C = 0.$$

For the solution of the first boundary-value problem it is necessary to calculate the value of $\frac{\partial G}{\partial n}$ on the circumference C.

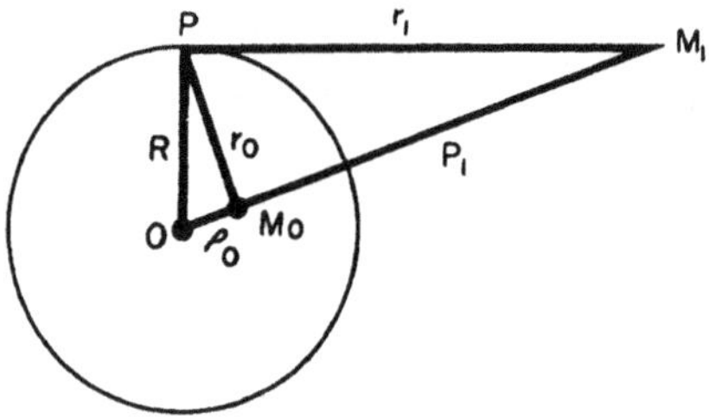

FIG. 52

The calculations are similar to the case of a sphere and give:

$$\left.\frac{\partial G}{\partial n}\right|_C = -\frac{1}{2\pi R}\,\frac{R^2 - \varrho_0^2}{r_0^2}.$$

Let (ϱ, θ) be the polar coordinates of point P, lying on the circumference, and (ϱ_0, θ_0) the coordinates of point M_0, then

$$r_0^2 = R^2 + \varrho_0^2 - 2R\varrho_0 \cos(\theta - \theta_0).$$

Substituting in the formula

$$u(\varrho_0, \theta_0) = \frac{1}{2\pi}\int_C u(P)\,\frac{R^2 - \varrho_0^2}{r_0^2}\,\frac{ds}{R}$$

this expression for r_0 and taking into consideration that

$$u(P)|_C = f(\theta) \text{ and } ds = Rd\theta,$$

we arrive at the expression for function $u(M_0)$

$$u(\varrho_0, \theta_0) = \frac{1}{2\pi}\int_0^{2\pi} \frac{R^2 - \varrho_0^2}{R^2 + \varrho_0^2 - 2R\varrho_0 \cos(\theta - \theta_0)}\, f(\theta)\, d\theta, \qquad (17)$$

called *Poisson's integral for a circle* (see page 344 (13)). The same formula except for a sign gives the solution of the external problem.

4. *Source function for a half-space*

The conception of the source function in (4) holds for an infinite semispace, if functions regular at infinity are considered (see § 2, sub-section 6). Let us find the source function for the semi-space $z > 0$. We place a unit charge at the point M_0 (x_0, y_0, z_0)

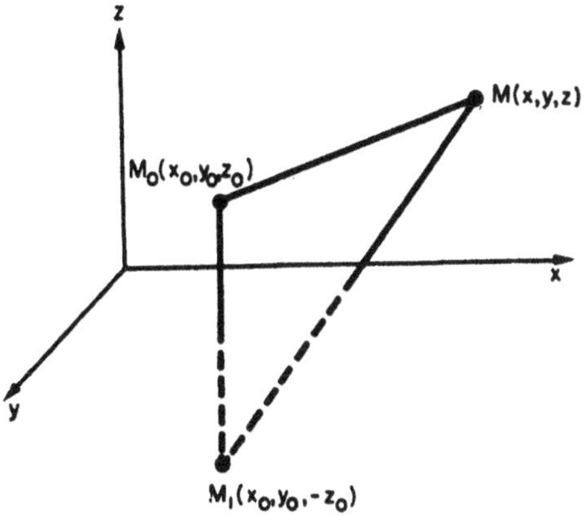

FIG. 53

which produces a field in the infinite semi-space, whose potential is given by the function

$$\frac{1}{4\pi}\frac{1}{r_{M_0M}}, \text{ where } r_{M_0M} = \sqrt{[(x-x_0)^2 + (y-y_0)^2 + (z-z_0)^2]}\,.$$

It is easy to see that the "induced field" v is the field of a negative unit charge, situated at the point M_1 $(x_0, y_0, -z_0)$ which is the mirror image of point M_0 on the plane $z = 0$ (Fig. 53). Function G, equal to

$$G(M, M_0) = \frac{1}{4\pi r_0} - \frac{1}{4\pi r_1},$$

where

$$r_0 = \overrightarrow{M_0 M} = \sqrt{[(x-x_0)^2 + (y-y_0)^2 + (z-z_0)^2]},$$

$$r_1 = \overrightarrow{M_1 M} = \sqrt{[(x-x_0)^2 + (y-y_0)^2 + (z+z_0)^2]},$$

reduces to zero at $z = 0$ and has the required singularity at point M_0.

Let us calculate $\left.\frac{\partial G}{\partial n}\right|_{z=0} = -\left.\frac{\partial G}{\partial z}\right|_{z=0}$. It is obvious that

$$\frac{\partial G}{\partial z} = \frac{1}{4\pi}\left[-\frac{z-z_0}{r_0^3} + \frac{z+z_0}{r_1^3}\right]$$

Assuming $z = 0$, we find:

$$\left.\frac{\partial G}{\partial n}\right|_{z=0} = -\left.\frac{\partial G}{\partial z}\right|_{z=0} = -\frac{z_0}{2\pi r_0^3}.$$

The solution of the first boundary-value problem is given by the relation

$$u(M_0) = \frac{1}{2\pi}\iint_{\Sigma_0} \frac{z_0}{r_{M_0P}^3} f(P)\, d\sigma_\varrho,$$

where Σ_0 is the plane $z = 0$, $f(P) = u|_{z=0}$, or

$$u(x_0, y_0, z_0) = \frac{1}{2\pi}\int_{-\infty}^{\infty}\int_{-\infty}^{\infty} \frac{z_0}{[(x-x_0)^2+(y-y_0)^2+z_0^2]^{3/2}} f(x,y)\,dx\,dy. \quad (18)$$

§ 5. Potential theory

The function

$$\frac{1}{r} = \frac{1}{\sqrt{[(x-\xi)^2+(y-\eta)^2+(z-\zeta)^2]}},$$

representing the potential of a field of unit mass (charge) situated at point $M_0(\xi, \eta, \zeta)$ is a solution of Laplace's equation, depending on the parameters ξ, η, ζ. Integrals of this function over these parameters are called *potential* functions and they have an intrinsic value from the point of view of direct application in physics, and also for the development of methods of solution of boundary-value problems.

1. *Volume potential*

Let a mass m_0 be situated at some point $M_0(\xi, \eta, \zeta)$. According to the law of universal gravitation a force of attraction acts on mass m, situated at point $M(x, y, z)$

$$\boldsymbol{F} = -\gamma\frac{mm_0}{r^3}\boldsymbol{r}, \quad (1)$$

where $\boldsymbol{r} = \overrightarrow{M_0 M}$, and γ is the gravitational constant. Let us choose a system of units so that $\gamma = 1$, and assuming $m = 1$, we obtain:

$$F = -\frac{m_0}{r^3}\boldsymbol{r},$$

Projections of this force on the coordinate axes are.

$$\left.\begin{aligned} X &= F\cos\alpha = -\frac{m_0}{r^3}(x-\xi), \\ Y &= F\cos\beta = -\frac{m_0}{r^3}(y-\eta), \\ Z &= F\cos\gamma = -\frac{m_0}{r^3}(z-\zeta), \end{aligned}\right\} \tag{2}$$

where α, β, and γ are angles, formed by vector $\boldsymbol{F}$ with the coordinate axes.

We introduce a function u, called the *potential of the force field* defined by the relation

$$\boldsymbol{F} = \operatorname{grad} u$$

or

$$X = \frac{\partial u}{\partial x}, \quad Y = \frac{\partial u}{\partial y}, \quad Z = \frac{\partial u}{\partial z}.$$

In our case

$$u = \frac{m_0}{r}.$$

The potential of a field of n point masses is expressed by the relation

$$u = \sum_{i=1}^{n} u_i = \sum_{i=1}^{n} \frac{m_i}{r_i}.$$

Let us proceed to the case of a continuous distribution of mass. A body T has mass density $\varrho(\xi, \eta, \zeta)$. Let us determine the potential of this body at a point $M(x, y, z)$. In order to do this we divide the body T into sufficiently small parts $\Delta\tau$. If we make the natural assumption that the effect of element $\Delta\tau$ is equivalent to the effect of this mass, concentrated at some "mean" point* of the volume $\Delta\tau$; then we obtain the

* More precisely, moreover, it is assumed that the action of some body T on a mass m at a point, lying outside a convex volume $\overline{T}$, containing this body, can be replaced by the action of some effective centre of the same mass m, lying inside $\overline{T}$.

following expression for the x-component of the force, acting at M:

$$\Delta X = -\frac{\varrho \Delta \tau}{r^3}(x - \xi),$$

where

$$r^2 = (x - \xi)^2 + (y - \eta)^2 + (z - \zeta)^2.$$

Integration over the total volume T gives the component of the total force of attraction at the point M

$$X = -\iiint_T \varrho \frac{x - \xi}{r^3} d\tau. \tag{3}$$

The potential at point M will be given by the relation

$$u(M) = \iiint_T \varrho \frac{1}{r} d\tau. \tag{4}$$

If the point M lies outside the body, then this can be proved directly by differentiation under the integral sign.* Similarly the derivatives of higher orders may be calculated. It is obvious that the potential $u(M)$ outside body T satisfies Laplace's equation.

Later, not aiming to give a theory for the most general conditions, we shall make use of these properties of potentials and develop theories with the condition that ϱ is a bounded function (assuming its integrability).

If point M lies inside region T, it is impossible to prove that $X = \frac{\partial u}{\partial x}$ without additional investigation. This will be given later.

2. *Plane problem. Logarithmic potential*

Let us consider a distribution of mass in space depending only on the two coordinates (x, y). Obviously the potential is independent of the z-coordinate, therefore it is sufficient to investigate the potential at the point (x, y) lying in the plane $z = 0$.

* Sufficient conditions for the differentiability of an integral of the form

$$f(M) = \int_T F(M, P)\varphi(P) d\tau_P$$

with respect to a parameter under the integral sign are the continuity of the derivative of the function $F(M, P)$ and the absolute integrability of $\varphi(P)$.

Let us determine the potential of a homogeneous infinite straight line L. We take the z-axis along this line. Let the linear density (i.e. the mass per unit length) equal μ. The force of

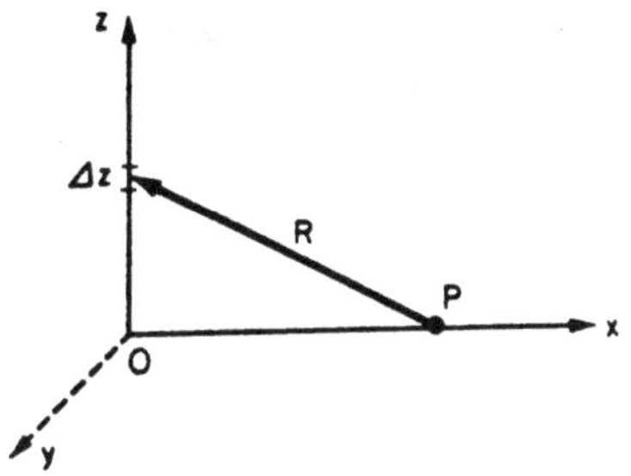

FIG. 54

attraction between element Δz and the point $P(x, 0)$ (Fig. 54) and its component along the x-axis equal, respectively,

$$\Delta F = -\frac{\mu \Delta z}{R^2} = -\frac{\mu \Delta z}{(x^2 + z^2)},$$

$$\Delta X = \Delta F \cos a = -\mu \Delta z \cdot \frac{x}{\sqrt{[(x^2 + z^2)^3]}}.$$

Hence

$$X = -\int_{-\infty}^{\infty} \mu x \frac{dz}{(x^2 + z^2)^{3/2}} = -\mu x^2 \frac{1}{x^3} \int_{-\frac{\pi}{2}}^{\frac{\pi}{2}} \cos a \, da = -\frac{2\mu}{x}$$

$$(z/r = \tan a).$$

If $P(x, y)$ is an arbitrary point, then the force of attraction at the point, due to the line L, will be directed along OP and have the value

$$F = -2\mu/\varrho,$$

where

$$\varrho = \sqrt{(x^2 + y^2)}.$$

The potential of this force is called the *logarithmic potential* and equals

$$V = 2\mu \ln l/\varrho, \tag{5}$$

as is easily verified by direct differentiation.

The logarithmic potential is a solution of Laplace's equation with two independent variables, possessing circular symmetry about the singularity at the point $\varrho = 0$ at which it tends to infinity.

Thus the potential of a homogeneous straight line gives a plane field and is expressed by relation (5). The representation (4) of the potential as an integral was derived only for a finite body T.* We note that in contrast to this potential the loga-

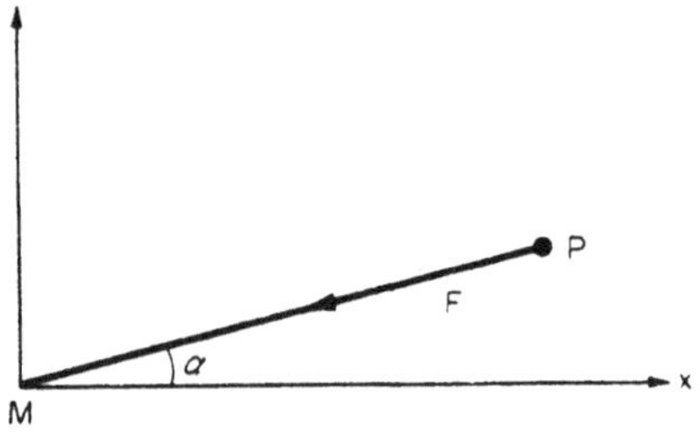

Fig. 55

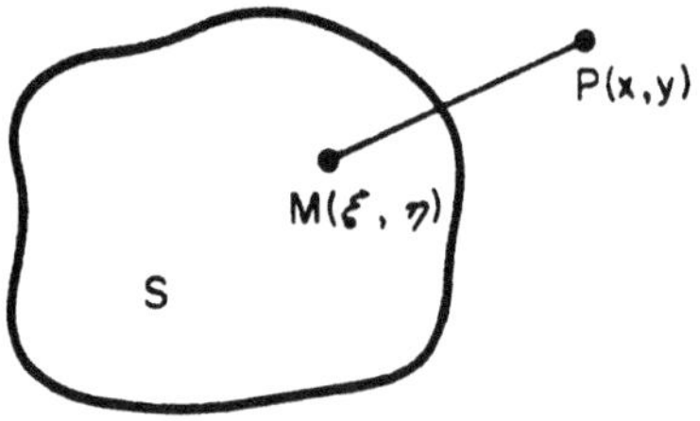

Fig. 56

rithmic potential does not reduce to zero at infinity, where it has a logarithmic singularity.

Let us calculate now the component of the force of attraction at points P (Fig. 55)

$$X = F\cos\alpha = -2\mu\frac{x}{\varrho^2} \qquad (\cos\alpha = x/\varrho),$$

$$Y = F\sin\alpha = -2\mu\frac{y}{\varrho^2} \qquad (\sin\alpha = y/\varrho).$$

If there are several points (infinite straight lines with a mass distribution along them), then because of the principle of superposition of force fields the potentials of the points (lines) will be additive.

* In the calculation of the potential of an infinite straight line it would be impossible to integrate the potentials of single elements directly using (4), because in this case a divergent integral is obtained. In fact, the potential of the element Δz equals

$$\Delta u = \mu\frac{\Delta z}{\sqrt{(\varrho^2 + z^2)}}.$$

Formal integration gives a divergent integral

$$u = \int_{-\infty}^{\infty} \mu\frac{dz}{\sqrt{(\varrho^2 + z^2)}}.$$

In the case of a region S with a continuously-distributed density $u(\xi, \eta)$ (Fig. 56) the components of the force of attraction at point P are expressed by the double integrals:

$$\left.\begin{aligned} X &= -2 \iint\limits_S \mu(\xi, \eta) \frac{x-\xi}{(x-\xi)^2+(y-\eta)^2} d\xi\, d\eta, \\ Y &= -2 \iint\limits_S \mu(\xi, \eta) \frac{y-\eta}{(y-\eta)^2+(x-\xi)^2} d\xi\, d\eta, \end{aligned}\right\} \tag{6}$$

and the potential will equal

$$u(x, y) = 2 \iint\limits_S \mu(\xi, \eta) \ln \frac{1}{\sqrt{[(x-\xi)^2+(y-\eta)^2]}} d\xi\, d\eta, \tag{7}$$

which it is easy to verify by differentiation at points lying outside S. If point P lies in the region S, then it is necessary to carry out an additional investigation.

3. *Improper integrals*

The potentials and components of the force of attraction are represented by integrals in which the functions under the integral sign tend to infinity at certain points in a region containing masses.

As is well known, if the function under the integral sign tends to infinity at some point of the region of integration, then it is impossible to define the integral as the limit of the integral sum. In fact, in this case the integral sum does not have a limit, since the term coming from an elemental volume, containing the singular point, can contribute any value to the sum depending upon the choice of the mean point of that volume. The integrals of such functions have to be defined as improper integrals.

In region T we are given a function $F(x, y, z)$ tending to infinity at some point $M_0\,(x_0, y_0, z_0)$. Let us consider the integral over a region $T - K_\varepsilon$, where K_ε is some neighbourhood of the point M_0 of diameter not exceeding ε.

If for an arbitrary contraction of the region K_{ε_n} to the point M_0 the series of integrals

$$I_n = \iiint\limits_{T-K_{\varepsilon_n}} F d\tau \qquad (\varepsilon_n \to 0)$$

has a limit, not depending on the choice of regions K_{ε_n}, then this limit is called the *improper integral* of function $F(x, y, z)$ over region T and is defined, as is usual

$$\iiint_T F d\tau .$$

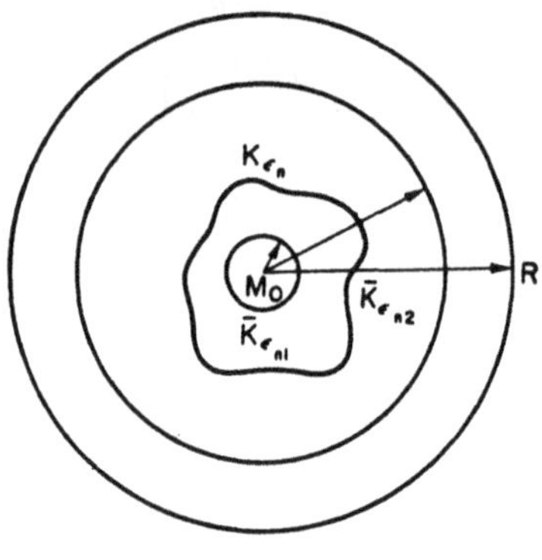

FIG. 57

If the limit $\bar{I}$ depends on the choice of the series of regions $\bar{K}_{\varepsilon_m}$, or even in certain cases does not exist, then the limit $\bar{I}$ is called a conditionally convergent improper integral. It is clear that in a consideration of a conditionally convergent improper integral $\bar{I}$ it is necessary to indicate the series of regions $\bar{K}_{\varepsilon_n}$, with respect to which the integral is defined.

We confine ourselves here to those cases, where the function under the integral sign has a singularity at an isolated point. We investigate the convergence of integrals of the type

$$\iiint_T \frac{C}{r^a} d\tau_M , \tag{8}$$

where C and $a > 0$ are constants,

$$r = r_{MM_0} = \sqrt{[(x_0 - \xi)^2 + (y_0 - \eta)^2 + (z_0 - \zeta)^2]} ,$$

M_0 is a point of region T. Without loss of generality it is possible to assume that T is a sphere of radius R with centre at point M_0. We take the regions K_{ε_n} to be spheres of radius ε_n with centres at the point M_0 and we search for the limit of the series of integrals

$$\iiint_{T-K_{\varepsilon_n}} \frac{C}{r^a} d\tau = \int_0^{2\pi} d\varphi \int_0^{\pi} \sin\theta \, d\theta \int_{\varepsilon_n}^{R} \frac{C}{r^{a-2}} dr = 2\pi \cdot 2C \int_{\varepsilon_n}^{R} \frac{dr}{r^{a-2}} =$$

$$= \begin{cases} 4\pi C \left[\dfrac{1}{3-a} r^{3-a}\right]_{\varepsilon_n}^{R}, & \text{if } a \neq 3 , \\ 4\pi C \, [\ln r]_{\varepsilon_n}^{R} & \text{if } a = 3 . \end{cases}$$

Transition to a limit for ε_n tending to zero shows that for $a < 3$ a limit exists, for $a \geqslant 3$ a limit does not exist.

Let us show that if the function $F(x, y, z)$ is positive and a limit

$$I_n = \iiint\limits_{T-\overline{K}_{\varepsilon_n}} F\,d\tau \qquad (\varepsilon_n \to 0),$$

exists, where K_{ε_n} is a sphere of radius ε_n with centre at point M, then a limit exists for any choice of the series of regions K_{ε_n}, containing the point M, and the value of this limit does not depend on the shape of the region K_{ε_n}. Any region K_{ε_n} can be included between the two spheres $K_{\varepsilon_{n_1}}$ and $K_{\varepsilon_{n_2}}$, whose radii ε_{n_1} and ε_{n_2} tend to zero together with ε_n (Fig. 57). Because of the positive nature of the function under the integral sign

$$\iiint\limits_{T-\overline{K}_{\varepsilon_{n_1}}} F\,d\tau \geqslant \iiint\limits_{T-K_{\varepsilon_n}} F\,d\tau \geqslant \iiint\limits_{T-\overline{K}_{\varepsilon_{n_2}}} F\,d\tau.$$

It follows that

$$\lim_{n\to\infty} \iiint\limits_{T-K_{\varepsilon_n}} F\,d\tau = \lim_{n\to\infty} \iiint\limits_{T-\overline{K}_{\varepsilon_n}} F\,d\tau = I,$$

since the limits of extreme integrals do exist and equal this quantity.

Thus, in the case of three independent variables the improper integral

$$\iiint\limits_{T} \frac{C}{r^a}\,d\tau \tag{8}$$

exists, if $a < 3$, and does not exist if $a \geqslant 3$.

For any number of independent variables the critical value a, fixing the boundaries of convergence of integrals of type (8), is equal to the number of dimensions; thus, for example, *for two independent variables, the integral*

$$\iint\limits_{\Sigma} \frac{C}{\varrho^a}\,d\sigma \quad \begin{array}{l}\text{converges if } a < 2,\\ \text{diverges if } a \geqslant 2.\end{array}$$

We now study the criterion for convergence of improper integrals. Let us prove that *for the convergence of an improper integral*

$$\iiint\limits_T F(x, y, z)\,dx\,dy\,dz \tag{9}$$

it is sufficient that there exists a function $\overline{F}(x, y, z)$, *for which the improper integral converges over region* T, *and such that the inequality*

$$|F(x, y, z)| < \overline{F}(x, y, z). \tag{10}$$

holds.

Let us consider some series of regions K_ε, containing the singular point M_0. Because of the convergence of the series of integrals $\overline{I}_n$ of the function $\overline{F}$ for any $\varepsilon > 0$ there exists $N(\varepsilon)$ such that

$$|\bar{I}_{n_1} - \bar{I}_{n_2}| = |\iiint\limits_{K_{\varepsilon_{n_1}} - K_{\varepsilon_{n_2}}} \overline{F}\,d\tau| < \varepsilon,$$

provided n_1, $n_2 > N(\varepsilon)$. Since $\overline{F}$ is a majorant function for $F(x, y, z)$ then it is possible to write:

$$|I_{n_1} - I_{n_2}| = |\iiint\limits_{K_{\varepsilon_{n_1}} - K_{\varepsilon_{n_2}}} F\,d\tau| \leqslant \iiint\limits_{K_{\varepsilon_{n_1}} - K_{\varepsilon_{n_2}}} |F|\,d\tau \leqslant \iiint\limits_{K_{\varepsilon_{n_1}} - K_{\varepsilon_{n_2}}} \overline{F}\,d\tau < \varepsilon, \tag{10'}$$

if n_1, $n_2 > N(\varepsilon)$. Fulfillment of condition (10′) by virtue of Cauchy's criterion of convergence is sufficient for the convergence of the series

$$I_n = \iiint\limits_{T - K_{\varepsilon_n}} F\,d\tau$$

to some limit

$$I = \lim_{n \to \infty} I_n = \iiint\limits_T F\,d\tau.$$

It is easy to see that this limit will not depend on the form of the region K_{ε_n}. Thus we have proved the existence of the improper integral (9).

If for a given function $F(x, y, z)$ it is possible to find a positive function $\overline{F}(x, y, z)$ such that $F(x, y, z) > \overline{F}$ and so that the improper integral of $\overline{F}$ over the region T diverges, then the improper integral (9) will, obviously, diverge.

Conclusion: *If for some function $F(M, P)$ tending to infinity for $P = M$, the following inequality holds*

$$|F(M,P)| < \frac{C}{r_{MP}^{\alpha}} \quad \alpha < 3,$$

then the improper integral over region T, containing the point M

$$\iiint_T F(M,P)\, d\tau_P$$

converges.

From the theory of integrals, depending on parameters, it is known that the continuity of a function under the integral sign with respect to parameters and an independent variable is a sufficient condition for the continuity of the same integral as a function of the parameters. For improper integrals the function under the integral sign is not continuous and therefore this criterion does not apply. Let us establish a criterion for the continuity of improper integrals, depending on parameters.

We consider the improper integral

$$V(M) = \int_T F(P, M) f(P)\, d\tau_P, \tag{11}$$

where $F(P, M)$ is a function, tending to infinity as $M \to P$ and continuous over M, and $f(P)$ is a bounded function.

Integral (11) *is said to be uniformly convergent at point M_0, if for any $\varepsilon > 0$ it is possible to assign $\delta(\varepsilon)$ such that the inequality*

$$|V_{\delta(\varepsilon)}(M)| = \left|\int_{T_{\delta(\varepsilon)}} F(P, M) f(P)\, d\tau_P\right| \leqslant \varepsilon$$

holds for any point M, whose distance from M_0 is less than $\delta(\varepsilon)$, and for any region $T_{\delta(\varepsilon)}$, containing the point M_0 and having diameter $d \leqslant \delta(\varepsilon)$.

Let us prove that the integral

$$V(M) = \int_T F(P, M) f(P)\, d\tau_P,$$

converging uniformly at the point M_0, is a continuous function at this point M_0. We should prove that for any ε it is possible to find $\delta(\varepsilon)$ such that

$$|V(M_0) - V(M)| < \varepsilon$$

for

$$|\overrightarrow{MM_0}| < \delta(\varepsilon).$$

Let us choose inside region T some region T_1, containing the point M_0 (Fig. 58) and let us separate the integral into two components

$$V = V_1 + V_2,$$

where integral V_1 corresponds to region T_1 and V_2 to region $T_2 = T - T_1$. Later we will specify the size of the region T_1. We consider the inequality

$$|V(M_0) - V(M)| \leqslant |V_2(M_0) - V_2(M)| + |V_1(M_0)| + |V_1(M)|$$

and show that each of the components, starting from the right, may be made less than $\varepsilon/3$ for sufficiently small $|\overrightarrow{M_0 M}|$.

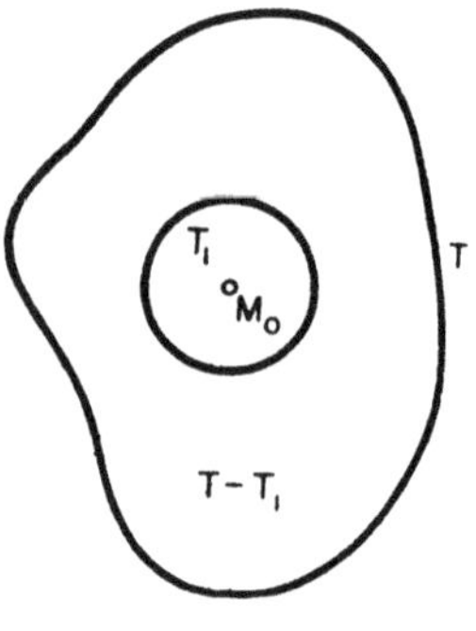

Fig. 58

Choosing region T_1 as a sphere of radius $\delta(\varepsilon/3)$, we have:

$$|V_1(M_0)| \leqslant \varepsilon/3 \text{ and } |V_1(M)| \leqslant \varepsilon/3,$$

if

$$|\overrightarrow{M_0M}| \leqslant \delta'(\varepsilon/3)$$

The existence of such a δ' follows from the condition of uniform convergence of the integral (11) at the point M_0. The choice of region T_1 determines the region T_2.

Since the point M_0 lies outside region T_2, then the integral V_2 is a continuous function at this point.

Hence there must exist $\delta''(\varepsilon/3)$ such that

$$|V_2(M_0) - V_2(M)| \leqslant \varepsilon/3$$

for

$$|\overrightarrow{M_0M}| \leqslant \delta''(\varepsilon/3).$$

Assuming

$$\delta(\varepsilon) = \min[\delta'(\varepsilon), \delta''(\varepsilon)],$$

we obtain:

$$|V(M) - V(M_0)| \leqslant \varepsilon \text{ when } |\overrightarrow{MM_0}| \leqslant \delta,$$

which implies the continuity of the uniformly converging integral.

We note that the results obtained are valid not only for integrals taken over a volume, but also for integrals over surfaces and lines. This result will be used by us later.

Let us consider the potential

$$V(M) = \iiint_T \frac{\varrho(P)}{r_{MP}} d\tau_P \tag{12}$$

and the components of the force of attraction

$$\left.\begin{aligned} X(M) &= -\iiint_T \frac{\varrho(P)}{r_{MP}^3} (x-\xi)\, d\tau_P; \\ Y(M) &= -\iiint_T \frac{\varrho(P)}{r_{MP}^3} (y-\eta)\, d\tau_P; \\ Z(M) &= -\iiint_T \frac{\varrho(P)}{r_{MP}^3} (z-\zeta)\, d\tau_P \end{aligned}\right\} \tag{13}$$

at points, lying inside the attracting body T. The improper integrals (12) and (13) are convergent if the density $\varrho(M)$ is bounded $|\varrho(M)| < C$. For the potential this is obvious, since

$$\frac{|\varrho|}{r} < \frac{C}{r^\alpha} \quad (\alpha = 1 < 3).$$

For the components of the force of attraction this follows from the inequality

$$\frac{|\varrho|}{r^2} \frac{|x-\xi|}{r} < \frac{C}{r^\alpha} \quad (\alpha = 2 < 3),$$

since $|x - \xi| < r$.

In order to illustrate the concept of uniform convergence of improper integrals let us show that integrals (12) and (13) are continuous functions.

In order to do this it is necessary to show that the integrals (12) and (13) are uniformly convergent at any point M_0.

Let us calculate the modulus of the integral*

$$\left|\iiint_{T_\delta} \frac{\varrho(P)}{r_{MP}} d\tau_P\right| \leqslant C \iiint_{K_\delta^{M_0}} \frac{d\tau_P}{r_{MP}},$$

* We note that integral (12) is obtained from integral (11) if $F(M, P) = 1/r_{MP}$, $f(P) = \varrho(P)$.

where $K_\delta^{M_0}$ is a sphere of radius δ with centre at point M_0, containing region T_δ. But calculation of this integral over region $K_\delta^{M_0}$ with centre at point M_0 is not easy. In order to estimate the second integral it is expedient to change to a spherical system of coordinates with centre at the point M. It is obvious that

$$C\iiint\limits_{K_\delta^{M_0}} \frac{d\tau_P}{r_{MP}} \leqslant C\left|\iiint\limits_{K_{2\delta}^{M}} \frac{d\tau_P}{r_{MP}}\right| = C\,8\,\pi\delta^2,$$

where $K_{2\delta}^M$ is a sphere of radius 2δ with centre at point M. If we are given some $\varepsilon > 0$ then by choosing

$$\delta(\varepsilon) = \sqrt{\frac{\varepsilon}{8\pi C}},$$

we have verified the uniform convergence of integral V.

Making a similar argument for the integral

$$X(M) = \iiint\limits_T \varrho(P)\frac{x-\xi}{r_{MP}^3}\,d\tau_P,$$

we obtain:

$$\left|\iiint\limits_{T_\delta} \varrho(P)\frac{x-\xi}{r_{MP}^3}\,d\tau_P\right| \leqslant C\left|\iiint\limits_{K_\delta^{M_0}} \frac{d\tau_P}{r_{MP}^2}\right| \leqslant$$

$$\leqslant C\left|\iiint\limits_{K_{2\delta}^{M}} \frac{d\tau_P}{r_{MP}^2}\right| = 8\,\pi\delta C \leqslant \varepsilon,$$

if

$$\delta \leqslant \delta(\varepsilon) = \frac{\varepsilon}{8\pi C}.$$

Thus, potential V and the components of the force of attraction X, Y, Z are continuous functions for all space.*

4. *First derivatives of the volume potential*

The functions, under the integral sign in the integrals

$$X(M) = -\iiint\limits_T \varrho(P)\frac{x-\xi}{r_{MP}^3}\,d\tau_P, \quad Y(M), \quad Z(M),$$

* The uniform convergence of integrals $V(M)$ and $X(M)$ is proved on the assumption that the density is bounded $|\varrho| < C$. Hence these integrals are also continuous at points of discontinuity of function ϱ, for instance on the boundary of a region, filled with masses.

are derivatives of the function, under the integral sign in the potential

$$V(M) = \iiint_T \frac{\varrho(P)}{r_{MP}}\, d\tau_P .$$

If differentiation under the integral sign is valid, for the function V, then

$$X = \frac{\partial V}{\partial x}, \quad Y = \frac{\partial V}{\partial y}, \quad Z = \frac{\partial V}{\partial z}, \tag{14}$$

i. e. V is a potential field, whose force components equal X, Y, Z.

If the point M lies outside region T, then the function

$$-\frac{x-\xi}{r_{MP}^3} = \frac{-(x-\xi)}{[(x-\xi)^2+(y-\eta)^2+(z-\zeta)^2]^{3/2}} = \frac{\partial}{\partial x}\frac{1}{r_{MP}}$$

is continuous with respect to both arguments $M(x, y, z)$ and $P(\xi, \eta, \zeta)$. Hence, in this case differentiation under the integral sign of V is valid.

Derivatives of higher order may also be calculated by differentiating under the integral sign everywhere outside the region T. Hence because of the lemma of Chapter III, § 2 it follows that the potential outside the attracting mass satisfies Laplace's equation

$$\nabla^2 V = 0 \quad \text{outside } T .$$

Let us prove that the derivatives of potential V may be calculated by differentiation under the integral sign in the case where point M lies inside body T.

In the proof we will use only the bounded nature of the function $\varrho(x, y, z)$, ($|\varrho(x, y, z)| < C$), not assuming its continuity from which it will follow that function $V(x, y, z)$ is differentiable at points of the boundary, which may be points of discontinuity of the function $\varrho(x, y, z)$.

Let us show that for any ε it is possible to find $\delta(\varepsilon)$ such that

$$\left| \frac{V(x+\Delta x, y, z) - V(x, y, z)}{\Delta x} - X \right| < \varepsilon ,$$

if

$$|\Delta x| < \delta(\varepsilon) .$$

We confine the point M_0 in a sufficiently small sphere $K_{\delta^1}^{M_0}$, the dimensions of which we specify later, and separate V into the two components

$$V = V_1 + V_2,$$

where V_1 and V_2 correspond to the integration over volume $T_1 = K_{\delta^1}^{M_0}$, and the complementary volume $T_2 = T - K_{\delta^1}^{M_0}$. Then

$$\frac{V(x+\Delta x, y, z) - V(x, y, z)}{\Delta x} = \frac{V_1(x+\Delta x, y, z) - V_1(x, y, z)}{\Delta x} + \\ + \frac{V_2(x+\Delta x, y, z) - V_2(x, y, z)}{\Delta x}.$$

For any fixed dimensions of region T_1

$$\lim_{\Delta x \to \infty} \frac{V_2(x+\Delta x, y, z) - V_2(x, y, z)}{\Delta x} = X_2 = \iiint_{T_2} \varrho(\xi, \eta, \zeta) \frac{\partial}{\partial x}\left(\frac{1}{r}\right) d\tau,$$

since the point M_0 lies outside region T_2.

Let us evaluate the difference

$$\left| X - \frac{V(x+\Delta x, y, z) - V(x, y, z)}{\Delta x} \right| \leqslant \\ \leqslant \left| X_2 - \frac{V_2(x+\Delta x, y, z) - V_2(x, y, z)}{\Delta x} \right| + |X_1| + \\ + \left| \frac{V_1(x+\Delta x, y, z) - V_1(x, y, z)}{\Delta x} \right|$$

and let us show that each of the components may be made smaller than $\varepsilon/3$. In fact,

$$|X_1| = \left| \iiint_{T_1} \varrho \frac{x-\xi}{r^3} d\tau \right| < C \int_0^{\delta^1} \int_0^{2\pi} \int_0^{\pi} \frac{r^2 \sin\vartheta \, d\vartheta \, d\varphi \, dr}{r^2} = \\ = 4\pi C \delta^1 < \frac{\varepsilon}{3}, \tag{15}$$

since $\left|\frac{x-\xi}{r}\right| < 1$ and $|\varrho| < C$. Let us consider the last component

$$|S| = \left| \frac{V_1(x+\Delta x, y, z) - V_1(x, y, z)}{\Delta x} \right| = \\ = \left| \frac{1}{\Delta x} \iiint_{T_1} \varrho \left(\frac{1}{r_1} - \frac{1}{r} \right) d\tau \right| = \left| \frac{1}{\Delta x} \iiint_{T_1} \varrho \frac{r - r_1}{r r_1} d\tau \right|,$$

where

$$r_1 = \sqrt{\{[(x + \Delta x) - \xi]^2 + (y - \eta)^2 + (z - \zeta)^2\}};$$

$$r = \sqrt{[(x - \xi)^2 + (y - \eta)^2 + (z - \zeta)^2]}.$$

The sides of the triangle $M_0 M M_1$ equal r, r_1, and $|\Delta x|$. Hence it follows that

$$|r - r_1| \leqslant |\Delta x|.$$

Therefore

$$|S| \leqslant C \iiint_{T_1} \frac{d\tau}{r r_1} \leqslant C \frac{1}{2} \left\{ \iiint_{T_1} \frac{d\tau}{r_1^2} + \iiint_{T_1} \frac{d\tau}{r^2} \right\},$$

since for any numbers a and b

$$ab \leqslant \frac{1}{2}(a^2 + b^2).$$

In addition

$$\iiint_{T_1} \frac{d\tau}{r^2} = 4\pi\delta^1$$

and

$$\iiint_{T_1} \frac{d\tau}{r_1^2} \leq \iiint_{K_{2\delta^1}^{M_1}} \frac{d\tau}{r_1^2} = 8\pi\delta^2,$$

where $K_{2\delta^1}^{M_1}$ is a sphere of radius $2\delta^1$ with centre at the point M_1,

By a suitable choice of δ^1 it is possible to satisfy the inequality

$$|S| < \frac{C}{2} 12\pi\delta^1 = 6\pi C \delta^1 < \frac{\varepsilon}{3}. \tag{16}$$

Selecting δ^1 from relation (16) we satisfy both inequalities (15) and (16). We fix region $T_1 = K_{\delta^1}^{M_0}$ and thus region $T_2 = T - T_1$.

Relation (14) applied to the selected region T_2 indicates that for any ε it is possible to find δ^2 such that

$$\left| \frac{V_2(x + \Delta x, y, z) - V_2(x, y, z)}{\Delta x} - X_2 \right| < \frac{\varepsilon}{3},$$

provided $|\Delta x| < \delta^2$. Choosing, finally, $\delta = \min[\delta^1, \delta^2]$, we obtain:

$$\left| \frac{V(x + \Delta x, y, z) - V(x, y, z)}{\Delta x} - X \right| < \varepsilon, \text{ if } |\Delta x| < \delta.$$

Thus we have proved that

$$\frac{\partial V}{\partial x} = X. \tag{17}$$

Formulae

$$\frac{\partial V}{\partial y} = Y \text{ and } \frac{\partial V}{\partial z} = z$$

do not require separate proof.

Thus we have proved that differentiation under the integral sign is valid and that the components of the force field X, Y, Z are the components of grad V.

5. *Second derivatives of the volume potential*

The improper integral

$$\iiint_T \varrho(P) \frac{\partial^2}{\partial x^2}\left(\frac{1}{r_{MP}}\right) d\tau_P = -\iiint_T \varrho\left(\frac{1}{r^3} - 3\frac{(x-\xi)^2}{r^5}\right) d\tau \quad (18)$$

does not converge absolutely for internal points P of a body T. In this case the majorant for the function under the integral sign has the form

$$\frac{C}{r^a} \text{ where } a = 3.$$

Let us establish formulae for calculating the second derivatives of the potential V inside T on the assumption of the continuity and the differentiability of the density $\varrho(x, y, z)$ in the neighbourhood of the points under investigation. In particular the discussion will not apply to the boundary points, where a discontinuity of the density occurs.

Let us represent the potential V as a sum of two components

$$V = V_1 + V_2,$$

relating to the regions T_1 and T_2, where $T_1 = K_\delta^{M_0}$ is a sphere of radius δ with centre at the point M_0 under consideration, inside which function ϱ is differentiable.

The second derivative of V_2 may be calculated by differentiation under the integral sign, since point M_0 lies outside the region T_2.

$$\frac{\partial^2 V_2}{\partial x^2} = \frac{\partial}{\partial x}\left(\frac{\partial V_2}{\partial n}\right) = \iiint_{T_2} \varrho(\xi, \eta, \zeta) \frac{\partial^2}{\partial x^2}\left(\frac{1}{r}\right) d\tau.$$

The first derivative of V_1 with respect to x equals

$$\frac{\partial V_1}{\partial x} = \iiint_{T_1} \varrho \frac{\partial}{\partial x}\left(\frac{1}{r}\right) d\tau = - \iiint_{T_1} \varrho \frac{\partial}{\partial \xi}\left(\frac{1}{r}\right) d\tau, \qquad (19)$$

since

$$\frac{\partial}{\partial x}\left(\frac{1}{r}\right) = - \frac{\partial}{\partial \xi}\left(\frac{1}{r}\right).$$

Let us transform integral (19), using Ostrogradskii's relation

$$\frac{\partial V_1}{\partial x} = - \iiint_{T_1} \varrho \frac{\partial}{\partial \xi}\left(\frac{1}{r}\right) d\tau = - \iiint_{T_1} \left[\frac{\partial}{\partial \xi}\left(\varrho \frac{1}{r}\right) - \frac{1}{r}\frac{\partial \varrho}{\partial \xi}\right] d\tau =$$

$$= - \iint_{\Sigma_\delta^{M_0}} \frac{\varrho}{r} \cos \alpha \, d\sigma + \iiint_{T_1} \frac{1}{r} \frac{\partial \varrho}{\partial \xi} d\tau,$$

where $\Sigma_\delta^{M_0}$ is the surface of a sphere, bounding volume T_1, and α is the angle between the outer normal to the surface $\Sigma_\delta^{M_0}$ and the x-axis. The first component is a differentiable function at the point M_0, since M_0 lies inside $\Sigma_\delta^{M_0}$. The second component in the neighbourhood of point M_0 is also a differentiable function, since function ϱ has a derivative in T_1. Hence it follows that at the point M_0 a second derivative of the function V_1 exists. Let us calculate it:

$$\frac{\partial}{\partial x}\left(\frac{\partial V_1}{\delta x}\right) = - \iint_{\Sigma_\delta^{M_0}} \varrho \frac{\partial}{\partial x}\left(\frac{1}{r}\right) \cos \alpha \, d\sigma + \iiint_{T_1} \frac{\partial}{\partial x}\left(\frac{1}{r}\right) \frac{\partial \varrho}{\partial \xi} d\tau.$$

The following relation holds for the second component at point M_0:

$$\left| \iiint_{T_1} \frac{\partial}{\partial x}\left(\frac{1}{r}\right) \frac{\partial \varrho}{\partial \xi} d\tau \right| < C_1 \iiint_{T_1} \frac{d\tau}{r^2} = C_1 \, 4 \pi \delta \qquad (20)$$

if

$$\left| \frac{\partial \varrho}{\partial \xi} \right| < C_1.$$

Applying the mean value theorem to the surface integral, we obtain:

$$- \iint_{\Sigma_\delta^{M_0}} \varrho \frac{\partial}{\partial x}\left(\frac{1}{r}\right) \cos \alpha \, d\sigma = - \iint_{\Sigma_\delta^{M_0}} \varrho \frac{\cos^2 \alpha}{r^2} d\sigma = - \varrho^* \frac{4\pi}{3}.$$

Here ϱ^* is the value of the density at some point of $\Sigma_\delta^{M_0}$,

$$-\frac{\partial}{\partial x}\left(\frac{1}{r}\right)=\frac{x-\xi}{r^3}=-\frac{1}{r^2}\cos\alpha$$

and, moreover,

$$\iint\limits_{\Sigma_\delta^{M_0}}\frac{\cos^2\alpha}{r^2}\,d\sigma=\frac{1}{3}\iint\limits_{\Sigma_\delta^{M_0}}\frac{1}{r^2}(\cos^2\alpha+\cos^2\beta+\cos^2\gamma)\,d\sigma=\frac{4}{3}\pi.$$

Transition to a limit as $\delta\to 0$ gives:

$$\lim_{\delta\to 0}\frac{\partial^2 V_1}{\partial x^2}=\lim_{\delta\to 0}\Big[-\iint\limits_{\Sigma_\delta^{M_0}}\varrho\frac{\partial}{\partial x}\left(\frac{1}{r}\right)\cos\alpha\,d\sigma\Big]=-\frac{4\pi}{3}\varrho(M_0). \tag{21}$$

The equation

$$\frac{\partial^2 V}{\partial x^2}=\frac{\partial^2 V_1}{\partial x^2}+\frac{\partial^2 V_2}{\partial x^2}$$

is true for any δ and the left-hand side of it is not dependent on δ, therefore

$$\frac{\partial^2 V}{\partial x^2}=\lim_{\delta\to 0}\left(\frac{\partial^2 V_1}{\partial x^2}+\frac{\partial^2 V_2}{\partial x^2}\right)=-\frac{4\pi}{3}\varrho(M)+\lim_{\delta\to 0}\iiint\limits_{T_2}\varrho\frac{\partial^2}{\partial x^2}\left(\frac{1}{r}\right)d\tau. \tag{22}$$

From the existence of a second derivative $\frac{\partial^2 V}{\partial x^2}$, proved above, there follows the existence of the limit

$$\lim_{\delta\to 0}\iiint\limits_{T_2}\varrho\frac{\partial^2}{\partial x^2}\left(\frac{1}{r}\right)d\tau=\overline{\iiint\limits_{T}}\varrho\frac{\partial^2}{\partial x^2}\left(\frac{1}{r}\right)d\tau. \tag{23}$$

The latter integral is obtained by a special limiting transition, where the neighbourhoods of the point M_0 are spheres,† and is distinguished by a line above the integral in (23). Variation in the shape of these neighbourhoods can alter the value of the limit. The integral (23) must be considered as conditionally convergent. Thus,

$$\frac{\partial^2 V}{\partial x^2}(M_0)=\overline{\iiint\limits_{T}}\varrho\frac{\partial^2}{\partial x^2}\left(\frac{1}{r}\right)d\tau-\frac{4\pi}{3}\varrho(M_0). \tag{24}$$

† The limit (23) may be considered as the *principal value* of the integral.

It follows that calculation of the second derivatives of the potential by formal differentiation under the integral sign leads to an incorrect result.

Similar expressions are obtained for the derivatives $\frac{\partial^2 V}{\partial y^2}$ and $\frac{\partial^2 V}{\partial^2 z^2}$. Substituting the values of all three derivatives in the expression for the Laplacian operator, we find:

$$\nabla^2 V = \frac{\partial^2 V}{\partial x^2} + \frac{\partial^2 V}{\partial y^2} + \frac{\partial^2 V}{\partial z^2} =$$
$$= \overline{\iiint_T} \varrho \left[\frac{\partial^2}{\partial x^2}\left(\frac{1}{r}\right) + \frac{\partial^2}{\partial y^2}\left(\frac{1}{r}\right) + \frac{\partial^2}{\partial z^2}\left(\frac{1}{r}\right)\right] d\tau - 4\pi\varrho(M_0) =$$
$$= -4\pi\varrho(M_0), \tag{25}$$

since $1/r$ is a harmonic function.*

Thus, the volume potential satisfies Poisson's equation

$$\nabla^2 V = -4\pi\varrho \text{ inside the body}$$

and Laplace's equation

$$\nabla^2 V = 0 \text{ outside the body.}$$

The inhomogeneous equation

$$\nabla^2 u = -f \tag{25'}$$

has a particular solution

$$u_0 = \frac{1}{4\pi}\iiint_T \frac{f\, d\tau}{r}.$$

if f is differentiable inside some region T.

It follows that the solution of the boundary-value problem for the inhomogeneous equation (25′) may be reduced to the solution of a similar boundary-value problem for Laplace's equation $\nabla^2 v = 0$, if the unknown function is represented in the sum $u = u_0 + v$.

* Formula (25) was established on the assumption of the differentiability of function ϱ, which is a sufficient condition. There are other less stringent conditions, but the condition of continuity of $\varrho(M)$ is insufficient, since there exist examples of continuous funtions $\varrho(M)$, for which the volume potential does not have second derivatives.

6. *Surface potentials*

As Green's fundamental theorem proves

$$U(M) = \frac{1}{4\pi} \iint_{\Sigma} \left[\frac{1}{r_{MP}} \frac{\partial u}{\partial n} - u \frac{\partial}{\partial n} \left(\frac{1}{r} \right) \right] d\sigma_P,$$

any harmonic function can be represented by means of integrals, which are surface potentials.

Let us consider the field, produced by masses, distributed on a surface and let us determine the potential of this field. The

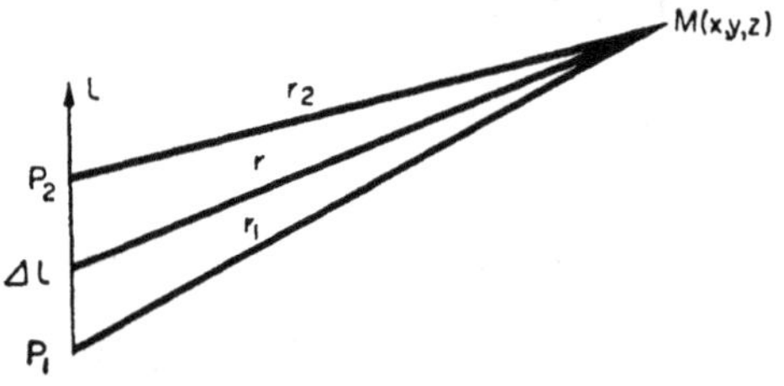

FIG. 59

surface density $\mu(P)$ at the point P of the surface Σ is the limit of the ratio of mass, in some element $d\sigma$ of the surface Σ, containing the point P, to the area of the element as $d\sigma$ contracts to the point P.* The potential of these masses may be described by the surface integral

$$V(M) = \iint_{\Sigma} \frac{\mu(P)}{r_{MP}} d\sigma_P. \tag{26}$$

This is called the *potential of a single layer.*

Another type of surface potential is the potential of a double layer. We consider a dipole, produced by two masses $-m$ and $+m$, situated at the points P_1 and P_2 a distance Δl apart (Fig. 59).

* If the masses with volume density ϱ are situated in some layer of thickness h near the surface Σ and the field is investigated at distances, great compared with $h (h/r \ll 1)$, then the effect of the thickness of the surface, generally speaking, is negligible. Therefore instead of the volume potential with density ϱ it is useful to consider the surface potential with surface density $\mu = \varrho h$.

The product $m \cdot \Delta l = N$ is called the *moment* of the dipole. The potential at some point $M(x, y, z)$ equals

$$V = \frac{m}{r_2} - \frac{m}{r_1} = m\left(\frac{1}{r_2} - \frac{1}{r_1}\right) = N\frac{1}{\Delta l}\left(\frac{1}{r_2} - \frac{1}{r_1}\right),$$

where r_1 and r_2 are the distances of point M from the points P_1 and P_2.

If Δl is small in comparison with the distance to the point

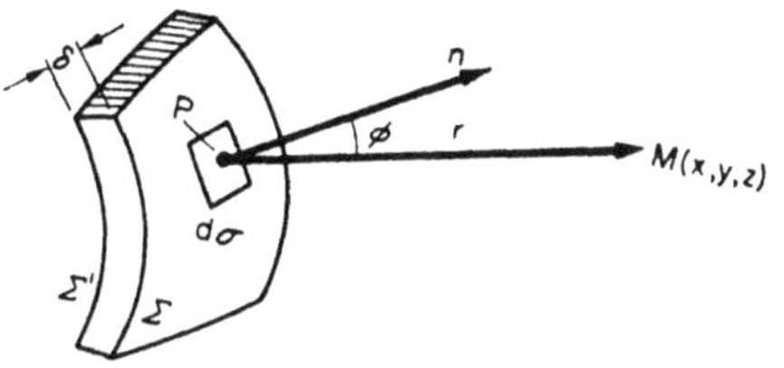

FIG. 60

M, $\left(\frac{\Delta l}{r_1} \ll 1\right)$, then using the mean value theorem, one may write:

$$V = N\frac{d}{dl}\left(\frac{1}{r}\right), \quad r = \sqrt{[(x-\xi)^2 + (y-\eta)^2 + (z-\zeta)^2]},$$

where the vector $\boldsymbol{l}$ is directed from the negative to the positive mass and r is the distance from the point $M(x, y, z)$ to some intermediate point $P(\xi, \eta, \zeta)$ of the segment Δl.

Let us calculate the derivative in the direction l

$$\frac{d}{dl}\left(\frac{1}{r}\right) = \frac{1}{r^2}\cos(\boldsymbol{r}, \boldsymbol{l}) = \frac{\cos\varphi}{r^2},$$

where vector $\boldsymbol{r}$ is directed from the dipole to the fixed point M, and φ is the angle between vector $\boldsymbol{l}$ and vector $\boldsymbol{r}$. Thus, the potential of the dipole is

$$V(M) = N\frac{\cos\varphi}{r^2}, \tag{27}$$

where N is the dipole moment.

Let masses be distributed on the two surfaces Σ and Σ' (Fig. 60), a small distance δ apart, in such a way that the mass of each element of surface Σ' is equal in magnitude but opposite in sign to the mass of a corresponding element of the surface Σ. Let us denote by $\boldsymbol{n}$ the common normal to surfaces Σ and Σ', directed from the negative to the positive masses. Passing to a limit as $\delta \to 0$, we obtain the double layer as a combination of two single layers with opposite densities, at a small distance from one another. If ν is the surface density of the moment, then the moment of an element of surface $d\sigma_P$ will equal

$$dN = \nu d\sigma_P .$$

We get the following expression for the potential of an element $d\sigma$ at the point $M(x, y, z)$:

$$\nu \frac{1}{dn_P} \left(\frac{1}{r_{MP}}\right) d\sigma_P = \nu (P) \frac{\cos \varphi_1}{r_{MP}} d\sigma_P ,$$

where $\varphi_1 = (\widehat{\boldsymbol{n}\, \overrightarrow{PM}})$.

We call the integral

$$W (M) = - \iint\limits_{\Sigma} \frac{d}{dn_P} \left(\frac{1}{r_{MP}}\right) \nu (P)\, d\sigma_P \tag{28}$$

the *potential of a double layer*. This definition, obviously, corresponds to the case where the outer side of the surface is positive, and the inner negative.

It is obvious that

$$W = \iint\limits_{\Sigma} \frac{\cos \varphi}{r_{MP}^2} \nu (P)\, d\sigma_P ,$$

where φ is the angle between the inner normal and the direction of the point of the surface P to the fixed point M.

The potentials of a single and double layer in the case of two independent variables have the form

$$V = \int\limits_{C} \mu (P) \ln \frac{1}{r_{MP}} ds , \tag{29}$$

$$W = - \int\limits_{C} \nu (P) \frac{d}{dn_P} \left(\ln \frac{1}{r_{MP}}\right) ds = \int\limits_{C} \frac{\cos \varphi}{r_{MP}} \nu (P)\, ds , \tag{30}$$

where C is some line, μ is the linear density of the single layer, ν is the density of the moment of a linear double layer, φ is the angle between the inner normal to the line C and the direction to the fixed point M.

If the point $M(x, y, z)$ lies outside the surface Σ, then the function under the integral sign and its derivatives with respect to x, y, z of any order in the relations

$$V(M) = \iint_{\Sigma} \mu(P) \frac{1}{r_{MP}} d\sigma_P ,$$

$$W(M) = -\iint_{\Sigma} \nu(P) \frac{d}{dn_P} \left(\frac{1}{r_{MP}}\right) d\sigma_P$$

are continuous with respect to variables x, y, z. Therefore at points, lying outside the surface Σ, the derivatives of surface potentials may be calculated by differentiating under the integral sign. Hence because of the principle of superposition it follows that surface potentials satisfy Laplace's equation everywhere outside the attracting masses. Functions (29) and (30), obviously, satisfy Laplace's equation in two dimensions.

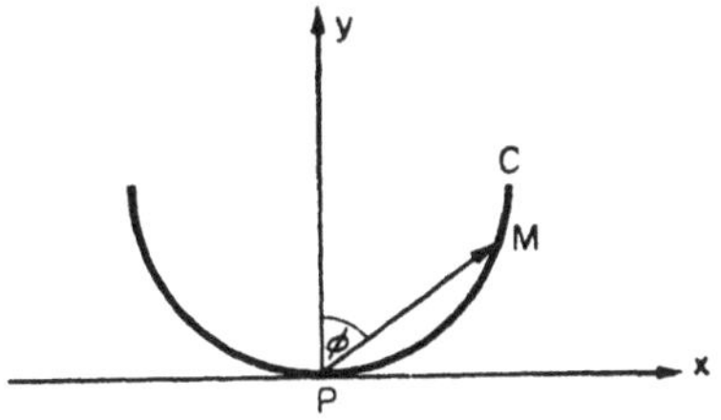

FIG. 61

Surface potentials at points of the surface Σ are represented by improper integrals. Let us show that if the surface has a continuous curvature, then the potential of a double layer exists at points of this surface. We give the proof for two dimensions:

$$W = \int_C \frac{\cos \varphi}{r} \nu \, ds .$$

We consider a curve on the plane (x, y) and choose the origin of coordinates at point P. We choose the x-axis along the tangent, and the y-axis along the normal at this point (Fig. 61). The equation of the curve in the neighbourhood of point P is

$$y = y(x) .$$

The curve has, by assumption, a continuous curvature, i. e. $y(x)$ has a continuous second derivative. Therefore

$$y(x) = y(0) + xy'(0) + \frac{x^2}{2} y''(\vartheta x) \quad (0 < \vartheta < 1) ,$$

or because of the choice of coordinate axes

$$y(x) = \frac{1}{2} x^2 y''(\vartheta x) .$$

Hence we have:

$$r = \sqrt{(x^2 + y^2)} = \sqrt{\left\{x^2 + x^4 \left[\frac{y''(\vartheta x)}{2}\right]^2\right\}} = x\sqrt{\left\{1 + x^2\left[\frac{y''(\vartheta x)}{2}\right]^2\right\}},$$

$$\cos\varphi = \frac{y}{r} = \frac{xy''(\vartheta x)}{2\sqrt{\left\{1 + x^2\left[\frac{y''(\vartheta x)}{2}\right]^2\right\}}}$$

and

$$\frac{\cos\varphi}{r} = \frac{y''(\vartheta x)}{2\left\{1 + x^2\left[\frac{y''(\vartheta x)}{2}\right]^2\right\}} .$$

From the expression for the curvature

$$K = \frac{y''}{(1 + y'^2)^{3/2}}$$

it follows

$$y''(0) = K(P) .$$

Therefore

$$\lim_{MP \to 0} \frac{\cos\varphi}{r} = \frac{1}{2} K(P) ,$$

which proves the continuity of $(\cos\varphi)/r$ along the curve, and thus the existence of a double layer potential at points of the curve C for the bounded function ν.

The potential of a double layer also exists at points of the surface in three dimensions, because the function

$$\cos\varphi/r^2$$

has an integrable singularity of order $1/r$. The existence of a potential of a single layer certainly exists.

7. *Surfaces and Lyapunov's curves*

The assumption of finiteness of the curvature appears unnecessary for the existence of surface potentials.

The potentials of a single and double layer at points of the surface $\sum$ are improper integrals. Let us show that these integrals converge for a given class of surfaces called Lyapunov surfaces, if the moment density is bounded $|\nu(P)| < C$, where C is some constant.

The surface is called a *Lyapunov surface*, if the following conditions are fulfilled:

1. At every point of surface Σ there exists a unique normal (tangent surface).

2. There exists a number $d > 0$ such that straight lines, parallel to the normal at some point P of the surface Σ intersect a part Σ'_P of the surface Σ, lying inside a sphere of radius d with centre P, not more than once. These parts of the surface Σ'_P are called the *Lyapunov neighbourhoods*.

3. The angle $\gamma(P, P') = (\widehat{n_P, n_{P'}})$, formed by the normals at points P and P', satisfies the following relation:

$$\gamma(P, P') < Ar^\delta, \tag{31}$$

where r is the distance between the points P and P', A is some constant and $0 < \delta \leqslant 1$.

Let P_0 be some point of the surface Σ. We choose a cartesian system of coordinates, locating the origin of coordinates at point P_0 and directing the z-axis along the outer normal. The plane (x, y) therefore coincides with the tangent surface. Because of condition 2. there exists ϱ_0 such that the equation of the surface Σ can be represented in the form

$$z = f(x, y)^{*)} \tag{32}$$

for

$$\varrho = \sqrt{(x^2 + y^2)} < \varrho_0. \tag{33}$$

Let us denote by Σ'_{P_0} the neighbourhood of point P_0 on the surface Σ, defined by (32) and (33). Let us establish some results for $f(x, y)$ and its derivatives.

From the existence of a normal at every point of the surface (condition 1) it follows that $f(x, y)$ is differentiable. The direction cosines of the normal (outer) are given by the relations:

$$\cos\alpha = \frac{z_x}{\sqrt{(1 + z_x^2 + z_y^2)}};$$

$$\cos\beta = \frac{z_y}{\sqrt{(1 + z_x^2 + z_y^2)}}; \quad \cos\gamma = \frac{1}{\sqrt{(1 + z_x^2 + z_y^2)}}.$$

* We note that if $f(x, y)$ has continuous second derivatives in the neighbourhood of point P_0, then the surface $z = f(x, y)$ satisfies the Lyapunov relations. Thus, surfaces with continuous curvature are Lyapunov surfaces.

Because of the choice of our coordinate system $z_x(P_0) = 0$, $z_y(P_0) = 0$. We shall assume that the surface is so small (ϱ_0 is so small), that

$$1 \geqslant \cos\gamma = \frac{1}{\sqrt{(1 + z_x^2 + z_y^2)}} > \frac{1}{2}. \tag{34}$$

Let us denote by $\boldsymbol{n}'_P$ the projection of vector $\boldsymbol{n}_P$ on the plane (x, y) and by α', β' the angles, formed by vector $\boldsymbol{n}'_P$ with the x and y axes. It is obvious that

$$\cos\alpha = \sin\gamma\cos\alpha', \quad \cos\beta = \sin\gamma\sin\alpha'.$$

Because of condition 3

$$\sin\gamma < \gamma < Ar^{\delta}_{PP_0},$$

then

$$|\cos\alpha| < Ar^{\delta}_{PP_0}, \quad |\cos\beta| < Ar^{\delta}_{PP_0}, \tag{35}$$

and since $z_x = \dfrac{\cos\alpha}{\cos\gamma}$, $z_y = \dfrac{\cos\beta}{\cos\gamma}$, where $\dfrac{1}{\cos\gamma} < 2$, then

$$|z_x| < 2\,Ar^{\delta}_{PP_0}, \quad |z_y| < 2\,Ar^{\delta}_{PP_0}.$$

Making use of Taylor's theorem for the function $z = f(x, y)$ in the neighbourhood of point $P_0\,(0, 0)$, we have:

$$z(x, y) = z\,(0, 0) + xz_x\,(\bar{x}, \bar{y}) + yz_y\,(\bar{x}, \bar{y}),$$

where

$$0 \leqslant \bar{x} \leqslant x, \quad 0 \leqslant \bar{y} \leqslant y,$$

from which it follows that

$$|z\,(x, y)| < 4\,Ar^{1+\delta}_{PP_0}. \tag{36}$$

The results (34), (36) enable us to prove that at points, lying on the surface Σ, the potential of a double layer

$$W\,(M) = \iint\limits_{\Sigma} \frac{\cos\varphi}{r^2_{MP}}\, v\,(P)\, d\sigma_P \tag{28}$$

is a convergent improper integral, if Σ is a Lyapunov surface. Let $M = P_0$ be a point of the surface Σ. Choosing a coordinate system, as was indicated above, let us represent the equation of the surface Σ in the neighbourhood of the point P_0 by the function

$$z = f\,(x, y),$$

satisfying relations (34) and (36). We calculate $\cos\varphi$, where φ is the angle between the direction of the inner normal at point $P(\xi, \eta, \zeta)$ and the direction $\overrightarrow{PP_0}$. It is easy to see that

$$|\cos\varphi| = \left|\frac{\xi}{r}\cos\alpha + \frac{\eta}{r}\cos\beta + \frac{\zeta}{r}\cos\gamma\right| \leqslant |\cos\alpha| +$$

$$+ |\cos\beta| + \frac{|\zeta|}{r} \leqslant Ar^{\delta}_{PP_0} + Ar^{\delta}_{PP_0} + 4Ar^{\delta}_{PP_0} = 6Ar^{\delta}_{PP_0}$$

and

$$\left|\frac{\cos\varphi}{r^2}\right| \leqslant 6A\frac{1}{r^{2-\delta}} \quad (0 < \delta \leqslant 1). \tag{37}$$

Now we represent W in the form of a sum of two integrals

$$W = W_1 + W_2,$$

where W_1 is the integral over surface Σ'_{P_0}, containing the singular point P_0, and integral W_2 is taken over the remaining part of the surface $\Sigma - \Sigma'_{P_0}$. Since the function under the integral sign in the integral W_2 nowhere tends to infinity, the convergence of the integral W_1 is sufficient to ensure the convergence of the integral W. Since

$$d\sigma = \frac{d\xi\, d\eta}{\cos\gamma} = \frac{\varrho\, d\varrho\, d\theta}{\cos\gamma},$$

where $\varrho = \sqrt{(\xi^2 + \eta^2)}$, θ are polar coordinates on the plane (xy), then transformation of the variables in integral (28) gives:

$$W_1 = \iint\limits_{\Sigma_{P_0}} \frac{\cos\varphi}{r^2_{PP_0}}\nu(P)\, d\sigma_P = \int_0^{\varrho_0}\int_0^{2\pi} \frac{\cos\varphi}{r^2_{PP_0}}\nu(P)\frac{1}{\cos\gamma}\varrho\, d\varrho\, d\theta.$$

Because of (34), (36) and (37) the function under the integral sign satisfies the relation:

$$\left|\nu(P)\frac{\cos\varphi}{r^2}\frac{1}{\cos\gamma}\right| \leqslant \bar{F} = \frac{12AC}{\varrho^{2-\delta}},$$

since $\varrho < r$.

Such a form of the majorant function $\bar{F}$ ensures the convergence of an improper integral in the case of two independent variables (see sub-section 3).

It is easy to establish that for a Lyapunov surface the potential of a single layer

$$V(M) = \iint_{\Sigma} \frac{1}{r_{MP}} \mu(P)\, d\sigma_P$$

also converges at points of the surface. One should note that this convergence occurs for surfaces of a more extensive class.

In the case of two independent variables the potentials of single and double layers converge at points of the curve [see (29) and (30)], if these potentials are considered as Lyapunov curves, defined by conditions, similar to conditions 1—3 for Lyapunov surfaces.

8. *Discontinuity of the potential of a double layer*

Let us show that the potential of a double layer at some point P_0 lying on the surface Σ is a discontinuous function, which obeys the relations

$$\left.\begin{aligned} W_i(P_0) &= W(P_0) + 2\pi\nu(P_0), \\ W_e(P_0) &= W(P_0) - 2\pi\nu(P_0), \end{aligned}\right\} \tag{38}$$

where $W_i\ (P_0)$ is the limiting value of the potential of a double layer approaching the point P_0 from the inner side, and $W_e\ (P_0)$ is the limiting value from the outer side of the surface.*

In the case of two independent variables the appropiate formulae are

$$\left.\begin{aligned} W_i(P_0) &= W(P_0) + \pi\nu(P_0), \\ W_e(P_0) &= W(P_0) - \pi\nu(P_0). \end{aligned}\right\} \tag{39}$$

For simplicity we limit ourselves to a proof of these formulae for two independent variables.

The potential of a double layer in this case is expressed by the integral

$$W(M) = \int_C \frac{\cos\varphi}{r_{MP}} \nu(P)\, ds_P .$$

* If Σ is an open surface, then the inner side may be conditionally determined by deciding which normal at point P_0 is called "inner" and which "outer". One should realize that in the case of an open surface the potential of a double layer is defined only for two-sided surfaces.

Let us consider some element of the curve ds, whose extremities are the points P and P_1. We draw through P an arc of a circle of radius MP with centre at point M which intersects the line MP_1 at point Q, then correct to the lowest order we can write (Fig. 62):

$$ds \cos \varphi = d\sigma\,, \quad \frac{d\sigma}{r} = d\omega\,,$$

where $ds = \breve{PP_1}$, $d\sigma = \breve{PQ}$, $d\omega$ is the angle subtended by the curve ds at the point M. The sign of $d\omega$ is the same as that of $\cos \varphi$, so that: $d\omega > 0$, if φ (the angle between the inner normal at point P and the vector $\overrightarrow{PM}$) is less than $\pi/2$ and $d\omega < 0$, if $\varphi > \pi/2$. If $d\omega > 0$, i.e. $\varphi < \pi/2$, then MP joins M to the "inner" side of curve C; for $d\omega < 0$, $\varphi > \pi/2$, MP joins M to the "outer" side of the curve. It follows that the angle subtended at M by some curve P_1P_2 equals the angle P_1MP_2 swept out by the line MP as P moves along the curve from P_1 to P_2.

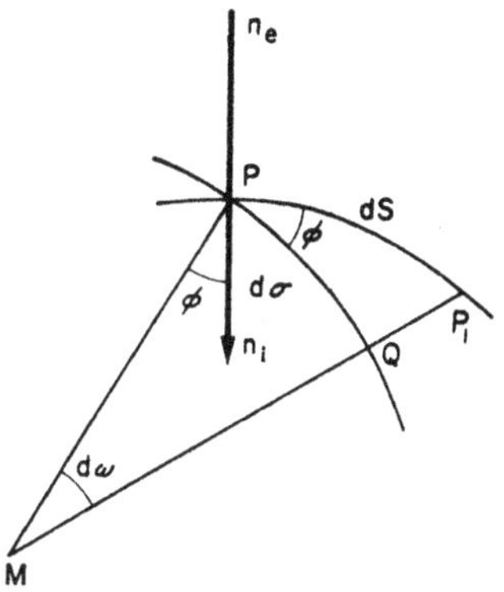

FIG. 62

Let us consider the potential of a double layer W° with constant density $\nu = \nu_0 = \text{const.}$ on a closed curve C. The line MP sweeps out an angle

$$\Omega = \begin{cases} 2\pi, \text{ if the point } M \text{ lies inside the curve } C, \\ \pi, \text{ if the point } M \text{ lies on the curve } C, \\ 0, \text{ if the point } M \text{ lies outside the curve } C, \end{cases}$$

as the point P passes around the curve C. Hence we obtain for the potential W°:

$$W^\circ = \nu_0 \Omega = \begin{cases} 2\pi\nu_0, \text{ if point } M \text{ lies inside curve } C, \\ \pi\nu_0, \text{ if point } M \text{ lies on curve } C, \\ 0, \text{ if point } M \text{ lies outside curve } C. \end{cases}$$

Thus, the potential of a constant density double layer is a partly-continuous function in which

$$\left.\begin{aligned} W_i^0 &= W_C^0 + \pi\nu_0, \\ W_e^0 &= W_C^0 - \pi\nu_0, \end{aligned}\right\} \tag{41}$$

where W_i°, W_C°, W_e° are the values of the potential inside, on and outside the curve C.

Similarly in the case of three independent variables we have:

$$\frac{d\sigma \cos \varphi}{r^2} = d\omega, \tag{42}$$

where $d\omega$ is the solid angle, subtended by the element $d\sigma$ of the closed surface Σ. Let $d\sigma'$ be an element of spherical surface, obtained by the intersection of a sphere, described by the radius MP from the point M, with a cone, having apex at the point M and resting on an element of surface $d\sigma$. The element of surface $d\sigma' = d\sigma \cos \varphi$. Hence formula (42) follows. The observation made above concerning the sign of $d\omega$ still holds good, leading us to the relations

$$W^\circ = \nu_0 \Omega = \begin{cases} 4\pi\nu_0, & \text{if point } M \text{ lies inside surface } \Sigma, \\ 2\pi\nu_0, & \text{if point } M \text{ lies on surface } \Sigma, \\ 0, & \text{if point } M \text{ lies outside surface } \Sigma, \end{cases}$$

and also to the relations

$$\left. \begin{aligned} W_i^0 &= W_\Sigma^0 + 2\pi\nu_0, \\ W_e^0 &= W_\Sigma^0 - 2\pi\nu_0, \end{aligned} \right\} \tag{41'}$$

where W_i^0, W_e^0 are values of the potential W° inside and on the outside of surface Σ, and W_Σ^0 is the value of W° on Σ.

We consider now the potential of a double layer with variable density and prove that at points of continuity of density formulae, similar to (41) and (41′) hold.

Let P_0 be a point of the surface Σ, at which $\nu(P)$ is continuous. We introduce the potential of a double layer W° with constant density $\nu_0 = \nu(P)$ and consider the function

$$I(M) = W(M) - W^0(M) = \iint_\Sigma [\nu(P) - \nu_0] \frac{\cos \varphi}{r_{MP}^2} d\sigma_P.$$

Let us prove that the function I is continuous at point P_0. In order to do this it is sufficient to prove the uniform convergence of the integral $I(M)$ at the point P_0. We choose some number $\varepsilon > 0$. From the continuity of the function $\nu(P)$ at point P_0 it follows that for any number $\eta > 0$ it is possible to find Σ_1, a neighbourhood of point P_0 on surface Σ such that

$$|\nu(P) - \nu(P_0)| < \eta,$$

if $P \subset \Sigma_1$. We write the integral I in the form of a sum

$$I = I_1 + I_2,$$

where I, is the contribution from Σ_1, and I_2 from $\Sigma_2 = \Sigma - \Sigma_1$. From the definition of Σ_1 it follows:

$$|I_1| < \eta B_\Sigma,$$

where B_Σ is a constant, given by the relation

$$\iint_\Sigma \frac{|\cos\varphi|}{r_{MP}^2} d\sigma_P \leqslant B_\Sigma \tag{43}$$

for all possible positions of point M, and not depending on the choice of surface Σ_1. Details concerning this constant will be given below.

Choosing $\eta = \varepsilon/B_\Sigma$, we satisfy ourselves that for any $\varepsilon > 0$ it is possible to find Σ_1, containing P_0 such that for any position

$$|I_1(M)| < \varepsilon$$

of point M. Hence the uniform convergence of integral $I(M)$ at point P_0 follows, and also its continuity at this point.

If $W_i\ (P_0)$ and $W_e\ (P_0)$ are the limits of the potential $W(M)$ as $M \to P_0$ from the inner and outer side of the surface Σ then

$$W_i(P_0) = W_i^0\ (P_0) + I\ (P_0) = W^0\ (P_0) + I\ (P_0) + 2\pi\nu_0 = \\ = W\ (P_0) + 2\pi\nu\ (P_0)$$

and similarly

$$W_e(P_0) = W\ (P_0) - 2\pi\nu\ (P_0)\,.$$

The validity of (43) is established.

The proof outlined above is valid for surfaces, satisfying the condition (43). For a convex surface where each line from the point M intersects the surface not more than twice, $B_\Sigma \leqslant 8\pi$; for surfaces consisting of a finite number of convex parts, B_Σ is also bounded. Thus, our proof holds for a very wide class of surfaces.

All the arguments developed above hold good for functions of two independent variables. In this case (41) takes the form

$$W_i(P_0) = W(P_0) + \pi\nu(P_0);$$

$$W_e(P_0) - W(P_0) - \pi\nu(P_0).$$

9. *Properties of the potential of a single layer*

In contrast to the potential of a double layer the potential of a single layer

$$V(M) = \iint_{\Sigma} \frac{1}{r_{MP}} \mu(P)\, d\sigma_P \tag{26}$$

is continuous at points of the surface Σ. In order to satisfy ourselves of this, it is sufficient to establish the uniform convergence of integral $V(M)$ at points of the surface Σ.

In fact, let P_0 be some point of the surface Σ. We represent the potential V in the form of a sum

$$V = \iint_{\Sigma_1} \frac{1}{r_{MP}} \mu(P)\, d\sigma_P + \iint_{\Sigma_2} \frac{1}{r_{MP}} \mu(P)\, d\sigma_P = V_1 + V_2,$$

where Σ_1 is a sufficiently small part of the surface Σ, enclosed by a sphere of radius δ with centre at the point P_0. We define the value of δ more exactly later.

Let us consider a coordinate system with origin at the point P_0, the z-axis of which is directed along the outer normal at P_0. Let $M(x, y, z)$ be an arbitrary point, such that $MP_0 < \delta$. Let us denote by Σ_1' the projection of Σ_1 on the plane (x, y) and by $K_{2\delta}^{M'}$ a circle of radius 2δ with centre at the point $M'(x, y, 0)$, entirely containing the region Σ_1'. Assuming the bounded nature of the function

$$|\mu(P)| < A$$

and taking into consideration that

$$d\sigma = \frac{d\sigma'}{\cos\gamma} = \frac{d\xi\, d\eta}{\cos\gamma}$$

and

$$r = \sqrt{[(x-\xi)^2 + (y-\eta)^2 + (z-\zeta)^2]} \geqslant \sqrt{[(x-\xi)^2 + (y-\eta)^2]} = \varrho,$$

we obtain:

$$V_1(M) < A \iint\limits_{\Sigma_1} \frac{d\sigma}{r_{MP}} = A \iint\limits_{\Sigma_1'} \frac{d\sigma'/\cos\gamma}{\sqrt{[(x-\xi)^2+(y-\eta)^2+(z-\zeta)^2]}} \leqslant$$

$$\leqslant 2A \iint\limits_{\Sigma_1'} \frac{d\xi\, d\eta}{\sqrt{[(x-\xi)^2+(y-\eta)^2]}} \leqslant 2A \iint\limits_{K_{2\delta}^{M'}} \frac{d\xi\, d\eta}{\sqrt{[(x-\xi)^2+(y-\eta)^2]}},$$

if δ is so small that $\cos\gamma > 1/2$.

Let us introduce the polar system of coordinates (ϱ, φ) in the plane (x, y) with origin at the point M'. Then it is possible to write

$$V_1(M) < 2A \iint\limits_{K_{2\delta}^{M'}} \frac{d\xi\, d\eta}{\sqrt{[(x-\xi)^2+(y-\eta)^2]}} = 2A \int_0^{2\delta}\int_0^{2\pi} \frac{\varrho\, d\varrho\, d\varphi}{\varrho} = 8A\pi\delta.$$

Choosing $\delta = \varepsilon/8\pi A$, we have:

$$V_1(M) < \varepsilon,$$

if $MP_0 < \delta$. Hence $V(M)$ converges uniformly at each point $P_0 \subset \Sigma$ and is a continuous function at this point.

Normal derivatives of the potential of a single layer on the surface Σ have discontinuities of a type similar to the potential of a double layer.

Outer and inner normal derivatives of function V, dV/dn_e and dV/dn_i are defined in the following way. Let P_0 be some point of Σ. From the point P_0 we draw the z-axis which can be directed either along the outer, or along the inner normal.

We consider the derivative dV/dz at some point M of the z-axis and denote by $(dV/dz)_i$ and $(dV/dz)_e$ the limits of the derivative dV/dz as M tends to the point P_0, from the inner or outer side of the surface Σ. If the z-axis is directed along the outer (inner) normal, then these values are called the inner and outer limiting values of the derivative with respect to the outer (inner) normal at the point P_0.*

* The limit of the difference relation $\dfrac{V(M) - V(P_0)}{MP_0}$ as $M \to P_0$ equals the limit from without for the derivative with respect to the outer normal or the limit from within for the derivative with respect to the inner normal, depending from which side M approaches P_0.

Let us investigate the discontinuity of the inner normal derivative of the potential of a single layer on Σ. The derivative dV/dz at the point M of the z-axis, directed along the inner normal, equals

$$\frac{dV}{dz}(M) = \iint_{\Sigma} \mu(P) \frac{\partial}{\partial n}\left(\frac{1}{r_{MP}}\right) d\sigma_P = \iint_{\Sigma} \frac{\cos\psi}{r_{MP}^2} \mu(P)\, d\sigma_P, \qquad (44)$$

where ψ is the angle between the z-axis and the vector MP. Let us draw (Fig. 63) the normal PQ at P and the straight line PN, parallel to the z-axis (normal at point P_0), and denote by θ the angle NPQ, equal to the angle between the normals at points P and P_0.* The expression for the potential of a double layer $W(M)$ contains the factor $(\cos\varphi)/r^2$, where $\varphi = \measuredangle MPQ$. Since the angle MPN equals $\pi - \psi$, then

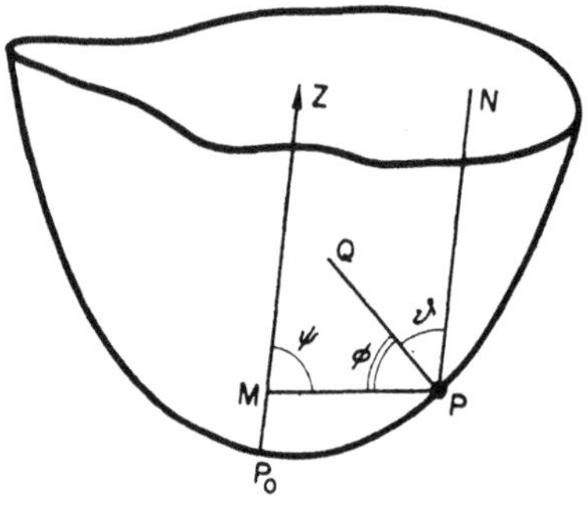

FIG. 63

$$\cos(\pi - \psi) = \cos\varphi\cos\theta + \sin\varphi\sin\theta\cos\Omega = -\cos\psi,$$

where Ω is the azimuthal angle of the side PQ**. Hence it follows that

$$\frac{\partial V}{\partial z}(M) = -\iint (\mu\cos\theta)\frac{\cos\varphi}{r^2} d\sigma - \iint_{\Sigma} \mu\sin\theta\cos\Omega\frac{\sin\varphi}{r^2} d\sigma =$$
$$= -W_1 - I(M), \qquad (45)$$

* It is obvious that θ and $\sin\theta$ tend to zero as $P \to P_0$. If the surface has a finite curvature in the neighbourhood of the point P_0, i.e. if its equation can be written in the form

$$z = f(x, y),$$

where $f(x, y)$ has second derivatives, then $\sin\theta$ will be a differentiable function of x, y and consequently

$$\sin\theta < Ar$$

(for a Lyapunov surface $\sin\theta < Ar^{\delta}$).

** If the direction PQ is taken as the axes of a new spherical system then this formula coincides with formula (13) on page 357.

where $W_1(M)$ is the potential of a double layer with density $\mu_1 = \mu \cos\theta$, having discontinuities on the surface Σ. It is obvious that the integral $I(M)$ is a function continuous at the point P_0, since $I(M)$ converges uniformly at this point (see the footnote, page 393).

Returning to formula (45), we see:

$$\left.\begin{aligned}\left(\frac{\partial V}{\partial z}\right)_i &= -W_1(P_0) - 2\pi\mu_1(P_0) - I(P_0),\\ \left(\frac{\partial V}{\partial z}\right)_e &= -W_1(P_0) + 2\pi\mu_1(P_0) - I(P_0).\end{aligned}\right\} \tag{46}$$

We note

$$\begin{aligned}\left(\frac{\partial V}{\partial z}\right)_0 &= -W_1(P_0) - I(P_0) = \\ &= \left[-\iint\limits_{\Sigma}(\mu\cos\theta)\frac{\cos\varphi}{r^2}\,d\sigma - \iint\limits_{\Sigma}\mu\sin\theta\cos\Omega\frac{\sin\varphi}{r^2}\,d\sigma\right]_{M=P_0} = \\ &= \iint\limits_{\Sigma}\mu\frac{\cos\psi_0}{r^2_{P_0P}}\,d\sigma\,,\end{aligned}$$

where ψ_0 is the angle between the z-axis and the vector $\overrightarrow{P_0P}$.

Observing that $\mu_1(P_0) = \mu(P_0)$, we find:

$$\left.\begin{aligned}\left(\frac{\partial V}{\partial n_i}\right)_i &= \left(\frac{\partial V}{\partial n_i}\right)_0 - 2\pi\mu(P_0),\\ \left(\frac{\partial V}{\partial n_i}\right)_e &= \left(\frac{\partial V}{\partial n_i}\right)_0 + 2\pi\mu(P_0),\end{aligned}\right\} \tag{47}$$

since by definition the z-axis is directed along the inner normal. If the z-axis is directed along the outer normal, then the sign of $\cos\psi$ changes and we obtain:

$$\left.\begin{aligned}\left(\frac{\partial V}{\partial n_e}\right)_i &= \left(\frac{\partial V}{\partial n_e}\right)_0 + 2\pi\mu(P_0),\\ \left(\frac{\partial V}{\partial n_e}\right)_e &= \left(\frac{\partial V}{\partial n_e}\right)_0 - 2\pi\mu(P_0).\end{aligned}\right\} \tag{48}$$

For the case of two dimensions similar formulae occur with 2π replaced by π.

10. *Application of surface potentials to the solution of boundary-value problems*

Surface potentials give a simple analytical device for solving boundary-value problems. Let us consider the internal boundary-value problems for some contour C:

Find the function u, harmonic in region T, bounded by contour C and satisfying on C the boundary conditions

$$u|_C = f \quad \textit{the first boundary-value problem}$$

$$\left.\frac{\partial u}{\partial n}\right|_C = f \quad \textit{the second boundary-value problem.}$$

Similar conditions define the external boundary-value problem.*

We shall look for a solution of the internal first boundary-value problem in the form of the potential of a double layer

$$W(M) = \int_C \frac{\cos \varphi}{r_{MP}} \nu(P)\, ds_P = -\int_C \frac{d}{dn_P}\left(\ln \frac{1}{r_{MP}}\right) \nu(P)\, ds_P.$$

For any choice of $\nu(P)$ the function $W(M)$ satisfies Laplace's equation inside C. Function $W(M)$ is discontinuous on the contour C. To fullfil the boundary condition, it is obviously necessary that

$$W_i(P_0) = f(P_0).$$

Remembering formula (41) we obtain an equation for $\nu(P)$

$$\pi\nu(P_0) + \int_C \frac{\cos \varphi}{r_{P_0P}} \nu(P)\, ds_P = f(P_0). \tag{49}$$

If we denote by s_0, s the arcs of the contour C, corresponding to points P_0 and P, then equation (49) may be written in the form

$$\pi\nu(s_0) + \int_0^L K(s_0, s)\, \nu(s)\, ds = f(s_0), \tag{50}$$

* In the statement of both internal and external second boundary-value problems, we will assume the normal to be internal in the boundary condition.

where L is the length of the contour C and

$$K(s_0, s) = -\frac{d}{dn_P}\left(\ln\frac{1}{r_{PP_0}}\right) = \frac{\cos\varphi}{r_{PP_0}} \tag{51}$$

is the kernel of this integral equation, which is an integral equation of the Fredholm type of the second kind.*
A similar equation is obtained for the external problem

$$-\pi\nu(s_0) + \int_0^L K(s_0, s)\,\nu(s)\,ds = f(s_0). \tag{52}$$

For the second boundary-value problem the equations

$$-\pi\mu(s_0) + \int_0^L K_1(s_0, s)\,\mu(s)\,ds = f(s_0) \quad \text{(internal problem)}, \tag{53}$$

$$\pi\mu(s_0) + \int_0^L K_1(k_0, s)\,\mu(s)\,ds = f(s_0) \quad \text{(external problem)}, \tag{54}$$

are obtained, where

$$K_1(s_0, s) = \frac{\partial}{\partial n_{P_0}}\left(\ln\frac{1}{r_{PP_0}}\right) = \frac{\cos\psi_0}{r_{PP_0}}\text{**}, \tag{55}$$

if its solution is sought in the form of the potential of a single layer

$$u(M) = \int_C \ln\frac{1}{r_{MP}}\,\mu(P)\,ds_P.$$

Problems connected with the solubility of these equations will be considered in sub-section 11 of the present section.

* Integral equations containing integrals with constant limits are called *Fredholm* equations:

$$\int_a^b K(x, s)\,\varphi(s)\,ds = f(x) \text{ — first type,}$$

$$\varphi(x) + \int_a^b K(x, s)\,\varphi(s)\,ds = f(x) \text{ — second type.}$$

** It is easily seen that $K(s_0, s) = K_1(s, s_0)$. The kernel K_1 is called the *adjoint* of K and their corresponding equations are called adjoint integral equations.

Let us consider boundary-value problems for some of the simplest regions, for which the appropriate integral equations are easily solved.

1. The first boundary-value problem for a circle. If the contour

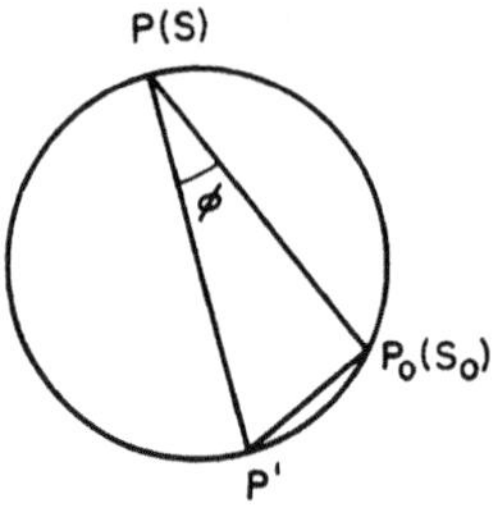

FIG. 64

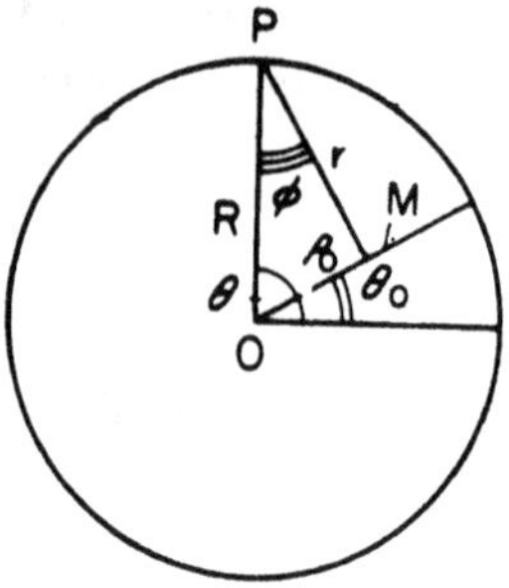

FIG. 65

C is the circumference of radius R, then the inner normal at P is directed along the diameter and

$$\frac{\cos \varphi}{r} = \frac{1}{2R},$$

since φ is the angle P_0PP' (Fig. 64). The integral equation for ν takes the form

$$\nu(s_0) + \frac{1}{\pi} \int_C \frac{1}{2R} \nu(s)\, ds = \frac{1}{\pi} f(s_0). \tag{56}$$

It is easy to see that its solution is the function

$$\nu(s) = \frac{1}{\pi} f(s) + A, \tag{57}$$

where A is some constant, to be determined. Substituting the assumed form of the solution in (56) we have:

$$\frac{1}{\pi} f(s_0) + A + \frac{1}{\pi} \int_C \frac{1}{2R} \left(\frac{1}{\pi} f(s) + A \right) ds = \frac{1}{\pi} f(s_0),$$

from which we find an expression for the constant A

$$A = -\frac{1}{4\pi^2 R} \int_C f(s)\, ds$$

Thus,

$$\nu(s) = \frac{1}{\pi} f(s) - \frac{1}{4\pi^2 R} \int\limits_C f(s)\, ds \tag{58}$$

is the solution of the integral equation.

The corresponding potential of the double layer equals

$$W(M) = \int\limits_C \frac{\cos\varphi}{r_{MP}} \nu(P)\, ds_P = \int\limits_C \frac{\cos\varphi}{r} \left[\frac{1}{\pi} f(s) - \frac{1}{4\pi^2 R} \int\limits_C f(s)\, ds\right] ds.$$

Let us transform the right-hand side of the preceding formula, assuming that M lies inside C:

$$W = \frac{1}{\pi} \int\limits_C \frac{\cos\varphi}{r} f(s)\, ds - \left(\frac{1}{4\pi^2 R} \int\limits_C f(s)\, ds\right) \int\limits_C \frac{\cos\varphi}{r}\, ds =$$

$$= \frac{1}{\pi} \int\limits_C \frac{\cos\varphi}{r} f(s)\, ds - \left(\frac{1}{4\pi^2 R} \int\limits_C f(s)\, ds\right) \cdot 2\pi =$$

$$= \frac{1}{\pi} \int\limits_C \left(\frac{\cos\varphi}{r} - \frac{1}{2R}\right) f(s)\, ds\,. \tag{59}$$

From Δ OPM (Fig. 65) we see that

$$K = \frac{\cos\varphi}{r} - \frac{1}{2R} = \frac{2R\cos\varphi - r}{2Rr} = \frac{2Rr\cos\varphi - r^2}{2Rr^2} =$$

$$= \frac{R^2 - \varrho_0^2}{2R\,[R^2 + \varrho_0^2 - 2R\varrho_0 \cos(\theta - \theta_0)]}, \tag{60}$$

since

$$\varrho_0^2 = R^2 + r^2 - 2Rr\cos\varphi\,.$$

Substituting expression (60) for K in (59) we obtain Poisson's integral

$$u = W(\varrho_0, \theta_0) = \frac{1}{2\pi} \int\limits_0^{2\pi} \frac{(R^2 - \varrho_0^2) f(\theta)\, d\theta}{R^2 + \varrho_0^2 - 2R\varrho_0 \cos(\theta - \theta_0)}, \tag{61}$$

giving the solution of the first boundary-value problem for a circle.

The arguments developed in this section show that any continuous function f, (61) defines a harmonic function, continuously adjoining the boundary values of f.

If f is partly-continuous, then because of the property of the potential of a double layer, the function M is also continuous at all points of the continuity of f. From the bounded nature of function f

$$|f| < C$$

the bounded nature of function (61) follows

$$|W(\varrho_0, \theta_0)| < C \frac{1}{2\pi} \int_0^{2\pi} \frac{(R^2 - \varrho_0^2)}{R^2 + \varrho_0^2 - 2R\varrho_0 \cos(\theta - \theta_0)} d\theta = C,$$

since*

$$\frac{1}{2\pi} \int_0^{2\pi} \frac{R^2 - \varrho_0^2}{R^2 + \varrho_0^2 - 2R\varrho_0 \cos(\theta - \theta_0)} d\theta = 1. \tag{62}$$

2. The first boundary-value problem for semi-space: find the harmonic function, continuous everywhere in region $z \geqslant 0$, which assumes on the boundary $z = 0$ the given value $f(x, y)$ and which reduces to zero at infinity.

We shall search for a solution of this problem in the form of a potential of a double layer

$$W(x, y, z) = \int\int_{-\infty}^{+\infty} \frac{\cos\varphi}{r^2} \nu(\xi, \eta)\, d\xi\, d\eta, \quad r^2 = (x - \xi)^2 + (y - \eta)^2 + z^2.$$

In this case

$$\frac{\cos\varphi}{r^2} = \frac{z}{r^3} = \frac{z}{[(x-\xi)^2 + (y-\eta)^2 + z^2]^{3/2}}$$

and the kernel of the integral equation

$$\frac{1}{2\pi}\left(\frac{\cos\varphi}{r^2}\right)_{z=0} = 0.$$

Thus, the surface density of the double layer

$$\nu(P) = \frac{1}{2\pi} f(P)$$

and potential is

$$u(x, y, z) = \frac{1}{2\pi} \int\int_{-\infty}^{+\infty} \frac{z}{[(x-\xi)^2 + (y-\eta)^2 + z^2]^{3/2}} f(\xi, \eta)\, d\xi\, d\eta.$$

* Equation (62) follows from the fact that the left-hand side represents the solution of the first boundary-value problem for $f = 1$.

It is easy to show that $u(x, y, z)$ tends uniformly to zero as $r = \sqrt{(x^2 + y^2 + z^2)} \to \infty$, if the function f possesses this property.

11. *Integral equations, corresponding to boundary-value problems*

In the solution of boundary-value problems for Laplace's equation in terms of the potentials of single and double layers we have arrived at Fredholm integral equations of second kind (50).

The conditions of solvability of Fredholm integral equations of second kind with a continuous kernel and bounded (integrable) right-hand part are similar to the conditions of solvability of a system of linear algebraic equations (to which they converge if the integral replaces the integral sum). Fredholm's first theorem states the following.

*An inhomogeneous integral equation of second kind has a unique solution, if the corresponding homogeneous equation has only a zero solution.**

* For curves with finite curvature Fredholm's theory is directly applicable, since the kernel of the integral equation (50) is continuous. Fredholm's theory is applicable also in those cases where one of the iterated kernels is continuous

$$K^{(n+1)}(P_1, P_2) = \iint_{\Sigma} K^{(1)}(P_1, M)\, K^{(n)}(M, P_2)\, d\sigma_M ,$$

$$K^{(1)}(P, M) = K(P, M).$$

Let us prove that if Σ is a Lyapunov surface then the iterated kernels of our equation, starting from a certain number, are continuous. As we saw, for Lyapunov surfaces

$$\left|\frac{\cos\varphi}{r^2}\right| < \frac{C}{r^{2-\delta}} .$$

Repeated kernels can be represented in the form

$$K_{1,2}(P_1 P_2) = \iint_{\Sigma} K_1(P_1, M)\, K_2(M, P_2)\, d\sigma_M .$$

Let us show that if

$$|K_i| < \frac{C_i}{r_i^{2-a_i}} \qquad (r_i = P_i M;\ a_i > 0;\ i = 1, 2),$$

then

$$|K_{1,2}| < \frac{C}{r^{2-a_1-a_2}}, \quad \text{if } a_1 + a_2 < 2,\ r = P_1 P_2.$$

Let us prove, that the integral equation (50) has a unique solution.

We confine ourselves to a consideration of convex contours, whose boundaries do not contain linear sections. In this case the kernel of equation (50) $K(P_0, P)$ is positive, since

$$K(P_0, P)\, ds_P = d\omega,$$

where $d\omega$ is the angle subtended by the curve ds_P at P_0.

Let us consider the first boundary-value problem for an internal region. The homogeneous equation, corresponding to equation (50) has the form

$$\pi\nu(s_0) + \int_0^L K(s_0, s)\,\nu(s)\,ds = 0. \tag{63}$$

As we saw (see sub-section 8) there is a relation

$$\int_0^L K(s_0, s)\,ds = \pi,$$

It is obvious that it is sufficient to establish these results for cases where the point P_2 lies in the Lyapunov neighbourhood Σ_0 of point P_1, where instead of an integral over Σ_0 it is possible to consider an integral over the projection S_0 of this neighbourhood on the tangent plane at point P_1, This is so because

$$1 \geqslant \frac{\varrho(P, M)}{r(P, M)} \geqslant B > 0$$

(where $\varrho(P, M)$ is the distance between the projections of the points P and M on the tangent plane, B is some constant), and also because of the relation between an element of surface $d\sigma$ and its projection dS: $d\sigma = dS/\cos\gamma$, where according to (34) $\cos\gamma > {}^1/_2$.

For a plane region the lemma is valid:

$$\text{if } |K_i| < \frac{C_i}{r_i^{2-\alpha_i}}, \text{ then}$$

$$|I| = \left|\iint_{S_0} K_1(P_1, M)\, K_2(M, P_2)\, dx\, dy\right| < \frac{C}{r^{2-\alpha_1-\alpha_2}}.$$

Let us denote by R the diameter of region S_0. Let us separate the integral I into the two integrals: I_1 taken over the circle G_1 of radius $2r$ with centre at the point P_1 and I_2 taken over the remaining region G_2 (Fig. 66). Since for point M lying in G_2

$$2 \geqslant \frac{r_1}{r_2} \geqslant \frac{2}{3} \qquad \left(\begin{array}{l} r_1 \leqslant r_2 + r \leqslant 2r_2, \\ r_2 \leqslant r_1 + r \leqslant r_1 + \dfrac{r_1}{2} = \dfrac{3r_1}{2} \end{array}\right),$$

which enables the homogeneous equation (63) to be written in the form

$$\int_0^L [\nu(s_0) + \nu(s)]\, K(s_0, s)\, ds = 0. \tag{64}$$

Let $P_0^*(s_0^*)$ be a point of contour C, at which $|\nu(s)|$ reaches its maximum value. Hence it follows that the sum $\nu(s_0^*) + \nu(s)$ has the same sign. Then, assuming in (64) $s_0 = s_0^*$ and making use of the fact that $K(s_0, s) \geqslant 0$ we obtain the equality

$$\nu(s_0^*) + \nu(s) = 0 \text{ or } \nu(s) = -\nu(s_0^*),$$

which contradicts the continuity at point s_0^*, unless $\nu(s^*) \neq 0$.

Consequently the homogeneous equation (63) has only a zero solution, and thus the inhomogeneous equation has a unique solution for any function f.†

---

then for the integral I_2 we obtain the relation

$$|I_2| < 4C_1C_2 \left| \int_0^{2\pi}\int_{2r}^{R} \frac{1}{r_1^{4-\alpha_1-\alpha_2}}\, r_1\, dr_1\, d\varphi \right| < \begin{cases} \dfrac{C_3}{r^{2-\alpha_1-\alpha_2}}, & \alpha_1 + \alpha_2 < 2. \\ C_3 R^{\alpha_1+\alpha_2-1}, & \alpha_1 + \alpha_2 > 2. \end{cases}$$

Making a change of the variables in the integral over G_1

$$x = rx',$$
$$y = ry',$$

we obtain:

$$|I_1| < \left| \frac{1}{r^{2-\alpha_1-\alpha_2}} \iint_{G_1'} \frac{C_1 C_2}{r_1'^{2-\alpha_1}\, r_2'^{2-\alpha_2}}\, dx'\, dy' \right|.$$

In the latter integral, taken over the circle G_1' with radius equal to $2r$, r_1' is the distance from the centre, r_2' from the middle of the radius. As a result this integral converges, and it does not depend on the position of the point P_2 i.e. on r.

Hence

$$|I_1| < \frac{C_4}{r^{2-\alpha_1-\alpha_2}}.$$

Putting $C_3 + C_4 = C$, we obtain the required inequality

$$|I| < \begin{cases} \dfrac{C}{r^{2-\alpha_1-\alpha_2}}, & \alpha_1 + \alpha_2 < 2, \\ CR^{\alpha_1+\alpha_2-2}, & \alpha_1 + \alpha_2 > 2. \end{cases}$$

Hence it follows that, beginning with some number, the integrals, form ing repeated kernels, are bounded and converge uniformly, i.e. they are continuous functions of their arguments.

† In the presence of linear sections of the boundary the discussions are rather complicated, but the result still holds.

The external second boundary-value problem, as we saw (see sub-section 10) reduces to the integral equation

$$\pi\mu(s_0) + \int_0^L K_1(s_0, s)\,\mu(s)\,ds = f(s_0), \tag{54}$$

the kernel of which $K_1(s_0, s)$ is related to the kernel $K(s_0, s)$, i.e. $K_1(s_0, s) = K(s, s_0)$.

Fredholm's second theorem states:

The number of linearly-independent non-zero solutions of a certain homogeneous integral equation is equal to the number of linearly-independent solutions of the adjoint equation.

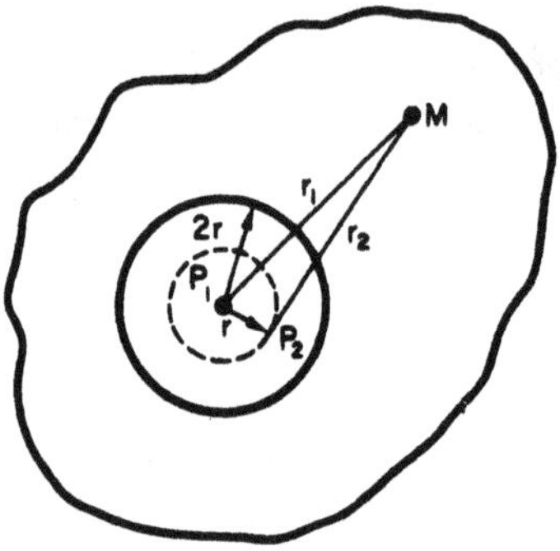

FIG. 66

From this theorem it follows that equation (54) has a unique solution. The exterior first boundary-value problem corresponds to the equation

$$-\pi\nu(s_0) + \int_0^L K(s_0, s)\,\nu(s)\,ds = f(s_0). \tag{52}$$

The homogeneous equation ($f = 0$) can therefore be reduced to the form

$$\int_0^L [\nu(s_0) - \nu(s)]\,K(s_0, s)\,ds = 0. \tag{65}$$

Denoting by s_0^* the point, at which $|\nu(s)|$ reaches a maximum value, we obtain from (65)

$$\nu(s^*) = \nu(s).$$

Hence it follows that

$$\nu(s) = \text{const} = \nu_0$$

is a unique non-zero solution of the homogeneous equation. Because of Fredholm's second theorem the conjugate homogeneous equation will have a unique solution.

Fredholm's third theorem states that *if a certain homogeneous integral equation*

$$\varphi(x) = \int_a^b K(x, s)\,\varphi(s)\,ds$$

has k linearly-independent solutions $\varphi_i(x)$ $(i = 1, 2, \ldots, k)$, then the adjoint inhomogeneous equation

$$\psi(x) = \int_a^b K(s, x)\,\psi(s)\,ds + f(x)$$

has a solution, only if

$$\int_a^b f(x)\,\varphi_i(x)\,dx = 0\,, \quad (i = 1, 2, \ldots, k).$$

Applying Fredholm's third theorem to equation (53), corresponding to the interior second boundary-value problem, we obtain the condition of solvability of this problem

$$\int_0^L f(s)\,ds = 0, \tag{66}$$

which we have already met in § 1.

The condition of solvability of the exterior first boundary-value problem is

$$\int_0^L f(s)\,h(s)\,ds = 0, \tag{67}$$

where $h(s)$ is the solution of the homogeneous equation, corresponding to (53). It is easy to explain the physical significance of this function.

Let a cylindrical conductor, with a cross-section bounded by a contour S, be charged to some potential V_0. In the conductor all the charge lies on the surface. We denote the density of the surface charges by $\bar{h}(s)$. The potential, produced by these surface charges, is the potential of a single layer with density $\bar{h}(s)$ and is expressed by formula (29). The normal derivatives inside it equal zero, since inside the conductor $V = \text{const}$. Therefore the function $\bar{h}(s)$ satisfies the homogeneous equation (53) and is proportional to the function $h(s)$, defined above. Thus we have an explanation of the physical significance of this function.

Thus, the integral equations, to which the boundary-value problems reduce, are always capable of solution for the interior first and exterior second boundary-value problems and are solvable with conditions (66) and (67) for the interior second and exterior first boundary-value problems. We will not discuss the question of the solvability of boundary-value problems in any further detail.

§ 6. Method of finite differences

1. *Concept of the method of finite differences for Laplace's equation*

In those cases, where one can not obtain an analytical expression of the solution, or use numerical methods for solution of integral equations, we can reduce the differential equation to an equation with finite differences.

In the case of a function of one argument $y = f(x)$ in place of the second derivative it is possible to consider the difference relation

$$\frac{\Delta^2 f}{h^2} = \frac{1}{h}\left[\frac{\Delta^+ f}{h} - \frac{\Delta^- f}{h}\right] = \frac{[f(x+h) - f(x)] - [f(x) - f(x-h)]}{h^2} =$$

$$= \frac{f(x+h) + f(x-h) - 2f(x)}{h^2}, \tag{1}$$

where $\Delta^+ f = f(x+h) - f(x)$, $\Delta^- f = f(x) - f(x-h)$ are first differences (right and left). Second differences for the case of two variables are

$$\Delta^2_{xx} u = u(x+h, y) + u(x-h, y) - 2u(x, y),$$

$$\Delta^2_{yy} u = u(x, y+h) + u(x, y-h) - 2u(x, y),$$

The difference relation for the Laplacian operator, obviously, has the form

$$\frac{\bar{\Delta}_h u}{h^2} =$$

$$= \frac{u(x+h, y) + u(x, y+h) + u(x-h, y) + u(x, y-h) - 4u(x, y)}{h^2}. \tag{2}$$

By replacing the differential equation by differences, we pass from a continuous change of the argument to a discrete one.

Let some positive number h be given and let us plot in the plane (x, y) a net, consisting of two systems of mutually perpendicular straight lines, separated from one another by a distance h (Fig. 67).

We shall consider the value of the function only at the nodal points of this net.

We consider the first boundary-value problem for Laplace's equation in some region S, bounded by curve C, on which the continuous bounded function f is given.

In the method of finite differences the given region S is replaced by the network region S_h, approximating to the region S. We shall assume that region S_h consists of the squares of our net, lying entirely inside S (as an alternative definition of the region S_h we could take a combination of all the squares of the net, having even one common point in S). Let us denote the broken line, bounding region S_h, by C_h. It is obvious that the distance from every nodal point of the curve C_h to the curve C does not exceed $h\sqrt{2}$.

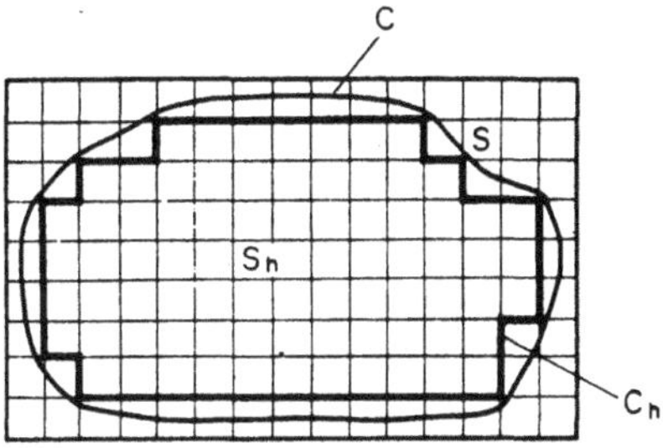

FIG. 67

We define the boundary function f_h at the nodal points of C_h by assuming it equal to the value of function f at the nearest point of the boundary C (or at one of the nearest points, if there are several).*

Let us consider the first boundary-value problem for Laplace's difference equation in region S_h:

Find the function, satisfying the difference equation at the nodal points M_{ik} (x_i, y_k) inside S_h

$$u^{(h)}_{i+1,k} + u^{(h)}_{i,k+1} + u^{(h)}_{i-1,k} + u^{(h)}_{i,k-1} - 4\,u^{(h)}_{i,k} = 0, \tag{3}$$

where

$$u^{(h)}_{i,k} = u\,(x_i, y_k),$$

and taking a value equal to f_h on C_h.

The solution of this problem involves searching for the values of the network function $u^{(h)}_{ik}$ at the internal nodal points M_{ik} of region S_h.

* The degree of uncertainty in the choice may be made as small as desired, because of the continuity of function f, if h is sufficiently small.

At each internal nodal point M_{ik} the difference equation (3) should be satisfied. Thus, for the determination of the network function we obtain a system of linear algebraic equations, the number of which equals the number of unknowns.

Let us prove that *the system of difference equations has a unique solution.*

In order to do this it is sufficient to satisfy ourselves that the corresponding homogeneous system has only a trivial solution.

The difference system is a system of inhomogeneous equations, since the values u_{ik} at the nodal points of C_h are given and equal f_h. Transition to a homogeneous system is equivalent to the case where the boundary function $f_h = 0$. Let us prove that in this case the solution of the difference system equals zero for all nodal points of region S_h.

Let some $u_{i_1,k_1} \neq 0$; for definiteness we shall assume that

$$u_{i_1,k_1} > 0.$$

Let u_{i_0,k_0} be the maximum value of our network function, so that

$$u_{i,k} \leqslant u_{i_0,k_0}$$

at all points M_{ik} of the region S_h. The equation

$$u_{i_0,k_0} = \frac{u_{i_0-1,k_0} + u_{i_0,k_0-1} + u_{i_0+1,k_0} + u_{i_0,k_0+1}}{4}$$

can hold in this case, only if

$$u_{i_0-1,k_0} = u_{i_0,k_0-1} = u_{i_0+1,k_0} = u_{i_0,k_0+1} = u_{i_0,k_0}.$$

Carrying out in succession similar arguments for u_{i_0+1,k_0}, u_{i_0+2,k_0} etc. we arrive at some boundary point, which leads us to a contradiction, since the boundary values equal zero. The assumption $u_{i_0,k_0} < 0$ also leads to a contradiction.

Hence it follows that

$$u_{i,k} \equiv 0$$

at all internal nodal points of region S, i.e. that the system of homogeneous difference equations has only a trivial solution. Thus we have proved the uniqueness theorem for the solution of the first boundary-value problem for Laplace's difference equation.

Solving the difference equations, we obtain a network function which represents an approximate solution of the original problem for Laplace's equation.

2. *Method of successive approximations for the solution of difference equations*

The method of finite differences involves the replacement of the boundary-value problem for the differential equation $\nabla^2 u = 0$ by the corresponding boundary-value problem for the difference equation. To justify this method it is necessary to satisfy ourselves that for sufficiently small h the function u_h differs from u by as small a quantity as desired, u being the exact solution of the equation $\nabla^2 u = 0$. We will not give this proof.

Let us discuss more fully the methods of solution of the difference equation. The solution of the boundary-value problem by a difference method leads to the solution of a system of algebraic equations with a large number of unknowns, often several hundred and sometimes even thousands. The solution of this system using determinants would produce excessive technical difficulty. The method of successive approximations is much more convenient.

The method of successive approximations for a system of linear algebraic equations consists of the following.

Let us write down the system of equations in the form

$$\left.\begin{array}{l} u_1 = f_1 - (a_{12}u_2 + \dots + a_{1n}u_n), \\ u_2 = f_2 - (a_{21}u_1 + \dots + a_{2n}n_n), \\ \dots\dots\dots\dots\dots\dots\dots \\ \dots\dots\dots\dots\dots\dots\dots \\ u_n = f_n - (a_{n1}u_1 + \dots + a_{n,n-1}u_{n-1}). \end{array}\right\} \quad (4)$$

Choosing as a zero approximation the arbitrary numbers $u_1^{(0)}, u_2^{(0)}, \dots, u_n^{(0)}$ and substituting them in the right-hand part of equations (4), we find the first approximations $u_1^{(1)}, u_2^{(1)}, \dots, u_n^{(1)}$.

Continuing this process further we define the $(k + 1)$th approximation by the formulae

$$u_1^{(k+1)} = f_1 - (a_{12}u_2^{(k)} + \ldots + a_{1n}u_n^{(k)}),$$
$$u_1^{(k+1)} = f_2 - (a_{22}u_1^{(k)} + \ldots + a_{2n}u_n^{(k)}),$$
$$\cdots\cdots\cdots\cdots\cdots\cdots$$
$$\cdots\cdots\cdots\cdots\cdots\cdots$$
$$u_n^{(k+1)} = f_n - (a_{n1}u_1^{(k)} + \ldots + a_{n,n-1}u_{n-1}^{(k)}).$$

If the successive approximations $\{u_n^{(k)}\}$ converge to a limit

$$\lim_{k\to\infty} u_n^{(k)} = u_n,$$

then these limiting values are the solutions of system (4).

The method of successive approximations is not applicable to every system of equations.

Let us prove that the first boundary-value problem for a system of difference equations $\overline{\nabla}^2{}_h u = 0$ can be solved by the method of successive approximations. The system of algebraic equations, corresponding to $\overline{\nabla}^2{}_h u = 0$ has the form

$$u_{ik}^{(h)} = \frac{u_{i+1,k}^{(h)} + u_{i,k+1}^{(h)} + u_{i-1,k}^{(h)} + u_{i,k-1}^{(h)}}{4}.$$

We shall carry out a successive approximation, starting from the boundary, where the boundary values are given and working inwards. Let us call $C_h^{(1)}$ the set of nodal points of region S_h, removed from the nodal points C_h by a distance h. In calculating the nth approximation at the points $C_h^{(1)}$, the values of the function f_h, given at C_h will be used directly. Let us call, $C_h^{(2)}$ the set of nodal points of region S_h inside $C_h^{(1)}$ and separated from $C_h^{(1)}$ by a distance h. In the calculation of the nth approximation on $C_h^{(2)}$ the values already obtained $u_{ik}^{(h)}$ on $C_h^{(1)}$ will be used. Similarly "zones" $C_h^{(3)}$, $C_h^{(4)}$ are determined. As a result, each of the nodal points of region S_h belongs to some $C_h^{(i)}$ $(i = 1, 2, \ldots N)$ where N is the last number of the zone $C_h^{(N)}$.

Let u_{ik} be the exact solution of the system of difference equations, $u_{ik}^{(n)}$ the nth approximation of this system.

We consider the difference

$$v_{ik}^{(n)} = u_{ik} - u_{ik}^{(n)},$$

equal to zero on C_h and satisfying the equations

$$v_{ik}^{(n)} = \frac{v_{i+1,k}^{(n-1)} + v_{i,k+1}^{(n-1)} + v_{i-1,k}^{(n-1)} + v_{i,k-1}^{(n-1)}}{4}.$$

inside S_h and prove that the successive approximation $\{v_{ik}^{(n)}\}$ converges to zero

$$\lim_{n\to\infty} v_{ik}^{(n)} = 0.$$

We assume $\max v_{ik}^{(n)} = A_n$, and evaluate the $(n+1)$th approximation. It is obvious that

$$v_{ik}^{(n+1)} \leqslant \frac{3}{4} A_n \quad \text{on} \quad C_h^{(1)},$$

since in this case at least one of the components in (5) equals zero. Hence it follows that

$$v_{ik}^{(n+1)} \leqslant \left(1 - \frac{1}{4^2}\right) A_n \quad \text{on} \quad C_h^{(2)}$$

and in general

$$v_{ik}^{(n+1)} \leqslant \left(1 - \frac{1}{4^s}\right) A_n \quad \text{on} \quad C_h^{(s)}.$$

For the last zone $C_h^{(N)}$ we obtain:

$$v_{ik}^{(n+1)} \leqslant \left(1 - \frac{1}{4^N}\right) A_n \quad \text{on} \quad C_h^{(N)}.$$

Hence we conclude that

$$A_{n+1} \leqslant a A_n, \quad a = 1 - \frac{1}{4^N},$$

i.e. $\lim\limits_{n\to\infty} A_n = 0$.

Denoting $\min v_{ik} = B_n$ in a similar manner we can show that

$$\lim_{n\to\infty} B_n = 0$$

and, therefore

$$\lim_{n\to\infty} v_{ik}^{(n)} = 0, \quad \text{i.e.} \quad \lim_{n\to\infty} u_{ik}^{(n)} = u_{ik}$$

for all points $M_{ik}\ (x_i, y_k)$ of region S_h.

The convergence holds for any choice of zeroth approximation, but the rate of convergence depends strongly on its choice.

3. *Modelling methods*

Mathematical machines are often used to solve systems of difference equations. These machines are constructed by utilizing the similarity existing between different physical phenomena, which obey the same differential equations.

For the solution of Laplace's equation (and also for some more

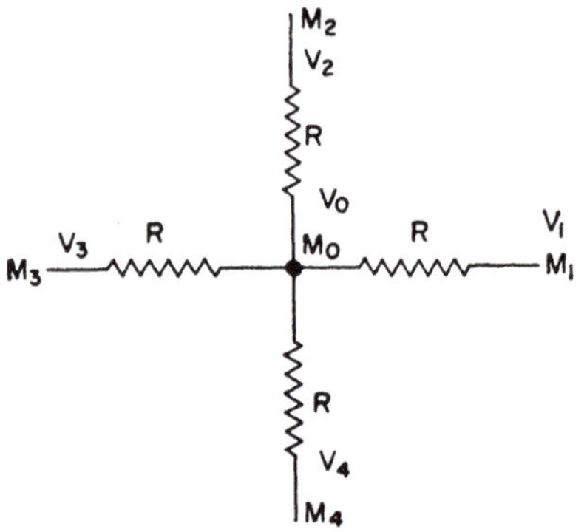

FIG. 68

complex equations) *electro-integrators* are extensively employed. We consider one of the simplest electrical systems for solving Laplace's equation.

We take a net of identical ohmic resistances, one of the nodes of which is shown in Fig. 68. We denote by V_i the potential at point M_i, and by j_i the current in the section M_0M_i. Using Ohm's law

$$j_i = \frac{V_i - V_0}{R} \; (i = 1, 2, 3, 4)$$

and Kirchhoff's law

$$j_1 + j_2 + j_3 + j_4 = 0,$$

we obtain

$$V_0 = \frac{V_1 + V_2 + V_3 + V_4}{4}.$$

Thus, the potential at any nodal point of such an electrical system equals the arithmetical mean of the potentials at four adjacent nodes. This relation, similar to equation (5) of the method of finite differences is the basis for an electric model of Laplace's equation.

The simplest electro-integrator consists of a panel in which there are sockets joined by identical resistances. Suppose we have to solve the first boundary-value problem for some region S of the plane (x, y), bounded by the contour C.

We consider a net with interval h and plot the region S_h with boundary C_h by the method indicated in the preceding sub-section. Voltages given at the nodes of the boundary C_h by means of a special device correspond to the boundary conditions of the first boundary-value problem. The distribution of voltages at other points gives an approximate solution of the problem under consideration.

In some systems of electro-integrators the resistances between nodal points can vary. This enables us to solve equations with variable coefficients of the form

$$\frac{\partial}{\partial x}\left(k_1 \frac{\partial u}{\partial x}\right) + \frac{\partial}{\partial y}\left(k_2 \frac{\partial u}{\partial y}\right) = 0,$$

where

$$k_1 = k_1(x, y), \quad k_2 = k_2(x, y).$$

Problems on Chapter IV.

1. Find the function u, harmonic inside a circle of radius a and taking on the circumference C the values

$$\text{(a) } u|_C = A\cos\varphi;$$

$$\text{(b) } u|_C = A + B\sin\varphi.$$

2. Solve Laplace's equation $\nabla^2 u = 0$ inside the rectangle $0 \leqslant x \leqslant a$, $0 \leqslant y \leqslant b$ for the boundary conditions

$$u|_{x=0} = f_1(y), \ u|_{y=0} = f_2(x), \ u|_{x=a} = 0, \ u|_{y=b} = 0.$$

Prove that the formulae obtained give the solution of the problem for an arbitrary partly-continuous function, given on the boundary.

Solve the problem for the particular case

$$f_1(y) = Ay(b-y), \ f_2(x) = B\cos\frac{\pi}{2a}x, \ f_3 = f_1 = 0.$$

3. Solve the equation $\nabla^2 u = 1$ for a circle of radius a for the boundary condition $u|_{r=a} = 0$.

4. Solve the equation $\nabla^2 u = Axy$ for a circle of radius a with centre at the point $(0, 0)$ for the boundary condition $u|_{r=a} = 0$.

5. Solve the equation $\nabla^2 u = A + B(x^2 - y^2)$ for the ring $a \leqslant \varrho \leqslant b$, if

$$u\Big|_{\varrho=a} = A_1, \quad \frac{\partial u}{\partial \varrho}\Big|_{\varrho=b} = 0.$$

The origin of coordinates is at the centre of the ring.

6. Plot the source function for Laplace's equation (the first boundary-value problem): (a) for a semi-circle, (b) for a ring, (c) for a layer ($0 \leqslant z \leqslant l$).

7. Find the harmonic function inside the ring $a \leqslant \varrho \leqslant b$, satisfying the following boundary conditions:

$$u\,|_{\varrho=a} = f_1(\varphi), \qquad u\,|_{\varrho=b} = f_2(\varphi).$$

8. Find the solution of Laplace's equation $\nabla^2 u = 0$ in the semi-plane $y \geqslant 0$ with the boundary condition

$$u(x, 0) = \begin{cases} 0 & \text{where } x < 0, \\ u_0 & \text{where } x > 0. \end{cases}$$

9. Find the function $u(\varrho, \varphi)$, harmonic inside the circular sector $\varrho \leqslant a$, $0 \leqslant \varphi \leqslant \varphi_0$ for the boundary conditions:

(a) $u\,|_{\varphi=0} = q_1$, $u\,|_{\varphi=\varphi_0} = q_1$, $u\,|_{\varrho=a} = q_2$, where q_1 and q_2 are constants;

(b) $u\,|_{\varphi=0} = u\,|_{\varphi=\varphi_0} = 0$, $u\,|_{\varrho=a} = f(\varphi)$.

10. Solve by the method of finite differences the first boundary-value problem for the equation $\nabla^2 u = 0$ inside the rectangle $0 \leqslant x \leqslant a$ $0 \leqslant y \leqslant b$, dividing each of its sides into eight equal parts, if the boundary conditions have the form

$$u\,|_{x=0} = \frac{y}{b}\left(1 - \frac{y}{b}\right), \quad u\,|_{y=b} = \frac{x}{a} \sin \frac{\pi}{a} x, \quad u\,|_{x=a} = u\,|_{y=0} = 0.$$

Compare with the analytical solution.

11. Find the potential of a sphere with constant density $\varrho = \varrho_0$. *Hint.* Solve the equation $\nabla^2 u = 0$ outside the sphere and $\nabla^2 u = 4\pi\varrho_0$ inside the sphere and join the solutions on the surface of the sphere.

12. Find the potential of a single layer, distributed with the constant density $\nu = \nu_0$ on the sphere.

Hint. Look for the solution of the equation $\nabla^2 u = 0$ outside and inside the sphere and use the conditions of discontinuity of the derivative of the potential of a single layer.

13. Solve the first boundary-value problem for a bounded circular cylinder ($\varrho \leqslant a$, $0 \leqslant z \leqslant l$):

(a) zero boundary conditions of the first or second kind are given at the ends of the cylinder while on the sides

$$u\,|_{\varrho=a} = f(z);$$

(b) on the sides and on one end of the cylinder zero boundary conditions are given (of first or second kind) and on the second end of the cylinder

$$u = f(\varrho),$$

for instance $f(\varrho) = A\ \varrho(1 - \varrho/a)$.

14. Solve the inhomogeneous equation

$$\nabla^2 u = -f$$

in an infinite cylindrical region for zero boundary conditions (of first or second kind) and construct the source function.

15. Find the function, harmonic inside a sphere, equal to u_1 on one half of the sphere and u_2 on the second half of the sphere.

16. Find by analysis into spherical functions the density of surface charges induced on a conducting sphere by a point charge.

17. Solve the problem of a polarized dielectric sphere in the field of a point charge.

18. Calculate the gravitational potential of a plane disc. Compare with the asymptotic form of a gravitational potential at large distances.

19. Calculate the magnetic potential of a circular current.

20. Solve the problem of the perturbation of a uniform electric field by a perfectly conducting sphere. Solve the problem for a dielectric sphere.

APPENDICES TO CHAPTER IV

I. Asymptotic expression for the volume potential

In an investigation of the volume potential

$$V(M) = \iiint_T \frac{\varrho(P)\, d\tau_P}{d}, \quad \text{where } d = r_{MP} \tag{1}$$

at great distances from a body one usually assumes a potential equal to m/R, where m is the mass of the body T, R is the distance of its centre of gravity from the point of observation. Let us determine a more exact asymptotic expression for V.*

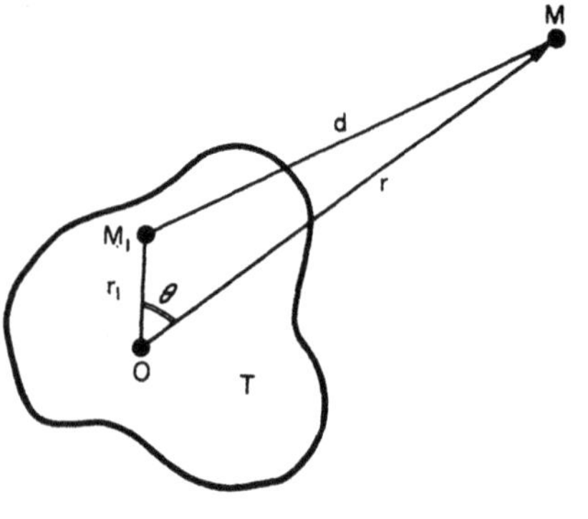

FIG. 69

Let $\sum$ be a sphere with centre at the origin entirely containing the body T. Outside this sphere the potential will be a harmonic function.

The distance from the point of observation $M(x, y, z)$ to the variable points inside the body $M(x_1, y_1, z_1)$ (Fig. 69), over which the integration is performed, equals

$$d = \sqrt{(r^2 + r_1^2 - 2rr_1 \cos\theta)} \quad (r = OM, r_1 = OM_1), \tag{2}$$

from which

$$\frac{1}{d} = \frac{1}{r} \frac{1}{\sqrt{(1 + a^2 - 2a\mu)}}; \quad a = \frac{r_1}{r}; \quad \mu = \cos\theta. \tag{3}$$

* V. I. Smirnov, *Course in Higher Mathematics*, vol. III, Gostekhizdat, 1949.

Since $r_1 < r$, $\alpha < 1$ and we can make the expansion (see the supplement, part II, § 1)

$$\frac{1}{d} = \frac{1}{r} \sum_{n=0}^{\infty} \alpha^n P_n(\mu), \tag{4}$$

where $P_n(\mu)$ is a Legendre polynomial of nth order (see Supplement II, § 1). Substituting this expression in (1) and taking into account that $1/r$ does not depend on the variables of integration we obtain:

$$\begin{aligned} V(M) &= \frac{1}{r} \iiint_T \varrho \sum_{n=0}^{\infty} \alpha^n P_n(\mu)\, d\tau = V_1 + V_2 + V_3 + \ldots = \\ &= \frac{1}{r} \iiint_T \varrho d\tau + \frac{1}{r^2} \iiint_T \varrho r_1 P_1(\mu)\, d\tau + \\ &\quad + \frac{1}{r^3} \iiint_T \varrho r_1^2 P_2(\mu)\, d\tau + \ldots \end{aligned} \tag{5}$$

The first term equals m/r, where m is the mass of the whole body, and gives us a first approximation for the calculation of the potential for large r.

Let us proceed to a calculation of the following terms of expansion (5). The expression under the integral sign in the second term is

$$\varrho P_1(\mu)\, r_1 = \varrho \mu r_1 = \varrho r_1 \cos\theta = \frac{\varrho x x_1 + \varrho y y_1 + \varrho z z_1}{r}.$$

The values x, y, z and r do not depend on the variables of integration and may be taken outside the integral sign. Thus the second term of the expansion of the potential takes the form

$$\begin{aligned} \frac{1}{r^2} \iiint_T \varrho r_1 P_1(\mu)\, d\tau &= \frac{1}{r^3}(M_1 x + M_2 y + M_3 z) = \\ &= \frac{M}{r^3}(x\bar{x} + y\bar{y} + z\bar{z}), \end{aligned}$$

where

$$M_1 = \iiint_T \varrho x_1\, d\tau = M\bar{x}, \quad M_2 = \iiint_T \varrho y_1\, d\tau = M\bar{y};$$

$$M_3 = \iiint_T \varrho z_1\, d\tau = M\bar{z}$$

are moments of first order, $\bar{x}, \bar{y}, \bar{z}$ are the coordinates of the centre of gravity. Thus, the second term decreases as $1/r^2$. If the origin of coordinates is located at the centre of gravity ($\bar{x} = 0$, $\bar{y} = 0$, $\bar{z} = 0$), then $V_2 = 0$.

In the third term of the expansion we transform the expression under the integral sign

$$\varrho r_1^2 P_2(\mu) = \varrho r_1^2 \frac{3\mu^2 - 1}{2} = \varrho r_1^2 \frac{3(xx_1 + yy_1 + zz_1)^2 - r_1^2 r^2}{2r_1^2 r^2} =$$

$$= \frac{\varrho}{2r^2}[3(x_1 x + y_1 y + z_1 z)^2 - r_1^2 r^2] .$$

Introducing the symbol

$$M_{ik} = \iiint_T \varrho x_i x_k \, d\tau \quad (x = x_1;\ y = x_2;\ z = x_3),$$

we arrive at the following expression for V_3:

$$V_3 = \frac{1}{r^3} \iiint_T \varrho r_1^2 P_2(\mu) \, d\tau =$$

$$= \frac{1}{2r^5}\{x^2[3M_{11} - (M_{11} + M_{22} + M_{33})] +$$

$$+ y^2[3M_{22} - (M_{11} + M_{22} + M_{33})] +$$

$$+ z^2[3M_{33} - (M_{11} + M_{22} + M_{33})] + 2 \cdot 3xyM_{12} +$$

$$+ 2 \cdot 3xzM_{13} + 2 \cdot 3yzM_{23}\} .$$

The polynomial in brackets is a harmonic polynomial, since it can be written in the form

$$V_3 = \frac{1}{2r^5}\{(x^2 - y^2)[M_{11} - M_{22}] + (z^2 - x^2)[M_{11} - M_{33}] +$$

$$+ (y^2 - z^2)[M_{22} - M_{33}] + 6[xyM_{12} + xzM_{13} + yzM_{23}]\},$$

where each component satisfies Laplace's equation. The coefficients in the brackets are related to the *moments of inertia*. The moment of inertia of a body T about the x-axis equals

$$A = \iiint_T \varrho (y_1^2 + z_1^2) \, d\tau = M_{22} + M_{33} .$$

Similarly the moments of inertia about the y and z-axes equal

$$B = M_{33} + M_{11};\ C = M_{11} + M_{22} .$$

Hence it follows that

$$M_{11} - M_{22} = B - A\,;\; M_{11} - M_{33} = C - A\,;\; M_{22} - M_{33} = C - B.$$

As a result we arrive at the asymptotic expression for the potential

$$V \cong \frac{m}{r} + \frac{m}{r^3}(x\bar{x} + y\bar{y} + z\bar{z}) + \frac{1}{2r^5}\{(x^2 - y^2)(B - A) + (y^2 - z^2)(C - B) + (z^2 - x^2)(A - C) + 6(xyM_{12} + yzM_{23} + zxM_{31})\}, \tag{6}$$

which is valid up to terms of order $1/r^6$.

Expression (6) is simplified if we locate the origin of coordinates at the centre of gravity, and the coordinate axes are directed along the principal axes of inertia:

$$V \cong \frac{m}{r} + \frac{1}{2r^5}\{(x^2 - y^2)(B - A) + (y^2 - z^2)(C - B) + (z^2 - x^2)(A - C)\}. \tag{7}$$

The asymptotic representation of the potential obtained allows us to answer a series of questions about the converse problem of potential theory, consisting of the determination of the characteristics of a body by its potential (or by any derivative of it). In fact determining the coefficients of expansion (6) it is possible to find the mass, the coordinates of the centre of gravity and the moments of inertia of the body.

II. Problems of electrostatics

In problems of electrostatics the solution of Maxwell's equations reduces to a search for one scalar function the potential φ, connected to the field strength by the relation

$$\boldsymbol{E} = -\operatorname{grad}\varphi.$$

Using Maxwell's equation

$$\operatorname{div}\boldsymbol{E} = -4\pi\varrho,$$

we obtain:

$$\nabla^2\varphi = -4\pi\varrho.$$

Thus, the potential satisfies Poisson's equation at those points of space where electric charges occur, and Laplace's equation at those points where there are no charges.

1. *The main problem of electrostatics* is the search for the field, produced by a system of charges on given conductors. There are two different statements of this problem possible.

I. The potentials of conductors are given and it is required to find the field outside the conductors and the density of the charges on the conductors. The mathematical problem is formulated as follows.

It is required to find the function φ, satisfying Laplace's equation $\Delta \varphi = 0$ everywhere outside the given system of conductors, reducing to zero at infinity and taking the given values φ_i on the surfaces of the conductors

$$\varphi \,|_{S_i} = \varphi_i.$$

Thus, in this case we arrive at the first boundary-value problem for Laplace's equation. The uniqueness of its solution follows from the general theory.

II. The total charges on the conductors are given. It is required to determine the potential of the conductors, the distribution of charges over their surfaces and the field outside the conductors. The solution of this problem reduces to a search for *the function φ, satisfying Laplace's equation*

$$\nabla^2 \varphi = 0$$

outside the given system of conductors, reducing to zero at infinity, taking constant values on the surfaces of the conductors

$$\varphi \,|_{S_i} = \text{const}$$

and satisfying the integral relation on the surfaces of the conductors

$$\oint_{S_i} \frac{\partial \varphi}{\partial n}\, d\sigma = -4\pi e_i,$$

where e_i is the total charge on the ith conductor.

2. The uniqueness of the solution of the second problem does not follow from the general theory, but may easily be proved.

Let us assume that there exists two solutions φ_1 and φ_2 of problem II. Then their difference

$$\varphi' = \varphi_1 - \varphi_2$$

will satisfy the equation

$$\nabla^2\varphi' = 0$$

and the relations

$$\varphi' |_{S_i} = \text{const},$$

$$\oint_{S_i} \frac{\partial \varphi'}{\partial n}\, d\sigma = 0,$$

$$\varphi' |_\infty = 0.$$

Let us enclose two given conductors inside a sphere Σ_R of sufficiently large radius R and let us apply Green's first theorem to function φ' in the region T_R, bounded by the sphere Σ_R and by the surfaces of the conductors S_i:

$$\int_{T_R} (\nabla\varphi')^2\, d\tau = \int_{\Sigma_R} \varphi' \frac{\partial \varphi'}{\partial n}\, d\sigma + \sum_{i=1}^{n} \int_{S_i} \varphi' \frac{\partial \varphi'}{\partial n}\, d\sigma.$$

Because of the condition at infinity* and on the surfaces we obtain:

$$\lim_{R\to\infty} \int_{T_R} (\nabla\varphi')^2\, d\tau = 0\,,$$

from which on account of the positive nature of the expression under the integral sign it follows:

$$\nabla\varphi' = 0$$

or

$$\varphi' = \text{const}$$

everywhere in the region under consideration. Remembering the value at infinity

$$\varphi'|_\infty = 0,$$

we obtain:

$$\varphi' \equiv 0,$$

which proves the uniqueness of the solution.

* From the condition $\varphi' \,|\, \infty = 0$ it follows that the function φ' is regular at infinity (see page 329), hence

$$\int_{\Sigma_R} \varphi' \frac{\partial \varphi'}{\partial n}\, d\sigma \to 0 \text{ where } R \to \infty.$$

3. From the uniqueness of the solution of the boundary-value problem for Laplace's equation it follows that *the potential of an isolated conductor is directly proportional to its charge*

$$e/\varphi = C.$$

If an isolated conductor carries charges e or $e' = me$, then the corresponding potentials φ or φ' should satisfy the equations

$$\nabla^2\varphi = 0; \quad \nabla^2\varphi' = 0$$

and the boundary conditions

$$-\frac{1}{4\pi}\oint\limits_S \frac{\partial\varphi}{\partial n}\,d\sigma = e; \qquad -\frac{1}{4\pi}\oint\limits_S \frac{\partial\varphi'}{\partial n}\,d\sigma = me,$$

from which it follows that $\varphi' - m\varphi = 0$, i.e. $\varphi'/\varphi = e'/e$.

Hence we obtain:

$$e'/\varphi' = e/\varphi = C = \text{const.}$$

This constant C is called the *capacity* of the isolated conductor. It does not depend on the charge of the conductor, and is determined by the shape and dimensions of the latter. Thus, for an isolated conductor

$$e = C\varphi.$$

The capacity of an isolated conductor is numerically equal to that charge, which gives the conductor a potential equal to 1. If the conductor is not isolated then its potential depends on the other conductors. For the system of conductors there are relations

$$\begin{aligned}
e_1 &= C_{11}\varphi_1 + C_{12}(\varphi_2 - \varphi_1) + \ldots + C_{1n}(\varphi_n - \varphi_1),\\
e_2 &= C_{21}(\varphi_1 - \varphi_2) + C_{22}\varphi_2 + \ldots + C_{2n}(\varphi_n - \varphi_2),\\
&\cdots\cdots\cdots\cdots\cdots\cdots\cdots\cdots\\
&\cdots\cdots\cdots\cdots\cdots\cdots\cdots\cdots\\
e_n &= C_{n1}(\varphi_1 - \varphi_n) + C_{n2}(\varphi_2 - \varphi_n) + \ldots + C_{nn}\varphi_n,
\end{aligned}$$

where e_i and φ_i are the charge and potential of the ith conductor. The value C_{ik} has the sense of the mutual capacitance of the ith conductor with respect to the kth conductor. It can be defined as that charge which must be imparted to the ith conductor in order that all the conductors, except the kth, would have zero potential and the kth conductor the potential 1.

4. It is easy to show that *the matrix of the coefficients* C_{ik} *is symmetrical*, i.e. the relation

$$C_{ik} = C_{ki}.$$

holds. For simplicity we consider the case of two conductors although in the case of n conductors the proof remains the same.

Let there be given two conductors a and b. Then determination of the coefficients C_{ab} and C_{ba} reduces to a determination of functions $u^{(1)}$ and $u^{(2)}$, satisfying the equations $\nabla^2 u^{(1)} = 0$ and $\nabla^2 u^{(2)} = 0$ and the boundary conditions

$$u^{(1)}|_{S_a} = 0; \quad u^{(1)}|_{S_b} = 1; \quad u^{(1)}|_{\infty} = 0,$$

$$-\frac{1}{4\pi}\oint_{S_a} \frac{\partial u^{(1)}}{\partial n} d\sigma = e_a^{(1)} = C_{ab};$$

$$u^{(2)}|_{S_a} = 1; \quad u^{(2)}|_{S_b} = 0; \quad u^{(2)}|_{\infty} = 0,$$

$$-\frac{1}{4\pi}\oint_{S_b} \frac{\partial u^{(2)}}{\partial n} d\sigma = e_b^{(2)} = C_{ba}.$$

Let us describe a sphere Σ_R of sufficiently large radius R, containing both conductors a and b and apply Green's theorem to the functions $u^{(1)}$ and $u^{(2)}$ in the region between the surface Σ_R and the surfaces of the conductors S_a and S_b

$$\int_{T_R} (u^{(1)} \nabla^2 u^{(2)} - u^{(2)} \nabla^2 u^{(1)})\, d\tau = \int_{\Sigma_R + S_a + S_b} \left(u^{(1)} \frac{\partial u^{(2)}}{\partial n} - u^{(2)} \frac{\partial u^{(1)}}{\partial n}\right) d\sigma.$$

The integral on the left-hand side of this equation equals zero. Using the boundary conditions and the conditions at infinity we obtain:

$$\int_{S_b} \frac{\partial u^{(2)}}{\partial n} d\sigma - \int_{S_a} \frac{\partial u^{(1)}}{\partial n} d\sigma = 0$$

or

$$C_{ab} = C_{ba},$$

which it was required to prove.

5. We proceed to concrete examples.

Let us consider the problem of the field of a charged sphere. The surface of the conducting sphere of radius a is at a potential φ_0. Solving problem 1 we see that the field and density of the charges on the surface of the sphere are given by the expressions

$$\varphi = \frac{\varphi_0}{r} a \text{ and } \sigma = \frac{\varphi_0}{4\pi a}.$$

If in place of the potential of the surface of the sphere φ_0 we are given its total charge e_0 then

$$\varphi_0 = \frac{e_0}{a}, \quad \sigma = \frac{e_0}{4\pi a^2}, \quad \varphi = \frac{e_0}{r} \quad (r > a).$$

The capacity of the sphere is

$$C = a,$$

i. e. in absolute units the capacity of an isolated sphere is numerically equal to its radius.

As a second example we consider the problem of a spherical condenser (a system of two concentric conducting spheres).

Let the internal sphere of radius r_1 have a given potential V_0, and the external sphere of radius r_2 be earthed. Then determination of the field inside the condenser reduces to a search for the function φ, satisfying the equation

$$\nabla^2\varphi = 0$$

and the conditions

$$\varphi|_{r_1} = V_0, \quad \varphi|_{r_2} = 0.$$

It is easy to show that

$$\varphi = \frac{r_1 r_2}{r_2 - r_1} V_0 \left(\frac{1}{r} - \frac{1}{r_2}\right),$$

and the capacity of the spherical condenser equals

$$C = \frac{r_1 r_2}{r_2 - r_1}.$$

A more complex problem is the determination of the potential of two charged non-concentric spheres. This problem is solved by the method of images, but the analytical solution is rather cumbersome and we will not give it here.*

* See Frank and Mises, *Differential and Integral Equations of Mathematical Physics*, vol. II, p. 713, 1937.

By means of an inversion* it is possible to transform a sphere into a plane. Let us show that by means of an inversion the problem of the determination of the potential of a plane in the presence of a sphere reduces to the problem of a system of two concentric conducting spheres.

Instead of a plane and a sphere, obviously, it is sufficient to consider a straight line E and a circle K with centre O of radius ϱ

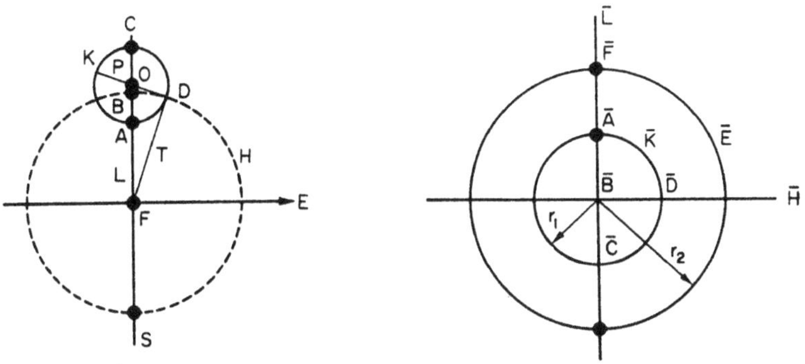

FIG. 70

(Fig. 70). Let us drop from O a perpendicular $OF = l$ on to the straight line E and draw through the point F the tangent FD to the circle K; we describe the circle H with radius FD and centre F intersecting the extension of the perpendicular OF in the point S.

We take the point S as the centre of inversion.

In this transformation the straight line $L = SO$, intersecting circle K at the point A, and the circle change into the mutually perpendicular straight lines $\bar{L}$ and $\bar{H}$, and the straight line E and circle K into the circles $\bar{E}$ and $\bar{K}$.

Since in the inversion transformation the property of orthogonality is retained, the circles $\bar{E}$ and $\bar{K}$ must be orthogonal to the straight lines $\bar{L}$ and $\bar{H}$. This is possible only if the centres of the circles $\bar{E}$ and $\bar{K}$ coincide with the point $\bar{B}$ of intersection of the straight lines $\bar{L}$ and $\bar{H}$.

* See the same.

Thus, the given transformation changes a straight line and circle into a pair of concentric circles.

Simple calculations allow us to find the radii of the circles $\overline{K}$ and $\overline{E}$:

$$r_1 = \frac{\sqrt{(l^2-\varrho^2)}-(l-\varrho)}{2\sqrt{(l^2-\varrho^2)}\,[(l-\varrho)+\sqrt{(l^2-\varrho^2)}]}\,; \quad r_2 = \frac{1}{2\sqrt{(l^2-\varrho^2)}},$$

from which

$$C = \frac{1}{2\ln\dfrac{\sqrt{(l^2-\varrho^2)}+(l-\varrho)}{\sqrt{(l^2-\varrho^2)}-(l-\varrho)}}.$$

6. As an example of a two-dimensional problem we consider a cylindrical condenser, formed by two infinitely long coaxial cylinders, on one of which an electric charge is uniformly distributed. It is obvious that the solution of the problem is the same for all planes, perpendicular to the axis of the cylinder. Therefore the problem may be considered in a plane and in place of the total charge we have a charge per unit length x.

If the outer cylinder of radius r_2 is earthed, and on the inner of radius r_1 there is given a charge x then the potential of the field in the condenser is given by the expression

$$\varphi = 2\varkappa \ln\frac{r_2}{r_1}.$$

and the capacity per unit length equals

$$C = \frac{1}{2\ln\dfrac{r_2}{r_1}}.$$

The example considered enables us to solve the more complex problem of the capacity of an infinitly long cylindrical conductor situated above a conducting plane. It is clear that this problem may be solved by an inversion transformation.

III. The fundamental problem of electrical prospecting

Electrical methods are extensively employed in a study of the inhomogeneity of the earth's crust with a view to searching for minerals. The fundamental idea of electrical prospecting is the following. By means of earthed electrodes, the current from a battery is conducted through the earth. On the surface of the earth the potentials of the field of constant current thus created are mapped. By means of controls on the surface the underground structure is determined. Methods of determining the underground structures (interpretation of the observations) are based on a mathematical solution of the appropriate problems.

The potential of a field of constant current in a homogeneous medium satisfies Laplace's equation

$$\nabla^2 V = 0 \qquad (z > 0) \tag{1}$$

with the additional relation

$$\left.\frac{\partial V}{\partial z}\right|_{z=0} = 0, \tag{2}$$

which indicates that the vertical component of the density of current along the surface $z = 0$ equals zero, since the semi-space $z < 0$ (air) is non-conducting.

Let us consider a point electrode on the boundary of the semi-space at point A. It is obvious that the potential of the field will equal

$$V = \frac{I\varrho}{2\pi R}, \tag{3}$$

where R is the distance from the source A, ϱ is the specific resistance of the medium, and I is the intensity of the current. This result differs from that for an infinite medium by the coefficient 2 because of condition (2).

Measuring the difference of the potentials at points M and N, lying on a straight line through A we obtain:

$$V(M) - V(N) = \frac{\partial V}{\partial r}\Delta r,$$

where Δr is the distance between the points N and M.

Assuming that the points M and N are sufficiently near one another, we obtain:

$$\frac{V(M) - V(N)}{\Delta r} \cong \left|\frac{\partial V}{\partial r}\right| \cong \frac{I\varrho}{2\pi r^2},$$

where r is the distance of point O (the centre of the receiving circuit MN) from the feeding electrode. The current intensity I

in the feeding circuit is known. Hence for the resistance of homogeneous semi-space we obtain:

$$\varrho = \frac{2\pi r^2}{I} \cdot \left| \frac{\partial V}{\partial r} \right| . \tag{4}$$

If the medium is inhomogeneous then the value ϱ, defined by (4) is called the impedance and is denoted by ϱ_k; ϱ_k will not be a constant value.

Let us consider the problem of vertical electrical probing, where layers of the earth's crust occur horizontally and their resistance depends only on the depth

$$\varrho = \varrho(z).$$

In this case the impedance will be a function of the distance $r = AO$. The question of interpretation of the results of vertical electrical probings consists of determining the function $\varrho(z)$ which gives the "electrical profile" of the medium from the known values

$$\varrho_k(r).$$

Let us consider in detail the problem of a two-layer medium, where a homogeneous layer of thickness l and resistance ϱ_0 rests on a homogeneous medium with resistance ϱ_1,

$$\varrho(z) = \begin{cases} \varrho_0 \text{ where } 0 \leqslant z < l, \\ \varrho_1 \text{ where } l < z. \end{cases}$$

It is obvious that at small distances $r \ll l$ the impedance ϱ_k equals ϱ_0, since the effect of a lower medium will be small. At large distances $(r \gg l)$ ϱ_k will equal ϱ_1.

The problem reduces, thus, to finding the solution of Laplace's equation V_0 for the layer $0 < z < l$ and V_1 for the semi-space $z > l$. For $z = l$ the conditions of continuity of the potential

$$V_0\big|_{z=l} = V_1\big|_{z=l} \tag{5}$$

must be fulfilled, and the continuity of the normal components of the current density

$$\frac{1}{\varrho_0}\frac{\partial V_0}{\partial z}\bigg|_{z=l} = \frac{1}{\varrho_1}\frac{\partial V_1}{\partial z}\bigg|_{z=l} . \tag{6}$$

For $z = 0$ the potential V_0 must satisfy condition (2), and at point A, which we choose as the origin of cylindrical coordinates (φ, r, z), the potential V_0 must have a singularity of type (3)

$$V_0 = \frac{\varrho_0 I}{2\pi} \frac{1}{\sqrt{(z^2 + r^2)}} + v_0, \tag{7}$$

where v_0 is a bounded function.

Function V_1 must be bounded at infinity. Functions v_0 and V_1 satisfy equation (1) which because of the cylindrical symmetry of the problem reduces to

$$\frac{\partial^2 V}{\partial r^2} + \frac{1}{r}\frac{\partial V}{\partial r} + \frac{\partial^2 V}{\partial z^2} = 0.$$

The method of separation of variables gives for V two types of solutions, bounded for $r = 0$:

$$e^{\pm \lambda z} J_0(\lambda r),$$

where J_0 is the Bessel function of zero order (see Supplement, part I, § 1) and λ is a separation parameter. We shall seek a solution in the form

$$V_0(r, z) = \frac{\varrho_0 I}{2\pi} \cdot \frac{1}{\sqrt{(r^2 + z^2)}} + \int_0^\infty (A_0 e^{-\lambda z} + B_0 e^{\lambda z}) J_0(\lambda r)\, d\lambda,$$

$$V_1(r, z) = \int_0^\infty (A_1 e^{-\lambda z} + B_1 e^{\lambda z}) J_0(\lambda r)\, d\lambda,$$

where A_0, B_0, A_1, B_1 are functions of λ. Condition (2) gives the relation between A_0 and B_0. Let us calculate

$$\frac{\partial V_0}{\partial z} = -\frac{\varrho_0 I}{2\pi} \cdot \frac{z}{(z^2 + r^2)^{3/2}} + \int_0^\infty (-\lambda A_0 e^{-\lambda z} + \lambda B_0 e^{\lambda z}) J_0(\lambda r)\, d\lambda.$$

Condition (2) takes the form

$$\int_0^\infty (B_0 - A_0) J_0(\lambda r) \lambda d\lambda = 0$$

for arbitrary r, from which

$$B_0 = A_0$$

From the condition of the bounded nature of V_1 as $z \to \infty$ it follows that

$$B_1 = 0.$$

Thus.

$$V_1(r, z) = \int_0^\infty A_1 e^{-\lambda z} J_0(\lambda r)\, d\lambda$$

and

$$V_0(r, z) = \int_0^\infty [q e^{-\lambda z} + A_0 (e^{-\lambda z} + e^{\lambda z})] J_0(\lambda r)\, d\lambda,$$

where we have made use of the formula

$$\frac{1}{\sqrt{(r^2 + z^2)}} = \int_0^\infty J_0(\lambda r)\, e^{-\lambda z}\, d\lambda$$

(see Supplement, part I, § 4) and we defined $\varrho_0 I / 2\pi = q$.

The remaining constants A_0 and A_1 are determined from relations (5) and (6) for $z = 1$, which reduce to a system of algebraic equations.

$$A_0 (e^{-2\lambda l} + 1) - A_1 e^{-2\lambda l} = -q e^{-2\lambda l},$$

$$\frac{1}{\varrho_0} A_0 (e^{-2\lambda l} - 1) - \frac{1}{\varrho_1} A_1 e^{-2\lambda l} = -\frac{q}{\varrho_0} e^{-2\lambda l},$$

from which it follows that

$$A_0 = q \frac{(\varrho_1 - \varrho_0) e^{-2\lambda l}}{(\varrho_1 + \varrho_0) - (\varrho_1 - \varrho_0) e^{-2\lambda l}},$$

and the solution V_0 for the upper layer is given by the relation

$$V_0(r, z) = \frac{I \varrho_0}{2\pi} \int_0^\infty \left[e^{-\lambda z} + \frac{k e^{-2\lambda l}}{1 - k e^{-2\lambda l}} (e^{-\lambda z} + e^{\lambda z}) \right] J_0(\lambda r)\, d\lambda, \quad (9)$$

where we put

$$\frac{\varrho_1 - \varrho_0}{\varrho_1 + \varrho_0} = k.$$

Let us transform the expression for V_0. Since $|k| < 1$, it is possible to write:

$$\frac{k e^{-2\lambda l}}{1 - k e^{-2\lambda l}} = \sum_{n=1}^\infty k^n \cdot e^{-2\lambda l n}$$

and

$$V_0(r, z) = \frac{I \varrho_0}{2\pi} \sum_{n=0}^{\infty} \int_0^{\infty} k^n e^{-\lambda(2nl+z)} J_0(\lambda r)\, d\lambda +$$

$$+ \frac{I \varrho_0}{2\pi} \sum_{n=1}^{\infty} \int_0^{\infty} k^n e^{-\lambda(2nl-z)} J_0(\lambda r)\, d\lambda. \tag{9'}$$

Hence using (8) we obtain:

$$V_0(r, z) = \frac{I \varrho_0}{2\pi} \left(\sum_{n=1}^{\infty} k^n \frac{1}{\sqrt{[r^2 + (z - 2nl)^2]}} + \right.$$

$$\left. + \sum_{n=0}^{\infty} k^n \frac{1}{\sqrt{[r^2 + (z + 2nl)^2]}} \right). \tag{10}$$

This expression for the solution (9) may be written down at once, if the problem is solved by the method of images. Assuming $z = 0$ we obtain the distribution of potential on the earth's surface

$$V_0(r, 0) = \frac{I \varrho_0}{2\pi} \left[\frac{1}{r} + 2 \sum_{n=0}^{\infty} \frac{k^n}{\sqrt{[r^2 + (2nl)^2]}} \right], \tag{11}$$

from which

$$\frac{\partial V_0}{\partial r} = -\frac{I \varrho_0}{2\pi} \left[\frac{1}{r^2} + 2 \sum_{n=1}^{\infty} \frac{k^n r}{[r^2 + (2nl)^2]^{3/2}} \right],$$

and for ϱ_k by formula (4) we have:

$$\varrho_k = \varrho_0 \left[1 + 2 \sum_{n=1}^{\infty} \frac{k^n r^3}{r^2 + (2nl)^2]^{3/2}} \right] =$$

$$= \varrho_0 \left[1 + 2 \sum_{n=1}^{\infty} \frac{k^n \left(\frac{\xi}{2}\right)^3}{\left[\left(\frac{\xi}{2}\right)^2 + n^2\right]^{3/2}} \right] = \varrho_0 f(\xi), \tag{12}$$

where $\xi = r/l$, $f(\xi)$ denotes the expression in brackets. For $r \ll l$ we have

$$\varrho_k \cong \varrho_0 .$$

In order to evaluate the behaviour of ϱ_k for large r, in (12) we let $r \to \infty$ ($\xi \to \infty$). The limit of the nth term of the sum will equal k^n, from which it follows that

$$\lim_{r \to \infty} \varrho_k = \varrho_0 \left(1 + 2 \sum_{n=1}^{\infty} k^n \right) = \varrho_0 \left(1 + \frac{2k}{1-k} \right) =$$

$$= \varrho_0 \frac{1+k}{1-k} = \varrho_0 \frac{\varrho_1 + \varrho_0 + (\varrho_1 - \varrho_0)}{\varrho_1 + \varrho_0 - (\varrho_1 - \varrho_0)} = \varrho_1 .$$

Comparing the experimental curve with the curve determined by (12), we can determine ϱ_0 from the values ϱ_k for small values of r, and ϱ_1 from the values ϱ_k for large values of r. The thickness of the upper conducting layer l is determined by trial. It equals that value of l for which the empirical curve agrees most closely with the curve calculated from (12).

In the case of many conducting layers the curves for ϱ_k are calculated similarly. The nature of the electrical profile of a medium is determined by fitting a theoretical curve to the empirical one. With a large number of layers the technique of interpretation is very complicated.

We note that for different conductivity profiles $\varrho_1(z) \neq \varrho_2(z)$ the corresponding impedances are also different

$$\varrho_k^{(1)}(r) \neq \varrho_k^{(2)}(r);$$

hence the problem of determining the electrical profile by the impedance from a mathematical view point has a unique solution.*

Problems, similar to the problem of electrical prospecting occur in various fields of physics and engineering.

Problems of determination of the magnetic field in an inhomogeneous medium occur, for instance, in the magnetic detection of defects. For the determination of a defect, for example the determination of the presence of a cavity under the surface, the metallic piece is placed between the poles of a magnet and the magnetic field on the surface is measured. By the perturbation of the magnetic field we try to determine the presence of a defect, and also, if possible, the size of the defect, its depth etc. Modelling methods are widely used for the solution of this type of problem.**

In fact, let us consider potential fields in inhomogeneous media of different physical nature (for example, a steady temperature field, a magnetic field in an inhomogeneous medium, an electro-

* A. N. Tikhonov, Concerning the uniqueness of the solution of the problem of electrical prospecting, *Dokl. Akad. Nauk. SSR* 69, N° 6, 797, 1949.

** A. V. Luk'yanov, Concerning the electrolytic model of three-dimensional problems, *Dokl. Akad. Nauk SSR* 75, N° 5, 1950.

static field, the velocity field of a fluid for filtration). The potential functions of these fields $u(x, y, z)$ in any homogeneous region satisfies Laplace's equation $\nabla^2 u = 0$. On the boundary of the regions G_1 and G_2 with different coefficients of heat conduction, magnetic permeability, etc., the following relation is fulfilled

$$k_1 \frac{\partial u^{(1)}}{\partial n} = k_2 \frac{\partial u^{(2)}}{\partial n},$$

where k_1 and k_2 are the appropriate physical constants.

On the boundaries of similar geometric regions for the different physical fields we may be given numerically equal values of the potentials or of their normal derivatives. Let us assume that physical inhomogeneities of these regions are geometrically similar and equally situated; the relationships of the physical constants (thermal conductivity, magnetic permeability, etc.) of any pair of corresponding inhomogeneities are also similar. Then the numerical values of the potentials of these fields at corresponding internal points are also equal, since they are the solution of the same mathematical problem, having a unique solution.

Direct measurements of temperature, magnetic and other fields are considerably more difficult than the measurement of the field of a steady current in an electrolytic tank. Therefore it is best to replace the direct measurements of these fields by measurements in an electrolytic tank.

IV. Determination of vector fields

Together with scalar problems, many questions in electrodynamics and hydrodynamics problems occur concerning the determination of vector fields with a given curl and divergence.

Let us prove that the vector field $\boldsymbol{A}$ is well defined inside some domain G, bounded by the surface S, if the curl and divergence of the field are given inside G:

$$\operatorname{curl} \boldsymbol{A} = \boldsymbol{B}, \tag{1}$$

$$\operatorname{div} \boldsymbol{A} = C \tag{2}$$

and the normal component of vector A is given on the boundary S

$$A_n|_S = f(M). \tag{3}$$

We note that the functions $\boldsymbol{B}, C$ and f cannot be arbitrary. The relations

$$\operatorname{div} \boldsymbol{B} = 0, \tag{4}$$

$$\iint_S f(M)\, dS = \iiint_G C\, d\tau. \tag{5}$$

must be fulfilled. Function f will be assumed continuous on the surface S, functions $\boldsymbol{B}$ and C continuous in G together with their derivatives and the surface S such that the second boundary-value problem with continuous boundary values is capable of solution.

The problem will be solved in several stages. Let us find the vector $\boldsymbol{A}_1$, satisfying the relations

$$\operatorname{curl} \boldsymbol{A}_1 = 0, \tag{6}$$

$$\operatorname{div} \boldsymbol{A}_1 = C. \tag{7}$$

From (6) it follows that

$$\boldsymbol{A}_1 = \operatorname{grad} \varphi. \tag{8}$$

Taking the function φ as

$$\varphi(P) = -\frac{1}{4\pi} \iiint_G \frac{C(Q)}{r_{PQ}}\, d\tau_Q, \tag{9}$$

we satisfy equation (7). Let us determine now a vector $\boldsymbol{A}_2$ so that

$$\operatorname{curl} \boldsymbol{A}_2 = \boldsymbol{B}, \tag{10}$$

$$\operatorname{div} \boldsymbol{A}_2 = 0. \tag{11}$$

Assuming

$$\boldsymbol{A}_2 = \operatorname{curl} \boldsymbol{\psi}, \tag{12}$$

we satisfy equation (11). The vector $\boldsymbol{\psi}$ may be chosen so that

$$\operatorname{div} \boldsymbol{\psi} = 0, \tag{13}$$

If (13) is not fulfilled, then, without violating condition (12) we may assume

$$\boldsymbol{\psi}^* = \boldsymbol{\psi} + \operatorname{grad} \chi, \tag{14}$$

where function χ is determined from the relation $\operatorname{div} \boldsymbol{\psi}^* = 0$, i.e.

$$\Delta\chi = -\operatorname{div} \boldsymbol{\psi}. \tag{15}$$

Let us return to the determination of vector $\boldsymbol{A}_2$. From equations (10) and (12) it follows that

$$\Delta\boldsymbol{\psi} = -\boldsymbol{B}. \tag{16}$$

Therefore assuming

$$\boldsymbol{\psi}(P) = \frac{1}{4\pi} \iiint_G \frac{\boldsymbol{B}(Q)}{r_{PQ}} d\tau_Q, \tag{17}$$

we satisfy equation (10). It is obvious that the vector $\boldsymbol{A}_1 + \boldsymbol{A}_2$ satisfies the relations

$$\operatorname{curl} (\boldsymbol{A}_1 + \boldsymbol{A}_2) = \boldsymbol{B}, \tag{18}$$

$$\operatorname{div} (\boldsymbol{A}_1 + \boldsymbol{A}_2) = C, \tag{19}$$

In order to find the vector $\boldsymbol{A}$ it remains for us to satisfy the boundary relation (2). In order to do this let us find vector $\boldsymbol{A}_3$, satisfying the relations inside G:

$$\operatorname{curl} \boldsymbol{A}_3 = 0, \tag{20}$$

$$\operatorname{div} \boldsymbol{A}_3 = 0, \tag{21}$$

and on S

$$A_n|_S = f(M) - A_{1n}|_S - A_{2n}|_S = f^*(M). \tag{22}$$

It is obvious that the function $f^*(M)$ is well defined. From equation (20) it follows that

$$\boldsymbol{A}_3 = \operatorname{grad} \theta.$$

Substituting this value of $\boldsymbol{A}_3$ in (21) we obtain inside G:

$$\Delta\theta = 0; \tag{23}$$

condition (22) gives

$$\left.\frac{\partial\theta}{\partial n}\right|_S = f^*(M), \tag{24}$$

i.e. we obtain the second boundary-value problem for the determination of function θ. Therefore vector $\boldsymbol{A}_3$ is well defined.

Thus, we have proved that problem (1)–(3) has a unique solution

$$\boldsymbol{A} = \boldsymbol{A}_1 + \boldsymbol{A}_2 + \boldsymbol{A}_3.$$

V. Application of the method of conformal mapping to electrostatics

1. For the solution of two-dimensional electrostatic problems one often uses the theory of functions of a complex variable. Let us consider, for example, the following problem of electrostatics:

Find the electric field of several charged conductors, the potentials of which equal $u_1, u_2, \ldots$

Such a problem, as is well known (see Appendix II), leads to the equation

$$\nabla^2 u = 0 \tag{1}$$

with boundary conditions

$$u\,|_{S_i} = u_i, \tag{2}$$

where the surface of the conductor with number i is denoted by S_i. If the field can be assumed plane, and invariant, for instance along the z-axis, then equation (1) and the boundary conditions take the form:

$$\frac{\partial^2 u}{\partial x^2} + \frac{\partial^2 u}{\partial y^2} = 0, \tag{3}$$

$$u\,|_{C_i} = u_i, \tag{4}$$

where C_i is the contour, bounding region S_i.

We shall search for the potential u as the imaginary part of some analytic function

$$f(z) = v(x, y) + iu(x, y) \qquad (z = x + iy), \tag{5}$$

which satisfies (3) because of the Cauchy-Riemann relations

$$v_x = u_y, \quad v_y = -u_x \tag{6}$$

and

$$v_x v_y + u_x u_y = 0. \tag{7}$$

From the boundary condition (4) it follows that $f(z)$ has a constant imaginary part on the contours C_i, bounding our conductors.

Returning to relation (6) we observe that

$$v(x, y) = \text{const} \tag{8}$$

represents the equation of a family of lines of force* while the equation

$$u(x, y) = \text{const} \tag{9}$$

by virtue of relation (7) determines a family of equipotential lines.

Thus, for the solution of the problem it is sufficient to find the conformal transformation

$$w = f(z),$$

changing the plane of the complex variable

$$z = x + iy$$

into the plane

$$w = v + iu,$$

for which the boundaries of the conductors change into the straight lines

$$u = \text{const}$$

or

$$\operatorname{Im} w = \text{const}.$$

If such a function $w = f(z)$ is known, then the unknown potential is found from the formula

$$u = u(x, y) = \operatorname{Im} f(z).$$

Knowing the potential, it is possible to calculate the electric field

$$E_x = -\frac{\partial u}{\partial x}, \quad E_y = -\frac{\partial u}{\partial y} \tag{10}$$

and the density of the surface charges per unit length along the z-axis:

$$\sigma = \frac{1}{4\pi}\sqrt{(E_x^2 + E_y^2)} = \frac{1}{4\pi}\sqrt{\left[\left(\frac{\partial u}{\partial x}\right)^2 + \left(\frac{du}{dy}\right)^2\right]},$$

which because of the Cauchy–Riemann relations is

$$\sigma = \frac{1}{4\pi}|f'(z)|. \tag{11}$$

* In fact, the equation of the lines of force is $dx/u_x = dy/u_y$. Substituting u_x and u_y according to relation (6) by $-v_y$ and v_x, we obtain $v_x\,dx + v_y\,dy = dv = 0$ or $v(x, y) = \text{const}$.

2. *The field of a semi-infinite plane condenser.* Let us find the field of a condenser, formed by infinitely thin metallic plates $y = -d/2$ and $y = d/2$, extending in the region $x < 0$. Not dwelling on the derivation of the conformal transformation, changing the region, described in Fig. 71 into a strip $|\text{Im } w| \leqslant \pi$, we apply it directly to the solution of the specified problem.*

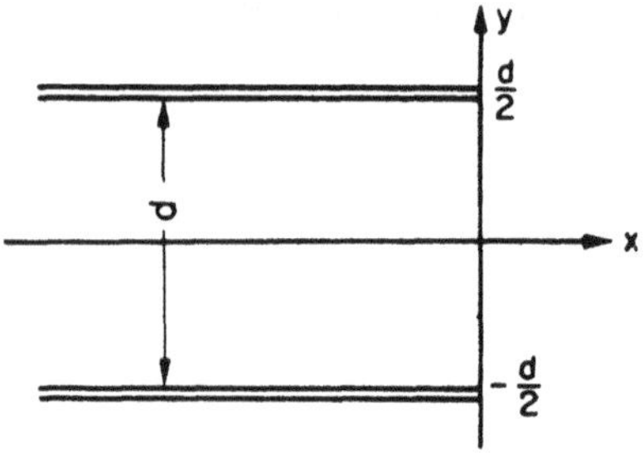

FIG. 71

The transformation

$$z = \frac{d}{2\pi}(w + e^w)$$

$$(w = \varphi + i\psi) \qquad (12)$$

changes the plane $z = x + iy$ with two sections ($y = \pm d/2$, $x < 0$) into the layer $|\psi| \leqslant \pi$ of the plane $w = \varphi + i\psi$ (Fig. 72). As a complex potential let us choose the function

$$\frac{u_0}{2\pi} w, \qquad (13)$$

where the difference in potential between the plates of a condenser is denoted by u_0, so that the potential of the electric field is expressed by the function

$$u(x, y) = \frac{u_0}{2\pi} \psi, \qquad (14)$$

where ψ is connected to x and y by the relations

$$\left.\begin{aligned} x &= \frac{d}{2\pi}(\varphi + e^{\varphi} \cos \psi), \\ y &= \frac{d}{2\pi}(\psi + e^{\varphi} \sin \psi), \end{aligned}\right\} \qquad (15)$$

In Fig. 73 the equipotentials and lines of force of a semi-infinite plane condenser are plotted.

Let us proceed to an investigation of the field of the condenser. From (15) it is seen that as $\varphi \to -\infty$

$$x \approx \frac{d}{2\pi}\varphi, \qquad y \approx \frac{d}{2\pi}\psi, \qquad (16)$$

* See Frank and Mises, *Differential and Integral Equations of Mathematical Physics*, vol. II, chap. XV, § 5, 1937.

i.e. inside the condenser, far from its edge, the field is plane and as $\varphi \to \infty$

$$\varrho = \sqrt{(x^2 + y^2)} \approx \frac{d}{2\pi} e^{\varphi}, \qquad \theta = \arctan \frac{y}{x} \approx \psi, \tag{17}$$

i.e. outside the condenser, at large distances from its edge, the equipotential lines are circles.

If instead of w we introduce the complex potential $f = u_0 w/2\pi$

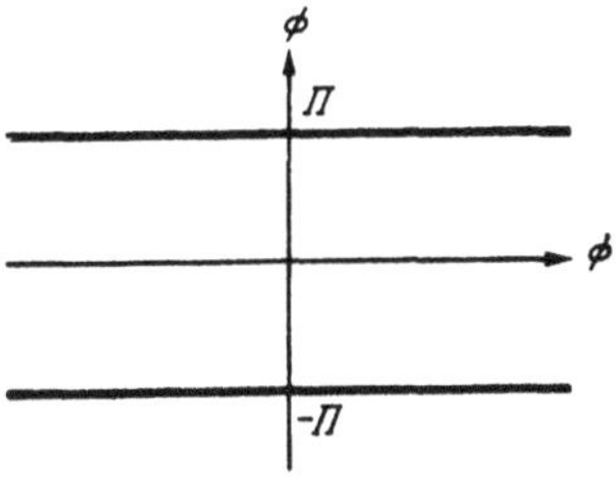

FIG. 72

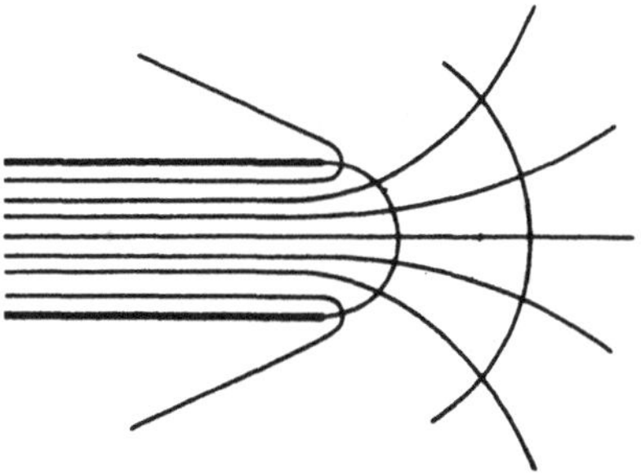

FIG. 73

so that $w = 2\pi f/u_0$, then the relation between z and $f(z)$ is given by the equation

$$z = d\left(\frac{f}{u_0} + \frac{1}{2\pi} e^{\frac{2\pi f}{u_0}}\right)$$

from which it follows:

$$\frac{dz}{df} = \frac{d}{u_0}\left(1 + e^{\frac{2\pi f}{u_0}}\right),$$

and for $f = u_0/2\pi \cdot (\varphi \pm \pi i)$ we obtain:

$$\frac{dz}{df} = \frac{d}{u_0}(1 - e^{\varphi}) \quad \text{or} \quad f'(z) = \frac{u_0}{d(1 - e^{\varphi})}.$$

Assuming $u_0 = 1$, we obtain the following relation for the charge density σ:

$$\sigma = \frac{|f'(z)|}{4\pi} = \frac{1}{4\pi d\,|1 - e^{\varphi}|}. \tag{18}$$

Hence it follows that as $\varphi \to -\infty$, $\sigma \approx 1/4\pi d$, and as $\varphi \to \infty$, $\sigma \approx 1/(4\pi\, d e^{\varphi})$, i. e. in this case the charge density decreases on the outer side of the plates as $1/\varrho$.

From (18) it is seen that for $\varphi = 0$ (at the edge of the condenser) $\sigma = \infty$. In fact, the end of the plane plate has an infinite curvature and in order to charge it to some potential it is necessary to place an infinite charge on it.

The range of problems, solvable by the method of conformal transformation, is very extensive. By means of it the problem of the end of a thick walled plane condenser can successfully be solved, as also a series of problems, concerning the effect of bends in the condenser etc. A disadvantage of the method is that the conformal transformation may be applied only to plane problems, reducible to the two-dimensional equation $\nabla_2^2 u = 0$.

VI. Application of the method of conformal mapping to hydrodynamics

1. The solution of problems of the motion of a solid body in a fluid involve the boundary relations on the surface of the body.

In the case of an ideal fluid the boundary relation requires that the component v_n of the velocity of the fluid normal to the surface of the body should equal the normal component of the velocity of motion of the body.

If the body is fixed, then the boundary relation takes the simple form

$$v_n = 0$$

on the surface of the body.

If the flow under consideration is potential, i.e.

$$v = \operatorname{grad} \varphi,$$

then the boundary conditions take the form

$$\left.\frac{\partial \varphi}{\partial n}\right|_S = 0 \text{ in the case of a rigid body,}$$

$$\left.\frac{\partial \varphi}{\partial n}\right|_S = u_n \text{ in the case of a body, moving with velocity } u.$$

As is well known from hydrodynamics, the velocity potential for an incompressible fluid satisfies the equation

$$\nabla^2 \varphi = 0.$$

Thus, the problem of the potential flow of an incompressible ideal fluid around a rigid body reduces to the solution of Laplace's equation

$$\nabla^2\varphi = 0$$

with the additional boundary condition on the surface of a streamline body

$$\left.\frac{\partial\varphi}{\partial n}\right|_S = u_n ,$$

i.e. to the solution of the second boundary value problem for Laplace's equation.

If the motion is plane, then the solution of the problem can be obtained by means of the theory of functions of a complex variable.

In the case of plane motion of an incompressible fluid the equation of continuity gives:

$$\frac{\partial v_x}{\partial x} = \frac{\partial(-v_y)}{\partial y} . \tag{1}$$

Let us write the equation of the lines of flow

$$\frac{dx}{v_x} = \frac{dy}{v_y}$$

in the form

$$v_x\,dy - v_y\,dx = 0 \tag{2}$$

and let us introduce the function ψ by means of the relations

$$v_x = \frac{\partial\psi}{\partial y}\,;\quad v_y = -\frac{\partial\psi}{\partial x} .$$

Then from equation (1) it follows that the left-hand side of (2) is the complete differential of function ψ:

$$v_x\,dy - v_y\,dx = d\psi .$$

The single parametric family of curves

$$\psi(x, y) = C$$

represents lines of flow of the incompressible fluid.

If there exists a velocity potential, then the equality curl $\boldsymbol{v} = 0$ is equivalent to the equation

$$\nabla^2\psi = 0 .$$

From the expressions for v_x and v_y it follows:

$$\frac{\partial \varphi}{\partial x} = \frac{\partial \psi}{\partial y},$$

$$\frac{\partial \varphi}{\partial y} = -\frac{\partial \psi}{\partial x},$$

i. e. the functions φ and ψ satisfy the Cauchy–Riemann relations. Hence the function of the complex variable

$$w(z) = \varphi(x, y) + i\psi(x, y)$$

is analytic.

Thus, any potential plane motion of a fluid corresponds to a given analytic function of a complex variable, and, conversely, any analytic function is connected with a given kinematic picture of the motion of a fluid (more precisely, with two pictures, since the functions φ and ψ can change roles).

Let us consider concrete examples of the application of the theory of analytic functions to the solution of problems concerning the plane flow of fluid around bodies.

2. *Flow around a circular cylinder.* We consider flow of fluid around a circular cylinder of radius $r = a$, when the flow has a constant velocity u at infinity. In the case of steady motion it is possible to consider the equivalent problem of motion of a cylinder with constant velocity u relative to the fluid.

We take a fixed system of coordinates with the axis Ox parallel to the velocity of motion of the cylinder.

Obviously the boundary condition

$$\frac{\partial \psi}{\partial s} = u\frac{\partial y}{\partial s},$$

is fulfilled on the surface of the body moving in the fluid, where ds is an element of an arc on the rim of the body.

In the case of uniform motion with velocity u this condition can be integrated over the surface of the body, and we obtain:

$$\psi = uy + C$$

on the surface of the body.

Thus, our problem reduces to the solution of the equation

$$\nabla^2\psi = 0$$

with boundary conditions:

(1) $\psi = uy + C$ on the surface of the cylinder,

(2) $\frac{\partial \varphi}{\partial x}$ and $\frac{\partial \varphi}{\partial y}$ tend to zero at infinity.

The latter condition indicates that the function

$$\frac{dw}{dz} = \frac{\partial \varphi}{\partial y} + i \frac{\partial \varphi}{\partial x} = v_x - iv_y$$

is a well-defined analytic function outside the circle C, reducing to zero at infinity. This allows us to represent the function w in the form

$$w = C_1 \ln z - \frac{C_2}{z} - \frac{C_3}{z^2} + \dots$$

Assuming

$$C_k = A_k + iB_k ,$$

we determine the constants A_k and B_k from the boundary relation

$$\psi = ua \sin \theta + C ,$$

converting to the polar coordinates $z = ae^{i\theta}$.

For the constants we obtain the expressions

$$A_1 = 0 ; \; A_2 = ua^2 ; \; B_2 = 0 ; \; A_3 = B_3 = 0 ;$$

$$B_1 = -\frac{\Gamma}{2\pi} .$$

Hence

$$w = \frac{\Gamma}{2\pi i} \ln z - u \frac{a^2}{z} ;$$

$$\varphi = \frac{\Gamma}{2\pi} \theta - u \cos \theta \frac{a^2}{r} ;$$

$$\psi = -\frac{\Gamma}{2\pi} \ln r + u \sin \theta \frac{a^2}{r} .$$

The first term in the expression for w implies a circulation around the cylinder of intensity Γ. In the simplest case of no circulation we obtain

$$w = -u \frac{a^2}{z} .$$

The complex potential for the current, flowing around a stationary cylinder and having a velocity u at infinity has the form

$$w = uz + \frac{ua^2}{z} + \frac{\Gamma}{2\pi i} \ln z .$$

3. *The flow around plates.* The results obtained for the flow around a circular cylinder enable us to solve problems of flow around arbitrary contours by using the method of conformal transformation. Let us consider its application in the problem of the flow around plates.

We consider plane flow around an infinitely long plate of width $2a$, situated on the axis Ox (Fig. 74)

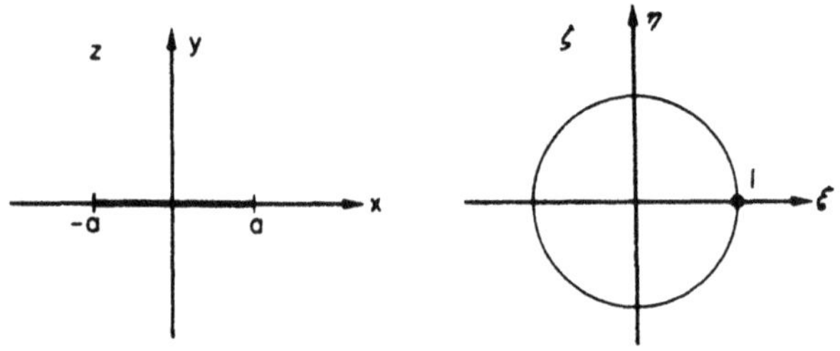

FIG. 74

The velocity has components u and v at infinity. By means of the analytic function

$$z = \frac{a}{2}\left(\zeta + \frac{1}{\zeta}\right) = f(\zeta)$$

it is possible to establish a reciprocal correspondence between the region around the plate in the plane z and the region outside a circle of unit radius in the plane ζ. In addition the point $z = \infty$ will correspond to the point $\zeta = \infty$, and

$$\frac{dz}{d\zeta} = \frac{a}{2} > 0 \text{ where } \zeta = \infty .$$

We observe that it changes the boundary condition at infinity. For the complex potential

$$w(z) = \varphi + i\,\psi$$

we have

$$\left(\frac{dw}{dz}\right)_{z=\infty} = u - iv = \bar{v}_\infty$$

the conjugate value of the complex velocity.

Let us find the value of the complex velocity of the corresponding current on the plane ζ

$$w(\zeta) = w[f(z)]; \quad \frac{dw}{d\zeta} = \frac{dw}{dz}\cdot\frac{dz}{d\zeta},$$

from which

$$\left(\frac{dw}{d\zeta}\right)_{\zeta=\infty} = k\bar{v}_\infty \quad \left(k = \frac{a}{2}\right).$$

Thus, the corresponding current represents the flow around a cylinder of unit radius having a complex velocity kv_∞ at infinity. For such a motion the complex potential has the form

$$w(\zeta) = k\bar{v}_\infty \zeta + \frac{kv_\infty}{\zeta} + \frac{\Gamma}{2\pi i}\ln\zeta.$$

From the relation $z = f(\zeta)$ it follows:

$$\zeta = \frac{z + \sqrt{(z^2 - a^2)}}{a}; \quad \frac{1}{\zeta} = \frac{z - \sqrt{(z^2 - a^2)}}{a}.$$

Using these relations we obtain the complex potential of a fluid, flowing around a plate,

$$w(z) = uz - iv\sqrt{(z^2 - a^2)} + \frac{\Gamma}{2\pi i}\ln\left(\frac{z + \sqrt{(z^2 - a^2)}}{a}\right).$$

In the case of no circulation this expression takes the form

$$w(z) = uz - iv\sqrt{(z^2 - a^2)}.$$

From the relations obtained we see that the velocity is infinite at the ends of the plate. Under real conditions this, naturally, does not occur. Our results are explained by the fact that we assume the fluid to be ideal. Applying Bernoulli's theorem we can find an expression for the force acting on a streamline in the fluid.

Aerodynamic wing theory deals with the study of the forces, with which air acts on the wings of an aeroplane moving in it. In the development of this theory, the principal role belongs to soviet scientists, in the first place to N. E. Zhukovskii and S. A. Chaplygin. In the simplest case of no circulation around a cylinder we obtain a paradoxical result that the flow does not exert any force on the cylinder. In the case of superposition of a circulation around the cylinder on the forward flow there is a force, acting on the cylinder perpendicular to the direction of the velocity of the flow at infinity.

The theory of analytic functions can be used only in the case of plane motion. In three-dimensional problems one has use other methods to solve problems of fluid flow around a solid body. Let us consider the simplest case of a sphere moving in an infinite fluid with constant velocity. The problem requires the solution of the equation

$$\nabla^2\varphi = 0$$

with the boundary condition

$$\left.\frac{\partial\varphi}{\partial n}\right|_{r=a} = u \cos \theta$$

on the surface of the sphere and

$$\frac{\partial\varphi}{\partial x} = \frac{\partial\varphi}{\partial y} = \frac{\partial\varphi}{\partial z} = 0$$

at infinity.

The solution has the form

$$\varphi = A\boldsymbol{u}\cdot \text{grad}\ \frac{1}{r}.$$

Using the boundary condition we obtain:

$$\varphi = -\frac{a^3\,\boldsymbol{u}\cdot\boldsymbol{r}}{2\,r^3},$$

which gives the solution of the problem.

In all the cases treated we have assumed the fluid to be ideal. For a viscous fluid the boundary conditions are changed. The condition of adhesion should be fulfilled on the surface of the body namely: at points of a solid boundary the velocity of the fluid should be equal to the velocity of the corresponding point of the boundary.

Problems of the flow around bodies by a viscous fluid introduce great mathematical difficulties. In the development of this field of hydrodynamics, theories of the boundary layer have played an important part.

VII. The biharmonic equation

In Appendix I to Chapter II the equation of the transverse vibrations of a rod was obtained

$$\frac{\partial^2 u}{\partial t^2} + a^2 \frac{\partial^4 u}{\partial x^4} = 0. \tag{1}$$

The problem of the vibrations of a thin plate, clamped at the edges leads to a similar equation*

$$\frac{\partial^2 u}{\partial t^2} + a^2 \left(\frac{\partial^4 u}{\partial x^4} + \frac{\partial^4 u}{\partial y^4} + 2 \frac{\partial^4 u}{\partial x^2 \partial y^2} \right) = 0 \text{ or } \frac{\partial^2 u}{\partial t^2} + a^2 \nabla^2 \nabla^2 u = 0 \tag{2}$$

with the boundary conditions

$$u = 0 \text{ and } \frac{\partial u}{\partial n} = 0 \text{ on the edge} \tag{3}$$

In addition, the function u should satisfy the initial conditions

$$u(x, y, 0) = \varphi(x, y); \quad \frac{\partial u}{\partial t}(x, y, 0) = \psi(x, y). \tag{4}$$

If an external force distributed with density $f(x, y)$ acts on the plate, then the static deflection of the plate clamped at the edge will be determined from the equation

$$\nabla^2 \nabla^2 u = f \tag{5}$$

with the boundary conditions

$$u = 0 \text{ and } \frac{\partial u}{\partial n} = 0. \tag{3}$$

The equation

$$\nabla^2 \nabla^2 u = 0 \tag{5'}$$

is called *biharmonic*, and its solutions, having derivatives up to 4th order inclusive, are called *biharmonic functions*.

The main boundary value problem for the biharmonic equation is given in the following manner:

Find the function $u(x, y)$, continuous together with the first derivative in the closed region $S + C$, having derivatives up to

* V. I. Smirnov, *Course in Higher Mathematics*, vol. III, Gostekhizdat, 1949.

the fourth order in S, satisfying equation (5) or (5′) inside S and the boundary conditions on C

$$u|_C = g(s); \qquad \frac{\partial u}{\partial n}\bigg|_C = h(s), \tag{6}$$

where $g(s)$ and $h(s)$ are continuous functions of the arc s of contour C.

In the solution of problems (2)–(4) by the method of separation of variables we assume solutions

$$u(x, y, t) = v(x, y)\, T(t). \tag{7}$$

Substituting this expression in (2) and separating the variables, we arrive at the eigen-value problem

$$\nabla^2\nabla^2 v - \lambda v = 0 \tag{8}$$

with boundary conditions

$$v = 0, \frac{\partial v}{\partial n} = 0 \quad \text{on} \quad C. \tag{9}$$

1. Uniqueness of the solution.

Let us prove that *the biharmonic equation*

$$\nabla^2\nabla^2 u = 0$$

with the boundary conditions

$$u|_C = g(s), \qquad \frac{\partial u}{\partial n}\bigg|_C = h(s) \tag{3′}$$

has a unique solution.

Let there exist two solutions u_1 and u_2. We consider their difference

$$v = u_1 - u_2.$$

The function v satisfies the biharmonic equation (5) and the homogeneous boundary conditions

$$v|_C = 0, \qquad \frac{\partial v}{\partial n}\bigg|_C = 0.$$

Applying Green's theorem

$$\int\limits_G (\nabla^2\varphi \cdot \psi - \varphi\, \nabla^2\psi)\, dS = \int\limits_C \left(\psi\, \frac{\partial \varphi}{\partial n} - \varphi\, \frac{\partial \varphi}{\partial n}\right) ds$$

to the functions $\varphi = v$, $\psi = \nabla^2 v$, we obtain:

$$\int_G (\Delta v)^2 \, dS = 0 ,$$

from which

$$\nabla^2 v = 0 .$$

Taking into consideration that $v|_c = 0$ we obtain

$$v \equiv 0 \quad \text{and} \quad u_1 \equiv u_2 .$$

Hence the biharmonic function is well defined by the boundary conditions (3).

2. *Representation of biharmonic functions by harmonic functions*

Let us prove the following theorem.

If u_1 and u_2 are two functions harmonic in some region G then the function $u = xu_1 + u_2$ is biharmonic in region G.

For the proof we make use of the identity

$$\nabla^2 (\varphi\psi) = \varphi \nabla^2 \psi + \psi \nabla^2 \varphi + 2 \left(\frac{\partial \varphi}{\partial x} \frac{\partial \psi}{\partial x} + \frac{\partial \varphi}{\partial y} \frac{\partial \psi}{\partial y} \right) . \tag{10}$$

Assuming

$$\varphi = x , \quad \psi = u_1 ,$$

we find:

$$\nabla^2 (xu_1) = 2 \frac{\partial u_1}{\partial x} . \tag{11}$$

Applying the operator ∇^2 once more, taking into account that $\nabla^2 \nabla^2 u_2 = 0$, we obtain:

$$\nabla^2 \nabla^2 (xu_1 + u_2) = 0 .$$

If region G is such that every straight line, parallel to the x-axis, intersects its boundary in not more than two points, then the converse theorem holds.

For every biharmonic function u given in region G there exist harmonic functions u_1 and u_2 such that

$$u = xu_1 + u_2 .$$

To prove this statement it is sufficient to show that there is a function u_1, satisfying the two conditions

$$\nabla^2 u_1 = 0 \tag{12}$$

and

$$\nabla^2(u - xu_1) = 0\,. \tag{13}$$

From (13) and (11) it follows:

$$\nabla^2 u = \nabla^2 (xu_1) = 2\frac{\partial u_1}{\partial x}\,. \tag{14}$$

Equation (14) is satisfied by the function

$$\bar{u}_1(x, y) = \int_{x_0}^{x} \frac{1}{2} \nabla^2 u(\xi, y)\, d\xi.$$

Since

$$\frac{\partial}{\partial x} \nabla^2 \bar{u}_1 = \nabla^2 \frac{\partial}{\partial x} \bar{u}_1 = \frac{1}{2} \nabla^2 \nabla^2 u = 0\,,$$

then $\Delta \bar{u}_1$ depends only on y:

$$\nabla^2 \bar{u}_1 = v(y)\,.$$

Let us define a function $\bar{\bar{u}}_1(y)$ so that

$$\nabla^2 \bar{\bar{u}}_1 = \frac{\partial^2 \bar{\bar{u}}_1}{\partial y^2} = -v(y)\,,$$

and let us assume

$$u_1 = \bar{u}_1 + \bar{\bar{u}}_1\,.$$

This function, obviously, will satisfy both conditions (12) and (13).

Let us consider other ways of representing biharmonic functions. We assume that the origin of coordinates is chosen inside region G and that any ray, emerging from the origin, intersects the boundary of region G at one point. Then *any function u biharmonic in G can be represented by means of the two harmonic functions u_1 and u_2 in the form*

$$u = (r^2 - r_0^2)\, u_1 + u_2\,. \tag{15}$$

Here $r^2 = x^2 + y^2$, and r_0 is a constant.

This is proved similarly by means of identity (10) and the relations

$$\nabla^2 r^2 = 4; \qquad \frac{\partial u_1}{\partial r} = \frac{\partial u_1}{\partial x}\frac{\partial x}{\partial r} + \frac{\partial u_1}{\partial y}\frac{\partial y}{\partial r}\,.$$

3. *Solution of the biharmonic equation for a circle.* Let us consider a circle of radius r_0 with centre at the origin of coordinates and let us search for the biharmonic function, satisfying the boundary conditions (6) at $r = r_0$. As was shown above, the unknown function can be represented in the form of a sum

$$u = (r^2 - r_0^2)\, u_1 + u_2\,, \tag{15}$$

where u_1 and u_2 are harmonic functions. From the boundary conditions we find:

$$u_2 \big|_{r=r_0} = g\,. \tag{16}$$

Hence u_2 is the solution of the first boundary value problem for Laplace's equation and can be represented by means of Poisson's integral

$$u_2 = \frac{1}{2\pi} \int_0^{2\pi} \frac{(r_0^2 - r^2)\, g\, da}{r^2 + r_0^2 - 2\, r r_0 \cos (a - \theta)}\,. \tag{17}$$

From the second boundary condition we obtain:

$$2\, r_0 u_1 + \frac{\partial u_2}{\partial r}\bigg|_{r=r_0} = h\,. \tag{18}$$

It is easy to satisfy ourselves by direct differentiation that the function

$$2\, r_0 u_1 + \frac{r}{r_0} \frac{\partial u_2}{\partial r} \tag{19}$$

satisfies Laplace's equation and therefore can be expressed by Poisson's integral

$$2\, r_0 u_1 + \frac{r}{r_0} \frac{\partial u_2}{\partial r} = \frac{1}{2\,\pi} \int_0^{2\pi} \frac{(r_0^2 - r^2)\, h\, da}{r^2 + r_0^2 - 2\, r r_0 \cos (a - \theta)}\,. \tag{20}$$

Having differentiated (17) with respect to r and substituting the value $\frac{\partial u_2}{\partial r}$ in (20) we find u_1. Substituting for u_1 and u_2 in formula (15) we obtain:

$$u = \frac{1}{2\,\pi\, r_0} (r^2 - r_0^2)^2 \left[\frac{1}{2} \int_0^{2\pi} \frac{-\, h\, da}{r^2 + r_0^2 - 2\, r r_0 \cos (a - \theta)} + \right.$$

$$\left. + \int_0^{2\pi} \frac{g\, [r_0 - r \cos (a - \theta)]\, da}{[r^2 + r_0^2 - 2\, r r_0 \cos (a - \theta)]^2} \right].$$

CHAPTER V

WAVE PROPAGATION IN SPACE

§ 1. Problem with initial conditions. Method of averaging

1. *The method of averaging*

In the present section we shall consider the wave equation in space

$$\nabla^2 u - \frac{1}{a^2}\frac{\partial^2 u}{\partial t^2} = -f \tag{1}$$

and shall search for its solution for the following initial conditions:

$$\left.\begin{aligned} u(x, y, z, 0) &= \varphi(x, y, z), \\ u_t(x, y, z, 0) &= \psi(x, y, z). \end{aligned}\right\} \tag{2}$$

This problem is important in the theory of the propagation of sound, in the theory of the propagation of electromagnetic fields and in other parts of physics. In the present section for simplicity we shall assume $f = 0$.

In order to determine the function u at some point M_0 let us introduce the new function

$$\bar{u}(r, t) = \frac{1}{4\pi r^2}\iint_{S_r^{M_0}} u\, dS = \frac{1}{4\pi}\iint_{S_r^{M_0}} u\, d\Omega, \tag{3}$$

where $S_1^{M_0}$ is a sphere about the point $M_0(x_0, y_0, z_0)$ of radius r, and $d\Omega = dS/r^2$ is the element of solid angle, corresponding to dS. Obviously, $\bar{u}(r, t)$ is an average value of the function u on the sphere $S_r^{M_0}$ of radius r with centre at M_0. We note at once that

$$u(M_0, t_0) = \bar{u}(0, t_0).$$

Let us show that $\bar{u}$ satisfies the homogeneous equation (1). The Laplacian operator $\nabla^2 u$ in a spherical system of coordinates equals

$$\nabla^2 u = \frac{1}{r^2}\frac{\partial}{\partial r}\left(r^2\frac{\partial u}{\partial r}\right) + \frac{1}{r^2 \sin\theta}\frac{\partial}{\partial\theta}\left(\sin\theta\frac{\partial u}{\partial\theta}\right) + \frac{1}{r^2\sin^2\theta}\frac{\partial^2 u}{\partial\varphi^2}.$$

Since $\bar{u}$ does not depend on θ or φ the equation

$$\frac{1}{r^2}\frac{\partial}{\partial r}\left(r^2\frac{\partial \bar{u}}{\partial r}\right)-\frac{1}{a^2}\frac{\partial^2 \bar{u}}{\partial t^2}=0$$

should be satisfied. Let us prove this. Integration of (1) with $f=0$ over the volume $T_r^{M_0}$, bounded by the sphere $S_r^{M_0}$, gives:

$$\frac{1}{a^2}\frac{\partial^2}{\partial t^2}\iiint\limits_{T_r^{M_0}} u\,d\tau=\iiint\limits_{T_r^{M_0}} \nabla^2 u\,d\tau=\iint\limits_{S_r^{M_0}}\frac{\partial u}{\partial n}\,dS\,.$$

Taking into account that the normal to $S_r^{M_0}$ is directed along the radius and that an element of surface equals $dS=r^2d\Omega$, we find:

$$\iint\limits_{S_r^{M_0}}\frac{\partial u}{\partial n}\,dS=\iint\limits_{S_r^{M_0}}\frac{\partial u}{\partial r}\,r^2\,d\Omega=r^2\frac{\partial}{\partial r}\Big[\iint\limits_{S_r^{M_0}} u\,d\Omega\Big]\,.$$

Hence, substituting in the left-hand side the expression $d\tau=$ $=r^2dr\,d\Omega$ for the element of volume, we obtain:

$$\frac{1}{a^2}\frac{\partial^2}{\partial t^2}\int_0^r r'^2\,dr'\Big[\iint\limits_{S_{r'}^{M_0}} u\,d\Omega\Big]=r^2\frac{\partial}{\partial r}\Big[\iint\limits_{S_r^{M_0}} u\,d\Omega\Big]\,.$$

By differentiating the latter equation with respect to r and dividing it by r^2, we can show that the equation for $\bar{u}$

$$\frac{1}{a^2}\frac{\partial^2\bar{u}}{\partial t^2}=\frac{1}{r^2}\frac{\partial}{\partial r}\left(r^2\frac{\partial\bar{u}}{\partial r}\right)$$

is valid. We note the following property of the Laplacian operator: for any function $U=U(r,t)$, not depending on φ or θ, the Laplacian operator takes the form:

$$\nabla^2 U=\frac{1}{r^2}\frac{\partial}{\partial r}\left(r^2\frac{\partial U}{\partial r}\right)=\frac{1}{r}\frac{\partial^2(rU)}{\partial r^2}\,.$$

Making use of this property we see that for any spherically symmetrical solution $\bar{u}$ of the wave equation in space the substitution $v=r\bar{u}$ leads to the one-dimensional wave equation

$$\frac{\partial^2 v}{\partial r^2}=\frac{1}{a^2}\frac{\partial^2 v}{\partial t^2}\,.$$

The general solution of this equation has the form

$$v = r\bar{u} = f_1\left(t - \frac{r}{a}\right) + f_2\left(t + \frac{r}{a}\right).$$

For $r = 0$ because of the bounded nature of $\bar{u}$ we find that

$$0 = f_1(t) + f_2(t) \text{ where } f_1(t) = -f_2(t) = -f(t)$$

for any value of t, i.e.

$$v = r\bar{u} = f\left(t + \frac{r}{a}\right) - f\left(t - \frac{r}{a}\right).$$

Differentiation of the latter relation with respect to r gives:

$$\bar{u} + r\frac{\partial \bar{u}}{\partial r} = \frac{1}{a}\left[f'\left(t - \frac{r}{a}\right) + f'\left(t + \frac{r}{a}\right)\right],$$

from which at $r = 0$ and $t = t_0$ it follows:

$$\bar{u}(0, t_0) = u(M_0, t_0) = \frac{2}{a}f'(t_0).$$

For the solution of the original problem with initial conditions it remains to express the function $f(t)$ in terms of the original functions φ and ψ. Adding together the equations

$$\frac{\partial}{\partial r}(r\bar{u}) = \frac{1}{a}\left[f'\left(t - \frac{r}{a}\right) + f'\left(t + \frac{r}{a}\right)\right],$$

$$\frac{1}{a}\frac{\partial}{\partial t}(r\bar{u}) = \frac{1}{a}\left[-f'\left(t - \frac{r}{a}\right) + f'\left(t + \frac{r}{a}\right)\right],$$

we find:

$$\frac{\partial}{\partial r}(r\bar{u}) + \frac{1}{a}\frac{\partial}{\partial t}(r\bar{u}) + \frac{2}{a}f'\left(t + \frac{r}{a}\right).$$

Assuming $t = 0$ and $r = at_0$, we obtain:

$$u(M_0, t_0) = \left[\frac{\partial}{\partial r}(r\bar{u}) + \frac{1}{a}\frac{\partial}{\partial t}(r\bar{u})\right]_{r=at_0,\ t=0}$$

or after substitution of $\bar{u}$ by its value

$$u(M_0, t_0) = \frac{1}{4\pi}\left[\frac{\partial}{\partial r}\iint\limits_{S_r^{M_0}} ru\,d\Omega + \frac{1}{a}\iint\limits_{S_r^{M_0}} r\frac{\partial u}{\partial t}\,d\Omega\right]_{r=at_0,\ t=0}.$$

Substituting here the given values and omitting the index 0 in M_0, t_0 we arrive at the following relation:

$$u(M, t) = \frac{1}{4\pi a}\left[\frac{\partial}{\partial t}\iint\limits_{S_{at}^M} \frac{\varphi}{r}\, dS + \iint\limits_{S_{at}^M} \frac{\psi}{r}\, dS\right], \tag{4}$$

called usually *Poisson's relation*. This relation is obtained on the assumption of the existence of a solution of the problem. Thus the discussions outlined above prove the uniqueness of the solution (see Chapter II, § 2). It is possible by a direct check to satisfy ourselves that this relation gives the solution of the problem with very general conditions on φ and ψ.

2. *Method of descent*

Formula (4) obtained in the preceding sub-section solves the homogeneous wave equation with initial conditions, which are, generally speaking, arbitrary functions of the variables x, y, and z. If the initial functions φ and ψ do not depend on z, obviously, function u, given by (4), also will not depend on the variable z. Therefore this function will satisfy the equation

$$u_{xx} + u_{yy} - \frac{1}{a^2} u_{tt} = 0$$

and the initial conditions

$$u(x, y, 0) = \varphi(x, y),$$
$$u_t(x, y, 0) = \psi(x, y).$$

Thus the formula, giving the solution of the three-dimensional problem, enables us to solve the problem for a plane.

In (4) integration proceeds over the sphere S_{at}^M. Because of the independence of the initial data on z integration over the upper half of the sphere can be replaced by an integration over the circle Σ_{at}^M, obtained by the intersection of the sphere S_{at}^M with the plane x, y (Fig. 75). The element of surface dS is connected with the plane $d\sigma$ by the relation

$$d\sigma = dS \cos\gamma,$$

where

$$\cos\gamma = \frac{\sqrt{[(at)^2 - \varrho^2]}}{at} = \frac{\sqrt{[(at)^2 - (x-\xi)^2 - (y-\eta)^2]}}{at}.$$

The same contribution occurs on integrating over the lower

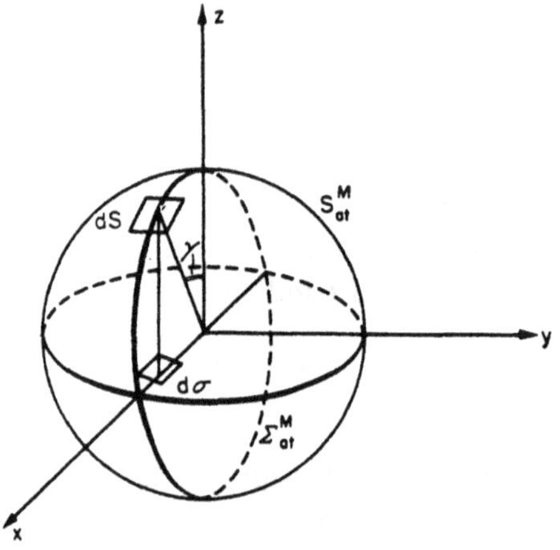

FIG. 75

half of the sphere; therefore the integral over the circle should be taken twice.

As a result we arrive at the relation

$$u(M,t) = u(x,y,t) = \frac{1}{2\pi a}\left[\frac{\partial}{\partial t}\iint\limits_{\Sigma_{at}^M} \frac{\varphi(\xi,\eta)\,d\xi\,d\eta}{\sqrt{[(at)^2 - (x-\xi)^2 - (y-\eta)^2]}} + \right.$$

$$\left. + \iint\limits_{\Sigma_{at}^M} \frac{\psi(\xi,\eta)\,d\xi\,d\eta}{\sqrt{[(at)^2 - (x-\xi)^2 - (y-\eta)^2]}}\right], \quad (5)$$

in which integration is performed over the interior circle of radius at with centre at the point (x, y).

Similarly if the initial functions φ and ψ depend only on one variable x, then (4) enables us to find $u(x, t)$, which is the solution of the equation

$$u_{xx} - \frac{1}{a^2}u_{tt} = 0$$

with initial data

$$u(x,0) = \varphi(x),$$

$$u_t(x,0) = \psi(x).$$

In order to do this let us introduce a spherical system of coordinates, having directed the polar axis along the x-axis. The element of surface dS is expressed in the following way:

$$dS = r^2 \sin \theta \, d\theta \, d\varphi = - r \, d\varphi \, d\xi \,,$$

since

$$\xi = r \cos \theta, \ d\xi = - r \sin\theta \, d\theta \,.$$

Integrating in Poisson's relation with respect to φ, we obtain:

$$u(x, t) = \frac{1}{2a} \left[\frac{\partial}{\partial t} \int_{x-at}^{x+at} \varphi(\xi) \, d\xi + \int_{x-at}^{x+at} \psi(\xi) \, d\xi \right].$$

Performing in the first integral a differentiation with respect to t, we arrive at D'Alembert's relation known from Chapter II, § 2

$$u(x, t) = \frac{\varphi(x + at) + \varphi(x - at)}{2} + \frac{1}{2a} \int_{x-at}^{x+at} \psi(\xi) \, d\xi \,. \tag{6}$$

The wave equations for three, two, and a one-dimensional argument respectively are often called the equations of spherical, cylindrical and plane waves. This terminology completely fits the methods applied above, called the *method* of *descent*, since for the solution of the wave equation in a plane and on a straight line we have proceeded from a spatial problem, as though "descending" to a smaller number of variables. The solutions obtained for two and one variable are characteristic of cylindrical and plane waves.

We apply this method not only to the wave equation, but also to other types of equations and it enables us in a series of cases to derive from formulae, giving the solution of the equation for many variables, the solution of the problem for the equation with a smaller number of independent variables.

3. *Physical interpretation*

Formulae (4) and (5) give an opportunity of explaining the physical picture of the propagation of spherical and cylindrical waves. Let us begin with the case of three variables, for which the physical nature of the process of propagation differs essentially from the case of two spatial variables.

We confine ourselves to an investigation of the propagation of a local disturbance, where the initial condition (functions $\varphi > 0$ and $\psi > 0$) differs from zero only in some bounded region T_0. Let us consider firstly the change of state $u(M_0, t)$ at the point M_0, lying outside region T_0. The state u at point M_0 at time t is determined from (4) by the initial conditions at points lying on the sphere $S_{at}^{M_0}$ of radius at with centre at M_0. Function $u(M_0, t)$ differs from zero only in that case where the sphere $S_{at}^{M_0}$ intersects the region of the initial values of T_0. Let d and D be the distances from point M_0 to the nearest and the most distant points of region T_0 (Fig. 76).

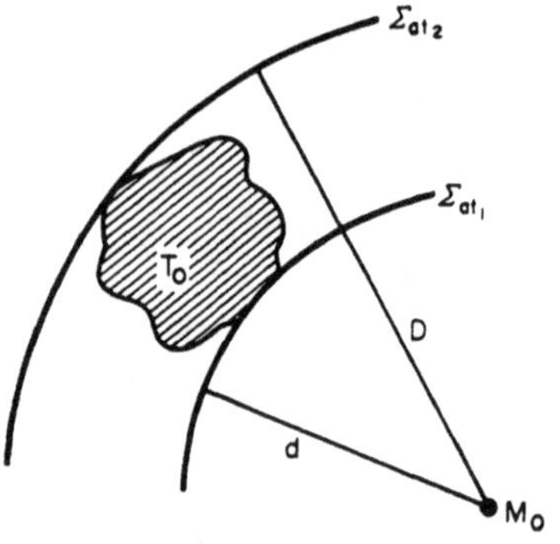

FIG. 76

Obviously, if t is sufficiently small ($t < t_1 = d/a$), then the sphere $S_{at}^{M_0}$ does not intersect region T_0, the surface integrals in (4) equal zero and the disturbance has not reached the point M_0. Starting at time $t_1 = d/a$, the sphere $S_{at}^{M_0}$ ($t_1 < t < t_2$) will intersect region T_0 up to time $t_2 = D/a$; the surface integrals in (4), generally speaking, differ from zero: the point M_0 exists in an excited state. With further increase of t the sphere $S_{at}^{M_0}$ will include the region T_0 inside itself, the surface integrals equal zero: the disturbance is past the point M_0. Thus, in the propagation of a local disturbance in three-dimensional space the phenomenon of "diffusion" is absent.

Let us consider now the instantaneous three-dimensional picture of the disturbance $u(M, t_0)$ at some time t_0. The points M, occurring in an excited state, are characterized by the fact that the spheres $S_{at_0}^{M}$ intersect the region of the initial disturbances T_0. In other words, this indicates that the geometric position of points W, at which the disturbance differs from zero, consists of the points M, occurring on the spheres $S_{at_0}^{P}$ of radius at_0 with centres at the point P of region T_0. The envelope of the spheres $S_{at_0}^{P}$ will be the boundaries of a region W. The outer curve is called the leading front, the inner the rear front of the propagating wave. In Fig. 77 the leading and rear fronts of the wave are given (1 and 2) for the case where region T_0 is a sphere of radius R_0.

Thus, an initial disturbance, localized in space, induces at every point M_0 of space an effect, localized in time.

We proceed to the case of two variables. Let the initial disturbance be given in region S_0 in the plane x, y. Let us consider the change of state $u(M_0, t)$ at point M_0, lying outside S_0. The state $u(M_0, t)$ at point M_0 at time t is defined according to (5) by the initial values at the points P, inside the circle $\Sigma_{at_0}^{M_0}$ of radius at_0 with centre at M_0. For times $t < t_1 = d/a$ (d is the distance from M_0 to the nearest point of region S_0) the function $u(M_0,t)=0$, i.e. the disturbance has not arrived at the point M_0. If $t > t_1$, then $u(M_0, t) \neq 0$. This means that, starting at a time $t = t_1$, a disturbance develops at the point M_0, which at first increases, and then, decreases to zero (as $t \to \infty$). This phenomenon, called diffusion, shows the difference between the plane and the three-dimensional case. The effect of the initial disturbances, localized in the plane, is not localized with respect to time and is characterized by a reaction continuing for a long time. Huygen's principle does not hold.

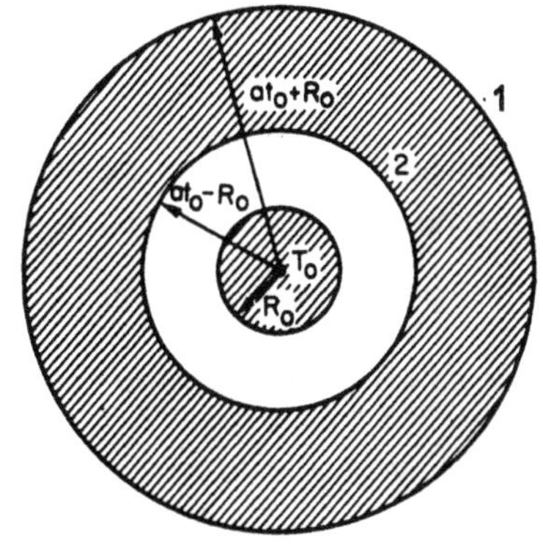

FIG. 77

An instantaneous picture of the disturbances in a plane has sharply outlined leading fronts, but does not have the rear front.

4. *The method of images*

The problem with initial conditions for the wave equation in regions, bounded by planes, can be solved by the method of images.

Let us consider the problem of the semi-space $z > 0$.

Find the solution of the wave equation

$$\nabla^2 u = \frac{1}{a^2} u_{tt} ,$$

satisfying the initial conditions

$$\left. \begin{aligned} u(x, y, z, 0) &= \varphi(x, y, z) , \\ u_t(x, y, z, 0) &= \psi(x, y, z) \end{aligned} \right\} \quad (z \geqslant 0)$$

and the boundary condition

$$u\Big|_{z=0} = 0 \quad \text{or} \quad \frac{\partial u}{\partial z}\Big|_{z=0} = 0 .$$

The solution of this problem is given by (4) if we extend the initial conditions as an odd function over all space (for $u|_{z=0}=0$)

$$\varphi(x, y, z) = -\varphi(x, y, -z); \quad \psi(x, y, z) = -\psi(x, y, -z)$$

or as an even function $\left(\text{for } \frac{\partial u}{\partial z}\Big|_{z=0} = 0\right)$

$$\varphi(x, y, z) = \varphi(x, y, -z); \quad \psi(x, y, z) = \psi(x, y, -z) .$$

Let us verify that for odd functions φ and ψ the boundary relation $u|_{z=0} = 0$ is automatically fulfilled. In fact

$$u(P, t) = u(x, y, 0, t) =$$
$$= \frac{1}{4\pi a}\left[\frac{\partial}{\partial t}\iint\limits_{S_{at}^P} \frac{\varphi(\xi, \eta, \zeta)}{at}\, ds + \iint\limits_{S_{at}^P} \frac{\psi(\xi, \eta, \zeta)}{dt}\, ds\right] = 0 ,$$

since the surface integrals with respect to spheres with centres at points of the plane $z = 0$ equal zero for the odd functions φ and ψ.

Similarly the problem of a plane layer $0 \leqslant z \leqslant l$ can be solved for boundary conditions of the first and second kind

$$u = 0 \quad \text{when} \quad z = 0 \quad \text{and} \quad z = l$$

or

$$\frac{\partial u}{\partial z} = 0 \quad \text{when} \quad z = 0 \quad \text{and} \quad z = l$$

and appropriate initial conditions.

Formula (4) gives the solution of the problem at once, if the initial conditions continue oddly (or evenly) with respect to the planes $z = 0$ and $z = l$. The initial functions φ and ψ specified in this way will be periodic with respect to the variable z with a period of $2l$.

If in the layer $0 < z < l$ the initial functions φ and ψ are local functions, differing from zero in region T_0, then the extended functions will differ from zero in a series of regions T_n, obtained from T_0 by means of mirror images. The function $u(M, t)$ for any M and t is represented as the sum of a finite number of components, given by the disturbances in T_n (see in Chapter II, § 2, sub-section 5). Physically for a finite interval of time a finite number of image waves are produced from the walls $z = 0$ and $z = l$. The problem of a parallelepiped can be solved similarly.

§ 2. Integral relation

1. *Derivation of the integral relation*

In solving the wave equation for a string

$$u_{xx} - \frac{1}{a^2} u_{tt} = - f$$

by the method of propagation of waves we made use of the idea of the *characteristic angle*. To solve the wave equation in a plane or in space

$$\Delta u - \frac{1}{a^2} u_{tt} = - f, \tag{1}$$

we consider the surface

$$\frac{1}{a} r_{MM_0} = | t - t_0 |,$$

called the *characteristic cone* for the point M_0 at time t_0. The set of points of the "phase" space (M, t), at which a signal arrives, propagating with velocity a from the point M_0 at time t_0, is given by the equation

$$\frac{1}{a} r_{MM_0} = t - t_0 \qquad (t > t_0)$$

and is the upper part of the characteristic cone of the point M_0. Similarly, a signal, emitted from the point M at time t, arrives at the point M_0 at time t_0, if

$$\frac{1}{a} r_{MM_0} = t_0 - t \qquad (t < t_0).$$

The set of such points (M, t) define the lower portion of the characteristic cone (Fig. 78).

For determination of the function $u(M, t)$ representing the solution of equation (1) at the point (M_0, t_0), let us introduce

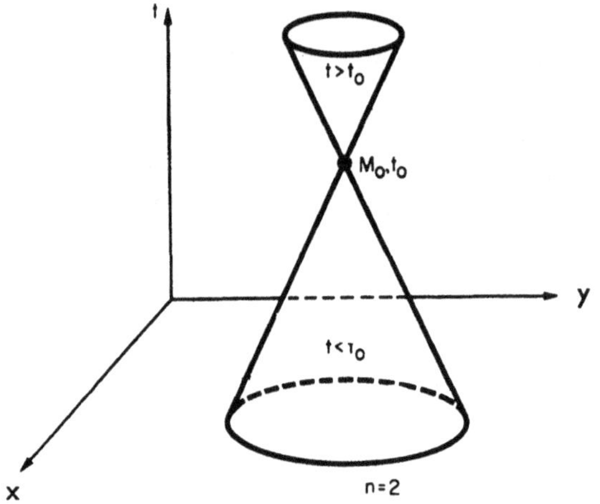

FIG. 78

in place of the time t a local time t^* of the point M_0 by the relation

$$t^* = t - \left(t_0 - \frac{r_{MM_0}}{a}\right).$$

Using a spherical system of coordinates (r, θ, φ), with origin at the point M_0, we have a new system of variables

$$r^* = r, \quad \theta^* = \theta, \quad \varphi^* = \varphi, \quad t^* = t - \left(t_0 - \frac{r}{a}\right).$$

Let us determine the equation which the function

$$u(r, \theta, \varphi, t) = u\left(r^*, \theta^*, \varphi^*, t^* + t_0 - \frac{r}{a}\right) = U(r^*, \theta^*, \pi^*, t^*).$$

satisfies. The Laplacian operator in a spherical system of coordinates has the form

$$\nabla^2 u = \frac{\partial^2 u}{\partial r^2} + \frac{2}{r}\frac{\partial u}{\partial r} + \frac{1}{r^2 \sin\theta}\frac{\partial}{\partial\theta}\left(\sin\theta\frac{\partial u}{\partial\theta}\right) + \frac{1}{r^2\sin^2\theta}\frac{\partial^2 u}{\partial\varphi^2}.$$

Let us express the derivatives of function u in terms of the derivatives of a new function U:

$$u_r = U_{r*} + \frac{1}{a} U_{t*},$$

$$u_{rr} = U_{r*r*} + \frac{2}{a} U_{r*t*} + U_{t*t*}$$

$$u_\theta = U_{\theta*}; \quad u_{\theta\theta} = U_{\theta*\theta*},$$

$$u_\varphi = U_\varphi \; ; \quad u_{\varphi\varphi} = U_{\varphi*\varphi*},$$

$$u_t = U_t^*; \quad u_{tt} = U_{t*t*}.$$

Equation (1) transforms to the equation

$$\Delta U = -\frac{2}{ar^*}\frac{\partial}{\partial r^*}(r^* U_{t*}) - F(r^*, \theta^*, \varphi^*, t^*), \tag{2}$$

where

$$F(r^*, \theta^*, \varphi^*, t^*) = f(r, \theta, \varphi, t).$$

Let the point $M_0(x_0, y_0, z_0)$ lie inside some region T, bounded by the surface S. Considering (2) as Laplace's inhomogeneous equation in which t^* acts as a parameter, we use Green's fundamental theorem (Chapter IV). Let us apply it to region T, assuming in it $t^* = 0$:†

$$4\pi U(M_0, 0) = \iint_S \left[\frac{1}{r^*}\frac{\partial U}{\partial n} - U\frac{\partial}{\partial n}\left(\frac{1}{r}\right)\right] dS +$$
$$+ \iiint_T \frac{2}{ar^{*2}}\frac{\partial}{\partial r^*}\left(r^* \frac{\partial U}{\partial t^*}\right) d\tau + \iiint_T \frac{F}{r^*} d\tau.$$

The point M_0 is a singular point of the spherical system of coordinates. Therefore the volume integrals are naturally considered as the limits as $\varepsilon \to 0$ of the corresponding integrals, taken over the volume $T - T_\varepsilon$, where T_ε is a sphere of radius ε with centre at the point M_0. Let us transform the volume integral

$$I_\varepsilon = \iiint_{T-T_\varepsilon} \frac{2}{ar^{*2}}\frac{\partial}{\partial r^*}\left(r^* \frac{\partial U}{\partial t^*}\right) d\tau =$$
$$= \iiint_{T-T_\varepsilon} \frac{2}{a}\frac{\partial}{\partial r^*}\left(r^* \frac{\partial U}{\partial t^*}\right) \sin\theta^* \, dr^* \, d\theta^* \, d\varphi^*.$$

† By virtue of the condition assumed in Chapter IV the signs in the formula correspond to the outer normal.

Integrating with respect to the variable r^*, we obtain:

$$I_\varepsilon = \iint_S \frac{2}{ar^*} \frac{\partial U}{\partial t^*} \cos (\widehat{n, r^*}) \, dS - \iint_{S_\varepsilon} \frac{2}{ar^*} \frac{\partial U}{\partial t^*} \, dS,$$

since

$$dS_n = dS \cos (\widehat{n, r^*}) = r^{*2} \sin \theta^* \, d\theta^* \, d\varphi^*.$$

The second component in I_ε tends to zero as $\varepsilon \to 0$, since the surface area S_ε equals $4\pi \varepsilon^2$. Thus the limit of I_ε as $\varepsilon \to 0$ equals

$$I_0 = \iiint_T \frac{2}{ar^{*2}} \frac{\partial}{\partial r^*} \left(r^* \frac{\partial U}{\partial t^*} \right) d\tau = \iint_S \frac{2}{ar^*} \frac{\partial U}{\partial t^*} \frac{dr^*}{dn} \, dS \, \dagger),$$

since

$$\cos (\widehat{n, r^*}) = \frac{dr^*}{dn},$$

which gives:

$$4\pi U (M_0, 0) =$$

$$= \iint_S \left[\frac{1}{r^*} \frac{\partial U}{\partial n} - U \frac{\partial}{\partial n} \left(\frac{1}{r^*} \right) + \frac{2}{ar^*} \frac{\partial U}{\partial t^*} \frac{dr^*}{dn} \right] dS + \iiint_T \frac{F}{r^*} \, d\tau.$$

Let us return to the old variables and the function u:

$$u (M, t) = U (M, t^*) \qquad \left(t = t^* + t_0 - \frac{r_{M_0M}}{a} \right),$$

so that

$$U (M_0, 0) = u (M_0, t_0),$$

and also

$$\frac{\partial u}{\partial n} = \frac{\partial U}{\partial n} + \frac{1}{a} \frac{\partial U}{\partial t^*} \frac{dr}{dn}.$$

As a result we obtain the following *integral* relation for $u(M_0, t_0)$:

$$u (M_0, t_0) = \frac{1}{4\pi} \iint_S \left\{ \frac{1}{r} \left[\frac{\partial u}{\partial n} \right] - [u] \frac{\partial}{\partial n} \left(\frac{1}{r} \right) + \frac{1}{ar} \left[\frac{\partial u}{\partial t} \right] \frac{dr}{dn} \right\} dS_M +$$

$$+ \frac{1}{4\pi} \iiint_T \frac{[f]}{r} \, d\tau_M, \qquad (3)$$

† In this transformation we make use of the fact that $d\tau = (r^*)^2 \, d\Omega \, dr$, we integrate with respect to r^* and then substitute $d\Omega = dS/(r^*)^2$.

often called *Kirchhoff's relation.*† Here the square brackets show that the value of the functions must be taken for $t^* = 0$, i.e. for $t = t_0 - \frac{r_{MM_0}}{a}$, so that $[f] = f\left(M, t - \frac{r_{MM_0}}{a}\right)$.

2. *Consequences of the integral relation*

Formula (3) may be applied to the solution of a whole series of problems. As a first example let us consider the following problem: *find the solution of the wave equation*

$$\nabla^2 u - \frac{1}{a^2} u_{tt} = 0$$

in infinite space, with the initial conditions

$$u\,|_{t=0} = \varphi(x, y, z),$$

$$u_t\,|_{t=0} = \psi(x, y, z).$$

The lower portion of the characteristic "cone" $r = a(t_0 - t)$ at the point M_0, t_0 intersects the region of space-time $t = 0$ in a sphere $S_{at_0}^{M_0}$ $(r = at_0)$ of radius at_0 with centre at M_0. Let us use (3), assuming $S = S_{at_0}^{M_0}$ in it. For any function $v(M, t)$ the values $[v]$ on the sphere $S_{at_0}^{M_0}$ have the form

$$[v] = v\left(M, t_0 - \frac{r_{M_0M}}{a}\right) = v(M, 0).$$

Therefore

$$[u]_{S_{at_0}^{M_0}} = u(M_0, 0) = \varphi(M_0),$$

$$\left[\frac{\partial u}{\partial t}\right]_{S_{at_0}^{M_0}} = \frac{\partial u}{\partial t}(M_0, 0) = \psi(M_0),$$

$$\left[\frac{\partial u}{\partial t}\right]_{S_{at_0}^{M_0}} = \frac{\partial u}{\partial r}(M_0, 0) = \frac{\partial \varphi}{\partial r}(M_0).$$

† Kirchhoff's relation was generalized by S. L. Sobolev for the equation of hyperbolic type with an even number of variables. By means of this Kirchhoff–Sobolev relation the solution of problems with initial conditions can be formed for the standard equation (S. L. Sobolev, *Dokl. Akad. Nauk SSSR*, 1933).

Further

$$\frac{1}{r}\left[\frac{du}{dn}\right]-[u]\frac{\partial}{dn}\left(\frac{1}{r}\right)=\frac{1}{r}\frac{\partial\varphi}{\partial r}-\varphi\frac{\partial}{\partial r}\left(\frac{1}{r}\right)=\frac{1}{r^2}\frac{\partial}{\partial r}(r\varphi).$$

Substituting this expression in (3), we find:

$$u(M_0, t_0)=\frac{1}{4\pi}\iint\limits_{S_{at_0}^{M_0}}\left\{\frac{1}{r^2}\frac{\partial}{\partial r}(r\varphi)+\frac{1}{a}\psi\right\}dS=$$

$$=\frac{1}{4\pi}\left[\frac{\partial}{\partial r}\iint\limits_{S_{at_0}^{M_0}} r\varphi\, d\Omega+\frac{1}{a}\iint\limits_{S_{at_0}^{M_0}} r\psi\, d\Omega\right]_{r=at_0}, \tag{4}$$

from which, omitting index 0 in M_0 and t_0, we obtain Poisson's relation

$$u(M,t)=\frac{1}{4\pi a}\left[\frac{\partial}{\partial t}\iint\limits_{S_{at}^{M}}\frac{\varphi}{r}\,dS+\iint\limits_{S_{at}^{M}}\frac{\psi}{r}\,dS\right].$$

As a second example let us consider the solution of the inhomogeneous wave equation with zero initial conditions. Choosing as before $S = S_{at_0}^{M_0}$, we see that the surface integral in (3) reduces to zero, and we obtain:

$$u(M_0, t_0)=\frac{1}{4\pi}\iiint\limits_{T_{at_0}^{M_0}}\frac{[f]}{r}\,d\tau_M, \tag{5}$$

where $T_{at_0}^{M_0}$ is a sphere of radius at_0 with centre at M_0. Let us investigate in more detail that case where the right-hand side is a periodic function of time

$$f(M,t)=f_0(M)\,e^{i\omega t},$$

where ω is the given vibration frequency. From (5) we find:

$$u(M_0, t_0)=e^{i\omega t_0}\frac{1}{4\pi}\iiint\limits_{T_{at_0}^{M_0}} f_0(M)\frac{e^{-ikr_{MM_0}}}{r_{MM_0}}\,d\tau_M \quad \left(k=\frac{\omega}{a}\right). \tag{6}$$

Let $f_0(M)$ be a local function, i. e. a function, differing from zero only inside some region T. If M_0 lies outside region T and the distance from M_0 to the nearest point of region T equals d, then the integral over T_{at_0} equals zero for $t_0 < d/a$. For such times the disturbance has not reached the point M_0. If the distance from M_0 to the most removed point of region T equals D, then for time $t_0 > D/a$ the integral on the left-hand side is constant and becomes the integral prevailing over the entire region T. Thus, at any point M_0, beginning at time $t_0 = D/a$, periodic oscillations occur with an amplitude

$$v(M_0) = \frac{1}{4\pi} \iiint_T f_0(M) \frac{e^{-ikr_{MM_0}}}{r_{MM_0}} d\tau_M, \tag{7}$$

so that

$$u(M_0, t_0) = v(M_0) e^{i\omega t_0}.$$

Direct substitution of expression (6) for u (for $t_0 > D/a$) in the wave equation shows that $v(M)$ must satisfy the equation

$$\Delta v + k^2 v = -f_0(M) \quad \left(k > \frac{\omega}{a}\right), \tag{8}$$

which we will later call the *wave equation* (see Chapter VII).

Let us consider (3) for the case of steady oscillations when

$$u(M, t) = v(M) e^{i\omega t},$$

where $v(M)$ is the amplitude of the oscillations, satisfying the wave equation (8).

In this case we have:

$$[u] = u\left(M, t - \frac{r_{MM_0}}{a}\right) = v(M) e^{i(\omega t - kr)},$$

$$\left[\frac{\partial u}{\partial n}\right] = \frac{\partial v}{\partial n} e^{i(\omega t - kr)},$$

$$\frac{1}{a}\left[\frac{\partial u}{\partial r}\right] = ikv(M) e^{i(\omega t - kr)},$$

$$|f| = f_0(M) e^{i(\omega t - kr)}.$$

Substituting this expression in (3) we arrive at the integral relation for the wave equation

$$v(M_0) = \frac{1}{4\pi} \iint_S \left[\frac{e^{-ikr}}{r} \frac{\partial v}{\partial n} - v \frac{\partial}{\partial n}\left(\frac{e^{-ikr}}{r}\right)\right] dS_M +$$

$$+ \frac{1}{4\pi} \iiint_T f_0(M) \frac{e^{-ikr}}{r} d\tau_M \quad (r = r_{MM_0}), \tag{9}$$

which is often called *Kirchhoff's* relation.

For $k = 0$ ($\omega = 0$, the static case) (9) becomes Green's fundamental relation (Chapter IV, § 2) for Laplace's inhomogeneous equation.

§ 3. Vibrations in finite volumes

1. *General scheme of the method of separation of variables. Standing waves*

The problem of vibrations in finite volumes consists of the following: *find the solution of the equation*

$$\operatorname{div}(k \operatorname{grad} u) - q(M)\, u = \varrho(M)\, u_{tt}, \tag{1}$$

$M = M(x, y, z)$, *inside some volume* T, *bounded in the closed surface* S, *satisfying the additional conditions*

$$\left.\begin{aligned} u(M, 0) &= \varphi(M), \\ u_t(M, 0) &= \psi(M) \end{aligned}\right\} \text{ inside } T, \tag{2}$$

$$u|_S = 0 \text{ when } t > 0. \tag{3}$$

In the case of a homogeneous medium (k = const., ϱ = const.) for $q = 0$ equation (1) takes the form

$$\Delta u = \frac{1}{a^2} u_{tt} - f \quad \left(a^2 = \frac{k}{\varrho};\ f = \frac{F}{k}\right).$$

Problems of a similar type are met with in investigations of the vibrations of a membrane (the case of two independent geometric variables), the acoustic vibrations of a gas, electromagnetic processes in non-conducting media. Special importance is attached to problems connected with the generation of electromagnetic oscillations in closed cavity resonators (klystrons, magnetrons, etc.).

We note that the homogeneity of the boundary condition (3) is not an important restriction. In fact the case

$$u|_S = \mu, \tag{3'}$$

where μ is an arbitrary function of the point P of surface S and the time t, readily reduces to the case of a homogeneous boundary condition by the method given in § 3, Chapter II, for a single variable.

We shall search for a solution $u(M, t)$ of the homogeneous equation (1) with conditions (2) and (3) by the method of separation of variables. We confine ourselves to an account of the formal scheme of solution. With this object let us consider the main auxiliary problem (see § 3, Chapter II):

Find the non-trivial solution of the homogeneous equation

$$\operatorname{div}(k \operatorname{grad} u) - qu = \varrho u_{tt} \tag{1*}$$

with the boundary conditions

$$u|_S = 0, \tag{3}$$

representable in the form of the product

$$u(M, t) = v(M)\, T(t). \tag{4}$$

Substituting the assumed form of the solution (4) in (1) and expanding, as is usual, we arrive at the following equations for the functions $v(M)$ and $T(t)$:

$$\operatorname{div}(k \operatorname{grad} v) - qv + \lambda\varrho v = 0; \tag{5}$$

$$v|_S = 0;$$

$$T'' + \lambda T = 0. \tag{6}$$

Problem (5) is an eigen-value problem. We have to find the eigen-values of λ, which correspond to non-trivial solutions (eigen-functions) of the homogeneous equation with homogeneous boundary conditions. Let us dwell at greater length on this problem (similar to § 3, Chapter II). In our case the eigen-value equation is a partial differential equation and as a result it is difficult to find the eigen-functions for an arbitrary region T. Later (sub-sections 2 and 3) we give examples of regions T, for which it is possible to find the eigen-functions, although the introduction of a new class of special functions is required. Here we consider the general properties of eigen-functions and eigen-values and introduce the formal scheme of the method of separation of variables. Let us enumerate these properties.

1. There exists a countably large number of eigen-values $\lambda_1 < \lambda_2 < \ldots < \lambda_n < \ldots$, with corresponding eigen-functions

$$v_1(x, y, z), \qquad v_2(x, y, z), \ldots, v_n(x, y, z), \ldots$$

The eigen-values λ_n increase indefinitely with n; $\lambda_n \to \infty$ as $n \to \infty$.

2. For $q \geqslant 0$ all eigen-values λ are positive:

$$\lambda_n > 0 .$$

3. The eigen-functions $\{v_n\}$ are orthogonal among themselves with weight $\varrho(x, y, z)$ in the region T:

$$\int_T v_m(M)\, v_n(M)\, \varrho(M)\, d\tau_M = 0 \qquad (m \neq n), \tag{7}$$

$$M = M(x, y, z); \quad d\tau_M = dx\, dy\, dz .$$

4. Expansion theorem. *An arbitrary function $F(M)$, twice continuously differentiable and satisfying the boundary condition*

$$F = 0 \text{ on } S,$$

is expandable in a uniformly and absolutely convergent series with respect to the eigen-functions $\{v_n(M)\}$:

$$F(M) = \sum_{n=1}^{\infty} F_n v_n(M),$$

where F_n are the expansion coefficients.

Proof of the properties 1 and 4 involve the theory of integral equations. Let us prove the properties 2 and 3 which do not require a special mathematical device. First we prove the orthogonality of the eigen-functions $\{v_n\}$ (property 3). Let $v_n(M)$ and $v_m(M)$ be two eigen-functions, corresponding to the different eigen-values λ_m and λ_n:

$$\frac{\partial}{\partial x}\left(k\frac{\partial v_m}{\partial x}\right) + \frac{\partial}{\partial y}\left(k\frac{\partial v_m}{\partial y}\right) + \frac{\partial}{\partial z}\left(k\frac{\partial v_m}{\partial z}\right) - q v_m + \lambda_m \varrho\, v_m = 0,$$

$$\frac{\partial}{\partial x}\left(k\frac{\partial v_n}{\partial x}\right) + \frac{\partial}{\partial y}\left(k\frac{\partial v_n}{\partial y}\right) + \frac{\partial}{\partial z}\left(k\frac{\partial v_n}{\partial z}\right) - q v_n + \lambda_n \varrho\, v_n = 0,$$

where $v_m = 0$ and $v_n = 0$ on S. Multiplying the first equation by $v_n(M)$ and subtracting from it the second equation, multiplied by $v_m(M)$, we find:

$$\iiint_T \{v_n \operatorname{div}(k \operatorname{grad} v_m) - v_m \operatorname{div}(k \operatorname{grad} v_n)\}\, d\tau +$$
$$+ (\lambda_m - \lambda_n) \iiint_T v_n v_m \varrho\, d\tau = 0 .$$

Hence after transformations, similar to those which are employed in the derivation of Green's second theorem* we obtain:

$$\iint_S \left(v_n k \frac{\partial v_m}{\partial \nu} - v_m k \frac{\partial v_n}{\partial \nu}\right) d\sigma + (\lambda_m - \lambda_n) \iiint_T v_m v_n \varrho \, d\tau = 0 .$$

The boundary conditions give $v_m = 0$ and $v_n = 0$ on S, so that

$$(\lambda_m - \lambda_n) \iiint_T v_m v_n \varrho \, d\tau = 0 ,$$

from which it follows that for $\lambda_m \neq \lambda_n$

$$\iiint_T v_m v_n \varrho \, d\tau = 0 \qquad (m \neq n) ,$$

i.e. eigen-functions, corresponding to different eigen-values are mutually orthogonal with weight $\varrho(M)$.

In the investigation of a similar boundary-value problem for one independent variable

$$X'' + \lambda \varrho X = 0 ,$$

$$X(0) = 0 ,$$

$$X(l) = 0$$

it was proved that to each eigen value there corresponds one normalized eigen-function. For two and three variables this result does not hold. Examples of eigen-functions in a rectangle and a circle, considered below (sub-sections 2 and 3), show that the same eigen-value can correspond to several eigen-functions. However it follows from the theory of integral equations, that each eigen-value has only a finite number of corresponding linearly independent eigen-functions. A certain eigen-value λ_n corresponds, to a system of linearly independent functions $v_n^{(1)}, v_n^{(2)}, \ldots, v_n^{(m)}$. It is obvious that any linear combination of these functions

$$\bar{v}_n = \sum_{i=1}^{m} a_i v_n^{(i)}$$

* In this formula the normal derivatives of eigenfunctions on surface S enter. For proof of this formula for Lyapunov type surfaces see V. I. Smirnov, *Course in Higher Mathematics*, vol. IV.

is also an eigen-function for the same eigen-value λ_n. Making use of the well-known method of *orthogonalization** it is possible to form the functions $\bar{v}_n^{(1)} \ldots \bar{v}_n^{(m)}$, which are linear combinations of the original functions and mutually orthogonal. Thus, if the eigen-functions corresponding to some λ_n, are not mutually orthogonal, then we can orthogonalize them and derive a new system of eigen-functions, mutually orthogonal and corresponding to the same λ_n.

A series of such systems of eigen-functions for different λ_n produces an orthogonal system of eigen-functions of the boundary-value problem under consideration

$$\operatorname{div}(k \operatorname{grad} v) - qv + \lambda \varrho v = 0\,,$$

$$v = 0 \text{ on } S\,.$$

The integral

$$N(v_n) = \iiint_T v_n^2 \varrho \, d\tau > 0$$

is called the *norm* of the eigen-function. Multiplying each function v_n by $\dfrac{1}{\sqrt{N(v_n)}}$, we obtain a system of eigen-functions, normalized to unity.

We can prove that the eigen-values are positive (property 2) by using Green's first theorem

$$\iiint_T (k \operatorname{grad} v_n)^2 \, d\tau =$$

$$= -\iiint_T v_n \operatorname{div}(k \operatorname{grad} v_n) \, d\tau + \iint_S v_n k \frac{\partial v_n}{\partial n} \, d\sigma =$$

$$= -\iiint_T q v_n^2 \, d\tau + \lambda_n \iiint_T v_n^2 \varrho \, d\tau\,.$$

Hence for $q \geqslant 0$ the eigen-values λ_n are positive.

Later we shall make use of the expansion theorem (property 4), assuming the proof holds for the appropriate class of integral equations. Let

$$F(M) = \sum_{n=1}^{\infty} F_n v_n(M)\,.$$

* See, for instance, V. I. Smirnov, *Course in Higher Mathematics*, vol. IV.

Hence in the usual way, using the condition of orthogonality (7), we find the expansion coefficients

$$F_n = \frac{\iiint\limits_T F(M)\, v_n(M)\, \varrho\, d\tau}{\iiint\limits_T v_n^2 \varrho\, d\tau} . \tag{8}$$

Let us return now to the partial differential equation. The solution of the equation

$$T_n'' + \lambda_n T_n = 0$$

has the form

$$T_n(t) = A_n \cos \sqrt{(\lambda_n)}\, t + B_n \sin \sqrt{(\lambda_n)}\, t,$$

so that the solution of our fundamental auxiliary problem will be the product

$$u_n(M, t) = T_n(t)\, v_n(M) = (A_n \cos \sqrt{[\lambda_n]}\, t + B_n \sin \sqrt{[\lambda_n]}\, t)\, v_n(M) .$$

We seek general solution of the original problem with initial data in the form of the sum

$$u(M, t) = \sum_{n=1}^{\infty} u_n(M, t) = \\ = \sum_{n=1}^{\infty} (A_n \cos \sqrt{[\lambda_n]} t + B_n \sin \sqrt{[\lambda_n]}\, t)\, v_n(M) . \tag{9}$$

Satisfying the initial conditions (2)

$$u(M, 0) = \varphi(M) = \sum_{n=1}^{\infty} A_n v_n(M) ,$$

$$u_t(M, 0) = \psi(M) = \sum_{n=1}^{\infty} B_n \sqrt{(\lambda_n)}\, v_n(M)$$

and making use of the expansion theorem 4 we find:

$$A_n = \varphi_n , \quad B_n \sqrt{\lambda_n} = \psi_n ,$$

where φ_n and ψ_n are Fourier coefficients of the functions $\varphi(M)$ and $\psi(M)$ in their expansions with respect to the orthogonal system with weight $\varrho(M)$ of the functions $v_n(M)$. In this way we complete the formal construction of the solution of the original problem.

A physical interpretation of the solution obtained is completely analogous to the case of a single variable. The particular solutions

$$u_n(M, t) = (A_n \cos \sqrt{[\lambda_n]}\, t + B_n \sin \sqrt{[\lambda_n]}\, t)\, v_n(M)$$

represent by themselves *standing waves* which may exist inside a bounded volume T.

"Profiles" of the standing waves, defined by the functions $v_n(M)$, for different times differ only by a factor of proportionality. Lines or surfaces (for two or three variables respectively), along which $v_n(M) = 0$, are called *nodal lines* (surfaces) of the standing waves $v_n(M)$. Points at which $v_n(M)$ reaches relative maxima or minima are called the *anti-nodes* of this standing wave.* The general solution is represented in the form of an infinite sum of such standing waves. The possibility of representing the general solution in the form of a sum of components of similar type implies the possibility of representing an arbitrary vibration in the form of a superposition of standing waves.**

Thus, the problem of the vibration of membranes or volumes reduces essentially to finding the appropriate eigen-functions. In sub-sections, 2 and 3 we consider the vibrations of rectangular or circular membranes, and derive formulae for the eigen-values and eigen-functions. As was already noted above, finding the eigen-functions in explicit analytical form involves great difficulties for regions of more complex form. In the case of arbitrary regions approximate methods can be used to find eigen-functions. There exist different approximate methods, based on the use of integral equations, variational principles, finite differences.

2. *Vibrations of a rectangular membrane*

The vibrations of a plane homogeneous membrane are described by the wave equation

$$u_{tt} = a^2 \Delta u . \tag{10}$$

A rectangular membrane with sides b_1 and b_2 lies in the (x, y) plane. It is secured at the sides and excited by an initial displace-

*If vibrations are excited in a membrane sprinkled with sand, then the sand will be thrown from an anti-node to the nodal lines, reproducing the nodal lines of the eigen-functions.

** The basis of the method of separation for the case of many variables is contained in the work of O. A. Ladyzhenskii (*Dokl. Akad. Nauk SSSR* 85, 3, 1952).

ment and an initial velocity. In order to find $u(x, y, t)$ we must solve the wave equation

$$\frac{\partial^2 u}{\partial t^2} = a^2 \left(\frac{\partial^2 u}{\partial x^2} + \frac{\partial^2 u}{\partial y^2}\right) \tag{10'}$$

for the given initial

$$\left.\begin{aligned} u(x, y, 0) &= \varphi(x, y), \\ \frac{\partial u}{\partial t}(x, y, 0) &= \psi(x, y) \end{aligned}\right\} \tag{11}$$

and boundary conditions

$$u(0, y, t) = 0, \quad u(b_1, y, t) = 0, \tag{12}$$

$$u(x, 0, t) = 0, \quad u(x, b_2, t) = 0. \tag{13}$$

We find the solution by the method of separation of variables assuming

$$u(x, y, t) = v(x, y)\, T(t). \tag{14}$$

Substituting (14) in (10) and separating the variables we obtain for $T(t)$ the equation

$$T'' + a^2 \lambda T = 0, \tag{15}$$

and for $v(x, y)$ the following boundary-value problem:

$$\left.\begin{aligned} v_{xx} + v_{yy} + \lambda v &= 0; \\ v(0, y) = 0, \quad v(b_1, y) &= 0; \\ v(x, 0) = 0, \quad v(x, b_2) &= 0. \end{aligned}\right\} \tag{16}$$

Thus the eigen-value problem reduces to the solution of the homogeneous equation with homogeneous boundary conditions. We shall solve this problem by the method of separation of variables, assuming

$$v(x, y) = X(x)\, Y(y).$$

Carrying out a separation of the variables, we obtain the following one-dimensional eigen-value problems:

$$\left.\begin{aligned} X'' + \nu X &= 0, \\ X(0) = 0, \; X(b_1) &= 0; \end{aligned}\right\} \tag{17}$$

$$\left.\begin{aligned} Y'' + \mu Y &= 0, \\ Y(0) = 0, \; Y(b_2) &= 0, \end{aligned}\right\} \tag{18}$$

where ν and μ are constants of the separation of variables connected by the relationship $\mu + \nu = \lambda$. The boundary conditions for $X(x)$ and $Y(y)$ follows from the corresponding conditions for function v. For example from

$$v(0, y) = X(0) Y(y) = 0$$

it follows $X(0) = 0$, since $Y(y) \neq 0$ (we have only non-trivial solutions).

We have already met problems, similar to (17) and (18) in the study of the vibrations of a string. The solutions of equations (17) and (18) have the form

$$X_n(x) = \sin\frac{n\pi}{b_1} x, \quad Y_m(y) = \sin\frac{m\pi}{b_2} y;$$

$$\nu_n = \left(\frac{n\pi}{b_1}\right)^2; \qquad \mu_m = \left(\frac{m\pi}{b_2}\right)^2.$$

The eigen-values

$$\lambda_{n,m} = \left(\frac{n\pi}{b_1}\right)^2 + \left(\frac{m\pi}{b_2}\right)^2$$

have corresponding eigen-functions

$$v_{n,m} = A_{n,m} \sin\frac{n\pi}{b_1} x \sin\frac{m\pi}{b^2} y,$$

where $A_{n,m}$ is some constant factor. We choose this so that the norm of function $v_{n,m}$ with weight 1 equals unity

$$\int_0^{b_1}\int_0^{b_2} v_{n,m}^2 \, dx \, dy = A_{n,m}^2 \int_0^{b_1} \sin^2\frac{n\pi}{b_1} x dx \int_0^{b_2} \sin^2\frac{m\pi}{b_2} y dy = 1.$$

Hence

$$A_{n,m} = \sqrt{\frac{4}{b_1 b_2}}.$$

The orthogonality of the functions $\{v_{n,m}\}$ has been proved. Hence, the functions

$$v_{n,m}(x, y) = \sqrt{\left(\frac{4}{b_1 b_2}\right)} \sin\frac{n\pi}{b_1} x \sin\frac{m\pi}{b_2} y \tag{19}$$

form an orthogonal system of eigen-functions of the rectangular membrane.

The number of eigen-values, belonging to $\lambda_{n,m}$ (multiple of $\lambda_{n,m}$) depends on the number of whole-number solutions n and m of the equation

$$\left(\frac{n\pi}{b_1}\right)^2 + \left(\frac{m\pi}{b_2}\right)^2 = \lambda_{n,m}\,.$$

The system of eigen-functions $v_{n,m}$ found is such that any function $F(x, y)$, twofold continuously differentiable and satisfy-

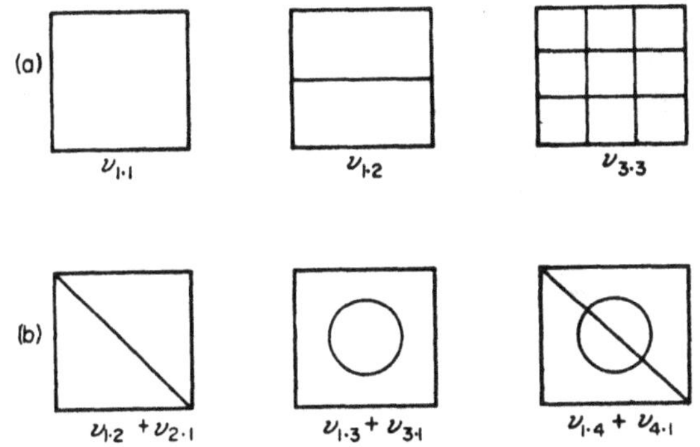

FIG. 79

ing a boundary condition, can be expanded in an absolutely and uniformly convergent series with respect to $v_{n,m}$. This statement can be proved from the theory of multiple Fourier series.

Let us show that the system (19) contains all the eigen-functions of our problem. We assume that there exists an eigen-function u_0 belonging to the eigen-value λ_0. Since the function u_0 is orthogonal to all the eigen-functions, belonging to the other values of λ, then in its expansion in system (19) a finite number of terms only is considered, corresponding to eigen-functions, belonging to the eigen-value $\lambda_{n,m} = \lambda_0$. Therefore u_0 is a linear combination only of those functions (19) which correspond to $\lambda_{n,m} = \lambda_0$. Thus all functions of a rectangular membrane are given by (19).

Returning to the original problem for equation (10) we see that the particular solutions

$$u_{n,m} = v_{n,m}\,(\overline{B}_{n,m} \cos \sqrt{[\lambda_{n,m}]}\, at + \overline{\overline{B}}_{n,m} \sin \sqrt{[\lambda_{n,m}]}\, at)$$

represent standing waves, whose profile is given by the eigen-functions $v_{n,m}$. The geometrical positions of points inside the rectangle at which the eigen-functions reduce to zero, are called nodal lines. Let us consider for simplicity the square with side b $(b_1 = b_2)$ The nodal lines of the function

$$v_{n,m} = \frac{2}{b} \sin \frac{n\pi}{b} x \sin \frac{n\pi}{b} y$$

are lines, parallel to the axes of coordinates (Fig. 79a).

For multiple eigen-values a linear combination of eigen-functions also will be an eigen-function. Its nodal lines can have a very complex form (Fig. 79b).

The required solution of equation (10) for the additional conditions (11)–(13) has the form

$$u(x, y, t) = \sum_{m=1}^{\infty} \sum_{n=1}^{\infty} (\overline{B}_{n,m} \cos \sqrt{[\lambda_{n,m}]}\, at + \overline{\overline{B}}_{n,m} \sin \sqrt{[\lambda_{n,m}]}\, at)\, v_{n,m}(x, y),$$

where $v_{n,m}$ is given by (19) and the coefficients $\overline{B}_{n,m}$ and $\overline{\overline{B}}_{n,m}$ equal:

$$\overline{B}_{n,m} = \int_0^{b_1}\int_0^{b_2} \varphi(x, y)\, v_{n,m}(x, y)\, dx\, dy = \sqrt{\left(\frac{4}{b_1 b_2}\right)} \int_0^{b_1}\int_0^{b_2} \varphi(x, y) \sin \frac{n\pi}{b_1} x \sin \frac{m\pi}{b_2} y\, dx\, dy,$$

$$\overline{\overline{B}}_{n,m} = \frac{1}{\sqrt{(a^2\lambda_{n,m})}} \sqrt{\left(\frac{4}{b_2 b_2}\right.} \int_0^{b_1}\int_0^{b_2} \psi(x, y) \sin \frac{n\pi}{b_1} x \sin \frac{m\pi}{b_2} y\, dx\, dy.$$

3. *Vibrations of a circular membrane*

In an investigation of the vibrations of a circular membrane it is useful to change to polar coordinates. Then the wave equation is written in the form

$$\frac{1}{r}\frac{\partial}{\partial r}\left(r \frac{\partial u}{\partial r}\right) + \frac{1}{r^2}\frac{\partial^2 u}{\partial \theta^2} = \frac{1}{a^2}\frac{\partial^2 u}{\partial t^2}. \tag{20}$$

We shall search for a solution of this equation with the given initial conditions

$$\left.\begin{aligned} u(r, \theta, 0) &= f_1(r, \theta), \\ u_t(r, \theta, 0) &= f_2(r, \theta) \end{aligned}\right\} \tag{21}$$

and the boundary condition

$$u\,(r_0, \theta, t) = 0 \tag{22}$$

(the edge of the membrane of radius r_0 is fixed). As in the case of a rectangular membrane, we resort to separation of the variables. Having assumed

$$u\,(r, \theta, t) = v\,(r, \theta)\,T\,(t)\,,$$

we obtain the equation for $T(t)$

$$T'' + a^2 \lambda T = 0\,,$$

$$T = C_1 \cos \sqrt{(a^2 \lambda)}\, t + C_2 \sin \sqrt{(a^2 \lambda)}\, t$$

and the following eigen-value problem for the function $v(r, \theta)$:

$$\left.\begin{aligned}&\frac{1}{r}\frac{\partial}{\partial r}\left(r\frac{\partial v}{\partial r}\right) + \frac{1}{r^2}\frac{\partial^2 v}{\partial \theta^2} + \lambda v = 0 \quad (0 < r < r_0)\,,\\ &|v(0, \theta)| < \infty \text{ (bounded at the origin)},\\ &v(r_0, \theta) = 0 \text{ (fixed edge)},\\ &v(r, \theta) = v(r, \theta + 2\pi) \text{ (single valued)}.\end{aligned}\right\}$$

The function v must be a single valued and differentiable function; since θ is a cyclic coordinate we must demand that v is periodic in θ with period 2π, i.e. $v(r, \theta + 2\pi) = v(r, \theta)$.

Let us assume

$$v\,(r, \theta) = R\,(r)\,\Theta\,(\theta)\,.$$

Substituting the assumed form of the solution in our equation and dividing by $R\,\Theta$, we obtain:

$$\frac{r\dfrac{d}{dr}\left(r\dfrac{dR}{dr}\right)}{R} + \frac{\Theta''}{\Theta} + \lambda r^2 = 0\,.$$

Hence we arrive at the equations

$$\Theta'' + \mu^2 \Theta = 0;$$

$$\Theta\,(\theta) = \Theta\,(\theta + 2\,\pi); \quad \Theta'\,(\theta) = \Theta'\,(\theta + 2\,\pi)\,;$$

$$\frac{1}{r}\frac{d}{dr}\left(r\frac{dR}{dr}\right) + \left(\lambda - \frac{\mu^2}{r^2}\right) R = 0\,,$$

$$R\,(r_0) = 0\,, \quad |R\,(0)| < \infty\,.$$

Non-trivial periodic solutions for $\Theta(\theta)$ exist only for $\mu^2 = n^2$ (n an integer) and have the form

$$\Theta_n(\theta) = D_{1n} \cos n\theta + D_{2n} \sin n\theta .$$

We note that the eigen-value n^2 belongs to two linearly-independent eigen-functions $\cos n\theta$ and $\sin n\theta$.

In order to determine $R(r)$ we have to solve the equation

$$\frac{d^2 R}{dr^2} + \frac{1}{r}\frac{dR}{dr} + \left(\lambda - \frac{n^2}{r^2}\right) R = 0 \tag{23}$$

with the homogeneous boundary conditions

$$R(r_0) = 0 ,$$
$$|R(0)| < \infty . \tag{24}$$

The second condition imposed on $R(r)$, a restriction for $r = 0$, is connected with the fact that $r = 0$ is a singular point of the equation; for boundary conditions at singular points it is sufficient to require that the solution be bounded (see the Supplement, p. 630).

Introducing a new variable

$$x = \sqrt{\lambda} \cdot r$$

and putting

$$R(r) = R\left(\frac{x}{\sqrt{\lambda}}\right) = y(x) ,$$

we obtain Bessel's equation of the nth order (see Supplement, part I)

$$\frac{d^2 y}{dx^2} + \frac{1}{x}\frac{dy}{dx} + \left(1 - \frac{n^2}{x^2}\right) y = 0 \tag{25}$$

with the boundary conditions

$$y(x_0) = 0 \quad (x_0 = \sqrt{\lambda} \cdot r_0) , \tag{26}$$
$$|y(0)| < \infty .$$

The general solution of this equation is

$$y(x) = d_1 J_n(x) + d_2 N_n(x) , \tag{27}$$

where $J_n(x)$ is the Bessel function, N_n is a Neumann function of nth order (see Supplement, part I). From the second condition it follows that $d_2 = 0$. The first condition gives:

$$J_n(\sqrt{\lambda} \cdot r_0) = 0 \text{ or } J_n(\mu) = 0 \tag{28}$$
$$(\mu = \sqrt{\lambda} \cdot r_0) .$$

If $\mu_m^{(n)}$ is the mth root of the equation $J_n(\mu) = 0$, then

$$\lambda_{n,m} = \left(\frac{\mu_m^{(n)}}{r_0}\right)^2. \tag{29}$$

The eigen-function belonging to this eigen-value is

$$R_{n,m} = y(\sqrt{\lambda} \cdot r) = J_n\left(\frac{\mu_m^{(n)}}{r_0} r\right). \tag{30}$$

We note the following properties of the eigen-functions (30):*

1. The eigen-functions $R(r)$, belonging to different eigen-values λ, are orthogonal with weight r:

$$\int_0^{r_0} r R_{nm_1}(r) R_{nm_2}(r)\, dr = 0 \qquad (m_1 \neq m_2)$$

or

$$\int_0^{r_0} r J_n\left(\frac{\mu_{m_1}^{(n)}}{r_0} r\right) J_n\left(\frac{\mu_{m_2}^{(n)}}{r_0} r\right) dr = 0. \tag{31}$$

2. The norm of these functions equals

$$\int_0^{r_0} r J_n^2\left(\frac{\mu_m^{(n)}}{r_0} r\right) dr = \frac{r_0^2}{2} [J'_n(\mu_m^{(n)})]^2. \tag{31'}$$

In particular, the norm of the function $J_0\left(\frac{\mu_m^{(0)}}{r_0} r\right)$ equals

$$\int_0^{r_0} r \left[J_0\left(\frac{\mu_m^{(0)}}{r_0} r\right)\right]^2 dr = \frac{r_0^2}{2} [J_1(\mu_m^{(0)})]^2. \tag{32}$$

3. Any function $f(r)$ continuous in the interval $(0, r_0)$, having partly-continuous first and second derivatives and satisfying the boundary conditions of the problem can be expanded in an absolutely and uniformly convergent series

$$f(r) = \sum_{m=1}^{\infty} f_m J_n\left(\frac{\mu_m^{(n)}}{r_0} r\right), \tag{33}$$

* See supplement, Part I, § 1.

where the expansion coefficients are given by the relation

$$f_m = \frac{\int\limits_0^{r_0} r f(r)\, J_n\left(\frac{\mu_m^{(n)}}{r_0} r\right) dr}{\frac{r_0^2}{2}\left[J_n'(\mu_m^{(n)})\right]^2}. \tag{34}$$

Returning to the eigen-value problem for a circular membrane, we obtain for the eigen-value

$$\lambda_{n,m} = \left(\frac{\mu_m^{(n)}}{r_0}\right)^2$$

two eigen-functions

$$\bar{v}_{n,m} = J_n\left(\frac{\mu_m^{(n)}}{r_0} r\right)\cos n\theta, \quad \bar{\bar{v}}_{n\cdot m} = J_n\left(\frac{\mu_m^{(n)}}{r_0} r\right)\sin n\theta. \tag{35}$$

Forming a linear combination of them we obtain:

$$v_{n,m}(r,\theta) = J_n\left(\frac{\mu_m^{(n)}}{r_0} r\right)(A_{n,m}\cos n\theta + B_{n,m}\sin n\theta). \tag{36}$$

Let us calculate the norm of the eigen-functions $v_{n,m}$. In passing we verify the orthogonality of the eigen-functions. In order to simplify the calculation let us confine ourselves to the eigen-functions $\bar{v}_{n,m}$:

$$\int\limits_0^{2\pi}\int\limits_0^{r_0} \bar{v}_{n_1,m_1}\bar{v}_{n_2 m_2}\, r\, dr\, d\theta =$$

$$= \int\limits_0^{r_0} J_{n_1}\left(\frac{\mu_{m_1}^{(n_1)}}{r_0} r\right) J_{n_2}\left(\frac{\mu_{m_2}^{(n_2)}}{r_0} r\right) r\, dr \int\limits_0^{2\pi} \cos n_1\theta \cos n_2\theta\, d\theta =$$

$$= \begin{cases} 0 \quad \text{where} \quad n_1 \neq n_2, \\ 0 \quad \text{where} \quad n_1 = n_2, \quad m_1 \neq m_2, \\ \frac{r_0^2}{2}\left[J_n'(\mu_m^{(n)})\right]^2 \pi \text{ where } n_1 = n_2 = n \neq 0 \text{ and } m_1 = m_2 = m, \\ \frac{r_0^2}{2}\left[J_0'(\mu_m^{(n)})\right]^2 2\pi \text{ where } n_1 = n_2 = 0 \quad \text{ and } m_1 = m_2 = m. \end{cases} \tag{37}$$

Similar conditions hold for the function

$$\bar{\bar{v}}_{n,m} = J_n\left(\frac{\mu_m^{(n)}}{r_0} r\right)\sin n\theta.$$

The expression for the norm can be written in the form

$$\iint_{\Sigma} v_{n,m}^2\, d\sigma = \frac{r_0^2}{2}\pi\varepsilon_n\left[J'_n(\mu_m^{(n)})\right]^2, \qquad (d\sigma = r\,dr\,d\theta) \tag{38}$$

where

$$\varepsilon_n = \begin{cases} 2 & \text{where} \quad n = 0, \\ 1 & \text{where} \quad n \neq 0. \end{cases}$$

From the general theory it follows that

any continuous function $F(r, \theta)$ with continuous first and second derivatives, satisfying the boundary conditions of the problem, can be expanded in an absolutely uniformly convergent series

$$F(r, \theta) = \sum_{n,m}\left(A_{n,m}\bar{v}_{n,m}(r, \theta) + B_{n,m}\bar{\bar{v}}_{n,m}(r, \theta)\right) \tag{39}$$

in the eigen-functions of the problem.

The expansion coefficients are calculated from the relations

$$\left.\begin{aligned} A_{n,m} &= \frac{\displaystyle\int_0^{2\pi}\int_0^{r_0} F(r, \theta)\, J_n\left(\frac{\mu_m^{(n)}}{r_0} r\right)\cos n\theta\, r\,dr\,d\theta}{\dfrac{\pi r_0^2}{2}\varepsilon_n\left[J'_n(\mu_m^{(n)})\right]^2}, \\ B_{n,m} &= \frac{\displaystyle\int_0^{2\pi}\int_0^{r_0} F(r, \theta)\, J_n\left(\frac{\mu_m^{(n)}}{r_0} r\right)\sin n\theta\, r\,dr\,d\theta}{\dfrac{\pi r_0^2}{8}\varepsilon_n\left[J'_n \mu_m^{(n)})\right]^2}. \end{aligned}\right\} \tag{40}$$

We note that the function

$$v_{0,m}(r, \theta) = J_0\left(\frac{\mu_m^{(0)}}{r_0} r\right)$$

does not depend on θ.

If the given functions F depends only on r, $F = F(r)$, then the series, representing the expansion of F, will contain only the functions $v_{0,m}$:

$$F(r) = \sum_{m=1}^{\infty} A_{0,m} J_0\left(\frac{\mu_m^{(0)}}{r_0} r\right), \tag{41}$$

where

$$A_{0,m} = \frac{\int_0^{r_0} F(r) J_0\left(\frac{\mu_m^{(0)}}{r_0} r\right) r\,dr}{\frac{r_0^2}{2} J_1^2(\mu_m^{(0)})} \tag{42}$$

All the remaining coefficients $A_{n,m}$ and $B_{n,m}$ equal zero.

Returning to the original problem of the vibration of a membrane with a given initial oscillation and initial velocity, we can write its solution in the form

$$u(r, \theta, t) = \sum_{n, m=0}^{\infty} \bar{v}_{n,m}(r, \theta)\left(A_{n,m} \cos \frac{a\mu_m^{(n)}}{r_0} t + B_{n,m} \sin \frac{a\mu_m^{(n)}}{r_0} t\right) +$$
$$+ \sum_{n, m=0}^{\infty} \bar{\bar{v}}_{n,m}(r, \theta)\left(C_{n,m} \cos \frac{a\mu_m^{(n)}}{r_0} t + D_{n,m} \sin \frac{a\mu_m^{(n)}}{r_0} t\right).$$

The coefficients $A_{n,m}$, $B_{n,m}$, $C_{n,m}$, $D_{n,m}$ are determined from the initial conditions

$$u(r, \theta, 0) = \sum_{n, m=0}^{\infty} (A_{n,m} \bar{v}_{n,m} + C_{n,m} \bar{\bar{v}}) = f_1(r, \theta),$$

$$u_t(r, \theta, 0) = \sum_{n, m=0}^{\infty} (B_{n,m} \bar{v}_{n,m} + D_{n,m} \bar{\bar{v}}_{n,m}) \frac{a\mu_m^{(n)}}{r_0} = f_2(r, \theta)$$

by the relations

$$A_{n,m} = \frac{\int_0^{2\pi} \int_\pi^{r_0} \varphi(r, \theta) J_n(\sqrt{[\lambda_{n,m}]}\, r) \cos n\theta\, r\,dr\,d\theta}{\frac{\pi r_0^2}{2} \varepsilon_n [J'n(\sqrt{[\lambda_{n,m}]}\, r)]^2}$$

$$C_{n,m} = \frac{\int_0^{2\pi} \int_0^{r^\bullet} \varphi(r, \theta) J_n(\sqrt{[\lambda_{n,m}]}\, r) \sin n\theta\, r\,dr\,d\theta}{\frac{\pi r_0^2}{2} \varepsilon_n [J'n(\sqrt{\{\lambda_{n,m}\}}\, r)]^2}.$$

Similar relations hold for $a\sqrt{(\lambda_{n,m})}\, B_{n,m}$ and $a\sqrt{(\lambda_{n,m})}\, D_{n,m}$ respectively.

Problems on Chapter V

1. Solve the wave equation $u_{tt} = a^2 \nabla^2 u$ in 3 dimensions assuming that the initial velocity everywhere equals zero, and the initial displacements $u|_{t=0} = \varphi$ have the form

$$(a)\quad \varphi = \begin{cases} 1 \text{ inside a circular region,} \\ 0 \text{ outside the circular region.} \end{cases}$$

or

$$(b)\quad \varphi = \begin{cases} A \cos \dfrac{\pi}{2r_0} r \text{ inside a circle of radius } r_0, \\ 0 \qquad\qquad \text{outside a circle of radius } r_0. \end{cases}$$

2. Solve the problem of the vibration of the semi-space $z > 0$ for a homogeneous boundary condition of first or second kind, if there is

(a) an initial local disturbance, i. e. an initial velocity and an initial displacement in some region T_0;

(b) a force, acting according to an arbitrary law.

3. Solve the same problem for the layer $-l \leqslant z \leqslant l$.

4. Solve the wave equation in a wedge, whose angle equals $\pi/2$ or in general π/n (n a whole number), if there are homogeneous boundary conditions of first or second kind and also an initial velocity and an initial displacement.

5. Derive the analogue of the integral relation (3) for the equation

$$u_{tt} = a^2 \nabla^2 u + cu, \quad \text{where } c = \text{const.}$$

Consider the case $c < 0$ and find the solution of the problem with initial data for the inhomogeneous equation. Give a physical interpretation of the results obtained.

6. Find the function $u(\varrho, \varphi, t)$, describing the vibrations of a membrane under the action of an impulse K, concentrated

(a) at the centre of a circular membrane,

(b) at an arbitrary point of a circular membrane,

(c) at an arbitrary point of a rectangular membrane.

7. Find the characteristic frequencies and eigenfunctions of membranes, having the form

(a) of a semi-circle,

(b) of a ring,

(c) of a circular sector,

(d) of a ring sector.

Consider the first and second boundary value problems.

8. Find the steady oscillations of a circular membrane (the membrane of a microphone) under the action of a periodic force, distributed over the membrane with constant density $f = A \sin \omega t$. Solve the same problem, if $f = A(1 - r^2/c^2) \sin \omega t$, where c is the radius of the membrane.

9. Derive the equation of the propagation of sound in a medium, moving with constant velocity. Transform the equation obtained, changing to a system of coordinates, moving along with the medium.

APPENDICES TO CHAPTER V

I. Reduction of the equations of the theory of elasticity to the wave equation

The theory of elasticity aims to investigate the deformations and motions arising in elastic bodies by mathematical methods. Strains and motions originating from the action of external forces can be characterized by means of a *displacement vector* $\boldsymbol{u}$ at every point, whose projection on the coordinate axes x, y, z we shall denote by $u(x, y, z, t)$, $v(x, y, z, t)$, $w(x, y, z, t)$. These displacements arise in an elastic body under the action of internal forces (stresses), which form a symmetrical stress tensor

$$\begin{pmatrix} \sigma_x & \tau_{xy} & \tau_{xz} \\ \tau_{yx} & \sigma_y & \tau_{yz} \\ \tau_{zx} & \tau_{zy} & \sigma_z \end{pmatrix},$$

where $\sigma_x, \tau_{xy}, \tau_{xz}$ are the components of force (stress), acting on unit area of a surface element, perpendicular to the x-axis; similarly $\tau_{yx}, \sigma_y, \tau_{yz}$ and $\tau_{zx}, \tau_{zy}, \tau_z$ are the components of stresses, acting on unit area of surface elements, perpendicular to the y and z axes. The components $\sigma_x, \sigma_y, \sigma_z$ are called *normal stresses*, and τ_{xy} τ_{xz}, etc., are called the *shear stresses*. Considering an element of volume and writing down its equation of motion we obtain:

$$\left.\begin{aligned} \varrho \frac{\partial^2 u}{\partial t^2} &= \frac{\partial \sigma_x}{\partial x} + \frac{\partial \tau_{xy}}{\partial y} + \frac{\partial \tau_{xz}}{\partial z} + X, \\ \varrho \frac{\partial^2 v}{\partial t^2} &= \frac{\partial \tau_{yx}}{\partial x} + \frac{\partial \sigma_y}{\partial y} + \frac{\partial \tau_{yz}}{\partial z} + Y, \\ \varrho \frac{\partial^2 w}{\partial t^2} &= \frac{\partial \tau_{zx}}{\partial x} + \frac{\partial \tau_{zy}}{dy} + \frac{\partial \sigma_z}{\partial z} + Z, \end{aligned}\right\} \tag{1}$$

where ϱ is the volume density at the point (x, y, z), X, Y, Z are the components of the external volume forces. The values of stresses, arising from strains, are given by Hooke's law, which is written in the following form:

$$\left.\begin{aligned}\sigma_x &= 2G\left\{\varepsilon_x + \frac{\theta}{m-2}\right\}, & \tau_{xy} &= G\gamma_{xy},\\ \sigma_y &= 2G\left\{\varepsilon_y + \frac{\theta}{m-2}\right\}, & \tau_{yz} &= G\gamma_{yz},\\ \sigma_z &= 2G\left\{\varepsilon_z + \frac{\theta}{m-2}\right\}, & \tau_{zx} &= G\gamma_{zx}.\end{aligned}\right\} \tag{2}$$

The quantities

$$\left.\begin{aligned}\varepsilon_x = \frac{\partial u}{\partial x}, \quad \varepsilon_y = \frac{\partial v}{\partial y}, \quad \varepsilon_z = \frac{\partial w}{\partial z},\\ \gamma_{xy} = \frac{\partial u}{\partial y} + \frac{\partial v}{\partial x}, \quad \gamma_{yz} = \frac{\partial v}{\partial z} + \frac{\partial w}{\partial y}, \quad \gamma_{zx} = \frac{\partial w}{\partial x} + \frac{\partial u}{\partial z}\end{aligned}\right\} \tag{3}$$

form a symmetrical tensor of the strains

$$\begin{pmatrix}\varepsilon_x & \gamma_{xy} & \gamma_{xz}\\ \gamma_{yx} & \varepsilon_y & \gamma_{yz}\\ \gamma_{zx} & \gamma_{zy} & \varepsilon_z\end{pmatrix}.$$

In (2) we used the following symbols: G the shear *modulus*, m the inverse of Poissons ratio, $\theta = \operatorname{div} \boldsymbol{u} = \frac{\partial u}{\partial x} + \frac{\partial v}{\partial y} + \frac{\partial w}{\partial z}$.

Boundary conditions should be added to (1) and (2) (at the boundary, for example, displacements u, v, w, or surface forces, etc., are given), which we will not consider here.

Equations (1) and (2) form a complete system of differential equations for the stresses and strains. Substituting the expressions for the stresses from (2) in the equation of motion (1) and taking into account relations (3), we obtain a set of equations for the displacements

$$\left.\begin{aligned}\varrho\frac{\partial^2 u}{\partial t^2} &= G\left\{\nabla^2 u + \frac{m}{m-2}\frac{\partial\theta}{\partial x}\right\} + X,\\ \varrho\frac{\partial^2 v}{\partial t^2} &= G\left\{\nabla^2 v + \frac{m}{m-2}\frac{\partial\theta}{\partial y}\right\} + Y,\\ \varrho\frac{\partial^2 w}{\partial t^2} &= G\left\{\nabla^2 w + \frac{m}{m-2}\frac{\partial\theta}{\partial z}\right\} + Z.\end{aligned}\right\} \tag{4}$$

One often introduces in place of the constants G and m the so-called *constants of Lamé* λ and μ, connected by the relations

$$\mu = G, \quad \lambda = \frac{2}{m-2} G.$$

This allows us to write the preceding set of equations in the form of a single vector equation

$$\varrho \frac{\partial^2 \boldsymbol{u}}{\partial t^2} = (\lambda + 2\mu) \operatorname{grad} \operatorname{div} \boldsymbol{u} - \mu \operatorname{curl} \operatorname{curl} \boldsymbol{u} + \boldsymbol{F}, \qquad (5)$$

where $\boldsymbol{u}$ is a displacement vector with components u, v, and w, F is a vector of the volume forces with components X, Y, Z.

We show that equations (5) can be reduced to the wave equations for the appropriate form of the chosen functions.*

The arbitrary vector $\boldsymbol{F}$ may always be described in the form of a sum

$$\boldsymbol{F} = \operatorname{grad} U + \operatorname{curl} \boldsymbol{L},$$

where U is a scalar, and $\boldsymbol{L}$ a vector potential.

Let us assume

$$\boldsymbol{u} = \operatorname{grad} \boldsymbol{\Phi} + \operatorname{curl} \boldsymbol{A},$$

where

$$\varrho \frac{\partial^2 \boldsymbol{\Phi}}{\partial t^2} = (\lambda + 2\mu) \nabla^2 \boldsymbol{\Phi} + U,$$

$$\varrho \frac{\partial^2 \boldsymbol{A}}{\partial t^2} = \mu \nabla^2 \boldsymbol{A} + \boldsymbol{L}.$$

It is easy to verify by direct substitution that the vector $\boldsymbol{u}$ defined in this way in fact satisfies the equations of elasticity (4).

If the volume forces are absent, then we obtain for the potentials $\boldsymbol{\Phi}$ and $\boldsymbol{A}$ the homogeneous wave equations

$$\varrho \frac{\partial^2 \boldsymbol{\Phi}}{\partial t^2} = (\lambda + 2\mu) \nabla^2 \boldsymbol{\Phi},$$

$$\varrho \frac{\partial^2 \boldsymbol{A}}{\partial t^2} = \mu \nabla^2 \boldsymbol{A}.$$

The wave equation for the vector potential $\boldsymbol{A}$ in certain cases (for example in a cartesian system of coordinates) breaks down

* S. L. Sobolev, *Some Problems in the Theory of Wave Propagation*; Frank and Mises, *Differential and Integral Equations of Mathematical Physics*, chap. XII, Gostekhizdat, 1937.

into three scalar equations. However the question of the reduction of the equations of elasticity to separate scalar wave equations cannot be considered until the end without boundary conditions which can connect different components and thus represent considerable difficulty for complete separation of the equations.

II. Equations of the electromagnetic field

1. *Equations of the electromagnetic field and boundary conditions*

The electromagnetic field is characterized by the vectors $\boldsymbol{E}$ and $\boldsymbol{H}$ of the strengths of the electric and magnetic fields and by the vectors $\boldsymbol{D}$ and $\boldsymbol{B}$ of the electric and magnetic induction. The complete set of Maxwell's equations, connecting these values, has the form

$$\operatorname{curl} \boldsymbol{H} = \frac{1}{c}\frac{\partial \boldsymbol{D}}{\partial t} + \frac{4\pi}{c}\boldsymbol{j} + \frac{4\pi}{c}\boldsymbol{j}^{(e)}, \tag{1}$$

$$\operatorname{curl} \boldsymbol{E} = -\frac{1}{c}\frac{\partial \boldsymbol{B}}{\partial t}, \tag{2}$$

$$\operatorname{div} \boldsymbol{B} = 0, \tag{3}$$

$$\operatorname{div} \boldsymbol{D} = 4\pi\varrho . \tag{4}$$

where $\boldsymbol{j}$ is the volume density of the conduction currents, $\boldsymbol{j}^{(e)}$ is the density of currents, arising from the action of the external e. m. f., ϱ is the volume density of charges, c the velocity of light in a vacuum. Later we shall often assume $\boldsymbol{j}^{(e)} = 0$.

To these equations should be added the so-called mathematical field equations

$$\boldsymbol{D} = \varepsilon \boldsymbol{E}, \tag{5}$$

$$\boldsymbol{B} = \mu \boldsymbol{H}, \tag{6}$$

$$\boldsymbol{j} = \sigma \boldsymbol{E}, \tag{7}$$

where ε is the dielectric constant, μ is the magnetic permeability, σ the conductivity of the medium. Later we shall assume the medium to be homogeneous and isotropic. In this case $\varepsilon = \text{const.}$ $\mu = \text{const.}$, $\sigma = \text{const.}$ We will often consider electromagnetic processes in a vacuum, where $\varepsilon = \mu = 1$, $\sigma = 0$, with the condition of the absence of charges and currents. In this case Maxwell's equations are simplified:

$$\operatorname{curl} \boldsymbol{H} = \frac{1}{c} \frac{\partial \boldsymbol{E}}{\partial t},$$

$$\operatorname{curl} \boldsymbol{E} = -\frac{1}{c} \frac{\partial \boldsymbol{H}}{\partial t},$$

$$\operatorname{div} \boldsymbol{H} = 0,$$

$$\operatorname{div} \boldsymbol{E} = 0.$$

Equations (1) and (4) are related, since ϱ and $\boldsymbol{j}$ satisfy the relation

$$\frac{\partial \varrho}{\partial t} + \operatorname{div} \boldsymbol{j} = 0,$$

expressing the law of the conservation of electric charge.

Laws of the electromagnetic field, expressed in differential form by the equations (1)–(4), can be expressed in integral form

$$\oint_C H_s\, ds = \frac{4\pi}{c} \iint_\Sigma i_n\, d\sigma, \tag{1'}$$

$$\oint_C E_s\, ds = -\frac{1}{c} \frac{d}{dt} \iint_\Sigma B_n\, d\sigma = -\frac{1}{c} \frac{d\Phi}{dt}, \tag{2'}$$

where

$$\boldsymbol{i} = \boldsymbol{j}_d + \boldsymbol{j} = \frac{1}{4\pi} \frac{\partial \boldsymbol{D}}{\partial t} + \boldsymbol{j} \tag{8}$$

is the total current, $\boldsymbol{j}_d = \frac{1}{4\pi} \frac{\partial \boldsymbol{D}}{\partial t}$ is the displacement current. Integration is performed over the contour C and over the surface Σ lying on this contour; $\Phi = \iint_\Sigma B_n d\sigma$ is the induction threading the contour C. Denoting by T a certain closed volume, and by Σ the surface confining it, in place of (3) and (4) we have

$$\iint_\Sigma B_n\, d\sigma = 0, \tag{3'}$$

$$\iint_\Sigma D_n\, d\sigma = 4\pi \iiint_T \varrho d\tau = 4\pi e, \tag{4'}$$

where e is the total charge inside the volume T.

Equations (1')–(4') have a simple physical meaning and are the mathematical expression of fundamental experimental facts, serving as the basis of the development of Maxwell's equations. Thus, equation (1′) is the generalization of the well-known Biot and Savart law, equation (2′) expresses Faraday's law of electromagnetic induction, equation (4′) can be directly obtained from Coulomb's law. Equation (3′) is a result of the closed nature of the lines of force of the magnetic field.

If the medium is inhomogeneous, then boundary conditions should be added to Maxwell's equation. At the boundary of separation of two different media (1) and (2) the following conditions must be fulfilled:

$$E_s^{(1)} = E_s^{(2)} \text{ (continuity of the tangential components of vector } \boldsymbol{E}\text{)}, \tag{9}$$

$$H_s^{(1)} = H_s^{(2)} \text{ (continuity of the tangential components of vector } \boldsymbol{H}\text{)} \tag{10}$$

and also

$$B_{n_1}^{(1)} = B_{n_2}^{(2)} \text{ (continuity of the normal components of vector } \boldsymbol{B}\text{)} \tag{11}$$

$$D_{n_1}^{(1)} - D_{n_2}^{(2)} = 4\pi\nu\,, \text{ or } \varepsilon_1 E_{n_1}^{(1)} - \varepsilon_2 E_{n_2}^{(2)} = 4\pi\nu \tag{12}$$

where $\boldsymbol{n}_1$ and $\boldsymbol{n}_2$ are the normals to the surface of separation of the two media, in which $\boldsymbol{n}_1$ is directed into the first medium, and $\boldsymbol{n}_2$ into the second medium, ν is the surface density of the charges. These conditions are easily obtained from equations (1′)–(4′).

Maxwell's equations together with the boundary conditions enable us to determine the electromagnetic field in space from a given initial state of the field. Moreover for a complete determination of the field it is sufficient to use conditions (9) and (10) of the continuity of the tangential components of the field.

If the electromagnetic process is static, i. e. does not change with time, then Maxwell's equations take the form

$$\begin{aligned}
&\operatorname{curl} \boldsymbol{E} = 0\,,\\
&\operatorname{div} \varepsilon\boldsymbol{E} = 4\pi\varrho\,,\\
&\operatorname{curl} \boldsymbol{H} = \frac{4\pi}{c}\sigma\boldsymbol{E} + \frac{4\pi}{c}\boldsymbol{j}^{(e)},\\
&\operatorname{div} \mu\boldsymbol{H} = 0\,.
\end{aligned}$$

If, moreover, the medium is non-conducting, i. e. $\sigma = 0$, then we obtain two independent sets of equations for the electric and magnetic fields

$$\left.\begin{aligned}\operatorname{curl} \boldsymbol{E} &= 0\,,\\ \operatorname{div} \varepsilon \boldsymbol{E} &= 4\pi\varrho\end{aligned}\right\} \quad \text{for the electric field}$$

$$\left.\begin{aligned}\operatorname{curl} \boldsymbol{H} &= \frac{4\pi}{c}\boldsymbol{j}^{(e)},\\ \operatorname{div} \mu \boldsymbol{H} &= 0\end{aligned}\right\} \quad \text{for the magnetic field.}$$

The equations of electrostatics were considered by us in Chapter IV and in the appendices to Chapter IV.

In the case of a homogeneous medium it is easy to obtain the equations for each of the vectors $\boldsymbol{E}$ and $\boldsymbol{H}$ separately. We assume that $\varrho = 0, \boldsymbol{j}^{(e)} = 0$.

Applying the operator curl to equation (1) we have:

$$\operatorname{curl}\operatorname{curl} \boldsymbol{H} = \frac{\varepsilon}{c}\frac{\partial}{\partial t}\operatorname{curl} \boldsymbol{E} + \frac{4\pi}{c}\sigma \operatorname{curl} \boldsymbol{E}\,,$$

from which because of equation (2) and the relation $\operatorname{curl}\operatorname{curl} \boldsymbol{H} = \operatorname{grad}\operatorname{div} \boldsymbol{H} - \nabla^2 \boldsymbol{H}$ we obtain:

$$\operatorname{grad}\operatorname{div} \boldsymbol{H} - \nabla^2 \boldsymbol{H} = -\frac{\varepsilon\mu}{c^2}\frac{\partial^2 \boldsymbol{H}}{\partial t^2} - \frac{4\pi\sigma\mu}{c^2}\frac{\partial \boldsymbol{H}}{\partial t}$$

or

$$\nabla^2 \boldsymbol{H} = \frac{1}{a^2}\frac{\partial^2 \boldsymbol{H}}{\partial t^2} + \frac{4\pi\sigma\mu}{c^2}\frac{\partial \boldsymbol{H}}{\partial t} \quad \left(a^2 = \frac{c^2}{\varepsilon\mu}\right), \tag{13}$$

since

$$\operatorname{div} \boldsymbol{H} = 0\,.$$

Similarly the equation for $\boldsymbol{E}$ is derived

$$\nabla^2 \boldsymbol{E} = \frac{1}{a^2}\frac{\partial^2 \boldsymbol{E}}{\partial t^2} + \frac{4\pi\sigma\mu}{c^2}\frac{\partial \boldsymbol{E}}{\partial t}\,. \tag{14}$$

In particular, the components E_x, E_y, E_z and H_x, H_y, H_z will satisfy equation (13) or (14)

$$\nabla^2 u = \frac{1}{a^2}\frac{\partial^2 u}{\partial t^2} + \frac{4\pi\sigma\mu}{c^2}\frac{\partial u}{\partial t}\,, \tag{15}$$

where u is one of the components E_x, E_y, E_z or H_x, H_y, H_z.

The form of this equation is given by the properties of the medium and by the nature of the process. If the medium is non-conducting ($\sigma = 0$) we obtain the ordinary wave equation

$$\nabla^2 u = \frac{1}{a^2} \frac{\partial^2 u}{\partial t^2}, \tag{16}$$

i. e. electromagnetic processes are propagated in a non-conducting medium without attenuation with a velocity $a = \frac{c}{\sqrt{(\varepsilon\mu)}}$ and, in particular, in vacuum with the velocity of light c.

If the medium possesses high conductivity and displacement currents can be neglected in comparison with conduction currents, then we shall have an equation of parabolic type

$$\nabla^2 u = \frac{4\pi\sigma\mu}{c^2} \frac{\partial u}{\partial t}. \tag{17}$$

In the general case when the conduction currents and displacement currents are of the same order, equation (15) is an equation of hyperbolic type, describing the process of propagation with attenuation, caused by the dissipation of energy due to conduction.

For steady processes, for example in problems of diffraction

$$u = v(x, y, z)\, e^{i\omega t},$$

we arrive at an equation of elliptic type

$$\nabla^2 v + (k^2 + iq^2)\, v = 0, \tag{18}$$

where

$$k^2 = \frac{\omega^2}{a^2}; \quad q^2 = \frac{4\pi\sigma\mu\omega}{c^2}. \tag{19}$$

Static fields, as was already noted in Chapter IV, are described by Laplace's equation.

2. *Potentials of the electromagnetic field*

In order to determine the electromagnetic field we must find six values, which are the components of the vectors $\boldsymbol{E}$ and $\boldsymbol{H}$. In a number of cases, however, one can reduce this problem to finding four, and sometimes a smaller number of values. With this aim let us introduce the field potentials vector $\boldsymbol{A}$, scalar φ in the following way. We consider Maxwell's equation in a homogeneous space, for instance, in vacuum. From equation (3)

$$\operatorname{div} \boldsymbol{H} = 0 \tag{3}$$

it follows that vector $\boldsymbol{H}$ is solenoidal and therefore can be described by means of another vector $\boldsymbol{A}$ in the form

$$\boldsymbol{H} = \operatorname{curl} \boldsymbol{A}. \tag{20}$$

Substituting this expression in equation (2)

$$\operatorname{curl} \boldsymbol{E} = -\frac{1}{c}\frac{\partial \boldsymbol{H}}{\partial t}.$$

we obtain:

$$\operatorname{curl}\left[\boldsymbol{E} + \frac{1}{c}\frac{\partial \boldsymbol{A}}{\partial t}\right] = 0,$$

i. e. the vector $\boldsymbol{E} + \frac{1}{c}\frac{\partial \boldsymbol{A}}{\partial t}$ is irrotational and therefore can be described in the form of the gradient of some scalar function φ

$$\boldsymbol{E} + \frac{1}{c}\frac{\partial \boldsymbol{A}}{\partial t} = -\operatorname{grad}\varphi, \tag{21}$$

from which it follows

$$\boldsymbol{E} = -\operatorname{grad}\varphi - \frac{1}{c}\frac{\partial \boldsymbol{A}}{\partial t}.$$

The vector potential $\boldsymbol{A}$ and scalar potential φ introduced thus are not uniquely defined. In fact, from (20) and (21) it is seen that we obtain the same field if we replace $\boldsymbol{A}$ and φ by the potentials

$$\boldsymbol{A}' = \boldsymbol{A} + \operatorname{grad} F, \qquad \varphi' = \varphi - \frac{1}{c}\frac{\partial F}{\partial t},$$

where F is an arbitrary function. In order to remove this uncertainty we impose on the potentials $\boldsymbol{A}$ and φ the additional condition

$$\operatorname{div}\boldsymbol{A} + \frac{1}{c}\frac{\partial \varphi}{\partial t} = 0, \tag{22}$$

often called the Lorentz condition. We show that if this condition is fulfilled the potentials $\boldsymbol{A}$ and φ satisfy the equations

$$\nabla^2\varphi - \frac{1}{c^2}\frac{\partial^2\varphi}{\partial t^2} = -4\pi\varrho, \tag{23}$$

$$\nabla^2\boldsymbol{A} - \frac{1}{c^2}\frac{\partial^2\boldsymbol{A}}{\partial t^2} = -\frac{4\pi}{c}\boldsymbol{j}, \tag{24}$$

where ϱ and $\boldsymbol{j}$ are the densities of the given charges and currents.

Substituting expressions (20) and (21) in equation (1)

$$\operatorname{curl} \boldsymbol{H} = \frac{1}{c}\frac{\partial \boldsymbol{E}}{\partial t} + \frac{4\pi}{c}\boldsymbol{j} \tag{1}$$

and using the vector identity

$$\operatorname{curl}\operatorname{curl} \boldsymbol{A} = \operatorname{grad}\operatorname{div} \boldsymbol{A} - \nabla^2 \boldsymbol{A},$$

we have

$$\operatorname{grad}\left(\operatorname{div}\boldsymbol{A} + \frac{1}{c}\frac{\partial\varphi}{\partial t}\right) - \nabla^2\boldsymbol{A} = -\frac{1}{c^2}\frac{\partial^2\boldsymbol{A}}{\partial t^2} + \frac{4\pi}{c}\boldsymbol{j},$$

from which because of conditions (22) equation (24) follows. Substituting next expression (21) in Maxwell's fourth equation

$$\operatorname{div}\boldsymbol{E} = 4\pi\varrho \tag{4}$$

and taking condition (22) into account, we obtain equation (23) for φ.

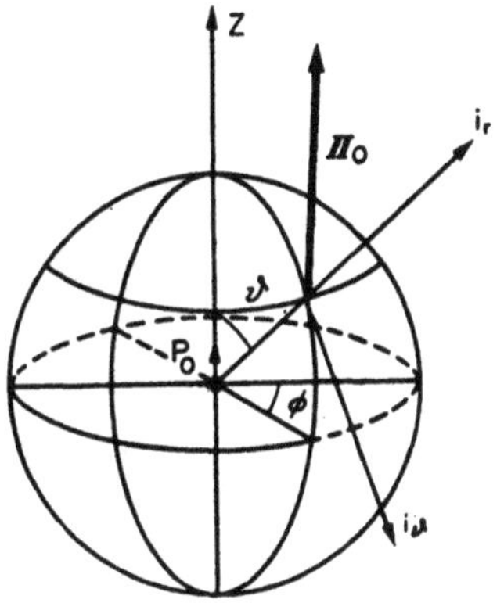

Fig. 80

In the case of a homogeneous conducting medium ($\sigma \neq 0$) the potentials are introduced by means of the relations

$$\boldsymbol{B} = \operatorname{curl}\boldsymbol{A},$$
$$\boldsymbol{E} = -\operatorname{grad}\varphi - \frac{1}{c}\frac{\partial \boldsymbol{A}}{\partial t}. \tag{25}$$

$\boldsymbol{A}$ and φ are connected to each other by the relation

$$\operatorname{div}\boldsymbol{A} + \frac{\varepsilon\mu}{c}\frac{\partial\varphi}{\partial t} + \frac{4\pi\mu\sigma}{c}\varphi = 0 \tag{26}$$

and satisfy the equations

$$\nabla^2\boldsymbol{A} - \frac{\varepsilon\mu}{c^2}\frac{\partial^2\boldsymbol{A}}{\partial t^2} - \frac{4\pi\mu\sigma}{c}\frac{\partial\boldsymbol{A}}{\partial t} = -\frac{4\pi\mu}{c}\boldsymbol{j}^{(e)}, \tag{27}$$

$$\nabla^2\varphi - \frac{\varepsilon\mu}{c^2}\frac{\partial^2\varphi}{\partial t^2} - \frac{4\pi\mu\sigma}{c}\frac{\partial\varphi}{\partial t} = -\frac{4\pi\varrho}{\varepsilon}, \tag{28}$$

which correspond to equations (23) and (24).

If there are no free charges ($\varrho \equiv 0$), the potential $\varphi = 0$ and the vector fields can be expressed by the one vector-potential $\boldsymbol{A}$, satisfying the additional condition

$$\operatorname{div}\boldsymbol{A} = 0.$$

In a number of cases the electromagnetic fields can be described by means of the vector potential, of which only one component differs from zero.

Some examples will be considered later (see also Appendix I to Chapter VII).

3. *The electromagnetic field of an oscillator*

1. In the theory of the emission of electromagnetic waves one often uses the idea of the oscillator or vibrator. This idea is closely related to the idea of linear currents. The oscillator represents a linear current of infinitely small length.

Let us consider a linear current L, whose intensity varies with time. In the simplest model it is assumed that the current intensity is constant over the length of the conductor.

The current, constant over the length of the conductor, is combined with the presence of charges, varying with time at its ends. By analogy with the electrostatic dipole, representing a combination of the two charges $+e$ and $-e$, the oscillator can be described by a moment

$$\boldsymbol{p}(t) = e(t)\,\boldsymbol{l}. \tag{1}$$

The current intensity in the oscillator, obviously, equals

$$J(t) = \dot{e}(t),$$

so that

$$\frac{d\boldsymbol{p}}{dt} = \boldsymbol{J}(t)\,\boldsymbol{l}. \tag{2}$$

The product $J(t)\,\boldsymbol{l} = \boldsymbol{J}_0(t)$ is called the moment of current

2. Let us find the electromagnetic field, excited by an oscillator of moment

$$\boldsymbol{p} = \boldsymbol{p}_0 f(t) \tag{3}$$

in infinite space, assuming that $\sigma = 0$, $\varepsilon = 1$, $\mu = 1$ (vacuum).

We consider the problem of the excitation of an electromagnetic field by a linear current L, the limiting case of which is the oscillator.

Outside the current L the electromagnetic field is determined from Maxwell's equations

$$\left.\begin{aligned} \operatorname{curl} \boldsymbol{H} &= \frac{1}{c}\frac{\partial \boldsymbol{E}}{\partial t}, & \operatorname{div} \boldsymbol{H} &= 0, \\ \operatorname{curl} \boldsymbol{E} &= -\frac{1}{c}\frac{\partial \boldsymbol{H}}{\partial t}, & \operatorname{div} \boldsymbol{E} &= 0. \end{aligned}\right\} \tag{4}$$

Along the path of the current L the field vectors $\boldsymbol{H}$ and $\boldsymbol{E}$ must have a singularity, characterized by the fact that a circuit over an infinitely small neighbourhood of K_ε, encircling the direction of the current L, has the following value:

$$\lim_{\varepsilon \to 0} \oint_{K_\varepsilon} H_s \, ds = \frac{4\pi}{c} J, \tag{5}$$

where $J = J(t)$ is the current intensity.

From this relation it follows that the component H_s in the magnetic field has a singularity of the type

$$H_s \approx \frac{2J}{c\varrho}, \tag{6}$$

where $\varrho = M_0 P$, M_0 is a point along the path L, P is a point on K_ε $(\varrho = \varepsilon)$.

For the field to be well defined it is necessary to add initial conditions, which we assume to be zero.

It is useful to introduce the potentials $\boldsymbol{A}$ and φ, by which as we saw (see p. 493), the field is expressed in the following manner:

$$\left.\begin{aligned} \boldsymbol{H} &= \operatorname{curl} \boldsymbol{A} \\ \boldsymbol{E} &= -\frac{1}{c}\frac{\partial \boldsymbol{A}}{\partial t} - \operatorname{grad} \varphi, \end{aligned}\right\} \tag{7}$$

where

$$\operatorname{div} \boldsymbol{A} + \frac{1}{c}\frac{\partial \varphi}{\partial t} = 0. \tag{8}$$

The vector potential outside the current L satisfies the homogeneous wave equation

$$\nabla^2 \boldsymbol{A} - \frac{1}{c^2}\frac{\partial^2 \boldsymbol{A}}{\partial t^2} = 0. \tag{9}$$

Let us introduce the cartesian system of coordinates, directing the z-axis along the current $\boldsymbol{L}$. We assume

$$\boldsymbol{A} = A(x, y, z)\, z^0,$$

where z^0 is a unit vector along the z-axis.

The function $A(x, y, z)$, obviously, satisfies the homogeneous wave equation outside the path of the current L

$$\nabla^2 A - \frac{1}{c^2}\frac{\partial^2 A}{\partial t^2} = 0 . \tag{10}$$

In order to find the singularity of vector $\boldsymbol{A}$ at the current we use condition (6). From equation (7) it follows that

$$H_s = \frac{\partial A}{\partial \varrho} .$$

Using condition (6) we find that the function $\boldsymbol{A}(x, y, z)$ at L should have a singularity of the form

$$A \approx \frac{2J}{c}\ln \varrho . \tag{11}$$

We shall search for the potential A in the form

$$A(P, t) = \int_L A_1(P, M, t)\, dl_M \qquad (P = P(x, y, z)) , \tag{12}$$

where $\boldsymbol{A}_0(P, M, t) = A_1(P, M, t)\, d\boldsymbol{l}_M$ is the vector-potential of an elementary current of the oscillator, whose moment equals $\boldsymbol{J}_0 = J d\boldsymbol{l}$.

In order that the potential $\boldsymbol{A}$ has the necessary singularity the function $A_0(P, M, t)$ must have a singularity of the form

$$A_0(P, M, t) \approx \frac{J_0(t)}{c r_{PM}} . \tag{13}$$

In fact, assuming that A_0 has the specified singularity, and calculating from (12) the value of A in the neighbourhood of the current L $(0 < z < l)$, we obtain:

$$A \approx J \int_0^l \frac{1}{c r_{MP}}\, d\zeta = \frac{J}{c}\int_0^l \frac{d\zeta}{\sqrt{[\varrho^2 + (z-\zeta)^2]}} =$$

$$= \frac{J}{c}\ln\left[z - \zeta + \sqrt{\{\varrho^2 + (z-\zeta)^2\}}\,\Big|_0^l\right] = \frac{J}{c}\ln\frac{z - l + \sqrt{[\varrho^2 + (z-l)^2]}}{z + \sqrt{(\varrho^2 + z^2)}} =$$

$$= \frac{J}{c}\ln\frac{(l-z)\left[-1 + 1 + \dfrac{\varrho^2}{2(l-z)^2} + \dots\right]}{z + \sqrt{(\varrho^2 + z^2)}} = \frac{2J}{c}\ln\varrho + \dots ,$$

where the terms, not tending to infinity at $\varrho = 0$ are omitted.

3. Thus, the function $A_0\,(P, M, t)$ must satisfy with respect to the variables $P(x, y, z)$, t, the wave equation

$$\nabla^2 A_0 - \frac{1}{c^2}\frac{\partial^2 A_0}{\partial t^2} = 0 \tag{14}$$

everywhere, except at the point M, and at this point it must have a singularity of the form

$$A_0 \approx \frac{J_0(t)}{c r_{MP}}\,. \tag{15}$$

The initial conditions are zero.

The solution of this problem, as we saw in Chapter V, is represented in the form of a retarded potential

$$A_0(M, P, t) = \frac{J_0\left(t - \dfrac{r_{MP}}{c}\right)}{c r_{MP}}\,. \tag{16}$$

As was noted above, the moment of the current equals

$$\boldsymbol{J}_0\,(t) = J\,(t)\,d\boldsymbol{l} = \frac{d\boldsymbol{p}}{dt} = \boldsymbol{p}_0 \mathring{f}\,(t)\,. \tag{17}$$

Thus, the vector-potential of an oscillator can also be represented in the form

$$\boldsymbol{A}_1 = \frac{\boldsymbol{p}_0 \mathring{f}\left(t - \dfrac{r}{c}\right)}{c r}\,. \tag{18}$$

Often in place of the vector-potential one uses the polarization potential or the Hertz vector $\boldsymbol{\Pi}$, assuming

$$\boldsymbol{A} = \frac{1}{c}\frac{\partial \Pi}{\partial t}\,. \tag{19}$$

The vector $\boldsymbol{\Pi}$ also satisfies the equation

$$\nabla^2 \boldsymbol{\Pi} - \frac{1}{c^2}\frac{\partial^2 \boldsymbol{\Pi}}{\partial t^2} = 0 \tag{20}$$

and is related to the scalar potential by the relation

$$\varphi = -\operatorname{div} \boldsymbol{\Pi} \tag{21}$$

The field vectors $\boldsymbol{E}$ and $\boldsymbol{H}$ are expressed in terms of the polarization potential by the relations

$$\boldsymbol{E} = \operatorname{grad}\operatorname{div} \boldsymbol{\Pi} - \frac{1}{c^2}\frac{\partial^2 \boldsymbol{\Pi}}{\partial t^2} = \operatorname{curl}\operatorname{curl} \boldsymbol{\Pi}\,, \tag{22}$$

$$\boldsymbol{H} = \frac{1}{c}\operatorname{curl}\frac{\partial \boldsymbol{\Pi}}{\partial t}\,. \tag{23}$$

Taking relation (18) into account, we obtain the following expression for the polarization potential for an oscillator

$$\boldsymbol{\Pi}_0 = \frac{\boldsymbol{p}_0 f\left(t - \frac{r}{c}\right)}{r}$$

or

$$\boldsymbol{\Pi}_0 = \frac{\boldsymbol{p}\left(t - \frac{r}{c}\right)}{r} \tag{24}$$

4. In order to calculate the fields $\boldsymbol{E}$ and $\boldsymbol{H}$ let us change to a spherical system of coordinates (r, ϑ, φ), at the origin of coordinates of which we locate an oscillator and let us direct the z-axis ($\vartheta = 0$) along the vector $\boldsymbol{p}_0$ (Fig. 80). Let us denote by $\boldsymbol{i}_r$, $\boldsymbol{i}_\vartheta$, $\boldsymbol{i}_\varphi$, the unit vectors of the spherical system of coordinates.

Vector $\boldsymbol{\Pi}_0$, parallel to vector $\boldsymbol{p}$, can be represented in the form

$$\left.\begin{aligned} \boldsymbol{\Pi}_0 &= \Pi_0 \cos\vartheta\, \boldsymbol{i}_r - \Pi_0 \sin\vartheta\, \boldsymbol{i}_\vartheta, \\ \Pi_0 &= |\boldsymbol{\Pi}_0|. \end{aligned}\right\} \tag{25}$$

Substituting expression (25) in (22) and (23), using the expression of the differential operator curl $\boldsymbol{F}$ in a spherical system of coordinates

$$\operatorname{curl} \boldsymbol{F} = \frac{1}{r\sin\vartheta}\left[\frac{\partial}{\partial\vartheta}(\sin\vartheta\, F_\varphi) - \frac{\partial F_\vartheta}{\partial\varphi}\right]\boldsymbol{i}_r + $$
$$+ \frac{1}{r}\left[\frac{1}{\sin\vartheta}\frac{\partial F_r}{\partial\varphi} - \frac{\partial}{\partial r}(r F_\varphi)\right]\boldsymbol{i}_\vartheta + \frac{1}{r}\left[\frac{\partial}{\partial r}(r F_\vartheta) - \frac{\partial F_r}{\partial\vartheta}\right]\boldsymbol{i}_\varphi$$

and taking into consideration that $\boldsymbol{\Pi}_0 = \boldsymbol{\Pi}_0(r, t)$ we obtain:

$$\left.\begin{aligned} E_r &= -\frac{2\cos\vartheta}{r}\frac{\partial\Pi_0}{\partial r}, \\ E_\vartheta &= \frac{\sin\vartheta}{r}\frac{\partial}{\partial r}\left(r\frac{\partial\Pi_0}{\partial r}\right), E_\varphi = 0, \end{aligned}\right\} \tag{26}$$

$$\left.\begin{aligned} H_r &= 0, \quad H_\vartheta = 0, \\ H_\varphi &= -\sin\vartheta\frac{1}{c}\frac{\partial^2\Pi_0}{\partial r\,\partial t}. \end{aligned}\right\} \tag{27}$$

From equations (26) and (27) it follows that the electric and magnetic fields of the oscillator are mutually perpendicular.

5. Let us consider the special cases, where the dipole moment periodically depends on the time

$$\boldsymbol{p}(t) = \boldsymbol{p}_0 e^{-i\omega t}.$$

In this case (26) and (27) give:

$$\left.\begin{aligned} E_r &= 2\cos\vartheta\left(\frac{1}{r^2} + \frac{ik}{r}\right)\Pi_0, \\ E_\vartheta &= \sin\vartheta\left(\frac{1}{r^2} - \frac{ik}{r} - k^2\right)\Pi_0 \qquad \left(k = \frac{\omega}{c}\right). \\ H_\varphi &= ik\sin\vartheta\left(ik - \frac{1}{r}\right)\Pi_0, \end{aligned}\right\} \tag{28}$$

where

$$\Pi_0 = p_0 \frac{e^{ikr}}{r} e^{-i\omega t}. \tag{29}$$

Proceeding from (28) it is easy to determine the singularity in the field of an oscillator. At distances, small in comparison with a wavelength $\lambda = (2\pi/k)$ $(kr \ll 1)$ in (28) it is possible to confine ourselves to one term. Formulae obtained from this for the strength of the electric field correspond to the field of a static dipole, whose electric moment p equals the instantaneous value of the moment of the oscillator $p(t)$. For the magnetic field strength we obtain an expression, corresponding to the Biot–Savart law. At large distances from the dipole $R \gg \lambda$ $(kr \gg 1)$ in (28) it is possible to neglect all terms of higher order than $1/r$. From this we obtain:

$$E_r = 0, \qquad E_\vartheta = H_\varphi = -k^2 \sin\vartheta\, \Pi_0, \tag{30}$$

i. e. the field becomes transverse relative to the direction of propagation. Regions where the field of emission becomes transverse are called *wave zones of the oscillator*. In order to calculate the *energy flow* across the surface of a sphere of radius R with centre in the oscillator it is necessary to calculate the Umov–Poynting vector

$$S = \frac{c}{4\pi}\left|[\boldsymbol{E}\boldsymbol{H}]\right| = \frac{c}{4\pi} EH$$

and integrate this expression over the sphere.

From (29) and (30) it follows that in a wave zone the real part of the vectors H_φ and E_ϑ is given by the expression

$$H_\varphi = E_\vartheta = -\frac{\omega^2 \sin\vartheta}{c^2 r} p_0 \cos\omega\left(t - \frac{r}{c}\right),$$

from which

$$S = \frac{p_0^2 \omega^4}{4\pi} \frac{\sin^2\vartheta}{R^2 c^3} \cos^2\omega\left(t - \frac{r}{c}\right).$$

The energy flow across a sphere of radius R in the time of one complete period $T = 2\pi/\omega$ will be given by the expression

$$\Sigma = \frac{1}{T}\int_0^T dt \int_0^\pi \int_0^{2\pi} S R^2 \sin\vartheta \, d\vartheta \, d\varphi = \frac{4p_0^2\omega^4}{3c^3} \frac{1}{T}\int_0^T \cos^2\omega\left(t - \frac{r}{c}\right) dt$$

or

$$\Sigma = \frac{2p_0^2\omega^4}{3c^3}.$$

The energy emitted by a harmonic oscillator is proportional to the fourth power of the frequency

$$\Sigma \sim \omega^4$$

or

$$\Sigma \sim \frac{1}{\lambda^4},$$

where λ is the wavelength.

CHAPTER VI

HEAT CONDUCTION IN SPACE

§ 1. Heat conduction in infinite space

1. *Temperature propagation function*

In Chapter III it was shown that the process of heat propagation in a homogeneous isotropic space is given by the equation of heat conduction

$$u_t = a^2 \nabla^2 u \qquad \left(a^2 = \frac{k}{c\varrho}\right), \tag{1}$$

where $u(M, t)$ is the temperature of the point $M(x, y, z)$ at time t, ϱ is the density, c the coefficient of specific heat, $k = \text{const.}$ and $a^2 = k/c\varrho$ are the coefficients of heat conduction and thermal conductivity. Equation (1) also permits a diffusion interpretation. In this case u is the concentration of the diffusing substance, $a^2 = D$ is the diffusion coefficient.

Let us consider the following problem in infinite space:

Find the solution of the inhomogeneous equation of heat conduction

$$u_t = a^2 \nabla^2 u + f/c\varrho \tag{2}$$

(f is the density of the heat sources) with the initial condition

$$u(x, y, z, 0) = \varphi(x, y, z). \tag{3}$$

The solution of this problem can be represented in the form of the sum

$$u = u_1 + u_2,$$

where u_1 is the solution of the homogeneous equation (1) with inhomogeneous initial conditions, u_2 is the solution of the inhomogeneous equation (2) with zero initial conditions. In an investigation of corresponding one-dimensional problems we saw that their solutions are determined in terms of a source function.

Let us find the source function for the equation of heat conduction in infinite space.

First we prove the following lemma, which will be used later.

If the solution of the equation $\nabla^2 u - \frac{1}{a^2} u_t = 0$ *depends only on* r *and* t, *then the function* $v = ru$ *satisfies the equation*

$$\frac{\partial^2 v}{\partial r^2} = \frac{1}{a^2} \frac{\partial v}{\partial t}. \tag{4}$$

In fact, writing down the Laplacian operator in a spherical system of coordinates, we see that the function $u = u(r, t)$ satisfies the equation

$$\frac{\partial^2 u}{\partial r^2} + \frac{2}{r} \frac{\partial u}{\partial r} - \frac{1}{a^2} \frac{\partial u}{\partial t} = 0 \text{ or } \frac{1}{r} \frac{\partial^2 (ru)}{\partial r^2} - \frac{1}{a^2} \frac{\partial u}{\partial t} = 0;$$

assuming next $ru = v$, we obtain the equation (4).* for v. Let a continuously acting heat source of constant magnitude q be located at the origin of coordinates, and let the initial temperature equal zero in the rest of space.

$$u(r, 0) = 0 \text{ where } r \neq 0.$$

It is obvious that in this case the temperature u is a function only of r and t.

The presence of a heat source at $r = 0$ indicates that the heat flow per unit time across the sphere S_ε (with centre at $r = 0$ and radius ε) as $\varepsilon \to 0$ equals q, i. e.

$$\lim_{\varepsilon \to 0} \left[- \iint_{S_\varepsilon} k \frac{\partial u}{\partial n} d\sigma \right] = q.$$

Since the normal derivative $\frac{\partial u}{\partial n} = \frac{\partial u}{\partial r}$ because of the symmetry of the surface S_ε, then

$$- k \frac{\partial u}{\partial r} \cdot 4 \pi r^2 \Big|_{r=\varepsilon} \to q \text{ where } \varepsilon \to 0,$$

* Compare with sub-section 1, § 1, of Chapter V.

which indicates a singularity of the form $-\frac{q}{4\pi kr^2}$ in the derivative $\frac{\partial u}{\partial r}$ for $r = 0$. Hence, the function itself at $r = 0$ must have a singularity of the form

$$u \sim \frac{q}{4\pi kr},$$

so that the product $ru = v$ remains bounded at $r = 0$.

The function v, defined by the conditions

$$\frac{\partial^2 v}{\partial r^2} = \frac{1}{a^2}\frac{\partial v}{\partial t},$$

$$v(0, t) = \frac{q}{4\pi k} = v_0, \tag{5}$$

$$v(r, 0) = 0,$$

is given by the formula

$$v(r, t) = v_0\left[1 - \Phi\left(\frac{r}{2\sqrt{(a^2 t)}}\right)\right] = q\frac{1}{4\pi k}\frac{2}{\sqrt{\pi}}\int\limits_{\frac{r}{2\sqrt{(a^2 t)}}}^{\infty} e^{-\alpha^2}\, d\alpha$$

[see formula (35), Chapter III, § 3]. Hence, the solution of the problem on the conduction of heat from a continuously acting source of magnitude q, situated at the origin of coordinates ($r = 0$) has the form

$$u(r, t) = q\,U(r, t) = q\frac{1}{4\pi k}\frac{1}{r}\frac{2}{\sqrt{\pi}}\int\limits_{\frac{r}{2\sqrt{(a^2 t)}}}^{\infty} e^{-\alpha^2}\, d\alpha, \tag{6}$$

where $U(r, t)$ is the temperature, corresponding to unit source ($q = 1$).

In order to proceed to the case of an instantaneous source, let us consider a source of magnitude q, situated at the point (ξ, η, ζ) and continuously acting during an interval of time τ.

Such a source is equivalent to two sources of magnitude $+q$ and $-q$, the first beginning at $t = 0$, the second at $t = \tau$. The temperature distribution is expressed by the relation

$$u_\tau(r, t) = q\left[U(r, t) - U(r, t - \tau)\right].$$

After an interval of time τ an amount of heat $Q = q\tau$ is liberated, therefore

$$u_\tau(r, t) = \frac{Q}{\tau}\left[U(r, t) - U(r, t - \tau)\right].$$

Passing to a limit as $\tau \to 0$ and assuming Q constant, we find:

$$u_0(r, t) = \lim_{\tau\to 0} u_\tau(r, t) = Q\frac{\partial U}{\partial t} = \frac{Q}{4\pi kr}\frac{2}{\sqrt{\pi}} e^{-\frac{r^2}{4a^2t}} \frac{r}{4a^2\sqrt{(a^2t^3)}}\, a^2$$

or

$$u_0(r, t) = \frac{Q}{c\varrho} G(x, y, z, t; \xi, \eta, \zeta),$$

where

$$G(x, y, z, t; \xi, \eta, \zeta) = \left(\frac{1}{2\sqrt{(\pi a^2 t)}}\right)^3 e^{-\frac{(x-\xi)^2+(y-\eta)^2+(z-\zeta)^2}{4a^2t}} \tag{7}$$

The function $G(x, y, z, t;\ \xi, \eta, \zeta)$ is the *temperature propagation function* of an instantaneous heat source. It represents the temperature at the point x, y, z at time t, produced by a point source of magnitude $Q = c\varrho$, situated at time $t = 0$ at the point (ξ, η, ζ).

It is easy to verify that

$$\iiint_{-\infty}^{\infty} G(x, y, z, t;\ \xi, \eta, \zeta)\, d\xi\, d\eta\, d\zeta = 1. \tag{8}$$

In fact, the triple integral (8) can be represented in the form of a product of three integrals, each of which equals unity:

$$\int_{-\infty}^{\infty} \frac{1}{2\sqrt{(\pi a^2 t)}} e^{-\frac{(x-\xi)^2}{4a^2t}} d\xi = \frac{1}{\sqrt{\pi}} \int_{-\infty}^{\infty} e^{-\alpha^2} d\alpha = 1 \quad \left(\alpha = \frac{\xi - x}{2\sqrt{(a^2 t)}}\right).$$

From (7) it is seen that the source function G possesses a *symmetry property*

$$G(x, y, z, t;\ \xi, \eta, \zeta) = G(\xi, \eta, \zeta, t;\ x, y, z),$$

which is the expression of the reciprocity principle: the effect at the point (x, y, z) of a source, existing at point (ξ, η, ζ) equals the effect at the point (ξ, η, ζ) of a similar source, situated at the point (x, y, z). But such symmetry does not occur with respect to the variable t, which expresses the irreversibility of heat conduction processes with time.

Let us determine the form of the propagation function G in the case of two dimensions. Let an infinite linear source be situated on the straight line, parallel to the z-axis and passing through the point (ξ, η), let us denote by $\bar{Q} = \text{const.}$ the magnitude of the source, per unit length. The propagation function G_2 of such a source will not depend on z and is fully described by its values in the plane (x, y). Let us calculate the function G_2. The quantity of heat liberated in the element $d\zeta$ is

$$dQ = \bar{Q} d\zeta ;$$

and the distribution of temperature in space is given by the integral

$$\bar{u} = \int_{-\infty}^{\infty} \frac{\bar{Q} d\zeta}{c\varrho} (x, y, z, t ; \xi, \eta, \zeta) .$$

Calculating the integral

$$\int_{-\infty}^{\infty} e^{-\frac{(z-\zeta)^2}{4a^2t}} d\zeta = 2\sqrt{(a^2 t)} \int_{-\infty}^{\infty} e^{-\alpha^2} d\alpha = 2\sqrt{(\pi a^2 t)} \quad \left(\alpha = \frac{\zeta - z}{2\sqrt{(a^2 t)}}\right),$$

we obtain:

$$\bar{u} = \frac{\bar{Q}}{c\varrho} G_2 ,$$

where

$$G_2 (x, y, t ; \xi, \eta) = \left(\frac{1}{2\sqrt{(\pi a^2 t)}}\right)^2 e^{-\frac{(x-\xi)^2+(y-\eta)^2}{4a^2t}} \qquad (8')$$

Comparing this function with (7) we see their similarity of structure.

In a similar way it is possible to obtain an expression for the source function in the one-dimensional case. Considering an infinite plane source with constant density $\bar{\bar{Q}}$, we obtain:

$$\bar{\bar{u}} = \int_{-\infty}^{\infty}\int_{-\infty}^{\infty} \frac{\bar{\bar{Q}} d\eta d\zeta}{c\varrho} G (x, y, z, t ; \xi, \eta, \zeta) =$$

$$= \frac{\bar{\bar{Q}}}{c\varrho} \frac{1}{2\sqrt{(\pi a^2 t)}} e^{-\frac{(x-\xi)^2}{4a^2t}} = \frac{\bar{\bar{Q}}}{c\varrho} G_1 (x, t, \xi) ,$$

where

$$G_1(x, t; \xi) = \frac{1}{2\sqrt{(\pi a^2 t)}} e^{-\frac{(x-\xi)^2}{4a^2t}}$$

is the source function for one-dimension.

In Chapter III graphs were given, showing the behaviour of the source function $G(x, t; \xi)$. The qualitative character of the source function, given in Chapter III, holds for the three-dimensional case.

2. *Conduction of heat in infinite space*

Let us use the source function, obtained in the previous section, to solve the problem of the propagation of an initial temperature distribution in infinite space.

We are required to find the solution of the equation

$$u_t = a^2 \nabla^2 u, \tag{1}$$

satisfying the initial condition

$$u(x, y, z, 0) = \varphi(x, y, z). \tag{3}$$

The initial temperature state can be described as a superposition of the effects of instantaneous sources, producing the initial temperature. Let us consider an element of volume $d\xi\, d\eta\, d\zeta$, containing the point ξ, η, ζ. In order to give the initial temperature $\varphi(\xi, \eta, \zeta)$ it is necessary to inject an amount of heat $dQ = c\, \varrho\varphi(\xi, \eta, \zeta)\, d\xi\, d\eta\, d\zeta$ into the volume $d\xi\, d\eta\, d\zeta$.

This concentrated amount of heat produces a temperature at the point (x, y, z) at time t

$$\frac{dQ}{c\varrho} G(x, y, z, t; \xi, \eta, \zeta) = G(x, y, z, t; \xi, \eta, \zeta)\, \varphi(\xi, \eta, \zeta)\, d\xi\, d\eta\, d\zeta. \tag{9}$$

Because of the principle of superposition the solution of our problem can be represented by the integration of (9) over all space

$$u(x, y, z, t) = \int\!\!\int\!\!\int_{-\infty}^{\infty} G(x, y, z, t; \xi, \eta, \zeta)\, \varphi(\xi, \eta, \zeta)\, d\xi\, d\eta\, d\zeta. \tag{10}$$

Formula (10) is derived by intuition. We have not shown the limits of its applicability nor its validity.

Let us prove that

if the function φ is piecewise-continuous and bounded, $|\varphi| < A$, then the function u, given by the expression

$$u(x, y, z, t) = \left(\frac{1}{2\sqrt{(\pi a^2 t)}}\right)^3 \int\int\int_{-\infty}^{\infty} e^{-\frac{(x-\xi)^2+(y-\eta)^2+(z-\zeta)^2}{4a^2 t}} \varphi(\xi, \eta, \zeta)\, d\xi\, d\eta\, d\zeta, \tag{10'}$$

(1) *is bounded for all space: $|u| < A$;*

(2) *is the solution of the equation of heat conduction for $t > 0$;*

(3) *for $t = 0$ is continuous at points of continuity of function φ, and satisfies the condition $u(x, y, z, 0) = \varphi(x, y, z)$.*

In order to prove that (10') satisfies equation (1), we make use of a lemma (see § 3, Chapter III).

If $U(x, y, z, t; \xi)$ for any value of the parameter ξ is the solution of the equation $\mathscr{L}(u) = 0$, where $\mathscr{L}(u)$ is a linear differential operator, then the function

$$u(x, y, z, t) = \int U(x, y, z, t; \xi)\, \varphi(\xi)\, d\xi$$

will also be a solution of the equation $\mathscr{L}(u) = 0$, if the derivatives of u, entering into the operator $\mathscr{L}(u)$, can be calculated by differentiation under the integral sign.

In our case $U = G$ satisfies the equation of heat conduction for any ξ, η, ζ. As is well known, differentiation with respect to a parameter under the sign of an improper integral is possible if: (1) the derivative with respect to a parameter of the function under the integral sign is continuous and (2) the integral, obtained after formal differentiation, converges uniformly.

Performing a formal differentiation of integral (10') with respect to x, we obtain:

$$\int\int\int_{-\infty}^{\infty} \left(-\frac{x-\xi}{2a^2 t}\right) e^{-\frac{(x-\xi)^2+(y-\eta)^2+(z-\zeta)^2}{4a^2 t}} \varphi(\xi, \eta, \zeta)\, d\xi\, d\eta\, d\zeta. \tag{11}$$

The function under the integral sign is continuous for $\bar{t} > t > 0$ and the presence of the factor $e^{-\frac{(x-\xi)^2+(y-\eta)^2+(z-\zeta)^2}{4a^2 t}}$ ensures the uniform convergence, if φ is bounded: $|\varphi| < A$. We obtain similar results for repeated differentiation with respect to x and for differentiation with respect to t; the same applies to differentiation with respect to y and z. Thus, G satisfies all the conditions of the lemma for $t > 0$. Hence, u satisfies the equation of heat conduction for $t > 0$.

The bounded nature of u, given by (10′), which we rewrite in the form

$$u(M, t) = \int\int_{-\infty}^{\infty}\int G(M, M', t)\,\varphi(M')\,d\tau_{M'}$$

$$(M = M(x, y, z),\ \ M' = M'(\xi, \eta, \zeta)),$$

is established directly, if equation (8) is considered:

$$|u| < A\int\int_{-\infty}^{\infty}\int G\,d\tau = A. \tag{12}$$

Let us proceed to a proof of the continuity of $u(x, y, z, t)$ for $t = 0$.

Let us consider the point $M_0\,(x_0, y_0, z_0)$, the point of continuity of function φ, and let us prove that for any $\varepsilon > 0$ there exists $\delta(\varepsilon)$ such that

$$|u(M, t) - \varphi(M_0)| < \varepsilon \text{ where } |\overline{MM_0}| < \delta(\varepsilon) \text{ and } t < \delta(\varepsilon). \tag{13}$$

We consider an auxiliary region T_1, containing the point M_0; its dimensions will be determined below; we denote by T_2 the remaining part of space. Taking into consideration the equations

$$u(M, t) = \int\int_{T_1}\int G(M, M', t)\,\varphi(M')\,d\tau_{M'} + \\ + \int\int_{T_2}\int G(M, M', t)\,\varphi(M')\,d\tau_{M'},$$

$$\varphi(M_0) = \int\int_{T_1}\int G(M, M', t)\,\varphi(M_0)\,d\tau_{M'} + \\ + \varphi(M_0)\int\int_{T_2}\int G(M, M', t)\,d\tau_{M'},$$

and also the positive nature of $G(M, M', t)$, we have:

$$|U(M, t) - \varphi(M_0)| \leqslant J_1 + J_2, \tag{14}$$

where

$$J_1 = \int\int_{T_1}\int G(M, M', t)\,|\varphi(M') - \varphi(M_0)|\,d\tau_{M'}. \tag{15}$$

$$J_2 = 2A\int\int_{T_2}\int G(M, M'', t)\,d\tau_{M''}. \tag{16}$$

From the continuity of φ at the point M_0 it follows: whatever $\eta > 0$ may be, there exists $\delta'(\eta) > 0$ such that

$$|\varphi(M') - \varphi(M_0)| < \eta, \quad \text{when} \quad |\overline{M'M_0}| < \delta'(\eta).$$

Hence, if the diameter of region T_1 does not exceed $\delta'(\varepsilon/3)$ then

$$J_1 < \frac{\varepsilon}{3}\iiint_{T_1} G\, d\tau_{M'} < \frac{\varepsilon}{3}\iiint_{-\infty}^{\infty} G\, d\tau = \frac{\varepsilon}{3}. \tag{17}$$

Let us consider the choice of region T_1 in more detail. As T_1 we choose a sphere with centre at $M(x, y, z)$, which is convenient in order to determine the integral J_2. Relation (17) for the integral J_0 retains validity, if the radius of this sphere is chosen equal to

$$\varrho_0 = \frac{1}{2}\delta'\left(\frac{\varepsilon}{3}\right) \text{ and if } |\overline{MM_0}| < \varrho_0.$$

Converting to a spherical system of coordinates with centre at the point M, we obtain:

$$\iiint_{T_1} G\, d\tau = \left(\frac{1}{2\sqrt{(\pi a^2 t)}}\right)^3 \int_0^{\varrho_0} e^{-\frac{r^2}{4a^2 t}} r^2\, dr =$$

$$= \frac{4}{\sqrt{\pi}} \int_0^{\frac{\varrho_0}{2\sqrt{(a^2 t)}}} \alpha^2 e^{-\alpha^2}\, d\alpha \xrightarrow[t\to 0]{} \frac{4}{\sqrt{\pi}}\int_0^{\infty} \alpha^2 e^{-\alpha^2}\, d\alpha = 1 \quad \left(\alpha = \frac{r}{2\sqrt{(a^2 t)}}\right),$$

since

$$\int_0^{\infty} \alpha^2 e^{-\alpha^2}\, d\alpha = -\frac{1}{2}\alpha e^{-\alpha^2}\Big|_0^{\infty} + \frac{1}{2}\int_0^{\infty} e^{-\alpha^2}\, d\alpha = \frac{\sqrt{\pi}}{4}.$$

Thus,

$$\iiint_{T_2} G\, d\tau = 1 - \iiint_{T_1} G\, d\tau \to 0 \text{ where } t \to 0,$$

i. e. for any $\varepsilon > 0$ it is possible to assign $\delta''(\varepsilon)$ such that

$$\iiint_{T_2} G\, d\tau < \varepsilon/3A$$

and, therefore,

$$J_2 < 2\varepsilon/3 \tag{18}$$

provided $t < \delta''(\varepsilon)$.

Choosing the smaller of the quantities $1/2\ \delta'(\varepsilon/3)$ and $\delta''(\varepsilon)$ and calling it $\delta(\varepsilon)$, we have the inequality

$$|u(M,t) - \varphi(M_0)| < \varepsilon \text{ where } |\overline{MM_0}| < \delta(\varepsilon) \text{ and } t < \delta(\varepsilon)\,, \quad (13)$$

which proves the continuity of $u(M, t)$ at $t = 0$ for any point M_0 of continuity of function $\varphi(M)$.

Let us pass now to the solution of the inhomogeneous equation

$$u_t = a^2 \nabla^2 u + f/c\varrho$$

with the zero initial condition

$$u(x, y, z, 0) = 0\,.$$

We consider the point (ξ, η, ζ) at time $\tau < t$. The amount of heat, liberated in element $d\xi\, d\eta\, d\zeta$ in a time $d\tau$ and equal to

$$dQ = f\, d\xi\, d\eta\, d\zeta\, d\tau\,,$$

produces at the point (x, y, z) at time t a temperature

$$\frac{1}{c\varrho} G(x, y, z, t;\ \xi, \eta, \zeta, \tau)\, f(\xi, \eta, \zeta, \tau)\, d\xi\, d\eta\, d\zeta\, d\tau\,.$$

Using the principle of superposition, we can write the solution of the given problem in the form

$$u(x, y, z, t) = \\ = \int_0^t \int \int_{-\infty}^{\infty} \int \frac{1}{c\varrho} G(x, y, z, t;\ \xi, \eta, \zeta, \tau)\, f(\xi, \eta, \zeta, \tau)\, d\xi\, d\eta\, d\zeta\, d\tau\,. \quad (19)$$

We will not consider a proof of this relation or a discussion of the conditions of its applicability.

Problems for semi-space with homogeneous boundary conditions of the first and second kind are solved by the method of images.

§ 2. Heat conduction in finite bodies

1. *Outline of the method of separation of variables*

We have considered heat propagation in infinite space. In an investigation of heat propagation in a finite body it is necessary to add boundary conditions and initial condition. In the simplest cases these are boundary conditions of the first, second or third kind.

Let us consider the simplest problem with a homogeneous boundary condition of the first kind:

$$\left.\begin{aligned} u_t &= a^2 \nabla^2 u \\ u(x, y, z, 0) &= \varphi(x, y, z) \\ u|_\Sigma &= 0, \end{aligned}\right\} \quad (1)$$

The solution of this problem can be found by the usual method of separation of the variables, given for the equation $u_{tt} = a^2 \nabla^2 u$ in Chapter V, § 2; the application of this method to our problem proceeds entirely analogously.

Let us consider the auxiliary problem:

$$u_t - a^2 \nabla^2 u = 0,$$

$$u|_\Sigma = 0 \quad (2)$$

$$u(M, t) = v(M) T(t).$$

Separating the variables in the usual way, we arrive at the following relations, defining the functions $v(M)$ and $T(t)$:

$$\left.\begin{aligned} \nabla^2 v + \lambda v = 0 \ \text{ in } T, \\ v = 0 \ \text{ on } \Sigma \end{aligned}\right\} \quad (3)$$

and

$$T' + a^2 \lambda T = 0. \quad (4)$$

For function v we have the eigen-value problem, which we met in the investigation of vibrations of finite volumes (see Chapter V, § 3, sub-section 1).

Let $\lambda_1, \lambda_2, \ldots, \lambda_n, \ldots$ be the eigen-values, and $v_1, v_2, \ldots, v_n, \ldots$ be the eigen-functions of problem (3). The functions $\{v_n\}$ form an orthogonal system.

The corresponding functions $T_n(t)$ have the form

$$T_n(t) = C_n e^{-a^2\lambda_n t},$$

and the auxiliary problem has the non-trivial solution

$$u_n(M, t) = C_n v_n(M) e^{-a^2\lambda_n t}. \tag{5}$$

The general solution of the original problem can be represented in the form

$$u(M, t) = \sum_{n=1}^{\infty} C_n e^{-a^2\lambda_n t} v_n(M). \tag{6}$$

Satisfying the initial condition

$$u(M, 0) = \varphi(M) = \sum_{n=1}^{\infty} C_n v_n(M), \tag{7}$$

we find the coefficients

$$C_n = \frac{\int\limits_T \varphi(M') v_n(M') d\tau_{M'}}{N_n},$$

where

$N_n = \int\limits_T v_n^2(M') d\tau_{M'}$ — are normalization coefficients of v_n.

Function (6) gives the solution of the problem.

The equation

$$u_t - a^2 \nabla^2 u = f(M, t) \qquad (f = F/c\varrho) \tag{8}$$

with homogeneous boundary and zero initial conditions can also be solved by the method of separation of variables.

Assuming, as is usual,

$$u(M, t) = \sum_{n=1}^{\infty} T_n(t) v_n(M) \tag{9}$$

and expanding the function $f(M, t)$ in the eigen-functions $v_n(M)$

$$f(M, t) = \sum_{n=1}^{\infty} f_n(t) v_n(M), \quad f_n(t) = \frac{1}{N_n} \int\limits_T f(M', t) v_n(M') d\tau_{M'}, \tag{10}$$

we obtain for $T_n(t)$ the equation

$$T_n + a^2 \lambda_n T_n = f_n(t) \tag{11}$$

with initial condition $T_n(0) = 0$, if $u(M, 0) = 0$, the solution of which has the form

$$T_n(t) = \int_0^t e^{-a^2\lambda_n(t-\tau)} f_n(\tau)\, d\tau. \tag{12}$$

Hence we obtain:

$$u(M, t) = \int_0^t \int_T \left\{ \sum_{n=1}^{\infty} e^{-a^2\lambda_n(t-\tau)} \frac{v_n(M)\, v_n(M')}{N_n} \right\} f(M', \tau)\, d\tau_{M'}\, d\tau. \tag{13}$$

The expression in brackets, obviously, corresponds to the source function of an instantaneous source of magnitude $Q = c\varrho$, located at point M' at time τ,

$$G(M, t, M', \tau) = \sum_{n=1}^{\infty} \frac{v_n(M)\, v_n(M')}{N_n} e^{-a^2\lambda_n(t-\tau)}. \tag{14}$$

The solution of the first boundary value problem $\bar{u}$ for the equation of heat conduction with inhomogeneous boundary conditions $\bar{u}|_\Sigma = \mu$ readily reduces to the solution u of the inhomogeneous equation with homogeneous boundary conditions $u|_\Sigma = 0$, if it is assumed

$$\bar{u} = u + \Phi, \tag{15}$$

where Φ is an arbitrary (sufficiently smooth) function, taking the value μ on Σ (see Chapter III, § 2). Very often cases of constant boundary values occur, $\mu_0 = \text{const.}$, leading to the problem with homogeneous boundary conditions, if the function

$$\bar{u} = u + \mu_0 \qquad (\Phi = \text{const} = \mu_0),$$

is introduced, representing the deviation from the fixed solution.

Thus, the main difficulty in solving the problem of heat propagation in a finite region lies in finding the eigen-functions and eigen-values for the given region.

The form of solution (6), obtained by the method of separation of variables, is suitable for investigating a stage of the process for large t. In fact, the eigen-values λ_n for any region increase rapidly with n. Therefore for $t > 0$ the series rapidly converges and, starting at a certain time, the first term different from zero predominates over the sum of the remaining terms

$$u(M, t) \approx C_1 v_1(M)\, e^{-a^2\lambda_1 t}. \tag{16}$$

This corresponds to the physical fact that, independently of the initial distribution, starting at some time, a "regular" temperature distribution is established in the body which has a "profile" invariant with time and the amplitude decreasing exponentially with time. This fact is the basis of non-steady methods of determining the coefficients of heat conduction. In fact, measuring the temperature of a body at an arbitrary point M_0, we find that

$$\ln |u(M_0, t)| \approx -a^2 \lambda_1 t + \ln |C_1 v_1(M)| . \tag{17}$$

A graph of this function is, for large times, a straight line with slope $-a^2 \lambda_1$. Knowing the value of λ, depending on the shape of the region, it is possible to find the coefficient of heat conduction.

2. *The cooling of a circular cylinder*

Let us consider the problem of the cooling of an infinitely long cylinder of radius r_0, having a given initial temperature, if the temperature on the surface is maintained equal to zero. We assume that the initial temperature does not depend on z (the z-axis is directed along the axis of the cylinder). Then, obviously, later the temperature will not depend on z and changes only over a cross-section S of the cylinder. Choosing in this section a polar system of coordinates with the pole at the centre of the circle S, we arrive at the problem of the determination of the function $u(r, \varphi, t)$ satisfying the equation

$$\frac{\partial^2 u}{\partial r^2} + \frac{1}{r}\frac{\partial u}{\partial r} + \frac{1}{r^2}\frac{\partial^2 u}{\partial \varphi^2} = \frac{1}{a^2}\frac{\partial u}{\partial t}, \tag{18}$$

initial condition

$$u(r, \varphi, 0) = \Phi(r, \varphi) \tag{19}$$

and boundary condition

$$u(r_0, \varphi, t) = 0 . \tag{20}$$

As we saw, the solution of a problem of such a kind can be represented in the form

$$u = \sum C e^{-a^2 \lambda t} v(M), \tag{21}$$

where summation extends over all the eigen-functions of the problem

$$\frac{\partial^2 v}{\partial r^2} + \frac{1}{r}\frac{\partial v}{\partial r} + \frac{1}{r^2}\frac{\partial^2 v}{\partial \varphi^2} + \lambda v = 0\,, \qquad (22)$$

$$v(r_0, \varphi) = 0\,.$$

This eigen-value problem was considered by us in the investigation of the vibrations of a circular membrane (see Chapter V, § 3). To each eigen-value

$$\lambda_{mn} = \left(\frac{\mu_m^{(n)}}{r_0}\right)^2 \qquad (23)$$

there corresponds the two eigen-functions

$$\bar{v}_{nm} = J_n\left(\frac{\mu_m^{(n)}}{r_0} r\right)\cos n\varphi$$

and

$$\bar{\bar{v}}_{nm} = J_n\left(\frac{\mu_m^{(n)}}{r_0} r\right)\sin n\varphi\,, \qquad (24)$$

whose norms equal

$$N_{nm} = \frac{\varepsilon_n r_0^2}{2}\left[J_n'(\mu_m^{(n)})\right]^2, \qquad \varepsilon_n = \begin{cases} 1; & n \neq 0; \\ 2; & n = 0\,, \end{cases} \qquad (25)$$

where $\mu_m^{(n)}$ is the mth root of the equation

$$J_n(\mu) = 0\,. \qquad (26)$$

Using the expression for v and λ, we obtain:

$$u(r, \varphi, t) = \\ = \sum_{n=0}^{\infty}\sum_{m=1}^{\infty}(\bar{C}_{mn}\cos n\varphi + \bar{\bar{C}}_{mn}\sin n\varphi)\, J_n\left(\frac{\mu_m^{(n)}}{r_0} r\right) e^{-a^2\left(\frac{\mu_m^{(n)}}{r_0}\right)^2 t} \qquad (27)$$

where the coefficients $\bar{C}_{nm}$ and $\bar{\bar{C}}_{nm}$ are given by the initial functions

$$\left.\begin{aligned} \bar{C}_{nm} &= \frac{\displaystyle\int_0^{r_0}\int_0^{2\pi} \Phi(r,\varphi) J_n\left(\frac{\mu_m^{(n)}}{r_0} r\right)\cos n\varphi\, r\, d\varphi\, dr}{\displaystyle\frac{\pi r_0^2}{8}\,\varepsilon_n\left[J_n'(\mu_m^{(n)})\right]^2}\,, \\ \bar{\bar{C}}_{nm} &= \frac{\displaystyle\int_0^{r_0}\int_0^{2\pi} \Phi(r,\varphi) J_n\left(\frac{\mu_m^{(n)}}{r_0} r\right)\sin n\varphi\, r\, d\varphi\, dr}{\displaystyle\frac{\pi r_0^2}{2}\left[J_n'(\mu_m^{(n)})\right]^2}\,, \\ \varepsilon_n &= \begin{cases} 1; & n \neq 0; \\ 2; & n = 0. \end{cases} \end{aligned}\right\} \qquad (28)$$

If the initial temperature Φ depends only on r then the double series (27) is replaced by the single series

$$u(r,t)=\sum_{m=1}^{\infty} C_m J_0\left(\frac{\mu_m^{(0)}}{r_0}r\right) e^{-a^2\left(\frac{\mu_m}{r_0}\right)^2 t}, \tag{29}$$

where

$$C_m=\frac{2\int_0^{r_0}\Phi(r)J_0\left(\frac{\mu_m^{(0)}}{r_0}r\right)r\,dr}{r_0^2[J_1(\mu_m^{(0)})]^2} \qquad (J_1=-J_0'), \tag{30}$$

and $\mu_m^{(0)}$ is the mth root of the equation $J_0(\mu)=0$.

Let us consider in more detail the problem of the uniform cooling of a heated cylinder with zero temperature on the surface. If the initial temperature

$$u(r,0)=\Phi=u_0,$$

the

$$C_m=\frac{2u_0\int_0^{r_0}J_0\left(\frac{\mu_m^{(0)}}{r_0}r\right)r\,dr}{r_0^2[J_1(\mu_m^{(0)})]^2}=\frac{2u_0\frac{r_0}{\mu_m^0}\left[rJ_1\left(\frac{\mu_m^{(0)}}{r_0}r\right)\right]_0^{r_0}}{r_0^2[J_1(\mu_m^{(0)})]^2}=\frac{2u_0}{\mu_m^{(0)}J_1(\mu_m^{(0)})}, \tag{30'}$$

since $aJ_0(a)=[aJ_1(a)]'$. Thus we obtain:

$$\frac{u(r,t)}{u_0}=\sum_{m=1}^{\infty}\frac{2J_0(\mu_m^{(0)}\varrho)}{\mu_m^{(0)}J_1(\mu_m^{(0)})}e^{-(\mu_m^{(0)})^2\theta} \qquad \left(\varrho=\frac{r}{r_0},\quad \theta=\frac{a^2t}{r_0^2}\right). \tag{31}$$

In the tables of cylindrical functions (see p. 753, Table 3) the numerical values of both the roots $\mu_m^{(0)}$ and $J_1(\mu_m^{(0)})$ are given. In particular,

$$\mu_1^{(0)}=2.40,\quad J_1(\mu_1^{(0)})=0.52,$$

$$\mu_2^{(0)}=5.52,\quad J_1(\mu_2^{(0)})=-0.34.$$

Series (31) converges rapidly and for large t it is possible to consider only the first term of this series. In particular, on the axis of the cylinder

$$\frac{u(r,t)}{u_0}=\frac{2}{2.40\times 0.52}1.60\,e^{-(2.40)^2\theta}=1\cdot 60\,e^{-5.76\,\theta}\quad\left(\theta=\frac{a^2t}{r_0}\right). \tag{32}$$

3. *Determination of critical sizes*

In Chapter III it was shown that the process of diffusion of an unstable gas, whose rate of decomposition is proportional to the concentration, leads to the equation

$$u_t = a^2 \nabla^2 u + \beta u \qquad (\beta < 0). \tag{33'}$$

Greatest interest lies in diffusion processes in the presence of *chain reactions*. Chain reactions are characterized by the fact that particles of a diffusing substance, entering into reaction with the surrounding medium, "multiply". Thus, for instance, in the collision of neutrons with "active" nuclei of uranium the reaction of the division of nuclei results, accompanied by the production of new neutrons, the number of which is greater than unity. These neutrons in their turn enter into reaction with a generation of new neutrons, etc. Thus the multiplication of neutrons arises, characterized by a chain reaction.

Considering the process described in "the diffusion approximation" we arrive at the following equation:

$$u_t = a^2 \nabla^2 u + \beta u \qquad (\beta > 0), \tag{33''}$$

since the chain reaction is equivalent to the presence of sources in a diffusing substance (neutrons), proportional to the concentration (density of neutrons).

Let us consider the following problem:

Find the solution of the equation

$$u_t = a^2 \nabla^2 u + \beta u \text{ inside } T, \tag{33}$$

satisfying the initial condition

$$u(M, 0) = \varphi(M) \tag{34}$$

and the boundary condition

$$u|_\Sigma = 0. \tag{35}$$

By means of the substitution

$$u(M, t) = \bar{u}(M, t)\, e^{\beta t} \tag{36}$$

equation (33) changes into equation (1); the initial and boundary conditions moreover remain invariant. Thus, the unknown function u has the form

$$u(M, t) = \sum_{\infty=1}^{\infty} C_n e^{(\beta - a^2\lambda_n)t} v_n(M), \qquad (37)$$

where C_n is given by the initial functions in (10). In the case $\beta < 0$ (diffusion with decomposition) the exponents of series (37) are greater than the corresponding exponents of series (6). This means that in the presence of decomposition, decrease in concentration takes place more rapidly compared with the case of pure diffusion ($\beta = 0$). In the case $\beta > 0$ (diffusion with multiplication), if even one of the exponents $\beta - a^2 \lambda > 0$, i. e. $\beta > a^2 \lambda_1$, then in time there will be, generally speaking ($C_1 \neq 0$), an increase in concentration according to an exponential law (chain reaction). The value β is a characteristic of the substance (multiplication coefficient), and λ_1 essentially depends on the shape and dimensions of the region. We shall say that a certain region T_{cr} has for a given β *critical dimensions* if $\lambda_1 = \beta/a^2$. Let us determine the critical dimensions for a infinite layer, cylinder and sphere.

1. *The layer* $0 \leqslant x \leqslant l$. Assuming the problem to be one-dimensional, we have (see Chapter II, § 2):

$$\lambda_n = (\pi n/l)^2 \quad \text{and} \quad \lambda_1 = \pi^2/l^2.$$

The critical thickness of the layer l_{cr}, for which the process of avalanche increase in concentration u takes place, is given by the relation

$$l_{cr} = \frac{a\pi}{\sqrt{\beta}} \approx \frac{3.14\, a}{\sqrt{\beta}} \qquad (\beta > 0). \qquad (38)$$

2. *Infinite cylinder.* Assuming the problem to be a plane one, we see that the least value of λ corresponds to an eigen-function, possessing radial symmetry, and equals

$$\lambda_1^{(0)} = \left(\frac{\mu_1^{(0)}}{r_0}\right) a \qquad (\mu_2^{(0)} = 2.4048).$$

Hence for the critical diameter we obtain the relation

$$d_{cr} = \frac{2\mu_1^{(0)} a}{\sqrt{\beta}} \approx \frac{4.80\, a}{\sqrt{\beta}}. \qquad (39)$$

3. *Sphere.* The least value of λ corresponds to an eigenfunction, possessing spherical symmetry, and equals

$$\lambda_1 = (\pi/R)^2,$$

whence for the critical diameter D_{cr} we obtain the relation

$$D_{cr} = \frac{2\,\pi\,a}{\sqrt{\beta}} \approx \frac{6.28\,a}{\sqrt{\beta}}. \tag{40}$$

§ 3. Boundary value problems for regions with moving boundaries

1. *Green's theorem for the equation of heat conduction and the source function*

For the equation of heat conduction it is possible to set boundary-value problems for regions with boundaries, varying with time.

For simplicity we shall consider the problem for the equation with one space variable

$$\mathscr{L}(u) = a^2 u_{xx} - u_t = 0, \tag{1}$$

although everything stated below can be transformed to the case of many variables.

Let us consider the region $BAEF$, bounded by the sides AB and EF ($t = \text{const.}$) and by the curves, defined by the equations

$$x = \chi_1(t) \quad (\text{curve } AE)$$

and

$$x = \chi_2(t) \quad (\text{curve } BF)\ (\text{Fig. } 81).$$

The first boundary-value problem for this region consists of the determination of the solution of the equation of heat conduction (1), satisfying the initial and boundary conditions:

$$\left. \begin{array}{c} u = \varphi(x) \text{ on } AB, \\ u\,|_{x=\chi_1(t)} = \mu_1(t),\ u\,|_{x=\chi_2(t)} = \mu_2(t). \end{array} \right\} \tag{2}$$

From the maximum value principle it immediately follows that this problem cannot have more than one continuous solution. Similarly other boundary value problems can be given.

Let us establish Green's theorem for equation (1) and the integral representation of the solution of this problem.

We consider the operator

$$\mathscr{M}(v) = a^2 \frac{\partial^2 v}{\partial x^2} + \frac{dv}{dt}; \tag{3}$$

integrating the expression

$$\psi\,\mathscr{L}(\varphi) - \varphi\,\mathscr{M}(\psi)$$

over some region $PABQ$ (Fig. 82), where $\varphi(x, t)$ and $\psi(x, t)$ are arbitrary functions, differentiable a sufficient number of times, we obtain after a simple transformation

$$\int\int [\psi \mathscr{L}(\varphi) - \varphi \mathscr{M}(\psi)]\, dx\, dt = \int\left[\varphi\psi\, dx + a^2\left(\psi\frac{\partial\varphi}{\partial x} - \varphi\frac{\partial\psi}{\partial x}\right) dt,\right]$$

where the integral on the right is taken over the closed contour $PABQ$.

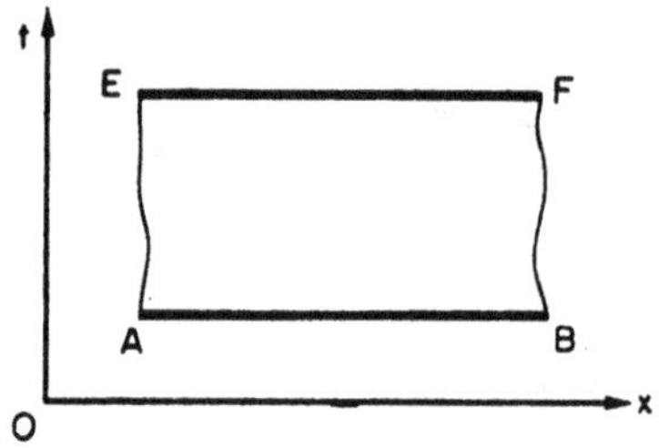

FIG. 81

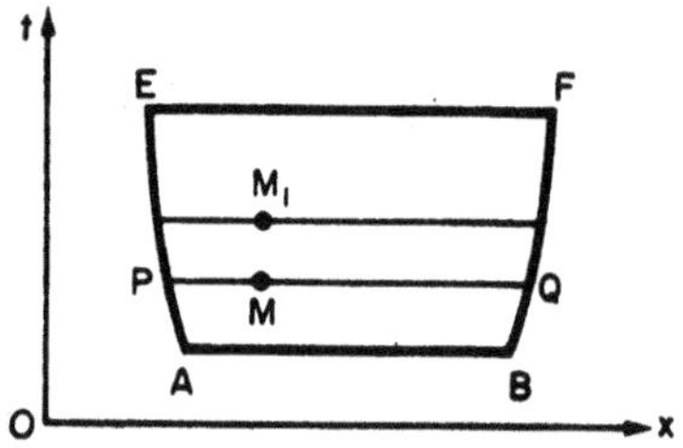

FIG. 82

If $\mathscr{L}(\varphi) = 0$ and $\mathscr{M}(\psi) = 0$, then writing the right-hand part more fully, we obtain:

$$\int_{PQ} \varphi\psi\, dx = \int_{AB} \varphi\psi\, dx + \int_{BQ} \left[\varphi\psi\, dx + a^2\left(\psi\frac{\partial\varphi}{\partial x} - \varphi\frac{\partial\psi}{\partial x}\right) dt\right] -$$

$$- \int_{AP} \left[\varphi\psi\, dx + a^2\left(\psi\frac{\partial\varphi}{\partial x} - \varphi\frac{\partial\psi}{\partial x}\right) dt\right]. \qquad (4)$$

Let $\varphi(x, t) = u(x, t)$ be any solution of the equation of heat conduction $L(u) = 0$, and $\psi = G_0(x, t, \xi, \tau)$ the source function for this equation in an infinite region

$$G_0(x, t, \xi, \tau) = \frac{1}{2\sqrt{[\pi a^2 (t - \tau)]}}\, e^{-\frac{(x-\xi)^2}{4a^2(t-\tau)}}, \qquad (5)$$

often called the *fundamental solution of the equation of heat conduction*. The function $G_0(x, t, \xi, \tau)$ satisfies the equation $\mathscr{L}(G_0) = 0$ with respect to the variables x, t and the related equation $\mathscr{M}(G_0) = 0$ with respect to the variables ξ, τ.

Let $M(x, t)$ be some fixed point inside the region $EAFB$, at which we wish to determine the value of the function $u(x, t)$, and M_1 be a point with coordinates x, $t + h$, where $h > 0$. Drawing the line PQ through the point M replacing x by ξ, t by τ in (4) and applying this to region $ABPQ$ (Fig. 82) and the functions

$$\varphi = u(x, t) \text{ and } \psi = G_0(x, t, + h, \xi, \tau), \qquad (6)$$

we have:

$$\int_{PQ} \frac{e^{-\frac{(x-\xi)^2}{4a^2h}}}{2\sqrt{(\pi a^2 h)}} u(\xi, t)\, d\xi =$$

$$= \int_{PABQ} u(\xi, \tau)\, G_0(x, t+h, \xi, \tau)\, d\xi + a^2 \left(G_0 \frac{du}{\partial \xi} - u \frac{\partial G_0}{\partial \xi}\right) d\tau. \qquad (7)$$

Passing to a limit as $h \to 0$ and taking into consideration the continuity with respect to h of the function $G(x, t+h, \xi, \tau)$ and $\frac{\partial G_0}{\partial \xi}$ on $PABQ$, and also the relation

$$\lim_{h \to 0} \int_{PQ} \frac{e^{-\frac{(x-\xi)^2}{4a^2h}}}{2\sqrt{(\pi a^2 h)}} u(\xi, t)\, d\xi = u(x, t)\,*\,, \qquad (8)$$

if (x, t) lies in the segment PQ we obtain the fundamental integral relation

$$u(x, t) = \int_{PABQ} u(\xi, \tau)\, G_0(x, t, \xi, \tau)\, d\xi + \int_{BQ+PA} a^2 \left(G_0 \frac{\partial u}{\partial \xi} - u \frac{\partial G_0}{\partial \xi}\right) d\tau, \qquad (9)$$

giving a representation of the general solution of the equation of heat conduction. Let us rewrite it once more in more detail

$$u(x, t) = \int_{PABQ} \frac{e^{-\frac{(x-\xi)^2}{4a^2(t-\tau)}}}{2\sqrt{[\pi a^2 (t-\tau)]}} u(\xi, \tau)\, d\xi + a^2 \int_{BQ+PA} \frac{e^{-\frac{(x-\xi)^2}{4a^2(t-\tau)}}}{2\sqrt{[\pi a^2 (t-\tau)]}} \frac{\partial u}{\partial \xi}\, d\tau -$$

$$- a^2 \int_{BQ+PA} u(\xi, \tau) \frac{\partial}{\partial \xi} \left(\frac{e^{-\frac{(x-\xi)^2}{4a^2(t-\tau)}}}{2\sqrt{[\pi a^2 (t-\tau)]}} \right) d\tau\,. \qquad (9')$$

This relation does not give the solution of boundary-value problems, since for calculation of the right-hand part it is necessary to know the value not only of u but also of $\frac{\partial u}{\partial \xi}$ along the curves AE and BF.

By means of a transformation, similar to that which was performed for Laplace's equation in the introduction of the source function, it is possible to eliminate $\frac{\partial u}{\partial \xi}$ from this relation.

Let v be some solution of the conjugate equation $\mathscr{M}(v) = 0$, reducing to zero on PQ, and u be the solution of the equation of heat conduction $\mathscr{L}(u) = 0$.

Applying (4) to the functions v and u for the region $PABQ$, we obtain:

$$0 = \int_{PABQ} \left[u(\xi, \tau)\, v(\xi, \tau)\, d\xi + a^2 \left(v \frac{\partial u}{\partial \xi} - u \frac{\partial v}{\partial \xi} \right) d\tau \right]. \qquad (10)$$

* See Chapter III, § 3.

Subtracting (10) from (9) we have:

$$u(x, t) = \int\limits_{PABQ} \left[u(\xi, \tau)\, G(x, t, \xi, \tau)\, d\xi + a^2 \left(G \frac{\partial u}{\partial \xi} - u \frac{\partial G}{\partial \xi} \right) d\tau \right], \tag{11}$$

where

$$G(x, t, \xi, \tau) = G_0(x, t, \xi, \tau) - v\,.$$

If v is chosen so that

$$G = 0 \text{ on } PA \text{ and } BQ\,,$$

then we obtain an integral representation for $u(x, t)$ in the form

$$u(x, t) = \int\limits_{AB} u(\xi, \tau)\, G(x, t, \xi, \tau)\, d\xi + a^2 \int\limits_{AP} u \frac{\partial G}{\partial \xi}\, d\tau - a^2 \int\limits_{BQ} u \frac{\partial G}{\partial \xi}\, d\tau\,. \tag{13}$$

Formula (13) gives the solution of the boundary-value problem, in terms of which the values of u on AP and BQ are given.

Let us consider further the function G. It is determined by (12), where $v(\xi, \tau)$ is characterized by the following conditions:

(1°) $v(\xi, \tau)$ is defined in the region $PABQ$ and for $\tau < t$ satisfies the conjugate equation $\mathcal{M}(v) = 0$.

(2°) $v = 0$ on PQ, i.e. where $\tau = t$.

(3°) $v(\xi, \tau) = -G_0(x, t, \xi, \tau)$ on PA and QB.

By virtue of these conditions, v depends on the parameters x, t, so that $v = v(x, t, \xi, \tau)$ and for its determination it is necessary to solve the boundary-value problem for the equation $\mathcal{M}(v) = 0$, which is equivalent to solving the boundary-value problem of type (2) for the equation $\mathcal{L}(u) = 0$, as is easily verified by change in sign of τ. Thus, in representing $u(x, t)$ by means of (11) which gives the solution of the boundary-value problem (2) the main difficulty consists of finding the function $v(x, t, \xi, \tau)$.

Let us consider the function $\bar{v}(x, t, \xi, \tau)$, defined by the conditions:

(1°) $\bar{v}(x, t, \xi, \tau)$ is defined in the region $PABQ$ for $t > \tau$ and, as a function of the variables x, t satisfies the equation of heat conduction $\mathcal{L}(v) = 0$.

(2°) $\bar{v} = 0$ on AB, i.e. where $t = \tau$.

(3°) $\bar{v} = -G_0$ on AP, and BQ.

Let us prove that $v(x, t, \xi, \tau) = \bar{v}(x, t, \xi, \tau)$.

We consider the function $\bar{G}(x, t, \xi, \tau) = G_0 + \bar{v}$. It is obvious that for any solution $\bar{u}$ of the equation $\mathcal{M} = 0$ a relation, similar to (9), holds,

$$\bar{u}(\xi, \tau) = \int\limits_{BQPA} \bar{u}\, G_0\, dx + a^2 \left(\bar{u} \frac{\partial G_0}{\partial x} - G_0 \frac{\partial \bar{u}}{\partial x} \right) dt\,, \tag{9'}$$

and also a relation, similar to (13),

$$\bar{u}(\xi, \tau) = \int\limits_{PQ} \bar{u}\, \bar{G}\, dx + a^2 \int\limits_{BQ} \bar{u} \frac{\partial \bar{G}}{\partial x}\, dt - a^2 \int\limits_{AP} \bar{u} \frac{\partial \bar{G}_0}{\partial x}\, dt\,. \tag{13'}$$

These relations can be derived from (9) and (13) by a change in sign of τ, since by this the equation $\mathscr{M} = 0$ becomes the equation $\mathscr{L} = 0$,

Applying (13) to the region $PQRS$ (Fig. 83), where RS is a segment of the straight line, corresponding to the ordinate θ, where $t > \theta > \tau$ and to the solution $u(x, t) = \bar{G}(x, t, \xi, \tau)$, continuous in this region, of the equation $\mathscr{L}(u) = 0$, we obtain:

$$\bar{G}(x, t, \xi, \tau) = \int_{RS} \bar{G}(x', \theta, \xi, \tau)\, G(x, t, x', \theta)\, dx',$$

since the integrals over RP and SQ by virtue of (3°) equal zero.

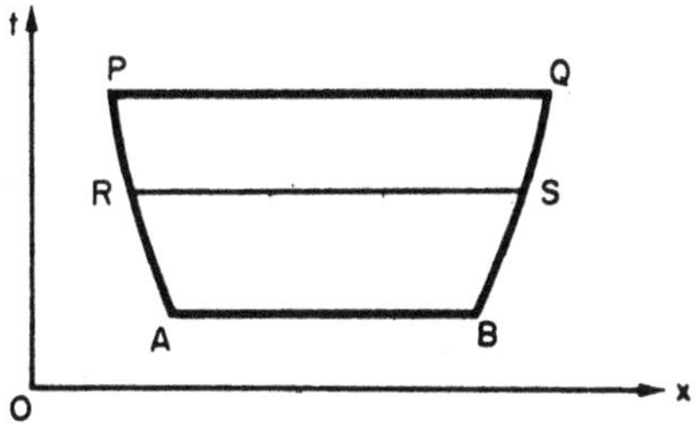

FIG. 83

Applying the similar relation (13′) to the region $ARSB$ and to the solution $u(\xi, \eta) = G(x, t, \xi, \eta)$, continuous in this region, of the equation $\mathscr{M} = 0$, we obtain:

$$G(x, t, \xi, \eta) = \int_{RS} G(x, t, x', \theta)\, \bar{G}(x', \theta, \xi, \tau)\, dx',$$

since the integrals over $BS = AR$ equal zero. Comparison of these formulae shows that

$$G(x, t, \xi, \tau) \equiv \bar{G}(x, t, \xi, \tau).$$

This equation proves that the function G, considered as a function of x, t, has a singularity at $t = \tau$ and $x = \xi$, characteristic of the source function equal to zero for $t = \tau$ and $x \neq \xi$, and satisfies the equation $\mathscr{L} = 0$ inside $APQB$ and reduces to zero on AP and BQ. Such a function of course is the *source function* of an *instantaneous source* for the equation of heat conduction in the region $APQB$.

Thus, *any solution of the equation of heat conduction can be represented by (13) in terms of the source function.*

If the inhomogeneous equation $\mathscr{L}(u) = f(x, t)$ is given, the one should add to the right-hand part in (13) the component

$$\int\!\!\int_{S} G(x, t, \xi, \tau)\, f(\xi, \tau)\, d\xi\, d\tau .$$

2. *Solution of the boundary-value problem.*

Formula (13) obtained above gives the solution of the boundary value problem for a finite segment with moving ends. If the ends of the segment AB are fixed, then the curves AE and BF are replaced by segments of straight lines, parallel to the t-axis. The region S in this case has the form of a rectangle with sides, parallel to the coordinate axes. From the general relation (11) it is possible by a limiting transition to derive Poisson's relation, which gives the solution of the equation of heat conduction with a given initial condition along an infinite straight line.

We assume that in part of the strip, bounded by the two characteristics $t = 0$ and $t = \delta$, passing through the points A and E, the functions u_x satisfy the inequalities

$$u(x, t)\, e^{-Kx^2} < N \text{ and } \frac{\partial u}{\partial x} e^{-Kx^2} < N. \tag{A}$$

where $K > 0$ and $N > 0$ are numbers. We replace the curve BQ (Fig. 84) by a segment of the straight line $x = l$, where l is a positive number, which later we shall increase indefinitely. In addition we shall proceed from formula (9), which we rewrite in the form

$$u(x, t) = \int\limits_{PABQ} \frac{e^{-\frac{(x-\xi)^2}{4a^2(t-\tau)}}}{2\sqrt{[\pi a^2(t-\tau)]}} \left[u(\xi, \tau)\, d\xi + a^2 \frac{\partial u}{d\xi} d\tau - u(\xi, \tau) \frac{x-\xi}{2(t-\tau)} d\tau\right].$$

Let us consider the integral over the segment BQ

$$\int\limits_{BQ} \frac{e^{-\frac{(x-l)^2}{4a^2(t-\tau)}}}{2\sqrt{[\pi a^2(t-\tau)]}} \left[a^2 \left(\frac{\partial u}{\partial \xi}\right)_{\xi=l} - u(l, \tau) \frac{x-l}{2(t-\tau)}\right] d\tau =$$

$$= \int\limits_0^t a^2 \left(\frac{\partial u}{\partial \xi}\right)_{\xi=l} \frac{e^{-\frac{(x-l)^2}{4a^2(t-\tau)}}}{2\sqrt{[\pi a^2(t-\tau)]}} d\tau -$$

$$- \int\limits_0^t u(l, \tau) \frac{e^{-\frac{(x-l)^2}{4a^2(t-\tau)}} (x-l)\, d\tau}{2\sqrt{[\pi a^2(t-\tau)]}(t-\tau)} = I_1 + I_2$$

and let us show that it tends to zero as $l \to \infty$.

We evaluate the integral $|I_1|$ for large values of l

$$|I_1| \leqslant \frac{Na^2}{2\sqrt{(\pi a^2)}} e^{Kl^2 - \frac{l^2}{16a^2\delta}} \int\limits_0^t \frac{dt}{\sqrt{(t-\tau)}}$$

$$\left(\text{if } x < \frac{l}{2} \text{ and } (t-\tau) < \delta\right).$$

It is obvious that $|I_1| \to 0$ as $l \to \infty$, since M is a fixed number, and δ can be chosen as small as desired, for instance, so that

$$K < \frac{1}{16a^2\delta}.$$

Similarly we can prove that $|I_2| \to 0$ as $\mathscr{L} \to \infty$.

If the function $u(x, t)$ and its derivative $\frac{\partial u}{\partial x}$ satisfy inequality (A) also for negative x, then it is possible to take the segment of the straight line $x = -l$ for the curve AE and, repeating the arguments outlined above, to verify that for a limiting transition the integral over PA in

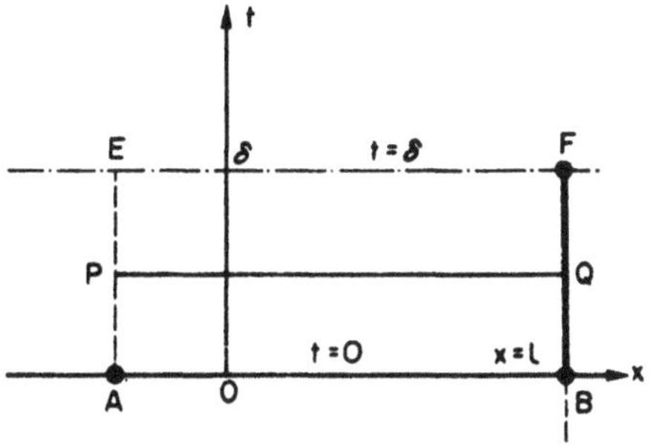

FIG. 84

(9) tends to zero. As a result we arrive at Poisson's relation* derived in Chapter III, § 3

$$u(x, t) = \frac{1}{2\sqrt{(\pi a^2)}} \int_{-\infty}^{+\infty} \frac{e^{-\frac{(x-\xi)^2}{4a^2 t}}}{\sqrt{t}} u(\xi, 0)\, d\xi .$$

Considering a semi-infinite region and assuming that the inequality (A) is fulfilled for the source function $G(x, t, \xi, \tau)$ by means of similar considerations we find:

$$u(x, t) = -\int_{PA} a^2 \mu(\tau) \frac{\partial G}{\partial \xi}\Big|_{\xi = x_A} d\tau + \int_{x_A}^{\infty} \varphi(\xi)\, G(x, t, \xi, 0)\, d\xi , \qquad (14)$$

where

$$\mu(t) = u(x_A, t) \text{ and } \varphi(x) = u(x, 0) .$$

As is easily verified, the source function for a semi-infinite straight line $x \geqslant 0$ may be derived by the method of images and equals

$$G(x, t, \xi, \tau) = \frac{1}{2\sqrt{[\pi a^2 (t-\tau)]}} \left[e^{-\frac{(x-\xi)^2}{4a^2(t-\tau)}} - e^{-\frac{(x+\xi)^2}{4a^2(t-\tau)}} \right],$$

and since it is representable in the form (12), satisfies the equation of heat conduction with respect to the variables x, t and reduces to zero at $x = 0$:

$$G(0, t, \xi, \tau) = 0 .$$

* It is impossible to consider the deductions as a result of this formula, since we have established them by the reduction of formula (9).

Let us calculate the derivative

$$\left.\frac{\partial G}{\partial \xi}\right|_{\xi=0} = \frac{x}{2\sqrt{\pi}\,[a^2(t-\tau)]^{3/2}}\, e^{-\frac{x^2}{4a^2(t-\tau)}}$$

and, substituting its value in (14), we obtain the relation

$$u(x, t) = \frac{1}{2\sqrt{\pi}} \int_0^\infty \frac{1}{\sqrt{(a^2 t)}} \left[e^{-\frac{(x-\xi)^2}{4a^2 t}} - e^{-\frac{(x+\xi)^2}{4a^2 t}}\right] \varphi(\xi)\, d\xi +$$

$$+ \frac{1}{2\sqrt{\pi}} \int_0^t \frac{a^2 x}{[a^2(t-\tau)]^{3/2}}\, e^{-\frac{x^2}{4a^2(t-\tau)}} \mu(\tau)\, d\tau\,, \tag{15}$$

which gives the function $u(x, t)$, satisfying the equation

$$u_t = a^2 u_{xx} \qquad (0 < x < \infty, t > 0)$$

and the additional conditions

$$\begin{aligned} u(x, 0) &= \varphi(x)\,, \\ u(0, t) &= \mu(t) \end{aligned} \qquad (0 < x < \infty)\,.$$

3. *The source function for a segment*

The solution of the equation of heat conduction in the finite segment $0 < x < l$ is given by formula (11), which after replacement of the curves PA and BQ by the segments of straight lines and displacement of the origin of coordinates to the point A, takes the form

$$u(x, t) = a^2 \int_0^t \left.\frac{\partial G}{\partial \xi}\right|_{\xi=0} \mu_1(\tau)\, d\tau -$$

$$- a^2 \int_0^t \left.\frac{\partial G}{\partial \xi}\right|_{\xi=l} \mu_2(\tau)\, d\tau + \int_0^l G(x, t, \xi, 0)\, \varphi(\xi)\, d\xi\,,$$

where

$$\mu_1(t) = u(0, t),\ \mu_2(t) = u(l, t),\ \varphi(x) = u(x, 0)\,.$$

The source function $G(x, t, \xi, \tau)$ for a segment can be formed by the method of images. Locating the positive sources at the points $2nl + \xi$ and the negative sources at the points $2nl - \xi$, we represent the source function by means of the series

$$u(x, t) = \sum_{n=-\infty}^{\infty}{}' \left[G_0(x, t, 2nl + \xi, \tau) - G_0(x, t, 2nl - \xi, \tau)\right], \tag{16}$$

where

$$G_0(x, t, \xi, \tau) = \frac{1}{2\sqrt{[\pi a^2(t-\tau)]}}\, e^{-\frac{(x-\xi)^2}{4a^2(t-\tau)}}$$

is the source function for a finite straight line.

The convergence of the series and also the fulfillment of the boundary conditions

$$G|_{x=0} = 0 \text{ and } G|_{x=l} = 0$$

are proved without difficulty.

In § 2, Chapter III, another expression for the source function was obtained

$$G(x, t, \xi, \tau) = \frac{2}{l} \sum_{n=1}^{\infty} e^{-\frac{n^2\pi^2}{l^2} a^2 (t-\tau)} \sin \frac{\pi n}{l} x \sin \frac{\pi n}{l} \xi . \tag{17}$$

Let us prove the equivalence of the two descriptions.

Formula (17) can be considered as the expansion of $G(x, t, \xi, \tau)$ in a Fourier sine series over the segment $(0, l)$

$$G(x, t, \xi, \tau) = \sum_{n=1}^{\infty} G_n(x, t, \tau) \sin \frac{\pi n}{l} \xi . \tag{18}$$

Let us calculate the Fourier coefficients G_n of function G, defined by series (16),

$$G_n = \frac{2}{l} \int_0^l G(x, t, \xi, \tau) \sin \frac{\pi n}{l} \xi \, d\xi =$$

$$= \frac{2}{l} \sum_{n=-\infty}^{\infty} \left\{ \int_0^l G_0(x, t, 2nl + \xi_1, \tau) \sin \frac{\pi n}{l} \xi_1 \, d\xi_1 - \right.$$

$$\left. - \int_0^l G_0(x, t, 2nl - \xi_2, \tau) \sin \frac{\pi n}{l} \xi_2 \, d\xi_2 \right\} .$$

Introducing new variables of integration

$$\xi' = 2nl + \xi_1 \text{ and } \xi'' = 2nl - \xi_2 ,$$

we obtain:

$$G_n = \frac{2}{l} \sum_{n=-\infty}^{\infty} \left\{ \int_{2nl}^{(2n+1)l} G_0(x, t, \xi', \tau) \sin \frac{\pi n}{l} \xi' \, d\xi' + \right.$$

$$+ \int_{(2n-1)l}^{2nl} G_0(x, t, \xi'', \tau) \sin \frac{\pi n}{l} \xi'' \, d\xi'' ,$$

from which it follows that

$$G_n = \frac{2}{l} \int_{-\infty}^{+\infty} G_0(x, t, \xi, \tau) \sin \frac{\pi n}{l} \xi \, d\xi =$$

$$= \frac{2}{l} \int_{-\infty}^{+\infty} \frac{1}{2\sqrt{[\pi a^2 (t-\tau)]}} e^{-\frac{(x-\xi)^2}{4a^2(t-\tau)}} \sin \frac{\pi n}{l} \xi \, d\xi .$$

We introduce the variable

$$\lambda = \frac{\xi - x}{2\sqrt{[a^2(t-\tau)]}}.$$

Then

$$d\lambda = \frac{d\xi}{2\sqrt{[a^2(t-\tau)]}}; \quad \sin\frac{\pi n}{l}\xi = \sin\frac{\pi n}{l}(x + 2\sqrt{[a^2(t-\tau)]}\,\lambda) =$$

$$= \sin\frac{\pi n}{l}x\cos\frac{2\pi n}{l}\sqrt{[a^2(t-\tau)]}\,\lambda + \cos\frac{\pi n}{l}x\sin\frac{2\pi n}{l}\sqrt{[a^2(t-\tau)]}\,\lambda;$$

$$G_n = \frac{2}{l}\sin\frac{\pi n}{l}\frac{1}{2\sqrt{\pi}}\int_{-\infty}^{+\infty} e^{-\lambda^2}\cos\frac{2\pi n}{l}\sqrt{[a^2(t-\tau)]}\,\lambda\,d\lambda +$$

$$+ \frac{2}{l}\cos\frac{\pi n}{l}x\int_{-\infty}^{+\infty} e^{-\lambda^2}\sin\frac{2\pi n}{l}\sqrt{[a^2(t-\tau)]}\,\lambda\,d\lambda.$$

The second integral equals zero, since the function under the integral sign is odd with respect to the origin of coordinates.

The first integral is a particular case of the integral

$$I(\alpha, \beta) = \int_{-\infty}^{+\infty} e^{-\alpha\lambda^2}\cos\beta\lambda\,d\lambda,$$

evaluated in § 3, Chapter III, and equal to

$$I(\alpha, \beta) = \frac{1}{2}\sqrt{\frac{\pi}{\alpha}}\cdot e^{-\frac{\beta^2}{4\alpha^2}}.$$

In our case

$$\alpha = 1, \quad \beta = \frac{2\pi n}{l}\sqrt{[a^2(t-\tau)]},$$

so that

$$I = \frac{\sqrt{\pi}}{2}e^{-\frac{\pi^2 n^2}{l^2}a^2(t-\tau)}$$

and

$$G_n = \frac{2}{l}e^{-\frac{\pi^2 n^2}{l^2}a^2(t-\tau)}\sin\frac{\pi n}{l}x.$$

Substituting the expression found for the Fourier coefficients A_n in (18), we at once obtain the second representation (17) for the source function G. Thus, the equivalence of the two different representations (16) and (17) is proved.

§ 4. Thermal potentials

1. *Properties of thermal potentials of a single and of a double layer*

As we saw, any solution of the equation of heat conduction may be represented in the form

$$u(x, t) = \int_{AB} G_0 u\,d\xi - \int_{AP} G_0 u\,d\xi + \int_{BQ} G_0 u\,d\xi + a^2\int_{BQ+PA}\left(G_0\frac{\partial u}{\partial\xi} - u\frac{\partial G_0}{\partial\xi}\right)d\tau.$$

Let us investigate the individual components of this sum and prove that each of them separately satisfies the equation of heat conduction. In fact the first component is Poisson's integral, for which this was already proved.

Let us prove that for interior points of the region $PABQ$ the equation is satisfied by the integrals

$$V = a^2 \int_{AP} G_0 \nu d\tau = \frac{a^2}{2\sqrt{\pi}} \int_0^t \frac{1}{\sqrt{[a^2(t-\tau)]}} e^{-\frac{[x-\chi_1(\tau)]^2}{4a^2(t-\tau)}} \nu(\tau)\, d\tau$$

$$(\xi = \chi_1(\tau)),$$

$$W = 2a^2 \int_{AP} \frac{\partial G_0}{\partial \xi} \mu d\tau = \frac{1}{2a\sqrt{\pi}} \int_0^t \frac{x-\chi_1(\tau)}{(t-\tau)^{3/2}} e^{-\frac{[x-\chi_1(\tau)]^2}{4a^2(t-\tau)}} \mu(\tau)\, d\tau .$$

The derivatives of these functions are evaluated by differentiating under the integral sign, since differentiation with respect to the limit t of integration gives zero. For example,

$$G_0(x, t, \chi_1(\tau), \tau)\, \mu(\tau) \Big|_{\tau=t} = \frac{1}{2\sqrt{\pi}} \frac{1}{\sqrt{[a^2(t-\tau)]}} e^{-\frac{[x-y_1(\tau)]^2}{4a^2(t-\tau)}} \mu(\tau) \Big|_{\tau=t} = 0$$

since $x \neq \chi_1(t)$. Thus, differentiation with respect to the parameters x, t refers to the function G which is a solution of the equation of heat conduction. Investigation of the remaining components proceeds similarly.

Let us consider now the behaviour of the functions V, W along the curve $AP(x = \chi_1(t)$. It is obvious that the integral V is continuous as the point (x, t) moves through the curve AP, since this integral converges uniformly (see Chapter IV, § 5). We will prove that W has a discontinuity at the curve AP, for which

$$W|_{x=\chi_1(t)+0} = W|_{x=\chi_1(t)} + \mu(t),$$

$$W|_{x=\chi_1(t)-0} = W|_{x=\chi_1(t)} - \mu(t).$$

This proof will be given on the assumption of the differentiability of $\chi_1(t)$ and $\mu(t)$.

Let us consider W for the constant density $\mu(t) = \mu_0$

$$W^0(x, t) = \frac{1}{2a\sqrt{\pi}} \int_{t_0}^t \frac{x-\chi_1(\tau)}{(t-\tau)^{3/2}} e^{-\frac{[x-\chi_1(\tau)]^2}{4a^2(t-\tau)}} \mu_0\, d\tau$$

and the auxiliary integral

$$\widetilde{V}^0(x, t) = \frac{1}{2a\sqrt{\pi}} \int_{t_0}^t \frac{2\chi_1'(\tau)}{\sqrt{(t-\tau)}} e^{-\frac{[x-\chi_1(\tau)]^2}{4a^2(t-\tau)}} \mu_0\, d\tau,$$

which is, by virtue of the observation made above, a continuous function at points of the curve AP.

The difference $W^0 - \tilde{V}_0$ is calculated directly

$$W^0(x,t) - \tilde{V}^0(x,t) = \frac{1}{2a\sqrt{\pi}} \int_{t_0}^{t} e^{-\frac{[x-\chi_1(\tau)]^2}{4a^2(t-\tau)}} \left[\frac{x-\chi_1(\tau)}{(t-\tau)^{3/2}} - \frac{2\chi_1'(\tau)}{\sqrt{(t-\tau)}}\right] \mu_0 \, d\tau =$$

$$= \mu_0 \frac{2}{\sqrt{\pi}} \int_{\frac{x-\chi_1(t_0)}{2\sqrt{[a^2(t-t)_0]}}}^{a_0} e^{-a^2} da \qquad \left(a = \frac{x-\chi_1(\tau)}{2\sqrt{[a^2(t-\tau)]}}\right),$$

where

$$a_0 = +\infty, \text{ if } x > \chi_1(t),$$

$$a_0 = 0, \qquad \text{if } x = \chi_1(t) = x_0,$$

$$a_0 = -\infty, \text{ if } x < \chi_1(t).$$

As $x \to x_0 \pm 0$ we obtain:

$$[W^0(x_0 \pm 0, t) - W^0(x_0, t)] - [\tilde{V}(x_0, \pm 0, t) - \tilde{V}(x_0 t)] =$$

$$= \mu_0 \frac{2}{\sqrt{\pi}} \int_0^{\pm\infty} e^{-a^2} da = \pm \mu_0 .$$

Because of the continuity of $\tilde{V}$ we have:

$$\tilde{V}(x_0 \pm 0, t) - \tilde{V}(x_0, t) = 0 .$$

Thus,

$$W^0(x_0 \pm 0, t) = W^0(x_0, t) \pm \mu_0.$$

If $\mu(t)$ is not constant, then

$$W(x,t) = W^0(x,t) - \psi(x,t) ,$$

where

$$\psi(x,t) = \frac{a^2}{2\sqrt{\pi}} \int_{t_0}^{t} \frac{x-\chi_1(\tau)}{[a^2(t-\tau)]^{3/2}} e^{-\frac{[x-\chi_1(\tau)]^2}{4a^2(t-\tau)}} [\mu(t) - \mu(\tau)] \, d\tau .$$

Because of the differentiability of $\mu(t)$ this integral has the same singularity at $\tau = t$ as V, converges uniformly and is a continuous function on the curve AP. Thus, the limit $W(x,t)$ for $x = x_0 \pm 0$ equals

$$W(x_0 \pm 0, t) = W^0(x_0, t) \pm \mu ,$$

which it was required to prove.

It is easily verified that the derivative $\frac{\partial V}{\partial x}(x, t)$, like $W(x, t)$ is discontinuous at $x = x_0$. This derivative equals

$$\frac{\partial V}{\partial x} = -\frac{1}{2a\sqrt{\pi}}\frac{1}{2}\int_0^t \frac{x - \chi_1(\tau)}{(t-\tau)^{3/2}} e^{-\frac{[x-\chi_1(\tau)]^2}{4a^2(t-\tau)}} \nu(\tau)\,d\tau$$

and equals $W(x, t)$ with density

$$\mu(t) = \frac{\nu(t)}{2}.$$

Hence it follows that

$$\frac{\partial V}{\partial x}(x_0 \pm 0, t) = \frac{\partial V}{\partial x}(x_0, t) \pm \frac{\nu(t)}{2}$$

where the integral

$$\frac{\partial V}{\partial x}(x_0, t) = -\frac{1}{2a\sqrt{\pi}}\frac{1}{2}\int_0^t \frac{x_0 - \chi_1(\tau)}{(t-\tau)^{3/2}} e^{-\frac{[x_0-\chi_1(\tau)]^2}{4a^2(t-\tau)}} \nu(\tau)\,d\tau$$

equals the semi-sum of the derivatives of V at the point x_0 from the right and left

$$\frac{1}{2}\left[\frac{\partial V}{\partial x}(x_0 + 0, t) + \frac{\partial V}{\partial x}(x_0 - 0, t)\right].$$

We note that the function $V(x, t)$ does not have a derivative at the point x_0.

With this we conclude the investigation of the potentials along AP. The properties of the potentials along the curve BQ are entirely similar.

2. *Solution of boundary-value problems*

Heat potentials are a convenient analytical device for solving boundary-value problems.

Let us consider in the first place the first boundary-value problem for a semi-bounded region $x \geqslant \chi_1(t)$:

Find the solution of the equation

$$u_t = a^2 u_{xx} \text{ where } x \geqslant \chi_1(t), t \geqslant t_0,$$

satisfying the conditions

$$u(x, t_0) = \varphi(x), \qquad x \geqslant \chi_1(t_0);$$
$$u[\chi_1(t); t] = \mu(t), \qquad t \geqslant t_0.$$

Without restriction it is possible to assume that $\varphi(x) = 0$, since, forming the difference between $u(x, t)$ and an arbitrary solution of the equation of heat conduction $v(x, t)$, satisfying the same initial condition, we obtain a new function, for which $\varphi(x) = 0$, and with the same boundary value as before.

Assuming that reduction to the zero initial condition is already made, we represent the solution in the form

$$u(x, t) = \frac{1}{2a^2} W(x, t) = \int_0^t \frac{\partial G_0}{\partial \xi}(x, t, \chi_1(\tau), \tau) \bar{\mu}(\tau) \, d\tau =$$

$$= \frac{1}{4\sqrt{\pi}} \int_0^t \frac{x - \chi_1(t)}{[a^2(t-\tau)]^{3/2}} e^{-\frac{[x-\chi_1(\tau)]^2}{4a^2(t-\tau)}} \bar{\mu}(\tau) \, d\tau ;$$

this function satisfies the equation for $x > \chi_1(t)$, is bounded at infinity and has a zero initial value for any choice of $\bar{\mu}(t)$. For $x = \chi_1(t)$ it is discontinuous and its limiting value for $x = \chi_1(t) + 0$ should equal $\mu(t)$

$$\frac{\bar{\mu}(t)}{2a^2} + \frac{1}{4\sqrt{\pi}} \int_0^t \frac{\chi_1(t) - \chi_1(\tau)}{[a^2(t-\tau)]^{3/2}} e^{-\frac{[\chi_1(t)-\chi_1(\tau)]^2}{4a^2(t-\tau)}} \bar{\mu}(\tau) \, d\tau = \mu(t) .$$

This relation is an integral equation of the *Volterra type* of the second kind for finding the function $\bar{\mu}(\tau)$, giving the solution $u(x, t)$. A solution always exists by the general theory, if the curve $x = \chi_1(t)$ is defined by a differentiable function.

This equation is particularly simple, if the boundary of our region is fixed: $\chi_1(t) = x_0$. In this case the integral reduces to zero and

$$\bar{\mu}(t) = 2a^2 \mu(t) ,$$

so that the solution has the form

$$u(x, t) = \frac{1}{2a\sqrt{\pi}} \int_0^t \frac{x - x_0}{(t-\tau)^{3/2}} e^{-\frac{(x-x_0)^2}{4a^2(t-\tau)}} \mu(\tau) \, d\tau .$$

We have already met this formula twice (see Chapter III, § 3 and Chapter VI, § 3, sub-section 3), but here we have proved that this function satisfies the equation and additional conditions.

The second and third boundary-value problems are solved similarly by means of potentials. Let us consider the boundary-value problem for a bounded region, taking the additional conditions in the form

$$u(x, 0) = \varphi(x) , \qquad \chi_1(0) < x < \chi_2(0) ,$$

$$u[\chi_1(t); t] = \mu_1(t) , \quad u[\chi_2(t); t] = \mu_2(t) \qquad (t > 0) .$$

Assuming that the initial value is reduced to zero: $\varphi(x) = 0$; we represent the solution in the form

$$u(x, t) = \frac{1}{2a^2}(W_1 + W_2) =$$

$$= \int_0^t \frac{\partial G_0}{\partial \xi}(x, t \ \chi_1(\tau), \tau) \bar{\mu}_1(t) \, d\tau + \int_0^t \frac{\partial G_0}{\partial \xi}(x, t, \chi_2(\tau), \tau) \bar{\mu}_2(\tau) \, d\tau.$$

This function satisfies the equation and the zero initial condition for any choice of $\overline{\mu}_1(t)$ and $\overline{\mu}_2(t)$. It is discontinuous for $x = \chi_1(t)$ and $x = \chi_2(t)$ and its limiting values for $x = \chi_1(t) + 0$ and $x = \chi_2(t) + 0$ must equal $\mu_1(t)$ and $\mu_2(t)$, which gives the system of equations

$$\frac{\overline{\mu}_1(t)}{2a^2} + \frac{1}{4\sqrt{\pi}} \int_0^t \frac{\chi_1(t) - \chi_1(\tau)}{[a^2(t-\tau)]^{3/2}} e^{-\frac{[\chi_1(t)-\chi_1(\tau)]^2}{4a^2(t-\tau)}} \overline{\mu}_1(\tau)\,d\tau +$$

$$+ \frac{1}{4\sqrt{\pi}} \int_0^t \frac{\chi_1(t) - \chi_2(\tau)}{[a^2(t-\tau)]^{3/2}} e^{-\frac{[\chi_1(t)-\chi_2(\tau)]^2}{4a^2(t-\tau)}} \overline{\mu}_2(\tau)\,d\tau = \mu_1(t):$$

$$-\frac{\overline{\mu}^2(t)}{2a^2} + \frac{1}{4\sqrt{\pi}} \int_0^t \frac{\chi_2(t) - \chi_1(\tau)}{[a^2(t-\tau)]^{3/2}} e^{-\frac{[\chi_2(t)-\chi_1(\tau)]^2}{4a^2(t-\tau)}} \overline{\mu}_1(\tau)\,d\tau +$$

$$+ \frac{1}{4\sqrt{\pi}} \int_0^t \frac{\chi_2(t) - \chi_2(\tau)}{[a_2(t-\tau)]^{3/2}} e^{-\frac{[\chi_2(t)-\chi_2(\tau)]^2}{4a^2(t-\tau)}} \overline{\mu}_2(\tau)\,d\tau = \mu_2(t).$$

This system is a system of integral equations of the Volterra type, and always has a solution.

Problems on Chapter VI

1. A sphere of radius R_0 is filled with a gas of concentration μ_0 at the initial time; outside the sphere the concentration equals zero. Find the function u, describing the process of diffusion of the gas into the region outside the sphere. Solve the same problem for semi-space in the presence of a boundary $z = 0$ impermeable to gas.
2. Solve the problem of the heating of a sphere of radius R_0, if the initial temperature equals zero, and a constant temperature is maintained on the boundary.
3. Find the temperature of a sphere, at whose surface a heat exchange occurs with a medium of zero temperature, if the initial temperature is constant and equal to u_0.
4. A homogenous rigid body is bounded by two concentric spheres of radii a and $2a$. The inner surface of the body is thermally insulated, and on the outer surface a heat exchange with a medium of zero temperature occurs. Find the temperature distribution in the body at time t, if the initial temperature of the body equals u_0.
5. Derive the diffusion equation in a medium, moving with constant velocity. Write down the expression for the source function in infinite space.
6. Examine the steady diffusion problem in a moving medium, assuming the velocity of motion constant and neglecting diffusion along the direction of motion of the medium (the problem of gaseous attack). Write down the source function for semi-space, assuming that the plane $z = 0$ is impermeable to gas.

7. Form the source function for a layer, bounded by the planes $z = 0$ and $z = e$ and also for a wedge with an inclination π/n (n a whole number) for zero boundary conditions. Investigate the solution.

8. Find the propagation function for an instantaneous source of heat of magnitude Q, uniformly distributed on the surface of a sphere of radius a.

9. Solve the problem of the heating of an infinite cylinder, whose initial temperature equals zero, when a constant temperature is maintained on the surface. Using the tables of Bessel functions, find the temperature profile (take ten points on the radius) and the mean temperature over a cross-section for large times. Draw the appropriate graphs.

10. Examine the problem of the magnetization of an infinite cylinder by a constant magnetic field, parallel to the axis of the cylinder. Use the Bessel function tables to calculate the value of the induction current through a cross-section of the cylinder.

11. Form the source function for an infinite cylindrical region of arbitrary cross-section with boundary conditions of the first kind. Examine the special case of a cross-section of circular shape.

Hint. Represent the solution in the form

$$u(M, z, t) = \sum_{n=1}^{\infty} u_n(z, t)\, \psi_n(M),$$

where $\psi_u(M)$ is the eigen-function of a cross-section of the cylinder.

APPENDICES TO CHAPTER VI

I. Diffusion of a cloud

Let us consider the diffusion process of a gaseous cloud, produced by the explosion of a shell.

In the explosion a certain amount of smoke Q is liberated which forms a cloud. The cloud at first grows, then it brightens at the edge, its dark opaque part decreases, the whole cloud brightens, it starts to "melt" and, finally, fades away. This picture is seen especially clearly on a bright day against the background of a blue sky.

The process of propagation of a smoky cloud can be treated as a diffusion process of smoke from an instantaneous point source of magnitude Q into infinite space. Such a process of diffusion has not a molecular but a turbulent character; it corresponds to some effective coefficient of turbulent diffusion D. We do not consider here the initial dispersion, and also the practically negligible effect of the earth. On these assumptions the concentration of the smoke is given by the relation

$$u(x, y, z, t) = Q \left(\frac{1}{2\sqrt{\pi Dt}}\right)^3 e^{-\frac{x^2+y^2+z^2}{4Dt}} \qquad (D = a^2),$$

if the origin of coordinates is situated at the point of explosion of the shell.

Let us consider the question of the visibility of the cloud. The time, at which the cloud is fully "grown" depends on the absorption of light in the atmosphere and on the threshold of sensitivity of the measuring instrument (the eye, photofilm, etc.).

As is well-known, the intensity of light, passing through homogeneous layers of gas, approximately equals

$$I = I_0 e^{-\alpha l},$$

where I_0 is the initial intensity of light, $a = a_0 u$ is the absorption coefficient, proportional to the concentration of the absorbing gas ($a_0 =$ const.), u is the concentration of the gas in a layer, l the thickness of the layer.

If there are two layers of thickness l_1 and l_2 with different concentrations of gas u_1 and u_2 then

$$I = I_0 e^{-a_0 u_1 l_1} e^{-a_0 u_2 l_2} = I_0 e^{-a_0(u_1 l_1 + u_2 l_2)}.$$

Hence it is obvious that the intensity of light, passing through the cloud with a continuously changing concentration of smoke will be given by the relation

$$I = I_0 e^{-a_0 \int u\, dl}.$$

The visibility of the cloud is given by the ratio I/I_0 depending on the value $\int udl$.

Let δ be the threshold of sensitivity of the control instrument; then for

$$\frac{I_0 - I}{I_0} < \delta \quad \text{or} \quad \frac{I}{I_0} > 1 - \delta$$

the cloud becomes invisible; for

$$\frac{I_0 - I}{I_0} > 1 - \delta \quad \text{or} \quad \frac{I}{I_0} < \delta$$

the cloud appears completely opaque. If

$$\delta < \frac{I}{I_0} < 1 - \delta,$$

then the cloud appears to an observer partially transparent. The degree of transparency depends on the value of the ratio

$$\frac{I}{I_0} = e^{-a_0 \int u\, dl},$$

i.e. on the value of the integral $\int u\, dl$.

Let us direct the z-axis along the line of vision and assume that the observer is at infinity. We consider the projection of the cloud on the xy plane. In order to determine the visibility of different parts of the cloud, corresponding to the points (x, y) we evaluate the integral

$$\int u\,dl = \int_{-\infty}^{\infty} u(x, y, z, t)\,dz = Q\left(\frac{1}{2\sqrt{(\pi Dt)}}\right)^3 \int_{-\infty}^{\infty} e^{-\frac{x^2+y^2+z^2}{4Dt}}\,dz =$$

$$= Q\left(\frac{1}{2\sqrt{(\pi Dt)}}\right)^3 e^{-\frac{x^2+y^2}{4Dt}}.$$

If the amount of smoke in the line of vision is small

$$\int u\,dl < \frac{\delta}{a_0}, \quad \text{or} \quad \frac{I}{I_0} > 1 - \delta,$$

and the corresponding part is completely transparent. If the amount of smoke in the line vision is large

$$\int u\,dl > \frac{\Delta}{a_0}, \quad \text{or} \quad \frac{I}{I_0} < e^{-\Delta} = \delta,$$

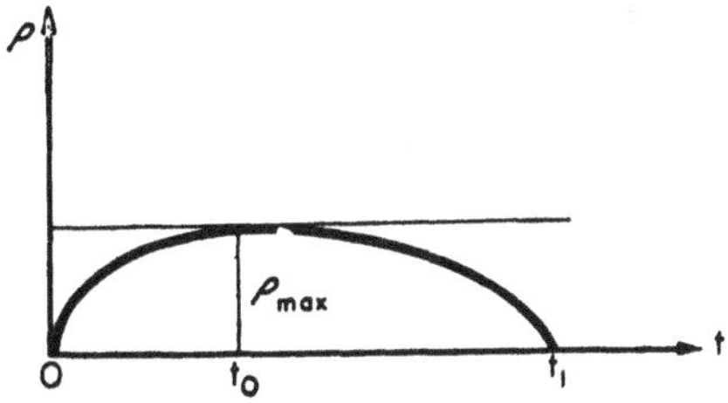

FIG. 95

i.e. for a suitable choice of $\Delta = \ln 1/\delta$ the corresponding part of the cloud is completely opaque. For

$$\frac{\delta}{a_0} \leqslant \int u\,dl < \frac{\Delta}{a_0}$$

the relation

$$a_0 \int u\,dl = \delta \text{ or } a_0 Q\left(\frac{1}{2\sqrt{(\pi Dt)}}\right)^2 e^{-\frac{\varrho^2}{4Dt}} = \delta \qquad (\varrho^2 = x^2 + y^2)$$

defines the boundary of the cloud, at the limits of which it becomes invisible. The radius of the cloud, obviously, equals

$$\varrho = 2\sqrt{\left(-Dt \ln \frac{\delta\, 4\pi Dt}{Q a_0}\right)} \qquad \text{(Fig. 85)}.$$

For small values or t the radius of the cloud (ϱ) is small and grows with t; for

$$t = t_0 = \frac{a_0 Q}{4\pi e \delta D}$$

ϱ reaches a maximum

$$\varrho_{\max} = 2\sqrt{(Dt_0)} = \sqrt{\left(\frac{a_0 Q}{\pi e \delta}\right)},$$

for $t > t_0$ the radius of the cloud ϱ decreases and for

$$t_1 = \frac{Q\,a_0}{\delta\, 4\,\pi\, D}$$

reduces to zero (the cloud vanishes).

Observing the process of dissolution of a cloud it is possible to determine the coefficient of turbulent diffusion in the clear atmosphere D (for instance, from the relation for t_1 or t_0).

II. The demagnetization of a cylinder by a coil

Let us consider the problem of the demagnetization of a cylinder by a coil. Such a problem arises in connection with the theory of the ballistic galvanometer.*

In the switching on or switching off of a magnetic field through a coil an induced current arises. In an exact treatment of the problem it is necessary to take into account the reaction of this current on the field inside the cylinder.

However this braking action of the coil is usually disregarded and one solves the problem with simplified boundary conditions.

Let us deal, first of all, with such a simplified statement of the problem. We consider an infinite cylinder of radius R, on whose surface is wound a conducting coil. The cylinder is in a homogeneous magnetic field H_0, parallel to the axis of the cylinder O_z. At time $t = 0$ the field is switched off.

Inside the cylinder, obviously, the equation

$$\nabla^2 H = \frac{1}{a^2}\frac{\partial H}{\partial t} \qquad (H = H_z) \tag{1}$$

will be satisfied, where

$$a^2 = \frac{c^2}{4\,\pi\,\mu\,\sigma}.$$

Because of the axial symmetry of the field

$$H_z = H\,(r, t)$$

equation (1) can be rewritten in the form

$$\frac{\partial^2 H}{\partial r^2} + \frac{1}{r}\frac{\partial H}{\partial r} = \frac{1}{a^2}\frac{\partial H}{\partial t}. \tag{1'}$$

* B. A. Vvedenskii, *Russian Journal of the Physical-Chemistry Society* 55, I, 1923.

If one neglects the effect of the induced current in the coil in the process of demagnetization of the cylinder, the boundary condition on its surface will have the form

$$H(R, t) = 0 \qquad (t > 0) . \tag{2}$$

The solution of equation (1′) with the boundary condition (2) is easily obtained by the method of separation of variables

$$H(r, t) = \sum_{k=1}^{\infty} \frac{2 H_0}{\mu_k^{(0)} J_1(\mu_k^{(0)})} e^{-\left(\frac{\mu_k^2}{R}\right) a^2 t} J_0\left(\mu_k^{(0)} \frac{r}{R}\right) . \tag{3}$$

Here J_0 and J_1, are Bessel functions of zero and first order, $\mu_k^{(0)}$ is the kth root of the equation

$$J_0(\mu) = 0 . \tag{4}$$

Since a^2 is very large, then for sufficiently large t it is possible to consider the first term in (3) (the usual method)

$$H(r, t,) \cong 1{\cdot}60 \times H_0 e^{-5{\cdot}77 \frac{a^2}{R^2} t} J_0\left(2{\cdot}4 \frac{r}{R}\right) . \tag{5}$$

Hence for the induced current we obtain:

$$\Phi(t) = 2\pi \int_0^R \mu H(r, t)\, r\, dr = \frac{4}{\mu_1^2} \Phi_0 e^{-\mu_1^2 \frac{a^2}{R^2} t} , \tag{6}$$

where Φ_0 is the initial current (for $t = 0$).

Formula (6) is used for practical calculations in measurements by means of the ballistic galvanometer.

In order to determine the region of applicability of this formula, one should solve the original problem, taking into account the braking action of the coil.*

The induced e. m. f. in the coil, as is well-known, equals

$$\mathscr{E}_1^{\text{ind}} = \oint_L E_l\, dl .$$

* This problem was solved by B. N. Nikitin *(Zh. Tekh. Fiz.*, in print).

Let us transform the contour integral, using for this purpose Maxwell's second equation and equation (1)

$$\mathscr{E}^{\text{ind}} = \iint\limits_S \operatorname{curl}_z \boldsymbol{E}\, dS = -\frac{\mu}{c} \iint\limits_S \frac{\partial H}{\partial t}\, ds =$$

$$= -\frac{c}{4\pi\sigma} \iint\limits_S \Delta H dS = -\frac{c}{4\pi\sigma} \oint\limits_L \frac{\partial H}{\partial \nu}\, dl$$

or

$$\mathscr{E}^{\text{ind}} = -\frac{cR}{2\sigma} \frac{dH}{d\nu}(R)\,. \tag{7}$$

Here S is the cylinder cross-section, L is the contour, bounding S, ν is the normal to the contour L.

The boundary conditions on the surface of the cylinder are written in the form

$$H_{z_1} - H_{z_2} = \frac{4\pi}{c} nJ,$$

where J is the induced current in the coil, n is the number of turns per unit length of the cylinder. Hence, taking into account that $H_{z_2} = 0$,

$$J = \frac{\mathscr{E}^{\text{ind}}}{\varrho l}\,,$$

where ϱ is the linear resistance of the coil, l is the length of one turn, we obtain:

$$H_{z_1} = H(R, t) = \frac{4\pi}{c} n \frac{\mathscr{E}^{\text{ind}}}{\varrho l}\,. \tag{8}$$

Comparing relations (7) and (8), we finally arrive at the boundary condition

$$H(R, t) + \frac{n}{\varrho\sigma} H_r(R, t) = 0\,.$$

Thus, we must solve the equation

$$H_{rr} + \frac{1}{r} H_r = \frac{1}{a^2} H_t \tag{9}$$

with the additional conditions

$$H(r, 0) = H_0\,,$$

$$H(R, t) + \alpha H_r(R, t) = 0 \qquad (\alpha = n/\varrho\sigma)\,.$$

The solution will be sought by the method of separation of variables, assuming

$$H(r, t) = X(r)\, T(t)\,.$$

For the functions $X(r)$ and $T(t)$ we obtain the relations

$$X'' + \frac{1}{r} X' + \lambda^2 X = 0 ;$$

$$X(R) + aX'(R) = 0 \qquad (X(0) < \infty), \tag{10}$$

$$T' + \lambda^2 a^2 T = 0 , \tag{11}$$

where λ^2 is a separation parameter.

From the second equation we find at once:

$$T(t) = e^{-a^2\lambda^2 t}.$$

The particular solutions of equation (10) are the functions $J_0(\lambda r)$ and $N_0(\lambda r)$ (see Supplement, Part I), but $J_0(\lambda r)$ only satisfies the boundary condition at $r = 0$. Therefore

$$X(r) = AJ_0(\lambda r) .$$

The boundary condition at $r = R$ gives the eigen-value equation

$$J_0(\lambda R) + a \frac{dJ_0(\lambda R)}{dR} = 0$$

or

$$J_0(y) - \beta y J_1(y) = 0 ,$$

where

$$\beta = \frac{a}{R}, \quad y = \lambda R . \tag{12}$$

The roots of this equation can be found either graphically or by the expansion of the Bessel functions in a power series in $y = \lambda R$.

Let us denote by y_k the roots of equation (12), so that

$$\lambda_k = \frac{y_k}{R} .$$

The general solution of our problem will have the form

$$H(r, t) = \sum_{k=1}^{\infty} A_k J_0\left(y_k \frac{r}{R}\right) e^{-\left(\frac{y_k}{R}\right)^2 a^2 t} \tag{13}$$

We determine the coefficients A_k from the initial condition

$$A_k = \frac{\int_0^R H_0 J_0\left(r \frac{y_k}{R}\right) r\, dr}{\int_0^R J_0^2\left(r \frac{y_k}{R}\right) r\, dr} = \frac{2 H_0 J_1(y_k)}{y_k [J_0^2(y_k) + J_1^2(y_k)]} . \tag{14}$$

The terms of series (13) rapidly decrease, since $a^2 = c^2/4\pi\mu\sigma$ is large ($\sim 10^{13} - 10^{14}$). Therefore to a sufficient degree of accuracy it is possible to confine oneself to the first term

$$H(r, t) \cong \frac{2 H_0 J_1(y_1) J_0\left(y_1 \frac{r}{R}\right)}{y_1 [J_0^2(y_1) + J_1^2(y_1)]} e^{-y_1^2 \frac{a^2}{R^2} t}, \tag{15}$$

which leads to the following expression for the current:

$$\Phi(t) \cong \Phi_0 \frac{4 J_1^2(y_1)}{y_1^2 \cdot [J_0^2(y_1) + J_1^2(y_1)]} e^{-y_1^2 \frac{a^2}{R^2}} \tag{16}$$

where

$\Phi_0 = N H_0 \pi R^2$ (N is the total number of turns in the coil).

Calculations lead to the following relations for the current for different values of the parameter β:

$$\Phi(t) = \begin{cases} 0{\cdot}804\,\Phi_0\, e^{-4{\cdot}75\,\theta} & \text{where } \beta = 0.1, \\ 0{\cdot}872\,\Phi_0\, e^{-3{\cdot}96\,\theta} & \text{where } \beta = 0.2, \quad \theta = \dfrac{a^2}{R^2} t, \\ 0{\cdot}912\,\Phi_0\, e^{-3{\cdot}35\,\theta} & \text{where } \beta = 0.3, \end{cases}$$

permitting us to determine the decay of the field in the cylinder from the braking action of the current. With increase in β, i.e. with increase of the current in the coil, the rate of decrease of current diminishes. For $\beta = 0$ we naturally arrive at expression (6) for the current, which is, thus, a lowest approximation.

In the theory of the ballistic galvanometer it is important to know the time τ of the fall of the current from Φ_0 to values, given by the sensitivity of the galvanometer, which characterizes the inertia of the instrument. Let γ be the relative sensitivity of the galvanometer i.e. the galvanometer can record only the values $\Phi \geqslant \gamma\Phi_0$. The value τ, can be found, by substituting in the formula

$$\Phi(t) = a\,\Phi_0\, e^{-bt},$$

$$\Phi = \gamma\Phi_0 \text{ at the time } t = \tau.$$

The table given below contains values of τ for different β, among them $\beta = 0$, with $\gamma = 10^{-3}$

	R	$\frac{a^2}{R^2}$	$\tau(\beta=0)$	$\tau(\beta=0.1)$	$\tau(\beta=0.2)$	$\tau(\beta=0.3)$
$\mu = 500$, $\sigma = 10^5 \frac{1}{\Omega\, cm}$	1 *cm*	1.59	0.71	0.888	1.15	1.28
$\mu = 2000$ $\sigma = 1.3 \times 10^4$	1 *cm*	5.12	0.217	0.272	0.330	0.405

The values of τ, obtained by this method enable us to interpret the results from the search coils of the ballistic galvanometer more exactly.

III. Method of finite differences for the equation of heat conduction

In order to find the approximate solutions of the equation of heat conduction one often makes use of the method of finite differences.

Let us consider the first boundary-value problem for the equation of heat conduction:

$$\frac{\partial u}{\partial t} = a^2 \frac{\partial^2 u}{\partial x^2} \qquad (0 < x < l,\ 0 < t < T), \tag{1}$$

$$u(x, 0) = \varphi(x), \tag{2}$$

$$\left.\begin{aligned} u(0, t) &= \mu_1(t), \\ u(l, t) &= \mu_2(t). \end{aligned}\right\} \tag{3}$$

Dividing the segment $(0, l)$ into n equal parts by the points $x_0 = 0$, $x_i = ih, \ldots, x_r = l$ $(h = l/n)$ and replacing the derivative with respect to x by the second difference relation, we obtain a differential-difference system for the functions $u_i(t)$:

$$\frac{du_i}{dt} = a^2 \frac{u_{i+1} + u_{i-1} - 2u_i}{h^2}, \tag{1'}$$

$$u_i(0) = \varphi(x_i), \tag{2'}$$

$$\left.\begin{aligned} u_0(t) &= \mu_1(t), \\ u_n(t) &= \mu_2(t), \end{aligned}\right\} \tag{3'}$$

representing a system of $(n - 1)$ ordinary differential equations for $(n - 1)$ unknown functions. The initial conditions determine the solution of this system.

Solving the system (1′–3′) by some numerical method we obtain a system of functions $u_i(t)$, which we shall consider as the approximate values of the function $u(x, t)$ of the solution of the equation of heat conduction at the points x_i. The approximate values of the function $u(x, t)$ at points x, different from x_i, and also the approximate values of the derivatives can be obtained by interpolation.

For this method to be valid it is necessary to prove that for sufficiently small h the functions $u_i(t)$ give the values of the function $u(x, t)$ to any degree of accuracy.

THEOREM. The solution of the first boundary-value problem for the equation of heat conduction with continuous initial and boundary conditions ($\varphi(0) = \mu_1(0)$, $\varphi(l) = \mu_2(0)$) can be approximated by the solution of the system (1′) to any degree of accuracy, i.e. whatever $\varepsilon > 0$ may be, there exists $h(\varepsilon)$ such that

$$|u(x, t) - u_i(t)| < \varepsilon$$

for all i simultaneously, if in system (1′) $h < h(\varepsilon)$.

First of all let us prove several lemmae.

LEMMA 1. If the function $u(x, t)$ is the solution of the first boundary-value problem for the equation of heat conduction, continuous together with its second derivatives with respect to x in the closed region $0 \leqslant x \leqslant l$, $0 \leqslant t \leqslant T$, then the functions $\bar{u}_i(t) = u(x_i, t)$ satisfy the system of equations

$$\frac{d\bar{u}_i}{dt} - a^2 \frac{\bar{u}_{i+1} + \bar{u}_{i-1} - 2\bar{u}_i}{h^2} = Q_i(t), \tag{4}$$

where $Q_i(t)$ is uniformly bounded by the quantity $Q_0(h)$, where $Q_0(h) \to 0$ as $h \to 0$.

Let us substitute the function $u(x, t)$ in the system (1′). Making use of the obvious equalities

$$\bar{u}_{i+1}(t) = u(x_i + h, t) = u(x_i, t) + hu_x(x_i, t) + \frac{h^2}{2} u_{xx}(x_i + \theta_1 h, t)$$
$$(0 \leqslant \theta_1 \leqslant 1),$$

$$\bar{u}_{i-1}(t) = u(x_i - h, t) = u(x_i, t) - hu_x(x_i, t) + \frac{h^2}{2} u_{xx}(x_i - \theta_2 h, t)$$
$$(0 \leqslant \theta_2 \leqslant 1)$$

and equation (1), we obtain:

$$Q_i(t) = \frac{a^2}{2}[u_{xx}(x_i + \theta_1 h, t) + u_{xx}(x_i - \theta_2 h, t) - 2u_{xx}(x_i, t)]. \quad (5)$$

By virtue of the closed nature of the region under consideration $(0 \leqslant x \leqslant l,\ 0 \leqslant t \leqslant T)$, the uniform continuity in this region follows from the assumption of the continuity of the function $u_{xx}(x, t)$, i.e. for any $\eta > 0$ it is possible to assign $h(\eta)$ such that

$$|u_{xx}(x + \theta h, t) - u_{xx}(x, t)| < \eta$$

for any x and t. Hence it follows that

$$|Q_i(t)| \leqslant Q_0 = a^2 \eta \qquad (\text{if } h < h(\eta)),$$

which proves the lemma.

Note. If the function $u(x, t)$ has continuous fourth derivatives, then

$$Q_0 \leqslant \frac{a^2 h^2}{12} U^{(IV)}, \quad (6)$$

where $U^{(IV)}$ is the maximum value of $u^{(IV)}(x, t)$ in the region $(0 \leqslant x \leqslant l,\ 0 \leqslant t \leqslant T)$. This follows immediately from the equalities

$$u(x_i + h, t) = u(x_i, t) + hu_x(x_i, t) + \frac{h^2}{2} u_{xx}(x_i, t) +$$
$$+ \frac{h^3}{3!} u_{xxx}(x_i, t) + \frac{h^4}{4!} u_x^{(IV)}(x_i + \theta_1 h, t),$$

$$u(x_i - h, t) = u(x_i, t) - hu_x(x_i, t) + \frac{h^2}{2} u_{xx}(x_i t) -$$
$$- \frac{h^3}{3!} u_{xxx}(x_i, t) + \frac{h^4}{4!} u^{(IV)}(x_i - \theta_2 h, t).$$

LEMMA 2. The minimum value of the solution of the system of equations

$$\frac{dw_i}{dt} - a^2 \frac{w_{i+1} + w_{i-1} - 2w_i}{h^2} = Q_i(t), \tag{7}$$

where

$$Q_i(t) \geq 0,$$

cannot be less than the minimum of the initial values or the minimum of the functions $w_0(t)$ or $w_n(t)$.

In fact, let the system of functions $w_i(t)$ have a minimum value for $i = i_0$, $t = t_0$ $(t_0 > 0)$, $w_i(t) \geqslant w_{i_0}(t_0)$.

Let us consider the equation for w_{i_0} with $t = t_0$. It is obvious that

$$\left.\frac{dw_{i_0}}{dt}\right|_{t=t_0} \leqslant 0$$

(the sign $<$ can occur only for $t_0 = T$) and

$$a^2 \frac{w_{i_0+1} + w_{i_0-1} - 2w_{i_0}}{h^2} \geqslant 0,$$

which at once leads to a contradiction, if $Q_{i_0}(t_0) > 0$. If $Q_{i_0}(t_0) = 0$, then $w_{i_0+1}(t_0) = w_{i_0=1}(t_0) = w_{i_0}(t_0)$.

Repeating these arguments for $i_0 + 1$, etc., we arrive at a contradiction, if even one of $Q_{i_0+k}(t_0) > 0$. If all

$$Q_{i_0+k}(t_0) = 0,$$

then

$$w_{i_0}(t_0) = w_{i_0+1}(t_0) = \ldots = w_{i_0+k}(t_0) = \ldots = w_n(t_0),$$

from which it follows that the minimum value of the system of functions $w_i(t)$ $(i = 1, \ldots, n-1)$ cannot be less than the minimum of the function $w_n(t)$. A similar lemma, occurs for the maximum values of the solution of system (7), if

$$Q_i(t) \leqslant 0,$$

Conclusion 1. If the system of functions $\{\bar{v}_i(t)\}$ $(i = 1, \ldots, n-1)$ satisfies the system of equations

$$\frac{d\bar{v}_i}{dt} - a^2 \frac{\bar{v}_{i+1} + \bar{v}_{i-1} - 2\bar{v}_i}{h^2} = \bar{Q}_i(t),$$

and the system of functions $\{\bar{v}_i(t)\}$ $(i = 1, 2, \ldots, n-1)$ satisfies the system of equations

$$\frac{d\bar{\bar{v}}_i}{dt} - a^2 \frac{\bar{\bar{v}}_{i+1} + \bar{\bar{v}}_{i-1} - 2\bar{\bar{v}}_i}{h^2} = \bar{\bar{Q}}_i(t).$$

where

$$Q_i(t) \leqslant \bar{\bar{Q}}_i(t); \quad \bar{v}_0(t) \leqslant \bar{\bar{v}}_0(t); \quad \bar{v}_n(t) \leqslant \bar{\bar{v}}_n(t); \quad \bar{v}_i(0) \leqslant \bar{\bar{v}}_i(0),$$

then

$$\bar{v}_i(t) \leqslant \bar{\bar{v}}_i(t) \quad (i = 1, 2, \ldots, n-1).$$

The proof follows directly from lemma 2 in the application to the function $w_i(t) = \bar{\bar{v}}_i(t) - \bar{v}_i(t)$.

Conclusion 2. The solution of the homogeneous system (7) satisfies the relation

$$|w_i(t)| < \varepsilon,$$

if

$$|w_i(0)| < \varepsilon \text{ and } |w_0(t)| < \varepsilon, \quad |w_n(t)| < \varepsilon.$$

The proof follows immediately from lemma 2.

LEMMA 3. If the function $u(x, t)$ is the solution of the first boundary value problem for the equation of heat conduction, continuous together with its second derivatives with respect to x in the closed region $0 \leqslant x \leqslant l$, $0 \leqslant t \leqslant T$, then for sufficiently small h it can approximate as accurately as desired to the solution of the finite-difference system (1′).

Let us consider the difference of the functions $\bar{u}_i(t) = u(x_i, t)$ and $u_i(t)$

$$\bar{v}_i(t) = \bar{u}_i(t) - u_i(t).$$

These functions satisfy the system of equations

$$\frac{d\bar{v}_i}{dt} - a^2 \frac{\bar{v}_{i+1} + \bar{v}_{i-1} - 2\bar{v}_i}{h^2} = Q_i(t),$$

$$\bar{v}_i(0) = 0, \quad \bar{v}_0(t) = 0, \quad \bar{v}_n(t) = 0,$$

where

$$|Q_i(t)| \lesssim Q_0(h) \quad (Q_0(h) \underset{h \to 0}{\to} 0).$$

Let us prove that

$$|\bar{v}_i(t)| \leqslant Q_0 t \qquad (i = 1, 2, \ldots, n-1).$$

Let us consider the auxiliary system of functions

$$\bar{\bar{v}}_i(t) = Q \cdot t \quad \text{for all} \quad i = 0, 1, \ldots, n.$$

This system of functions satisfies the equations

$$\frac{d\bar{\bar{v}}_i}{dt} - a^2 \frac{\bar{\bar{v}}_{i+1} + \bar{\bar{v}}_{i-1} - 2\bar{\bar{v}}_i}{h^2} = Q_0.$$

with the additional conditions

$$\bar{\bar{v}}_i(0) = 0, \quad \bar{\bar{v}}_0(t) = Q_0 t; \quad \bar{\bar{v}}_n(t) = Q_0 t.$$

By virtue of conclusion 1 of lemma 2 the relation

$$\bar{v}_i(t) \leqslant \bar{\bar{v}}(t) = Q_0 t$$

holds. Similarly, considering the system of functions $-\bar{\bar{v}}_i(t)$, we obtain

$$\bar{v}_i(t) \geqslant -\bar{\bar{v}}_i(t) = -Q_0 t.$$

Hence it follows that

$$|\bar{v}_i(t)| \leqslant Q_0 t \quad (i = 1, 2, \ldots, n-1), \tag{8}$$

and also the proof of lemma 3, since

$$Q_0(h) \to 0 \text{ where } h \to 0.$$

Let us determine some properties of the solution of the equation of heat conduction, which will be necessary to us later.

LEMMA 4. If the function $u(x, t)$ is a solution of the first boundary-value problem for the equation of heat conduction, in which the initial function $\varphi(x)$ is continuous together with its derivatives up to the second order in the segment $0 \leqslant x \leqslant l$, and the boundary values $\mu_1(t)$ and $\mu_2(t)$ are continuous together with derivatives of the first order and the conditions of matching

$$(\varphi(0) = \mu_1(0), \quad \varphi(l) = \mu_2(0), \quad \varphi''(0) = \mu_1'(0), \quad \varphi''(l) = \mu_2'(l)),$$

are fulfilled, then the function $u(x, t)$ possesses second continuous derivatives with respect to x in the closed region $0 \leqslant x \leqslant l$, $0 \leqslant t \leqslant T$.

Let us consider the function $z_1(x, t)$, satisfying the equation of heat conduction and the additional conditions

$$z_1(x, 0) = a^2 \varphi''(x),$$
$$z_1(0, t) = \mu_1'(t),$$
$$z_1(l, t) = \mu_2'(t).$$

Let us form the function

$$z(x, t) = \varphi(x) + \int_0^t z_1(x, \tau)\, d\tau.$$

The boundary and initial values of this function respectively equal

$$z(x, 0) = \varphi(x),$$
$$z(0, t) = \mu_1(0) + \int_0^t \mu_1'(\tau)\, d\tau = \mu_1(t) \qquad (\varphi(0) = \mu_1(0)),$$
$$z(l, t) = \mu_2(t).$$

Calculating the derivatives

$$\frac{\partial z}{\partial t} = z_1(x, t),$$

$$\frac{\partial^2 z}{\partial x^2} = \varphi''(x) + \int_0^t \frac{\partial^2 z_1}{\partial x^2}(x, \tau)\, d\tau =$$
$$= \frac{1}{a^2} z_1(x, 0) + \frac{1}{a^2} \int_0^t \frac{\partial z_1}{\partial \tau}(x, \tau)\, d\tau =$$
$$= \frac{1}{a^2} z_1(x, t),$$

we see that for $t > 0$ the function $z(x, t)$ satisfies the equation of heat conduction. Thus,

$$z(x, t) \equiv u(x, t),$$

which proves the continuity of the function $u(x, t)$ together with its derivatives up to the second order inclusive.

Conclusion. The solution of the first boundary-value problem for the equation of heat conduction for sufficiently small h approximates to the solution of the differential-difference system to any degree of accuracy, if the initial function $\varphi(x)$ is continuous together with its second derivatives, and the boundary values are continuous together with the first derivatives, and the conditions of matching are fulfilled.

Let us turn now to a proof of the main theorem. In order to do this it is sufficient to show that the additional conditions of differentiability of the functions $\varphi(x)$, $\mu_1(t)$ and $\mu_2(t)$ in the preceding conclusion of Lemma 4 are unnecessary. Let us consider the function $u(x, t)$ which is the solution of the first boundary-value problem. with continuous initial and boundary conditions:

$$u(x, 0) = \varphi(x), \quad u(0, t) = \mu_1(t), \quad u(l, t) = \mu_2(t)$$
$$(\varphi(0) = \mu_1(0), \quad \varphi(l) = \mu_2(0)).$$

Let us consider some functions: $\hat{\varphi}(x)$ continuous together with its second derivatives and $\hat{\mu}_1(t)$ and $\hat{\mu}_2(t)$ continuous together with the first derivatives, satisfying the conditions of matching

$$\hat{\varphi}(0) = \hat{\mu}_1(0); \quad \hat{\varphi}(l) = \hat{\mu}_2(0),$$
$$\hat{\varphi}''(0) = \hat{\mu}_1'(0); \quad \hat{\varphi}''(l) = \hat{\mu}_2'(0),$$

and approximating the initial and boundary values of our function correct to $\varepsilon/3$*

$$|\hat{\varphi}(x) - \varphi(x)| < \varepsilon/3,$$
$$|\hat{\mu}_1(t) - \mu_1(t)| < \varepsilon/3,$$
$$|\hat{\mu}_2(t) - \mu_2(t)| < \varepsilon/3.$$

Let us consider the function $\hat{u}(x, t)$, which is the solution of the corresponding boundary-value problem. Because of the principle of maximum value

$$|\hat{u}(x, t) - u(x, t)| < \varepsilon/3.$$

Similarly, because of the principle of a maximum for the differential-difference system (Conclusion 2, Lemma 2) the inequality

$$|\hat{u}_i(t) - u_i(t)| < \varepsilon/3,$$

* Such functions can be formed in very many ways. They can be obtained, for instance, by replacing the functions φ and ψ by discontinuous lines and smoothing these discontinuities at corner points so that the conditions of differentiability and matching are fulfilled.

holds, where the functions $\hat{u}_i(t)$ and $u_i(t)$ are solutions of the corresponding differential-difference systems with additional conditions

$$\hat{u}_i(0) = \hat{\varphi}(x_i)\ ; \quad \hat{u}_0(t) = \hat{\mu}_1(t)\ ; \quad \hat{u}_n(t) = \hat{\mu}_2(t)\ ;$$

$$u_i(0) = \varphi(x_i)\ ; \quad u_0(t) = \mu_1(t)\ ; \quad u_n(t) = \mu_2(t)\ .$$

Because of the conclusion of Lemma 4, the relation

$$|\hat{u}(x, t) - \hat{u}_i(t)| < \varepsilon/3\ ;$$

holds, if h is chosen sufficiently small (see formula (8)), so that

$$Q_0 < \varepsilon/3T\ .$$

Comparing the inequalities obtained, we obtain

$$|u(x, t) - u_i(t)| \leqslant |u(x, t) - \hat{u}(x, t)| + |\hat{u}(x, t) - \hat{u}_i(t)| + \\ + |\hat{u}_i(t) - u_i(t)| < \varepsilon\ ,$$

which proves the theorem.

Replacing the derivatives with respect to t in the system of equations (1′) by the differences

$$\frac{u_i^{j+1} - u_i^j}{k}\ ,$$

where $k = \Delta t$ is the increment in t, $u_i^j = u(x_i, t_j)$, we obtain a system of finite-difference equations

$$u_i^{j+1} - u_i^j = \gamma a^2 (u_{i+1}^j + u_{i-1}^j - 2u_i^j) \qquad (\gamma = k/h^2)$$

$$\begin{pmatrix} i = 1, 2, \ldots, n-1, \\ j = 0, 1, \ldots, N-1, N = T/k \end{pmatrix}$$

or

$$u_i^{j+1} = (1 - 2a^2\gamma) u_i^j + a^2\gamma (u_{i+1}^j + u_{i-1}^j) \tag{1″}$$

with the additional conditions

$$u_i^0 = \varphi(x_i), \qquad u_0^j = \mu_1(t_j), \qquad u_n^j = \mu_2(t_j). \tag{2″}$$

If $\gamma < 1/2a^2$, then all the coefficients in equation (1′′) are positive and the maximum value principle holds for the finite-difference system.

In fact, assuming that $u_i^{j_0+1}$ is the maximum value of the solution of the problem $(0 < i_0 < n,\ j_0 > 0)$, and noting that all the coefficients of equation (1'') are positive and their sum is always equal to unity, we deduce that the equation can be fulfilled only for the conditions of equality

$$u_{i_0}^{j_0+1} = u_{i_0}^{j_0} = u_{i_0-1}^{j_0} = u_{i_0+1}^{j_0+1}.$$

Pursuing these arguments further, we arrive at the conclusion that the maximum values be reached either at $i = 0$, $i = n$ or at $j = 0$.

Repeating all the arguments outlined above for the differential-difference system, we easily verify that the solution of the finite-difference system (1'') approximates to the solution of the boundary value problem for the equation of heat conduction to any degree of accuracy.

The finite-difference system (1'') is obtained by numerical integration of the differential-difference system (1') by Euler's method. Using other methods of numerical integration, it is possible to obtain other types of finite-difference systems, corresponding to the equation of heat conduction.

Approximate solutions of the equation of heat conduction can be obtained also, replacing only the derivatives with respect to t by the finite differences and reducing the determination of the approximate solution to a step-by-step solution of the boundary-value problems

$$\frac{u^{j+1}(x) - u^j(x)}{k} = a_2 \frac{d^2 u^{j+1}(x)}{dx^2}$$

or

$$a^2 \frac{d^2 u^{j+1}}{dx^2} - \frac{1}{k} u^{j+1} x = -\frac{1}{k} u^j(x)$$

with the additional conditions

$$u^0(x) = \varphi(x); \qquad u^j(0) \mu_1(t_j); \qquad u^j(l) = \mu_2(t_j).$$

For a fuller account of the solution of the equation of heat conduction by this method see Smirnov, *Course in Higher Mathematics*, vol. IV.

The solution of the homogeneous equation of heat conduction with constant coefficients was considered above. Similar results may be obtained for the inhomogeneous equation and also for the linear equation of parabolic type with variable coefficients,* assuming the continuity of the coefficients and their differentiability with respect to t. Differentiability with respect to t is necessary for the proof of the analogue of Lemma 4. In the case, where the coefficients of the equation depend only on x, the proof remains the same.

* In addition it is assumed that the corresponding boundary-value problem is capable of solution for continuous initial and boundary conditions.

CHAPTER VII

EQUATIONS OF ELLIPTIC TYPE
(CONTINUATION)

§ 1. Fundamental problems leading to the equation $\nabla^2 v + cv = 0$

1. *Steady vibrations*

A very extensive class of problems, connected with *steady vibrations* (mechanical, acoustic, electromagnetic, etc.) leads to the so-called *wave equation*

$$\nabla^2 v + k^2 v = 0 \qquad (k^2 = c > 0)\,. \tag{1}$$

Let us consider as an example a membrane S, fixed at its boundary C and vibrating under the action of forces periodic with time. The appropriate equation has the form

$$\nabla_2^2 \bar{u} = \frac{1}{a^2}\,\overline{u_{tt}} - F_0(x, y)\,\cos\,\omega t. \tag{2}$$

In an investigation of periodic processes it is convenient to use complex functions, replacing (2) by the equation

$$\nabla_2^2 u = \frac{1}{a^2}\,u_{tt} - F_0(x, y)\,e^{i\omega t}\,. \tag{3}$$

The function $\bar{u}$, obviously, is the real part of function u in (3).
We shall look for the steady vibrations, having the form

$$u = ve^{i\omega t} \tag{4}$$

For the amplitude of the steady vibrations v we obtain the following equation:

$$\nabla_2^2 v + k^2 v = -\,F_0(x, y) \qquad k = \omega/c\,, \tag{5}$$

to which it is necessary to add the boundary condition

$$v\,|_C = 0 \tag{6}$$

if the contour of the membrane C is not fixed, and performs periodic oscillations with the same frequency ω

$$u|_C = f_0 \, e^{i\omega t}, \tag{6'}$$

then there is an inhomogeneous boundary condition for the function v on the contour C,

$$v|_C = f_0 . \tag{6''}$$

As has already been noted, problems on steady vibrations occur also in acoustics and in the theory of the electromagnetic field. Moreover, problems on steady vibrations often occur in an inhomogeneous medium, in particular, in a partly-homogeneous medium (where, for instance, there are isolated regions in space, disturbing the homogeneity). Problems in diffraction theory, which we consider below, are related to this class.

2. *Diffusion of a gas in the presence of decay and with chain reactions*

In the diffusion of certain gases (for instances, the emanation from radium) a decay of the molecules of the diffusing gas takes place. The rate of decay is usually proportional to the concentration of the gas. In writing the diffusion equation this is equivalent to the presence of negative sources of gas. In the case of a steady diffusion process we arrive at the equation

$$D \nabla^2 v + cv = 0 \qquad (c < 0), \tag{7}$$

where D is the diffusion coefficient. As was shown in Chapter VI, § 3, sub-section 3, great interest lies in the case $c > 0$, corresponding to diffusion in the presence of chain reactions, leading to the multiplication of diffusing particles. In the steady case we obtain the equation

$$\nabla^2 v + cv = 0 \qquad (c > 0),$$

since the chain reaction is equivalent to the presence of sources of a diffusing substance, proportional to the concentration $v(x, y, z)$.

3. *Diffusion in a moving medium*

In Chapter IV the problem concerning the diffusion of a gas in a stationary medium was considered. Let us consider the problem of the diffusion of a gas in a moving medium, whose velocity at the point $M(x, y, z)$ has components $\vartheta_1(x, y, z)$, $\vartheta_2(x, y, z)$, $\vartheta_3(x, y, z)$. The amount of gas, flowing across an element of area $d\sigma$ at the point $M(x, y, z)$ equals

$$dQ = -D\boldsymbol{n}\cdot \operatorname{grad} u\, d\sigma + u\boldsymbol{\vartheta}\cdot\boldsymbol{n}\, d\sigma\,,$$

where $u(x, y, z)$ is the concentration of gas in unit volume, $\boldsymbol{n}$ is the unit vector, normal to the area $d\sigma$, D is the diffusion coefficient at the point (x, y, z), $\vartheta(x, y, z)$ is the velocity vector, of the current.

Forming the equation for the conservation of matter for a certain volume T with boundary Σ, we obtain:

$$\int_{\Sigma} [-D\boldsymbol{n}\cdot \operatorname{grad}\ u + u\boldsymbol{\vartheta}\cdot\boldsymbol{n}]\, d\sigma = 0\,.$$

Let us transform the surface integral into a volume integral, using Ostrogradskii's relation

$$\int_{T} [\operatorname{div}\ (D\ \operatorname{grad}\ u) - \operatorname{div}\ (u\boldsymbol{\vartheta})]\, d\tau = 0\,.$$

Hence because of the arbitrary nature of the volume T the diffusion equation follows

$$\operatorname{div}(D\ \operatorname{grad}\ u) - \operatorname{div}(u\boldsymbol{\vartheta}) = 0 \tag{8}$$

or in scalar form

$$\frac{\partial}{\partial x}\left(D\frac{\partial u}{\partial x}\right) + \frac{\partial}{dy}\left(D\frac{\partial u}{\partial y}\right) + \frac{\partial}{\partial z}\left(D\frac{\partial u}{\partial z}\right) - $$
$$- \frac{\partial}{\partial x}(u\vartheta_1) - \frac{\partial}{\partial y}(u\vartheta_2) - \frac{\partial}{\partial z}(u\vartheta_3) = 0\,. \tag{8'}$$

The problem of heat propagation in a moving medium leads to similar equations.

Let us consider the following example. In the semi-space $z \geqslant 0$ there is an air current with constant velocity u_0, directed along the x-axis. Assuming the diffusion coefficient constant, we obtain from (8) the equation

$$D \nabla^2 u - u_0 \frac{\partial u}{\partial x} = 0,$$

which is the simplest version of the equation of *gas attack.*

Assuming

$$u = ve^{\mu x}$$

and choosing

$$\mu = \frac{u_0}{2D},$$

we obtain an equation for $v(x, y, z)$

$$\nabla^2 v + cv = 0,$$

where

$$c = -\frac{u_0^2}{4D} < 0.$$

4. *Formulation of interior boundary-value problems for the equation $\Delta v + cv = 0$.*

As was shown in Chapter I during the investigation of the canonical form of equations with constant coefficients, any equation of elliptic type with constant coefficients can be reduced to the form

$$\nabla^2 v + cv = 0. \tag{1}$$

The properties of the solution of equation (1) depend essentially on the sign of the coefficient c, which is physically obvious, if we think of a diffusion interpretation of this equation.

Let us consider the question of the uniqueness of the solution of the first boundary-value problem for equation (1). For the equation $\nabla^2 v + cv = 0$ with $c < 0$ there is a maximum value principle occuring in the following form.

The solution $v(M)$ of the equation $\nabla^2 v + cv = 0$ $(c < 0)$, defined inside some region T with boundary $\sum$, cannot reach positive maximum (and negative minimum) values at interior points of region T.

In fact, we assume that at some point M_0, lying inside T, the function $v(M)$ reaches a positive maximum value $[v(M_0) > 0]$. Then at the point M_0.

$$\frac{\partial^2 v}{\partial x^2} \leqslant 0, \quad \frac{\partial^2 v}{\partial y^2} \leqslant 0, \quad \frac{\partial^2 v}{\partial z^2} \leqslant 0$$

and, therefore, $\nabla^2 v \leqslant 0$, which is in contradiction to the negative sign of the coefficient c and the positive sign of $v(M_0)$.*

The uniqueness of the solution of the first boundary-value problem for equation (1) follows from the maximum value principle.

There can exist only one solution of the equation $\nabla^2 v + cv = 0$ $(c < 0)$, *defined and continuous in the closed region* $T + \Sigma$, *taking given values on the boundary* Σ,

$$v|_\Sigma = f.$$

In fact, assuming the existence of two different solutions v_1 and v_2, considering their difference $v_1 - v_2$ and carrying out the discussions in the manner outlined above (see Chapters III and IV) we arrive at a contradiction of the maximum value principle.

If $c = 0$, then we obtain the first boundary-value problem for Laplace's equation, the uniqueness of whose solution has been proved.

If $c > 0$, then the solution is not unique. When considering the eigen-value problem

$$\nabla^2 v + \lambda v = 0, \quad v|_\Sigma = 0,$$

in Chapter V we showed by examples the existence of non-trivial solutions (eigen-functions) for $\lambda > 0$. It is obvious that the question of the multiplicity or uniqueness of the solution of the first boundary-value problem is equivalent to the question of whether c coincides with one of the eigen-values λ_n of the region T under consideration.

§ 2. The source function

1. *The source function*

Potential theory, developed in Chapter IV for Laplace's equation, may be extended to the equation $\nabla^2 v + cv = 0$. In order to form the source function we consider the solution

* Compare with the proof of the principle of maximum value for the equation of heat conduction.

v_0, depending only on r. The Laplacian operator for the function $v_0(r)$ in a spherical system of coordinates has the form

$$\frac{1}{r^2}\frac{d}{dr}\left(r^2\frac{dv_0}{dr}\right)=\frac{1}{r}\frac{d^2(rv_0)}{dr^2},$$

which leads to the ordinary differential equation

$$\frac{d^2w}{dr^2}+cw=0 \quad (w=v_0 r).$$

Introducing the symbol $c = k^2$ for $c > 0$ and $c = -\varkappa^2$ for $c < 0$, we obtain:

$$\frac{d^2w}{dr^2}+k^2 w=0 \qquad (c>0), \tag{1}$$

$$\frac{d^2w}{dr^2}-\varkappa^2 w=0 \qquad (c<0). \tag{1'}$$

From equation (1) we find:

$$w=C_1 e^{ikr}+C_2 e^{-ikr} \tag{2}$$

and, consequently,

$$v_0=C_1\frac{e^{ikr}}{r}+C_2\frac{e^{-ikr}}{r} \tag{3}$$

In the case of real k we obtain two linearly-independent solutions $\frac{e^{ikr}}{r}$ and $\frac{e^{-ikr}}{r}$, to which there correspond the real linearly-independent solutions

$$\frac{\cos kr}{r} \quad \text{and} \quad \frac{\sin kr}{r}.$$

For $c < 0$ $(c = -\varkappa^2)$, using equation (1′), we obtain two real linearly-independent solutions

$$\frac{e^{-\varkappa r}}{r} \quad \text{and} \quad \frac{e^{\varkappa r}}{r} \qquad (\varkappa > 0). \tag{4}$$

The functions

$$\frac{e^{\pm ikr}}{r}\,(c>0) \quad \text{and} \quad \frac{e^{\pm \varkappa r}}{r} \qquad (c<0)$$

have a discontinuity at $r = 0$, tending to infinity as $1/r$. The source function for Laplace's equation $(c = 0)$ has a similar type of singularity, proportional to $1/r$.

Let us consider the behaviour of these functions at infinity. The case $c < 0$ corresponds to a process, accompanied by absorption (see the diffusion equation (6), § 1). One of the solutions $\frac{e^{-\varkappa r}}{r}$ tends exponentially to zero at infinity, which in terms of the diffusion problem means a decrease in concentration, caused by absorption. This decrease is greater, the greater the coefficient $|c| = \varkappa^2$, describing the intensity of the absorption. The second solution increases exponentially to infinity and has no physical significance for the problem in an infinite region (it could be interpreted as the presence of sources at infinity).

The case $c = k^2 > 0$ corresponds to steady wave processes (see § 1, sub-section 1). The function v represents the amplitude of the function

$$u(M, t) = v(M) e^{i\omega t},$$

satisfying the wave equation (§ 1).

One of the main solutions of equation (1)

$$v_0(r) \frac{e^{-ikr}}{r}$$

corresponds to the vibration process

$$u_0(r, t) = \frac{e^{i(\omega t - kr)}}{r},$$

which has the character of a spherical wave, *diverging* from the source at the point $r = 0$. The second solution

$$v_0(r) = \frac{e^{ikr}}{r}$$

corresponds to a vibration process

$$u_0(r, t) = \frac{e^{i(\omega t + kr)}}{r},$$

having the character of a spherical wave, coming from infinity to the point $r = 0$ (*converging* wave). It is obvious that this solution has no direct physical significance in an investigation of processes, produced by a point source in infinite space.

We note that the function $v(M)$ may be considered as the amplitude of vibrations of the type $e^{i\omega t}$ or $e^{-i\omega t}$. We have taken a time factor of the first type. In the second case the diverging wave has the form

$$u_0(r, t) = \frac{e^{-i(\omega t - kr)}}{r},$$

i.e. the second solution corresponds to

$$v_0(r) = \frac{e^{ikr}}{r}.$$

The first solution

$$v_0(r) = \frac{e^{-ikr}}{r}$$

has no physical significance.

2. *Integral representation of the solution*

For equation (1), § 1 with $c \neq 0$, it is possible to write down formulae, similar to Green's formulae, which were established for Laplace's equation. Introducing the symbol

$$\mathscr{L}(u) = \nabla^2 u + cu, \tag{5}$$

we at once obtain the relation

$$\int_T (u\mathscr{L}(v) - v\mathscr{L}(u))\, d\tau = \int_\Sigma \left(u\frac{\partial v}{\partial \nu} - v\frac{\partial u}{\partial \nu}\right) d\sigma, \tag{6}$$

which is analogous to Green's second theorem (see Chapter IV, § 2). Substituting here one of the "functions of a point source" in place of v, for instance $\frac{e^{-\varkappa r}}{r}$, and repeating literally all the discussions of Chapter IV, § 2, we arrive at the analogous fundamental Green's theorem

$$u(M_0) = -\frac{1}{4\pi}\int_\Sigma \left[u\frac{\partial}{\partial \nu}\left(\frac{e^{-\varkappa r}}{r}\right) - \frac{e^{-\varkappa r}}{r}\frac{\partial u}{\partial \nu}\right] d\sigma_M +$$

$$+ \frac{1}{4\pi}\int_T f(M)\frac{e^{-\varkappa r}}{r}\, d\tau_M \quad (r = r_{MM_0}), \tag{7}$$

where $u(M)$ is the solution of the inhomogeneous equation $\mathscr{L}(u) = -f(M)$.

For the case $c = k^2$ we have

$$u(M_0) = -\frac{1}{4\pi}\int_{\Sigma}\left[u\frac{\partial}{\partial\nu}\left(\frac{e^{-ikr}}{r}\right) - \frac{e^{-ikr}}{r}\frac{\partial u}{\partial\nu}\right]d\sigma + \\ + \frac{1}{4\pi}\int_{T} f(M)\frac{e^{-ikr}}{r}d\tau_M, \tag{7'}$$

which was obtained in Chapter V from Kirchhoff's relation.

Let us introduce the idea of the source function of the equation $\mathscr{L}(u) = 0$ for the region T whith boundary $\sum$. Let $v(M)$ be a solution of the equation $\mathscr{L}(v) = 0$, regular everywhere in T. Formula (6) gives

$$0 = -\int_{\Sigma}\left(u\frac{\partial v}{\partial\nu} - v\frac{\partial u}{d\nu}\right)d\sigma + \int_{T} fv\,d\tau. \tag{8}$$

Adding (8) and (7) we obtain:

$$u(M_0) = -\int_{\Sigma}\left[u\frac{\partial}{\partial\nu}\left(\frac{e^{-\varkappa r}}{4\pi r} + v\right) - \left(\frac{e^{-\varkappa r}}{4\pi r} + v\right)\frac{\partial u}{\partial\nu}\right]d\sigma_M + \\ + \int_{T}\left(\frac{e^{-\varkappa r}}{4\pi r} + v\right)f(M)\,d\tau_M \qquad (r = r_{MM_0}). \tag{9}$$

This relation is valid for an arbitrary solution $v(M)$ of the equation $\nabla^2 v - \varkappa^2 v = 0$, regular in the region T. Using a special choice of v, we obtain:

$$u(M_0) = -\int_{\Sigma} u(M)\frac{\partial G(M_0 M)}{\partial\nu}d\sigma_M + \int_{T} G(M_0, M)f(M)\,d\tau_M, \tag{10}$$

where

$$G(M_0, M) = \frac{e^{-\varkappa r}}{4\pi r} + v \tag{11}$$

is the source function, possessing the following properties:

(1) $G(M, M_0)$ tends to infinity for $M = M_0$ as $1/4\pi r$, which follows from (11);

(2) $G(M, M_0)$ satisfies the equation $L(u) = 0$ everywhere in T, except the point M_0;

(3) $G(P, M_0) = 0$ at points P, lying on the boundary $\sum$.

The question of the existence of a source function is connected with the question of the existence of a function v, satisfying the equation

$$\mathscr{L}(v) = 0 \text{ in } T$$

and the boundary condition

$$u = -\frac{e^{-\varkappa r}}{4\pi r} \text{ on } \Sigma.$$

It is obvious that the function $G(M, M_0)$ is well defined for any region, and gives a unique solution of the first boundary-value problem. In particular, for $c = -\varkappa^2 < 0$ this function is defined for any region. In the simplest cases it is possible to determine the source function in explicit form, by use of a method, similar to the method of electrostatic images.*

Thus, for instance, for the semi-space $z > 0$ the source function has the form

$$G(M, M_0) = \frac{e^{-\varkappa r}}{4\pi r} - \frac{e^{-\varkappa r_1}}{4\pi r_1}, \tag{12}$$

$$r = r_{MM_0} = \sqrt{[(x-x_0)^2 + (y-y_0)^2 + (z-z_0)^2]},$$

$$r_1 = r_{MM_1} = \sqrt{[(x-x_0)^2 + (y-y_0)^2 + (z+z_0)^2]},$$

where $M_1(x_0, y_0, -z_0)$ is the image in the plane $z = 0$ of the point $M_0(x_0, y_0, z_0)$.

We will not consider here the question of the applicability of the preceding formulae for an infinite region, which, however, can be established without difficulty in the case $c < 0$. Problems for infinite regions with $c > 0$ are related to the "radiation principle" and will be considered in the following section.

For the source function $G(M, M_0)$, defined for an arbitrary region T, there exists the "*reciprocity principle*", expressed by the equation

$$G(M, M_0) = G(M_0, M).$$

Proof of this relationship is a literal repetition of the corresponding proof for the case of Laplace's equation (Chapter IV, § 2).

* The method of electrostatic images is inapplicable for a sphere with $c \neq 0$.

In the case of two independent variables the equation for the function $v_0(r)$ has the form

$$\frac{1}{r}\frac{d}{dr}\left(r\frac{dv_0}{dr}\right)+k^2v_0=0 \quad \text{or} \quad \frac{d^2v_0}{dr^2}+\frac{1}{r}\frac{dv_0}{dr}+k^2v_0=0,$$

i. e. *Bessel's equation of zero order*, the general solution of which may be written in the following way (see Supplement I)

$$v_0(r)=C_1H_0^{(1)}(kr)+C_2H_0^{(2)}(kr),$$

where $H_0^{(1)}(kr)$ and $H_0^{(2)}(kr)$ are *Hankel functions of zero order of the first and second kind.*

The functions $H_0^{(1)}(kr)$ and $H_0^{(2)}(kr)$ have a logarithmic singularity at $r \doteq 0$:

$$\left.\begin{aligned} H_0^{(1)}(\varrho) &= \frac{-2i}{\pi}\ln\frac{1}{\varrho}+\ldots, \\ H_0^{(2)}(\varrho) &= \frac{2i}{\pi}\ln\frac{1}{\varrho}+\ldots \end{aligned}\right\} \quad (\varrho=kr),$$

where higher terms are finite for $\varrho=0$. At infinity (as $\varrho\to\infty$) the behaviour of the Hankel functions is determined by the asymptotic relations

$$H_0^{(1)}(\varrho)=\sqrt{\left(\frac{2}{\pi\varrho}\right)}\,e^{i\left(\varrho-\frac{\pi}{4}\right)}+\ldots,$$

$$H_0^{(2)}(\varrho)=\sqrt{\left(\frac{2}{\pi\varrho}\right)}\,e^{-i\left(\varrho-\frac{\pi}{4}\right)}+\ldots.$$

Thus, the equation $\nabla_2^2 v+k^2v=0$ has two fundamental solutions

$$v_0(r)=\begin{cases} H_0^{(1)}(kr), \\ H_0^{(2)}(kr), \end{cases}$$

having a logarithmic singularity and corresponding to the functions e^{ikr}/r and e^{-ikr}/r in three dimensions.

Choice of either fundamental function depends on the form of the radiation condition at infinity (see § 3, subsection 4). If the time dependence is taken in the form $e^{i\omega t}$, then the function $H_0^{(2)}(kr)$ defines a divergent cylindrical wave. For a time dependence $e^{-i\omega t}$ the function $H_0^{(1)}(kr)$ defines a divergent wave.

If $c = -\varkappa^2 < 0$, the the linearly-independent solutions of the equation

$$\frac{d^2 v_0}{dr^2} + \frac{1}{r}\frac{dv_0}{dr} - \varkappa^2 v_0 = 0$$

are cylindrical functions of imaginary argument

$$I_0(\varkappa r) \quad \text{and} \quad K_0(\varkappa r).$$

The first of these functions $I_0(\varkappa r)$ is bounded at $r = 0$ and increases exponentially as $r \to \infty$; the function $K_0(\varkappa r)$ has a logarithmic singularity at the point $r = 0$

$$K_0(\varrho) = \ln\frac{1}{\varrho} + \ldots$$

and thus is the fundamental solution we require. At infinity it decreases according to the law

$$K_0(\varrho) = \sqrt{\left(\frac{\pi}{2\varrho}\right)}\, e^{-\varrho} + \ldots$$

We will not consider Green's theorems in detail nor the idea of the source function G in the case of two independent variables, since an account of this would be a repetition of the preceding.

2. *Potentials*

In Chapter IV potentials were considered for the equation $\nabla^2 u = 0$. Potentials of a similar type can be formed for the equation $\nabla^2 u - \varkappa^2 u = 0$.

We shall call the *volume potential* (for the equation $\nabla^2 u - \varkappa^2 u = 0$) the integral

$$V(M) = \int_T \varrho(P) \frac{e^{-\varkappa r}}{r}\, d\tau_P,$$

$$r = r_{MP} = \sqrt{[(x-\xi)^2 + (y-\eta)^2 + (z-\zeta)^2]},\; d\tau_P = d\xi\, d\eta\, d\zeta, \quad (13)$$

where $\varrho(P)$ is the density of the potential.

Let us describe the basic properties of the volume potential. Proofs are carried out by analogy with Chapter IV.

1. Outside region T the function $V(M)$ satisfies the equation

$$\Delta V - \varkappa^2 V = 0.$$

2. Inside region T the integral (13) converges, and the integrals, resulting from formal differentiation of $V(M)$ under the integral sign, also converge

$$\int_T \varrho(\xi, \eta, \zeta) \frac{\partial}{\partial x}\left[\frac{e^{-\varkappa r}}{r}\right] d\xi\, d\eta\, d\zeta \text{ etc.}$$

3. The function $V(x, y, z)$ is differentiable, and its first derivatives can be evaluated by differentiation under the integral sign

$$\frac{\partial V}{\partial x} = \int \varrho(\xi, \eta, \zeta) \frac{\partial}{\partial x}\left[\frac{e^{-\varkappa r}}{r}\right] d\xi\, d\eta\, d\zeta \text{ etc.}$$

The proof of the differentiability of $V(x, y, z)$ requires only that ϱ be bounded. Hence, in particular, (x, y, z) is differentiable at points of the surface Σ, bounding the region T, where, as a rule there is a discontinuity of the density $\varrho(M)$.

4. At interior points of region T, in the neighbourhood of which the density ϱ is differentiable, second derivatives of the volume potential V exist, and the potential V satisfies the equation

$$\nabla^2 V - \varkappa^2 V = -4\pi\varrho(M).$$

5. First derivatives of the volume potential are represented by uniformly converging integrals assuming that ϱ is uniformly bounded. Therefore first derivatives are continucus functions for all space, including points of the surface Σ.

Volume potentials enable us to represent the solution of the boundary value problem for the inhomogeneous equation $\nabla^2 u - \varkappa^2 u = -f$ in the form of the sum

$$u(M) = V(M) + u_1(M),$$

where $V(M)$ is the volume potential with density $\varrho = f/4\pi$, $u_1(M)$ is the solution of the boundary-value problem for the homogeneous equation $\nabla^2 u_1 - \varkappa^2 u_1 = 0$.

Let us proceed to a survey of the properties of potentials of a single and double layer. Let us call the *potential of a double layer* the integral

$$W(M) = \int_\Sigma \mu(P) \frac{\partial}{\partial \nu_P}\left[\frac{e^{-\varkappa r}}{r}\right] d\sigma_P \quad (r = r_{MP}), \tag{14}$$

where $\mu(P)$ is the surface density of the potential W.

Let us enumerate the main properties of the potential of a double layer, referring for their proof to Chapter IV, § 5.

1. Outside the surface Σ the potential of a double layer everywhere satisfies the homogeneous equation $\nabla^2 W - \varkappa^2 W = 0$.

2. The potential of a double layer converges at points of the boundary, if Σ belongs to the class of Lyapunov surfaces.

3. Function W is discontinuous at points of the surface Σ and the relations

$$W_i(M_0) = W(M_0) + 2\pi\mu(M_0),$$
$$W_e(M_0) = W(M_0) - 2\pi\mu(M_0)$$

hold. Here $W_i(M_0)$ is the limiting value of the function $W(M)$ as M tends to M_0 inside region T, $W_e(M_0)$ is the limiting value of $W(M)$ as M tends to M_0 on the outside of T. *The potential of a single layer*, defined by the surface integral

$$V(M) = \int_\Sigma \varrho(P) \frac{e^{-\varkappa r}}{r} d\sigma_P \qquad (r = r_{MP}), \tag{15}$$

possesses the following properties:

1. Outside surface Σ the potential of the single layer everywhere satisfies the homogeneous equation $\Delta V - \varkappa^2 V = 0$.

2. The integral converges uniformly on Σ and determines a function $V(M)$, continuous over all space.

3. The normal derivatives of the potential of a single layer for a Lyapunov class of surface satisfy the relations (see (48), § 5. Chapter IV)

$$\left(\frac{\partial V}{\partial \nu}\right)_i = U_0 + 2\pi\varrho(M_0),$$
$$\left(\frac{\partial V}{\partial \nu}\right)_e = U_0 - 2\pi\varrho(M_0),$$

where

$$\left(\frac{\partial V}{\partial \nu}\right)_i \quad \text{or} \quad \left(\frac{\partial V}{\partial \nu}\right)_e$$

are the limiting values for the normal derivative from inside and from outside Σ respectively at the point M_0 on surface Σ (ν is the outer normal)

$$U_0(M_0) = \int_\Sigma \varrho(P) \frac{\partial}{\partial \nu}\left[\frac{e^{-xr}}{r}\right] d\sigma_P \qquad (r = r_{M_0P}).$$

The surface potentials enable us, for a very large class of surfaces (for instance, surfaces of the Lyapunov type) to reduce boundary-value problems to integral equations.

Let us consider the first interior boundary-value problem for the equation $\nabla^2 u - \varkappa^2 u = 0$ with the boundary value $u|_\Sigma = f$. We assume that the unknown function can be represented in the form of the potential of a double layer

$$u(M) = W(M) = \int_\Sigma \mu(P) \frac{\partial}{\partial \nu_P} \left[\frac{e^{-\varkappa r}}{r}\right] d\sigma_P, \tag{14}$$

which, as was noted above, satisfies the homogeneous equation $\nabla^2 u - \varkappa^2 u = 0$ inside T. Requiring that the boundary condition $u|_\Sigma = f$ be satisfied, we arrive at the following integral equation for the determination of μ:

$$2\pi \mu(M) + \int_\Sigma \mu(P) \frac{\partial}{\partial \nu_P} \left[\frac{e^{-\varkappa r}}{r}\right] d\sigma_P = f(M),$$

which is a *Fredholm* linear integral equation of the second kind. We will not consider here the question of the existence and uniqueness of the solution of this integral equation.

The method of finite differences can be applied to the equation $\nabla^2 u - \varkappa^2 u = 0$, as for Laplace's equation.

§ 3. Problems for an infinite region Radiation principle

1. *The equation* $\nabla^2 v + cv = -f$ *in infinite space*

Let us consider the solution of the inhomogeneous equation

$$\nabla^2 v + cv = -f \tag{1}$$

in infinite space. For simplicity we shall assume that f differs from zero inside some bounded region (local function). The nature of the solution of this equation essentially depends on the sign of the coefficients c. First we consider the case $c = -\varkappa^2 < 0$. The solution of the equation $\nabla^2 v - \varkappa^2 v = -f$ may be represented by volume potentials

$$v_1(M) = \int_T f(P) \frac{e^{-\varkappa r}}{4\pi r} d\tau_P \quad \text{and} \quad v_2(M) = \int_T f(P) \frac{e^{\varkappa r}}{4\pi r} d\tau_P.$$

Thus, the solution of equation (1) without additional conditions at infinity is not well defined. We shall search, in analogy to the exterior problem for Laplace's equation, for the solution of equation (1), tending to zero at infinity. This condition is satisfied by $v_1(M)$ but is not satisfied by $v_2(M)$.

Let us prove the following uniqueness theorem:

The equation

$$\nabla^2 v - \varkappa^2 v = -f$$

*cannot have more than one solution, tending to zero at infinity.**

We assume that there exist two different solutions of the problem $\bar{v}(M)$ and $\bar{\bar{v}}(M)$ and consider their difference $w = \bar{v} - \bar{\bar{v}}$. We assume, there exists a point M_0 such that $w(M_0) = A \neq 0$. For definiteness we assume $A > 0$. Because $w(M) \to 0$ at infinity, it is possible to find R_0 such that for $r > R_0$ the function $w < A/2$. Hence it follows that the point M_0 lies inside T_{R_0}, a sphere of radius R_0 and that the function $w(M)$ reaches its maximum value inside T_{R_0}. Thus, we arrive at a contradiction of the maximum value principle, which holds for our equation (see § 2, sub-section 1). The uniqueness theorem is proved.

Now we consider the case $c = k^2 > 0$.

The functions

$$v_1(M) = \int_T f(P) \frac{e^{-ikr}}{4\pi r} d\tau_P \quad \text{and} \quad v_2(M) = \int_T f(P) \frac{e^{ikr}}{4\pi r} d\tau_P$$

are the solutions of equation (1) as before. However in this case both functions decrease at infinity. Hence we need additional conditions at infinity, in order to determine a unique solution of equation (1). These conditions will be discussed in sub-sections 2, 3 and 4 of the present section.

2. *Principle of vanishing absorption*

The problem of forced vibrations with damping reduces to the equation

$$\nabla^2 u = \frac{1}{a^2} u_{tt} + \beta u_t - F(M, t) \qquad (\beta > 0). \tag{2}$$

* By the term "function, tending to zero at infinity" we understand the following: whatever ε may be, there exists $r(\varepsilon)$ such that for any point $M(r, \theta, \varphi)$, for which $r > r(\varepsilon)$, $|u(M)| < \varepsilon$, i. e. we assume the uniform convergence to zero as $r \to \infty$.

We assume that the function $F(M, t)$ is periodic in time, i.e. $F(M, t) = f(M)e^{i\omega t}$. In this case equation (2) has a periodic solution of the form

$$u(M, t) = v(M)e^{i\omega t}. \tag{3}$$

The function $v(M)$, obviously, satisfies the equation

$$\nabla^2 v + cv = -f(M), \tag{4}$$

where $c = k^2 - i\beta\omega$ is a complex quantity.

We shall call equation (9) with a complex value of the coefficient c the *equation with a complex absorption of the first* (Im $c < 0$) *or second type* (Im $c > 0$), depending on the sign of the imaginary part of c, which corresponds to a time dependence $e^{i\omega t}$ (first type) or $e^{-i\omega t}$ (second type).

The fundamental solutions of this equation, depending only on r have the form

$$\bar{v}_0(r) = \frac{e^{-iqr}}{r} \quad \text{and} \quad \bar{\bar{v}}_0(r) = \frac{e^{+iqr}}{r},$$

where

$$q = \sqrt{\left[\left(\frac{\omega}{a}\right)^2 - i\beta\omega\right]} =$$

$$= \sqrt{\left[\frac{\sqrt{(k^4 + \beta^2\omega^2)} + k^2}{2}\right]} - i\sqrt{\left[\frac{\sqrt{(k^4 + a^2\omega^2)^2} - k^2}{2}\right]} = q_0 - iq_1\,. \tag{5}$$

Let us choose the signs of the roots so that $q_1 > 0$. Therefore

$$\bar{v}_0(r) = \frac{e^{-iq_0 r}}{r} e^{-q_1 r}, \qquad \bar{\bar{v}}_0(r) = \frac{e^{iq_0 r}}{r} e^{q_1 r}.$$

The boundary condition at infinity is satisfied only by function $\bar{v}_0(r)$; function $\bar{\bar{v}}_0(r)$ increases indefinitely as $r \to \infty$ and therefore has no direct physical significance.

The volume potential

$$\bar{v}(M) = \int_T f(P) \frac{e^{-iq_0 r}}{4\pi r} e^{-q_0 r}\, d\tau_P \qquad (r = r_{MP}) \tag{6}$$

represents a unique solution of equation (9), reducing to zero at infinity. The limit $\bar{v}(M)$ for $\beta \to 0$ equals

$$v(M) = \lim_{\beta=0} \bar{v}(M) = \int_T f(P) \frac{e^{-ikr}}{4\pi r}\, d\tau_P \qquad (r = r_{MP}),$$

since for $\beta \to 0$ we have: $q_0 \to k$ and $q_1 \to 0$. For the time dependence $e^{i\omega t}$ chosen by us the quantity $q_0 > 0$, since the sign of q_0 is connected to the sign of q_1 by the relation $2q_0 q_1 = = \beta\omega$ (since $q^2 = c$).

If the dependence on time is taken in the form $e^{-i\omega t}$ (Im $c > 0$), then a positive value of q will correspond to $q_0 < 0$ and the the limit of q_0 for $\beta \to 0$ equals $-k$.

Thus, the additional condition, enabling us to separate the solution of the wave equation

$$\Delta v + k^2 v = -f,$$

corresponding to a divergent wave, is the requirement that the function $v(M)$ should be the limit of the bounded solution of the wave equation with complex absorption of the first kind with an infinitely small imaginary part of the complex absorption.*

3. *Principle of limiting amplitude*

In this section we give an alternative to the principle of vanishing absortion. The wave equation

$$\nabla^2 v + k^2 v = -f \tag{7}$$

is most often encountered in an investigation of steady vibrations, produced by periodic forces (see § 1. sub-section 1).

Let us consider the wave equation with a periodic right-hand side

$$\Delta u - \frac{1}{a^2} u_{tt} = -F \qquad (F = fe^{i\omega t}). \tag{8}$$

In order to find the solution certain initial conditions must be added to the equation, for example zero conditions:

$$\left.\begin{aligned} u(M, 0) &= 0, \\ u_t(M, 0) &= 0. \end{aligned}\right\} \tag{9}$$

The function $u(M, t)$ at an initial stage of the process will not be strictly periodic. However, in the course of time periodic

* See A. G. Sveshnikov: Principle of radiation, *Dokl. Akad. Nauk. SSR.* 73, 5. 1950.

oscillations will be established in the system with the frequency of the constraining force, i.e. the solution $u(M, t)$ has the form

$$u(M, t) = v(M)\, e^{i\omega t}, \tag{10}$$

$v(M)$ represents the limiting amplitude of the oscillations, i.e. $v(M) = \lim_{t\to\infty} ue^{-i\omega t}$, and satisfies the equation

$$\nabla^2 v + k^2 v = -f \qquad (k = \omega/a).$$

The requirement that $v(M)$ should be the limiting amplitude of the oscillations with zero initial data, represents an additional condition, which must be added to the wave equation to determine a unique solution.

Thus, we arrive at the following problem:

Find the solution of the wave equation $\nabla^2 v + k^2 v = -f$ *which is the limiting amplitude of the solution of the wave equation*

$$\nabla^2 u - \frac{1}{a^2} u_{tt} = -f(M)\, e^{i\omega t} \tag{8*}$$

with initial conditions

$$\left.\begin{aligned} u(M, 0) &= 0, \\ u_t(M, 0) &= 0. \end{aligned}\right\} \tag{9}$$

Let us represent the limiting amplitude in explicit form. In order to do this we find the solution of the wave equation (3*) with zero initial data, using the relation

$$u(M, t) = \frac{1}{4\pi} \int_{T_{at}^M} \frac{f(P)\, e^{i\omega\left(t - \frac{r}{a}\right)}}{r}\, d\tau_P \qquad (r = r_{MP}),$$

obtained in Chapter V (§ 2, (6)). Here T_{at}^M is a sphere of radius at with centre at the point M.

Let $f(P)$ be a local function, differing from zero only inside some bounded region T_0. Then for the limiting amplitude $v(M)$ we obtain the expression

$$v(M) = \lim_{t\to\infty} u(M, t)\, e^{-i\omega t} = \lim_{t\to\infty} \frac{1}{4\pi} \int_{T_{at}} \frac{e^{-ikr}}{r} f(P)\, d\tau_P =$$

$$= \frac{1}{4\pi} \int_{T_0} f(P) \frac{e^{-ikr}}{r}\, d\tau_P \qquad (r = r_{MP}).$$

Thus, the limiting amplitude is represented by the volume potential, given by the main solution $\frac{e^{-ikr}}{r}$, which corresponds to the divergent wave $\frac{e^{i(\omega t-kr)}}{r}$.

The principle of limiting amplitude leads mathematically to the same result as the principle of vanishing absorption. This is natural, since both these principles give a solution corresponding to a divergent wave.

4. *Radiation condition*

In the preceding sections general physical principles were considered, enabling us to define the solution of the wave equation, corresponding to divergent waves. However, these methods require the solution of auxiliary problems. Let us establish now an analytical condition, describing a divergent wave expressed directly in terms of the solution of the wave equation being investigated.

Plane waves, propagating along the x-axis, have the form

$\bar{u} = f\left(t - \frac{x}{a}\right)$ — the forward wave (travelling in the positive direction of the x-axis);

$\bar{\bar{u}} = f\left(t + \frac{x}{a}\right)$ — the backward wave (travelling in the negative direction of the x-axis).

The forward wave satisfies the equation

$$\frac{\partial \bar{u}}{\partial x} + \frac{1}{a}\frac{\partial \bar{u}}{\partial t} = 0,$$

the backward wave the equation.*

$$\frac{\partial \bar{\bar{u}}}{\partial x} - \frac{1}{a}\frac{\partial \bar{\bar{u}}}{\partial t} = 0.$$

For a steady state

$$u = v(M)\, e^{i\omega t}$$

these relations take the form

$$\frac{\partial \bar{v}}{\partial x} + ik\bar{v} = 0 \qquad \text{for the forward wave,} \tag{11}$$

$$(k = \omega/a)$$

$$\frac{\partial \bar{\bar{v}}}{\partial x} - ik\bar{\bar{v}} = 0 \qquad \text{for the backward wave.} \tag{12}$$

* These are partial differential equations of the first order, whose solutions have the form of a forward and backward wave.

Let us proceed now to the case of spherical waves. If a spherical wave is produced by sources, situated in a finite part of space, then at very large distances from the sources the spherical wave is similar to a plane wave, whose amplitude decreases as $1/r$. Hence it is natural to assume that a divergent spherical wave should satisfy the relation*

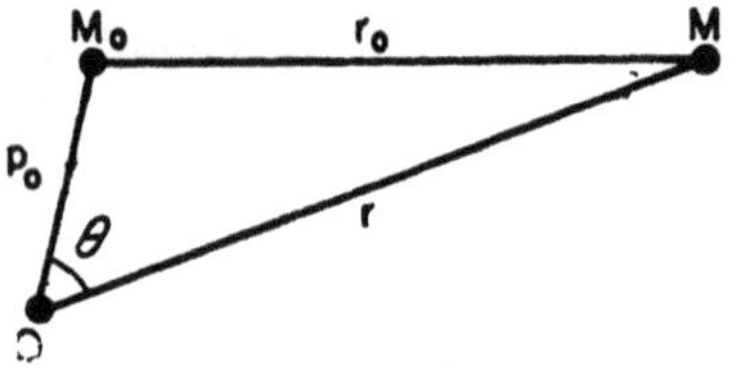

FIG. 86

$$\frac{\partial u}{\partial r} + \frac{1}{a}\frac{\partial u}{\partial t} = o\left(\frac{1}{r}\right); \qquad (13)$$

similarly for a converging spherical wave

$$\frac{\partial u}{\partial r} - \frac{1}{a}\frac{\partial u}{\partial t} = o\left(\frac{1}{r}\right). \qquad (14)$$

For the amplitude of the steady oscillations these conditions take the form

$$\frac{\partial v}{\partial r} + ikv = o\left(\frac{1}{r}\right) \text{ for diverging spherical waves,} \qquad (15)$$

$$\frac{\partial v}{\partial r} - ikv = o\left(\frac{1}{r}\right) \text{ for converging spherical waves.} \qquad (16)$$

We derived (15) and (16), assuming that at very large distances any divergent wave is similar to a plane wave, whose amplitude decreases as $1/r$. Let us verify this statement.

1. In the case of a point source, at the origin of coordinates, this statement is perfectly obvious, since the wave has the form

$$u(r, t) = \frac{e^{i(\omega t - kr)}}{r} = v_0(r)\, e^{i\omega t},$$

so that

$$\frac{\partial v_0}{\partial r} + ikv_0 = o\left(\frac{1}{r}\right).$$

2. Let a spherical wave be produced by a point source, situated at some point M_0. The amplitude of the spherical wave equals

$$v_0(M) = \frac{e^{-ikr_0}}{r_0},$$

* Later we make use of two symbols: $\varrho(\xi)$ a quantity decreasing with ξ as $\xi \to 0$, $o(\xi)$ a small quantity of a higher order than ξ as $\xi \to 0$.

where r_0 is the distance between the points M and M_0 equal to (Fig. 86)

$$r_0 = \sqrt{(r^2 + \varrho_0^2 - 2\, r\, \varrho_0 \cos\vartheta)}\,.$$

Let us evaluate the derivative

$$\frac{\partial r_0}{\partial r} = \frac{r - \varrho_0 \cos\vartheta}{r_0} \approx 1 + O\left(\frac{1}{r}\right).$$

By virtue of the result 1

$$\frac{\partial v_0}{\partial r_0} + ikv_0 = o\left(\frac{1}{r_0}\right).$$

Let us verify now the validity of (15):

$$\mathscr{L}(v_0) = \frac{\partial v_0}{\partial r} + ikv_0 = o\left(\frac{1}{r}\right). \tag{15'}$$

In fact,

$$\frac{\partial v_0}{\partial r} = \frac{\partial v_0}{\partial r_0}\frac{\partial r_0}{\partial r} = \frac{\partial v_0}{\partial r_0}\left(1 + O\left(\frac{1}{r}\right)\right) = \frac{\partial v_0}{\partial r_0} + o\left(\frac{1}{r}\right),$$

since

$$\frac{\partial v_0}{\partial r_0} O\left(\frac{1}{r}\right) = o\left(\frac{1}{r}\right).$$

Hence from 1 it follows:

$$\mathscr{L}(v_0) = \frac{\partial v_0}{\partial r_0} + ikv_0 + o\left(\frac{1}{r}\right) = o\left(\frac{1}{r}\right),$$

which it was required to prove.

3. Let us show that the volume potential

$$v(M) = \int_T f(P)\frac{e^{-ikr}}{4\pi r}\, d\tau_P \qquad (r = r_{MP}) \tag{16'}$$

satisfies condition (15). It is obvious that

$$\mathscr{L}(v) = \int_T f(P)\, \mathscr{L}\left(\frac{e^{-ikr}}{4\pi r}\right) d\tau_P = \int_T f(P)\, o\left(\frac{1}{r}\right) d\tau_P = o\left(\frac{1}{r}\right).$$

Formula (16′) represents the amplitude of a divergent wave, produced by sources, distributed arbitrarily inside a finite region of space T. We have seen that the function $v(M)$ satisfies the wave equation

$$\Delta v + k^2 v = -f(M)$$

and tends to zero as $1/r$ at infinity; moreover, it fulfills a condition at infinity

$$\frac{\partial v}{\partial r} + ikv = o\left(\frac{1}{r}\right),$$

which is the necessary additional condition.

Let us show that

there exists a unique solution of the wave equation

$$\nabla^2 v + k^2 v = -f(M),$$

where $f(M)$ is a local function, satisfying conditions at infinity

$$\left.\begin{aligned} v &= O\left(\frac{1}{r}\right) \\ \frac{\partial v}{\partial r} + ikv &= o\left(\frac{1}{r}\right). \end{aligned}\right\} \qquad (a)$$

Assuming the existence of two different solutions v_1 and v_2, we find that their difference

$$w = v_1 - v_2$$

satisfies the homogeneous equation and the condition (a). Let $\sum_R$ be a sphere of radius R, which later we expand to infinity. Making use of Green's fundamental theorem for the functions $w(M)$ and $v_0(M) = e^{-ikr}/4\pi r$, we have at the point M_0, lying inside $\sum$,

$$w(M_0) = \int_{\Sigma_R} \left(v_0 \frac{\partial w}{\partial r} - w \frac{\partial v_0}{\partial r}\right) d\sigma.$$

Condition (a) for $v_0(r)$ and $w(M)$ gives:

$$v_0 \frac{\partial w}{\partial r} - w \frac{\partial v_0}{\partial r} = v_0\left[-ikw + o\left(\frac{1}{r}\right)\right] - w\left[-ikv_0 + o\left(\frac{1}{r}\right)\right] =$$

$$= v_0\, o\left(\frac{1}{r}\right) - w o\left(\frac{1}{r}\right) = o\left(\frac{1}{r^2}\right).$$

Therefore

$$w(M_0) = \int_{\Sigma_R} o\left(\frac{1}{r^2}\right) d\sigma \to 0 \text{ where } R \to \infty,$$

from which the uniqueness of the solution follows, because of the arbitrary nature of the point M_0.

The conditions

$$\left.\begin{aligned} v &= O\left(\frac{1}{r}\right), \\ \frac{\partial v}{\partial r} + ikv &= o\left(\frac{1}{r}\right) \end{aligned}\right\} \qquad (a)$$

are often called the *radiation conditions* or the *Sommerfeld* conditions.

It should be noted that for infinite regions, not including all space, the radiation conditions can have a form, differing from the Sommerfeld conditions.

Thus, relation (a) representing the analytical form of the radiation conditions for infinite space, is not based on a physical principle, which can be extended to regions of more complex shape.

The radiation conditions, obtained by introducing an infinitely small complex absorption into the wave equation, were used first by V. S. Ignatovski.* This principle can be applied to infinite regions of different shape and more complex problems.

For two dimensional problems, described by the equation

$$\nabla_2^2 v + k^2 v = 0, \qquad (17)$$

the radiation conditions at infinity take the form

$$\left.\begin{aligned} v &= O\left(\frac{1}{\sqrt{r}}\right), \\ \lim_{r\to\infty} \sqrt{r}\left(\frac{\partial v}{\partial r} + ikv\right) &= 0. \end{aligned}\right\} \qquad (18)$$

The simplest solutions of this equation are Hankel functions of zero order

$H_0^{(1)}(kr)$ and $H_0^{(2)}(kr)$ (see Appendix I. § 3).

From the asymptotic formulae

$$H_\nu^{(1)}(kr) = \sqrt{\left(\frac{2}{\pi kr}\right)}\, e^{i\left(kr - \frac{\pi}{2}\nu - \frac{\pi}{4}\right)}\left[1 + O\left(\frac{1}{r}\right)\right],$$

$$H_\nu^{(2)}(kr) = \sqrt{\left(\frac{2}{\pi kr}\right)}\, e^{-i\left(kr - \frac{\pi}{2}\nu - \frac{\pi}{4}\right)}\left[1 + O\left(\frac{1}{r}\right)\right]$$

* V. S. Ignatovski, *Ann. de Physics 18.* 1905.

and the recurrence relations

$$\frac{dH_0^{(1)}}{dx} = -H_1^{(1)}(x), \quad \frac{dH_0^{(2)}}{dx} = -H_1^{(2)}(x)$$

we see that the radiation condition is satisfied only by the function $H_0^{(2)}(kr)$.

Thus, the function $H_0^{(2)}(kr)$ satisfies equation (11), the radiation conditions (12) and has a logarithmic singularity at $r = 0$. Therefore the function $H_0^{(2)}(kr)$, as has already been noted in § 2, is the required source function for the wave equation (7) in the case of two independent variables. The solution of the inhomogenous equation

$$\nabla_2^2 v + k^2 v = -f$$

is given by the relation

$$v(M) = -\frac{i}{4} \iint_S H_0^{(2)}(kr_{MP}) f(P) \, d\sigma_P,$$

where S is the region, in which the function f differs from zero.

§ 4. Problems of the mathematical theory of diffraction

1. *Statement of the problem*

The propagation of waves (electromagnetic, elastic, acoustic, etc.) is accompanied by a whole series of typical phenomena (diffraction, refraction, reflection, etc.). Problems, connected with these phenomena, can be studied by solving the wave equation in an inhomogeneous medium

$$\frac{\partial}{\partial x}\left(p \frac{\partial v}{\partial x}\right) + \frac{\partial}{\partial y}\left(p \frac{\partial v}{\partial y}\right) + \frac{\partial}{\partial z}\left(p \frac{\partial v}{\partial z}\right) + \varrho \omega^2 v = -\bar{f} \quad (p > 0), \quad (1)$$

where p and ϱ are parameters of the medium.

From the point of view of physical applications, greatest interest lies in the case of partly-constant parameters p and ϱ. The appropriate mathematical problem consists of the following. In infinite space there are a number of finite regions T_i with constant parameters p_i and ϱ_i; the part of space T_0, external to the regions T_i, is also homogeneous ($p_0 = \text{const.}$, $\varrho_0 = \text{const.}$). The wave equation inside each region T_i takes the usual form

$$\nabla^2 v_i + k_i^2 v_i = -f_i \text{ at } T_i \quad (i = 0, 1, \ldots, n), \tag{2}$$

where u_i is the value of the function u being sought inside T_i,

$$k_i^2 = \frac{\varrho_i \omega^2}{p_i}, \quad f_i = \frac{\bar{f}}{p_i}$$

in region T_i. On the surface Σ_i, bounding the region T_i* the differential equations are replaced by the matching conditions

$$\left.\begin{aligned} v_i &= v_0 \text{ on } \Sigma_i, \\ p_i \frac{\partial v_i}{\partial n} &= p_0 \frac{\partial v_0}{\partial n} \text{ on } \Sigma_i \quad (i = 1, 2, \ldots, n). \end{aligned}\right\} \tag{3}$$

At infinity the function v_0, which is the solution of the wave equation $\nabla^2 v + k_0^2 v = -f_0$ in T_0, should satisfy the radiation conditions

$$\left.\begin{aligned} \frac{\partial v_0}{\partial r} + ikv_0 &= o\left(\frac{1}{r}\right), \\ v_0(M) &= O\left(\frac{1}{r}\right). \end{aligned}\right\} \tag{4}$$

It will be shown that the matching condinitions and the radiation conditions are sufficient to determine the solution v uniquely. The problem given above is the simplest problem of the *mathematical theory of diffraction.*

2. *Uniqueness of the solution of the diffraction problem*

Let us prove that the problem of the mathematical theory of diffraction, formulated in sub-section 1, has a unique solution. For simplicity we shall assume that the homogeneity of the medium is disturbed only by one body T_1, bounded by the closed surface Σ, surrounded by the region T_0. In addition we make no assumption concerning the simply connected nature of region T_1.

Let us prove the following theorem:

There can exist only one function, satisfying :

* In addition we consider for simplicity the case where the inhomogeneities T_i have a common boundary only with the surrounding medium.

(a) the equations

$$\left.\begin{aligned}\mathscr{L}_0(v_0) &= \nabla^2 v_0 + k_0^2 v_0 = -f_0 \text{ in } T_0, \\ \mathscr{L}_1(v_1) &= \nabla^2 v_1 + k_1^2 v_1 = -f_1 \text{ in } T_1;\end{aligned}\right\} \tag{2'}$$

(b) *the matching conditions on the surface* Σ

$$\left.\begin{aligned} v_1 &= v_0, \\ p_1 \frac{\partial v_1}{\partial \nu} &= p_0 \frac{\partial v_0}{\partial \nu};\end{aligned}\right\} \tag{3}$$

(c) *the radiation conditions at infinity*

$$\left.\begin{aligned} v_0 &= O\left(\frac{1}{r}\right), \\ \frac{\partial v_0}{\partial r} + ikv_0 &= o\left(\frac{1}{r}\right).\end{aligned}\right\} \tag{4}$$

Assuming the existence of two different solutions

$$\bar{v} = \{\bar{v}_1, \bar{v}_0\} \text{ and } \bar{\bar{v}} = \{\bar{\bar{v}}_1, \bar{\bar{v}}_0\},$$

we show that their difference

$$w = \{w_1, w_0\},$$

where

$$w_1 = \bar{v}_1 - \bar{\bar{v}}_1, \quad w_0 = \bar{v}_0 - \bar{\bar{v}}_0$$

satisfies the homogenous equations and the preceding additional conditions

$$\mathscr{L}_0(w_0) = 0 \text{ at } T_0, \quad \mathscr{L}_1(w_1) = 0 \text{ at } T_1 \tag{2'}$$

$$w_1 = w_0, \quad p_1 \frac{\partial w_1}{\partial \nu} = p_0 \frac{\partial w_0}{\partial \nu} \text{ on } \Sigma_1 \tag{3'}$$

$$w_0 = O\left(\frac{1}{r}\right), \quad \frac{\partial w_0}{\partial r} + ikw_0 = o\left(\frac{1}{r}\right) \quad \text{as} \quad r \to \infty. \tag{4'}$$

The functions w_0^*, w_1^*, the complex-conjugates of functions w_0 and w_1 satisfy the homogenous equations (2′), conditions (3′) and radiation conditions

$$w_0^* = O\left(\frac{1}{r}\right), \qquad \frac{\partial w_0^*}{\partial r} - ikw_0^* = o\left(\frac{1}{r}\right). \tag{4''}$$

Let Σ_R be a sphere of sufficiently large radius R, encompassing the region T_1, and T_R be a region, bounded by the surfaces Σ_1 and Σ_R.

Applying Green's theorem to the functions w_1, w_1^* in region T_1 and to the functions w_0, w_0^* in region T_R, we obtain:

$$\int_{T_1} (w_1 \mathscr{L}_1(w_1^*) - w_1^* \mathscr{L}_1(w_1))\, d\tau = \int_{\Sigma_1} \left(w_1 \frac{\partial w_1^*}{\partial \nu_1} - w_1^* \frac{\partial w_1}{\partial \nu_1}\right) d\sigma,$$

$$\int_{T_R} [w_0 \mathscr{L}_0(w_0^*) - w_0^* \mathscr{L}_0(w_0)]\, d\tau =$$

$$= \int_{\Sigma_1} \left(w_0 \frac{\partial w_0^*}{\partial \nu_0} - w_0^* \frac{\partial w_0}{\partial \nu_0}\right) d\sigma + \int_{\Sigma_R} \left(w_0 \frac{\partial w_0^*}{\partial \nu} - w_0^* \frac{\partial w_0}{\partial \nu}\right) d\sigma = 0,$$

where ν_0 is the normal, external to region T_R, ν_1 is the normal, external to region T_1. Obviously $\frac{\partial}{\partial \nu_0} = -\frac{\partial}{\partial \nu_1}$ on Σ_1.

Multiplying the first equation by p_1, the second by p_0, adding them and using the matching conditions (3′), we find:

$$\int_{\Sigma_R} \left(v_0 \frac{\partial w_0^*}{\partial r} - w_0^* \frac{\partial w_0}{\partial r} \right) d\sigma = 0.$$

Calculating the derivatives from the radiation conditions

$$\frac{\partial w_0^*}{\partial r} = ikw_0^* + o\left(\frac{1}{r}\right), \qquad \frac{\partial w_0}{\partial r} = -ikw_0 + o\left(\frac{1}{r}\right),$$

we arrive at the following equation:

$$2ik \int_{\Sigma_R} w_0 w_0^* \, d\sigma + \int_{\Sigma_R} \left[w_0 \, o\left(\frac{1}{R}\right) - w_0^* \, o\left(\frac{1}{R}\right) \right] d\sigma = 0\,.$$

The second integral tends to zero as $R \to \infty$, therefore

$$\int_{\Sigma_R} w_0 w_0^* \, d\sigma = \int_{\Sigma_R} |\, R w_0 \,|^2 \, d\Omega \to 0, \;\; R \to \infty \qquad (d\Omega = \sin\theta \, d\theta \, d\varphi)\,. \tag{5}$$

In the Supplement, Part II, § 3, it is shown that the function

$$V_m(r, \theta, \varphi) = \zeta_m^{(2)} Y_m(\theta, \varphi)\,,$$

where

$$\zeta_m^{(2)}(\varrho) = \sqrt{\left(\frac{2}{\pi\varrho}\right)} H^{(2)}_{m+\frac{1}{2}}(\varrho) \qquad (\varrho = k_0 r)\,,$$

and $Y_m(\theta, \varphi)$ is a spherical function of m th order, satisfying the wave equation

$$\mathscr{L}_0(V_m) = \nabla^2 V_m + k_0^2 V_m = 0$$

and the radiation condition

$$\frac{\partial V_m}{\partial r} + ik_0 V_m = o\left(\frac{1}{r}\right).$$

Let us apply Green's theorem in the region T_R to the functions w_0 and V_m

$$0 = \int_{T_R} [w_0 \mathscr{L}(V_m) - V_m \mathscr{L}(w_0)] \, d\tau =$$

$$= \int_{\Sigma_1} \left(w_0 \frac{\partial V_m}{\partial \nu} - V_m \frac{\partial w_0}{\partial \nu} \right) d\sigma + \int_{\Sigma_R} \left(w_0 \frac{\partial V_m}{\partial \nu} - V_m \frac{\partial w_0}{\partial \nu} \right) d\sigma = I_1 + I_R\,.$$

The second component I_R by virtue of the radiation conditions tends to zero as $R \to \infty$ (see Theorem, § 3, sub-section 4). Since the first integral I_1 does not depend on R, it follows that $I_1 = 0$ and therefore $I_R = 0$ for any R, i.e.

$$\left. \frac{d\zeta_m^{(2)}(k_0 r)}{dr} \right|_{r=R} \cdot \int_{\Sigma_R} w_0 Y_m(\theta, \varphi) \, d\Omega - \zeta_m^{(2)} \Big|_{r=R} \cdot \int_{\Sigma_R} \frac{\partial w_0}{\partial r} Y_m(\theta, \varphi) \, d\Omega = 0.$$

If we denote

$$\int\limits_{\Sigma_r} w_0\, Y_m(\theta, \varphi)\, d\Omega = a_m(k_0 r),$$

then it is possible to write:

$$\zeta_m^{(2)\prime}(k_0 R)\, a_m(k_0 R) - a_m'(R)\, \zeta_m^{(2)}(k_0 R) = 0.$$

from which we find

$$a_m(r) = a_m \zeta_m^{(2)}(k_0 r),$$

where a_m is a constant factor.

The amplitude condition of spherical functions

$$\int\limits_{\Sigma_R} |R w_0|^2\, d\Omega = \sum_{m=0}^{\infty} R^2 a_m^2(k_0 R) \tag{6}$$

and formula (5) give:

$$R a_m(k_0 R) \to 0 \text{ where } R \to \infty.$$

However the asymptotic relation

$$\zeta_m^{(2)}(\varrho) \approx \frac{1}{\varrho} e^{-i\left(\varrho - \frac{m+1}{2}\pi\right)},$$

shows that the modulus of the product $r\,\zeta_m^{(2)}(k_0 r)$, remains greater than some positive number for large values of r; hence, $a_m = 0$, i.e. $a_m(r) \equiv 0$; hence because of (6) it follows that $w_0 = 0$ on the sphere Σ_{r_0}. Thus, if the sphere Σ_{r_0} of some radius r_0 encompasses the region T_1, then outside this sphere the function $w \equiv 0$. Hence because of the analytical nature* of the solution of the equation $\mathscr{L} = 0$ we deduce that the function $w_0 \equiv 0$ everywhere in the region T_1. Further, from the matching conditions it follows that on the surface Σ_1

$$w_1 = 0 \quad \text{and} \quad \frac{\partial w_1}{\partial \nu} = 0. \tag{7}$$

Green's fundamental theorem, applied in region T_1 to the function w_1, shows that

$$w_1(M) = \frac{1}{4\pi} \int\limits_{\Sigma_1} \left[\frac{e^{-ik_1 r}}{r} \frac{\partial w_1(P)}{\partial \nu_1} - w_1(P) \frac{\partial}{\partial \nu_1} \left(\frac{e^{-ik_1 r}}{r} \right) \right] d\sigma = 0, \tag{8}$$

where $r = r_{MP}$, at any point M of region T_1.

Thus we have verified that $w(M) = 0$ over all space; this proves the uniqueness theorem.

* The analytical nature of function w in region T_1 follows from (7), § 2, for the complex value $x = ik$ and for the surface Σ, lying entirely inside T_1.

3. *Diffraction by a sphere*

1. In practice an important class of solutions of the wave equation

$$\nabla^2 u - \frac{1}{a^2} u_{tt} = 0$$

is plane waves. A plane wave, propagating in any given direction, is a solution, depending on time and on one spatial coordinate, measured in the direction of propagation. For example, the plane wave, propagating along the x-axis, satisfies the equation with two independent variables

$$u_{xx} - \frac{1}{a^2} u_{tt} = 0$$

and has the form

$$u(x, t) = f\left(t - \frac{x}{a}\right).$$

In the case of a steady system, where the time dependence is determined by the factor $e^{i\omega t}$, the plane wave has the form

$$u(x, t) = Ae^{i(\omega t - kx)}, \tag{9}$$

where $k = \omega/a$ is the wave number, $|A|$ is the amplitude.

The plane wave, propagating in the direction $\boldsymbol{l}$, where $\boldsymbol{l}(l_x, l_y, l_z)$ is the unit vector, may be written in the following way:

$$u(x, y, z, t) = Ae^{i[\omega t - k(xl_x + yl_y + zl_z)]} = Ae^{i[\omega t - k\boldsymbol{l}.\boldsymbol{r}]}. \tag{10}$$

The functions

$$v(x) = Ae^{-ikx}, \qquad v(x, y, z) = Ae^{-ik\boldsymbol{l}.\boldsymbol{r}}, \tag{11}$$

which are the solutions of the wave equation

$$\nabla^2 v + k^2 v = 0. \tag{12}$$

are also usually called *plane waves*.

In the mathematical theory of diffraction one usually studies the excitations of a field in a homogenous medium, produced by the presence of impurities T_i, disturbing the homogeneity of the medium. Let $\bar{v}(M)$ be a field in a homogeneous medium, produced by given sources which we assume distributed outside the regions T_i $(i = 1, \ldots, n)$; and are sufficiently removed so that in the neighbourhood of the impurities $\bar{v}(M)$ can be represented by a plane wave,

$$\bar{v}(x, y, z) = Ae^{-ik\boldsymbol{l}.\boldsymbol{r}}. \tag{13}$$

The actual field v_0, occuring in region T_0 in the presence of inhomogeneities, may be represented as the sum

$$v_0(M) = w_0(M) + \bar{v}_0(M),$$

where $\bar{v}_0(M)$ is the "incident wave", $w_0(M)$ is the diffracted wave, representing the excitation of the external field $\bar{v}$ by the inhomogenities T_i.

We shall look for for the diffracted field $w_0(M)$ in the region T_0, and the "refracted field" inside T_i. Let us give the conditions, defining the unknown functions w_0 and v_i ($i = 1, 2, \ldots, n$):

(a) the functions w_0 and v_i satisfy the equations

$$\nabla^2 w_0 + k_0^2 w_0 = 0 \quad \text{in} \quad T_0, \tag{14}$$
$$\nabla^2 v_i + k_i^2 v_i = 0 \quad \text{in} \quad T_i \; (i = 1, 2, \ldots, n);$$

(b) at the boundaries of separation Σ_i of the regions T_i and T_0 the following matching conditions are fulfilled:

$$v_i = w_0 + \bar{v}_0 \quad \text{on} \quad \Sigma_i, \tag{15}$$

where $\bar{v}_0$ is a given function,

$$p_i \frac{\partial v_i}{\partial \nu} = p_0 \frac{\partial w_0}{\partial \nu} + f_i \quad \text{on} \quad \Sigma_i, \tag{16}$$

where $f_i = \varrho_0 \dfrac{\partial \bar{v}_0}{\partial \nu}$ is a given function;

(c) the diffracted wave $w_0(M)$ is at infinity a divergent spherical wave, i. e. it satisfies the radiation condition

$$w_0(M) = O\left(\frac{1}{r}\right),$$

$$\frac{\partial w_0}{\partial r} + ikw_0 = o\left(\frac{1}{r}\right).$$

2. Let us consider in more detail the diffraction of a plane wave by a sphere.* A plane wave travelling in the direction of the z-axis

$$\bar{v} = Ae^{-ikz} \tag{17}$$

falls on a sphere of radius R with centre at the origin of coordinates. We can expand the reflected and refracted field in a series of spherical functions: First we expand $\bar{v}_0$ and $f = p_0 \dfrac{\partial \bar{v}_0}{\partial r}$, appearing on the right-hand side of the matching conditions.

Let us assume $z = r \cos \theta$; then it is possible to use the following expansion of a plane wave in spherical functions:

$$e^{-ikr\cos\theta} = \sum_{m=0}^{\infty} (2m+1)(-i)^m \psi_m(kr) P_m(\cos\theta), \tag{18}$$

where

$$\psi_m(kr) = \sqrt{\left(\frac{\pi}{2kr}\right)} J_{m+\frac{1}{2}}(kr),$$

and $J_{m+1/2}(kr)$ is a Bessel function of the first kind of the $(m+1/2)$th order, $P_m(kr)$ is a Legendre polynomial of the m th order. In fact, the left-hand

* Similar methods are often used in quantum mechanics in problems on the scattering of particles.

side gives a solution of the wave equation, depending only on z. Any solution of the wave equation may be represented as the sum of the products of spherical functions and $\psi_m(kr)$. Since in our case the left-hand side of (18) possesses axial symmetry, then

$$e^{-ikr\cos\theta} = \sum_{\infty}^{m=0} C_m \psi_m(kr) P_m(\cos\theta), \tag{19}$$

where C_m are coefficients, indeterminate as yet. Utilizing the orthogonality of the Legendre polynomials and their norm (see Supplement, Part II), we obtain:

$$C_m \psi_m(\varrho) = \frac{2m+1}{2} \int_{-1}^{1} e^{-i\varrho\xi} P_m(\xi)\, d\xi \tag{20}$$

$$(\varrho = kr, \quad \xi = \cos\theta).$$

Let us determine the first term of the asymptotic representation for the integral on the right-hand side. Comparing it with the first term of the asymptotic expansion of function $\psi_m(\varrho)$ enables us to determine C_m. We integrate m times by parts, integrating each time $e^{-i\varrho\xi}$ and differentiating $P_m(\xi)$.

As a result we obtain an expansion of the integral in powers of $1/\varrho$ Retaining only the first term of the expansion, we have:

$$\int_{-1}^{+1} e^{-i\varrho\xi} P_m(\xi)\, d\xi \cong \frac{1}{-i\varrho}[e^{-i\varrho\xi} P_m(\xi)]_{-1}^{+1} =$$

$$= \frac{1}{-i\varrho}\left(e^{-i\varrho} P_m(1) - e^{i\varrho} P_m(-1)\right) = \frac{1}{-i\varrho}(e^{-i\varrho} - (-1)^m e^{i\varrho}) =$$

$$= \frac{1}{-i\varrho}(e^{-i\varrho} - e^{-im\pi} e^{i\varrho}) =$$

$$= \frac{e^{-im\frac{\pi}{2}}}{-i\varrho}\left[e^{-i\left(\varrho - m\frac{\pi}{2}\right)} - e^{i\left(\varrho - m\frac{\pi}{2}\right)}\right] = 2(-i)^m \frac{\sin\left(\varrho - \frac{m\pi}{2}\right)}{\varrho}.$$

On the other hand, as is well known (see Supplement, Part II, § 3, subsection 2),

$$\psi_m(\varrho) \cong \frac{\sin\left(\varrho - \frac{m\pi}{2}\right)}{\varrho}.$$

Comparing these expressions, we find from (20):

$$C_m = (2m+1)(-i)^m, \tag{21}$$

which proves formula (18).

From (17) it follows:

$$\left.\begin{aligned} \bar{v}_0 |_{r=R} &= \sum_{m=0}^{\infty} a_m P_m(\cos\theta); \\ a_m &= A(2m+1)(-i)^m \psi_m(k_0 R); \end{aligned}\right\} \tag{22}$$

$$\left.\begin{aligned} p_0 \frac{\partial \bar{v}_0}{\partial r}\bigg|_{r=R} &= \sum_{m=0}^{\infty} b_m P_m(\cos)\theta); \\ b_m &= A k_0 p_0 (2m+1)(-i)^m \psi_m'(k_0 R). \end{aligned}\right\} \tag{23}$$

The reflected and refracted fields are solutions of the wave equation and, like the incident field, possess axial symmetry. Therefore the functions v_1 and w_0 are taken in the form

$$v_1 = \sum_{m=0}^{\infty} \alpha_m \psi_m(k_1 r) P_m(\cos\theta), \tag{24}$$

$$w_0 = \sum_{m=0}^{\infty} \beta_m \zeta_m(k_0 r) P_m(\cos\theta), \tag{25}$$

$$\zeta_m(\varrho) = \sqrt{\left(\frac{\pi}{2\varrho}\right)} H^{(2)}_{m+\frac{1}{2}}(\varrho). \tag{26}$$

Let us proceed now to the determination of the expansion coefficients α_m and β_m. Utilizing the matching condition and comparing the coefficients with $P_m(\cos\theta)$, we obtain:

$$\alpha_m \psi_m(k_1 R) - \beta_m \zeta_m(k_0 R) = a_m = A(2m+1)(-i)^m \psi_m(k_0 R),$$

$$p_1 k_1 \alpha_m \psi_m'(k_1 R) - p_0 k_0 \beta_m \zeta_m'(k_0 R) =$$

$$= b_m = A k_0 p_0 (2m+1)(-i)^m \psi_m'(k_0 R),$$

from which

$$\alpha_m = A(2m+1)(-i)^m \frac{p_0 k_0 [\psi_m(k_0 R)\zeta_m'(k_0 R) - \zeta_m(k_0 R)\psi_m'(k_0 R)]}{p_0 k_0 \psi_m(k_1 R)\zeta_m'(k_0 R) - p_1 k_1 \psi_m'(k_1 R)\zeta_m(k_0 R)} \tag{27}$$

$$\beta_m = A(2m+1)(-i)^m \frac{p_1 k_1 \psi_m(k_0 R)\psi_m'(k_1 R) - p_0 k_0 \psi_m'(k_0 R)\psi_m(k_1 R)}{p_0 k_0 \psi_m(k_1 R)\zeta_m'(k_0 R) - p_1 k_1 \psi_m'(k_1 R)\zeta_m(k_0 R)} \tag{28}$$

3. Let us consider as an example the problem of the scattering of sound by a solid spherical obstacle. A plane sound wave, propagating in the direction of the z-axis is incident, on a perfectly rigid and stationary sphere of radius R with centre at the origin of coordinates. The sound pressure $p(x, y, z, t)$ (Chapter II, § 1), satisfies the wave equation

$$\frac{\partial^2 p}{\partial t^2} = a^2 \nabla^2 p, \quad a^2 = \gamma \frac{p_0}{\varrho_0},$$

where a is the velocity of sound, γ is the index of the adiabatic law, p_0 and ϱ_0 are the pressure and density of the medium in the unperturbed state.

The pressure in the incident plane wave is given by the function

$$\bar{p}_0 = Ae^{-i(\omega t - kz)} \qquad (k = \omega/a),$$

where A is a constant.

Considering a steady process

$$p(x, y, z, t) = p(x, y, z)\, e^{-i\omega t},$$

we obtain the wave equation for $p(x, y, z)$

$$\nabla^2 p + k^2 p = 0.$$

On the surface of the sphere S_R, because of its absolute rigidity, the normal component of the velocity u should equal zero. The projection of the velocity in the direction of the normal n is related to the pressure by the following equation:

$$\frac{\partial u_n}{\partial t} = -\frac{1}{\varrho}\frac{\partial p}{\partial n},$$

which in the steady case gives

$$u_n = \frac{1}{i\,\omega\varrho}\frac{\partial p}{\partial n}.$$

Hence we obtain the boundary condition

$$\left.\frac{\partial p}{\partial n}\right|_{S_R} = 0.$$

Assuming $p = \bar{p}_0 + w$, where $w(x, y, z)$ is the pressure of the scattered wave, we obtain the following relations for w

(a) the function $w(x, y, z)$ satisfies the wave equation

$$\nabla^2 w + k^2 w = 0;$$

(b) on the surface of the sphere S_R the boundary condition

$$\left.\frac{\partial w}{\partial n}\right|_{S_R} = -\left.\frac{\partial \bar{p}_0}{\partial n}\right|_{S_R}$$

is fulfilled;

(c) the scattered wave w is known at infinity as a divergent spherical wave, i.e. it satisfies the radiation condition for $r \to \infty$,

$$w(M) = O\left(\frac{1}{r}\right),$$

$$\frac{\partial w}{\partial r} + ikw = o\left(\frac{1}{r}\right).$$

It is readily seen that this problem is a particular case of the diffraction problem considered above and corresponds to a value of the parameter $p_1 = 0$.

Assuming in (25) and (28) $p_1 = 0$, we obtain:

$$\beta_m = -A(2m+1)(-i)^m \frac{\psi'_m(k_0 R)}{\zeta'_m(k_0 R)} \tag{29}$$

and

$$w = -A \sum_{m=0}^{\infty}{}' (2m+1)(-i)^m \frac{\psi'_m(k_0 R)}{\zeta'_m(k_0 R)} \zeta_m(k_0 r) P_m(\cos\theta). \quad (30)$$

If the wavelength is large in comparison with the dimensions of the sphere, i.e. $k_0 R \ll 1$, then in (29) it is possible to use the expansions of the functions $\psi_m(kr)$ and $\zeta_m(kr)$ in series, which follows from the expansion of the functions $J_{m+1/2}(kR)$ and $H^{(2)}_{m+1/2}(kR)$ in powers of the small argument kR (see Supplement, Part I, §§ 1 and 3):

$$\psi_0(kR) \cong \sqrt{\left(\frac{\pi}{2kR}\right)} \left(\frac{\left(\frac{1}{2}kR\right)^{\frac{1}{2}}}{\Gamma\left(\frac{3}{2}\right)} - \frac{\left(\frac{1}{2}kR\right)^{\frac{5}{2}}}{\Gamma\left(\frac{5}{2}\right)} \right), \quad \psi_1(kR) \cong \sqrt{\left(\frac{\pi}{2kR}\right)} \frac{\left(\frac{1}{2}kR\right)^{\frac{3}{2}}}{\Gamma\left(\frac{5}{3}\right)},$$

$$\zeta_0(kR) \cong \sqrt{\left(\frac{\pi}{2kr}\right)} \frac{\left(\frac{kR}{2}\right)^{-\frac{1}{2}}}{\Gamma\left(\frac{3}{3}\right)}, \quad \zeta_1(kR) \cong -i\sqrt{\left(\frac{\pi}{2kR}\right)} \frac{\left(\frac{kR}{2}\right)^{-\frac{3}{2}}}{\Gamma\left(-\frac{1}{2}\right)}.$$

Since

$$\Gamma\left(\frac{1}{2}\right) = \sqrt{\pi}, \quad \Gamma\left(\frac{3}{2}\right) = \frac{\sqrt{\pi}}{2}. \quad \Gamma\left(\frac{5}{2}\right) = \frac{3}{4}\sqrt{\pi}, \quad \Gamma\left(-\frac{1}{2}\right) = -2\sqrt{\pi},$$

then we obtain:

$$\psi_0(kR) \cong 1 - \frac{(kR)^2}{6}, \quad \psi_1(kR) \cong \frac{kR}{3},$$

$$\zeta_0(kR) \cong \frac{i}{kR}, \quad \zeta_1(kR) \cong \frac{i}{(kR)^2},$$

from which it follows:

$$\psi'_0(kR) \cong -\frac{kR}{3}, \quad \psi'_1(kR) \cong \frac{1}{3},$$

$$\zeta'_0(kR) \cong -\frac{i}{(kR)^2}, \quad \zeta'_1(kR) \cong -\frac{2i}{(kR)^3}.$$

Substituting the expressions determined for ψ'_m and ζ'_m in (29) we find:

$$\beta_0 = i\frac{A}{3}(kR)^3, \quad \beta_1 = -\frac{A}{2}(kR)^3.$$

We see that the next coefficients are proportional to $(kR)^5$, therefore for wave lengths ($kR \ll 1$) the perturbation w is approximately represented by the first two terms of series (30)

$$\left.\begin{aligned} & w \cong \beta_0 \zeta_0(kr) + \beta_1 \zeta_1(kr)\cos\theta \\ & [P_0(\cos\theta) = 1, \quad P_1(\cos\theta) = \cos\theta]. \end{aligned}\right\} \quad (31)$$

At large distances from the excited sphere ($kr \gg 1$) in the so-called "distant" or "wave" zone we have the asymptotic representations for the functions $\zeta_0(kr)$ and $\zeta_1(kr)$

$$\zeta_0(kr) \cong \frac{i}{kr} e^{-ikr}, \quad \zeta_1(kr) \cong -\frac{1}{kr} e^{-ikr}, \tag{32}$$

which follow from the asymptotic representations of the Hankel functions.

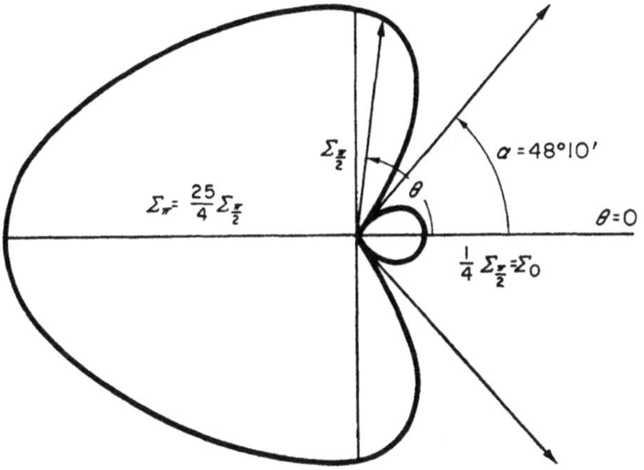

FIG. 87

Substituting in (31) expressions (32) for $\zeta_0(kr)$ and $\zeta_1(kr)$ and replacing β_0 and β_1 by their approximate values, we obtain:

$$w \cong -\frac{Ak^2 R^3}{3r}\left(1 - \frac{3}{2}\cos\theta\right) e^{-ikr}. \tag{33}$$

Let us turn now to a calculation of the intensity of the scattered wave; this is the mean value of the energy flow *(Umov vector)*, equal to the product of the excess sound pressure w and the velocity u, where u and w are the real parts of the appropriate expressions. In our case

$$\left.\begin{aligned} w &= w_0 \cos(\omega t - kr), \\ u &= u_0 \cos(\omega t - kr), \end{aligned}\right\} \tag{34}$$

where w_0 and u_0 are the appropriate amplitudes.

Let us evaluate the intensity of sound I in the wave zone, retaining the main terms of the asymptotic expansions

$$I = \frac{u_0 w_0}{2T} \int_0^T \cos^2(\omega t - kr)\, dt = \frac{u_0 w_0}{2} \qquad \left(T = \frac{2\pi}{\omega} - \text{period}\right).$$

From the equation of motion

$$\frac{\partial u}{\partial t} = -\frac{1}{\varrho}\frac{\partial w}{\partial r}$$

and formula (34) we find:

$$u_0 = \frac{w_0}{a\varrho}.$$

Thus

$$I = \frac{w_0^2}{2a\varrho} = \frac{A^2 k^4 R^6}{18a\varrho r^2}\left(1 - \frac{3}{2}\cos\theta\right)^2.$$

Denoting the power, scattered by the sphere in a cone $d\theta$, by

$$2\pi r^2 \Sigma(\theta) \sin\theta \, d\theta ,$$

we have:

$$\Sigma(\theta) = \frac{A^2 k^4 R^6}{18a\varrho}\left(1 - \frac{3}{2}\cos\theta\right)^2.$$

The polar diagram of the intensity of sound scattered from a sphere is given in Fig. 87. If

$$\cos\theta = +\,2/3, \quad \theta = \alpha = 48^\circ\, 10',$$

then there is no scattering in the direction $\theta = \alpha$.

Problems on Chapter VII

1. Find the source function of a stationary point source of gas, assuming that the gas decays in the process of diffusion. Solve the problem for diffusion in space and in a plane.

2. Solve the same problem in the semi-plane $y > 0$, assuming that at $y = 0$ the concentration equals zero.

3. (a) Solve the interior and exterior problems for the equation

$$\nabla^2 u - \varkappa^2 u = 0 ,$$

if on the sphere $r = r_0$ there is given the boundary condition $u|_{r=r_0} = A = A\cos\theta$.

In the case of the exterior problem state the conditions at infinity, ensuring the uniqueness of the solution.

Consider similar problems, assuming that

$$u|_{r=r_0} = F(\theta) .$$

(b) Solve similar problems for the equation with two independent variables, where the boundary conditions are given on a circle of radius r_0 and have the form

$$u|_{r=r_0} = A\cos\varphi$$

and, move generally,

$$u|_{r=r_0} = F(\varphi) .$$

4. Solve problems 3 (a), (b) for the equation

$$\nabla^2 u + k^2 u = 0 .$$

In the case of the interior problem, find the values of r_0 for which there is a unique solution (assume k given).

Formulate conditions for a solution for both two and three independent variables.

5. At a depth h under the surface of the earth there is a medium, in which a radioactive substance is distributed with constant density. Find the concentration of the radiation, assuming that its concentration at the surface equals zero.

6. Find the natural frequencies of a membrane, having the shape of a ring, whose radii equal a and b $(a < b)$, assuming that $v|_{r=a} = 0$ and $v|_{r=b} = 0$. Show that the limit of the first eigen-value as $a \to 0$ equals the first eigen-value of a circular membrane of radius b with a fixed boundary.

7. Find the natural oscillations and the frequencies for a cavity of cylindrical shape, assuming the walls of the cavity ideally conducting. Consider the same problem from an acoustic viewpoint.

Hint. In the case of electromagnetic oscillations introduce the polarization potential (see Appendix I to Chapter VII).

8. Determine the electromagnetic field of a point dipole in infinite space, assuming that the time variation of the field is proportional to $e^{i\omega t}$. Investigate the asymptotic behaviour of the solution at large distances (in the wave zone). Solve the same problem for a dipole, above a perfectly conducting surface (vertical dipole).

Hint. Introduce the polarization potential.

9. Set up the problem of propagation of electromagnetic waves inside an infinite cylindrical waveguide of arbitrary cross-section with ideally conducting walls. Consider the wave of electric type, of longest wavelength, propagating along a circular cylindrical waveguide. Find the field, and calculate the energy flow across a section, perpendicular to the arcis (see Appendix I to Chapter VII).

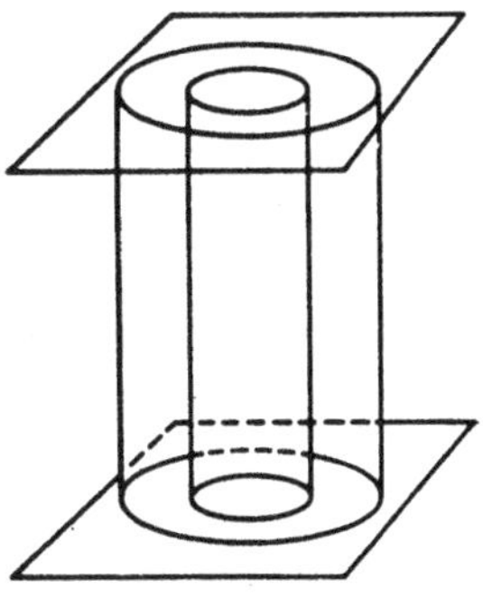

Fig. 88

10. Solve the inhomogeneous equation

$$\nabla^2 u + k^2 u = -f$$

in an infinite cylindrical region of circular cross-section, with homogeneous boundary conditions of first or second kind on the surface, and find the source function (see Appendix II to Chapter VII).

11. Find the source function in the case of the first boundary-value problem for the equation

$$\nabla^2 u + k^2 u = 0$$

(a) in the semi-space $z > 0$;
(b) in the semi-plane $y > 0$;
(c) inside the layer $-l \leqslant z \leqslant l$.

12. Solve the problem of the diffraction of a plane electromagnetic wave by an infinitely ideally-conducting cylinder. Solve the same problem in acoustics.

13. Find the characteristic electromagnetic oscillations of a spherical vibrator with ideally conducting walls. Consider the case of oscillations of *TE* and *TM* type (see Appendix II to Chapter VII).

14. Find the characteristic electromagnetic oscillations of a cavity consisting of a region enclosed between two coaxial cylindrical surfaces and two planes, perpendicular to the axis of the cylinders (Fig. 88).

Hint. For the polarization potential $\Pi_{n,m}$ use the relation, similar to (14) of Appendix II to Chapter VII.

APPENDICES TO CHAPTER VII

I. Waves in cylindrical guides

1. In the design of a certain kind of transmitter one has to solve the problem of the transmission of electromagnetic energy from the transmitter to the transmitting aerial or, conversely, from the aerial to the receiver. Problems of relaying the electromagnetic energy are also met in a number of other practical problems of modern radio-engineering.

Formerly this problem was satisfactorily solved by means of a two-wire line, consisting of two metal conductors, between which an electromagnetic wave is propagated. However, it appears that along with the disadvantages, inherent in general in transmission lines, such a two-wire line radiates electromagnetic energy, this radiation increasing with increase in frequency of the radiowave. Therefore such a form of transmission line becomes less suitable in the region of ultra-short waves.

Now other transmitting devices are employed in the technique of ultra-short (centimetre and micro) waves for the transmission of energy, hollow metal pipes *(wave guides)*, inside which the propagation of radiowaves takes place. Such transmitting devices, having small losses, are very suitable transmission lines.*

The mathematical theory of the propagation of radiowaves in pipes was begun by Rayleigh, who investigated the propagation of acoustic waves in pipes. There has been an intense development in the theory of waveguides, especially in the work of

* B. A. Vvedenskii and Y. G. Arenberg, *Waveguides*, part I, Gostekhizdat, 1946.

Soviet scientists. The properties of circular, rectangular and other forms of waveguides have been throughly investigated.

Let us first consider the properties of waveguides of arbitrary cross-section, and subsequently illustrate them by a number of concrete examples. Thus, we consider a cylindrical pipe, extending indefinitely along the z-axis. We shall assume the walls of the pipe to be ideally conducting. Let us denote by Σ the surface, S the cross-section of the pipe and by C the contour, bounding this section. We assume that: (1) the characteristics of the medium, filling such a waveguide, ε and μ, equal 1, $\sigma = 0$; (2) inside the waveguide, field sources are absent; (3) the field varies periodically according to the law $e^{-i\omega t}$.

Maxwell's equations in this case take the form

$$\left.\begin{aligned} &\operatorname{curl} \boldsymbol{H} = -ik\boldsymbol{E}, \\ &\operatorname{curl} \boldsymbol{E} = ik\boldsymbol{H}. \\ &\operatorname{div} \boldsymbol{H} = 0, \\ &\operatorname{div} \boldsymbol{E} = 0. \end{aligned} \quad \left(k = \frac{\omega}{c}\right) \right\} \tag{1}$$

Since the walls of the waveguide are ideally conducting, the tangential component E_t along the waveguide wall equals zero

$$E_t|_\Sigma = 0. \tag{2}$$

Let us show that *travelling electromagnetic waves can be propagated inside the waveguide*. We shall look for a solution of equation (1) in the form

$$\left.\begin{aligned} \boldsymbol{E} &= \operatorname{grad} \operatorname{div} \boldsymbol{\Pi} + k^2 \boldsymbol{\Pi}, \\ \boldsymbol{H} &= -ik \operatorname{curl} \boldsymbol{\Pi}, \end{aligned}\right\} \tag{3}$$

where $\boldsymbol{\Pi}$ is the polarization potential. Let us consider the case, where the vector $\boldsymbol{\Pi}$ has only one component, directed along the z-axis. In this case equations (1) after substitution of expressions (3) in them give:

$$\nabla^2 \Pi + k^2 \Pi = 0 \text{ or } \nabla_2^2 \Pi + \frac{\partial^2 \Pi}{\partial z^2} + k^2 \Pi = 0 \tag{4}$$

$$(\boldsymbol{\Pi} = \Pi \boldsymbol{i}_z).$$

Condition (2) will be fulfilled, if

$$\Pi|_\Sigma = 0. \tag{5}$$

We seek a solution in the form

$$\Pi(M, z) = \psi(M) f(z), \tag{6}$$

where M is a point, lying on the cross-section S. Substituting (6) in (4), we find that $\psi(M)$ is an eigen-function of the problem of the oscillations of a membrane, with a fixed boundary, i.e.

$$\left.\begin{array}{l} \nabla_2^2 \psi + \lambda\psi = 0 \quad \text{inside} \quad S, \\ \psi|_c = 0 . \end{array}\right\} \tag{7}$$

Here $\nabla_2^2 = \frac{\partial^2}{\partial x^2} + \frac{\partial^2}{\partial y^2}$ is the two-dimensional Laplacian operator.

Let us denote by $\{\lambda_n\}$ and $\{\psi_n\}$ the system of eigenvalues and eigen-functions of this problem. The particular solution of problem (4) has the form

$$\Pi_n(M, z) = \psi_n(M) f_n(z),$$

where the function $f_n(z)$ is determined from the equation

$$f_n'' + (k^2 - \lambda_n) f_n = 0 . \tag{8}$$

The general solution of equation (8) is

$$f_n(z) = A_n e^{i\gamma_n z} + B_n e^{-i\gamma_n z} \qquad (\gamma_n = \sqrt{[k^2 - \lambda_n]}) . \tag{9}$$

The term $A_n e^{i\gamma_n z}$ corresponds to a wave, travelling in the positive direction of the z-axis, and the second term in formula (9) to a wave, travelling in the opposite direction.

Considering only the wave, travelling in one direction, we assume

$$f_n(z) = A_n e^{i\gamma_n z},$$

then we obtain a solution in the form

$$\Pi_n(M, z) = A_n \psi_n(M) f_n(z), \tag{10}$$

where A_n is a constant, obtainable from the conditions of excitation of the fields.

Substituting expression (10) in formula (3) and restoring the factor $e^{-i\omega t}$ we find the components of the field in the form

$$\Pi_n(M, z) = F_n(M) e^{i(\gamma_n z - \omega t)}, \tag{11}$$

where F_n is a function, expressed in terms of the eigen-function of the membrane $\psi_n(M)$ or its derivative.

If $k^2 > \lambda_n$, then γ_n is real and expression (11) represents a travelling wave, propagating along the z-axis with a phase velocity

$$v = \frac{\omega}{\sqrt{(k^2 - \lambda_n)}} = \frac{c}{\sqrt{\left(1 - \frac{\lambda_n}{k^2}\right)}} > c\,.$$

The group velocity of the wave equals

$$u = \frac{c^2}{v} = c\sqrt{\left(1 - \frac{\lambda_n}{k^2}\right)} < c\,,$$

i.e. in a hollow waveguide dispersion occurs.

If $k^2 < \lambda_n$, then $\gamma_n = i\,\varkappa_n$ $(\varkappa_n > 0)$ and in place of (11) we obtain an attenuated wave

$$F_n(M)\, e^{-i\omega t - \varkappa_n z}, \tag{12}$$

propagating along the z-axis in the positive direction.

Since the characteristic frequencies λ_n of the membrane increase indefinitely with increase in n, then whatever the frequency ω may be, starting from some number $n = N$, we have:

$$k^2 < \lambda_n\,.$$

Hence it is possible to propagate only a finite number of travelling waves in the waveguide. If $k^2 < \lambda_1$ then not even one travelling wave can exist in the waveguide.

In order that in a waveguide of a given shape and dimensions one travelling wave can be propagated the condition

$$\lambda_1 < k^2 \text{ or } \Lambda < \frac{2\pi}{\sqrt{\lambda_1}}\,,$$

should be fulfilled, where Λ is the wavelength, propagating in the pipe.

For a waveguide of rectangular cross-section with sides a and b, we have:

$$\lambda_n = \lambda_{mn} = \pi^2 \left(\frac{m^2}{a^2} + \frac{n^2}{b^2}\right) \qquad \left(\begin{array}{c} m = 1, 2, \ldots \\ n = 1, 2 \ldots \end{array}\right), \tag{13}$$

and therefore, a travelling wave can exist only on condition

$$k > \pi \sqrt{\left(\frac{1}{a^2} + \frac{1}{b^2}\right)} \text{ or } \Lambda < \frac{2}{\sqrt{\left(\frac{1}{a^2} + \frac{1}{b^2}\right)}}\,. \tag{14}$$

The solutions of Maxwell's equations can also be fields with the z-component of the electric field equal to zero

$$E_z = 0\,. \tag{15}$$

Introducing the vector $\hat{\boldsymbol{\Pi}} = \hat{\Pi}\, i_z$ and assuming

$$\hat{\boldsymbol{E}} = ik \operatorname{curl} \hat{\boldsymbol{\Pi}}\,; \quad \hat{\boldsymbol{H}} = \operatorname{grad} \operatorname{div} \hat{\boldsymbol{\Pi}} + k^2 \hat{\boldsymbol{\Pi}} \tag{16}$$

$$(\hat{E}_z = 0),$$

we verify that the function $\Pi\,(M, z)$ should be determined from the equation

$$\nabla^2 \hat{\Pi} + k^2 \hat{\Pi} = 0 \;\text{ or }\; \nabla_2^2 \hat{\Pi} + \frac{\partial^2 \hat{\Pi}}{\partial z^2} + k^2 \hat{\Pi} = 0 \tag{17}$$

and the boundary condition

$$\frac{\partial \hat{\Pi}}{\partial \nu} = 0 \text{ on } \Sigma \tag{18}$$

Repeating the arguments deduced above, we find the solution of this problem

$$\hat{\Pi}_n = \hat{A}_n \hat{\psi}_n (M)\, e^{i\hat{\gamma}_n z} \quad (\hat{\gamma}_n = \sqrt{[k^2 - \hat{\lambda}_n]})\,, \tag{19}$$

to which there correspond solutions of Maxwell's equations of the form

$$\hat{F}_n (M)\, e^{i(\hat{\gamma}_n z - \omega t)}.$$

Here $\hat{\psi}_n\,(M)$ and $\hat{\lambda}_n$ denote the eigen-functions and eigen-values of a membrane S with free boundaries

$$\nabla_2^2 \hat{\psi}_n + \hat{\lambda}_n \hat{\psi}_n = 0 \text{ at } S\,,$$

$$\frac{\partial \hat{\psi}_n}{\partial \nu} = 0 \text{ on } C\,.$$

Thus, there can exist in the waveguide electromagnetic fields of two kinds $\{\bar{\boldsymbol{E}}, \bar{\boldsymbol{H}}\}$ and $\{\hat{\boldsymbol{E}}, \hat{\boldsymbol{H}}\}$, defined by formulae (3) and (16). The following terminology is adopted: we speak of electric waves (or TM waves), if $H_z = 0$, or of magnetic waves (TE type), if $E_z = 0$. We have verified that in the waveguide TE and TM waves can exist. It is possible to show that any field in the waveguide can be represented as a sum of TE and

TM fields. Hence it follows that an arbitrary field in the waveguide can be determined, if the two scalar functions $\Pi(M, z)$ and $\hat{\Pi}(M, z)$ are known.

2. Let us determine the energy, carried by the travelling wave, for instance of the TM type.

In order to do this we calculate the flux of the Umov–Poynting vector across the section S:

$$W_z = \frac{c}{8\pi} \iint_S [\boldsymbol{E}\boldsymbol{H}^*]_z \, dS, \tag{20}$$

where $\boldsymbol{H}^*$ is the complex-conjugate of the vector $\boldsymbol{H}$, S is a perpendicular section of the waveguide.

We introduce a rectangular system of coordinates, x, y, z. Then

$$W_z = \frac{c}{8\pi} \iint_S (E_x H_y^* - E_y H_x^*) \, dx \, dy. \tag{21}$$

Let us express the field components in terms of the polarization potential Π by the relations

$$E_x = \frac{\partial^2 \Pi}{\partial x \, \partial z}, \qquad E_y = \frac{\partial^2 \Pi}{\partial y \, \partial z},$$

$$H_x^* = ik \frac{\partial \Pi^*}{\partial y}, \qquad H_y^* = -ik \frac{\partial \Pi^*}{\partial x}$$

and substitute their values in (21)

$$W_z = -\frac{c}{8\pi} ik \iint_S \left(\frac{\partial^2 \Pi}{\partial x \, \partial z} \frac{\partial \Pi^*}{\partial x} + \frac{\partial^2 \Pi}{\partial y \, \partial z} \frac{\partial \Pi^*}{\partial y} \right) dx \, dy. \tag{22}$$

The function Π and its conjugate function Π^* according to (10) are represented in the form

$$\Pi(M, z) = A_n \psi_n(M) \, e^{i\gamma_n z},$$
$$\Pi^*(M, z) = A_n^* \psi_n(M) \, e^{-i\gamma_n z},$$

where ψ_n is the eigen-function which is zero on C $(\psi_n|_c = 0)$. Hence it follows that in place of (22) one may write

$$W_z = \frac{ck}{8\pi} \gamma_n |A_n|^2 \iint_S \left[\left(\frac{\partial \psi_n}{\partial x} \right)^2 + \left(\frac{\partial \psi_n}{\partial y} \right)^2 \right] dx \, dy =$$
$$= \frac{ck}{8\pi} \gamma_n |A_n|^2 \int_S (\nabla \psi_n)^2 \, dS.$$

Applying Green's first theorem

$$\iint\limits_S (\nabla\psi_n)^2\,dS = -\iint\limits_S \psi_n\,\nabla_2^2\psi_n\,dS + \int\limits_C \psi_n \frac{\partial\psi_n}{\partial\nu}\,ds = \lambda_n \iint\limits_S \psi_n^2\,dS = \lambda_n,$$

we obtain an expression for the energy flow of the travelling wave of number n

$$W_z = \frac{ck}{8\pi}\,|A_n|^2\,\gamma_n\,\lambda_n\,. \tag{23}$$

If several waves are propagated simultaneously, then W_z will equal the sum of components of type (23).

We pass now to the problem of exciting of electromagnetic fields in a waveguide.*

3. In a certain volume V_0 inside a waveguide Σ there are given currents $\boldsymbol{j}(M, z)e^{-i\omega t}$, varying harmonically with time. Let us determine the field, produced by these currents. By virtue of the principle of superposition of fields it is sufficient, to solve the problem of the excitation of the waveguide by an elementary dipole of arbitrary orientation.

In order to describe the method of solving the general problem let us consider the simpler case of the excitation of the waveguide by a linear current $I = I_0\,(z)e^{-i\omega t}$, of length L, parallel to the z-axis.

To determine the electromagnetic fields excited in the waveguide, it is necessary to make use of

1. Maxwell's equation (1).
2. The boundary conditions

$$E_{\tan} = 0 \text{ on } \Sigma\,.$$

3. The radiation conditions, in the form of the requirement of the absence of waves arriving from infinity.
4. The excitation condition, which we write in the form**

$$\oint\limits_{K_\varepsilon} H_s\,ds = \frac{4\pi}{c} I_0 \quad\text{or}\quad H_s \approx \frac{2I_0}{c\varrho}\,, \tag{24}$$

where K_ε is a circle of radius ε ($\varepsilon \to 0$), encompassing the path L, $\varrho = MM_0$, where M_0 is a point in the current, M is a point,

* A. A. Samarskii and A. N. Tikhonov, *Zh. Tekh. Fiz.* 27, copy 11 12, 1947.

** See Chapter V, Appendix II, Subsection 3.

on the circumference of K_ε. In other words, the electromagnetic field at the current should have a singularity of a given kind.

Let us go to the potential Π, using formula (3) for this. Let (M_0, ξ) be an arbitrary point on the current. We choose a cylindrical system of coordinates ϱ, φ, z with centre at the point (M_0, ξ) and let us calculate H_s, using equation (3)

$$H_s = -\frac{\partial \Pi}{\partial \varrho}.$$

Hence from (24) it follows that at the point (M_0, ξ) function Π should have a logarithmic singularity

$$\Pi \approx \frac{2I_0}{c} \ln \frac{1}{\varrho}. \tag{25}$$

Thus $\Pi(M, z)$ should satisfy the wave equation (4), the boundary condition $\Pi = 0$ on Σ, the radiation condition and the excitation condition (25).

We shall look for a solution of this problem in the form

$$\Pi = K \int_L \Pi_0 (M, M_0;\ z, \zeta)\, I_0(\zeta)\, d\zeta, \tag{26}$$

where $\Pi_0 (M, M_0;\ z, \zeta)$ is the source function, defined as the solution of the equation

$$\nabla^2 \Pi_0 + k^2 \Pi_0 = 0$$

in the variables (M, z) and (M_0, ζ), satisfying the boundary condition

$$\Pi_0 = 0 \text{ on } \Sigma,$$

and the radiation condition and having a singularity of the type $\frac{1}{4\pi} \frac{e^{ikr}}{r}$ when the arguments coincide, i.e. representable in the form of the sum

$$\Pi_0 (M, M_0;\ z, \zeta) = \frac{e^{ikr}}{4\pi r} + v (M, M_0;\ z, \zeta)$$

$$(r = \sqrt{[\varrho^2 + (z - \zeta)^2]},\ \varrho = MM_0),$$

where v is a regular function, determinable from the equation and the boundary condition

$$v = -\frac{e^{+ikr}}{4\pi r} \text{ on } \Sigma.$$

The function $\Pi(M, z)$, defined by formula (26), has a logarithmic singularity, and the excitation condition is fulfilled, if the normalizing factor is chosen to be

$$K = 4\pi/c .$$

Hence it follows that

$$\Pi(M, z) = \frac{4\pi}{c} \int_L \Pi_0 (M, M_0 ; z, \zeta) I_0 (\zeta) d\zeta .$$

In particular, for a short element of current of length l

$$\Pi(M, z) = \frac{4\pi}{c} I \cdot l \cdot \Pi_0 .$$

Therefore, Π_0 has the physical meaning of the *polarization potential*, corresponding to the excitation by an element of current, situated at the point (M_0, ζ) parallel to the axis of the waveguide.

Thus, the problem of determining the field in a waveguide is completely reduced to the problem of finding the source function Π_0 of the first boundary value problem for the equation $\nabla^2 u + k^2 u = 0$ in the waveguide.

In order to construct the source function, the method given in Chapter VI, § 2, may be applied. Let us consider the inhomogeneous equation

$$\nabla^2 u + k^2 u = - f(M, z), \tag{27}$$

where $f(M, z)$ is a given function with the boundary condition

$$u|_{\Sigma} = 0 .$$

We shall look for the function $u(M, z)$ in the form of the series

$$u(M, z) = \sum_{n=1}^{\infty} u_n (z) \psi_n (M) , \tag{28}$$

where $\psi_n (M)$ are the normalized eigen-functions of equations (7)

$$\nabla_2^2 \psi_n + \lambda_n \psi_n = 0, \quad \psi_n|_C = 0 . \tag{7}$$

Expanding $f(M, z)$ in the series

$$f(M, z) = \sum_{n=1}^{\infty} f_n (z) \psi_n (M) , \quad f_n (z) = \iint_S f(M', z) \psi_n (M') d\sigma_M , \tag{29}$$

and substituting expressions (28) and (29) in equation (27), we obtain the equation

$$u_n''(z) - p_n^2 u_n(z) = -f_n(z), p_n = \sqrt{(\lambda_n - k^2)}. \tag{30}$$

The solution of this equation is given by the relation

$$u_n(z) = \int_{-\infty}^{+\infty} \frac{e^{-p_n|z-\zeta|}}{2p_n} f_n(\zeta)\, d\zeta, \tag{31}$$

which by (29) may be written in the form

$$u_n(z) = \iint_S \int_{-\infty}^{+\infty} \frac{e^{-p_n|z-\zeta|}}{2p_n} f(M', \zeta)\, \psi_n(M')\, d\sigma_{M'}\, d\zeta. \tag{31'}$$

Substituting this expression in (28) and changing the order of summation and integration, we have:

$$u(M, z) = \iiint_T \Pi_0(M, M', z - \zeta)\, f(M', \zeta)\, d\sigma_{M'}\, d\zeta, \tag{32}$$

where

$$\Pi_0(M, M', z - \zeta) = \sum_{n=1}^{\infty} \frac{\psi_n(M)\, \psi_n(M')}{2p_n} e^{-p_n|z-\zeta|}. \tag{33}$$

The series $\Pi_0(M, M', z - \zeta)$ for $z \neq \zeta$ is uniformly and absolutely convergent by the definition for eigen-functions* and the presence of the exponential factor. The function

* For the eigen-functions $\psi_n(M)$ the inequality $|\psi_n(M)| \leqslant A\lambda_n$ holds, where A is a constant independent either of the point M, or of the suffix n. In fact, the boundary-value problem (7) is equivalent to the integral equation $\psi_n(M) = \lambda_n \iint_S G(M, M')\psi_n(M')\, d\delta M')$ where $G(M, M')$ is the source function for Laplace's equation $\nabla_2^2 u = 0$ with the boundary condition $u|_c = 0$. From this integral equation by virtue of Bunyakovskii's inequality it follows

$$|\psi_n| \leqslant \lambda_n | \sqrt{[\iint_S G^2(M, M')\, d\sigma_{M'} \iint_S \psi_n^2(M')\, d\sigma_{M'}]} \leqslant A|\lambda_n|,$$

since

$$\iint_S \psi_n^2(M')\, d\sigma_{M'} = 1; \qquad \iint_S G^2(M, M')\, d\sigma_{M'} \leqslant A^2.$$

By a similar method, inequalities for the derivatives are obtained

$$\left|\frac{\partial \psi_n}{\partial x}\right| \leqslant B\lambda_n^2, \qquad \left|\frac{\partial \psi_n}{\partial y}\right| \leqslant B\lambda_n^2.$$

$\Pi(M, M', z-\zeta)$ at the point ($M = M'$, $z = \zeta$) has a singularity of the kind $1/r$. We will not give a proof of this statement.* From what has been said above it follows that

$$G(M, M', z-\zeta) = \Pi_0(M, M', z-\zeta),$$

i. e. the source function Π_0 has the form

$$\Pi_0(M, M', z-\zeta) = \sum_{n=1}^{\infty} \frac{\psi_n(M)\,\psi_n(M')}{2\,p_n} e^{-p_n|z-\zeta|}.$$

From formula (33) it follows that the field in this case is represented as a superposition of waves of type (11) and (12). From the note on p. 598 it follows that series (33) will consist of a finite number of components of the form

$$B_n \psi_n(M)\, e^{i\gamma_n|z-\zeta|} \qquad \text{(travelling waves)}$$

$$(\gamma_n = \sqrt{[k^2 - \lambda_n]}\,,\ p_n = -i\gamma_n)$$

and of an infinite number of components of the form

$$B'_n \psi_n(M)\, e^{-p_n|z-\zeta|} \qquad \text{(attenuated waves)},$$

where

$$B'_n = \frac{\psi_n(M')}{2p_n},\ p_n = \sqrt{(\lambda_n - k^2)},\ \lambda_n > k^2.$$

In order to determine the fields it is necessary to make use of formulae (26) and (3).

The problem of the excitation of a waveguide by an elementary magnetic dipole, parallel to the z-axis (an infinitely small current loop with the electric current in the plane $S_{z=\zeta}$) leads us to a second source function

$$\hat{\Pi}_0(M, M'; z-\zeta) = \sum_{n=1}^{\infty} \frac{\hat{\psi}_n(M)\,\hat{\psi}_n(M')}{2\hat{p}_n} e^{-\hat{p}_n|z-\zeta|},$$

$$p_n = \sqrt{(\lambda_n - k^2)},$$

satisfying the boundary condition $\frac{\partial \hat{\Pi}_0}{\partial \nu} = 0$ on Σ. In addition $H_z = 0$; $\hat{\Pi} = \frac{4\pi}{c}\, kl\, \hat{\Pi}_0$ (kl is the moment of the magnetic dipole).

* See A. A. Samarskii and A. N. Tikhonov, *Zh. Tekh. Fiz.* 27, copy 11, 1947.

In a similar way it is possible to solve the problem of excitation by an arbitrarily orientated dipole (element of current). The corresponding functions Π will be given by a formula, similar to formula (33). In the case of surface and volume currents the functions Π are given by surface and volume integrals (by analogy with (26)).

Thus the problem of the excitation of any cylindrical waveguide by arbitrary given currents is completely solved. In order to make use of general formulae for a waveguide of given cross-section, it is sufficient to find the characteristic oscillations of a membrane, having the shape of a perpendicular section of the waveguide i.e. to solve the eigenvalue problem (7) or (19).

Expressions for the normalized eigen-functions of a rectangular membrane of sides a and b are

$$\psi_n(M) = \psi_{nm}(x, y) = \sqrt{\left(\frac{4}{ab}\right)} \sin\frac{\pi m}{a} x \sin\frac{\pi n}{b} y;$$

$$\hat{\psi}_n(M) = \hat{\psi}_{nm}(x, y) = \sqrt{\left(\frac{\varepsilon_m \varepsilon_n}{ab}\right)} \cos\frac{\pi m}{a} x \cos\frac{\pi n}{b} y$$

$$(\varepsilon_j = 2,\ j \neq 0;\ \varepsilon_0 = 1);\ \lambda_{mn} = \pi^2\left(\frac{m^2}{a^2} + \frac{n^2}{b^2}\right).$$

For a circular membrane of radius a we have:

$$\psi_n(M) = \psi_{mn}(r, \varphi) = \sqrt{\left(\frac{\varepsilon_n}{\pi a^2}\right)} \frac{J_n\left(\mu_{mn}\frac{r}{a}\right)}{J'_n(\mu_{mn})} \begin{matrix}\cos\\ \sin\end{matrix}\, n\varphi,$$

$$\hat{\psi}_n(M) = \hat{\psi}_{mn}(r, \varphi) = \sqrt{\left(\frac{\varepsilon_n}{\pi a^2}\right)} \frac{\hat{\mu}_{mn}}{\sqrt{(\hat{\mu}_{mn}^2 - n^2)}} \frac{J_n\left(\frac{\hat{\mu}_{mn}}{a} r\right)}{J_n(\hat{\mu}_{mn})} \begin{matrix}\cos\\ \sin\end{matrix}\, n\varphi,$$

where μ_{mn} is a root of the equation $J_n(\mu) = 0$; $\lambda_{mn} = \frac{\mu_{mn}^2}{a^2}$ $\hat{\mu}_{mn}$ is a root of the equation $J'_n(\mu) = 0$; $\hat{\lambda}_{mn} = \frac{\hat{\mu}_{mn}^2}{a^2}$.

II. Electromagnetic vibrations in hollow resonators

Recently there has been an extensive development of cavity *resonators* which are metal cavities filled with dielectric (in particular, air). In such cavities steady electromagnetic fields (standing waves), can exist.

In radio engineering oi ultra-short waves cavities of a very complicated shape are employed. The general problem of determining the characteristic vibrations of resonators of an arbitrary shape is extremely complex, but for cavities of the simplest shape a solution can be obtained in explicit form. Since the walls are made from highly conducting metal, in the calculation of the characteristic vibrations it is usual to assume the walls to be ideally conducting. Correction for the finite conductivity can be made, using Leontovich's boundary conditions. Later we shall assume that the walls of the vibrator are ideally conducting and all field quantities vary with time according to the lax $e^{-i\omega t}$

Without giving an exhaustive account of the theory of cavities, let us consider some general problems of the theory of these oscillatory systems.

1. *Characteristic oscillations of a cylindrical resonator*

The problem of determining the characteristic electromagnetic oscillations involves finding the non-trivial solutions of Maxwell's equations* or more exactly determining the characteristic frequencies ω, for which the system of homogeneous Maxwellian equations has non-trivial solutions, and finding the solutions.

Maxwell's equations in this case have the form

$$\left.\begin{aligned} \operatorname{curl} \boldsymbol{H} &= -ik\boldsymbol{E}, \\ \operatorname{curl} \boldsymbol{E} &= ik\boldsymbol{H}, \\ \operatorname{div} \boldsymbol{E} &= 0, \\ \operatorname{div} \boldsymbol{H} &= 0 \end{aligned} \quad \left(k = \frac{\omega}{c}\right) \right\} \tag{1}$$

inside the cavity T, with boundary conditions on the surface Σ

$$E_t = 0 \tag{2}$$

or

$$\frac{\partial H_\nu}{\partial \nu} = 0; \tag{3}$$

Both these conditions are equivalent.

Let us find the characteristic oscillations for a cavity, consisting of "a segment" of cylindrical waveguide of arbitrary

* The factor $e^{-i\omega t}$ is omitted everywhere.

section, bounded by two walls at $z = \pm l$ (the z-axis is parallel to the cylinder axis).

Just as in the cylindrical waveguide, oscillations of both electric type ($H_z = 0$) and magnetic type ($E_z = 0$) are possible.

For waves of electric type we assume

$$\left.\begin{aligned} \boldsymbol{E} &= \operatorname{grad}\operatorname{div}\boldsymbol{\Pi} + k^2\boldsymbol{\Pi}, \\ \boldsymbol{H} &= -ik\operatorname{curl}\boldsymbol{\Pi}, \end{aligned}\right\} \tag{4}$$

where $\boldsymbol{\Pi} = \Pi\,\boldsymbol{i}_z$ ($\boldsymbol{i}_z$ is the unit vector, directed along the z-axis) is the polarization vector-potential, whose component along the z-axis only differs from zero. From formula (4) we see that in this case $H_z = 0$.

The function Π, as is usual, satisfies the wave equation

$$\nabla^2\Pi + k^2\Pi = 0. \tag{5}$$

On the surface Σ let us choose a local rectangular system of coordinates $(\boldsymbol{s}, \boldsymbol{\nu}, \boldsymbol{i}_z)$, where $\boldsymbol{\nu}$ is the unit vector, directed along the normal to the surface, $\boldsymbol{s}$ is a tangent to the contour C, bounding the perpendicular section S of the cylindrical resonator.

By the boundary conditions (2) we have:

$$\left.\begin{aligned} E_s|_\Sigma &= \frac{\partial^2\Pi}{\partial s\,\partial z}\bigg|_\Sigma = 0, \\ E_z|_\Sigma &= \left(\frac{\partial^2\Pi}{\partial z^2} + k^2\Pi\right)\bigg|_\Sigma = 0. \end{aligned}\right\} \tag{6}$$

Both these equations will be satisfied if

$$\Pi|_\Sigma = 0. \tag{7}$$

At $z = \pm l$ from (2) we obtain the conditions

$$E_s|_{z=\pm l} = \frac{\partial^2\Pi}{\partial s\,\partial z}\bigg|_{z=\pm l} = 0,$$

$$E_\nu|_{z=\pm l} = \frac{\partial^2\Pi}{\partial \nu\,\partial z}\bigg|_{z=\pm l} = 0,$$

which are fulfilled if

$$\frac{\partial\Pi}{\partial z}\bigg|_{z=\pm l} = 0. \tag{8}$$

Thus, we arrive at the following boundary-value problem:

Find the non-trivial solutions of the wave equation

$$\nabla_2^2 \Pi + \frac{\partial^2 \Pi}{\partial z^2} + k^2 \Pi = 0 \tag{6'}$$

with the homogeneous boundary conditions

$$\Pi|_\Sigma = 0, \tag{7}$$

$$\left.\frac{\partial \Pi}{\partial z}\right|_{z=\pm l} = 0. \tag{8}$$

As in the case of the cylindrical waveguide (see p. 596), we put the solution in the form

$$\Pi(M, z) = \psi(M) f(z). \tag{9}$$

Substituting this expression in equation (6′) and using condition (7), we obtain for $\psi(M)$ the problem of characteristic oscillations of a fixed membrane, having the shape of a perpendicular section S,

$$\nabla_2^2 \psi + \lambda \psi = 0 \text{ in } S, \tag{10}$$

$$\psi = 0 \text{ on } C.$$

The function $f(z)$ satisfies the equation

$$f'' + (k^2 - \lambda) f = 0 \tag{12}$$

with the boundary condition

$$f'(\pm l) = 0, \tag{13}$$

resultant from condition (8).

One should realize that here, in contrast to the problem for waveguides, k^2 can not be arbitrary. We must find those values of k^2 for which the problem (6)–(8) has a non-trivial solution.

Solving equation (12) with conditions (13) we find the eigen-functions

$$f_m(z) = A_m \cos \frac{\pi m}{2l}(l - z),$$

corresponding to the eigen-values

$$\mu_m = \left(\frac{\pi m}{2l}\right)^2 \qquad (m = 0, 1, \ldots),$$

where

$$\mu_m = k_m^2 - \lambda.$$

The boundary value problem (10)-(11) gives a spectrum of eigen-values $\{\lambda_n\}$ with a corresponding system of normalized eigen-functions $\{\psi_n(M)\}$. Hence it follows that there can exists in the cavity only such oscillations, whose characteristic or resonant frequencies equal

$$\omega_{mn} = c\sqrt{(\lambda_n + \mu_m)}\,.$$

To these frequencies there corresponds a system of eigen-functions

$$\Pi_{n,m}(M, z) = \bar{A}_{n,m}\,\psi_n(M)\,\cos\frac{\pi m}{2l}(l - z) \tag{14}$$

or

$$\Pi_{n,m}(M, z) = A_{n,m}\,\psi_n(M)\,f_m(z)\,, \tag{14'}$$

where

$$f_m(z) = \sqrt{\left(\frac{\varepsilon_m}{2l}\right)}\cos\frac{\pi m}{2l}(l - z)\,, \quad \varepsilon_m = \begin{cases} 2; & m \neq 0\,; \\ 1; & m = 0\,, \end{cases}$$

are functions normalized to unity. The solution is given up to an amplitude factor $A_{n,m}$, which is found from the excitation conditions.

If the eigen-functions $\psi_n(M)$ are known, then it is possible to calculate the field components from formulae (4) and (5).

If the cross-section S of the resonator is a rectangle with sides a and b, then we have:

$$\psi_n(M) = \psi_{p,q}(x, y) = \sqrt{\left(\frac{4}{ab}\right)}\sin\frac{\pi p}{a}x\,\sin\frac{\pi q}{b}y$$

$$(p, q = 1, 2, 3, \ldots)\,,$$

$$\lambda_n = \lambda_{p,q} = \pi^2\left(\frac{p^2}{a^2} + \frac{q^2}{b^2}\right)\,,$$

$$\Pi_{n,m} = A_{m,p,q}\sqrt{\left(\frac{2\varepsilon_m}{abl}\right)}\sin\frac{\pi p}{a}x\,\sin\frac{\pi q}{b}y\,\cos\frac{\pi m}{2l}(l - z)\,.$$

In this case the minimum characteristic frequency

$$\omega_{0,1,1,} = c\sqrt{(\lambda_{1,1})} = c\pi\sqrt{\left(\frac{1}{a^2} + \frac{1}{b^2}\right)}$$

corresponds to a maximum permissible wavelength

$$\Lambda_0 = \frac{2}{\sqrt{\left(\frac{1}{a^2} + \frac{1}{b^2}\right)}}\,.$$

In particular, for $b = a$, the maximum wavelength

$$\Lambda_0 = a\sqrt{2}$$

equals the diagonal of the square. Therefore, in such a resonator only characteristic oscillations of frequency

$$\omega \geqslant \omega_{0,1,1}$$

are possible, or wavelengths

$$\Lambda \leqslant \Lambda_0 .$$

By complete analogy the characteristic oscillations of magnetic type ($E_z = 0$) are found. In this case we assume

$$\boldsymbol{E} = ik \operatorname{curl} \hat{\boldsymbol{\Pi}} ,$$

$$\boldsymbol{H} = \operatorname{grad} \operatorname{div} \hat{\boldsymbol{\Pi}} + k^2 \hat{\boldsymbol{\Pi}} ,$$

where

$$\hat{\boldsymbol{\Pi}} = \hat{\Pi} \boldsymbol{i}_z .$$

In order to determine $\hat{\Pi}(\boldsymbol{M}, z)$ we use equation (6) with the boundary conditions

$$\left.\frac{\partial \hat{\Pi}}{\partial \nu}\right|_{\Sigma} = 0 , \tag{7'}$$

$$\hat{\Pi}\,|_{z=\pm l} = 0 , \tag{8'}$$

solving which we find

$$\hat{\Pi}_{n,m} = \hat{A}_{n,m} \psi_m(M) \sin \frac{\pi m}{l} (l - z) . \tag{15}$$

In this case by $\hat{\psi}_n (M)$ one should understand the eigen-functions of the membrane S with the boundary condition $\frac{\partial \hat{\psi}}{\partial \nu} = 0$ on C.

2. *Electromagnetic energy of the natural vibrations*

Let us calculate the energy of the electric and magnetic fields in a standing wave in a cylindrical resonator.

For simplicity we confine ourselves to the case of a wave of electric type. Taking into account the dependence of $\boldsymbol{E}$ and $\boldsymbol{H}$ in formulae (4) and (5) on time according to the law $e^{-i\omega t}$ and considering only the real part, we obtain:

$$\left.\begin{aligned} E_x &= \frac{\partial^2 \Pi}{\partial z\, \partial x} \cos \omega t, \\ E_y &= \frac{\partial^2 \Pi}{\partial z\, \partial y} \cos \omega t, \\ E_z &= \left(\frac{\partial^2 \Pi}{\partial z^2} + k^2 \Pi\right) \cos \omega t, \end{aligned}\right\} \tag{16}$$

$$\left.\begin{aligned} H_x &= -k \frac{\partial \Pi}{\partial y} \sin \omega t, \\ H_y &= k \frac{\partial \Pi}{\partial x} \sin \omega t, \\ H_z &= 0. \end{aligned}\right\} \tag{17}$$

In order to calculate the energy of the electric and magnetic field we use the relations

$$\mathscr{E}_{el}(t) = \frac{c}{8\pi} \iiint_T \boldsymbol{E}^2\, d\tau, \tag{18}$$

$$\mathscr{E}_m(t) = \frac{c}{8\pi} \iiint_T \boldsymbol{H}^2\, d\tau, \tag{19}$$

where the integration is performed over the volume T of the resonator.

Substituting in (18) expressions (16) and using (14′), we have:*

$$\mathscr{E}_{el}(t) = \frac{A^2 c}{8\pi} \cos^2 \omega t \left\{ \iint_S \left[\left(\frac{\partial \psi}{\partial x}\right)^2 + \left(\frac{\partial \psi}{\partial y}\right)^2\right] d\sigma \int_{-l}^{l} [f'(z)]^2\, dz + \right.$$

$$\left. + \iint_S \psi^2\, d\sigma \int_{-}^{l} (f'' + k^2 f)^2\, dz \right\}.$$

Performing simple calculations, we obtain:

$$\int_{-l}^{l} [f'(z)]^2\, dz = ff' \big|_{-l}^{l} - \int_{-l}^{l} ff''\, dz = (k^2 - \lambda) \int_{-l}^{l} f^2\, dz = k^2 - \lambda, \tag{19′}$$

$$\int_{-l}^{l} (f'' + k^2 f)^2\, dz = \lambda^2 \int_{-l}^{l} f^2\, dz = \lambda^2, \tag{20}$$

* We temporarily omit the signs of m, n.

because of the normalization of the functions f

$$\int_{-l}^{l} f^2\,dz = 1\,. \tag{21}$$

In order to calculate the integrals over S we make use of Green's first theorem, the equation for ψ_n, the boundary conditions and the normalization condition

$$\int_S\int\left[\left(\frac{\partial\psi}{\partial x}\right)^2+\left(\frac{\partial\psi}{\partial y}\right)^2\right]dx\,dy=\int_S\int(\nabla_2\psi)^2\,d\sigma=$$

$$=-\int_S\int\psi\,\nabla_2^2\psi\,d\sigma+\int_C\psi\,\frac{\partial\psi}{\partial\nu}\,ds=\lambda\int_S\int\psi^2\,d\sigma=\lambda\,, \tag{22}$$

where ∇_2 is the operator "nabla" in the plane S, $\nabla_2^{\,2}$ is the two-dimensional Laplacian operator. As a result we obtain an expression for the energy of the electric field

$$\mathscr{E}_{el}(t)=\frac{A^2c}{8\pi}\,k^2\lambda\cos^2\omega t\,. \tag{23}$$

For the energy of the magnetic field because of (17), (19) and (14′) we have:

$$\mathscr{E}_m(t)=\frac{A^2ck^2}{8\pi}\int_S\int\left[\left(\frac{\partial\psi}{\partial x}\right)^2+\left(\frac{\partial\psi}{\partial y}\right)^2\right]dx\,dy\int_{-l}^{l}f^2\,dz\sin^2\omega t\,,$$

from which, taking into account (21) and (22) we find:

$$\mathscr{E}_m(t)=\frac{A^2ck^2}{8\pi}\,\lambda\sin^2\omega t\,. \tag{24}$$

The total energy of the electromagnetic field, obviously, does not vary with time:

$$\mathscr{E}=\mathscr{E}_{el}(t)+\mathscr{E}_m(t)=\frac{A^2ck^2}{8\pi}\,\lambda\,. \tag{25}$$

From formulae (23) and (24) we see that in the standing wave a reciprocal transformation of electric into magnetic energy takes place and, conversely, the mean value of the energy of the electric field

$$\overline{\mathscr{E}}_{el}=\frac{1}{2}\,\frac{A^2ck^2}{8\pi}\,\lambda=\frac{1}{2}\,\mathscr{E} \tag{26}$$

equals the mean energy of the magnetic field

$$\overline{\mathscr{E}}_m=\frac{1}{2}\,\frac{A^2ck^2}{8\pi}\,\lambda=\frac{1}{2}\,\mathscr{E}\,. \tag{27}$$

3. *Excitation of oscillations in the resonator*

In order to excite the field in the resonator by an external source it is necessary to introduce a coupling element through a hole in its case. This coupling element can be either a loop or a rod, acting as a small antenna. In order that the coupling element should not disturb the field in the resonator, it is necessary that its dimensions should be very much less than a wavelength. Other methods of exciting the resonator are possible, for instance by a beam of electrons, traversing the resonator cavity (through a gap in its walls).

The solution of the problem of the excitation of the resonator by an antenna, situated inside, or in the limiting case by an elementary dipole, requires the inclusion of the finite conductivity of the walls. Otherwise a steady process is impossible. Calculation in the case of walls of finite conductivity can be made by means of Leontovitch's conditions.

We consider here the problem of exciting a spherical resonator by a dipole, assuming a simple analytical solution.* At the centre of a sphere of radius r_0 there is a dipole, oscillating with a frequency ω and amplitude l, and directed along the z-axis. It is required to find the field inside the sphere, taking into account the finite conductivity of the walls.

In this case the fields $\boldsymbol{E}$ and $\boldsymbol{H}$ can be expressed by the function U:

$$\left.\begin{aligned} E_r &= \frac{i}{\varrho \sin\theta}\frac{\partial}{\partial\theta}\left(\sin\theta\frac{\partial U}{\partial\theta}\right), \\ E_\theta &= -\frac{i}{\varrho}\frac{\partial}{\partial\varrho}\left(\varrho\frac{\partial U}{\partial\theta}\right), \\ H_\varphi &= \frac{\partial U}{\partial\theta}. \end{aligned}\right\} \tag{28}$$

The remaining components E_φ, H_r, H_θ, equal zero.

Since the dipole is directed along the z-axis ($\theta = 0$) then the fields, obviously, should not depend on the angle φ.

The function U satisfies the equation

$$\frac{1}{\varrho^2}\frac{\partial}{\partial\varrho}\left(\varrho^2\frac{\partial U}{\partial\varrho}\right) + \frac{1}{\varrho^2\sin\theta}\frac{\partial}{\partial\theta}\left(\sin\theta\frac{\partial U}{\partial\theta}\right) + U = 0, \tag{29}$$

* See S. M. Rytov, *Dokl. Akad. Nauk SSR* 51, copy 2, 1946.

where $\varrho = kr$, where U has, as $\varrho \to 0$, a singularity of the form

$$\frac{ie^{ikr}}{r^2} = \frac{ik^2 e^{i\varrho}}{\varrho^2}. \tag{30}$$

On the surface of the sphere ($\varrho = \varrho_0$) Leontovicth's condition

$$E_\theta = aH_\varphi \tag{31}$$

should be fulfilled, where

$$a = \mu\, kd \sqrt{\frac{i}{2}} \qquad \left(d = \frac{c}{\sqrt{(2\pi\mu\sigma\omega)}}\right) \tag{32}$$

is the effective depth of the skin-layer.

From relations (31) and (28) the boundary condition for U follows

$$\left[\frac{\partial}{\partial\varrho}(\varrho U) - i\,\varrho_0\, aU\right]_{\varrho=\varrho_0} = 0$$

or

$$\varrho_0 \frac{\partial U}{\partial\varrho}\bigg|_{\varrho=\varrho_0} + (1 - i\,\varrho_0\, a)\, U\bigg|_{\varrho=\varrho_0} = 0\,. \tag{33}$$

The solution of (29), having the singularity (30) is the function

$$U = -k^2 \sqrt{\left(\frac{\pi}{2\varrho}\right)}\, [H^{(1)}_{3/2}(\varrho) + CJ_{3/2}(\varrho)]\, P_1(\cos\theta)\,,$$

where $P_1(\cos\theta)$ is a Legendre polynomial of first order, $H^{(1)}_{3/2}$ is a Hankel function of first kind, $J_{3/2}$ is Bessel's function

$$P_1(\cos\theta) = \cos\theta\,,$$

$$H^{(1)}_{3/2}(\varrho) = \sqrt{\left(\frac{2}{\pi\varrho}\right)}\, e^{i\varrho}\left(\frac{1}{i\varrho} - 1\right),$$

$$J_{3/2}(\varrho) = \sqrt{\left(\frac{2}{\pi\varrho}\right)}\left(\frac{\sin\varrho}{\varrho} - \cos\varrho\right).$$

The constant C is determined from the boundary condition (33)

$$C = -e^{i\varrho_0} \frac{1 - \frac{1}{\varrho_0^2} - \frac{i}{\varrho_0} + a\left(\frac{1}{i\varrho_0} - 1\right)}{i\left[\frac{\cos\varrho_0}{\varrho_0} + \left(1 - \frac{1}{\varrho_0^2}\right)\sin\varrho_0 - ia\left(\frac{\sin\varrho_0}{\varrho_0} - \cos\varrho_0\right)\right]}.$$

The solution obtained may be used to determine the value of the wall losses. The power absorbed in the walls,

$$Q = \frac{\mu\omega\, d}{16\,\pi}\int_0^\pi |H_\varphi|^2\, 2\,\pi\varrho_0^2 \sin\theta\, d\theta,$$

is calculated directly and equals

$$Q = \frac{\mu\omega\, k^4 d}{6} \frac{1}{|B - iaA|^2},$$

where

$$A = \frac{\sin \varrho_0}{\varrho_0} - \cos \varrho_0, \quad B = \frac{\cos \varrho_0}{\varrho_0} + \left(1 - \frac{1}{\varrho_0^2}\right) \sin \varrho_0 .$$

If the dipole is not situated at the centre of the sphere, then calculation of the fields is very complicated, but a solution may be obtained in the form of a series.*

III. Skin-effect

A variable current in contrast to a constant one is not distributed uniformly over a cross-section of a conductor, and has a greater density on its surface. This effect is called the *skin-effect.***

Let us consider, for simplicity, an infinite homogeneous cylindrical conductor ($\mu = \text{const.}$, $\sigma = \text{const.}$) on which a variable current flows. We shall assume that the total current $I = I_0 e^{i\omega t}$, flowing through a section of the conductor, is known.

Neglecting displacement currents in comparison with the conduction current*** and assuming the process to be steady, i. e. depending on time according to $e^{i\omega t}$, we obtain, after division by the factor $e^{i\omega t}$, Maxwell's equations in the form

$$\text{curl}\, \boldsymbol{H} = \frac{4\pi\sigma}{c} \boldsymbol{E}, \tag{1}$$

$$\text{curl}\, \boldsymbol{E} = - ik\mu \boldsymbol{H}, \tag{2}$$

$$\text{div}\; \boldsymbol{E} = 0, \tag{3}$$

$$\text{div}\; \boldsymbol{H} = 0, \tag{4}$$

* See V. Nikitin *(Zh. Tekh. Fiz.*, in print).

** I. E. Tamm, *Foundations of the Theory of Electricity*, Gostekhizdat 1946.

*** We note that inside conductors, in particular inside metals, the density of the displacement currents is negligibly small in comparison with the density of the conduction currents: $j_d \ll j = \sigma E$.

In our case the latter relation is equivalent to the requirement

$$\varepsilon\omega \ll \sigma .$$

For hard metals the conductivity $\sigma \approx 10^{17}$ abs. units, so that it is possible to neglect displacement currents for all frequencies employed in engineering.

where $k = \frac{\omega}{c}$. Equations (3) and (4) in the given case, obviously, follow from equations (1) and (2).

Let us introduce a cylindrical system of coordinates (r, θ, φ), so that the z-axis would coincide with the axis of the conductor. Then by virtue of the axial symmetry of the current all quantities may be assumed to depend only on the variable r.

Since in our case the vector $\boldsymbol{E}$ is directed along the z-axis, then from equations (1) and (2) we have:

$$\frac{1}{r}\frac{d}{dr}(rH_\varphi) = \frac{4\pi\sigma}{c}E_z, \tag{1'}$$

$$\frac{d}{dr}E_z = ik\mu H_\varphi. \tag{2'}$$

Eliminating H_φ, we find:

$$\frac{1}{r}\frac{d}{dr}\left(r\frac{dE_z}{dr}\right) = i\,\frac{4\pi\sigma\mu k}{c}E_z. \tag{5}$$

Let us introduce a boundary condition on the surface of the conductor when $r = R$. In order to do this we make use of the fact that the total current I_0, flowing through the cylinder, is known.

We write Maxwell's first equation (1) in integral form:

$$\oint_C H_s\,ds = \frac{4\pi}{c}I_0,$$

where C is the contour, encompassing the conductor, H_s is the tangential component of vector $\boldsymbol{H}$ on C. For such a contour we obtain:

$$\int_0^{2\pi} H_\varphi(R)\,d\varphi = \frac{4\pi}{cR}I_0$$

or

$$H_\varphi(R) = \frac{2I_0}{cR}. \tag{6}$$

Hence, using relation (2) we find:

$$\left.\frac{dE_z}{dr}\right|_{r=R} = \frac{2ik\mu}{cR}I_0. \tag{7}$$

Thus, we must solve Bessel's equation

$$E_z''(r) + \frac{1}{r} E_z'(r) + (a\sqrt{-i})^2 E_z(r) = 0 \tag{5'}$$

$$\left(a^2 = \frac{4\pi\sigma\mu\omega}{c^2} \right)$$

with the boundary condition

$$E_z'(R) = \frac{2ik\mu}{cR} I_0 \tag{7}$$

and the limiting condition at $r = 0$

$$E_z(0) < \infty. \tag{8}$$

The general solution of equation (5′) has the form

$$AJ_0(ar\sqrt{-i}) + BN_0(ar\sqrt{-i}), \tag{9}$$

where J_0 and N_0 are Bessel functions of the first and second kind (see Supplement, part I), A and B are constants, to be determined.

The function N_0 has a logarithmic singularity at $r = 0$. Therefore because of (8) $B = 0$ and, consequently,

$$E_z(r) = AJ_0(ar\sqrt{-i}). \tag{10}$$

We determine the coefficient A from the boundary condition (7):

$$A = \frac{2\sqrt{(-i)}\, k\mu I_0}{ac\, RJ_1(aR\sqrt{-i})}. \tag{11}$$

Hence for the current density

$$j = \sigma E_z$$

we obtain:

$$j(r) = \frac{Ia\sqrt{-i}}{2\pi RJ_1(aR\sqrt{-i})} J_0(ar\sqrt{-i}). \tag{12}$$

On the right-hand side of this formula is Bessel's function for a complex argument

$$x\sqrt{-i} = \frac{1-i}{\sqrt{2}} x.$$

Usually one uses for these functions the following symbols:

$$J_0(x\sqrt{-i}) = \mathrm{ber}_0\, x + i\, \mathrm{bei}_0\, x;$$

$$J_1(x\sqrt{-i}) = \mathrm{ber}_1\, x + i\, \mathrm{bei}_1\, x.$$

It is easy to find expressions for the real functions ber x and bei x, using the expansion of Bessel functions in series. For instance

$$J_0(x\sqrt{-i}) = J_0(xi\sqrt{i}) =$$

$$= 1 - \frac{\left(\frac{x}{2}\right)^2(-1)i}{(1!)^2} + \frac{\left(\frac{x}{2}\right)^4(-1)}{(2!)^2} - \frac{\left(\frac{x}{2}\right)^6 i}{(3!)^2} + \frac{\left(\frac{x}{2}\right)^8}{(4!)^2} - \ldots =$$

$$= \left\{1 - \frac{\left(\frac{x}{2}\right)^4}{(2!)^2} + \frac{\left(\frac{x}{2}\right)^8}{(4!)^2} - \ldots\right\} + i\left\{\frac{\left(\frac{x}{2}\right)^2}{(1!)^2} - \frac{\left(\frac{x}{2}\right)^6}{(3!)^2} + \ldots\right\},$$

from which we obtain:

$$\text{ber}_0\, x = 1 - \frac{\left(\frac{x}{2}\right)^4}{(2!)^2} + \frac{\left(\frac{x}{2}\right)^8}{(4!)^2} - \ldots; \tag{13}$$

$$\text{bei}_0\, x = \frac{\left(\frac{x}{2}\right)^2}{(1!)^2} - \frac{\left(\frac{x}{2}\right)^6}{(3!)^2} + \ldots . \tag{14}$$

It is easily verified in a similar manner that

$$\text{ber}_1\, x = \frac{1}{\sqrt{2}}\left\{\frac{x}{2} + \frac{\left(\frac{x}{2}\right)^3}{1!\,2!} - \frac{\left(\frac{x}{2}\right)^5}{2!\,3!} - \frac{\left(\frac{x}{2}\right)^7}{3!\,4!} + \ldots\right\}, \tag{15}$$

$$\text{bei}_1\, x = \frac{1}{\sqrt{2}}\left\{-\frac{x}{2} + \frac{\left(\frac{x}{2}\right)^3}{1!\,2!} + \frac{\left(\frac{x}{2}\right)^5}{2!\,3!} - \frac{\left(\frac{x}{2}\right)^7}{3!\,4!} - \ldots\right. . \tag{16}$$

In the appendices one also meets the derivatives

$$\text{ber}_0'\, x,\ \text{bei}_0'\, x,$$

where

$$J_1(x\sqrt{-i}) = \sqrt{-i}\cdot(\text{bei}_0'\, x - i\,\text{ber}_0'\, x). \tag{17}$$

Using the functions introduced, expression (12) for the current may be rewritten in the form

$$j(r) = \frac{I_0 a}{2\pi R}\,\frac{\text{bei}_0\, ar + i\,\text{ber}_0\, ar}{\text{bei}_0'\, aR - i\,\text{ber}_0'\, aR}.$$

or

$$j(r) = \frac{I_0 a}{2\pi R}\left\{\frac{\text{ber}_0\, ar\,\text{bei}_0'\, aR - \text{ber}_0\, ar\,\text{bei}_0'\, aR)}{(\text{bei}_0'\, aR)^2 + (\text{ber}_0'\, aR)^2} + \right.$$

$$\left. + i\,\frac{\text{bei}_0\, ar\,\text{bei}_0'\, aR + \text{ber}_0\, ar\,\text{ber}_0'\, aR)}{(\text{bei}_0'\, aR)^2 + (\text{ber}_0'\, aR)^2}\right\} \tag{18}$$

Evaluating the absolute value of this expression, we obtain:

$$|j(r)| = \frac{I_0 a}{2\pi R} \sqrt{\left[\frac{(\text{ber}_0\, ar)^2 + (\text{bei}_0\, ar)^2}{(\text{ber}_0'\, aR)^2 + (\text{bei}_0'\, aR)^2}\right]}\,. \tag{19}$$

The quantity, describing the distribution of current over a cross-section is the ratio

$$\frac{|j(r)|}{|j(R)|} = \sqrt{\left[\frac{(\text{ber}_0\, ar)^2 + (\text{bei}_0\, ar)^2}{(\text{ber}_0\, aR)^2 + \text{bei}_0\, aR)^2}\right]}\,. \tag{20}$$

Let us perform the calculation of the distribution of current over a cross-section for the frequencies $\omega_1 = 314$ (50 c/s), $\omega_2 = 314 \times 10^4$ (5×10^5 c/s). For the calculation it is more convenient to use the c. g. s. system.

All the calculations outlined above were performed in a gaussian symmetrical system. Therefore in a transition to a c. g. s. system one should take into account that $\mu_{\text{c. g. s.}} = \mu_{\text{gauss}}\, 1/c^2$. All the remaining quantities, appearing in formulae (12), (18), (19) and (20) coincide in both systems (gaussian and c. g. s.) Therefore in the c. g. s. system

$$a^2 = 4\,\pi\mu\sigma\omega\,.$$

For copper $\sigma = 57 \times 10^5$ c. g. s., therefore $a_1 = 0.4444$ (for ω_1), $a_2 = 44.44$ (for ω_2). Let us calculate the ratio of the moduli of the currents (20) for the low frequency $\omega_1 = 314$ for two values of r: $r = $ o and $r = 0.5R$. We assume in this R equal to unity.

Taking into account* that

$$\begin{aligned}
\text{ber}_0\, 0 &= 1,\\
\text{bei}_0\, 0 &= 0,\\
\text{ber}_0\, 0.222 &= 7 - 0.000036 + \ldots = 0.999964,\\
\text{bei}_0\, 0.222 &= 0.0123 - 0{,}000002 + \ldots = 0.012300,\\
\text{ber}_0\, 0.444 &= 1 - 0.00061 + \ldots = 0.9939,\\
\text{bei}_0\, 0.444 &= 0.493 - 0.0003 + \ldots = 0.4990,
\end{aligned}$$

we find that

$$\left.\frac{j(0)}{j(R)}\right|_{R=1} = 0.9994, \quad \left.\frac{j(0.5R)}{j(R)}\right|_{R=1} = 0.9999,$$

* See also Janke and Emde, *Tables of Special Functions*, Gostekhizdat, 1949.

i.e. at a low frequency the current is distributed over a cross-section approximately uniformly (skin-effect is absent).

Let us consider now the second case: $\omega_2 = 314 \times 10^4$. Since a is large, it is more convenient to proceed not from the expansions of the functions ber and bei in series, but from the asymptotic relations

$$J_0(ar\sqrt{-i}) = \frac{1}{\sqrt{(2\pi a r)}}\, e^{\frac{ar}{\sqrt{2}} - i\left(\frac{ar}{\sqrt{2}} - \frac{\pi}{8}\right)},$$

$$J_1(aR\sqrt{-i}) = \frac{1}{\sqrt{(2\pi a R)}}\, e^{\frac{aR}{\sqrt{2}} - i\left(\frac{aR}{\sqrt{2}} - \frac{\pi}{8}\right)},$$

from which we obtain, for the values $r = 0.9R$; $R = 1$:

$$\left.\frac{j(0.9\,R)}{j(R)}\right|_{R=1} = \sqrt{2}\, e^{-\frac{44}{\sqrt{2}}0.1} \approx 0.28\,.$$

This result indicates the extremely rapid decrease of current density inside the conductor at high frequencies. We note in conclusion that the skin effect is widely used in practice for the hardening of metals.

IV. Propagation of radiowaves over the surface of the earth

Problems, connected with the propagation of radiowaves both in free space and in the presence of surfaces of separation, have immense theoretical and practical value. A very large amount of work by soviet and foreign authors has been devoted to these problems.

We consider the problem of the effect of the earth on the propagation of radio waves emitted by a vertical dipole. In addition we shall assume the earth to be flat.*

There is a dipole at the point P_0, a distance h above the earth's surface, emitting periodic oscillations of frequency ω. We call the earth's plane the plane $z = 0$ and direct the z-axis along the dipole axis (Fig. 89). We assume that in the atmosphere ($z > 0$) $\varepsilon_0 = \mu_0 = 1$, $\sigma_0 = 0$. We assume further that the earth

* This problem was first solved by Sommerfeld in 1909. Sommerfeld's original solution contained an error, which was corrected by V. A. Fok.

($z < 0$) is characterized by a dielectric constant ε, conductivity σ, and the magnetic permeability μ may be taken equal to unity; we shall assume ε and σ to be constants.

Our problem is to find the field produced by the dipole. The process of propagating electromagnetic waves is described by Maxwell's equations.

As was shown in Appendix II to Chapter V, the solution of Maxwell's equations can be reduced to the solution of the wave equation for the polarization potential $\boldsymbol{\Pi}$:*

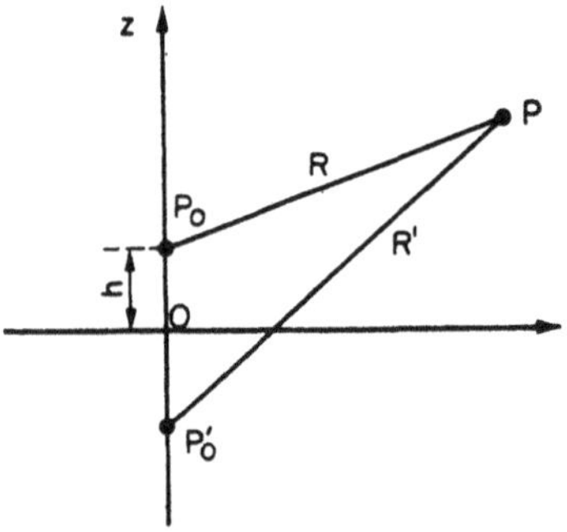

FIG. 89

$$\Delta \boldsymbol{\Pi} + k^2 \boldsymbol{\Pi} = 0\,, \tag{1}$$

where

$$k^2 = \begin{cases} k_0^2 = \dfrac{\omega^2}{c^2}\,; & z > 0\,; \\ k_3^2 = \dfrac{\varepsilon\omega^2 + i\,\sigma\omega}{c^2}\,; & z < 0\,. \end{cases}$$

The potential $\boldsymbol{\Pi}$ is connected to the field strength by the relations

$$\left.\begin{aligned} \boldsymbol{E} &= k^2 \boldsymbol{\Pi} + \operatorname{grad} \operatorname{div} \boldsymbol{\Pi}\,, \\ \boldsymbol{H} &= -\,i\,\frac{k^2}{k_0} \operatorname{curl} \boldsymbol{\Pi}. \end{aligned}\right\} \tag{2}$$

In our case the vector $\boldsymbol{\Pi}$ is parallel to the radiating dipole

$$\boldsymbol{\Pi} = (0, 0, \Pi_z)\,; \qquad \Pi_z = \Pi_z(r, z)\,. \tag{3}$$

Putting

$$n^2 = \varepsilon + i\,\frac{\sigma}{\omega}\,,$$

we obtain:

$$k_3^2 = n^2 k_0^2\,.$$

Relations (2) and (3) give:

$$E_r = \frac{\partial}{\partial r}\frac{\partial \Pi_0}{\partial z}\,;\ \ H_\varphi = -\,ik_0^2\,\frac{\partial \Pi_0}{\partial r}\,;\ \ E_\varphi = H_r = 0 \text{ where } z > 0, \tag{4}$$

$$E_r = \frac{\partial}{\partial r}\frac{\partial \Pi_3}{\partial z}\,;\ \ H_\varphi = -\,\frac{ik_3^2}{k_0}\frac{\partial \Pi_3}{\partial r}\,;\ \ E_\varphi = H_r = 0 \text{ where } z < 0, \tag{5}$$

* Since the process is considered steady, the time factor $e^{-i\omega t}$ may be omitted.

In order to obtain the boundary conditions for $z = 0$ we make use of the condition of the continuity of the tangential components of the field strengths. These conditions, as formulae (4) and (5) show, will be fulfilled if it is assumed:

$$\frac{\partial \Pi_0}{\partial z} = \frac{\partial \Pi_3}{\partial z}; \quad \Pi_0 = n^2 \Pi_3 \quad \text{where} \quad z = 0. \tag{6}$$

We shall look for the solution of equation (1) with boundary conditions (6) as a superposition of particular solutions of the form

$$J_0(\lambda r) e^{\pm \mu z} \qquad (k^2 = \lambda^2 + \mu^2).$$

For an infinite region in place of a discrete spectrum of the eigen-values λ, a continuous spectrum is obtained. Therefore solution Π may be sought in the form

$$\Pi = \int_0^\infty F(\lambda) J_0(\lambda r) e^{\pm \mu z} d\lambda; \tag{7}$$

the sign of μ should be chosen so that the convergence of the integral (7) is ensured. The remaining function $F(\lambda)$, undetermined so far, represents the amplitude factor of the oscillations.

Let us utilize the integral representation of the potential of a dipole in free space (see Supplement, part I, § 5)

$$\Pi_0 = \frac{e^{ikR}}{R} \int J_0(\lambda r) e^{-\mu|z|} \frac{\lambda\, d\lambda}{\mu}$$

$$(\mu = \sqrt{[\lambda^2 - k^2]}, \qquad R = \sqrt{[r^2 + z^2]}). \tag{8}$$

We consider two different regions:

(a) *Air* $(z > 0)$.

The field in this region will have the form

$$\Pi_0 = \Pi_p + \Pi_s,$$

where

$$\Pi_p = \frac{e^{ikR}}{R} \tag{9}$$

is the field potential of the primary excitation, produced by the same dipole, and Π_s is the field potential of the secondary excitation, produced by currents originating in the earth.

Utilizing (7), (8), and (9) we can write:

$$\left.\begin{aligned} \Pi_p &= \int_0^\infty J_0(\lambda r)\, e^{-\mu|z-h|} \frac{\lambda\, d\lambda}{\mu}, \\ \Pi_s &= \int_0^\infty F(\lambda)\, J_0(\lambda r)\, e^{-\mu(z+h)}\, d\lambda, \end{aligned}\right\} \tag{10}$$

where $F(\lambda)$ is a function undetermined for the time being.

(b) *The earth* $(z < 0)$.

In this region only secondary excitation occurs, which we can write in the form

$$\Pi_3 = \int_0^\infty F_3(\lambda)\, J_0(\lambda r)\, e^{\mu_3 z - \mu h}\, d\lambda, \tag{11}$$

where $\mu_3^2 = k_3^2 - \lambda^2$. Since $z < 0$, the sign of the exponent of the exponential ensures the convergence or the integral.

In order to determine $F(\lambda)$ and $F_3(\lambda)$ we use the boundary conditions (5) which give:

$$\left.\begin{aligned} &\int_0^\infty J_0(\lambda r)\, e^{-\mu h} [\lambda - \mu F(\lambda) - \mu_3 F_3(\lambda)]\, d\lambda = 0, \\ &\int_0^\infty J_0(\lambda r)\, e^{-\mu h} [\lambda + \mu F(\lambda) - n^2 \mu F_3(\lambda)]\, d\lambda = 0. \end{aligned}\right\} \tag{12}$$

Conditions (12) will be fulfilled, if we assume

$$\left.\begin{aligned} \mu F(\lambda) + \mu_3 F_3(\lambda) &= \lambda, \\ \mu F(\lambda) - n^2 \mu F_3(\lambda) &= -\lambda. \end{aligned}\right\} \tag{13}$$

Solving equation (13), we determine $F(\lambda)$ and $F_3(\lambda)$ in the form

$$\left.\begin{aligned} F(\lambda) &= \frac{\lambda}{\mu}\left(1 - \frac{2\mu_3}{n^2 \mu + \mu_3}\right); \\ F_3(\lambda) &= \frac{2\lambda}{n^2 \mu + \mu_3}. \end{aligned}\right\} \tag{14}$$

Substituting expression (14) obtained in formulae (10) and (11), we obtain the following expressions for the polarization of the field of a vertical dipole:

$$\left.\begin{aligned} \Pi_0 &= \int_0^\infty J_0(\lambda r)\, e^{-\mu|z-h|} \frac{\lambda\, d\lambda}{\mu} + \int_0^\infty J_0(\lambda r)\, e^{-\mu|z+h|} \frac{\lambda\, d\lambda}{\mu} - \\ &\qquad - 2\int_0^\infty J_0(\lambda r)\, e^{-\mu(z+h)} \frac{\mu^3}{n^2\mu+\mu_3} \frac{\lambda\, d\lambda}{\mu}; \\ \Pi_3 &= 2\int_0^\infty J_0(\lambda r)^{\mu_3 z-\mu h} \frac{\lambda\, d\lambda}{n^2\mu+\mu_3}. \end{aligned}\right\} \tag{15}$$

Denoting by $R = \sqrt{[r^2 + (z-h)^2]}$ the distance from the point of observation to the dipole, by $R' = \sqrt{[r^2 + (z+h)^2]}$ the distance from the point of observation to the mirror image of the dipole in the plane $z = 0$ and using (8) we rewrite the expression for the function Π in the form

$$\Pi_0 = \frac{e^{ikR}}{R} + \frac{e^{ikR'}}{R'} - 2\int_0^\infty J^0(\lambda r)\, e^{-\mu(z+h)} \frac{\mu_3}{n^2\mu+\mu_3} \frac{\lambda d\lambda}{\mu}. \tag{15'}$$

Let us consider some limiting cases.

(1) *Ideally conducting earth.* In this case $\sigma \to \infty$, and therefore, $|k_3|$ and $|n| \to \infty$. Moreover formulae (15) and (15') give:

$$\Pi_0 = \frac{e^{ikR}}{R} + \frac{e^{ikR'}}{R'}; \quad \Pi_3 = 0.$$

This result is readily obtained directly, solving the problem by the method of images.

(2) *A dipole in a homogeneous medium.* In this case $k_0 = k_3$; $n = 1$; $\mu = \mu_3$. Formulae (15) and (15') give:

$$\Pi_0 = \Pi_3 = \int_0^\infty J_0(\lambda r)\, e^{-\mu|z-h|} \frac{\lambda d\lambda}{\mu} = \frac{e^{ikR}}{R},$$

i. e. one primary excitation occurs, as should be.

The integral expressions (15) obtained are very complicated for study and for practical application. The expressions under the integral sign have branch points and poles. The method of approximate calculation of these integrals by means of deformation of the contour of integration was suggested by Sommerfeld. By this the following approximate formula for the field near the surface of the earth was obtained:

$$\Pi_0 = 2\frac{e^{ik_0 r}}{r}\left(1 + i\sqrt{(\pi\varrho)}\, e^{-\varrho} - 2\sqrt{\varrho}\, e^{-\varrho}\int_0^{\sqrt{\varrho}} e^{\alpha^2}\, d\alpha\right), \tag{16}$$

where the quantity ϱ is the so-called "numerical distance" connected to the pole p of the expression (15) under the integral sign by the relation

$$\varrho = i\,(k_0 - p)\, r. \tag{16}$$

Formula (16) agrees with formulae obtained by other means by a number of different authors (Weyl, Van der Pol, Fok).

SUPPLEMENT

Special functions

INTRODUCTION

§ 1. Equations for special functions

The method of separation of variables, as we verified in the solution of the simplest problems in Chapters II and III, leads to the eigen-value problem

$$\mathscr{L}[y] + \lambda \varrho y(x) = 0 \qquad (a < x < b), \tag{1}$$

where

$$\mathscr{L}[y] = \frac{d}{dx}\left[k(x)\frac{dy}{dx}\right] - q(x)y \qquad \left(k(x) > 0\right).$$

Besides the equation for the trigonometric functions

$$y'' + \lambda y = 0, \quad y(0) = y(l) = 0, \tag{2}$$

corresponding to the case $a = 0$, $b = 1$; $q = 0$, $k = \varrho = \text{const.}$ more complex boundary-value problems occur. Thus, for instance, the problem of the characteristic oscillations of a circular membrane (of radius r_0) leads to *Bessel's equation*

$$\frac{1}{x}\frac{d}{dx}\left(x\frac{dy}{dx}\right) + \left(\lambda - \frac{n^2}{x^2}\right)y = 0 \tag{3}$$

or

$$(xy')' - \frac{n^2}{x}y + \lambda xy = 0$$

$$\left[k(x) = x, \quad q(x) = \frac{n^2}{x}, \quad \varrho(x) = x, \quad a = 0, \quad b = r_0\right];$$

the problem of the characteristic oscillations of a sphere to *Legendre's equation*

$$[(1 - x^2)y']' + \lambda y = 0 \tag{4}$$

$$[k(x) = 1 - x^2, \quad q = 0, \quad \varrho = 1, \quad a = -1, \quad b = 1]$$

and to the *equation for the associated functions*

$$\left.\begin{array}{c}[(1-x^2)\,y']' - \dfrac{n^2}{1-x^2}\,y + \lambda y = 0 \\ \left[k\,(x) = 1 - x^2,\ \ \varrho = 1,\ \ q = \dfrac{n^2}{1-x^2},\ \ a = -1,\ \ b = 1\right].\end{array}\right\} \tag{5}$$

The characteristic feature of these equations is the fact that the coefficient $k(x)$ reduces to zero, at least at one end of the segment $[a, b]$.

Other equations occur, for example in problems of quantum mechanics, *Chebyshev-Hermite equations* ($k(x) = e^{-x^2}$, $\varrho = e^{-x^2}$, $q = 0$ $a = -\infty$, $b = \infty$), *Chebyshev-Laguerre equations* ($k(x) = xe^{-x}$, $\varrho(x) = e^{-x}$, $q = 0$, $a = 0$, $b = \infty$), and others. Boundary-value problems for these equations define a very important class of special function (cylindrical functions, spherical functions, Chebyshev-Hermite polynomials, Chebyshev-Laguerre polynomials, and others).

§ 2. Formulation of boundary-value problems in the case $k(a) = 0$

Let us proceed to an investigation of some general properties of the equation

$$\mathscr{L}[y] = \frac{d}{dx}\left[k\,(x)\,\frac{dy}{dx}\right] - q\,(x)\,y = 0, \tag{6}$$

in which $k(a) = 0$. We shall assume that $k(x)$ in the neighbourhood of the point a has the form $k(x) = (x - a)\,\varphi(x)$, where $\varphi(x)$ is a continuous function and $\varphi(a) \neq 0$.

If in this equation $q(x)$ is replaced by the function $q(x) - \lambda\varrho(x)$, then it is obvious that everything proved later will hold for the equation

$$\mathscr{L}\,[y] + \lambda\varrho y = 0. \tag{1}$$

Let us prove two lemmas concerning the solution of equation (6).

LEMMA 1. *If in equation (6) the coefficient $k(x)$ reduces to zero for $x = a$ ($k(x) = (x - a)\,\varphi(x)$, $\varphi(a) \neq 0$), and if one solution of this equation $y_1\,(x)$ remains finite for $x = a$, $[y_1\,(a) \neq \infty]$, then any other solution y_2 of equation* (6), *linearly-independent of $y_1\,(x)$, tends to infinity for $x = a$.*

If one solution y_1 of the linear differential equation (6) is known, then any other linearly-independent solution $y_2(x)$ can be represented as a quadrature. In fact, writing equation (6) for the functions $y_1(x)$ and $y_2(x)$, multiplying these equations by y_2 and y_1 respectively, and subtracting, we find:

$$y_2 \frac{d}{dx}\left(k \frac{dy_1}{dx}\right) - y_1 \frac{d}{dx}\left(k \frac{dy_2}{dx}\right) = \frac{d}{dx}\left[k\left(y_2 y_1' - y_1 y_2'\right)\right] = 0$$

Hence it follows that the *Wronskian* of the functions y_1 and y_2 equals $C/k(x)$:

$$y_1 \frac{dy_2}{dx} - y_2 \frac{dy_1}{dx} = \frac{C}{k(x)},$$

where C is a constant, not equal to zero, if y_1 and y_2 are linearly-independent. Dividing the latter equation by y_1^2, we find that

$$\frac{d}{dx}\left(\frac{y_2(x)}{y_1(x)}\right) = \frac{C}{ky_1^2}$$

or

$$y_2(x) = y_1(x)\left[\int_{x_0}^{x} \frac{C\,d\alpha}{k(\alpha)\, y_1^2(\alpha)} + C_1\right], \tag{7}$$

where C_1 is a constant of integration, depending on the choice of x_0.

The function $y_1(x)$ has a finite value at the point $x = a$, and could be zero there. We assume (see for more detail, Lemma 3) that

$$y_1(x) = (x-a)^n z_1(x), \qquad z_1(a) \neq 0, \; n \geqslant 0.$$

Let us choose x_0 in formula (7) so that

$$z_1(x) \neq 0, \qquad a < x \leqslant x_0$$

(or $y_1(x) \neq 0$, $a \leqslant x \leqslant x_0$, if $n = 0$). In this case $y_2(x)$ in the interval $a < x \leqslant x_0$ is represented by the integral

$$y_2(x) = (x-a)^n z_1(x)\left\{C_1 - \int_{x}^{x_0} \frac{1}{(x-a)^{2n+1}}\left[\frac{C}{\varphi(x)\, z_1^2(x)}\right] dx\right\}.$$

Applying the mean value theorem to the integral on the right-hand side, we find:

$$y_2(x) = (x-a)^n z_1(x)\left\{C_1 - A\int_{x}^{x_0} \frac{dx}{(x-a)^{2n+1}}\right\},$$

$$A = \frac{C}{\varphi(\bar{x})\, z_1^2(\bar{x})} \qquad (x < \bar{x} < x_0).$$

Evaluating the integral, we obtain for $y_2(x)$ the expression

$$y_2(x) = \begin{cases} (x-a)^n z_1(x)\left[C_1 + A\dfrac{(x_0-a)^{-2n}}{2n} - \right. \\ \qquad\qquad \left. - A\dfrac{(x-a)^{-2n}}{2n}\right] \text{where } n > 0\,, \\ y_1(x)\left[C_1 - A\ln(x_0-a) + A\ln(x-a)\right] \text{where } n = 0\,. \end{cases} \tag{8}$$

Thus, in both cases $\lim_{x\to a} y_2(x) = \pm\infty$, which proves the lemma; for $n > 0$ it follows from (8):

$$\lim_{x\to a}(x-a)^n y_2(x) = -\frac{C}{2n\varphi(a)z_1(a)}\,. \tag{9}$$

Formulae (8) and (9) enable us to formulate Lemma 1 more accurately.

If in the conditions of Lemma 1 the finite solution of equation (6) at the point $x = a$ differs from zero ($y_1(a) \neq 0$), then $y_2(x)$ has a logarithmic singularity at the point $x = a$

$$y_2(x) \sim \ln(x-a)\,.$$

If at the point $x = a$ $y_1(x)$ has a zero of order n ($y_1(x) = (x-a)^n z_1(x)$, $z_1(a) \neq 0$) then $y_2(x)$ has a pole of order n at the point $x = a$, $y_2(x) \sim (x-a)^{-n}$.

Let us prove a second lemma.

LEMMA 2. *If in equation (6) the coefficient $k(x)$ reduces to zero at $x = a$ ($k(x) = (x-a)\varphi(x)$, where $\varphi(a) \neq 0$), and the coefficient $q(x)$ is either bounded, or tends to $+\infty$ as $x \to a$, then the solution $y_1(x)$ of equation (6), bounded at the point $x = a$ satisfies the condition*

$$\lim_{x\to a} k(x)y_1'(x) = 0\,.$$

Let us consider firstly the case where $q(x)$ is bounded for $x \to a$. Integrating equation (6) from some value to x_1 ($x < x_1$), we obtain:

$$k(x)\frac{dy_1}{dx} = k(x)\frac{dy_1}{dx}\bigg|_{x=x_1} - \int_x^{x_1} q(\alpha)y(\alpha)\,d\alpha = Q(x) \quad (x < x_1)\,. \tag{10}$$

Hence it follows that $Q(x)$ is a continuous function ($a \leqslant x \leqslant x_1$). Passing to a limit as $x \to a$, we see that there exists a limit

$$C = \lim_{x\to a} k(x)y_1'(x) \qquad (C = Q(a))\,.$$

Let us prove that $C = Q(a) = 0$. Expressing function $y_1(x)$ by $Q(x)$, we find:

$$y_1(x) = y_1(x_2) - \int_x^{x_2} \frac{Q(\alpha)}{k(\alpha)}\, d\alpha =$$

$$= y_1(x_2) - \int_x^{x_2} \frac{Q(\alpha)}{(\alpha - a)\,\varphi(\alpha)}\, d\alpha \quad (a \leqslant x \leqslant x_2). \tag{11}$$

From this formula we see immediately that if $Q(a) \neq 0$, then as $x \to a$, $y_1(x) \to \infty$, which contradicts the boundary condition of $y_1(x)$ at $x = a$. Thus, the lemma is proved for the case of finite $q(x)$. Let us consider now the case where $q(x)$ tends to $+\infty$ as $x \to a$. Formula (10) still holds. Let $q(x)$ be positive in the interval $(a \leqslant x \leqslant x_1)$. It is readily verified that in this interval $y_1(x)$ is monotonic,* and has the same sign in some interval $(a \leqslant x \leqslant x_2)$ and has a limit $y_1(a)$ for $x \to a$. Hence it follows that in this interval $Q(x)$ is a monotonic function, having a finite or infinite limit at $x = a$. From (11), as before, it follows that $y_1(x) \to \infty$, if $Q(a) \neq 0$, which proves the lemma.

We note further from formula (10) and the condition $Q(x) \to 0$ as $x \to a$ that

$$k(x_1)\, y_1'(x_1) = \int_a^{x_1} q(\alpha)\, y_1(\alpha)\, d\alpha \tag{12}$$

both for the finite function $q(x)$ and for the unbounded $q(x)$ $(q(x) > 0)$.

LEMMA 3. *If $y_1(x)$ is a solution of equation (6), bounded at $x = a$, and $q(x)$ is a continuous function $(a \leqslant x \leqslant x_1)$, then $y_1(a) \neq 0$ and $y_1'(a) = \dfrac{q(a)\, y_1(a)}{\varphi(a)}$.*

If $q(x)$ has the form $q(x) = \dfrac{q_1(x)}{(x-a)}$, where $q_1(x)$ is a continuous function and $q_1(a) > 0$, then $y_1(a) = 0$.

* In fact assuming $\xi = \int_x^{x_1} \frac{dx}{k(x)}$, we can write equation (6) in the form $y'' - kqy = 0$ $(k > 0,\ q > 0)$. Hence it follows that y cannot have positive maximum (negative minimum) values since at the corresponding point $y'' \leqslant 0$ $(y'' \geqslant 0)$, and $y > 0$ $(y < 0)$.

The second part of the lemma follows directly from (12) because the integral

$$\int_a^x q(\alpha)\, y_1(\alpha)\, d\alpha = \int_a^x \frac{q_1(\alpha)}{(\alpha - a)}\, y_1(\alpha)\, d\alpha$$

converges, which is possible only for $y_1(a) = 0$.

If $q(x)$ is a continuous function in $(a \leqslant x \leqslant x_1)$, then because of the arbitrary nature of x_1 in (12) we obtain:

$$y_1'(x) = \frac{1}{\varphi(x)} \frac{1}{x - a} \int_a^x q(\alpha)\, y(\alpha)\, d\alpha\,. \tag{13}$$

Hence it follows that $y_1'(x)$ is bounded in some interval $(a \leqslant \leqslant x \leqslant x_2)$ and thus there exists a limit of $y_1(x)$ as $x \to a$. Passing to a limit as $x \to a$ in the latter formula we obtain:

$$\lim_{x \to a} y_1'(x) = \frac{q(a)\, y_1(a)}{\varphi(a)} = y_1'(a)\,.$$

From this relation it follows that if $y_1(a) = 0$, then $y_1'(a) = 0$. We prove further that $y_1(x) \equiv 0$ (it is impossible to utilize the uniqueness theorem directly because $x = a$ is a singular point of equation (6)). Let $|\, q(x)\, | \leqslant \bar{q}$ and $|\, \varphi(x)\, | \geqslant \underline{\varphi}$ for $(a \leqslant x \leqslant x_2)$.

Let us consider the interval $(a; a + h)$, where h is a sufficiently small number so that $h(\bar{q}/\underline{\varphi}) < 1$ and $a + h < x_2$. From (13) it follows that

$$y_1(x) = \int_a^x \left(\frac{1}{\varphi(\alpha_1)} \frac{1}{(\alpha_1 - a)} \int_a^{\alpha_1} q(\alpha_2)\, y(\alpha_2)\, d\alpha_2 \right) d\alpha_1 \quad (y_1(a) = 0)\,.$$

At $x = x_0$ the function $y_1(x)$ reaches a maximum value $y_1(x_0) = A$ in the interval $(a; a + h)$. Applying the mean value theorem to the latter relation, we obtain:

$$y_1(x_0) = A \leqslant \bar{A} q \frac{1}{\underline{\varphi}} (x_0 - a) < A\,,$$

which proves the statement.

LEMMA 4. *If the function* $q(x)$ *has the form* $q(x) = \frac{q_1(x)}{(x-a)}$ *where* $q_1(x) > 0$ *in the interval* $(a \leqslant x \leqslant x_1)$, *then* $y_1(x)$, *a solution of equation* (6) *bounded at* $x = a$, *has the form* $y_1(x) = (x-a)^\nu z(x)$, *where* $\nu = \sqrt{\frac{q_1(a)}{\varphi(a)}}$, *and* $z(x)$ *is a continuous and differentiable function in the interval* $(a \leqslant \leqslant x \leqslant x_1)$ *and* $z(a) \neq 0$.

Assuming $y_1(x) = (x-a)^\nu z(x)$, where ν is some number, we obtain from formula (12):

$$\nu k(x)(x-a)^{\nu-1} z(x) + k(x)(x-a)^\nu \frac{dz}{dx} = \int_a^x q(\alpha) y(\alpha) d\alpha . \quad (14)$$

Let us perform an integration by parts on the integral:

$$\int_{a+\varepsilon}^x q(\alpha) y(\alpha) d\alpha = \int_{a+\varepsilon}^x \left[\frac{q_1(\alpha)}{(\alpha-a)}(\alpha-a)^\nu\right] z(\alpha) d\alpha =$$

$$= z(\alpha) u(\alpha)\Big|_{a+\varepsilon}^x - \int_{a+\varepsilon}^x u(\alpha) z'(\alpha) d\alpha ,$$

where

$$u(\alpha) = \int_a^\alpha \frac{q_1(\alpha_1)}{(\alpha_1-a)}(\alpha_1-a)^\nu d\alpha_1 =$$

$$= \frac{q_1(a)}{\nu}(\alpha-a)^\nu + \int_a^\alpha \frac{q_1(\alpha)-q_1(v)}{(\alpha_1-a)}(\alpha_1-a)^\nu d\beta_1 =$$

$$= \frac{q_1(a)}{\nu}(\alpha-a)^\nu + u_1(\alpha) .$$

Because $z(\alpha)u(\alpha) = y_1(\alpha)\left(\frac{q_1(a)}{\nu} + \frac{u_1(\alpha)}{(\alpha-a)^\nu}\right)$ we deduce that $z(\alpha)$ $\underset{\alpha\to 0}{u(\alpha)} \to 0$, since $\frac{u_1(\alpha)}{(\alpha-a)^\nu}$ is a bounded function, and $y_1(\alpha) \to 0$ by Lemma 3.

Writing z' from formula (14) we find:

$$z'(x) = z(x)\left[-\frac{\nu}{(x-a)} + \frac{u(x)}{k(x-a)^\nu}\right] - \int_a^x u(\alpha) z'(\alpha) d\alpha =$$

$$= z(x) U(x) - \int_a^x u(\alpha) z'(\alpha) d\alpha . \quad (15)$$

Choosing $\nu = \sqrt{\left[\frac{q_1(a)}{\varphi(a)}\right]}$ we find that the function

$$U(x) = -\frac{\nu}{(x-a)} + \frac{u(x)}{k(x-a)^\nu} = \frac{u_1(x)}{k(x-a)^\nu}$$

is a continuous and differentiable function for $(a \leqslant x \leqslant x_1)$. Differentiating equation (15), we obtain:

$$z''(x) = [U(x) - u(x)]\, z'(x) + U'(x)\, z(x)\,. \tag{16}$$

The point $x = a$ is not a singular point of equation (16). Hence it follows that $z(x)$ is continuous and has a continuous derivative for $(a \leqslant x \leqslant x_1)$. Passing to a limit as $x \to a$ in equation (15) we find that

$$z'(a) = z(a)\, U(a)\,. \tag{17}$$

Formula (17) shows that $z(a) \neq 0$, since otherwise $z(a) = z'(a) = 0$ and thus $z(x) \equiv 0$.

We note that for the continuity of the function $U(x)$ at $x = a$ it is sufficient to assume the one-fold differentiability of function $q_1(x)$ and for the differentiability of function $U'(x)$ the differentiability of function $\varphi(x)$ and the twofold differentiability of function $q_1(x)$. However Lemma 4 can be proved, not by using the differential equation (16) but by considering equation (15) as an integral equation for $z(x)$. For this proof $U(x)$ need only be continuous.

These lemmae enable us to make the following statements on the formulation of boundary-value problems for the equations

$$\mathscr{L}[y] + \lambda \varrho y = 0 \text{ and } \mathscr{L}[y] = 0$$

in the interval $[a, b]$, at one or both ends of which $k(x)$ reduces to zero. If $k(a) = 0$, then for $x = a$, we shall impose the natural restriction that the solution be bounded. It is not necessary that it takes a given value at $x = a$.

Thus, we arrive at the following boundary-value problem: *find the eigen-values and eigen functions of the equation*

$$\mathscr{L}[y] = \frac{d}{dx}\left[k(x)\frac{dy}{dx}\right] - q(x)\, y = -\lambda \varrho y \qquad (a < x < b)\,,$$

where $k(x) > 0$ *for* $x > a$ *and* $k(a) = 0$ *with the boundary conditions*

$$y(b) = 0$$

and $y(x)$ *is bounded at* $x = a$.

If $k(a) = 0$ and $k(b) = 0$, then the solution must be bounded at both ends of the interval (a, b).

We state the general properties of the eigen-functions and eigen-values of the boundary-value problem:

1. There exists an infinite number of eigen-values $\lambda_1 \leqslant \lambda_2 \leqslant$, ..., with corresponding eigen-functions

$$y_1(x),\ y_2(x), \ldots,\ y_n(x), \ldots$$

2. For $q \geqslant 0$ all the eigenvalues are positive

$$\lambda_n \geqslant 0 .$$

3. The eigen-functions $y_n(x)$ and $y_m(x)$, corresponding to different eigen-values λ_n and λ_m are mutually orthogonal with weight $\varrho(x)$

$$\int_a^b y_n(x)\, y_m(x)\, \varrho(x)\, dx = 0 .$$

4. The expansion theorem holds: the function $f(x)$ with a continuous first and piece-wise continuous second derivative can be expanded in an absolutely and uniformly convergent series with respect to the eigenfunctions $y_n(x)$

$$f(x) = \sum_{n=1}^{\infty} f_n y_n(x), \quad \text{where } f_n = \frac{\int_a^b f(x)\, y_n(x)\, \varrho(x)\, dx}{\int_a^b y_n^2(x)\, \varrho(x)\, dx},$$

if $f(x)$ satisfies the boundary conditions of the problem. If $k(a) = 0$, then the boundary condition is

$$|f(a)| < \infty \quad \text{where} \quad q(a) < \infty ,$$

$$f(a) = 0 \quad \text{where} \quad q(x) = \frac{q_1(x)}{x-a} \qquad (q_1(a) \neq 0) .$$

Properties 2 and 3 are proved, as in Chapter II, § 3, by means of Green's theorem. In addition the bounded nature of $y_n(x)$ at $x = a$ is utilized, and also the equality proved above $\lim\limits_{x \to a} k(x)\, y_n'(x) = 0$, which ensures that the limit for $x = a$ equals zero. The properties 1 and 4 are usually proved by the theory of integral equations. We do not give these proofs here. In order that 1 and 4 should hold, it is sufficient that the function $k(x)$ should be continuous together with its derivative, and the function $q(x)$ should be either continuous, or have the form $\frac{q_1(x)}{x-a}$, where $q_1(x)$ is a continuous function. For the class of special function investigated below these conditions are fulfilled.

Our boundary-value problem is equivalent to the integral equation

$$y(x) = \lambda \int_a^b G(x, \xi)\, y(\xi)\, \varrho(\xi)\, d\xi ,$$

which by means of the substitution

$$\varphi(x) = \sqrt{[\varrho(x)]}\, y(x), \quad K(x, \xi) = \sqrt{[\varrho(x)]}\, G(x, \xi) \sqrt{[\varrho(\xi)]}$$

reduces to an integral equation with a symmetrical kernel $K(x, \xi)$:

$$\varphi(x) = \lambda \int_a^b K(x, \xi)\, \varphi(\xi)\, d\xi .$$

Here $G(x, \xi)$ is the source function for the equation $\mathscr{L}[y] = 0$ determined for the case $k(a) = 0$, $k(b) = 0$, $y(b) = 0$ by the conditions:

1. $G(x, \xi)$ is a continuous function of x for a fixed value of ξ.
2. The first derivative $\frac{\partial G}{\partial x}$ has a discontinuity at $x = \xi$:

$$\left.\frac{dG}{dx}\right|_{x=\xi-0}^{x=\xi+0} = -\frac{1}{k(\xi)}$$

or

$$G'(\xi + 0, \xi) - G'(\xi - 0, \xi) = -\frac{1}{k(\xi)} .$$

3. $\mathscr{L}[G] = 0$ everywhere, except $x = \xi$.
4. $G(x, \xi)$ satisfies the boundary conditions

$$G(a, \xi) < \infty, \quad G(b, \xi) = 0 .$$

PART I

CYLINDRICAL FUNCTIONS

§ 1. Cylindrical functions

In solving many problems of mathematical physics one comes across the ordinary differential equation

$$\left.\begin{aligned} &\frac{d^2 y}{dx^2} + \frac{1}{x}\frac{dy}{dx} + \left(1 - \frac{n^2}{x^2}\right) y = 0 \\ &\frac{1}{x}\frac{d}{dx}\left(x\frac{dy}{dx}\right) + \left(1 - \frac{n^2}{x^2}\right) y = 0 , \end{aligned}\right\} \tag{1}$$

called the *equation of cylindrical functions of nth order*. This equation is often also called *Bessel's equation of nth. order.*

Characteristic problems (see Chapters V, VI and VII), leading to cylindrical functions, are boundary-value problems for the equation

$$\nabla^2 u + k^2 u = 0 \tag{2}$$

outside or inside a circle (outside or inside a cylinder in the case of three independent variables). Introducing polar coordinates, we transform equation (2) to the form

$$\frac{1}{r}\frac{\partial}{\partial r}\left(r\frac{\partial u}{\partial r}\right) + \frac{1}{r^2}\frac{\partial^2 u}{\partial \varphi^2} + k^2 u = 0 . \tag{3}$$

Assuming $u = R\Phi$ and separating the variables in (3), we obtain:

$$\frac{1}{r}\frac{d}{dr}\left(r\frac{dR}{dr}\right) + \left(k^2 - \frac{\lambda}{r^2}\right) R = 0$$

and

$$\Phi'' + \lambda\Phi = 0 .$$

The condition of periodicity for $\Phi(\varphi)$ gives $\lambda = n^2$, where n is a whole number. Substituting $x = kr$, we get the equation for cylindrical functions

$$\frac{1}{x}\frac{d}{dx}\left(x\frac{dy}{dx}\right) + \left(1 - \frac{n^2}{x^2}\right) y = 0, \quad R(r) = y(kr)$$

or
$$y'' + \frac{1}{x} y' + \left(1 - \frac{n^2}{x^2}\right) y = 0 .$$

In the case of solutions of the wave equation (2), possessing radial (cylindrical) symmetry, we obtain Bessel's equation of zero order
$$\frac{1}{x}\frac{d}{dx}\left(x\frac{dy}{dx}\right) + y = 0 \quad \text{or} \quad y'' + \frac{1}{x} y' + y = 0 .$$

1. *Power series*

Bessel's equation of ν th order
$$y'' + \frac{1}{x} y' + \left(1 - \frac{\nu^2}{x^2}\right) y = 0 \tag{1}$$
or
$$x^2 y'' + xy' + (x^2 - \nu^2) y = 0 \tag{1'}$$
(ν is an arbitrary real or complex number, whose real part we may assume positive) has a singular point at $x = 0$. Therefore a solution $y(x)$ should be sought in the form of a power series*
$$y(x) = x^\sigma (a_0 + a_1 x + a_2 x^2 + \ldots + a_k x^k + \ldots), \tag{4}$$
beginning with x^σ, where σ is the characteristic index, to be determined. Substituting series (4) in equation (1′) and equating to zero the coefficients of x^σ, $x^{\sigma+1}$, ..., $x^{\sigma+k}$, we obtain an equation for determining σ and a set of equations for determining the coefficients a_k:
$$\left.\begin{array}{r} a_0 (\sigma^2 - \nu^2) = 0, \\ a_1 \left[(\sigma + 1)^2 - \nu^2\right] = 0, \\ a_2 \left[(\sigma + 2) - \nu^2\right] + a_0 = 0, \\ \cdots\cdots\cdots\cdots \\ \cdots\cdots\cdots\cdots \\ a_k \left[(\sigma + k)^2 - \nu^2\right] + a_{k-2} = 0 \end{array}\right\} \tag{5}$$
$$(k = 2, 3, \ldots) .$$

Since we can assume that $a_0 \neq 0$, then from the first equation of (5) it follows that
$$\sigma^2 - \nu^2 = 0 \quad \text{or} \quad \sigma = \pm \nu . \tag{6}$$

* See Stepanov, *Ordinary Differential Equations*, 1950.

Let us rewrite the kth equation of (5) $k > 1$ in the form

$$(\sigma + k + \nu)(\sigma + k - \nu) a_k + a_{k-2} = 0 .$$

Let us leave for the moment the case where $\sigma + \nu$ or $\sigma - \nu$ (and correspondingly -2ν or 2ν) equals a negative whole number. Then from the second equation of (5), because of (6), we have

$$a_1 = 0 . \tag{8}$$

Equation (7) gives a recurrence relation for determining a_k from a_{k-2}:

$$a_k = -\frac{a_{k-2}}{(\sigma + k + \nu)(\sigma + k - \nu)} . \tag{9}$$

Hence from (8) we deduce that all the odd coefficients equal zero. If ν is real, then for $\sigma = -\nu$ the solution tends to infinity at the point $x = 0$.

Let us consider the case $\sigma = \nu$. From (9) it follows that every even coefficient can be expressed in terms of the preceding one

$$a_{2m} = -a_{2m-2} \frac{1}{2^2 m (m + \nu)} . \tag{10}$$

Successive application of this formula enables us to find an expression for a_{2m} in terms of a_0

$$a_{2m} = (-1)^m \frac{a_0}{2^{2m} m! (\nu + 1)(\nu + 2) \ldots (\nu + m)} . \tag{11}$$

Let us use the property of the gamma-function $\Gamma(s)$*

$$\Gamma(s + 1) = s\Gamma(s) = \ldots = s(s - 1) \ldots (s - n)\Gamma(s - n) .$$

* The function $\Gamma(s)$, defined by the formula

$$\Gamma(s) = \int_0^\infty e^{-x} x^{s-1} dx \text{ where } s > 0 ,$$

is called a *gamma-function*. It is possible also to consider complex values of the variable s: $s = s_0 + is_1$, $s_0 > 0$. We give the main properties of gamma-functions:

(a) integration by parts in the relation for $\Gamma(s)$ gives:

$$\Gamma(s + 1) = \int_0^\infty e^{-x} x^s dx = - e^{-x} x^s \Big|_0^\infty + s \int_0^\infty e^{-x} x^{s-1} dx = s\Gamma(s)$$

or

$$\Gamma(s + 1) = s\Gamma(s) ;$$

(b) for $s = 1$ we immediately obtain:

$$\Gamma(1) = 1 ;$$

(c) if $s = n$ a whole number, then from (a) and (b) it follows:

$$\Gamma(n + 1) = n! ;$$

If S is a whole number, then

$$\Gamma(S+1) = S!$$

The coefficient a_0 up to now has remained arbitrary. If $\nu \neq -n$, where $n > 0$ is a whole number, then, assuming

$$a_0 = \frac{1}{2^{\nu}\,\Gamma(\nu+1)} \tag{12}$$

and utilizing the property of gamma-functions noted above, we obtain:

$$a_{2k} = (-1)^k \frac{1}{2^{2k+\nu}\,\Gamma(k+1)\,\Gamma(k+\nu+1)}. \tag{13}$$

If now $\sigma = -\nu$, $\nu \neq n$, where $n > 0$ is a whole number, then, assuming

$$a_0 = \frac{1}{2^{-\nu}\,\Gamma(-\nu+1)}, \tag{12'}$$

we obtain:

$$a_{2k} = (-1)^k \frac{1}{2^{2k-\nu}\,\Gamma(k+1)\,\Gamma(k-\nu+1)}. \tag{14}$$

Series (3), corresponding to $\sigma = \nu \geqslant 0$, with coefficients (12) and (13)

$$J_{\nu}(x) = \sum_{k=0}^{\infty} (-1)^k \frac{1}{\Gamma(k+1)\,\Gamma(k+\nu+1)} \left(\frac{x}{2}\right)^{2k+\nu} \tag{15}$$

(d) for $s = 1/2$

$$\Gamma(s) = \Gamma\left(\frac{1}{2}\right) = \int_0^{\infty} \frac{e^{-x}}{\sqrt{x}}\,dx = \int_0^{\infty} e^{-\xi^2}\,d\xi_2 = \sqrt{\pi}$$

or

$$\Gamma\left(\frac{1}{2}\right) = \sqrt{\pi};$$

(e) the functional relation $\Gamma(s+1) = s\,\Gamma(s)$ enables us to determine the gamma-function for negative values of s.

We note that

$$\Gamma(0) = \frac{\Gamma(s+1)}{s}\bigg|_{s\to 0} = \infty$$

and

$$\Gamma(-n) = \frac{\Gamma(-n+1)}{-n} = \ldots = (-1)^n \frac{\Gamma(0)}{n!} = \infty \text{ for all integral } n.$$

is called the *Bessel function of first kind of* ν*th order*. The series

$$J_{-\nu}(x) = \sum_{k=0}^{\infty} \frac{(-1)^k}{\Gamma(k+1)\,\Gamma(k-\nu+1)} \left(\frac{x}{2}\right)^{2k-\nu}, \tag{16}$$

corresponding to $\sigma = -\nu$, represents a second solution of equation (1), linearly independent of $J_\nu(x)$. Series (15) and (16), obviously, converge for all values of x.

Let us consider now the case where ν equals half an odd integer. Let $\nu^2 = (n + 1/2)^2$, where $n \geqslant 0$ is a whole number. Assuming $\sigma = n + 1/2$ in relations (5) we obtain:

$$2(n+1)\,a_1 = 0,$$

$$k(k+2n+1)\,a_k + a_{k-2} = 0 \qquad (k > 1),$$

so that

$$a_1 = 0,$$

$$a_k = -\frac{a_{k-2}}{k(k+2n+1)}.$$

Applying this formula in succession we find:

$$a_{2k} = \frac{(-1)^k a_0}{2\cdot 4\ldots(2k)\,(2n+3)(2n+5)\ldots(2n+2k+1)}.$$

Assuming here $\nu = n + 1/2$, we obtain formula (11).

If we put

$$a_0 = \frac{1}{2^{n+\frac{1}{2}}\,\Gamma\left(n+\frac{3}{2}\right)}.$$

we obtain formula (13):

Let

$$\sigma = -n - \frac{1}{2},$$

then equations (5) for a_k take the form

$$a_1 \cdot 1\,(-2n) = 0,$$

$$\cdots\cdots\cdots\cdots\cdots$$

$$\cdots\cdots\cdots\cdots\cdots$$

$$k(k-1-2n)\,a_k + a_{k-2} = 0.$$

As before all the coefficients $a_1, a_3, \ldots, a_{2n-1}$ equal zero, but for a_{2n+1} we obtain the equation $0 \cdot a_{2n+1} + a_{2n-1} = 0$, which is satisfied for any value of a_{2n+1}. For $k > n$ the coefficient a_{2k+1} is determined by the equation

$$a_{2k+1} = \frac{(-1)^{k-n}\,a_{2n+1}}{(2n+3)(2n+5)\ldots 2\cdot 4\ldots(2k-2n)}.$$

Assuming $a_{2n+1} = 0$, $a_0 = \dfrac{1}{2^{-n-\frac{1}{2}} \Gamma\left(\frac{1}{2} - n\right)}$, we obtain (14).

Thus, for $\nu = \pm (n + 1/2)$, a change in definition of the function $J_\nu(x)$ is not necessary. Formulae (15) and (16) remain valid.

We note that (16) defines $J_{-\nu}(x)$ only for fractional values of ν, since the definition of a_0 by formula (12) for negative integer $\nu = -n$ has no meaning. Let us investigate the behaviour of (16) as ν tends to an integer n. Since $\Gamma(k - n + 1) = \infty$ for $k \leqslant k_0 < n - 1$, the sum (16) begins with the values $k = k_0 + 1 = n$. Changing the index of summation $k = n + k'$ in (16) we obtain:

$$J_{-n}(x) = (-1)^n \sum_{k'=0}^{\infty} \frac{(-1)^{k'}}{\Gamma(k' + n + 1)\, \Gamma(k' + 1)} \left(\frac{x}{2}\right)^{2k'+n} = (-1)^n J_n(x),$$

since the summation begins with $k' = 0$.

We give as examples the series for Bessel functions of the first kind of zero ($n = 0$) and first ($n = 1$) orders:

$$J_0(x) = 1 - \left(\frac{x}{2}\right)^2 + \frac{1}{(2!)^2}\left(\frac{x}{2}\right)^4 - \frac{1}{(3!)^2}\left(\frac{x}{2}\right)^6 + \ldots,$$

$$J_1(x) = \frac{x}{2} - \frac{1}{2!}\left(\frac{x}{2}\right)^3 + \frac{1}{2!\,3!}\left(\frac{x}{2}\right)^5 - \ldots$$

The functions $J_0(x)$ and $J_1(x)$ occur most often in applications and there exist detailed tables.* On p. 751 graphs of $J_0(x)$ and $J_1(x)$ are given.

The functions $J_n(x)$ and $J_{-n}(x)$ (n an integer), as we saw, are linearly dependent

$$J_{-n}(x) = (-1)^n J_n(x).$$

For non-integral values of ν the functions $J_\nu(x)$ and $J_{-\nu}(x)$ are linearly independent. In fact, $J_\nu(x)$ has a zero, and $J_{-\nu}(x)$ a pole of νth order at the point $x = 0$. Thus, if ν is a non-integral number, then any solution $y_\nu(x)$ of Bessel's equation (1) can

* In all the tables of special functions there always exist tables of Bessel functions of the first kind (see, for instance, Janke and Emde, *Tables of Bessel Functions*, where $J_0(x)$ and $J_1(x)$ are given to five places for values of x in the interval from 0 to 14.9).

be represented as a linear combination of the functions $J_\nu(x)$ and $J_{-\nu}(x)$

$$y_\nu(x) = C_1 J_\nu(x) + C_2 J_{-\nu}(x).$$

If a bounded solution of equation (1) is required, then $C_2 = 0$ and

$$y_\nu(x) = C_1 J_\nu(x).$$

2. *Recurrence relations*

Let us establish the following relations, existing between Bessel functions of the first kind of different orders,

$$\frac{d}{dx}\left(\frac{J_\nu(x)}{x^\nu}\right) = -\frac{J_{\nu+1}(x)}{x^\nu}, \tag{17}$$

$$\frac{d}{dx}\left(x^\nu J_\nu(x)\right) = x^\nu J_{\nu-1}(x). \tag{18}$$

These relations are verified directly by differentiation of the series of Bessel functions. We show, for example, the validity of (17)

$$x^\nu \frac{d}{dx}\left(\frac{J_\nu(x)}{x^\nu}\right) = x^\nu \frac{1}{2!}\sum_{k=1}^{\infty}(-1)^k \frac{\frac{1}{2}\left(\frac{x}{2}\right)^{2k-1} 2k}{k!\,\Gamma(k+\nu+1)} =$$

$$= \sum_{k=1}^{\infty}(-1)^k \frac{1}{\Gamma(k)\,\Gamma(k+\nu+1)}\left(\frac{x}{2}\right)^{2k+(\nu-1)}.$$

In the latter sum k goes from 1 to ∞. Let us introduce a new index of summation $l = k - 1$, which will vary from 0 to ∞. Then we have:

$$x^\nu \frac{d}{dx}\left(\frac{J_\nu(x)}{x^\nu}\right) = -\sum_{l=1}^{\infty}(-1)^l \frac{1}{\Gamma(l+1)\,\Gamma[l+(\nu+1)+1]}\left(\frac{x}{2}\right)^{[2l+(\nu+1)]} =$$

$$= -J_{\nu+1}(x),$$

which proves formula (17). The validity of (18) is proved similarly.

We note two important particular cases of the recurrence relations. For $\nu = 0$ it follows from (17):

$$J_0'(x) = -J_1(x). \tag{19}$$

For the case $\nu = 1$ (18) gives:

$$[x J_1(x)]' = x J_0(x) \quad \text{or} \quad x J_1(x) = \int_0^x \xi J_0(\xi)\,d\xi. \tag{20}$$

Let us establish recurrence relations, connecting $J_\nu(x)$, $J_{\nu+1}^1(x)$ and $J_{\nu-1}(x)$. Performing a differentiation in (17) and (18) we obtain:

$$\frac{\nu J_\nu(x)}{x} - J'_\nu(x) = J_{\nu+1}(x), \tag{17'}$$

$$\frac{\nu J_\nu(x)}{x} + J'_\nu(x) = J_{\nu-1}(x). \tag{18'}$$

Adding and subtracting (17′) and (18′) we find the recurrence relations

$$\left.\begin{aligned} J_{\nu+1}(x) + J_{\nu-1}(x) &= \frac{2\nu}{x} J_\nu(x), \\ J_{\nu+1}(x) - J_{\nu-1}(x) &= -2J'_\nu(x), \end{aligned}\right\} \tag{21}$$

By means of (21) it is possible to evaluate $J_{\nu+1}(x)$, if $J_\nu(x)$ and $J_{\nu-1}(x)$ are known:

$$J_{\nu+1}(x) = -J_{\nu-1}(x) + \frac{2\nu J_\nu(x)}{x}. \tag{21'}$$

3. *Functions of half-integral order*

Let us determine the values of the functions $J_{\frac{1}{2}}(x)$ and $J_{-\frac{1}{2}}(x)$:

$$J_{\frac{1}{2}}(x) = \sum_{m=0}^{\infty} \frac{(-1)^m}{m!\,\Gamma\left(\frac{3}{2}+m\right)} \left(\frac{x}{2}\right)^{\frac{1}{2}+2m}, \tag{22}$$

$$J_{-\frac{1}{2}}(x) = \sum_{m=0}^{\infty} \frac{(-1)^m}{m!\,\Gamma\left(\frac{1}{2}+m\right)} \left(\frac{x}{2}\right)^{-\frac{1}{2}+2m} \tag{23}$$

Making use of a property of gamma-functions, we find:

$$\left.\begin{aligned} \Gamma\left(\frac{3}{2}+m\right) &= \frac{1\cdot 3\cdot 5\ldots(2m+1)}{2^{m+1}}\,\Gamma\left(\frac{1}{2}\right), \\ \Gamma\left(\frac{1}{2}+m\right) &= \frac{1\cdot 3\ldots(2m-1)}{2^m}\,\Gamma\left(\frac{1}{2}\right), \end{aligned}\right\} \tag{24}$$

where

$$\Gamma\left(\frac{1}{2}\right) = \sqrt{\pi}.$$

Substituting (24) in (22) and (23), we obtain:

$$J_{\frac{1}{2}}(x) = \sqrt{\left(\frac{2}{\pi x}\right)} \sum_{m=0}^{\infty} \frac{(-1)^m}{(2m+1)!} x^{2m+1}, \tag{25}$$

$$J_{-\frac{1}{2}}(x) = \sqrt{\left(\frac{2}{\pi x}\right)} \sum_{m=0}^{\infty} \frac{(-1)^m}{(2m)!} x^{2m}. \tag{26}$$

The sum in (25) is the expansion of $\sin x$, and the sum in (26) is an expansion of $\cos x$ in powers of x. Thus $J_{\frac{1}{2}}(x)$ and $J_{-\frac{1}{2}}(x)$ are expressed by the elementary functions

$$J_{\frac{1}{2}}(x) = \sqrt{\left(\frac{2}{\pi x}\right)} \sin x, \tag{27}$$

$$J_{-\frac{1}{2}}(x) = \sqrt{\left(\frac{2}{\pi x}\right)} \cos x. \tag{28}$$

Let us calculate the function $J_{n+\frac{1}{2}}(x)$, where n is an integer. From (21′) it follows:

$$J_{\frac{3}{2}}(x) = \frac{1}{x} J_{\frac{1}{2}}(x) - J_{-\frac{1}{2}}(x) = \sqrt{\left(\frac{2}{\pi x}\right)} \left(-\cos x + \frac{\sin x}{x}\right) =$$

$$= \sqrt{\left(\frac{2}{\pi x}\right)} \left[\sin\left(x - \frac{\pi}{2}\right) + \frac{1}{x} \cos\left(x - \frac{\pi}{2}\right)\right];$$

$$J_{\frac{5}{2}}(x) = \sqrt{\left(\frac{2}{\pi x}\right)} \left\{-\sin x + \frac{3}{x}\left[\sin\left(x - \frac{\pi}{2}\right) + \frac{1}{x}\cos\left(x - \frac{\pi}{2}\right)\right]\right\} =$$

$$= \sqrt{\left(\frac{2}{\pi x}\right)} \left\{\sin(x - \pi)\left(1 - \frac{3}{x^2}\right) + \cos(x - \pi) \cdot \frac{3}{x}\right\},$$

Applying (21′) in succession we find:

$$J_{n+\frac{1}{2}}(x) = \sqrt{\left(\frac{2}{\pi x}\right)} \left\{\sin\left(x - \frac{n\pi}{2}\right) P_n\left(\frac{1}{x}\right) + \cos\left(x - \frac{n\pi}{2}\right) Q_n\left(\frac{1}{x}\right)\right\}, \tag{29}$$

where $P_n(1/x)$ is a polynomial of degree n in $1/x$, and $Q_n(1/x)$ is a polynomial of degree $n-1$. We note that $P_n(0) = 1$, $Q_n(0) = 0$.

4. *Asymptotic form of cylindrical functions*

Let us show that any cylindrical function for large x can be represented in the form

$$y_\nu(x) = \gamma_\infty \frac{\sin(x + \delta_\infty)}{\sqrt{x}} + O\left(\frac{1}{x^{3/2}}\right). \tag{30}$$

where $\gamma_\infty \neq 0$, δ_∞ are constants, $O(1/x^{3/2})$ indicates terms of order not less than $1/x^{3/2}$.

If we put
$$y = \frac{v(x)}{\sqrt{x}}, \tag{31}$$

evaluate the derivatives $y' = -0.5x^{-3/2}v + x^{-1/2}v'$, $y'' = x^{-1/2}v'' - x^{-3/2}v' + 0.75x^{-5/2}v$ and substitute them in Bessel's equation, we obtain the equation

$$v'' + \left(1 - \frac{\nu^2 - \frac{1}{4}}{x^2}\right) v = 0, \tag{32}$$

which is a particular case of the equation

$$v'' + v + \varrho(x) v = 0, \tag{33}$$

where

$$\varrho(x) = O\left(\frac{1}{x^2}\right). \tag{34}$$

Let us assume

$$v = \gamma \sin(x + \delta), \qquad v' = \gamma \cos(x + \delta), \tag{35}$$

where $\gamma(x)$ and $\delta(x)$ are some functions of x, we must have $\gamma(x) \neq 0$ for all x, otherwise v and v' would simultaneously reduce to zero and $v(x)$ would be identically equal to zero. Utilizing (35) and (33) we have:

$$v' = \gamma \cos(x + \delta) = \gamma' \sin(x + \delta) + \gamma(\delta' + 1)\cos(x + \delta),$$
$$v'' = \gamma' \cos(x + \delta) - \gamma(\delta' + 1)\sin(x + \delta) = -(1 + \varrho)\gamma \sin(x + \delta).$$

Hence we find:

$$\delta' = \varrho \sin^2(x + \delta), \qquad \delta' = O\left(\frac{1}{x^2}\right), \tag{36}$$

$$\frac{\gamma'}{\gamma} = -\frac{\delta'}{\tan(x + \delta)} = \varrho \sin(x + \delta)\cos(x + \delta) = O\left(\frac{1}{x^2}\right). \tag{37}$$

Let us show that there exist definite limiting values of γ and δ as $x \to \infty$.

In fact

$$\delta(x) = \delta(a) - \int_x^a \delta'(s)\, ds,$$

from which, because of (36), it follows that there exists a limit $\lim_{a \to \infty} \delta(a) = \delta_\infty$ and

$$\delta(x) = \delta_\infty + O\left(\frac{1}{x}\right). \tag{38}$$

Similarly we find from (37)

$$\gamma(x) = \gamma_\infty \left(1 + O\left(\frac{1}{x}\right)\right), \tag{39}$$

in which $\gamma_\infty \neq 0$.

Thus, any solution of equation (33) and, therefore, of equation (32) as $x \to \infty$ has the form

$$v(x) = \gamma_\infty \sin(x + \delta_\infty) + O\left(\frac{1}{x}\right). \tag{40}$$

In this way the validity of the asymptotic relation (30) is established for any cylindrical function $y_\nu(x)$.

We note that there cannot exist two different cylindrical functions with the same asymptotic form. In fact, let $\bar{y}_\nu(x)$ and $\bar{\bar{y}}_\nu(x)$ be two different cylindrical functions, for which

$$\bar{\gamma}_\infty = \bar{\bar{\gamma}}_\infty, \qquad \bar{\delta}_\infty = \bar{\bar{\delta}}_\infty. \tag{41}$$

The difference of these functions

$$\tilde{y}_\nu(x) = \bar{y}_\nu(x) - \bar{\bar{y}}_\nu(x) \not\equiv 0$$

is also a cylindrical function, having, by (41), the following asymptotic form:

$$\tilde{y}_\nu(x) = O\left(\frac{1}{x^{3/2}}\right).$$

But this contradicts formula (30) for any cylindrical function $\tilde{y}_\nu(x)$.

Therefore, $\tilde{y}_\nu(x) \equiv 0$ and $\bar{y}_\nu(x) = \bar{\bar{y}}_\nu(x)$.

The values of the constants γ_∞ and δ_∞ can be found by further analyses, which gives

$$\gamma_\infty = \sqrt{\frac{2}{\pi}} \text{ for all } \nu.$$

In § 1, subsection 3, formula (29) for $\nu = n + 1/2$ was derived, from which it follows that

$$J_{n+\frac{1}{2}}(x) = \sqrt{\left(\frac{2}{\pi x}\right)} \sin\left(x - \frac{n\pi}{2}\right) + O\left(\frac{1}{x^{3/2}}\right). \tag{42}$$

In § 3 the asymptotic form for the function $J_\nu(x)$ will be derived with $\nu = n$, where n is an integer:

$$J_\nu(x) = \sqrt{\left(\frac{2}{\pi x}\right)} \cos\left(x - \frac{\pi}{2}\nu - \frac{\pi}{4}\right) + O\left(\frac{1}{x^{3/2}}\right). \qquad (43)$$

Analyses, based on representing the functions $J_\nu(x)$ and $J_{-\nu}(x)$ by means of contour integrals show that formula (43) holds for arbitrary ν, hence

$$J_{-\nu}(x) = \sqrt{\left(\frac{2}{\pi x}\right)} \cos\left(x + \frac{\pi}{2}\nu - \frac{\pi}{4}\right) + O\left(\frac{1}{x^{3/2}}\right). \qquad (44)$$

§ 2. Boundary-value problems for Bessel's equation

The simplest boundary-value problem for Bessel's equation in the segment $[0, r_0]$ is related to the problem of the characteristic oscillations of a circular membrane

$$\nabla_2^2 v + \lambda v = 0, \quad \nabla_2^2 v = \frac{1}{r}\frac{\partial}{\partial r}\left(r\frac{\partial v}{\partial r}\right) + \frac{1}{r_2}\frac{\partial^2 v}{\partial \varphi^2}, \qquad (1)$$

$$v(r, \varphi)\big|_{r=r_0} = 0. \qquad (2)$$

Assuming $v(r, \varphi) = R(r)\,\Phi(\varphi)$ and separating the variables, we obtain:

$$\Phi'' + \nu\Phi = 0, \qquad (3)$$

$$\frac{1}{r}\frac{d}{dr}\left(r\frac{dR}{dr}\right) + \left(\lambda - \frac{\nu}{r^2}\right)R = 0, \quad R(r_0) = 0. \qquad (4)$$

The condition of periodicity for $\Phi(\varphi)$ gives $\nu = n^2$, where n is an integer. Thus, the function $R(r)$ should be determined from Bessel's equation

$$\mathscr{L}[R] + \lambda r R = 0 \qquad \left(\mathscr{L}[R] = \frac{d}{dr}\left(r\frac{dR}{dr}\right) - \frac{n^2}{r}R\right) \qquad (5)$$

with the boundary condition

$$R(r_0) = 0 \qquad (6)$$

and by the condition that $R(r)$ should be bounded at the point $r = 0$

$$|R(0)| < \infty. \qquad (7)$$

Assuming

$$\left.\begin{aligned} x &= \sqrt{\lambda}\, r \\ y(x) &= R(r) = R\left(\frac{x}{\sqrt{\lambda}}\right), \end{aligned}\right\} \tag{8}$$

we get the equation

$$\frac{1}{x}\frac{d}{dx}\left(x\frac{dy}{dx}\right) + \left(1 - \frac{n^2}{x^2}\right) y = 0 \tag{9}$$

with the boundary conditions

$$y(\sqrt{\lambda}\, r_0) = 0\,, \tag{10}$$

$$|y(0)| < \infty\,. \tag{11}$$

Hence we find

$$y(x) = A J_n(x)\,. \tag{12}$$

By virtue of the boundary condition $y(r_0\sqrt{\lambda}) = 0$ we have:

$$J_n(\mu) = 0\,. \tag{13}$$

This transcendental equation has infinitely many real roots $\mu_1^{(n)}, \mu_2^{(n)}, \ldots, \mu_m^{(n)}, \ldots,$* i. e. equation (1) has infinitely many eigen-values

$$\lambda_m^{(n)} = \left(\frac{\mu_m^{(n)}}{r_0}\right)^2 \quad (m = 1,\ 2,\ \ldots)\,, \tag{14}$$

to which correspond the eigen-functions

$$R(r) = A J_n\left(\frac{\mu_m^{(n)}}{r_0}\, r\right) \tag{15}$$

of the boundary value-problem (1)–(2).

From the general theory of equations of the form $\mathscr{L}[y] + \lambda \varrho\, y = 0$, considered above, the orthogonality of the system of eigen-functions follows

$$\left\{J_n\left(\frac{\mu_m^{(n)}}{r_0}\, r\right)\right\}$$

with weight r:

$$\int_0^{r_0} J_n\left(\frac{\mu_m^{(n)}}{r_0}\, r\right) J_n\left(\frac{\mu_{m_2}^{(n)}}{r_0}\, r\right) r\, dr = 0 \text{ where } m_1 \neq m_2\,. \tag{16}$$

* On p. 753 tables of the roots of the equation $J_0(\mu) = 0$ are given, in particular the first root $\mu_1^{(0)} = 2.4048$.

Let us evaluate the norm of the eigen-functions $R_1(r) = J_n(a_1 r)$, where $a_1 = -\mu_m^{(n)}/r_0$. In order to do this let us consider the function $R_2(r) = J_n(a_2 r)$, where a_2 is an arbitrary parameter.

The functions $R_1(r)$ and $R_2(r)$ satisfy the equations

$$\frac{d}{dr}\left(r\frac{dR_1}{dr}\right) + \left(a_1^2 r - \frac{n^2}{r}\right) R_1 = 0,$$

$$\frac{d}{dr}\left(r\frac{dR_2}{dr}\right) + \left(a_2^2 r - \frac{n^2}{r}\right) R_2 = 0,$$

in which $R_1(r_0) = 0$ and $R_2(r)$ no longer satisfies this boundary condition. Subtracting the second equation from the first, after multiplying them by $R_2(r)$ and $R_1(r)$ respectively and integrating next with respect to r from 0 to r_0 we have:

$$(a_1^2 - a_2^2)\int_0^{r_0} rR_1(r)\,R_2(r)\,dr + [r(R_2R_1' - R_1R_2')]\Big|_0^{r_0} = 0,$$

from which we find:

$$\int_0^{r_0} R_1R_2 r\,dr = -\frac{r_0J_n(a_2r_0)\,a_1J_n'(a_1r_0) - r_0J_n(a_1r_0)\,a_2J_n'(a_2r_0)}{a_1^2 - a_2^2} =$$

$$= -\frac{r_0J_n(a_2r_0)\,a_1J_n'(a_1r_0)}{a_1^2 - a_2^2}.$$

Passing to a limit as $a_2 \to a_1$ and expanding the right-hand side, we obtain an expression for the norm

$$N = \int_0^{r_0} rR_1^2(r)\,dr = \frac{r_0^2}{8}\,[J_n'(a_1 r_0)]^2$$

or

$$\int_0^{r_0} J_n^2\left(\frac{\mu_m^{(n)}}{r_0}r\right) r\,dr = \frac{r_0^2}{2}\,[J_n'(\mu_m^{(n)})]^2. \qquad (17)$$

In particular, the norm of the function $J_0(\mu_m^{(0)}/r_0\, r)$ equals

$$\int_0^{r_0} J_0^2\left(\frac{\mu_m^{(0)}}{r_0}r\right) r\,dr = \frac{r_0^2}{2}\,J_1^2(\mu_m^{(0)}). \qquad (18)$$

The general properties of the eigen-functions of boundary-value problems and the note on p. 635, show that the expansion theorem holds:

Any twofold differentiable function $f(r)$, bounded at $r = 0$ and reducing to zero at $r = r_0$, can be expanded in an absolutely and uniformly convergent series

$$f(r) = \sum_{m=1}^{\infty} A_m J_n\left(\frac{\mu_m^{(n)}}{r_0} r\right),$$

where

$$A_m = \frac{\int\limits_0^{r_0} f(r) J_n\left(\frac{\mu_m^{(n)}}{r_0} r\right) r\,dr}{N_m},$$

$$N_m = \frac{r_0^2}{2}\left[J_n'\left(\mu_m^{(n)}\right)\right]^2.$$

The second boundary-value problem for Bessel's equation

$$\mathscr{L}(R) + \lambda r R = 0,$$
$$R'(r_0) = 0,$$
$$R(0) < \infty$$

is solved similarly. The eigen-functions and eigen-values will also be expressed by formulae (15) and (14), where by $\mu_m^{(n)}$ one should understand the mth root of the equation

$$J_n'(\mu) = 0.$$

The eigen-functions of the problem are mutually orthogonal with weight r [see (16)] and have a norm equal to

$$\int\limits_0^{r_0} J_n^2\left(\frac{\mu_m^{(n)} r}{r_0}\right) r\,dr = \frac{r_0^2}{2}\left[1 - \frac{n^2}{\left(\mu_m^{(n)}\right)^2}\right] J_n^2\left(\mu_m^{(n)}\right).$$

Similarly the third boundary-value problem is solved. In this case we obtain an equation of the form

$$J_n'(\mu) = h J_n(\mu).$$

for determining $\mu_m^{(n)}$.

§ 3. Various types of cylindrical functions

1. *Hankel functions*

Together with Bessel functions of the first kind, $J_\nu(x)$, other special forms of solution of Bessel's equation have value in applications. The *Hankel functions of first and second kind* $H_\nu^{(1)}(x)$ and $H_\nu^{(2)}(x)$ are complex-conjugate solutions of Bessel's equation. From the point of view of physical applications the main characteristic of Hankel functions is the asymptotic behaviour for large values of the argument

$$H_\nu^{(1)}(x) = \sqrt{\left(\frac{2}{\pi x}\right)}\, e^{i\left(x-\frac{\pi}{2}\nu-\frac{\pi}{4}\right)} + \ldots, \tag{1}$$

$$H_\nu^{(2)}(x) = \sqrt{\left(\frac{2}{\pi x}\right)}\, e^{-i\left(x-\frac{\pi}{2}\nu-\frac{\pi}{4}\right)} + \ldots, \tag{2}$$

Separating the real and imaginary parts, let us represent the Hankel function in the form

$$H_\nu^{(1)}(x) = J_\nu(x) + iN_\nu(x), \tag{3}$$

$$H_\nu^{(2)}(x) = J_\nu(x) - iN_\nu(x), \tag{4}$$

where the functions

$$J_\nu(x) = \frac{1}{2}\left[H_\nu^{(1)}(x) + H_\nu^{(2)}(x)\right], \tag{3'}$$

$$N_\nu(x) = \frac{1}{2i}\left[H_\nu^{(1)}(x) - H_\nu^{(2)}(x)\right] \tag{4'}$$

have an asymptotic form

$$J_\nu(x) = \sqrt{\left(\frac{2}{\pi x}\right)} \cos\left(x - \frac{\pi}{2}\nu - \frac{\pi}{4}\right) + \ldots, \tag{5}$$

$$N_\nu(x) = \sqrt{\left(\frac{2}{\pi x}\right)} \sin\left(x - \frac{\pi}{2}\nu - \frac{\pi}{4}\right) + \ldots, \tag{6}$$

which follows from (1) and (2).

As will be shown below (see sub-section 3), the function J_ν (x introduced here is a Bessel function of the first kind, considered in § 1. The imaginary part $N_\nu(x)$ of the Hankel function is called the *Neumann function* or the *cylindrical function of second kind of νth order*.

Formulae (3) and (4) determine a relation between Hankel, Bessel and Neumann functions, similar to the relation between the exponential functions of imaginary argument, sine and cosine. The asymptotic formulae (1), (2), (5) and (6) indicate the analogy.

In a study of solutions of the wave equation

$$u_{tt} = a^2 (u_{xx} + u_{yy})$$

we saw that the amplitude $v(x, y)$ of the steady vibrations

$$u(x, y, t) = v(x, y) e^{i\omega t}$$

satisfied the wave equation

$$v_{xx} + v_{yy} + k^2 v = \Delta v + k^2 v = 0 \quad \left(k^2 = \frac{\omega^2}{a^2}\right).$$

If the solution of the wave equation possesses radial symmetry $v(x, y) = v(r)$, then, as was noted in § 1, the function $v(kr)$ satisfies Bessel's equation of zero order.

Thus, the functions

$$H_0^{(1)} (kr) e^{i\omega t} = \sqrt{\left(\frac{2}{\pi kr}\right)} e^{i(\omega t + kr)} \frac{1}{\sqrt{i}} + \ldots \tag{7}$$

$$\left(\sqrt{i} = e^{i\frac{\pi}{4}}\right)$$

and

$$H_0^{(2)} (kr) e^{i\omega t} = \sqrt{\left(\frac{2}{\pi kr}\right)} e^{i(\omega t - kr)} \sqrt{i} + \ldots \tag{8}$$

are solutions of the wave equation, having the form of cylindrical waves. The function $H_0^{(2)} (kr) e^{i\omega t}$ corresponds to a diverging cylindrical wave, and the function $H_0^{(1)} (kr) e^{i\omega t}$ to a converging cylindrical wave.*

2. *Hankel and Neumann functions*

As was noted in sub-section 1, any solution of Bessel's equation of non-integral order ν can be expressed as a linear combination of J_ν or $J_{-\nu}$. Let us establish the connection between the functions $H_\nu^{(1)}$, $H_\nu^{(2)}$, N_ν and J_ν, $J_{-\nu}$.

* If the time factor $e^{-i\omega t}$ is taken, then $H_0^{(1)} (kr)e^{-i\omega t}$ corresponds to a diverging wave, and $H_0^{(2)} (kr)e^{-i\omega t}$ to a converging wave.

Since any solution of Bessel's equation for non-integral ν can be represented in the form of a linear combination of functions $J_\nu(x)$ and $J_{-\nu}(x)$, then

$$H_\nu^{(1)}(x) = C_1 J_\nu(x) + C_2 J_{-\nu}(x)\,, \tag{9}$$

where C_1 and C_2 are constants, to be determined. A similar equation holds for the main terms of the asymptotic expansions

$$\sqrt{\left(\frac{2}{\pi x}\right)} e^{i\left(x-\frac{\pi}{2}\nu-\frac{\pi}{4}\right)} =$$
$$= C_1 \sqrt{\left(\frac{2}{\pi x}\right)} \cos\left(x - \frac{\pi}{2}\nu - \frac{\pi}{4}\right) + C_2 \sqrt{\left(\frac{2}{\pi x}\right)} \cos\left(x + \frac{\pi}{2}\nu - \frac{\pi}{2}\right). \tag{10}$$

Let us transform the argument of the second component to the form $\left(x - \frac{\pi}{2}\nu - \frac{\pi}{4}\right)$

$$\cos\left(x + \frac{\pi}{2}\nu - \frac{\pi}{4}\right) = \cos\left[\left(x - \frac{\pi}{2}\nu - \frac{\pi}{4}\right) + \pi\nu\right] =$$
$$= \cos\left(x - \frac{\pi}{2}\nu - \frac{\pi}{4}\right)\cos \pi\nu - \sin\left(x - \frac{\pi}{2}\nu - \frac{\pi}{4}\right)\sin \pi\nu\,.$$

Dividing both sides of (10) by $\sqrt{\frac{2}{\pi x}}$ and using Euler's theorem for the left-hand side, we obtain:

$$\cos\left(x - \frac{\pi}{2}\nu - \frac{\pi}{4}\right) + i\sin\left(x - \frac{\pi}{2}\nu - \frac{\pi}{4}\right) =$$
$$= (C_1 + C_2\cos\pi\nu)\cos\left(x - \frac{\pi}{2}\nu - \frac{\pi}{4}\right) -$$
$$- C_2 \sin\pi\nu \sin\left(x - \frac{\pi}{2}\nu - \frac{\pi}{4}\right),$$

from which

$$C_1 + C_2 \cos\pi\nu = 1\,,$$
$$- C_2 \sin\pi\nu = i$$

or

$$\left.\begin{aligned} C_2 &= \frac{1}{i\sin\pi\nu}\,; \\ C_1 &= -\frac{\cos\pi\nu - i\sin\pi\nu}{i\sin\pi\nu} = -C_2 e^{-i\pi\nu}\,. \end{aligned}\right\} \tag{11}$$

Substituting (11) in (9) we find:

$$H_\nu^{(1)}(x) = -\frac{1}{i\sin\pi\nu}\left[J_\nu(x)e^{-i\pi\nu} - J_{-\nu}(x)\right]. \tag{12}$$

Similarly

$$H_\nu^{(2)}(x) = \frac{1}{i \sin \pi\nu}\left[J_\nu(x) e^{i\pi\nu} - J_{-\nu}(x)\right]. \tag{13}$$

Using (4′), defining $N_\nu(x)$, we obtain from (12) and (13):

$$N_\nu(x) = \frac{J_\nu(x)\cos\pi\nu - J_{-\nu}(x)}{\sin\pi\nu}. \tag{14}$$

Formulae (12), (13) and (14) are obtained by us for non-integral values of ν. For the integral value $\nu = n$ the Hankel and Neumann functions can be determined from (12), (13) and (14) by means of a limiting transition as $\nu \to n$. Passing to a limit as $\nu \to n$ in these formulae, we have:

$$H_n^{(1)}(x) = J_n(x) + i\frac{1}{\pi}\left[\left(\frac{\partial J_\nu}{\partial\nu}\right)_{\nu=n} - (-1)^n\left(\frac{\partial J_{-\nu}}{\partial\nu}\right)_{\nu=n}\right], \tag{12′}$$

$$H_n^{(2)}(x) = J_n(x) - i\frac{1}{\pi}\left[\left(\frac{\partial J_\nu}{\partial\nu}\right)_{\nu=n} - (-1)^n\left(\frac{\partial J_{-\nu}}{\partial\nu}\right)_{\nu=n}\right], \tag{13′}$$

$$N_n(x) = \frac{1}{\pi}\left[\left(\frac{\partial J_\nu}{\partial\nu}\right)_{\nu=n} - (-1)^n\left(\frac{\partial J_{-\nu}}{\partial\nu}\right)_{\nu=n}\right]. \tag{14′}$$

Using the representations of the functions J_ν and $J_{-\nu}$ as power series, it is possible to derive similar expansions for $N_\nu(x)$ and also $H_\nu^{(1)}(x)$ and $H_\nu^{(2)}(x)$.

Formulae (12) and (13) may be considered as an analytical definition of the Hankel functions. There are, however, other methods of introducing the Hankel functions. In § 6 we will give a definition of the Hankel functions as contour integrals.

If $\nu = n + 1/2$, then the Hankel and Neumann functions are expressed in finite form by elementary functions. In particular, for $\nu = 1/2$, we have:

$$N_{\frac{1}{2}}(x) = -J_{-\frac{1}{2}}(x) = -\sqrt{\left(\frac{2}{\pi x}\right)}\cos x = \sqrt{\left(\frac{2}{\pi x}\right)}\sin\left(x - \frac{\pi}{2}\right),$$

$$H_{\frac{1}{2}}^{(1)}(x) = J_{\frac{1}{2}}(x) + iN_{\frac{1}{2}}(x) = \sqrt{\left(\frac{2}{\pi x}\right)}\left[\cos\left(x - \frac{\pi}{2}\right) + \right.$$
$$\left. + i\sin\left(x - \frac{\pi}{2}\right)\right] = \sqrt{\left(\frac{2}{\pi x}\right)}\, e^{i\left(x - \frac{\pi}{2}\right)},$$

$$H_{\frac{1}{2}}^{(2)}(x) = J_{\frac{1}{2}}(x) - iN_{\frac{1}{2}}(x) = \sqrt{\left(\frac{2}{\pi x}\right)}\left[\cos\left(x - \frac{\pi}{2}\right) - \right.$$
$$\left. - i\sin\left(x - \frac{\pi}{2}\right)\right] = \sqrt{\left(\frac{2}{\pi x}\right)}\, e^{-i\left(x - \frac{\pi}{2}\right)}.$$

3. *Functions of an imaginary argument*

Cylindrical functions may be considered not only for real, but also for complex values of the argument. In the present section we consider Bessel functions of first kind for a purely imaginary argument.

Substituting in the series, defining $J_\nu(x)$, the value ix in place of x, we obtain:

$$J_\nu(ix) = i^\nu \sum_{k=0}^{\infty} \frac{(-1)^k i^{2k}}{\Gamma(k+1)\,\Gamma(k+\nu+1)} \left(\frac{x}{2}\right)^{2k+\nu} = i^\nu I_\nu(x)\,, \quad (15)$$

where

$$I_\nu(x) = \sum_{k=0}^{\infty} \frac{1}{\Gamma(k+1)\,\Gamma(k+\nu+1)} \left(\frac{x}{2}\right)^{2k+\nu} \quad (16)$$

is a real function, related to $J_\nu(ix)$ by the equation

$$I_\nu(x) = i^{-\nu} J_\nu(ix) \text{ or } I_\nu(x) = e^{-\frac{1}{2}\pi\nu i} J_\nu(ix)\,.$$

In particular, for $\nu = 0$

$$I_0(x) = J_0(ix) = 1 + \left(\frac{x}{2}\right)^2 + \frac{1}{(2!)^2}\left(\frac{x}{2}\right)^4 + \frac{1}{(3!)^2}\left(\frac{x}{2}\right)^6 + \ldots \quad (17)$$

From series (16) we see that $I_\nu(x)$ are monotonic increasing functions. Using the asymptotic relation (5), we see that

$$I_\nu(x) \approx \sqrt{\left(\frac{1}{2\pi x}\right)}\, e^x \quad (18)$$

should hold for large values of the argument x.

The Bessel functions for imaginary argument are solutions of the equation

$$y'' + \frac{1}{x} y' - \left(1 + \frac{\nu^2}{x^2}\right) y = 0 \quad (19)$$

and, in particular the function $I_0(x)$ satisfies the equation

$$y'' + \frac{1}{x} y' - y = 0\,. \quad (20)$$

Along with the functions $I_\nu(x)$ one considers the function $K_\nu(x)$, defined by the Hankel function for a purely imaginary argument

$$K_\nu(x) = \frac{1}{2} \pi i e^{\frac{1}{2}\pi\nu i} H_\nu^{(1)}(ix)\,. \quad (21)$$

Below we verify that $K_\nu(x)$ has real values for a real value of x. This will follow from the relation

$$K_\nu(x) = \frac{1}{2}\int_{-\infty}^{\infty} e^{-x\cosh\eta - \nu\eta}\, d\eta, \tag{22}$$

which will be derived in § 6. Using the asymptotic expression for $H_\nu^{(1)}$, we find:

$$K_\nu(x) = \sqrt{\left(\frac{\pi}{2x}\right)}\, e^{-x} + \ldots \tag{23}$$

Formulae (23) and (18) show that $K_\nu(x)$ decreases exponentially, and $I_\nu(x)$ increases exponentially as $x \to \infty$. Hence the linear independence of these functions follows, and also the possibility of representing any solution of (19) as a linear combination

$$y = AI_\nu(x) + BK_\nu(x).$$

In particular, if y is bounded at infinity, then $A = 0$ and $y = BK_\nu(x)$.

The function

$$K_0(x) = \int_0^\infty e^{-x\cosh\eta}\, d\eta.$$

is of great importance.

4. The function $K_0(x)$

Let us show that the following integral representation is valid for the function $K_0(x)$:

$$K_0(x) = \int_0^\infty e^{-x\cosh\xi}\, d\xi \qquad (x > 0). \tag{24}$$

We can show that the integral

$$F(x) = \int_0^\infty e^{-x\cosh\xi}\, d\xi \tag{24'}$$

satisfies the equation

$$\mathscr{L}(y) = y'' + \frac{1}{x}y' - y = 0. \tag{25}$$

In fact

$$\mathscr{L}(F) = \int_0^\infty e^{-x\cosh\xi}\left(\cosh^2\xi - \frac{1}{x}\cosh\xi - 1\right) d\xi =$$

$$= \int_0^\infty e^{-x\cosh\xi}\sinh^2\xi\, d\xi - \frac{1}{x}\int_0^\infty e^{-x\cosh\xi}\cosh\xi\, d\xi = S_1 - S_2.$$

Integrating the second term by parts, we obtain:

$$S_2 = \frac{1}{x}\int_0^\infty e^{-x \cosh \xi} \cosh \xi \, d\xi = \frac{\sinh \xi}{x} e^{-x \cosh \xi} \Big|_0^\infty +$$

$$+ \int_0^\infty e^{-x \cosh \xi} \sinh^2 \xi \, d\xi = S_1 ,$$

from which it follows:

$$\mathscr{L}(F) = 0 .$$

Assuming $\cosh \xi = \eta$, we transform the integral (24′) for $F(x)$ to the form

$$F(x) = \int_1^\infty \frac{e^{-xy}}{\sqrt{(\eta^2 - 1)}} d\eta .$$

Using this relation, it is possible to investigate the behaviour of the function $F(x)$ as $x \to \infty$. Making another change of variable

$$x(\eta - 1) = \xi$$

we obtain:

$$F(x) = \frac{e^{-x}}{\sqrt{x}} \int_0^\infty \frac{e^{-\xi}}{\sqrt{\left[\xi\left(\frac{\xi}{x} + 2\right)\right]}} d\xi = \frac{e^{-x}}{\sqrt{x}} F_1(x) .$$

For $x \to \infty$

$$\lim F_1(x) = \int_0^\infty \frac{e^{-\xi}}{\sqrt{(2\xi)}} d\xi = \frac{2}{\sqrt{2}} \int_0^\infty e^{-t^2} dt = \frac{\sqrt{\pi}}{\sqrt{2}} \qquad (t = \sqrt{\xi}) .$$

Therefore, for large values of x

$$F_1(x) = \frac{\sqrt{\pi}}{\sqrt{2}} (1 + \varepsilon) ,$$

where $\varepsilon \to 0$ as $x \to \infty$. Hence we obtain an asymptotic relation

$$F(x) = \sqrt{\left(\frac{\pi}{2x}\right)} e^{-x} + \ldots \tag{26}$$

The function $F(x)$ introduced by means of the integral (24′) is a solution of equation (25), bounded at infinity, therefore

$$F(x) = BK_0(x)$$

Comparison of the asymptotic formulae for $K_0(x)$ and $F_0(x)$ shows that $B = 1$ and therefore

$$K_0(x) = \int_0^\infty e^{-x \cosh \xi} d\xi \qquad (x > 0) . \tag{24}$$

Let us investigate the nature of the function $K_0(x)$ as $x \to 0$.

We write the integral

$$K_0(x) = F(x) = \int_0^\infty \frac{e^{-x\eta}}{\sqrt{(\eta^2 - 1)}} d\eta$$

in the form

$$K_0(x) = \int_x^\infty \frac{e^{-\lambda}}{\sqrt{(\lambda^2 - x^2)}} d\lambda \qquad (x\eta = \lambda).$$

Dividing this integral into three parts

$$K_0(x) = \int_x^A \frac{d\lambda}{\sqrt{(\lambda^2 - x^2)}} + \int_x^A \frac{(e^{-\lambda} - 1)\,d\lambda}{\sqrt{(\lambda^2 - x^2)}} + \int_A^\infty \frac{e^{-\lambda}\,d\lambda}{\sqrt{(\lambda^2 - x^2)}},$$

where A is some arbitrary constant, we see that the first component equals

$$\ln \frac{A + \sqrt{(A^2 - x^2)}}{x} = -\ln x + \ldots,$$

and the second and third components are bounded as $x \to 0$. Hence it follows that

$$K_0(x) = -\ln x + \ldots = \ln\frac{1}{x} + \ldots. \tag{27}$$

where the dots denote components, remaining finite for $x = 0$. Thus, the function $K_0(x)$ is a solution of equation (25), having a logarithmic singularity at the point $x = 0$ and decreasing exponentially as $x \to \infty$.

The following problem gives a physical interpretation of the function $K_0(x)$. At the origin of coordinates there is a fixed source of unstable gas of magnitude Q_0. A steady diffusion process accompanies the decay of the gas and is described by the equation

$$\Delta u - \varkappa^2 u = \frac{1}{r}\frac{\partial}{\partial r}\left(r\frac{\partial u}{\partial r}\right) + \frac{1}{r^2}\frac{\partial^2 u}{\partial \varphi^2} - \varkappa^2 u = 0 \tag{28}$$

$$\left(\varkappa^2 = \frac{\beta}{D^2}\right),$$

where β is the decay coefficient, and D is the diffusion coefficient. The source function of this equation possesses circular symmetry and, consequently, satisfies the equation

$$\frac{1}{x}\frac{d}{dx}\left(x\frac{du}{dx}\right) - u = 0 \qquad (x = \varkappa r);$$

moreover, the source function has a logarithmic singularity at the origin of coordinates and is bounded at infinity. Hence it follows that the source function is proportional to $K_0(\varkappa r)$:

$$\bar{G} = AK_0(\varkappa r). \tag{29}$$

In order to determine the factor A we make use of the known source strength

$$\lim_{\varepsilon \to 0} \int_{K_\varepsilon} \left(-D\frac{\partial u}{\partial r}\right) ds = Q_0, \tag{30}$$

where the integral on the left expresses the diffusion current through a circle K_ε of radius ε with centre at the source. Substituting in this condition in place of u the function $\bar{G} = AK_0(\varkappa r)$ and taking into account the logarithmic singularity of the function $K_0(x)$ at $x = 0$, we obtain:

$$\lim_{\varepsilon \to 0} \left\{ - \int_{K_\varepsilon} D \frac{\partial \bar{G}}{\partial r} dS \right\} = \lim_{\varepsilon \to 0} \left\{ D\, 2\pi\varepsilon\, A \frac{1}{\varepsilon} \right\} = 2\pi AD = Q_0,$$

Hence

$$A = \frac{Q_0}{2\pi D}$$

and

$$\bar{G} = \frac{Q_0}{2\pi D} K_0(\varkappa r). \tag{31}$$

The integral relation (24) for $K_0(x)$ can be derived from simple physical considerations.

Let us consider the non-steady problem of the diffusion of a gas with decay. At the origin of coordinates a source of constant magnitude Q_0 exists, acting from the time $t = 0$. We shall assume that at the initial time $t = 0$ the concentration of the gas everywhere equals zero. The concentration $u(x, y, t)$ must satisfy the equation

$$D \nabla^2 u - \beta u = u_t \tag{32}$$

and the appropriate additional conditions. Equation (39) by means of the substitution

$$u = \tilde{u} e^{-\beta t}$$

is transformed into the ordinary diffusion equation

$$D \nabla^2 \tilde{u} = \tilde{u}_t,$$

for which, we know the source function

$$\tilde{G} = \frac{1}{(2\sqrt{[\pi D(t-\tau)]})^2} e^{-\frac{r^2}{4D(t-\tau)}} \qquad (D = a^2).$$

Thus, the source function for an instantaneous point source for equation (32) equals

$$G = \frac{Q}{(2\sqrt{[\pi D(t-\tau)]})^2} e^{-\frac{r^2}{4D(t-\tau)} - \beta(t-\tau)}$$

The source function for a source of magnitude Q_0, continuously acting from $t = 0$ to the time t, is given by the relation

$$G = Q_0 \int_0^t \frac{1}{4\tau D(t-\tau)} e^{-\frac{r^2}{4D(t-\tau)} - \beta(t-\tau)} d\tau$$

Introducing a new variable

$$\theta = t - \tau ,$$

we obtain:

$$G = \frac{Q_0}{4\pi D} \int_0^t e^{-\frac{r^2}{4D\theta} - \beta\theta} \frac{d\theta}{\theta} .$$

The source function for the steady problem can be determined by a limiting transition for $t \to \infty$ in the preceding formula

$$\bar{G} = \lim_{t \to \infty} G = \frac{Q_0}{4\pi D} \int_0^\infty e^{-\frac{r^2}{4D}\frac{1}{\theta} - \beta\theta} \frac{d\theta}{\theta} .$$

Let us transform this integral by means of the substitution

$$\theta = Ce^{\xi}$$

where C is some constant,

$$\bar{G} = \frac{Q_0}{4\pi D} \int_{-\infty}^{\infty} e^{-\left[\frac{r^2}{4DC} e^{-\xi} + \beta C e^{\xi}\right]} d\xi .$$

If we determine C by the relation

$$\frac{r^2}{4DC} = \beta C ,$$

we find:

$$C = \frac{r}{2\sqrt{(\beta D)}} \text{ and } \frac{r^2}{4DC} = \beta C = \frac{r}{2}\sqrt{\frac{\beta}{D}} = \frac{\varkappa r}{2} \qquad \left(\varkappa^2 = \frac{\beta}{D}\right)$$

Hence it follows that the steady source function has the form

$$\bar{G} = \frac{Q_0}{4\pi D} \int_{-\infty}^{\infty} e^{-\varkappa r \cosh \xi} d\xi = \frac{Q_0}{2\pi D} \int_0^{\infty} e^{-\varkappa r \cosh \xi} d\xi = \frac{Q_0}{2\pi D} K_0(\varkappa r) .$$

Thus, the diffusion problem leads to an integral representation for the function $K_0(x)$.

§ 4. Integral representations. Asymptotic formulae

1. *Integral representations for functions of integral order*

We shall consider periodic solutions of the wave equation

$$u_{xx} + u_{yy} = \frac{1}{a^2} u_{tt} .$$

Assuming

$$u(x, y, t) = v(x, y)\, e^{i\omega t},$$

we obtain an equation for the amplitude of the vibrations $v(x, y)$

$$v_{xx} + v_{yy} + k^2 v = 0 \qquad \left(k = \frac{\omega}{a}\right), \tag{1}$$

which has a solution of the form

$$v = e^{\mp ikx} \text{ and } v = e^{\mp iky}, \tag{2}$$

corresponding to the plane waves

$$u = e^{i(\omega t \mp kx)} \text{ and } u = e^{i(\omega t \mp ky)}.$$

The choice of sign — or + determines a plane wave, travelling in the positive or negative direction of the axis. Later we shall omit the time factor $e^{i\omega t}$, calling the function (2) a plane wave.

The plane wave, propagating in the direction s, obviously, has the form

$$v = e^{-iksr} = e^{-ik(x \cos \alpha + y \sin \alpha)}$$

where α is the angle between the direction s and the x-axis.

Let us introduce polar coordinates, assuming

$$x = r \cos \varphi, \qquad y = \sin \varphi.$$

Then

$$v = e^{-ikr \cos (\varphi - \alpha)},$$

in which $\alpha = 0$ corresponds to a plane wave, travelling in the positive direction of the x-axis, $\alpha = \pi/2$ to a wave, travelling in the positive direction of the y-axis.

Let us determine an expansion of the plane wave, travelling along the y-axis

$$v = e^{-ikr \sin \varphi} \tag{3}$$

in a Fourier series in the variable φ

$$v = \sum_{n=-\infty}^{\infty} A_n(\varrho)\, e^{-in\varphi} \qquad (\varrho = kr), \tag{4}$$

where

$$A_n(\varrho) = \frac{1}{2\pi}\int_{-\pi}^{\pi} e^{-i\varrho\sin\varphi + in\varphi}\, d\varphi. \tag{5}$$

Let us prove that

$$A_n(\varrho) = J_n(\varrho). \tag{6}$$

We verify first that $A_n(\varrho)$ satisfies Bessel's equation (§ 1, Supplement I) which we write in the form

$$\varrho^2 \frac{d^2y}{d\varrho^2} + \varrho\frac{dy}{d\varrho} + (\varrho^2 - n^2)\, y = \mathscr{L}(y) = 0.$$

Let us evaluate

$$\mathscr{L}(A_n) = \frac{1}{2\pi}\int_{-\pi}^{\pi} e^{-i\varrho\sin\varphi + in\varphi}\,[-\varrho^2\sin^2\varphi - i\varrho\sin\varphi + \varrho^2 - n^2]\, d\varphi. \tag{7}$$

By means of a twofold integration by parts we transform the integral

$$n^2\int_{-\pi}^{\pi} e^{-i\varrho\sin\varphi + in\varphi}\, d\varphi = n\varrho\int_{-\pi}^{\pi} e^{-i\varrho\sin\varphi + in\varphi}\cos\varphi\, d\varphi =$$

$$= -\frac{\varrho}{i}\int_{-\pi}^{\pi} e^{-i\varrho\sin\varphi + in\varphi}\,[-i\varrho\cos^2\varphi - \sin\varphi]\, d\varphi =$$

$$= \int_{-\pi}^{\pi} e^{-i\varrho\sin\varphi + in\varphi}\,[\varrho^2 - \varrho^2\sin^2\varphi - i\varrho\sin\varphi]\, d\varphi. \tag{8}$$

We do not write the terms coming from the limits, since by the periodicity of the functions they equal zero. Substituting (8) in formula (7) for $\mathscr{L}(A_n)$ we verify that

$$\mathscr{L}(A_n) = 0. \tag{9}$$

From (5) we see that the function A_n is bounded at $\varrho = 0$. Hence from (9) it follows that $A_n(\varrho)$ is proportional to $J_n(\varrho)$

$$A_n(\varrho) = C_n J_n(\varrho), \tag{10}$$

where C_n is some constant. In order to calculate C_n we compare both sides of (10) with $\varrho = 0$. Noting that at $\varrho = 0$ $J_n(\varrho)$ has a zero of nth order and differentiating $J_n(\varrho)$ n times, we find from (12) § 1

$$\left[\frac{d^n}{d\varrho^n} J_n(\varrho)\right]_{\varrho=0} = \frac{1}{2^n}. \tag{11}$$

On the other hand, differentiating A_n, we find from (5):

$$\left[\frac{d^k}{d\varrho^k} A_n(\varrho)\right]_{\varrho=0} = \frac{1}{2\pi}(-i)^k \int_{-\pi}^{\pi} e^{in\varphi} \sin^k \varphi \, d\varphi =$$

$$= \frac{(-i)^k}{2\pi . 2^k i^k} \int_{-\pi}^{\pi} e^{in\varphi} (e^{i\varphi} - e^{-i\varphi})^k \, d\varphi.$$

Since

$$\frac{1}{2\pi} \int_{-\pi}^{\pi} e^{in\varphi} e^{ik\varphi} \, d\varphi = \begin{cases} 0 \text{ where } k \neq -n, \\ 1 \text{ where } k = -n, \end{cases}$$

then

$$\left[\frac{d^k}{d\varrho^k} A_n(\varrho)\right]_{\varrho=0} = \begin{cases} 0 \text{ where } k < n, \\ \dfrac{(-1)^n}{2^n} \cdot \dfrac{1}{2\pi} \displaystyle\int_{-\pi}^{\pi} e^{in\varphi} (-e^{-i\varphi})^n \, d\varphi = \dfrac{1}{2^n} \\ \qquad\qquad\qquad \text{where } k = n. \end{cases} \tag{12}$$

Comparing (11) and (12) we deduce that

$$C_n = 1.$$

Thus we have derived an integral representation for a Bessel function of first kind of integral order n:

$$J_n(\varrho) = \frac{1}{2\pi} \int_{-\pi}^{\pi} e^{-i\varrho \sin \varphi + in\varphi} \, d\varphi. \tag{13}$$

From (4) we have an expansion for the plane wave (3)

$$e^{-i\varrho \sin \varphi} = \sum_{n=-\infty}^{\infty} J_n(\varrho) e^{-in\varphi}.$$

Hence, in particular, it follows that

$$\cos(\varrho \sin \varphi) = J_0(\varrho) + 2J_2(\varrho) \cos 2\varphi + 2J_4(\varrho) \cos 4\varphi + \ldots.$$
$$\sin(\varrho \sin \varphi) = 2J_1(\varrho) \sin \varphi + 2J_3(\varrho) \sin 3\varphi + \ldots$$

Let us determine another integral relation for $J_n(\varrho)$. Substituting $\varphi = \psi - \pi/2$, we obtain from (13):

$$J_n(\varrho) = \frac{(-i)^n}{2\pi} \int_{-\pi}^{\pi} e^{i\varrho \cos\psi + in\psi}\, d\psi \,. \qquad (13')$$

Because of the periodicity of the functions under the integral signs in (13) and (13′) integration can be performed over any interval of length 2π. Formula (13′) may also be derived from the expansion

$$e^{i\varrho\cos\varphi} = \sum_{n=-\infty}^{\infty} i^n J_n(\varrho)\, e^{-in\varphi}.$$

Utilizing (13) and (13′) let us write down the integral relations for the function $J_0(\varrho)$:

$$J_0(\varrho) = \frac{1}{2\pi} \int_{-\pi}^{\pi} e^{-i\varrho\sin\varphi}\, d\varphi \,, \qquad (14)$$

$$J_0(\varrho) = \frac{1}{2\pi} \int_{-\pi}^{\pi} e^{i\varrho\cos\varphi}\, d\varphi \,. \qquad (14')$$

It should be noted that the integral representations (13) and (13′) are obtained only for functions of integral order. If the ν is non-integral, then the following relation holds for function $J_\nu(\varrho)$:

$$J_\nu(\varrho) = \int_{-\pi}^{\pi} \frac{1}{2\pi} e^{-i\varrho\sin\varphi + i\nu\varphi}\, d\varphi - \frac{\sin\nu\pi}{\pi} \int_0^{\infty} e^{-\varrho\sinh\xi - \nu\xi}\, d\xi \,, \qquad (15)$$

similar to formula (13) [compare with formula (7) § 6].

2. *Asymptotic relations*

Before proceeding to an investigation of the asymptotic behaviour of Bessel functions, let us prove the following lemma:

The function $Q(\varrho)$, defined by an integral of the type

$$Q(\varrho) = \frac{1}{\pi} \int_0^1 \frac{e^{\pm i\varrho\xi}}{\sqrt{(1-\xi)}} f(\xi)\, d\xi \,, \qquad (16)$$

where $f(\xi)$ is some continuous, twofold differentiable function, for large values of ϱ may be represented in the form

$$Q(\varrho) = \frac{e^{\pm i\left(\varrho - \frac{\pi}{4}\right)}}{\sqrt{(\pi\varrho)}} f(1)\left[1 + O\left(\frac{1}{\varrho}\right)\right], \tag{17}$$

where $O(1/\varrho) \to 0$ as $\varrho \to \infty$.

We give the proof, choosing the plus sign in the exponent.

The function under the integral sign in (16) has a singularity at $\xi = 1$. In order to remove this singularity let us introduce a new variable

$$\eta = 1 - \xi \qquad (\xi = 1 - \eta),$$

so that

$$Q(\varrho) = \frac{e^{i\varrho}}{\pi}\int_0^1 e^{-i\varrho\eta}\,\frac{f(1-\eta)}{\sqrt{\eta}}\,d\eta. \tag{18}$$

The second factor of the expression under the integral sign may be represented in the form

$$\frac{f(1-\eta)}{\sqrt{\eta}} = \frac{f(1)}{\sqrt{\eta}} + g(\eta), \tag{19}$$

where

$$g(\eta) = \frac{f(1-\eta) - f(1)}{\sqrt{\eta}} = \frac{f(1-\eta) - f(1)}{\eta}\sqrt{\eta} =$$

$$= g_1(\eta)\sqrt{\eta} \qquad (g(0) = 0), \tag{20}$$

in which $g_1(\eta)$ is a continuous and differentiable function of η. In particular, $g'(\eta)$ is absolutely integrable, i.e.

$$\int_0^1 |g'(\eta)|\,d\eta < M,$$

where M is some constant. Substituting (19) in (18), we obtain:

$$Q(\varrho) = \frac{e^{i\varrho}}{\pi}\left\{f(1)\int_0^1 \frac{e^{-i\varrho\eta}}{\sqrt{\eta}}\,d\eta + \int_0^1 e^{-i\varrho\eta}\,g(\eta)\,d\eta\right\} = \frac{e^{i\varrho}}{\pi}(q_1 + q_2), \tag{21}$$

where

$$q_1 = f(1)\int_0^1 \frac{e^{-i\varrho\eta}}{\sqrt{\eta}}\,d\eta, \quad q_2 = \int_0^1 e^{-i\varrho\eta}\,g(\eta)\,d\eta.$$

For q_1 we obtain an expression

$$q_1(\varrho) = f(1)\sqrt{\frac{\pi}{\varrho}}\,e^{-i\frac{\pi}{4}}\left[1 + O\left(\frac{1}{\varrho}\right)\right], \tag{22}$$

since

$$\int_0^{\varrho} \frac{e^{-i\xi}}{\sqrt{\xi}} d\xi = \int_0^{\infty} \frac{e^{-i\xi}}{\sqrt{\xi}} d\xi \left[1 + O\left(\frac{1}{\varrho}\right)\right] = \sqrt{\pi}\, e^{-i\frac{\pi}{4}}\left(1 + O\left(\frac{1}{\varrho}\right)\right].$$

In fact,

$$\int_0^{\infty} \frac{e^{-a\xi}}{\sqrt{\xi}} d\xi = \frac{2}{\sqrt{a}} \int_0^{\infty} e^{-t^2} dt = \frac{\sqrt{\pi}}{\sqrt{a}}.$$

Extending this formula to a complex value of a, the real part of which is positive, we find:*

$$\int_0^{\infty} \frac{e^{-i\xi}}{\sqrt{\xi}} d\xi = \frac{\sqrt{\pi}}{\sqrt{i}} = \sqrt{\pi} e^{-i\frac{\pi}{4}}$$

$$\left(\text{alternativly} \int_0^{\infty} \frac{e^{i\xi}}{\sqrt{\xi}} d\xi = \frac{\sqrt{\pi}}{\sqrt{-i}} = \sqrt{\pi} e^{i\frac{\pi}{4}}\right).$$

We transform the integral q_2 by means of integration by parts:

$$q^2 = \int_0^1 e^{-i\varrho\eta} g(\eta)\, d\eta = \frac{1}{-i\varrho} e^{-i\varrho\eta} g(\eta) \Big|_0^1 - \frac{1}{-i\varrho} \int_0^1 e^{-i\varrho\eta} g'(\eta)\, d\eta =$$

$$= \frac{1}{\sqrt{\varrho}} O\left(\frac{1}{\varrho}\right). \tag{23}$$

Substituting (22) and (23) in (21), we have:

$$Q(\varrho) = \frac{e^{i\left(\varrho - \frac{\pi}{4}\right)}}{\sqrt{(\pi\varrho)}} f(1)\left[1 + O\left(\frac{1}{\varrho}\right)\right], \tag{17'}$$

if the plus sign is taken in (16).

A similar relation is obtained, if a minus sign is taken in (16)

$$Q(\varrho) = \frac{e^{-i\left(\varrho - \frac{\pi}{4}\right)}}{\sqrt{(\pi\varrho)}} f(1)\left[1 + O\left(\frac{1}{\varrho}\right)\right]. \tag{17''}$$

The lemma is proved.

* See for more detail V. I. Smirnov, *Course in Higher Mathematics*, vol. III.

Let us apply this lemma to the derivation of the asymptotic relations for the Bessel function of the first kind of integral order. Introducing a new variable

$$\cos\psi = \xi,$$

we transform formula (13′):

$$J_n(\varrho) = \frac{(-i)^n}{2\pi}\int_{-\pi}^{\pi} e^{i\varrho\cos\psi}\cos n\psi\, d\psi = \frac{(-i)^n}{\pi}\int_{0}^{\pi} e^{i\varrho\cos\psi}\cos n\psi\, d\psi =$$

$$= \frac{(-i)^n}{\pi}\int_{-1}^{1} e^{i\varrho\xi}\frac{T_n(\xi)}{\sqrt{(1-\xi^2)}}\, d\xi, \tag{24}$$

where

$$T_n(\xi) = \cos(n \operatorname{arc}\cos\xi).$$

Since

$$\cos n\psi = \operatorname{Re}(\cos\psi + i\sin\psi)^n.$$

by Demoivre's theorem, is a polynomial of order n, then $T_n(\xi)$ is also a polynomial of order n in $\xi = \cos\psi$. The polynomials $T_n(\xi)$ are called *Chebyshev polynomials.*

Let us consider

$$T_n(-\xi) = \cos\left(n \operatorname{arc}\cos(-\xi)\right) = \cos(n\pi - n\operatorname{arc}\cos\xi) =$$

$$= \cos n\pi\cos(n\operatorname{arc}\cos\xi) + \sin n\pi\sin(n\operatorname{arc}\cos\xi) = (-1)^n T_n(\xi),$$

i.e.

$$T_n(-\xi) = (-1)^n T_n(\xi).$$

Representing the integral (24) as the sum of integrals from 0 to 1 and from —1 to 0 and utilizing these properties of Chebyshev polynomials, we obtain:

$$J_n(\varrho) = \frac{(-i)^n}{\pi}\int_0^1 \frac{e^{i\varrho\xi}}{\sqrt{(1-\xi)}} f(\xi)\, d\xi + \frac{i^n}{\pi}\int_0^1 \frac{e^{-i\varrho\xi}}{\sqrt{(1-\xi)}} f(\xi)\, d\xi, \tag{24′}$$

where

$f(\xi) = \dfrac{T_n(\xi)}{\sqrt{(1+\xi)}}$ is a continuous function, differentiable any number of times, for which $f(1) = \dfrac{1}{\sqrt{2}}$.

Applying the lemma proved above to each of the integrals (24′), we find from (17′) and (17″):

$$J_n(\varrho) = \frac{1}{\sqrt{2\pi\varrho}}\left[e^{i\left(\varrho - \frac{\pi}{4} - \frac{\pi}{2}n\right)} + e^{-i\left(\varrho - \frac{\pi}{4} - \frac{\pi}{2}\right)}\right]\left[1 + O\left(\frac{1}{\varrho}\right)\right] =$$

$$= \sqrt{\left(\frac{2}{\pi\varrho}\right)} \cos\left[\varrho - \frac{\pi}{2}\left(n + \frac{1}{2}\right)\right]\left[1 + O\left(\frac{1}{\varrho}\right)\right].$$

Thus, for a Bessel function of the first kind for large arguments the asymptotic relation

$$J_n(\varrho) = \sqrt{\left(\frac{2}{\pi\varrho}\right)} \cos\left[\varrho - \frac{\pi}{2}\left(n + \frac{1}{2}\right)\right]\left[1 + O\left(\frac{1}{\varrho}\right)\right] \tag{25}$$

holds. In particular, for $n = 0$

$$J_0(\varrho) = \sqrt{\left(\frac{2}{\pi\varrho}\right)} \cos\left(\varrho - \frac{\pi}{4}\right)\left[1 + O\left(\frac{1}{\varrho}\right)\right], \tag{25′}$$

and for $n = 1$

$$J_1(\varrho) = -\sqrt{\left(\frac{2}{\pi\varrho}\right)} \cos\left(\varrho + \frac{\pi}{4}\right)\left[1 + O\left(\frac{1}{\varrho}\right)\right]. \tag{25″}$$

In § 6 it will be shown that the following asymptotic relations are valid for Hankel functions:

$$H_n^{(1)}(\varrho) = \sqrt{\left(\frac{2}{\pi\varrho}\right)}\, e^{i\left[\varrho - \frac{\pi}{2}\left(n+\frac{1}{2}\right)\right]}\left[1 + O\left(\frac{1}{\varrho}\right)\right], \tag{26}$$

$$H_n^{(2)}(\varrho) = \sqrt{\left(\frac{2}{\pi\varrho}\right)}\, e^{-i\left[\varrho = \left(n+\frac{1}{2}\right)\right]}\left[1 + O\left(\frac{1}{\varrho}\right)\right]. \tag{27}$$

Using (26) and (27), and also the formulae

$$H_n^{(1)}(\varrho) = J_n(\varrho) + iN_n(\varrho) \qquad (3,\ \S\ 3)$$

and

$$H_n^{(2)}(\varrho) = J_n(\varrho) - iN_n(\varrho) \qquad (4,\ \S\ 3)$$

we find:

$$N_n(\varrho) = \sqrt{\left(\frac{2}{\pi\varrho}\right)} \sin\left[\varrho - \frac{\pi}{2}\left(n + \frac{1}{2}\right)\right]\left[1 + O\left(\frac{1}{\varrho}\right)\right]. \tag{28}$$

Formulae (3) and (4) § 3 along with the asymptotic relations for $H_n^{(1)}$, $H_n^{(2)}$ and J_n, N_n are similar structurally to Euler's relations for trigonometric functions:

$$e^{ix} = \cos x + i \sin x, \qquad e^{-ix} = \cos x - i \sin x.$$

§ 5. The Fourier–Bessel integral and some integrals containing Bessel functions

1. *Fourier–Bessel integral*

Let us find the expansion of the given function $f(r)$ as an integral in Bessel functions. The Fourier integrals for the function $f(x)$ and for the function $f(x, y)$ of two variables have the form

$$f(x) = \frac{1}{2\pi} \int_{-\infty}^{\infty} d\mu \int_{-\infty}^{\infty} f(\xi)\, e^{i\mu(x-\xi)}\, d\xi, \tag{1}$$

$$f(x, y) = \frac{1}{(2\pi)^2} \int_{-\infty}^{\infty} \int_{-\infty}^{\infty} d\mu\, d\mu' \int_{-\infty}^{\infty} \int_{-\infty}^{\infty} f(\xi, \eta)\, e^{i\mu(x-\xi)+i\mu'(y-\eta)}\, d\xi\, d\eta. \tag{2}$$

Introducing polar coordinates by means of the relations

$$x = r\cos\varphi, \qquad \xi = \varrho\cos\psi, \qquad \mu = \lambda\cos\xi,$$

$$y = r\sin\varphi\,; \qquad \eta = \varrho\sin\psi\,; \qquad \mu' = \sin\xi\,;$$

we obtain:

$$d\xi\, d\eta = \varrho\, d\varrho\, d\psi, \qquad d\mu\, d\mu' = \lambda\, d\lambda\, d\xi,$$

$$\mu x + \mu' y = \lambda r \cos(\xi - \varphi),$$

$$\mu\xi + \mu'\eta = \lambda\varrho\cos(\psi - \xi).$$

Assuming that $f(x, y)$ has the form

$$f(x, y) = f(r)\, e^{in\varphi}, \tag{3}$$

where n is an integer, and transforming the Fourier integral (2) by means of the relations written above, we find:

$$f(r)\, e^{in\varphi} = \int_0^{\infty}\int_0^{\infty} f(\varrho)\, \varrho\, d\varrho\, \lambda\, d\lambda \cdot \frac{1}{2\pi}\int_{-\pi}^{\pi} e^{i\lambda r\cos(\xi-\varphi)+in(\xi-\varphi)}\, d\xi \cdot e^{in\varphi} \times$$

$$\times \frac{1}{2\pi}\int_{-\pi}^{\pi} e^{-i\lambda\varrho\cos(\psi-\xi)+in(\psi-\xi)}\, d\psi. \tag{4}$$

Utilizing the relations

$$J_n(z) = \frac{1}{2\pi} \int_{-\pi}^{\pi} e^{iz \cos \xi + in\xi} e^{i\frac{\pi n}{2}} d\xi, \tag{5}$$

$$J_n(z) = \frac{1}{2\pi} \int_{-\pi}^{\pi} e^{-iz \cos \xi' + in\xi'} e^{i\frac{\pi n}{2}} d\xi' \quad (\xi = \pi + \xi'); \tag{6}$$

since the expressions under the integral signs in (5) and (6) are periodic functions of ξ and ξ' and can therefore be integrated over any interval of length 2π, it is possible to write:

$$\frac{1}{2\pi} \int_{-\pi}^{\pi} e^{iz \cos (\xi - \xi_0) + in(\xi - \xi_0)} d\xi = J_n(z)\, e^{i\frac{\pi n}{2}}, \tag{7}$$

$$\frac{1}{2\pi} \int_{-\pi}^{\pi} e^{-iz \cos (\xi' - \xi_0') + in(\xi' - \xi_0')} d\xi' = J_n(z)\, e^{-i\frac{\pi n}{2}}, \tag{8}$$

where ξ_0 and ξ_0' are arbitrary numbers. Substituting (7) and (8) in (4) and dividing both sides by $e^{in\varphi}$ we obtain the *Fourier–Bessel integral*:

$$f(r) = \int_0^\infty \int_0^\infty f(\varrho)\, J_n(\lambda\varrho)\, J_n(\lambda r)\, \lambda\, d\lambda\, \varrho\, d\varrho \tag{9}$$

or

$$f(r) = \int_0^\infty \varphi(\lambda)\, J_n(\lambda r)\, \lambda\, d\lambda,$$

where

$$\varphi(\lambda) = \int_0^\infty f(\varrho)\, J_n(\lambda\varrho)\, \varrho\, d\varrho.$$

In order that an expansion as a Fourier–Bessel integral may be possible, it is sufficient that the function $f(r)$, defined in the interval $(0, \infty)$, fulfill the following conditions:

(1) $f(r)$ is continuous in the interval $(0, \infty)$;

(2) $f(r)$ has a finite number of maxima and minima in any finite interval;

(3) the following integral converges:

$$\int \varrho\, |f(\varrho)|\, d\varrho.$$

We do not give a proof of this.

2. *Some integrals, containing Bessel functions*

In various applications, definite integrals containing Bessel functions often occur.

We consider the following integral as an example

$$B_1 = \int_0^\infty e^{-z\lambda} J_0(\varrho\lambda)\, d\lambda = \frac{1}{\sqrt{(\varrho^2 + z^2)}} \qquad (z > 0). \tag{10}$$

To prove this formula we replace J_0 by its integral representation [(14) § 4] and then change the order of integration:

$$B_1 = \int_0^\infty e^{-z\lambda} J_0(\varrho\lambda)\, d\lambda = \frac{1}{2\pi} \int_0^\infty e^{-z\lambda}\, d\lambda \int_{-\pi}^{\pi} e^{-i\varrho\lambda \sin\varphi}\, d\varphi =$$

$$= \frac{1}{2\pi} \int_{-\pi}^{\pi} d\varphi \int_0^\infty e^{-(z+i\varrho\sin\varphi)\lambda}\, d\lambda = \frac{1}{2\pi} \int_{-\pi}^{\pi} \frac{d\varphi}{z + i\varrho \sin\varphi} =$$

$$= \frac{1}{2\pi} \int_{-\pi}^{\pi} \frac{z\, d\varphi}{z^2 + \varrho^2 \sin^2\varphi} - \frac{i}{2\pi} \int_{-\pi}^{\pi} \frac{\varrho \sin\varphi\, d\varphi}{z^2 + \varrho^2 \sin^2\varphi} = \frac{1}{\pi} \int_0^{\pi} \frac{z\, d\varphi}{z^2 + \varrho^2 \sin^2\varphi}$$

since

$$\int_{-\pi}^{\pi} \frac{\sin\varphi\, d\varphi}{z^2 + \varrho^2 \sin^2\varphi} = 0$$

because the integrand is an odd function.

If we put $\tan\varphi = \xi$, and then $\sqrt{\left(\frac{z^2}{z^2 + \varrho^2}\right)}\, \xi = \eta$, we obtain:

$$B_1 = \frac{1}{\pi} \int_0^{\pi} \frac{z\, d\varphi}{z^2 + \varrho^2 \sin^2\varphi} = \frac{2z}{\pi} \int_0^{\frac{\pi}{2}} \frac{d\varphi}{z^2 + \varrho^2 \sin^2\varphi} =$$

$$= \frac{2z}{\pi} \int_0^\infty \frac{d\xi}{z^2(1 + \xi^2) + \varrho^2\xi^2} = \frac{2}{\pi \sqrt{(z^2 + \varrho^2)}} \int_0^\infty \frac{d\eta}{1 + \eta^2} = \frac{1}{\sqrt{(z^2 + \varrho^2)}};$$

in this way formula (10) is proved. Using (10), we find:

$$\int_0^\infty J_1(\varrho\lambda)\, e^{-z\lambda}\, d\lambda = \frac{1}{\varrho}\left(1 - \frac{z}{\sqrt{(z^2 + \varrho^2)}}\right). \tag{11}$$

Assuming $z = ia$ in formulae (10) and (11) and separating the real and imaginary parts, we obtain a series of results:

$$\left.\begin{aligned} &\int_0^\infty J_0(\varrho\lambda)\cos a\lambda\,d\lambda = \frac{1}{\sqrt{(\varrho^2 - a^2)}}, \\ &\int_0^\infty J_0(\varrho\lambda)\sin a\lambda\,d\lambda = 0, \\ &\int_0^\infty J_1(\varrho\lambda)\cos a\lambda\,d\lambda = 1/\varrho, \\ &\int_0^\infty J_1(\varrho\lambda)\sin a\lambda\,d\lambda = \frac{a}{\varrho\sqrt{(\varrho^2 - a^2)}} \end{aligned}\right\} \text{ if } \varrho > a; \qquad (12)$$

$$\left.\begin{aligned} &\int_0^\infty J_0(\varrho\lambda)\cos a\lambda\,d\lambda = 0, \\ &\int_0^\infty J_0(\varrho\lambda)\sin a\lambda\,d\lambda = \frac{1}{\sqrt{(a^2 - \varrho^2)}} \end{aligned}\right\} \text{ if a} > \varrho; \qquad (13)$$

$$\left.\begin{aligned} &\int_0^\infty J_1(\varrho\lambda)\cos a\lambda\,d\lambda = \frac{1}{\varrho}\left(1 - \frac{a}{\sqrt{(a^2 - \varrho^2)}}\right), \\ &\int_0^\infty J_1(\varrho\lambda)\sin a\lambda\,d\lambda = 0 \end{aligned}\right\} \text{ if } a > \varrho. \qquad (13')$$

Let us prove the second integral relation

$$B_2 = \int_0^\infty J_\nu(\lambda\varrho)\, e^{-t\lambda^2}\lambda^{\nu+1}\,d\lambda = \frac{1}{2t}\left(\frac{\varrho}{2t}\right)^\nu e^{-\frac{\varrho^2}{4t}}. \qquad (14)$$

We substitute a power series for J_ν and perform a term-by-term integration ($t > 0$!):

$$B_2 = \sum_{k=0}^\infty \frac{(-1)^k}{\Gamma(k+1)\,\Gamma(k+\nu+1)}\left(\frac{\varrho}{2}\right)^{2k+\nu}\int_0^\infty \lambda^{2k+2\nu+1}\,e^{-t\lambda^2}\,d\lambda.$$

Evaluating the auxiliary integral

$$\int_0^\infty \lambda^{2k+2\nu+1} e^{-t\lambda^2} d\lambda = \frac{1}{2\,t^{k+\nu+1}} \int_0^\infty e^{-\xi} \xi^{k+\nu} d\xi = \frac{1}{2\,t^{k+\nu+1}} \Gamma(k+\nu+1),$$

we obtain:

$$B_2 = \frac{1}{2t}\left(\frac{\varrho}{2t}\right)^\nu \sum_{k=0}^\infty \frac{(-1)^k}{k!}\left(\frac{\varrho^2}{4t}\right)^k = \frac{1}{2t}\left(\frac{\varrho}{2t}\right)^\nu e^{-\frac{\varrho^2}{4t}},$$

which was to be proved.

We note that evaluation of B_1 can be carried out similarly, expanding the Bessel function in a series and subsequently performing an integration term-by-term.

Let us consider the integral

$$C = \int_0^\infty J_0(\lambda\varrho)\, \frac{e^{-\sqrt{(\lambda^2-k^2)}|z|}}{\sqrt{(\lambda^2-k^2)}}\, \lambda\, d\lambda. \tag{15}$$

It is readily verified that it is a solution of the equation

$$\nabla^2 v + k^2 v = 0 \quad \left(\nabla^2 = \frac{\partial^2}{\partial x^2} + \frac{\partial^2}{\partial y^2} + \frac{\partial^2}{\partial z^2}\right).$$

The function

$$v_0 = \frac{e^{ikr}}{r} \quad (r = \sqrt{[\varrho^2 + z^2]})$$

also satisfies the wave equation

$$\nabla^2 v_0 + k^2 v_0 = \frac{1}{r}\frac{\partial^2}{\partial r^2}(r v_0) + k^2 v_0 = -k^2 \frac{e^{ikr}}{r} + k^2 v_0 = 0.$$

Let us expand the function $v_0(\varrho) = \dfrac{e^{ik\varrho}}{\varrho}$ as a Fourier–Bessel integral

$$\frac{e^{ik\varrho}}{\varrho} = \int_0^\infty F(\lambda)\, J_0(\varrho\lambda)\, \lambda\, d\lambda, \tag{16}$$

where

$$F(\lambda) = \int_0^\infty e^{ik\varrho} J_0(\lambda\varrho)\, d\varrho. \tag{17}$$

In order to evaluate the function $F(\lambda)$ we make use of formula (12)

$$F(\lambda) = \int_0^\infty J_0(\lambda\varrho)\,(\cos k\varrho + i\sin k\varrho)\,d\varrho =$$

$$= \begin{cases} \dfrac{1}{\sqrt{(\lambda^2 - k^2)}}, & \text{when } \lambda > k, \\[2ex] \dfrac{i}{\sqrt{(k^2-\lambda^2)}} = \dfrac{1}{\sqrt{(\lambda^2 - k^2)}}, & \text{when } k > \lambda. \end{cases}$$

Thus,

$$\frac{e^{ik\varrho}}{\varrho} = \int_0^\infty J_0(\varrho\lambda)\frac{\lambda\,d\lambda}{\sqrt{(\lambda^2-k^2)}}, \tag{18}$$

i.e. the function

$$v_0 = \frac{e^{ik\sqrt{(\varrho^2+z^2)}}}{\sqrt{(\varrho^2+z^2)}}$$

coincides with the integral $C(\varrho, z)$ at $z = 0$. Thus, both functions

$$v_0(\varrho, z) \quad \text{and} \quad C(\varrho, z)$$

are solutions of the wave equation, coincide at $z = 0$ and have the same singularity at the point $z = 0$, $\varrho = 0$.

Hence it follows that they are identically equal to one another, i.e.

$$\int_0^\infty J_0(\lambda\varrho)\frac{e^{-\sqrt{(\lambda^2-k^2)}|z|}}{\sqrt{(\lambda^2-k^2)}}\lambda\,d\lambda = \frac{e^{ik\sqrt{(\varrho^2+z_2)}}}{\sqrt{(\varrho^2+z^2)}}. \tag{19}$$

This formula was widely used by A. Sommerfeld and is often called the Sommerfeld relation.

§ 6. Representation of cylindrical functions by means of contour integrals

1. *Representation of cylindrical functions by means of contour integrals*

Often the solutions of Bessel's equation are represented by contour integrals of the type

$$\int e^{-ix\sin\varphi + i\nu\varphi}\,d\varphi = \int_C K(x,\varphi)\,\Phi(\varphi)\,d\varphi, \tag{1}$$

where C is some contour, lying in the plane of the complex variable φ, $K(x, \varphi) = e^{-ix \sin \varphi}$, $\Phi(\varphi) = e^{i\nu\varphi}$. The function (1) may be considered as a superposition of plane waves

$$e^{-i(x \sin \varphi + y \cos \varphi)},$$

propagating at different, among them complex, angles to the

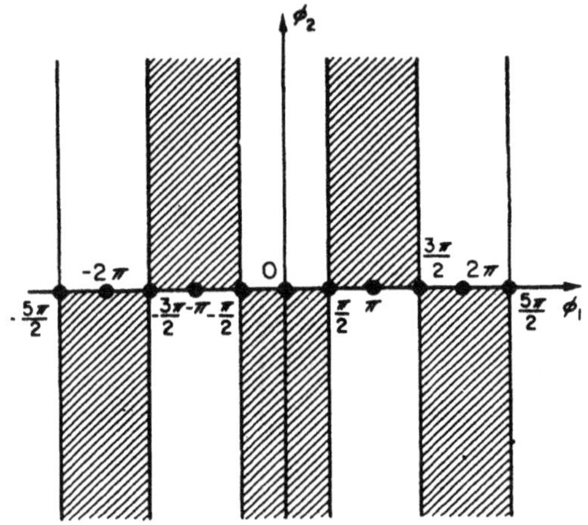

FIG. 90

y-axis, if their values at $y = 0$ (along the x-axis) are taken; $\Phi = e^{i\nu\varphi}$ is the amplitude factor of the plane wave.

Let us examine the question of the convergence of the integral (1). In Fig. 90 those parts of the plane of the complex variable $\varphi = \varphi_1 + i\,\varphi_2$ are shaded, for which the real part of the expression

$$ix \sin \varphi = xi \sin (\varphi_1 + i\varphi_2) = x\,(i \sin \varphi_1 \cosh \varphi_2 - \cos \varphi_1 \sinh \varphi_2) \quad (2)$$

is positive (for $x > 0$).

The Hankel functions of first and second kind respectively are usually defined by means of the integrals

$$H_\nu^{(1)}(x) = -\frac{1}{\pi} \int\limits_{C_1} e^{-ix \sin \varphi + i\nu\varphi}\, d\varphi \quad (3)$$

$$H_\nu^{(2)}(x) = -\frac{1}{\pi} \int\limits_{C_2} e^{-ix \sin \varphi + i\nu\varphi}\, d\varphi, \quad (4)$$

taken over the contours C_1 and C_2, where the contour C_1 consists of the section $(-i\infty, 0)$, $(0 - \pi)$, $(-\pi, -\pi + i\infty)$, and the contour C_2 of the sections $(\pi + i\infty, \pi)$, $(\pi, 0)$, $(0, -i\infty)$ (Fig. 91, paths I and II).

It is obvious that for $x > 0$, the integrals over C_1 and C_2 converge.

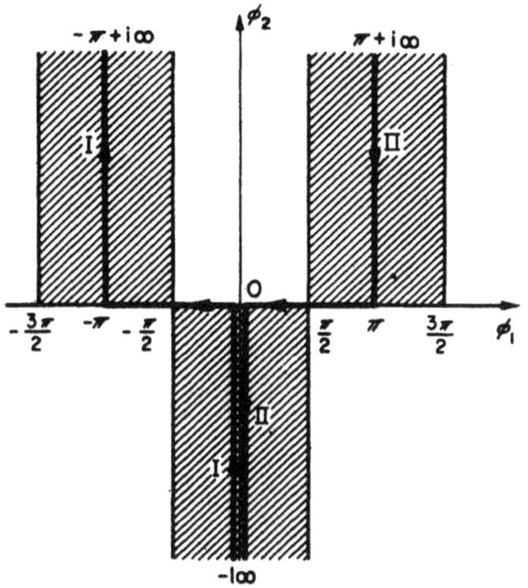

FIG. 91

Let us show that the functions (3) and (4) satisfy Bessel's equation

$$\mathscr{L}(y) = x^2 y'' + xy' + (x^2 - \nu^2)\, y = 0.$$

Taking into consideration that

$$K(x, \varphi) = e^{-x \sin \varphi}$$

satisfies the equation

$$x^2 K_{xx} + x K_x + x^2 K + K_{\varphi\varphi} = 0.$$

we find after twofold integration by parts

$$\mathscr{L}[H_\nu^{(1)}(x)] = -\frac{1}{\pi}\int_{C_1} \mathscr{L}(K)\, e^{i\nu\varphi}\, d\varphi = \frac{1}{\pi}\int_{C_1} [K_{\varphi\varphi} - \nu^2 K]\, \Phi\, d\varphi =$$

$$= \frac{1}{\pi}\int_{C_1} (\Phi'' + \nu^2 \Phi)\, K\, d\varphi + \frac{1}{\pi}\int_{C_1} \frac{\partial}{\partial\varphi}(K_\varphi \Phi - K\Phi')\, d\varphi.$$

The first integral on the right-hand side reduces to zero, by the particular choice of the function Φ, the second may be integrated and reduces to zero for $x > 0$ because of the choice of the path of integration; therefore, $H_\nu^{(1)}$, and also $H_\nu^{(2)}$ are solutions of Bessel's equation.

Similar discussions show that the integral over any contour C, the infinite branches of which lie in the shaded region, satisfies Bessel's equation. Later it will be proved that the integral taken over the contour $C_0 = C_1 + C_2$

$$J_\nu(x) = \frac{1}{2}[H_\nu^{(1)}(x) + H_\nu^{(2)}(x)] = -\frac{1}{2\pi}\int\limits_{C_0} e^{-ix \sin\varphi + i\nu\varphi}\, d\varphi \quad (5)$$

is a Bessel function of the first kind of ν th order. For integral values of $\nu = n$ the integrals over the infinite branches I and II (see Fig. 91) cancel because of the periodicity of the function under the integral sign, and the remaining integral

$$J_n(x) = \frac{1}{2\pi}\int\limits_{-\pi}^{\pi} e^{-ix \sin\varphi + in\varphi}\, d\varphi \quad (6)$$

agrees with the usual integral representation obtained in § 3.

For non-integral values ν there is a relation

$$J_\nu(x) = \frac{1}{2\pi}\int\limits_{-\pi}^{\pi} e^{-ix \sin\varphi + i\nu\varphi}\, d\varphi - \frac{\sin \nu\pi}{\pi}\int\limits_{0}^{\infty} e^{-x \sinh \xi - \nu\xi}\, d\xi, \quad (7)$$

resulting from formula (5). The integral (7) for $\nu > 0$ remains finite at $x = 0$; hence it follows that the function, defined by (7), agrees with the function defined by formula (15) § 1 up to a factor. Comparing coefficients for the low powers of x in the series expansion with respect to x, it is possible to show that these functions are identical.*

Let us show that the Hankel functions $H_\nu^{(k)}(x)$ $(k = 1, 2)$ satisfy recurrence relations, similar to formulae (21) for the functions $J_\nu(x)$, proved in § 1, sub-section 2.

* See for instance: A. Sommerfeld: *Partial Differential Equations of Physics*, 1950, Foreign Literature Publishing House, p. 430.

Because the paths of integration C_1 and C_2 in (3) and (4) do not depend on ν, it is possible to write

$$H^{(k)}_{\nu+1} + H^{(k)}_{\nu-1} = -\frac{2}{\pi}\int e^{-ix\sin\varphi+i\nu\varphi}\cos\varphi\, d\varphi =$$

$$= \frac{2}{\pi i x}\int \frac{\partial}{\partial\varphi}\left(e^{-ix\sin\varphi}\right)e^{i\nu\varphi}\, d\varphi\,,$$

$$H^{(k)}_{\nu+1} - H^{(k)}_{\nu-1} = -\frac{2i}{\pi}\int e^{-ix\sin\varphi+i\nu\varphi}\sin\varphi\, d\varphi =$$

$$= 2\frac{\partial}{\partial x}\frac{1}{\pi}\int e^{-ix\sin\varphi+i\nu\varphi}\, d\varphi = -2\frac{dH^{(k)}_{\nu}}{dx} \quad (k = 1, 2)\,.$$

Integrating by parts in the first equation, we obtain recurrence relations for the Hankel functions

$$\left.\begin{aligned} H^{(1)}_{\nu+1} + H^{(1)}_{\nu-1} &= \frac{2\nu}{x}H^{(1)}_{\nu}\,, \\ H^{(1)}_{\nu+1} + H^{(1)}_{\nu-1} &= -2\frac{dH^{(1)}_{\nu}}{dx} \end{aligned}\right\} \tag{8}$$

and similarly

$$\left.\begin{aligned} H^{(2)}_{\nu+1} + H^{(2)}_{\nu-1} &= \frac{2\nu}{x}H^{(2)}_{\nu}\,, \\ H^{(2)}_{\nu+1} - H^{(2)}_{\nu-1} &= -2\frac{dH^{(2)}_{\nu}}{dx}\,. \end{aligned}\right\} \tag{8'}$$

Hence, in particular, it follows that

$$\frac{dH^{(1)}_0(x)}{dx} = -H^{(1)}_1(x)\,, \qquad \frac{dH^{(2)}_0(x)}{dx} = -H^{(2)}_1(x)\,. \tag{9}$$

If $H^{(1)}_{\nu}(x)$ and $H^{(2)}_{\nu}(x)$ are considered as functions of the complex variable $x = x_1 + ix_2$, then formulae (3) and (4) determine these functions for those values of $x = x_1 + ix_2$, where the real part

$$\mathrm{Re}\,(-ix\sin\varphi) = \mathrm{Re}\left[(-ix_1 + x_2)\sin(\varphi_1 + i\varphi_2)\right] =$$
$$= (x_2\sin\varphi_1\cosh\varphi_2 + x_1\cos\varphi_1\sinh\varphi_2) < 0\,.$$

If $\varphi_1 = 0$ or $\varphi_1 = -\pi$, then integral (3) over the contour C_1 determines the function $H^{(1)}_{\nu}(x)$ for those values of x, for which the real part $\mathrm{Re}(x) = x_1 > 0$.

Let us denote by $C_{1\psi}$ the contour (Fig. 92), for which the vertical parts of the path C_1 have the abscissae $-\pi - \psi$ and ψ ($\psi < 0$) instead of $-\pi$ and 0.

In particular, $C_{10} = C_1$. By Cauchy's theorem replacement of the contour of integration C_1 in formula (3) by the contour $C_{1\psi}$ does not affect the value of the integral. In the integration over the contour $C_{1\psi}$ the integral (3) will converge for those values of $x = x_1 + ix_2$ for which

$$\text{Re}\,(-ix \sin \varphi) = x_2 \sin \varphi_1 \cosh \varphi_2 + x_1 \cos \varphi_1 \sinh \varphi_2 < 0$$

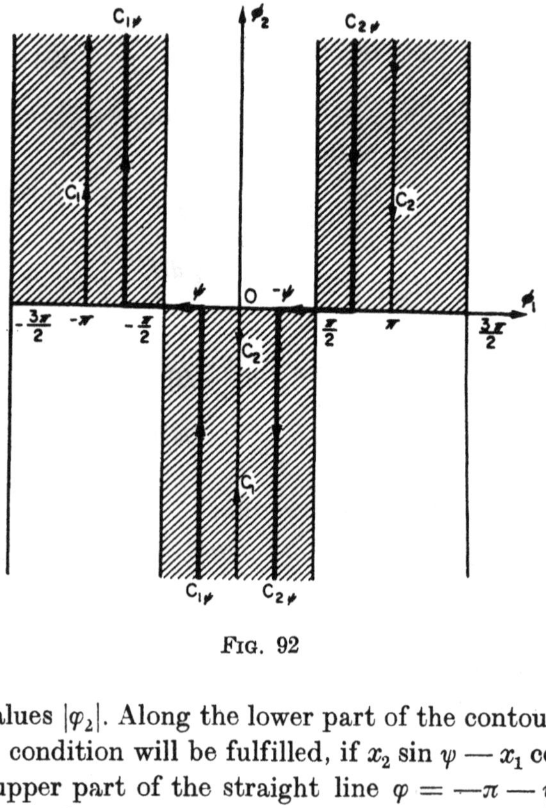

FIG. 92

for large values $|\varphi_2|$. Along the lower part of the contour $\varphi_1 = \psi$, $\varphi_2 < 0$ this condition will be fulfilled, if $x_2 \sin \psi - x_1 \cos \psi < 0$; along the upper part of the straight line $\varphi = -\pi - \psi$, $\varphi_2 > 0$ the condition of convergence is also equivalent to the inequality

$$x_2 \sin \psi - x_1 \cos \psi < 0 .$$

Thus, the integral over $C_{1\psi}$ converges if the inequality

$$x_2 \sin \psi - x_1 \cos \psi < 0$$

is fulfilled (or $x_1 \cos \psi_0 + x_2 \sin \psi_0 > 0$, where $\psi_0 = -\psi$).

If $0 \leqslant \psi_0 < \pi/2$, then the integrals (3) over $C_{1\psi}$ are equal for real values of $x = x_1$, but they determine the functions $H_\nu^{(1)}(x)$ in different regions of the plane of the complex variable x. Therefore, for a change in ψ the integrals (3) over a contour of the type $C_{1\psi}$ determines the analytic continuation of $H_\nu^{(1)}(x)$. In particular, assuming

$$\psi_0 = \pi/2 \quad \text{and} \quad \varphi = -\pi/2 + i\eta\,,$$

we obtain the relation

$$H_\nu^{(1)}(x) = -\frac{1}{\pi}\int_{C_{1\psi_0}} e^{-ix\sin\varphi + i\nu\varphi}\,d\varphi = \frac{e^{-\frac{1}{2}\pi\nu i}}{\pi i}\int_{-\infty}^{\infty} e^{ix\cosh\eta - \nu\eta}\,d\eta\,, \quad (3)$$

defining the function $H_\nu^{(1)}(x)$ in the semi-plane $x_2 = \mathrm{Im}(x) > 0$. Hence it is possible to derive an expression for the Hankel function for a purely imaginary argument

$$H_\nu^{(1)}(ix) = \frac{1}{\pi i}\,e^{-\frac{1}{2}\nu\pi i}\int_{-\infty}^{\infty} e^{-x\cosh\eta - \nu\eta}\,d\eta \quad (x > 0)\,. \qquad (3'')$$

The function $K_\nu(x)$ by definition (see § 3) is expressed in terms of the Hankel function of imaginary argument by the relation

$$K_\nu(x) = \frac{\pi}{2}\,ie^{i\nu\frac{\pi}{2}}\,H_\nu^{(1)}(ix)\,.$$

Utilizing the preceding formula, we obtain an integral representation of $K_\nu(x)$:

$$K_\nu(x) = \frac{1}{2}\int_{-\infty}^{\infty} e^{-x\cosh\eta - \nu\eta}\,d\eta\,,$$

showing that $K_\nu(x)$ is a real function of argument x. For $\nu = 0$ the relation

$$K_0(x) = \frac{1}{2}\int_{-\infty}^{\infty} e^{-x\cosh\eta}\,d\eta = \int_0^{\infty} e^{-x\cosh\eta}\,d\eta$$

was established directly (see § 4).

2. *The method of steepest descent. Asymptotic formulae*

The function under the integral sign

$$e^{-ix\sin\varphi+i\nu\varphi}$$

in formulae (3) and (4), defining the functions $H_\nu^{(1)}$ and $H_\nu^{(2)}$, does not have a singularity in the finite part of the plane of the

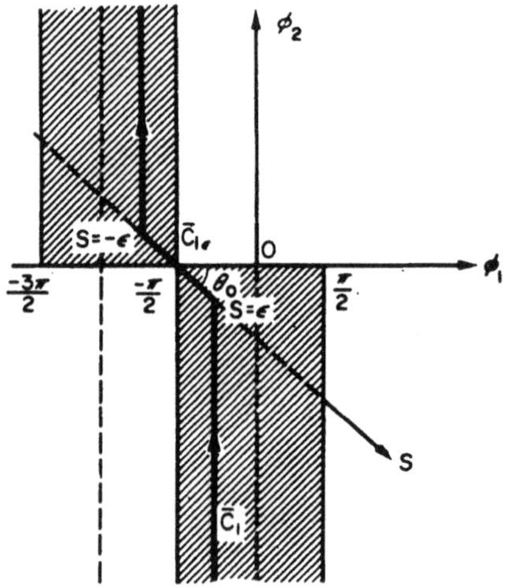

FIG. 93

complex variable φ. Therefore by Cauchy's theorem the contour of integration C_1 in the finite part of the plane can be arbitrarily deformed provided that the asymptotes of branches, extending to infinity, lie in the same shaded zones of the plane φ, as for the contour C_1 (C_2 respectively).

If the contour $\bar{C}_1$ lies entirely within the shaded region (Fig. 93), then at all points, where $\sin\varphi \neq 0$, because $\operatorname{Im}\sin\varphi < 0$, the function under the integral sign tends exponentially to zero for $x \to \infty$. If parts of the contour penetrate an unshaded region, then complicated interference effects occur in those parts of the integral.

To derive the asymptotic behaviour of the function $H_\nu^{(1)}(x)$ for large values of the argument x it is expedient to choose a contour $\bar{C}_1$ lying entirely within the shaded region. Such a contour must pass through the point $-\pi/2$, at which the real part

$$\mathrm{Re}\,(-i \sin \varphi) = \cos \varphi_1 \sinh \varphi_2$$

reduces to zero. For $x \to \infty$ the function under the integral sign outside a neighbourhood of this point tends uniformly to zero, therefore the main part of the integral over $\bar{C}_1$ for $x \to \infty$ comes from a small arc, containing the point $-\pi/2$. From this point of view $\bar{C}_1$ must be chosen so that

$$e^{-ix \sin \varphi}$$

decreases most rapidly with distance from the point $\varphi = -\pi/2$. Let us consider "the topography" of the function $e^{-ix \sin \varphi}$ in the neighbourhood of $\varphi = -\pi/2$.

We put

$$\varphi = -\frac{\pi}{2} + se^{i\theta}.$$

Then for small values of s we have:

$$- i \sin \varphi = i \cos (se^{i\theta}) \left(1 - \frac{s^2}{2} e^{2i\theta} + \ldots\right) =$$

$$= \frac{s^2}{2} \sin 2\theta + i\left(1 - \frac{s^2}{2} \cos 2\,\theta\right) + \ldots$$

The real part $s^2/2 \sin 2\theta$ has a saddle point at $s=0$; in the shaded zones this function is negative, in the unshaded zones positive and for $s = 0$ ($\varphi = -\pi/2$) it reduces to zero. The direction of steepest descent through this saddle point, corresponding to $\theta_0 = -\pi/4$, is the direction of most rapid descent (decrease) for the function $s^2/2 \sin 2\theta$ Hence it follows that for the modulus of the function $e^{-i \sin \varphi}$ the point $s = 0$ is a saddle point, and $\theta_0 = -\pi/4$ corresponds to the direction of most rapid descent.

Let us choose a contour $\bar{C}_1$ so that it contains the rectilinear section $\bar{C}_{1\varepsilon}$ $(-\varepsilon < s < \varepsilon)$, passing through the point $s = 0$ at an angle $\theta_0 = -\pi/4$, and its branches, extending to infinity, lie entirely within the shaded regions. Since the function under the integral sign in (3) decreases exponentially with distance from the point ($s = 0$) ($\varphi = -\pi/2$), it is possible to write:

$$H_\nu^{(1)}(x) = -\frac{1}{\pi}\int_{C_\varepsilon} e^{-ix\sin\varphi + i\nu\varphi}\,d\varphi \cong$$

$$\cong \frac{1}{\pi}\int_{-\varepsilon}^{\varepsilon} e^{x\left(-\frac{s^2}{2}+i\right)-i\nu\frac{\pi}{2}}\,ds\cdot e^{-\frac{i\pi}{4}},$$

since along $\bar{C}_{1\varepsilon}$

$$\varphi = -\frac{\pi}{2} + se^{-\frac{i\pi}{4}}, \quad d\varphi = e^{-\frac{i\pi}{4}}\,ds,$$

$$-i\sin\varphi \cong -\frac{s^2}{2} + i, \quad e^{i\nu\varphi} \approx e^{-i\nu\frac{\pi}{2}},$$

where s changes from ε to $-\varepsilon$. Introducing the symbols

$$\xi = s\sqrt{\frac{x}{2}}, \quad d\xi = \sqrt{\frac{x}{2}}\,ds,$$

we obtain:

$$H_\nu^{(1)}(x) = \frac{1}{\pi}\sqrt{\frac{2}{x}}\,e^{i\left(x-\frac{\pi}{2}\nu-\frac{\pi}{4}\right)}\int_{-\varepsilon\sqrt{\frac{x}{2}}}^{\varepsilon\sqrt{\frac{x}{2}}} e^{-\xi^2}\,d\xi. \tag{10}$$

If $x \to \infty$, then*

$$\int_{-\varepsilon\sqrt{\frac{x}{2}}}^{\varepsilon\sqrt{\frac{x}{2}}} e^{-\xi^2}\,d\xi \to \int_{-\infty}^{\infty} e^{-\xi^2}\,d\xi = \sqrt{\pi},$$

Hence from (10) the asymptotic relation follows

$$H_\nu^{(1)}(x) = \sqrt{\left(\frac{2}{\pi x}\right)}\,e^{i\left[x-\frac{\pi}{2}\left(\nu+\frac{1}{2}\right)\right]} + \ldots, \tag{11}$$

valid for large values of the argument x.

* The error, produced in replacing finite limits by infinite limits has an exponentially decreasing character, since

$$\int_z^{\infty} e^{-\xi^2}\,d\xi \approx \frac{e^{-z^2}}{2z}$$

Similarly

$$H_\nu^{(2)}(x) = \sqrt{\left(\frac{2}{\pi x}\right)}\, e^{-i\left[x-\frac{\pi}{2}\left(\nu+\frac{1}{2}\right)\right]} + \ldots \tag{12}$$

Equations (11) and (12) are approximate; the omitted terms are of order $x^{-3/2}$. The higher terms of the expansion may be derived, if higher power of s are taken in the expressions for $-i \sin \varphi$ and $e^{i\nu\varphi}$.

From the relations for $H_\nu^{(1)}$ and $H_\nu^{(2)}$ asymptotic formulae for the functions J_ν and N_ν follow:

$$J_\nu(x) = \frac{1}{2}\{H_\nu^{(1)}(x) + H_\nu^{(2)}(x)\} \approx$$

$$\approx \sqrt{\left(\frac{2}{\pi x}\right)} \cos\left[x - \frac{\pi}{2}\left(\nu + \frac{1}{2}\right)\right] + \ldots, \tag{13}$$

$$N_\nu(x) = \frac{1}{2i}\{H_\nu^{(1)}(x) - H_\nu^{(2)}(x)\} \approx$$

$$\approx \sqrt{\left(\frac{2}{\pi x}\right)} \sin\left[x - \frac{\pi}{2}\left(\nu + \frac{1}{2}\right)\right] + \ldots \tag{14}$$

We recall that formula (13) for J_ν was proved in § 3, sub-section 2 for integral $\nu = n$.

In order to prove the asymptotic formulae for all values of ν, we must verify that cylindrical functions, introduced as contour integrals, are identical to cylindrical functions, introduced as series. We will not give a proof of this statement.

---

for large values of z, which is readily verified, considering the relation

$$\int_z^\infty e^{-\xi^2}\, d\xi \Big/ \frac{e^{-z^2}}{2z}$$

and expanding the exponential.

PART II

SPHERICAL FUNCTIONS

Spherical functions were introduced in connection with investigations of the solutions of Laplace's equation, and, in particular, with potential theory. In § 1 we consider Legendre polynomials, which are used subsequently for the definition of spherical functions (§ 2). Spherical functions provide a very powerful tool for solving many problems of mathematical physics.

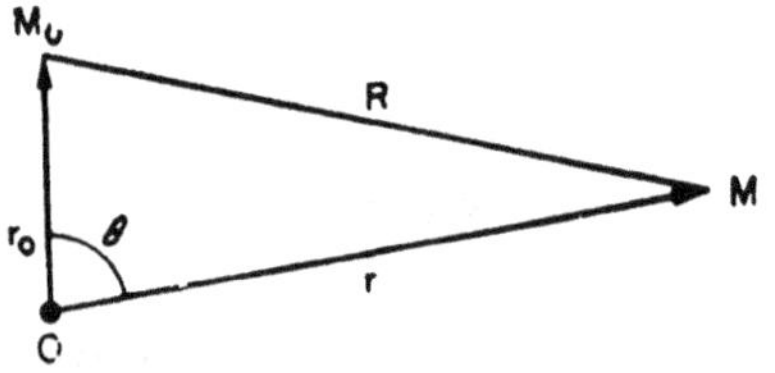

FIG. 94

§ 1. Legendre polynomials

1. *The generating function and Legendre polynomials*

Legendre polynomials are closely connected with the fundamental solution of Laplace's equation $1/R$, where R is the distance of the point M from the fixed point M_0. Let r and r_0 be the radii-vectors of the points M and M_0, and θ the angle between them (Fig. 94). Obviously, it is possible to write

$$\frac{1}{R} = \frac{1}{\sqrt{(r_0^2 + r^2 - 2rr_0 \cos\theta)}} = \begin{cases} \dfrac{1}{r_0}\,\dfrac{1}{\sqrt{(1+\varrho^2 - 2\varrho x)}} & \text{where } r < r_0. \\[2ex] \dfrac{1}{r}\,\dfrac{1}{\sqrt{(1+\varrho^2 - 2\varrho x)}} & \text{where } r > r_0, \end{cases} \tag{1}$$

where $x = \cos\theta$ $(-1 \leqslant x \leqslant 1)$ and $\varrho = r/r_0 < 1$ or $\varrho = r_0/r < 1$ (in both cases ϱ is less than unity).

The function

$$\Psi(\varrho, x) = \frac{1}{\sqrt{(1+\varrho^2-2\varrho x)}} \quad (0 < \varrho < 1, \ -1 \leqslant x \leqslant 1) \tag{2}$$

is called the *generating function* of the Legendre polynomials.

Let us expand the function $\Psi(\varrho, \varkappa)$ as a series in powers of ϱ:

$$\Psi(\varrho, x) = \sum_{n=0}^{\infty} P_n(x)\,\varrho^n . \tag{3}$$

The expansion coefficients of (3) $P_n(x)$ are polynomials of nth order and are called Legendre polynomials.

This expansion corresponds to the expansion of the potential

$$\frac{1}{R} = \begin{cases} \dfrac{1}{r_0} \sum\limits_{n=0}^{\infty} \left(\dfrac{r}{r_0}\right)^n P_n(\cos\theta) \text{ where } r < r_0, \\ \dfrac{1}{r} \sum\limits_{n=0}^{\infty} \left(\dfrac{r}{r_0}\right)^n P_n(\cos\theta) \text{ where } r > r_0 . \end{cases} \tag{4}$$

For sufficiently small values of ϱ considering the expansion

$$(1+\varrho^2-2\varrho x)^{-\frac{1}{2}} = 1 - \frac{1}{2}(\varrho^2 - 2\varrho x) + \frac{3}{8}(\varrho^2 - 2\varrho x)^2 + \ldots =$$

$$= 1 + \varrho x + \varrho^2\left(\frac{3}{2}x^2 - \frac{1}{2}\right) + \ldots + \varrho^n(a_0 x^n + \ldots) + \ldots =$$

$$= \sum_{n=0}^{\infty} P_n(x)\,\varrho^n , \tag{5}$$

it is readily verified that each of the functions $P_n(x)$ is a polynomial of order n. In fact, the terms containing ϱ^n are obtained from brackets of the form $(\varrho^2 - 2\varrho x)^k$ for $k \leqslant n$, where x^n appears in only one term with $k = n$. If $k < n$, then the coefficient in ϱ^n will contain x to a power less than n. Since the general term of the polynomial $(\varrho^2 - 2\varrho x)^k$ has the form $A_m \varrho^{2m}(\varrho x)^{k-m} =$

$A_m \varrho^{m+k} x^{k-m}$, where A_m is some coefficient, then even powers of x appear only with even powers of ϱ and odd powers only with odd powers of ϱ, therefore each polynomial $P_n(x)$ contains either only even, or only odd powers of x, depending on the evenness or oddness of n.

Hence, in particular, it follows that

$$P_n(-x) = (-1)^n P_n(x). \tag{6}$$

For $x = 1$ the generating function $\psi(\varrho, x)$ equals

$$\frac{1}{1-\varrho} = 1 + \varrho + \varrho^2 + \ldots + \varrho^n + \ldots = \sum_{n=0}^{\infty} P_n(1)\,\varrho^n, \tag{7}$$

i. e.

$$P_n(1) = 1 \quad \text{for all integral } n. \tag{8}$$

From formulae (6) and (8) it follows that

$$P_n(-1) = (-1)^n. \tag{9}$$

2. *Recurrence relation*

Differentiating $\Psi(\varrho, x)$ with respect to ϱ, we can show that

$$(1 - 2\varrho x + \varrho^2)\frac{\delta \Psi}{\delta \varrho} - (x - \varrho)\,\Psi = 0. \tag{10}$$

Let us write the left-hand side of this formula as a power series in ϱ, substituting the series (3) for Ψ and the series

$$\frac{\delta \Psi}{\delta \varrho} = P_1(x) + 2P_2(x)\,\varrho + 3P_3(x)\,\varrho^2 + \ldots$$

The coefficient of ϱ^n in the series obtained must equal zero for all x:

$$(n+1)\,P_{n+2}(x) - x\,(2n+1)\,P_n(x) + nP_{n-1}(x) = 0. \tag{11}$$

This identity is a *recurrence* relation, connecting three consecutive Legendre polynomials. It can be used to calculate $P_n(x)$ $(n > 2)$, if we remember that $P_0(x) = 1$, $P_1(x) = x$.

Let us evaluate the coefficient a_n of the highest power in x^n of the polynomial $P_n(x)$. From (11) we see that

$$a_{n+1} = \frac{2n+1}{n+1}\,a_n. \tag{12}$$

Substituting here $a_0 = 1$, successively determining $a_1 = 1$, $a_2 = 3/2$, $a_3 = 5/2$, etc., and utilizing the method of induction, we find:

$$a_n = \frac{1 \cdot 3 \cdot 5 \ldots (2n-1)}{n!}. \tag{13}$$

3. *Legendre's equation*

Let us show that $P_n(x)$ are bounded solutions of Legendre's equation

$$\frac{d}{dx}\left[(1-x^2)\frac{dy}{dx}\right] + \lambda y = 0$$

or

$$(1-x^2)\,y'' - 2\,xy' + \lambda y = 0$$

in the region $-1 < x < 1$ for $\lambda = n(n+1)$.

Differentiating the relation

$$\Psi(\varrho, x) = \sum_{n=0}^{\infty} P_n(x)\,\varrho^n$$

with respect to x, we obtain the equality

$$(1 - 2\,x\varrho + \varrho^2)\frac{\partial \Psi}{\partial x} - \varrho \Psi = 0. \tag{14}$$

Combining this with (10) we have:

$$\varrho \frac{\partial \Psi}{\partial \varrho} - (x - \varrho)\frac{\partial \Psi}{\partial x} = 0. \tag{15}$$

Hence we obtain a second recurrence relation

$$nP_n(x) - xP_n'(x) + P_{n-1}'(x) = 0. \tag{16}$$

Let us substitute in the identity

$$\varrho \frac{\partial}{\partial \varrho}(\varrho \Psi) = \varrho\left(\varrho \frac{\partial \Psi}{\partial \varrho} + \Psi\right) \tag{17}$$

the expression for $\varrho\psi$ and $\varrho \dfrac{\partial \Psi}{\partial \varrho}$, found from (14) and (15); this gives

$$\varrho \frac{\partial}{\partial \varrho}(\varrho \Psi) - (1 - \varrho x)\frac{\partial \Psi}{\partial x} = 0. \tag{18}$$

We expand the left-hand side in a power series with respect to ϱ and let us compare the zero coefficient with ϱ^n; this gives the third recurrence relation

$$nP_{n-1}(x) - P'_n(x) + xP'_{n-1}(x) = 0. \tag{19}$$

Let us eliminate $P_{n-1}(x)$ and $P'_{n-1}(x)$ from (16) and (19). In order to do this we substitute the expression for $P'_{n-1}(x)$ from (16) into (19), then differentiate the relation obtained with respect to x and again use (16). As a result we obtain the equation

$$(1 - x^2)P''_n(x) - 2xP'_n(x) + n(n+1)P_n(x) = 0. \tag{20}$$

Thus, Legendre polynomials are the eigen-functions of the following boundary-value problem:

Find those values of λ, for which there exist non-trivial solutions of Legendre's equation in the region $-1 < x < 1$

$$\frac{d}{dx}\left[(1 - x^2)\frac{dy}{dx}\right] + \lambda y = 0, \tag{21}$$

bounded at $x = \pm 1$ and satisfying the normalizing condition $P_n(1) = 1$, corresponding to the eigen-values $\lambda_n = n(n+1)$.

4. *Orthogonality of the Legendre polynomials*

Legendre's equation

$$\frac{d}{dx}\left[(1 - x^2)\frac{dy}{dx}\right] + \lambda y = 0 \tag{21}$$

is a particular case of the equation considered on p. 627

$$\frac{d}{dx}\left[k(x)\frac{dy}{dx}\right] - q(x)y + \lambda\varrho y = 0 \tag{22}$$

with $q = 0$, $\varrho = 1$, $k(x) = 1 - x^2$. Therefore the general theory developed for equation (22) is applicable to it. From this theory it follows that

(1) Legendre polynomials of different order are mutually orthogonal in the interval $(-1, 1)$:

$$\int_{-1}^{1} P_n(x)\,P_m x\,dx = 0 \quad \text{where} \quad m \neq n; \tag{23}$$

(2) the second linearly independent solution of Legendre's equation for $\lambda = n(n+1)$ tends to infinity at $x = \pm 1$ as $\log |1 \mp x|$.

The system of orthogonal polynomials is complete.* Hence it follows that Legendre's equation does not have non-trivial bounded solutions $P_n(x)$ for any $\lambda \neq n(n+1)$. In fact, if a solution $y_n(x)$ exists for $\lambda \neq n(n+1)$ it will be orthogonal to every $P_n(x)$. Hence because of the closed nature of the system of orthogonal polynomials $P_n(x)$ it follows that

$$y_n(x) \equiv 0.$$

5. *Norm of the Legendre polynomials*

Let us evaluate the norm

$$N_n = \int_{-1}^{1} P_n^2(x)\,dx. \tag{24}$$

Let us consider the polynomial

$$Q(x) = P_n(x) - \frac{a_n}{a_{n-1}} x P_{n-1}(x), \tag{25}$$

* The system of orthogonal functions $\{\varphi_n\}$ is said to be *closed* if there exists no continuous function, not equal identically to zero and orthogonal to all functions of the given system.

The system of orthogonal functions $\{\varphi_n\}$ is said to be *complete* in an interval (a, b), if any continuous function can be approximated in the mean to any degree of accuracy by means of a linear combination of functions $\{\varphi_n\}$. In other words, whatever $\varepsilon > 0$ may be, it is always possible to assign a linear combination of functions

$$S_n = c_1\varphi_1 + \ldots + c_n\varphi_n,$$

such that

$$\int_a^b [f(x) - S_n(x)]^2\,dx < \varepsilon.$$

For a complete system of functions $\{\varphi_n\}$ the relation

$$\int_a^b f^2(x)\,dx = \sum_{n=1}^{\infty} N_n f_n^2$$

holds, where f_n are Fourier coefficients of the function $f(x)$

$$\left(f_n = \frac{1}{N_n}\int_a^b f(\xi)\,\varphi_n(\xi)\,d\xi\right),$$

where a_n is the coefficient of x^n in $P_n(x)$, a_{n-1} is the coefficient of x^{n-1} in $P_{n-1}(x)$. The polynomial $Q(x)$ is a polynomial of order $n-2$ and may be represented in the form

$$Q(x) = \sum_{k=0}^{n=2}{}' A_k P_k(x),$$

where the coefficients A_k are determined in succession, beginning with $k = n - 1$, by comparing the coefficients of x^k.

Hence it follows that $Q(x)$ is orthogonal to $P_n(x)$:

$$\int_{-1}^{1} P_n(x) Q(x)\, dx = 0. \tag{26}$$

Therefore

$$N_n = \int_{-1}^{1} P_n(x) \left[\frac{a_n}{a_{n-1}} x P_{n-1}(x) + Q(x)\right] dx =$$

$$= \frac{a_n}{a_{n-1}} \int_{-1}^{1} x P_n(x) P_{n-1}(x)\, dx$$

expressing from the first recurrence relation

$$x P_n(x) = \frac{n+1}{2n+1} P_{n+1}(x) + \frac{n}{2n+1} P_{n+1}(x)$$

The closed nature is the result of the completeness. Let there be given some complete system of orthogonal functions $\{\varphi_n(x)\}$. We assume that there exists a continuous function $f(x) \not\equiv 0$, orthogonal to all $\varphi_n(x)$. Then by virtue of the completeness of the system of functions $\{\varphi_n\}$ the equality

$$\int_a^b f^2(x)\, dx = \sum_{n=1}^{\infty} N_n f_n^2 = 0,$$

must hold, since $f_n = 0$ by assumption. Hence it follows that $f \equiv 0$, which contradicts the given assumption, i.e. the system $\{\varphi_n(x)\}$ is closed.

The completeness and, thus, the closed nature of the system of orthogonal polynomials $\{P_n(x)\}$ are consequences of *Weierstrass's theorem* concerning the possibility of the uniform approximation of a continuous function by means of polynomials: *for any continuous function* $f(x)$, *given in the interval* (a, b) *and any* $\varepsilon > 0$, *there exists a polynomial* $Q_n(x)$ *such that*

$$|f(x) - Q_n(x)| < \varepsilon, \tag{A}$$

In fact, representing the polynomial $Q_n(x)$ as a linear combination of orthogonal polynomials $\{P_n(x)\}$ and utilizing the inequality (A), we obtain the condition of completeness of a system of orthogonal polynomials.

we have

$$N_n = \frac{n}{2n+1} \frac{a_n}{a_{n-1}} N_{n-1}. \tag{27}$$

Since, by formula (12),

$$\frac{a_n}{a_{n-1}} = \frac{2\,n-1}{n},$$

then

$$N_n = \frac{2n-1}{2n+1} N_{n-1}. \tag{28}$$

Taking into account that

$$P_0(x) = 1, \qquad N_0 = 2,$$

we determine from formula (28) in succession

$$N_2 = \frac{2}{3}, \qquad N_3 = \frac{2}{5}$$

and by induction

$$N_n = \frac{2}{2n+1}. \tag{29}$$

Thus, the Legendre polynomials $P_n(x)$ form an orthogonal system with norm $N_n = \dfrac{2}{2n+1}$:

$$\int_{-1}^{1} P_n(x) P_n(x)\,dx = \begin{cases} 0, & \text{if} \quad m \neq n, \\ \dfrac{2}{2n+1} & \text{if} \quad m = n. \end{cases} \tag{30}$$

6. *Differential formula for the Legendre polynomials*

Let us prove that the Legendre polynomial $P_n(x)$ may be represented in the form

$$P_n(x) = \frac{1}{2^n\,n!} \frac{d^n}{dx^n} [(x^2-1)^n]. \tag{31}$$

This differential relation is often called *Rodrigue's formula.*

In order to prove (31) it is sufficient to show that

(1) $\bar{P}_n(x) = \dfrac{1}{2^n\,n!} \dfrac{d^n}{dx^n} [(x^2-1)^n]$ is a solution of Legendre's equation;

(2) $\bar{P}_n(1) = 1$.

If we put

$$u = (x^2-1)^n$$

and evaluate

$$u' = 2nx(x^2-1)^{n-1},$$

we have

$$(x^2 - 1) u' - 2nxu = 0.$$

Differentiating this equation $(m + 1)$ times, we obtain

$$(x^2 - 1) u^{(m+2)} - (2n - 2m - 2) xu^{(m+1)} +$$
$$+ [m (m + 1) - 2n (m + 1)] u^{(m)} = 0.$$

Assuming $m = n$ and noting that the expression in square brackets equals $n(n + 1)$, we obtain:

$$(1 - x^2) u^{(n+2)} - 2xu^{n+1} + n (n + 1) u^{(n)} = 0,$$

i. e. the function

$$\bar{P}_n = \frac{1}{2^n n!} \frac{d^n u}{dx^n}$$

satisfies Legendre's equation

$$(1 - x^2) \bar{P}''_n - 2x\bar{P}'_n + n (n + 1) \bar{P}_n = 0.$$

Therefore

$$\bar{P}_n (x) = C_n P_n(x), \qquad (31')$$

where C is some constant.

We show now that $\bar{P}_n (1) = 1$.

Let us consider the derivative

$$\frac{d^m}{dx^m} [(x^2 - 1)^n] = \frac{d^m}{dx^m} [(x + 1)^n (x - 1)^n] = c_0 (x + 1)^{n-m} +$$
$$+ (x - 1)^n + c_1 (x + 1)^{n-m+1} (x - 1)^{n-1} + \ldots +$$
$$+ c_n (x + 1)^n (x - 1)^{n-m}.$$

If $m < n$, then at the points $x = \pm 1$ all the components reduce to zero:

$$\left\{\frac{d^m}{dx^m} [(x^2 - 1)^n]\right\}_{x = \pm 1} = 0 \qquad (m < n).$$

If $m = n$, then

$$\left\{\frac{d^n}{dx_n} [(x^2 - 1)^n]\right\}_{x=1} = 2^n n!,$$

from which it follows that

$$\bar{P}_n (1) = 1.$$

Puting $x = 1$ in equality (31') we obtain:

$$C_n = 1,$$

i.e.

$$\bar{P}_x (x) \equiv P_n (x).$$

In this way the relation

$$P_n(x) = \frac{1}{2^n\, n!} \frac{d^n}{dx^n} [(x^2 - 1)^n]$$

is established.

Using this relation it is possible to prove the following important properties of Legendre polynomials.

The Legendre polynomial $P_n(x)$ *has n zeros inside the interval* (—1, 1) *and its derivative of kth order* $d^k/dx^k\, P_n(x)$ $(k \leqslant n)$ *has* $n - k$ *zeros inside the interval* (—1, +1) *and does not reduce to zero at its ends.*

In fact, the function $(t^2 - 1)^n$ reduces to zero at the ends of the interval (—1, +1). Its first derivative reduces to zero at the ends of the interval and has only one zero inside the interval. The second derivative has, at least, two zeros inside the interval and reduces to zero at its ends (Fig. 95).

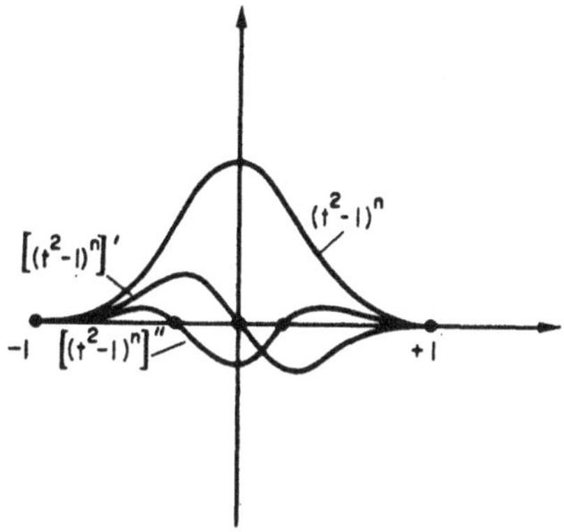

FIG. 95

Continuing this, we reach the conclusion that the nth derivative has, at least, n zeros in the interval (—1, +1) or more accurately, exactly n zeros, since it is determined by a polynomial of nth order. The first part of the statement is proved. The derivative $d/dx\, P_n(x)$ must have, at least, $n - 1$ zeros inside the interval; but since this derivative is a polynomial of $(n - 1)$th order, it has exactly $n - 1$ roots inside the interval. Similarly the function $d^k/dx^k\, P_n$ has $n - k$ roots inside the interval. The lemma is proved.

7. *Integral formulae and the bounds of the Legendre polynomials*

The Legendre polynomials $P_n(\mu)$ are bounded functions for all values of the argument $-1 \leqslant \mu \leqslant +1$.

Let us consider the integral

$$I = \int_0^\pi \frac{d\varphi}{a - ib\cos\varphi},$$

where a and b are constants, and evaluate it

$$I = a\int_0^\pi \frac{d\varphi}{a^2 + b^2 \cos\varphi} + ib\int_0^\pi \frac{\cos\varphi\, d\varphi}{a^2 + b^2 \cos^2\varphi} =$$

$$= a\int_0^\pi \frac{d\varphi}{a^2 + b^2 \cos^2\varphi} + ib\int_{-\frac{\pi}{2}}^{\frac{\pi}{2}} \frac{\sin\psi\, d\psi}{a^2 + b^2 \sin^2\psi} = I_1 + I_2 \quad (32)$$

$$\left(\psi = \varphi - \frac{\pi}{2}\right),$$

The integral I_2 equals zero, since the function under the integral sign is odd. Evaluation of the integral I_1 gives:

$$I_1 = \begin{cases} \dfrac{\pi}{\sqrt{(a^2 + b^2)}} & \text{where } a > 0, \\ -\dfrac{\pi}{\sqrt{(a^2 + b^2)}} & \text{where } a < 0. \end{cases} \quad (32')$$

If we put

$$a = 1 - x\mu ; \quad b = x\sqrt{(1 - \mu^2)} \quad (0 < x < 1 ; \ -1 \leq \mu \leq 1)$$

and use formula (32′) we find:

$$\frac{1}{\pi}\int \frac{d\varphi}{1 - x\,(\mu + i\sqrt{[1 - \mu^2]}\cos\varphi)} = \frac{1}{\sqrt{(1 + x^2 - 2x\mu)}}. \quad (33)$$

The right-hand side is the generating function, defining the Legendre polynomials

$$\frac{1}{\sqrt{(1 + x^2 - 2x\mu)}} = \sum_{n=0}^{\infty} x^n P_n(\mu).$$

We expand the function under the integral sign in the series

$$\frac{1}{1 - x\,(\mu + i\sqrt{[1 - \mu^2]}\cos\varphi)} = \sum_{n=0}^{\infty} x^n\,(\mu + i\sqrt{[1 - \mu^2]}\cos\varphi)^n. \quad (34)$$

Since

$$|x\,(\mu + i\sqrt{[1 - \mu^2]}\cos\varphi)| = x\sqrt{(\mu^2 \sin^2\varphi + \cos^2\varphi)} \leq x < 1,$$

then series (34) converges uniformly and it is possible to integrate it term wise:

$$\frac{1}{\pi}\int_0^\pi \frac{1}{1 - x(\mu + i\sqrt{[1-\mu^2]}\cos\varphi)}\,d\varphi =$$

$$\sum_{n=0}^\infty x^n \cdot \frac{1}{\pi}\int_0^\pi (\mu + i\sqrt{[1-\mu^2]}\cos\varphi)^n\,d\varphi$$

or by formulae (3) and (17)

$$\frac{1}{\sqrt{(1 + x^2 - 2x\mu)}}\sum_{n=0}^\infty x^n P_n(\mu) = \sum_{n=0}^\infty n^n \cdot \frac{1}{\pi} \times$$

$$\times \int_0^\pi (\mu + i\sqrt{[1-\mu^2]}\,\cos\,\varphi)^n\,d\varphi.$$

Hence we obtain a representation of the Legendre polynomials as an integral

$$P_n(\mu) = \frac{1}{\pi}\int_0^\pi (\mu + i\sqrt{[1-\mu^2]}\cos\varphi)^n\,d\varphi \quad (-1 \leq \mu \leq 1). \qquad (35)$$

If $-1 \leqslant \mu \leqslant 1$, then

$$|\mu + i\sqrt{(1-\mu^2)}\,\cos\,\varphi| = \sqrt{(\mu^2\,\sin^2\,\varphi + \cos^2\,\varphi)} \leqslant 1$$

and the inequality

$$|P_n(\mu)| \leq \int_0^\pi d\varphi = 1 \quad \text{where} \quad -1 \leqslant \mu \leqslant 1,$$

holds for $P_n(\mu)$. The equality sign occurs only for $\mu = \pm 1$.

Thus the finiteness of the Legendre polynomials

$$|P_n(\mu)| \leq 1 \qquad (36)$$

is proved over every region of variation of μ between -1 and $+1$.

8. *Associated functions*

Along with the Legendre polynomials $P_n(x)$ the *associated functions* $P_n^{(m)}(x)$ play an important part, being solutions, bounded in the interval $(-1, +1)$ of the equation

$$\frac{d}{dx}\left[(1-x^2)\cdot\frac{dz}{dx}\right] + \left(\lambda - \frac{m^2}{1-x^2}\right)z = 0 \qquad (37)$$

or

$$(1 - x^2) z'' - 2xz' + \left(\lambda - \frac{m^2}{1 - x^2}\right) z = 0. \tag{37}$$

The associated functions can be defined in terms of the Legendre polynomials

$$P_n^{(m)}(x) = (1 - x^2)^{\frac{m}{2}} \frac{d^m}{dx^m} P_n(x). \tag{38}$$

Let us prove this relation.

The equation of the associated functions (37′) is simplified by the substitution

$$z = (1 - x^2)^{\frac{m}{2}} Y(x)$$

to the form

$$(1 - x^2) Y'' - 2(m + 1) xY' + [\lambda - m(m + 1)] Y = 0. \tag{39}$$

We show that equation (39) is satisfied by the mth derivative of the Legendre polynomial. In fact, differentiating Legendre's equation (21) m times with respect to x and simplifying, we obtain:

$$(1 - x^2) \frac{d^{m+2} y}{dx^{m+2}} - 2(m + 1) x \frac{d^{m+1} y}{dx^{m+1}} + [\lambda - m(m + 1)] \frac{d^m y}{dx^m} = 0. \tag{40}$$

Comparing this result with (39), we see that a particular solution of equation (39) is the function

$$Y(x) - \frac{d^m y(x)}{dx^m},$$

and therefore, the expression

$$Z = (1 - x^2)^{\frac{m}{2}} Y(x) = (1 - x^2)^{\frac{m}{2}} \frac{d^m}{dx^m} y(x) \tag{41}$$

satisfies the equation of the associated functions and is bounded at $x = \pm 1$.

For

$$\lambda = n(n + 1) \quad \text{and} \quad y = P_n(x)$$

we obtain formula (41) for the associated function of mth order. From this formula we see that $P_n^{(m)}(x)$ differs from zero only for $m \leqslant n$.

The associated functions according to the general theorem on *p.* 635 *form an orthogonal system.* Let us evaluate the norm of the associated functions. In passing their orthogonality will be proved.

Let us multiply equation (40) by $(1-x^2)^m$ and write the result in the form

$$\frac{d}{dx}\left[(1-x^2)^{m+1}\frac{d^{m+1}}{dx^{m+1}}P_n\right]=-\left[\lambda-m(m+1)\right](1-x^2)^m\frac{d^mP}{dx^m}$$

or after the substitution $m'=m+1$

$$\frac{d}{dx}\left[(1-x^2)^{m'}\frac{d^{m'}P_n}{dx^{m'}}\right]=-\left[\lambda-m'(m'-1)\right](1-x^2)^{m'-1}\frac{d^{m'-1}P_n}{dx^{m'-1}}. \tag{42}$$

Let us introduce the symbol

$$L_{n,k}^m=\int_{-1}^{+1}P_n^{(m)}(x)\,P_k^{(m)}(x)\,dx=\int_{-1}^{+1}(1-x^2)^m\frac{d^mP_n}{dx^m}\frac{d^mP_k}{dx^m}\,dx\,.$$

Integrating the right-hand side by parts we obtain:

$$L_{n,k}^m=\left[\frac{d^{m-1}P_k}{dx^{m-1}}\frac{d^mP_n}{dx^m}(1-x^2)^m\right]_{-1}^1-$$
$$-\int_{-1}^{1}\frac{d^{m-1}P_k}{dx^{m-1}}\frac{d}{dx}\left[(1-x^2)^m\frac{d^mP_n}{dx^m}\right]dx\,.$$

The first term reduces to zero, and the integral term by virtue of the differential equation (42) is transformed to the form

$$L_{n,k}^m=\left[n(n+1)-m(m-1)\right]\int_{-1}^{+1}(1-x^2)^{m-1}\frac{d^{m-1}}{dx^{m-1}}P_n\frac{d^{m-1}}{dx^{m-1}}\times$$
$$\times P_k\,dx=(n+m)(n-m+1)\,L_{n,k}^{m-1}\,.$$

Repeating this argument we get the formula

$$L_{n,k}^m=(n+m)(n+m-1)\ldots(n+1)\,n\ldots(n-m+1)\,L_{n,k}^0=$$
$$=\frac{(n+m)!}{n!}\frac{n!}{(n-m)!}L_{n,k}^0=\frac{(n+m)!}{(n-m!)}L_{n,k}^0\,.$$

Hence, using equation (30)

$$L^0_{n,k} = \int_{-1}^{1} P_n(x)\, P_k(x)\, dx = \begin{cases} 0 & \text{where } k \neq n, \\ \dfrac{2}{2n+1} & \text{where } k = n, \end{cases}$$

finally we obtain:

$$\int_{-1}^{+1} P_n^{(m)}(x) P_k^{(m)}(x)\, dx = \begin{cases} 0 & \text{where } k \neq 0, \\ \dfrac{2}{2n+1} \dfrac{(n+m)!}{(n-m)!} & \text{where } k = n, \end{cases} \tag{43}$$

i.e. the associated functions are mutually orthogonal and have a norm

$$\int_{-1}^{+1} [P_n^{(m)}(x)]^2\, dx = \frac{2}{2n+1} \frac{(n+m)!}{(n-m)!}. \tag{43'}$$

9. *The closed nature of the system of associated functions*

Let us prove that the system of associated functions $\{P_n^{(m)}(x)\}$ completely exhausts all the bounded solutions of equation (37).

In fact, for $\lambda = n(n+1)$ the solution, linearly independent of $P_n^{(m)}(x)$, tends to infinity at $x = \pm 1$. The bounded solution for $\lambda \neq n(n+1)$ must be orthonogal to all $P_n^{(m)}(x)$.

In order to verify that there are no bounded solutions of equation (37), different from $P_n^{(m)}(x)$, it is sufficient to establish that the system of associated functions $\{P_n^{(m)}(x)\}$ is closed, i.e. that there does not exist any continuous function, not identically equal to zero, which would be orthogonal to all functions of the system.

LEMMA. *Any function $f(x)$, continuous in the segment $[-1, 1]$ and reducing to zero at its ends at $x = 1$ and $x = -1$, can be uniformly approximated to any degree of accuracy by a series in the associated functions of any order m.*

We note, first of all, that derivatives of the Legendre polynomials $d^m/dx^m\, P_n(x)$ are polynomials of order $n - m$. Since any polynomial in powers of x can be represented as a linear combination of these polynomials, then, by Weierstrass's theorem, any function $f(x)$, continuous in the segment $[-1, 1]$, can be uniformly approximated to any degree of accuracy by means of a linear combination of $d^m/dx^m\, P_n(x)$:

$$|\bar{f}(x) - \sum_{n=m}^{n_0} c_n \frac{d^m}{dx^m} P_n(x)| < \varepsilon, \text{ if } n_0 > N(\varepsilon). \tag{44}$$

Multiplying this inequality by $(1-x^2)^{m/2}$ we find that

$$|f_1(x) = \sum_{n=m}^{n_0} c_m P_n^{(m)}(x)| < \varepsilon, \text{ if } n_0 > N(\varepsilon), \tag{45}$$

where

$$f_1(x) = \bar{f}(x)(1-x^2)^{\frac{m}{2}}, \tag{46}$$

i.e. any function $f(x)$, representable in the form (46), where $\bar{f}(x)$ is a function, continuous in the segment [—1,1] can be uniformly approximated to any degree of accuracy by a linear combination of associated functions.

We shall say that the function $f_1(x)$ belongs to the class H_1, if it is continuous in the segment [—1,1] and is identically zero in a small neighbourhood of the points $x = -1$ and $x = 1$.

$$f_1(x) = 0 \text{ where } |1-\delta| \leq |x| \leq 1.$$

Since for every function $f_1(x)$ of the class H_1 the function

$$\bar{f}(x) = \frac{f_1(x)}{(1-x^2)^{\frac{m}{2}}}$$

is continuous in [—1,1], then in this way the lemma is proved for functions of the class H_1.

Let us consider some function $f(x)$, continuous in the segment [—1,1], reducing to zero at the ends. It is obvious that this function can be uniformly approximated by means of the function $f_1(x)$ from the class H_1 correct to $\varepsilon/2$:

$$|f(x) - f_1(x)| < \frac{\varepsilon}{2}.$$

Approximating $f_1(x)$ by a polynomial from the associated functions correct to $\varepsilon/2$

$$|f_1(x) - \Sigma_1(x)| < \frac{\varepsilon}{2}, \Sigma_1(x) = \sum_{n=m}^{n_0} c_n P_n^{(m)}(x),$$

we obtain the inequality

$$|f(x) - \Sigma_1(x)| < \varepsilon,$$

which proves the lemma.

Using this lemma the completeness of the system of associated functions is readily proved, and thus its closed nature.

We recall that the system of functions $\{\varphi_n(x)\}$ is called complete in some segment $[a, b]$, if any function $F(x)$, continuous in $[a, b]$, can be approximated in the mean to any degree of accuracy by means of a linear combination of these functions

$$\int_a^b [F(x) - \sum_{n=1}^{n_0} c_n \varphi_n(x)]^2 \, dx < \varepsilon, \text{ if } n_0 > N(\varepsilon).$$

It is obvious that any function, continuous in the segment $[-1,1]$, can be approximated in the mean to any degree of accuracy by means of a function $f(x)$, continuous in $[-1,1]$ and reducing to zero at $x = \pm 1$:

$$\int_{-1}^{1} [F(x) - f(x)]^2 \, dx < \varepsilon'.$$

Considering a linear combination of associated functions, uniformly approximating the functions $f(x)$

$$|f(x) - \Sigma_1(x)| < \varepsilon'',$$

and using the inequality

$$(a + b)^2 \leqslant 2(a^2 + b^2),$$

we obtain:

$$\int_{-1}^{1} [F(x) - \Sigma_1]^2 \, dx \leqslant 2 \int_{-1}^{1} [F(x) - f(x)]^2 \, dx + $$
$$+ 2 \int_{-1}^{1} [f(x) - \Sigma_1]^2 \, dx < \varepsilon$$
$$(\text{if } 2\varepsilon + 4(\varepsilon'')^2 \leqslant \varepsilon),$$

which proves the completeness, and thus the closed nature of the system of associated functions.

§ 2. Harmonic polynomials and spherical functions

1. *Harmonic polynomials*

The harmonic polynomial is a polynomial, satisfying Laplace's equation

$$\nabla^2 u = u_{xx} + u_{yy} + u_{zz} = 0 . \tag{1}$$

It is readily verified that the first two homogeneous harmonic polynomials have the form

$$u_1(x, y, z) = Ax + By + Cz ,$$

$$u_2(x, y, z) = Ax^2 + By^2 - (A + B) z^2 + Cxy + Dxz + Eyz ,$$

where A, B, C are arbitrary coefficients.

Let us find the number of linearly-independent homogeneous harmonic polynomials of order n

$$u_n = \sum_{p+q+r=n} a_{p,q,r}\, x^p y^q z^r . \tag{2}$$

A homogeneous polynomial of order n has $\frac{(n+1)(n+2)}{2}$ coefficients. In fact, the right-hand side of (2) may be represented in the form

$$a_{0,0,n} z^n + (a_{1,0,n-1} x + a_{0,1,n-1} y) z^{n-1} \ldots +$$
$$+ (a_{n-1,0,1} x^{n-1} + a_{n-2,1,1} x^{n-2} y + \ldots + a_{0,n-1,1} y^{n-1}) z +$$
$$+ (a_{n,0,0} x^n + a_{n-1,1,0} x^{n-1} y + \ldots + a_{0,n,0} y^n) z^0 .$$

For z^n there exists one coefficient, for z^{n-1} two, ..., for z we have n coefficients, and for z^0 the number of coefficients equals $(n+1)$, so that the general number of coefficients equals

$$1 + 2 + \ldots + n + (n+1) = \frac{(n+1)(n+2)}{2} . \tag{3}$$

Equation (1) imposes on the coefficient $\frac{n(n-1)}{2}$ linear homogeneous relations, since $\nabla^2 U_n$ is a homogeneous function of order $n - 2$. Thus, the polynomial must have not less than $\frac{(n+1)(n+2)}{2} - \frac{(n-1)n}{2} = (2n+1)$ independent coefficients. If the $\frac{(n-1)n}{2}$ relations specified were proved to be linearly-dependent, then the number of independent coefficients would be greater than $2n + 1$.

Let us show that only $2n + 1$ of the coefficients are linearly independent. The coefficients $a_{p,q,r}$ of a homogeneous polynomial can be represented in the form

$$a_{p,q,r} = \frac{1}{p!\,q!\,r!} \frac{\partial^n u_n}{\partial x^p \partial y^q \partial z^r} .$$

If u_n is a harmonic polynomial, then $a_{p,q,r}$ for $r \geqslant 2$ can be expressed by the coefficients $a_{p,q,0}$ and $a_{p,q,1}$, the number of which exactly equals $2n + 1$.

In fact,

$$\begin{aligned} a_{p,q,r} &= \frac{1}{p!\,q!\,r!} \frac{\partial^{n-2}}{\partial x^p \partial y^q \partial z^{r-2}} \left[\frac{\partial^2 u_n}{\partial z^2}\right] = \\ &= \frac{1}{p!\,q!\,r!} \frac{\partial^{n-2}}{\partial x^p \partial y^q \partial z^{r-2}} \left[-\frac{\partial^2 u_n}{\partial x^2} - \frac{\partial^2 u_n}{\partial y^2}\right] = \\ &= \beta_1 a_{p+2,q,r-2} + \beta_2 a_{p,q+2,r-2} . \end{aligned}$$

Treating the coefficients $a_{p+2,r-2}$ and $a_{p,q+2,r-2}$ similarly, we finally express $a_{p,q,r}$ in terms of coefficients of the type $a_{p,q,0}$ $(p + q = n)$ and $a_{p,q,1}$ $(p + q + 1 = n)$. The number of coefficients of the form $a_{p,q,0}$ equals $(n + 1)$ and $a_{p,q,r}$ n. Thus, the general number of linearly independent coefficients, and therefore of independent harmonic polynomials of nth order exactly equals $2n + 1$.

Homogeneous harmonic polynomials are called *spherical functions*.

2. *Spherical functions*

Spherical functions can most simply be introduced as solutions of Laplace's equation for a spherical region by the method of separation of variables.

We shall look for a solution of the equation

$$\nabla^2 u = \frac{1}{r^2}\frac{\partial}{\partial r}\left(r^2 \frac{\partial u}{\partial r}\right) + \frac{1}{r^2 \sin\theta}\frac{\partial}{\partial \theta}\left(\sin\theta \frac{\partial u}{\partial \theta}\right) + \frac{1}{r^2 \sin^2\theta}\frac{\partial^2 u}{\partial \varphi^2} = 0 , \tag{1}$$

assuming

$$u(r, \theta, \varphi) = R(r)\, Y(\theta, \varphi) .$$

We obtain Euler's equation for $R(r)$

$$r^2 R'' + 2r R' - \lambda R = 0 , \tag{4}$$

and the following equation for $Y(\theta, \varphi)$

$$\nabla^2_{\theta,\varphi} Y + \lambda Y = \frac{1}{\sin\theta}\frac{\partial}{\partial\theta}\left(\sin\theta\frac{\partial Y}{\partial\theta}\right) + \frac{1}{\sin^2\theta}\frac{\partial^2 Y}{\partial\varphi^2} + \lambda Y = 0, \quad (5)$$

with the additional conditions that Y be single valued and bounded over the whole sphere.

In particular, the function $Y(\theta, \varphi)$ satisfies the conditions

$$\left.\begin{aligned} &Y(\theta, \varphi + 2\pi) = Y(\theta, \varphi), \\ &|Y(0, \varphi)| < \infty, \quad |Y(\pi, \varphi)| < \infty. \end{aligned}\right\} \quad (5')$$

The bounded solutions of equation (5), possessing derivatives, continuous up to second order, are called *spherical functions.*

The equation for $Y(\theta, \varphi)$ may also be solved by the method of the separation of variables, assuming

$$Y(\theta, \varphi) = \Theta(\theta)\,\Phi(\varphi).$$

The function $\Phi(\varphi)$ satisfies the equation

$$\Phi'' + \mu\Phi = 0$$

and the condition of periodicity

$$\Phi(\varphi + 2\pi) = \Phi(\varphi).$$

The problem for $\Phi(\varphi)$ has a solution only for integral $\mu = m^2$ and the linearly-independent solutions are the functions $\cos m\varphi$ and $\sin m\varphi$. The function $\Theta(\theta)$ is determined from the equation

$$\frac{1}{\sin\theta}\frac{d}{d\theta}\left(\sin\theta\frac{d\Theta}{d\theta}\right) + \left(\lambda - \frac{\mu}{\sin^2\theta}\right)\Theta = 0$$

with the boundary condition at $\theta = 0$ and $\theta = \pi$. Introducing the variable

$$t = \cos\theta$$

and defining $X(t)|_{t=\cos\theta} = X(\cos\theta) = \Theta(\theta)$, we obtain the equation of the associated functions for $X(t)$.

$$\frac{d}{dt}\left[(1 - t^2)\frac{dX}{dt}\right] + \left(\lambda - \frac{m^2}{1 - t^2}\right)X = 0 \quad (-1 < t < 1). \quad (6)$$

Equation (6), as we have already seen in § 1, has bounded solutions only for $\lambda = n(n + 1)$

$$X(t)|_{t=\cos\theta} = P_n^{(m)}(t)|_{t=\cos\theta} = P_n^{(m)}(\cos\theta) = \Theta(\theta),$$

where $m \leqslant n$.

Let us write down the system of spherical functions of nth order. We use negative upper indices for those functions which contain $\cos k\varphi$, and positive for those functions which contain $\sin k\varphi$. Then we have:

$$\left.\begin{aligned}
&m=0 \qquad Y_n^{(0)} = P_n(\cos\theta)\,,\\
&m=1 \quad Y_n^{(-1)}(\theta,\varphi) = P_n^{(1)}(\cos\theta)\cos\varphi,\ Y_n^{(1)}(\theta,\varphi) =\\
&\qquad\qquad = P_n^{(1)}(\cos\theta)\sin\varphi\,,\\
&\cdots\cdots\cdots\cdots\cdots\cdots\cdots\cdots\\
&m=k \quad Y_n^{(-k)}(\theta,\varphi) = P_n^{(k)}(\cos\theta)\cos k\varphi,\ Y_n^{(k)}(\theta,\varphi) =\\
&\qquad\qquad = \varphi_n^{(k)}(\cos\theta)\sin k\varphi
\end{aligned}\right\} \quad (7)$$

$$(k = 1, 2, \ldots, n)\,.$$

The number of different spherical functions of nth order $Y_n^{(m)}$ equals $2n+1$. A linear combination of these $2n+1$ spherical functions (7)

$$Y_n(\theta,\varphi) = \sum_{m=0}^{n} (A_{nm}\cos m\varphi + B_{nm}\sin m\varphi)\, P_n^{(m)}\cos(\theta) \quad (7^*)$$

or

$$Y_n(\theta,\varphi) = \sum_{m=-n}^{n} C_{mn}\, Y_n^{(m)}(\theta,\varphi)\,,$$

where

$$C_{mn} = \begin{cases} A_{nm} & \text{where } m \leqslant 0\,,\\ B_{nm} & \text{where } m > 0 \end{cases}$$

is also a spherical function.

The functions $Y_n^{(0)} = P_n(\cos\theta)$ do not depend on φ and are called *zonal*. Since $P_n(t)$ by lemma § 1, sub-section 6 has exactly n zeros inside the interval $(-1, +1)$, then the sphere is divided into $(n+1)$ latitudinal zones, inside which the zonal function has the same sign.

Let us consider the behaviour of the function

$$Y_n^{(\pm k)} = \sin^k\theta\left[\frac{d^k}{dt^k}\,P_n(t)\right]_{t=\cos\theta} \begin{cases}\sin k\varphi\,,\\ \cos k\varphi\end{cases}$$

on the sphere. Since $\sin\theta$ reduces to zero at the poles, $\sin k\varphi$ or $\cos k\varphi$ reduces to zero at $2k$ meridians, and $d^k/dt^k\, P_n(t)$ at

$(n - k)$ latitudes, then the whole sphere is divided into cells, in which $Y_n^{(\pm k)}$ has a fixed sign (Fig. 96). The functions $Y^{(\pm k)}$ are called *tessoral.*

Let us return now to find the function R. We shall look for the function $R(r)$ in the form

$$R = r^{\sigma}.$$

Substituting in equation (4) we obtain a characteristic equation for the determination of σ:

$$\sigma(\sigma + 1) - n(n + 1) = 0,$$

from which we find two values of σ:

$$\sigma = n \quad \text{and} \quad \sigma = -(n + 1).$$

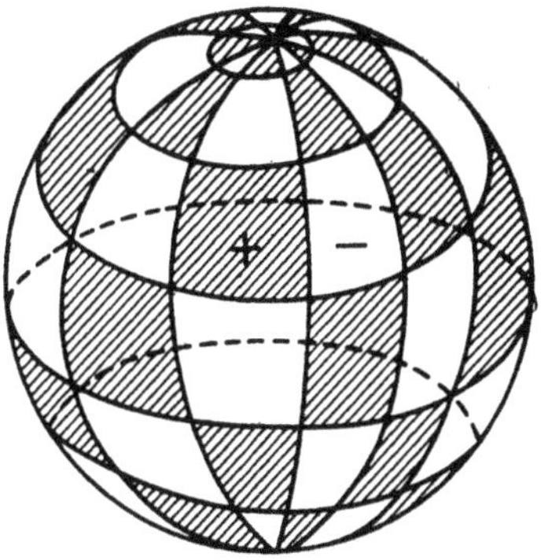

FIG. 96

Therefore, the particular solutions of Laplace's equation are the functions

$$r^n Y_n^{(k)}(\theta, \varphi), \tag{7'}$$

$$r^{-(n+1)} Y_n^{(k)}(\theta, \varphi), \tag{7''}$$

the first of which occur in the solution of interior problems, and the second in exterior problems.

We will show that these solutions of Laplace's equation are homogeneous polynomials of n th order. The general term, for instance, in formula (7′) can be written thus:

$$v = r^n \sin^k \theta \cos k\varphi \cos^{n-k-2q} \theta,$$

where q varies from 0 to $n - k/2$. The function v can be represented as the product of three polynomials:

$$v = u_1 \cdot u_2 \cdot u_3,$$

where

$$u_1 = r^k \sin^k \theta \cos k\varphi = \text{Re}\,[r \sin \theta e^{i\varphi}]^k = \text{Re}\,[(x + iy)^k],$$

$$u_2 = r^{n-k-2q} \cos^{n-k-2q} \theta = z^{n-k-2q},$$

$$u_3 = r^{2q} = (x^2 + y^2 + z^2)^q.$$

Hence it is evident that the function $r^n Y_n^{(k)}(\theta, \varphi)$ is a homogeneous harmonic polynomial of order $k + n - k - 2q + 2q = n$

It is obvious that spherical functions are values of the functions (7′) and (7″) on a sphere of radius unity.

3. *Orthogonality of the system of spherical functions*

Let us prove that the spherical functions, corresponding to different values of λ, are orthogonal on the surface of the sphere Σ. Let Y_1 and Y_2 satisfy the equations

$$\nabla^2_{\theta,\varphi} Y_1 + \lambda_1 Y_1 = 0; \quad \Delta_{\theta,\varphi} Y_2 + \lambda_2 Y_2 = 0, \tag{5'}$$

where

$$\nabla^2_{\theta,\varphi} = \frac{1}{\sin\theta}\frac{\partial}{\partial\theta}\left(\sin\theta\frac{\partial}{\partial\theta}\right) + \frac{1}{\sin^2\theta}\frac{\partial^2}{\partial\varphi^2}.$$

The formula

$$\iint_\Sigma Y_2 \Delta_{\theta,\varphi} Y_1 \, d\Omega = -\iint_\Sigma \left\{\frac{\partial Y_1}{\partial\theta}\frac{\partial Y_2}{\partial\theta} + \frac{1}{\sin^2\theta}\frac{\partial Y_1}{\partial\varphi}\cdot\frac{\partial Y_2}{\partial\varphi}\right\} d\Omega \tag{8}$$

$$(d\Omega = \sin\theta \, d\theta \, d\varphi),$$

may be proved by integration by parts.

On the surface of the sphere

$$\operatorname{grad} u = \frac{\partial u}{\partial\theta}\boldsymbol{i}_\theta + \frac{1}{\sin\theta}\frac{\partial u}{\partial\varphi}\boldsymbol{i}_\varphi,$$

$$\operatorname{div}\boldsymbol{A} = \frac{1}{\sin\theta}\left[\frac{\partial}{\partial\theta}(\sin\theta\, A_\theta) + \frac{\partial A_\varphi}{\partial\varphi}\right],$$

so that

$$\nabla^2_{\theta,\varphi} u = \operatorname{div}\operatorname{grad} u$$

and formula (8) can be written in the form

$$\iint_\Sigma Y_2 \Delta Y_1 \, d\Omega = -\iint_\Sigma \operatorname{grad} Y_1 \cdot \operatorname{grad} Y_2 \cdot d\Omega.$$

Changing the places of Y_1 and Y_2 in formula (8) and subtracting the formula obtained from formula (8), we have:

$$J = \iint_\Sigma \{Y_2 \Delta_{\theta,\varphi} Y_1 - Y_1 \Delta_{\theta,\varphi} Y_2\} d\Omega = 0. \tag{9}$$

Formulae (8) and (9) are Green's formulae for the operator of the spherical functions.

The orthogonality of the functions Y_1 and Y_2 readily follows from (9). In fact, using equations (5), we obtain from formula (9)

$$J = (\lambda_2 - \lambda_1) \iint_\Sigma Y_1 Y_2 \, d\Omega = 0 ,$$

from which for $\lambda_1 \neq \lambda_2$

$$\iint_\Sigma Y_1 Y_2 \, d\Omega = 0$$

or

$$\int_0^{2\pi} \int_0^{\pi} Y_1(\theta, \varphi) Y_2(\theta, \varphi) \sin \theta \, d\theta \, d\varphi = 0 .$$

Thus the orthogonality of the spherical functions, corresponding to different λ, is proved.

We have obtained above for $\lambda = n(n+1)$ a system of $2n+1$ spherical functions of *nth* order. Let us prove that these *spherical functions are mutually orthogonal on the sphere.*

Let $Y_n^{(k_1)}$ and $Y_n^{(k_2)}$ be two spherical functions. Integrating their product and using formulae (7) and (26) § 1, we obtain:

$$\iint_\Sigma Y_n^{(k_1)} Y_n^{(k_2)} \, d\Omega = \int_0^{2\pi} \int_0^{\pi} Y_n^{(k_1)}(\theta, \varphi) \, Y_n^{(k_2)}(\theta, \varphi) \sin \theta \, d\theta \, d\varphi =$$

$$= \int_0^{2\pi} \cos k_1 \varphi \cos k_2 \varphi \, d\varphi \int_0^{\pi} P_n^{(k_1)} \cos \theta) \, P_n^{(k_2)} \cos \theta) \sin \theta \, d\theta =$$

$$= \int_0^{2\pi} \cos k_1 \varphi \cos k_2 \varphi \, d\varphi \int_{-1}^{+1} P_n^{(k_1)}(t) \, P_n^{(k_2)}(t) \, dt =$$

$$= \left\{ \begin{array}{ll} 0 & \text{where} \quad k_1 \neq k_2 , \\ \dfrac{2\pi}{2n+1} \dfrac{(n+k)!}{(n-k)!} & \text{where} \quad k_1 = k_2 = k \neq 0 , \\ 2\pi \cdot \dfrac{2}{2n+1} & \text{where} \quad k_1 = k_2 = 0 , \end{array} \right\} \tag{8'}$$

i.e. the spherical functions, defined by formula (7), form an orthogonal system in the region $0 \leqslant \theta \leqslant \pi$, $0 \leqslant \varphi \leqslant 2\pi$ and have a norm, equal to

$$\int_0^{2\pi} \int_0^{\pi} [Y_n^{(k)}(0, \varphi)]^2 \sin \theta \, d\theta \, d\varphi = \frac{2}{2n+1} \pi \varepsilon_k \frac{(n+k)!}{(n-k)!} , \tag{8''}$$

where $\varepsilon_0 = 2$, $\varepsilon_k = 1$ for $k > 0$.

Assuming the possibility of expanding an arbitrary function $f(\theta, \varphi)$ in a series of spherical functions (the possibility of such an expansion for a twofold continuously differentiable function will be proved below, sub-section 5, allowing a termwise integration, we obtain

$$f(\theta, \varphi) = \sum_{n=0}^{\infty} \sum_{m=0}^{n} (A_{nm} \cos m\varphi + B_{nm} \sin m\varphi) P_n^{(m)} (\cos \theta) = \sum_{n=0}^{\infty} Y_n (\theta, \varphi),$$

where A_{nm} and B_{nm} are Fourier coefficients, given by the relations

$$\left.\begin{aligned} A_{nm} &= \frac{\int_0^{2\pi}\int_0^{\pi} f(\theta, \varphi) P_n^{(m)} \cos \theta) \cos m\varphi \sin \theta \, d\theta \, d\varphi}{N_{nm}}, \\ B_{nm} &= \frac{\int_0^{2\pi}\int_0^{\pi} f(\theta, \varphi) P_n^{(m)} (\cos \theta) \sin m\varphi \sin \theta \, d\theta \, d\varphi}{N_{nm}}, \end{aligned}\right\}$$

$$N_{nm} = \frac{2\pi\varepsilon_m}{2n+1} \frac{(n+m)!}{(n-m)!}, \qquad \varepsilon_m = \begin{cases} 2 \text{ where } m = 0, \\ 1 \text{ where } m > 0. \end{cases}$$

The general solution of Laplace's equation can be represented in the form

$$u(r, \theta, \varphi) = \sum_{n=0}^{\infty} \left(\frac{r}{a}\right)^n Y_n (\theta, \varphi)$$

for the interior boundary-value problem or

$$u(r, \theta, \varphi) = \sum_{n=0}^{\infty} \left(\frac{a}{r}\right)^{n+1} Y_n (\theta, \varphi)$$

for the exterior boundary-value problem.

If the boundary condition $u|_{\Sigma} = f(\theta, \varphi)$ is given on a sphere of radius a then we obtain:

$$f(\theta, \varphi) = \sum_{m=0}^{\infty} Y_n (\theta, \varphi),$$

4. *Completeness of a system of spherical functions*

Let us prove the completeness of the system of spherical functions, given by formula (7). We prove firstly that any function $f(\theta, \varphi)$ having continuous second derivatives, can be uniformly approximated by some finite series of the spherical functions.

Consider the expansion of such a function in a Fourier series

$$f(\theta, \varphi) = \sum_{m=0}^{\infty} [A_m(\theta) \cos m\varphi + B_m(\theta) \sin m\varphi].$$

Using the bounded nature of the second derivative, the coefficients A_m and B_m of this expansion are satisfy the inequalities

$$|A_m| < \frac{M}{m^2}; \quad |B_m| < \frac{M}{m^2},$$

where

$$M = \max |f_{\varphi\varphi}|.$$

Hence it follows that the remainder of the Fourier series satifies the uniform relation

$$\left| f - \sum_{m=0}^{m_0} [A_m(\theta) \cos m\varphi + B_m(\theta) \sin m\varphi] \right| =$$

$$= |R_m| < 2M \sum_{m=m_0}^{\infty} \frac{1}{m^2} < \varepsilon', \tag{10}$$

where $\varepsilon' > 0$ is an arbitrary number.

On the basis of sub-section 8, § 1, the Fourier coefficients $A_m(\theta)$ and $B_m(\theta)$, being continuous functions of θ, and reducing to zero for θ, equal to 0 and π, may be uniformly approximated by linear combinations of the associated functions of mth order

$$\left| A_m(\theta) - \sum_{k=0}^{n} a_k P_k^{(m)} \cos \theta) \right| < \frac{\varepsilon'}{2m_0 + 1},$$

$$\left| B_m(\theta) - \sum_{k=0}^{n} b_k P_k^{(m)} (\cos \theta) \right| < \frac{\varepsilon'}{2m_0 + 1}. \tag{11}$$

Then from the inequalities (10) and (11) it will follow:

$$|f(\theta, \varphi) - \sum_{m=0}^{m_0} \sum_{k=0}^{n} [a_k P_k^{(m)} (\cos \theta) \cos m\varphi +$$

$$+ b_k P_k^{(m)} (\cos \theta) \sin m\varphi]| < 2\varepsilon', \tag{12}$$

which proves the possibility of the uniform approximation of any twofold differentiable function $f(\theta, \varphi)$ by a finite series of the spherical functions. Hence it follows that any continuous function can be uniformly approximated by a finite series of the spherical functions, which proves the completeness of the system of functions, defined by formula (7). This system is closed because it is complete.

Thus, the equation of the spherical functions does not have bounded solutions for $\lambda \neq n(n+1)$ and any spherical function of nth order $(\lambda = n(n+1))$ is representable by formula (7*).

5. *Expansion in terms of spherical functions*

The spherical functions are eigen-functions of the equation

$$\frac{1}{\sin\theta}\frac{\partial}{\partial\theta}\left(\sin\theta\frac{\partial u}{\partial\theta}\right)+\frac{1}{\sin^2\theta}\frac{\partial^2 u}{\partial\varphi^2}+\lambda u=0 \text{ or } \nabla^2_{\theta,\varphi}u+\lambda u=0 \quad (13)$$

on the surface of the sphere Σ $(0 \leqslant \varphi \leqslant 2\pi,\ 0 \leqslant \theta \leqslant \pi)$ with additional boundary conditions.

In order to prove that an arbitrary twofold continuously differentiable function $f(\theta, \varphi)$ may be expanded in a series of spherical functions we form the integral equation corresponding to (13). With this in view let us find the source function of the equation

$$\nabla^2_{\theta,\varphi}u=\frac{1}{\sin\theta}\frac{\theta}{\partial\theta}\left(\sin\theta\frac{\partial u}{\partial\theta}\right)+\frac{1}{\sin^2\theta}\frac{\partial^2 u}{\partial\varphi^2}=0. \quad (14)$$

As was noted above

$$\nabla^2_{\theta,\varphi}u=(\text{div grad } u)_{\theta,\varphi} \quad (15)$$

on the surface of the sphere. Equation (14) may be considered as the equation for a steady distribution of temperature or of a steady electric current on the surface of the sphere.

From this point of view it is clear that it is impossible to find a solution of the homogeneous equation

$$\nabla^2_{\theta,\varphi}u=0 \quad (16)$$

with a singularity at one point only, since for the existence of a steady temperature it is necessary that total flow of heat from sources should equal zero.

We introduce a generalized source function, as the solution of the equation

$$\nabla^2_{\theta,\varphi} u = q \qquad (q = 1/4\pi), \tag{17}$$

regular everywhere, except at the pole $\theta = 0$, where it has a logarithmic singularity. The right-hand side of equation (17) represents the density of negative (sources) sinks uniformly distributed on the surface of the sphere so that

$$\iint_{\Sigma} q \, d\sigma = 1. \tag{18}$$

Assuming that the unknown source function u is a function only of the one variable θ, we obtain an ordinary differential equation for it, which has a solution:

$$u = -q \ln \sin \theta + c \ln \tan \frac{\theta}{2}. \tag{19}$$

Requiring that u should have a singularity only at $\theta = 0$ we obtain:

$$c = -q$$

and

$$u = -2q \ln \sin \frac{\theta}{2} - q \ln 2.$$

Since $u_1 = \text{const.}$ is a solution of the homogeneous equation. the source function G is determined up to an arbitrary constant. Therefore we can write:

$$G = -\frac{1}{2\pi} \ln \sin \frac{\theta}{2}. \tag{20}$$

If the source is located at some point M_0, then the source function has the form

$$G(M, M_0) = -\frac{1}{2\pi} \ln \sin \frac{\gamma_{MM_0}}{2}, \tag{21}$$

where γ_{MM_0} is the angular distance between the points $M_0(\theta_0, \varphi_0)$ and $M(\theta, \varphi)$.*

Let us solve the inhomogeneous equation

$$\nabla^2_{\theta,\varphi} u = \frac{1}{\sin \theta} \frac{\partial}{\partial \theta} \left(\sin \theta \frac{\partial u}{\partial \theta} \right) + \frac{1}{\sin^2 \theta} \frac{\partial^2 u}{\partial \varphi^2} = -F(\theta, \varphi). \tag{22}$$

* The angle γ is determined from the relation

$$\cos \gamma = \cos \theta \cos \theta_0 + \sin \theta \sin \theta_0 \cos (\varphi - \varphi_0).$$

This equation can have a solution, regular everywhere on Σ, only if

$$\iint_{\Sigma} F\,d\sigma = 0\,, \tag{23}$$

expressing the fact that the sum of the sources and sinks must equal zero. This result may be proved from Green's formulae for the operator $\Delta_{\theta\varphi}$, established in sub-section 3.

Let us show that any solution of equation (22), satisfying conditions (23), is representable in the form

$$u(M) = \iint_{\Sigma} G(M, P)\,F(P)\,d\sigma_P + A\,,$$

where A is some constant, and $G(M, P)$ is the source function, given by formula (21). Let M be some fixed point of the sphere, which by a change of coordinates we can take to be the point $\Theta = 0$, and let M_1 be the diametrically opposite point. The points M and M_1 are singular points of equation (22). We construct small circles K_ε^M and $K_\varepsilon^{M_1}$ at these points and consider the integral

$$I = \iint_{\Sigma_1 = \Sigma - K_\varepsilon^M - K_\varepsilon^{M_1}} (u\nabla^2 G - G\,\nabla^2 u)\,d\sigma.$$

Substituting for Δu and ΔG, we have:

$$I = \int_0^{2\omega}\int_\varepsilon^{\pi-\varepsilon}\left[u\frac{\partial}{\partial\theta}\left(\sin\theta\frac{\partial G}{\partial\theta}\right) - G\frac{\partial}{\partial\theta}\left(\sin\theta\frac{\partial u}{\partial\theta}\right)\right]d\theta\,d\varphi +$$

$$+\int_\varepsilon^{\pi-\varepsilon}\frac{d\theta}{\sin\theta}\int_0^{2\pi}\left[u\frac{\partial^2 G}{\partial\varphi^2} - D\frac{\partial^2 u}{\partial\varphi^2}\right]d\varphi$$

The expressions in square brackets are derivatives of

$$\sin\theta\left[\mu\frac{\partial G}{\partial\theta} - G\frac{\partial u}{\partial\theta}\right] \quad\text{and}\quad u\frac{\partial G}{\partial\varphi} - G\frac{\partial u}{\partial\gamma}\,,$$

hence we obtain the result

$$I = \int_0^{2\pi}\left[\sin\theta\left(u\frac{\partial G}{\partial\theta} - G\frac{\partial u}{\partial\theta}\right)\right]_\varepsilon^{\pi-\varepsilon} d\varphi.$$

Further, noting that

$$\frac{\partial G}{\partial \theta} = -\frac{1}{2\pi}\frac{\partial}{\partial \theta}\ln \sin\frac{\theta}{2} = -\frac{1}{4\pi}\cot\frac{\theta}{2},$$

we have:

$$I = \frac{1}{2\pi}\int_0^{2\pi}\left[\sin\frac{\theta}{2}\cdot\cos\frac{\theta}{2}\cot\frac{\Theta}{2}\mu\right]_{\varepsilon}^{\pi-\varepsilon} d\varphi -$$

$$-\frac{1}{2\pi}\left[\sin\theta \ln \sin\frac{\theta}{2}\int_0^{2\pi}\frac{\partial u}{\partial \theta}\,d\varphi\right]_{\varepsilon}^{\pi-\varepsilon} = I_1 + I_2.$$

Hence

$$\lim_{\varepsilon\to 0} I_1 = u(M) \text{ and } \lim_{\varepsilon\to 0} I_2 = 0.$$

Therefore,

$$u(M) = \iint_{\Sigma} G(M,P)\,F(P)\,d\sigma_P + A, \tag{24}$$

where

$$A = \frac{1}{4\pi}\iint_{\Sigma} u\,d\sigma \text{ — is a constant}$$

The solution of our problem is given correct to an additive constant. The solution which satisfies

$$\iint_{\Sigma} u\,d\sigma = 0,$$

is given by the relation

$$u(M) = \iint_{\Sigma} G(M,P)\,F(P)\,d\sigma_P.$$

Applying (24) to the equation for spherical functions $\nabla^2_{\theta,\varphi} u = -\lambda u$ we deduce:

the spherical functions, defined by formula (7), *are the linearly-independent eigenfunctions of the integral equation*

$$u(M) = \lambda\iint_{\Sigma} G(M,P)\,u(P)\,d\sigma_P.$$

with a symmetrical kernel $G(M, P)$, *defined by formula* (21).

The general theory of integral equations with a symmetrical kernel is applicable to this equation. Hence it follows that an arbitrary twofold differentiable function $f(\theta, \varphi)$ can be expanded in a uniformly and absolutely convergent series of spherical functions

$$f(\theta,\varphi) = \sum_{n=0}^{\infty} Y_n(\theta,\varphi) =$$

$$= \sum_{n=0}^{\infty} \sum_{m=0}^{\infty} (A_{nm} \cos m\varphi + B_{nm} \sin m\varphi) P_n^{(m)}(\cos\theta), \quad (25)$$

where

$$Y_n(\theta,\varphi) = \sum_{m=0}^{\infty} (A_{nm} \cos m\varphi + B_{nm} \sin m\varphi) P_n^{(m)}(\cos\theta) \quad (26)$$

and A_{nm} and B_{nm} are Fourier coefficients.

§ 3. Some examples of the application of spherical functions

1. *Polarization of a sphere in a homogeneous field*

Let us consider as an example of the application of spherical functions the problem of the polarization of a dielectric sphere in a homogeneous field.

In an electrostatic field in a homogeneous isotropic medium with dielectric constant ε, there is a sphere of radius a of a dielectric with constant ε_i (Fig. 97). We look for the potential of the field in the form of the sum

$$u = \begin{cases} u_1 = u_0 + v_1 & \text{outside the sphere,} \\ u_2 = u_0 + v_2 & \text{inside the sphere.} \end{cases}$$

where u_0 is the potential of the unperturbed (in the absence of a dielectric sphere) field, and v is the perturbation, produced by placing a sphere in the field. The potential u satisfies the equation

$$\nabla^2 u = 0$$

with the boundary conditions

$$u_1 = u_2 \qquad \text{on} \quad S,$$

$$\varepsilon_1 \frac{\partial u_1}{\partial n} = \varepsilon_2 \frac{\partial u_2}{\partial n} \quad \text{on} \quad S,$$

where S is the surface of the sphere, u_1 and u_2 are values of the function u outside and inside the sphere. Hence it follows that the potential v will be given by the conditions

$$\nabla^2 v = 0, \tag{1}$$

$$v_1 = v_2 \quad \text{on} \quad S, \tag{2}$$

$$\varepsilon_1 \frac{\partial v_1}{\partial n} - \varepsilon_2 \frac{\partial v_2}{\partial n} = -(\varepsilon_1 - \varepsilon_2) \frac{\partial u_0}{\partial n} \quad \text{on } S, \tag{3}$$

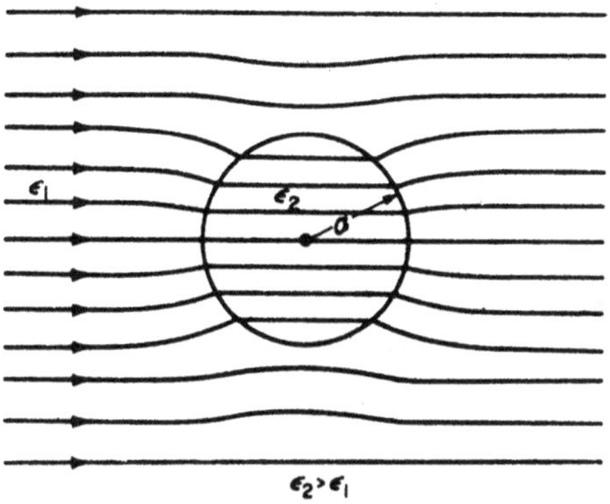

FIG. 97

since for the function u_0 we have:

$$\nabla^2 u_0 = 0,$$

$$(u_0)_1 = (u_0)_2 \quad \text{on} \quad S,$$

$$\left(\frac{\partial u_0}{\partial n}\right)_1 = \left(\frac{\partial u_0}{\partial n}\right)_2 \quad \text{on} \quad S.$$

On the right-hand side of (3) there is a well known function of θ and φ, which we expand in spherical functions:

$$\left.\frac{\partial u_0}{\partial n}\right|_S = \sum_{n=0}^{\infty} Y_n(\theta, \varphi).$$

Assuming

$$v_1 = \sum_{n=0}^{\infty} \left(\frac{a}{r}\right)^{n+1} \bar{Y}_n(\theta, \varphi): \qquad v_2 = \sum_{n=0}^{\infty} \left(\frac{r}{a}\right)^n \bar{\bar{Y}}_n$$

and using the boundary conditions (2) and (3) we obtain:

$$\bar{Y}_n = \bar{\bar{Y}}_n$$

and

$$\varepsilon_1 \sum_{n=0}^{\infty} \frac{-(n+1)}{r}\left(\frac{a}{r}\right)^{n+1} \bar{Y}_n \bigg|_{r=a} - \varepsilon_2 \sum_{n=0}^{\infty} \frac{n}{a}\left(\frac{r}{a}\right)^{n-1} \bar{Y}_n \bigg|_{r=a} =$$

$$= -(\varepsilon_1 - \varepsilon_2) \sum_{n=0}^{\infty} Y_n,$$

from which

$$\bar{Y}_n = Y_n \frac{(\varepsilon_1 - \varepsilon_2)\, a}{\varepsilon_1 (n+1) + \varepsilon_2 n}, \tag{4}$$

Now we consider special cases. The sphere is situated in a homogeneous parallel external field E^0, directed along the z-axis. The potential of this field is

$$u_0 = -E_0 z = -E_0 r \cos\theta,$$

so that

$$\frac{\partial u_0}{\partial n}\bigg|_S = \frac{\partial u_0}{dr}\bigg|_{r=a} = -E_0 \cos\theta = Y_1(\theta).$$

Formula (4) gives:

$$\bar{Y}_n = 0 \quad \text{where} \quad n \neq 1,$$

$$\bar{Y}_1 = -E_0 \cos\theta \frac{(\varepsilon_1 - \varepsilon_2)\, a}{2\varepsilon_1 + \varepsilon_2}.$$

For the potential of the perturbed field we have:

$$u_1 = -E_0 z \left[1 + \frac{\varepsilon_1 - \varepsilon_2}{2\varepsilon_1 + \varepsilon_2}\left(\frac{a}{r}\right)^3\right] \quad \text{outside the sphere } (r > a),$$

$$u_2 = -E_0 z \frac{3\varepsilon_1}{2\varepsilon_1 + \varepsilon_2} \quad \text{inside the sphere } (r < a),$$

from which it follows that

$$E_1 = -\frac{\partial u_1}{\partial z} = \left[1 - \frac{\varepsilon_1 - \varepsilon_2}{2\varepsilon_1 + \varepsilon_2}\frac{a^3}{r^3}\right] E_0,$$

$$E_2 = -\frac{\partial u_2}{\partial z} = \frac{3\varepsilon_1}{2\varepsilon_1 + \varepsilon_2} E_0,$$

i.e. the field inside the sphere is parallel and homogeneous.

By the same method it is possible to derive a solution of the problem of the polarization of a sphere in the presence of a point source, using the expansion of $1/r$ in spherical functions (see § 1).

Similar problems occur in an investigation of magnetic or thermal fields, and steady currents in the presence of a spherical inhomogeneity. For the thermal problem the coefficients of heat conduction will appear in the boundary condition (3) in place of ε_1 and ε_2, for the magnetic problem — the magnetic permeabilities μ_1 and μ_2, and for the last problem — the conductivities λ_1 and λ_2.

2. *Natural oscillations of a sphere*

Let us consider the problem of the natural oscillations of a sphere of radius r_0 with zero boundary conditions of the first kind. This problem reduces to a search for the eigen-values and eigen-functions of the equation

$$\nabla^2 v + \lambda v = 0 \tag{5}$$

with a boundary condition on the surface of the sphere

$$v = 0. \tag{6}$$

Placing the origin of a spherical system of coordinates at the centre of the sphere, we rewrite equation (5) in the form

$$\frac{1}{r^2}\frac{\partial}{dr}\left(r^2\frac{\partial v}{\partial r}\right) + \frac{1}{r^2}\nabla^2_{\theta,\varphi} v + \lambda v = 0, \tag{5'}$$

where

$$\nabla^2_{\theta,\varphi} v = \frac{1}{\sin\theta}\frac{\partial}{\partial v}\left(\sin\theta\frac{\partial v}{\partial\theta}\right) + \frac{1}{\sin^2\theta}\frac{\partial^2 v}{\partial\varphi^2}.$$

We shall look for a solution by the method of separation of variables, assuming

$$\theta(r, \theta, \varphi) = R(r)\,Y(\theta, \varphi). \tag{6'}$$

After substitution in equation (5) we obtain:

$$\frac{\frac{d}{dr}\left(r^2\frac{dR}{dr}\right)}{R} + \lambda r^2 + \frac{\nabla^2_{\theta,\varphi}Y}{Y} = 0, \tag{7}$$

from which it follows:

$$\nabla^2_{\theta,\varphi}Y + \mu Y = 0, \tag{8}$$

$$\frac{1}{r^2}\frac{d}{dr}\left(r^2\frac{dR}{dr}\right) + \left(\lambda - \frac{\mu}{r^2}\right)R = 0. \tag{9}$$

Solving equation (8) with natural boundary conditions at the poles of the sphere

$$|Y|_{\theta=0,\pi} < \infty \tag{10}$$

and conditions of periodicity with respect to φ: $Y(\theta, \varphi + 2\pi) = Y(\theta, \varphi)$ we obtain the eigen-values

$$\mu = n(n+1), \tag{11}$$

to each of which there corresponds $2n + 1$ spherical functions:

$$\left.\begin{aligned} Y_n^{(-j)}(\theta, \varphi) &= P_n^{(j)}(\cos\theta)\cos j\varphi, \\ Y_n^{(j)}(\theta, \varphi) &= P_n^{(j)}(\cos\theta)\sin i\,\varphi \qquad (j = 0,\ 1, 2, \ldots, n)) \end{aligned}\right\} \tag{12}$$

Let us turn now to equation (9). Taking into account equation (11) and the boundary conditions at $r = r_0$ and the natural boundary condition at $r = 0$, we obtain for the function $R(r)$ the following eigen-value problem:

$$\frac{1}{r^2}\frac{d}{dr}\left(r^2\frac{dR}{dr}\right) + \left(\lambda - \frac{n(n+1)}{r^2}\right)R = 0 \tag{9'}$$

$$R(r_0) = 0, \tag{13}$$

$$|R(0)| < \infty. \tag{14}$$

Making the substitution

$$R(r) = \frac{y(r)}{\sqrt{r}} \tag{15}$$

this equation reduces to Bessel's equation of order $(n + 1/2)$:

$$y'' + \frac{1}{r}y' + \left[\lambda - \frac{\left(n + \frac{1}{2}\right)^2}{r^2}\right]y = 0, \tag{16}$$

the general solution of which has the form (see Supplement, Part I, § 1)

$$y(r) = AJ_{n+1/2}(\sqrt{\lambda}r) + BN_{n+1/2}(\sqrt{\lambda}\,r). \tag{17}$$

From the boundary condition (14) it follows that

$$B = 0.$$

The boundary condition (13) gives:

$$AJ_{n+1/2}(\sqrt{\lambda}\cdot r_0) = 0.$$

Since we have non-trivial solutions of the equation, then $A \neq 0$, and, therefore,

$$J_{n+1/2}(\sqrt{\lambda} \cdot r_0) = 0.$$

Denoting the roots of the transcendental equation

$$J_{n+1/2}(\nu) = 0, \tag{18}$$

by $\nu_1^{(n)}, \nu_2^{(n)}, \ldots, \nu_m^{(n)}$, we find the eigen-values

$$\lambda_{m.n} = \left(\frac{\nu_m^{(n)}}{r_0}\right)^2. \tag{19}$$

Each eigen-value $\lambda_{n,m}$ has $2n + 1$ corresponding eigen-functions. Let us introduce the symbol

$$\psi_n(x) = \sqrt{\left(\frac{\pi}{2x}\right)} J_{n+1/2}(x). \tag{20}$$

Then the eigen-functions of equation (5) with the boundary condition (6) can be represented in the form

$$v_{n.m.j}(r, \theta, \varphi) = \psi_n\left(\frac{\nu_m^{(n)}}{r_0} r\right) Y_n^{(j)}(\theta, \varphi)$$

$$(j = -n, \ldots, -1, 0, 1, \ldots, n) \tag{21}$$

Let us consider now the first interior boundary-value problem for the wave equation

$$\nabla^2 v + k^2 v = 0 \tag{22}$$

with the boundary condition

$$v = f\,\theta, \varphi) \tag{23}$$

on the surface of a sphere of radius r_0.

From a previous account it is obvious that the solution of this problem can be represented in the form

$$v(r, \theta, \varphi) = \sum_{n=0}^{\infty} \sum_{j=-n}^{n} f_{nj} \frac{\psi_n(kr)}{\psi_n(kr_0)} Y_n^{(j)}(\theta, \varphi), \tag{24}$$

where f_{nj} are expansion coefficients of the function $f(\theta, \varphi)$ in a series of spherical functions $\{Y_n^{(j)}(\theta, \varphi)\}$

$$f(\theta, \varphi) = \sum_{n=0}^{\infty} \sum_{j=-n}^{n} f_{nj} Y_n^{(j)}(\theta, \varphi). \tag{25}$$

If k^2 coincides with one of the eigen-values

$$k^2 = \lambda_{m_0 n_0} = \left(\frac{\nu_{m_0}^{(n_0)}}{r_0}\right)^2 ,$$

then the boundary-value problem (22)–(23) has a solution only for some functions $f(\theta, \varphi)$. Formula (24) shows that a necessary and sufficient condition for the solvability of our boundary-value problem in this case is

$$f_{n_0 j} = 0$$

or

$$\int_0^\pi \int_0^{2\pi} f(\theta, \varphi)\, Y_{n_0}^{(j)}(\theta, \varphi) \sin\theta \, d\theta \, d\varphi = 0 .$$

If these conditions are fulfilled, then the solution is determined by (24), in which the components, corresponding to $n = n_0$ are absent. But this solution is definitely not unique, any linear combination of eigen-functions, corresponding to $k^2 = \lambda_{m_0 n_0}$ may be added to it.

3. *Exterior boundary-value problem for a sphere*

Let us consider the exterior first boundary-value problem for a sphere (see Chapter VII, § 3)

$$\nabla^2 v + k^2 v = 0 \quad (k^2 > 0) ,$$

$$v|_{r=r_0} = f(\theta, \varphi) ,$$

$$v = O\left(\frac{1}{r}\right) \quad \text{where} \quad r \to \infty ,$$

$$\lim_{r\to\infty} r\left(\frac{\partial v}{\partial r} + ikv\right) = 0 \quad \text{(radiation condition)}.$$

As was shown in § 3 of Chapter VII, this problem has a unique solution.

We expand the solution and the function $f(\theta, \varphi)$ in a series of spherical functions

$$v(r, \theta, \varphi) = \sum_{n=0}^{\infty} \sum_{j=-n}^{n} R_n(r)\, Y_n^{(j)}(\theta, \varphi) ,$$

$$f(\theta, \varphi) = \sum_{n=0}^{\infty} \sum_{j=-n}^{n} f_{nj}\, Y_n^{(j)}(\theta, \varphi) .$$

The expansion coefficients $R_n(r)$, satisfy the equation

$$R_n'' + \frac{1}{r} R_n' + \left(k^2 - \frac{n(n+1)}{r^2}\right) R_n = 0\,,$$

the boundary condition

$$R_n(r_0) = f_n$$

and the radiation condition for $r \to \infty$

$$R_n(r) = 0\left(\frac{1}{r}\right),$$

$$\lim_{r \to \infty} r\,(R_n' + ik\,R_n) = 0\,.$$

The general solution of this equation has the form (see subsection 2 and Supplement I, § 3)

$$R_n(r) = A_{n=n}\,\zeta_u^{(1)}(kr) + B_{n=n}\,{}_n^{(a')}(kr)\,,$$

where

$$\zeta_n^{(1)}(\varrho) = \sqrt{\left(\frac{\pi}{2\varrho}\right)}\, H_{n+1/2}^{(1)}(\varrho)\,,$$

$$\zeta_n^{(2)}(\varrho) = \sqrt{\left(\frac{\pi}{2\varrho}\right)}\, H_{n+1/2}^{(2)}(\varrho) \qquad (\varrho = kr)\,.$$

Using the asymptotic formulae for the Hankel functions $H_n^{(1)}(\varrho)$ and $H_n^{(2)}(\varrho)$ (see Supplement I, § 3):

$$H_n^{(1)}(\varrho) = \sqrt{\left(\frac{2}{\pi\varrho}\right)} e^{i\left[\varrho - \frac{\pi n}{2} - \frac{\pi}{4}\right]} + \ldots,$$

$$H_n^{(2)}(\varrho) = \sqrt{\left(\frac{2}{\pi\varrho}\right)} e^{-i\left[\varrho - \frac{\pi n}{2} - \frac{\pi}{4}\right]} + \ldots,$$

we obtain the following asymptotic relations for the function $\zeta_n^{(1)}$ and $\zeta_n^{(2)}$:

$$\zeta_n^{(1)}(kr) = \frac{e^{i\left[kr - \frac{\pi n}{2} - \frac{\pi}{4}\right]}}{r} + \ldots,$$

$$\zeta_n^{(2)}(kr) = \frac{e^{-i\left[kr - \frac{\pi n}{2} - \frac{\pi}{4}\right]}}{r} + \ldots\,.$$

Hence it is seen that the radiation condition is satisfied only by the function $\zeta_n^{(2)}$.

Therefore

$$A_n = 0 .$$

Utilizing the boundary condition for $r = r_0$ we find:

$$B_{nj} = \frac{f_{jn}}{\zeta_n^{(2)}(kr_0)} .$$

Thus, we obtain the function $v(r, \theta, \varphi)$ in the form

$$v(r, \theta, \varphi) = \sum_{n=0}^{\infty} \sum_{j=-n}^{n} \frac{f_{nj}\,\zeta_n^{(2)}(kr)}{\zeta_n^{(2)}(kr_0)} Y_n^{(j)}(\theta, \varphi) ,$$

where

$$f_{nj} = \frac{\int_0^{\pi}\int_0^{2\pi} f(\theta, \varphi)\, Y_n^{(j)}(\theta, \varphi) \sin\theta \, d\theta \, d\varphi}{N_{nj}} ,$$

and

$$N_{nj} = \int_0^{\pi}\int_0^{2\pi} [Y_n^{(j)}]^2 \sin\theta \, d\theta \, d\varphi =$$

$$= \frac{2\pi\varepsilon_j}{2n+1} \frac{(n+j)!}{(n-j)!} , \quad \varepsilon_j = \begin{cases} 2, i = 0, \\ 1, j > 0 \end{cases}$$

is the norm of the spherical function $Y_n^{(j)}(\theta, \varnothing)$

PART III

CHEBYSHEV-HERMITE AND CHEBYSHEV-LAGUERRE POLYNOMIALS

§ 1. Chebyshev-Hermite polynomials

The problem of the linear harmonic oscillator in quantum mechanics leads to the equation

$$\frac{d^2y}{dx^2} - 2x\frac{dy}{dx} + \lambda y = 0, \tag{1}$$

which may be written in the form

$$\frac{d}{dx}\left[e^{-x^2}\frac{dy}{dx}\right] + \lambda e^{-x^2}\,y = 0. \tag{2}$$

The Chebyshev–Hermite polynomials are defined as the eigen-functions of equation (2) on an infinite straight line $-\infty < x < \infty$ with the following boundary condition: the solution for $x \to \infty$ tends to infinity not faster than a finite power of x.

We shall look for a solution of this problem in the form of a power series

$$y = \sum_{n=0}^{\infty} a_n x^n. \tag{3}$$

Substituting expression (3) in equation (1) we have:

$$\sum_{n=0}^{\infty} \{a_{n+2}(n+2)(n+1) - 2na_n + \lambda a_n\}\,x^n = 0. \tag{3a}$$

Hence we obtain the recurrence relation for the coefficients

$$a_{n+2} = \frac{2n-\lambda}{(n+2)(n+1)}\,a_n. \tag{4}$$

From (4) it follows that for $a_1 = 0$ (3) contains only even, and for $a_0 = 0$ only odd powers of x. The coefficients a_0 and a_1 are arbitrary.

For $\lambda = 2n$ (1) has a solution in the form of a polynomial of order n

$$\bar{H}_n(x) = a_0 + a_2 x^2 + \ldots + a_n x^n \qquad (n - \text{even})$$

and

$$\bar{H}_n(x) = a_1 x + a_3 x^3 + \ldots + a_n x^n \qquad (n - \text{odd}) \tag{5}$$

which, up to a normalizing factor, is a *Chebyshev–Hermite polynomial* and is an eigen-function of the boundary-value problem under consideration.

If $\lambda \neq 2n$, then the power series will have an infinite number of coefficients, different from zero and having, with the exception, perhaps, of a finite number, the same sign. In this case there are terms $a_n x^n$ with n as large as desired, which exceed any power of x as $x \to \infty$: therefore, the power series does not satisfy the boundary condition and cannot be a solution of the boundary-value problem. Thus we have established that Chebyshev–Hermite polynomials give a unique solution of our problem.

It is possible to define the Chebyshev–Hermite polynomials $H_n(x)$, by a generating function

$$\Psi(x, t) = e^{2tx - t^2} \tag{6}$$

and considering its expansion in a Taylor's series with respect to the variable t

$$\Psi(x, t) = e^{x^2} e^{-(t-x)^2} = \sum_{n=0}^{\infty} H_n(x) \frac{t^n}{n!}. \tag{7}$$

It is readily verified that the expansion coefficients $H_n(x)$ are polynomials of n th order.

Let us show that the polynomials $H_n(x)$ coincide with the polynomials $\bar{H}_n(x)$, given by formula (5) up to a factor.

In fact, the relations

$$\frac{\partial \Psi}{\partial x} = 2t\Psi$$

and

$$\frac{\partial \Psi}{\partial t} + 2(t - x)\Psi = 0$$

give

$$H_n'(x) = 2n H_{n-1}(x) \tag{8}$$

and, respectively

$$H_{n+1} - 2\,x\,H_n + 2\,n\,H_{n-1} = 0\,. \tag{9}$$

From formulae (8), (9) it follows that $H_n(x)$ satisfies the equation

$$H''_n = 2\,x\,H'_n + 2\,n\,H_n = 0\,,$$

i.e. the Chebyshev–Hermite polynomials $H_n(x)$ coincide with the polynomials $\bar{H}_n(x)$ up to a factor of proportionality, which was undefined in formula (5).

Equation (7) leads to the following differential relation for *Chebyshev–Hermite polynomials* :

$$H_n(x) = \left(\frac{\partial^n \Psi}{\partial t^n}\right)_{t=0} = (-1)^n\, e^{x^2} \frac{d^n}{dx^n} e^{-x^2}\,. \tag{10}$$

Hence, in particular, it follows that $a_n = 2^n$.

Using this relation, we evaluate several polynomials $H_n(x)$:

$$H_0(x) = 1\,; \qquad H_3(x) = 8\,x^3 - 12\,x\,;$$
$$H_1(x) = 2\,x\,; \qquad H_4(x) = 16\,x^4 - 48\,x^2 + 12\,.$$
$$H_2(x) = 4\,x^2 - 2\,;$$

Let us prove that Chebyshev–Hermite polynomials for $-\infty < x < \infty$ form an orthogonal system with weight e^{-x^2}:

$$\int_{-\infty}^{\infty} H_m(x)\,H_n(x)\,e^{-x^2}\,dx = \begin{cases} 0, \text{ if } m \neq n\,, \\ 2^n\, n!\,\sqrt{\pi}, \text{ if } m = n\,. \end{cases} \tag{11}$$

In fact,

$$J = \int_{-\infty}^{\infty} H_m(x)\,H_n(x)\,e^{-x^2}\,dx = (-1)^n \int_{-\infty}^{\infty} H_m(x) \frac{d^n}{dx^n}(e^{-x^2})\,dx\,.$$

We assume for definiteness that $m \leqslant n$. Integrating by parts and using formula (8), and also the fact that at infinity the product of any polynomial and the function e^{-x^2} and all its derivatives of as high an order as desired reduce to zero, we obtain:

$$J = (-1)^{n-1}\, 2\,m \int_{-\infty}^{\infty} H_{m-1}(x) \frac{d^{n-1}}{dx^{n-1}}(e^{-x^2})\,dx = \ldots =$$

$$= (-1)^{n-m}\, 2^m m! \int_{-\infty}^{\infty} \frac{d^{n-m}}{dx^{n-m}}(e^{-x^2})\,dx\,.$$

If $m < n$, then

$$J = (-1)^{n-m} 2^m m! \frac{d^{n-m-1}}{dx^{n-m-1}} (e^{-x^2}) \Big|_{-\infty}^{\infty} = 0 .$$

If $m = n$, then

$$J = 2^n n! \int_{-\infty}^{\infty} e^{-x^2} dx = 2^n n! \sqrt{\pi} ,$$

since

$$\int_{-\infty}^{\infty} e^{-x^2} dx = \sqrt{\pi} .$$

Thus relation (11) is proved.

In applications one often utilizes the functions

$$\psi_n(x) = \frac{H_n(x) e^{-\frac{x^2}{2}}}{\sqrt{(2^n n! \sqrt{\pi})}} ,$$

forming an orthogonal and normalized system in the interval $-\infty < x < \infty$. These functions satisfy the equation

$$\psi_n'' + \left[(2n+1) - x^2\right] \psi_n = 0 .$$

§ 2. Chebyshev-Laguerre polynomials

1. Chebyshev–Laguerre polynomials may be defined as the solutions of the equation

$$xy'' + (1-x) y' + \lambda y = 0 \qquad (0 < x < \infty) \tag{1}$$

or in self-conjugate form

$$(xe^{-x} y')' + \lambda e^{-x} y = 0 \qquad (0 < x < \infty) \tag{1'}$$

with the following boundary conditions: that the solution should be bounded at $x = 0$ and tend to infinity not faster than a finite power of x as $x \to \infty$.

It is natural to look for a solution of equation (1) as a series in powers of x:

$$y = \sum_{n=0}^{\infty} a_n x^n . \tag{2}$$

Substituting this expression in equation (1) we have:

$$\sum_{n=0}^{\infty} [a_{n+1} (n+1) n + a_{n+1} (n+1) - n a_n + \lambda a_n] x^n = 0$$

or

$$\sum_{n=0}^{\infty} \left[a_{n+1}(n+1)^2 - a_n(n-\lambda)\right] x^n = 0,$$

from which we obtain a recurrence relation for the coefficients

$$a_{n+1} = \frac{n-\lambda}{(n+1)^2} a_n. \tag{3}$$

The coefficients a_0 is arbitrary. For $\lambda = n$ equation (1) has a solution in the form of a polynomial of n th order. Choosing a_0 so that the coefficient with highest power of x^n is equal to $(-1)^n$, we get the *Chebyshev–Laguerre polynomials* $L_n(x)$. For this to be so $a_0 = n!$

We give expressions for the first five polynomials $L_n(x)$:

$$L_0(x) = 1;$$

$$L_1(x) = -x + 1;$$

$$L_2(x) = x^2 - 4x + 2;$$

$$L_3(x) = -x^3 + 9x^2 - 18x + 6;$$

$$L_4(x) = x^4 - 16x^3 + 72x^2 - 96x + 24.$$

The Chebyshev–Laguerre polynomials have the following generating function:

$$\Psi(x, t) = \frac{e^{-\frac{xt}{1-t}}}{1-t}. \tag{4}$$

Expanding it in a series of powers of t, we obtain:

$$\Psi(x, t) = \frac{e^{-\frac{xt}{1-t}}}{1-t} = \sum_{n=0}^{\infty} \bar{L}_n(x) \frac{t^n}{n!}, \tag{5}$$

where $\bar{L}_n(x)$ are the expansion coefficients, equal to

$$\bar{L}_n(x) = \left[\frac{\partial^n \Psi(x, t)}{\partial t^n}\right]_{t=0}$$

or

$$\bar{L}_n(x) = e^x \frac{d^n}{dx^n}(x^n e^{-x}). \tag{6}$$

Thus $\bar{L}_n(x)$ is a polynomial of order n.

Let us prove that

$$\bar{L}_n(x) \equiv L_n(x) .$$

We will show that (1) the polynomials $\bar{L}_n(x)$ satisfy the Chebyshev–Laguerre equation, (2) the coefficient of x^n equals $\bar{a}_n = (-1)^n$.

Property (2) follows from formula (6). Let us prove property (1) We put

$$\bar{L}_n(x) = e^x \frac{d^n z}{dx^n},$$

where

$$z = x^n e^{-x} .$$

Evaluating $dz/dx = -z + nz/x$, we obtain an equation for z

$$xz' + (x - n) z = 0 .$$

Differentiating this identity $n + 1$ times, we obtain:

$$xz^{(n+2)} + (x + 1) z^{(n+1)} + (n + 1) z^{(n)} = 0 ,$$

i.e. the function

$$u = \frac{d^n z}{dx^n}$$

satisfies the equation

$$xu'' + (x + 1) u' + (n + 1) u = 0 . \tag{7}$$

The derivatives of $\bar{L}_n$ are

$$\bar{L}_n' = e^x (u + u') , \tag{8}$$

$$\bar{L}_n'' = e^x (u'' + 2u' + u) . \tag{9}$$

Using (8), (9) and subsequently (7), we find

$$x\bar{L}_n'' + (1 - x) \bar{L}_n' = e^x [xu'' + (1 + x) u' + u] = - n\bar{L}_n ,$$

i.e. $\bar{L}_n(x)$ satisfies the equation

$$x\bar{L}_n'' + (1 - x) \bar{L}_n' + n\bar{L}_n = 0 .$$

Thus we have proved that $\bar{L}_n(x) \equiv L_n(x)$ or

$$L_n(x) = e^x \frac{d^n}{dx^n} (e^{-x} x^n) . \tag{10}$$

Using equation (1) for $L_n(x)$, it is easy to show that the polynomials $L_n(x)$ and $L_m(x)$ for $m \neq n$ are mutually orthogonal with weight e^{-x}. But we will prove the orthogonality from the differential relation (10) since the norm of $L_n(x)$ will be evaluated at the same time.

We consider the integral

$$J = \int_0^\infty L_m(x) L_n(x) e^{-x} dx = \int_0^\infty L_m(x) \frac{d^n}{dx^n}(x^n e^{-x}) dx .$$

Let $m \leqslant n$. Integrating m times by parts and taking into account that owing to the presence of a factor of the form $x^k e^{-x}$ $(k > 0)$ all substitutions reduce to zero, we obtain:

$$J = (-1)^m \int_0^\infty \frac{d^m L_m}{dx^m} \frac{d^{n-m}}{dx^{n-m}}(x^n e^{-x}) dx .$$

If $m < n$, then, integrating once more by parts, and using the equality $\frac{d^{m+1}}{dx^{m+1}} L_m = 0$, we get that $J = 0$, i.e.

$$\int_0^\infty L_m(x) L_n(x) e^{-x} dx = 0 .$$

If $m = n$, then

$$J = (-1)^n \int_0^\infty (-1)^n n!\, x^n e^{-x} dx = n!\, \Gamma(n+1) = (n!)^2 .$$

Thus, we have proved that

$$\int_0^\infty L_m(x) L_n(x) e^{-x} dx = \begin{cases} 0, & \text{if } m \neq n, \\ (n!)^2, & \text{if } m = n, \end{cases} \tag{11}$$

i.e. the Chebyshev–Laguerre polynomials $L_n(x)$ form an orthogonal system with weight e^{-x} of eigen-functions of equation (1), corresponding to the eigen-values $\lambda = n$.

The normalized polynomials are

$$l_n(x) = \frac{L_n(x)}{n!} = \frac{1}{n!} e^x \frac{d^n}{dx^n}(x^n e^{-x}) . \tag{12}$$

There are orthogonal normalized functions corresponding to the Chebyshev–Laguerre polynomials

$$\psi_n = \frac{e^{-\frac{x}{2}}}{n!} L_n(x) , \tag{13}$$

which satisfy the equation

$$(x\psi')' + \left(\frac{1}{2} - \frac{x}{4}\right)\psi + \lambda\psi = 0 \tag{14}$$

with the boundary conditions: they are bounded at $x = 0$ and reduce to zero as $x \to \infty$.

$$S_n = x^{-\frac{1}{2}}\psi_n, \tag{15}$$

Functions often occur satisfying the self-conjugate differential equation

$$(x^2 S')' - \frac{x^2 - 2x - 1}{4} S + \lambda x S = 0 \tag{16}$$

and a boundary condition: that the solution vanishes for $x \to \infty$ In both cases the eigen-values are integral positive numbers

$$\lambda = n,$$

and the conditions of orthogonality have the form

$$\int_0^\infty \psi_n(x)\,\psi_m(x)\,dx = \begin{cases} 0 & \text{where} \quad m \neq n, \\ 1 & \text{where} \quad m = n; \end{cases} \tag{17}$$

$$\int_0^\infty S_m(x)\,S_n(x)\,x\,dx = 0 \quad \text{where} \quad m \neq n. \tag{18}$$

2. In investigating the motion of an electron in a coulomb field in wave mechanics the *generalized Chebyshev–Laguerre polynomials occur*

$$y(x) = Q_n^{(s)}(x) \qquad (s \leqslant n), \tag{19}$$

satisfying the differential equation

$$xy'' + (s + 1 - x)\,y' + \left(\lambda - \frac{s+1}{2}\right) y = 0, \tag{20}$$

which may be written in the following self-conjugate form:

$$(x^{s+1} e^{-x} y')' + x^s e^{-x} \left(\lambda - \frac{s+1}{2}\right) y = 0. \tag{21}$$

The boundary conditions are

$$\left.\begin{array}{ll} y(x) \text{ is bounded at} & x = 0, \\ y(x) \text{ is bounded by a polynomial in } x & \text{as } x \to \infty. \end{array}\right\} \tag{22}$$

Solving equation (20) in series

$$y = \sum_{n=0}^{\infty} a_n x^n , \tag{23}$$

we obtain the recurrence relation

$$a_{n+1} = \frac{n + \frac{s+1}{2} - \lambda}{n(n+s)} a_n . \tag{24}$$

If

$\lambda = n + \frac{s+1}{2}$, when n is a positive integer or half integer, then series (23) terminates and we obtain a solution in the form of a polynomial.

The coefficient a_0 is arbitrary and determines the normalization of the solution. Choosing a_0 so that the coefficients of the highest power of x^n equals $(-1)^n$

$$a_n = (-1)^n , \tag{26}$$

we obtain the generalized Chebyshev–Laguerre polynomials, which are a unique solution of equation (21), satisfying conditions (22) and (26).

The generating function of the polynomials $Q_n^{(s)}(x)$ is

$$\Psi(x, t) = \frac{e^{-\frac{xt}{1-t}}}{(1 - t^{s+t})} , \tag{27}$$

since the Maclaurin's series of the function $\psi(x, t)$ has the form

$$\Psi(x, t) = \sum_{n=0}^{\infty} Q_n^{(s)}(x) \frac{t^n}{n!} . \tag{28}$$

For $s = 0$ we obtain formula (5) for the Chebyshev–Laguerre polynomials.

Let us show that for $Q_n^{(s)}(x)$ the following differential relation is valid:

$$Q_n^{(s)}(x) = \frac{e^x}{x^s} \frac{d^n}{dx^n} (e^{-x} x^{n+s}) . \tag{29}$$

From the relation

$$T_n^{(s)}(x) = \frac{e^x}{x^s} \frac{d^n}{dx^n} (e^{-x} x^{n+s})$$

we see that T_n is a polynomial of order n, in which the coefficient in x^n equals $(-1)^n$.

Let us prove that the polynomials $T_n^{(s)}(x)$ satisfy equation (20). Introducing the function

$$z = x^{n+s} e^{-x},$$

we see that for it the relation

$$xz' + (x - n - s) z = 0 .$$

holds. Differentiating it $(n + 1)$ times with respect to x, we obtain:

$$xz^{(n+2)} + (x + 1 - s) z^{(n+1)} + (n + 1) z^{(n)} = 0,$$

i.e. the function

$$u = \frac{d^n z}{dx^n}$$

satisfies the equation

$$xu'' + (x + 1 - s) u' + (n + 1) u = 0 .$$

Evaluating the derivatives of $T_n^{(s)}(x)$ and taking into account the equation for u, we have:

$$x \frac{dT_n^{(s)}}{dx} = xT_n^{(s)} - sT_n^{(s)} + e^x x^{-s+1} u',$$

$$x \frac{d^2 T_n^{(s)}}{dx^2} = (x - s - 1) \frac{dT_n^{(s)}}{dx} + T_n^{(s)} + e^x e^{-s} [xu'' + (x - s + 1)u'] =$$

$$= (x - s - 1) \frac{dT_n^{(s)}}{dx} - nT_n^{(s)},$$

i.e. the polynomial $y = T_n(x)$ satisfies equation (20) for $\lambda = n + \frac{s+1}{2}$.

Hence it follows that

$$T_n^{(s)}(x) \equiv Q_n^{(s)}(x).$$

Let us show now that the generalized polynomials are orthogonal with weight $x^s e^{-x}$ in the region $0 < x < \infty$, and evaluate their norm.

We consider the integral

$$J = \int_0^\infty Q_m^{(s)} Q_n^{(s)} e^{-x} x^s \, dx = \int_0^\infty Q_m^{(s)} \frac{d^n}{dx^n} (e^{-x} x^{n+s}) \, dx.$$

Let $m \leqslant n$. Integrating m times by parts and noting that all the substitutons reduce to zero, we obtain:

$$J = (-1)^n \int_0^\infty \frac{d^m}{dx^m} Q_m^{(s)} \frac{d^{n-m}}{dx^{n-m}} (e^{-x} x^{n+s}) \, dx.$$

If $m < n$, then, integrating once more by parts, we derive by virtue of the equality $\frac{d^{m+1} Q_m^{(s)}}{dx^{m+1}} = 0$ that $J = 0$, i. e.

$$\int_0^\infty Q_m^{(s)}(x) \, Q_n^{(s)}(x) \, e^{-x} x^s \, dx = 0.$$

If $m = n$, then

$$J = n! \int_0^\infty e^{-x} x^{n+s} \, dx = n! \, \Gamma(n+s+1),$$

i.e.

$$\int_0^\infty [Q_n^{(s)}(x)]^2 e^{-x} x^s \, dx = n! \, \Gamma(n+s+1).$$

Thus,

$$\int_0^\infty Q_m^{(s)}(x) \, Q_n^{(s)}(x) \, e^{-x} x^s \, dx = \begin{cases} 0, \text{ if } m \neq n \quad (s > -1), \\ n! \, \Gamma(n+s+1), \text{ if } m = n. \end{cases} \tag{30}$$

To the generalized Chebyshev–Laguerre polynomials $Q_n^{(s)}(x)$ there correspond the orthogonal and normalized functions

$$\Phi_n^{(s)}(x) = \frac{x^{\frac{s}{2}} e^{-\frac{x}{2}} Q_n^{(s)}(x)}{\sqrt{[n! \Gamma(n+s+1)]}} \tag{31}$$

$$\left(\int_0^\infty \Phi_n^{(s)}(x) \, \Phi_m^{(s)}(x) \, dx = \begin{cases} 0 \text{ where } m \neq n, \\ 1 \text{ where } m = n \end{cases} \right)$$

which are solutions of the equation

$$\frac{d}{dx}\left(x \frac{d\Phi}{dx}\right) + \left(\lambda - \frac{x}{4} - \frac{s^2}{4x}\right) \Phi = 0 \tag{32}$$

with the corresponding eigenvalues

$$\lambda = n + \frac{s+1}{2}.$$

The boundary conditions are obvious from the preceding. Sometimes it is convenient to make use of the normalized polynomials

$$Q_n^{*(s)}(x) = \frac{Q_n^{(s)}(x)}{\sqrt{[n! \, \Gamma(n+s+1)]}}. \tag{33}$$

§ 3. Simplest problems for Schrödinger's equation*

1. *Schrödinger's equation*

In wave mechanics the behaviour of a particle, in a potential field, is described by Schrödinger's equation

$$i\hbar \frac{\partial \varphi}{\partial t} = -\frac{\hbar^2}{2\mu} \nabla^2 \varphi + U(x, y, z, t)\, \varphi, \tag{1}$$

where $\hbar = 1.05 \times 10^{-27}$ ergs/sec is Planck's constant, $i = \sqrt{-1}$, $\mu =$ the mass of the particle, U is its potential energy in the force field, $\psi = \psi(x, y, z, t)$ is the wave function.

If the forces do not depend on time, $U = U(x, y, z)$, then steady states are possible with a given energy, i.e. there exist solutions of the form

$$\psi = \psi^0(x, y, z)\, e^{-\frac{iE}{\hbar}t}, \tag{2}$$

where E is the total energy of the particle. Substituting this expression in equation (1), get Schrödinger's second equation

$$\nabla^2 \psi^0 + \frac{2\mu}{\hbar^2}(E - U)\, \psi^0 = 0, \tag{3}$$

in which E plays the part of an eigen-value, to be determined. Later in place of ψ^0 we will write ψ:

$$\nabla^2 \psi + \frac{2\mu}{\hbar^2}(E - U)\, \psi = 0. \tag{4}$$

In the case of the absence of a force field $U = 0$ equation (4) takes the form

$$\nabla^2 \psi + \frac{2\mu E}{\hbar^2}\, \psi = 0. \tag{5}$$

This equation is similar to the wave equation of classical physics

$$\nabla^2 \psi + k^2 \psi = 0, \tag{6}$$

where $k = \omega/c = 2\pi/\lambda$ is the wave number, λ is the wave-length. However this similarity is purely superficial because of

* The problems considered here for Schrödinger's equation give examples of the application of the Chebyshev-Hermite and the Chebyshev-Laguerre polynomials. The account given below does not claim to be a complete account of problems, connected with the Schrödinger equation. In a university course quantum mechanics is studied after a course in mathematical physics.

the difference in physical significance of the functions, appearing in equations (5) and (6).

In the Schrödinger equation direct physical significance is attached not to the function ψ itself, but to $|\psi|^2$, which is explained statistically: the expression $|\psi|^2\, dx\,dy\,dz$ denotes the probability of particle occupying the elementary volume $dxdydz$ at a point x, y, z of space.

In this connection, the normalization of the eigen-functions to unity, which we have repeatedly made use of earlier for mathematical simplicity, now acquires fundamental importance. The normalization condition.

$$\int\int\int |\psi|^2\, dx\, dy\, dz = 1 \tag{7}$$

indicates that a particle exists at some points of space and therefore the probability of finding the particle somewhere in space equals unity.

Let us consider some simple problems on Schrödinger's equation.

2. *The Harmonic oscillator*

Schrödinger's equation for the harmonic oscillator takes the form

$$\frac{h^2}{2\mu}\frac{d^2\psi}{dx^2} + (E - U)\,\psi = 0,$$

where $U = \frac{\mu\omega_0^2}{2}\,x^2$, ω_0 is the characteristic angular frequency of the oscillator. Where to find the steady states, i.e. the spectrum of eigenvalues of energy E and the corresponding eigen-functions ψ of the equation

$$\psi'' + \frac{2\mu}{h^2}\left(E - \frac{\mu\omega_0^2}{2}\,x^2\right)\psi = 0 \tag{8}$$

with the additional condition

$$\int_{-\infty}^{\infty} |\psi|^2\, dx = 1. \tag{9}$$

Introducing the symbols

$$\left.\begin{aligned} \lambda &= \frac{2E}{\hbar\omega_0}, \\ x_0 &= \sqrt{\frac{h}{\mu\omega_0}}. \\ \xi &= \frac{x}{x_0}, \end{aligned}\right\} \tag{10}$$

for the function $\psi = \psi(\xi)$, after obvious transformations we obtain the equation

$$\frac{d^2\psi}{d\xi^2} + (\lambda - \xi^2)\,\psi = 0 \tag{11}$$

with the additional normalization condition

$$\int_{-\infty}^{\infty} |\psi|^2\, d\xi = \frac{1}{x_0}. \tag{12}$$

This equation is solved by the functions (§ 1, p. 728)

$$\psi_n(\xi) = \frac{1}{\sqrt{x_0}} \frac{e^{-\frac{1}{2}\xi^2} H_n(\xi)}{\sqrt{(2^n n!\, \sqrt{\pi})}},$$

corresponding to the eigen-values

$$\lambda_n = 2n + 1.$$

Returning to the original symbols, we find:

$$\psi_n(x) = \frac{1}{\sqrt{x_0}} \frac{e^{-\frac{1}{2}\left(\frac{x}{x_0}\right)^2} H_n\left(\frac{x}{x_0}\right)}{\sqrt{(2^n n!\, \sqrt{\pi})}} \tag{13}$$

$$E_n = \hbar\omega_0\left(n + \frac{1}{2}\right) \qquad (n = 0,\ 1,\ 2, \ldots). \tag{14}$$

In classical mechanics the energy of the oscillator

$$E = \frac{p_x^2}{2\mu} + \frac{\mu\omega_0^2}{2} x^2,$$

where p_x is the momentum of the particle, can take a continuous series of values. From the point of view of quantum mechanics the energy of the oscillator, as formula (14) shows, can take only a discrete set of values E_n. It is said that the energy is quantized. The number n, defining the number of the quantum level, is called the principal quantum number. In the lowest quantum state with $n = 0$ the energy of the oscillator differs from zero and equals

$$E_0 = \frac{1}{2}\hbar\omega.$$

3. *Rotator*

Let us determine the eigen-values of the energy of a rotator with a free axis, i.e. of a particle rotating at a fixed distance about a fixed centre.

The potential energy U of the rotator has the same value for all positions of the particle and it is possible to put it equal to zero $U = 0$.

In a spherical system of coordinates (r, θ, φ) with origin at the fixed centre Schrödinger's equation for the rotator

$$\nabla^2\psi + \frac{2\mu}{\hbar^2}E\psi = 0$$

may be written in the form

$$\frac{1}{r^2\sin\theta}\frac{\partial}{\partial\theta}\left(\sin\theta\frac{\partial\psi}{\partial\theta}\right) + \frac{1}{r^2\sin^2\theta}\frac{\partial^2\psi}{\partial\varphi^2} + \frac{2\mu}{h^2}E\psi = 0. \qquad (15)$$

We have used the condition

$$\frac{\partial\psi}{\partial r} = 0.$$

Introducing in place of the mass μ the moment of inertia

$$I = \mu r^2,$$

we obtain

$$\frac{1}{\sin\theta}\frac{\partial}{\partial\theta}\left(\sin\theta\frac{\partial\psi}{\partial\theta}\right) + \frac{1}{\sin^2\theta}\frac{\partial^2\psi}{\partial^2\varphi} + \lambda\psi = 0$$

or

$$\nabla^2_{\theta,\varphi}\psi + \lambda\psi = 0, \qquad (16)$$

where

$$\lambda = \frac{2I}{h^2}E. \qquad (17)$$

Thus, we arrive at the eigen-value problem for the equation

$$\nabla^2_{\theta,\varphi}\psi + \lambda\psi = 0 \qquad (16)$$

with boundary conditions requiring that the solution be bounded at the points $\theta = 0$ and $\theta = \pi$ and a normalization condition

$$\int_0^{\pi}\int_0^{2\pi} |\psi|^2 \sin\theta \, d\theta \, d\varphi = 1. \tag{18}$$

The solutions of this problem are normalized spherical functions

$$\psi_{lm}(\theta, \varphi) = \sqrt{\left[\frac{2l+1)(l-m)!}{2\varepsilon_m \pi (l+m)!}\right]} Y_l^{(m)}(\theta, \varphi) \quad \left(\varepsilon_m = \begin{cases} 2 \text{ where } m = 0 \\ 1 \text{ where } m \neq 0 \end{cases}\right), \tag{19}$$

$$Y_l^{(m)}(\theta, \varphi) = P_l^{(m)}(\cos\theta) \begin{matrix} \cos m\varphi \\ \sin m\varphi \end{matrix} \quad (m = 0, \quad 1, \ldots, \quad l),$$

corresponding to the eigen-values

$$\lambda = l(l+1). \tag{20}$$

Replacing λ by its value according to formula (17), we obtain a relation for the quantum values of the energy of the rotator

$$E_{lm} = l(l+1)\frac{\hbar^2}{2I}. \tag{21}$$

4. *Motion of an electron in a coulomb field*

One of the simplest problems of atomic mechanics is the problem on the motion of an electron in the coulomb field of the nucleus, having great practical interest, since its solution gives not only the theory of the hydrogen spectrum, but also an approximate theory of atomic spectra with one valence electron (atoms similar to hydrogen), for example the sodium atom.

In the hydrogen atom the electron moves in the electrostatic field of the nucleus (proton), so that the potential energy $U(x, y, z)$ equals

$$U = -\frac{e^2}{r}, \tag{22}$$

where r is the distance of the electron from the nucleus, $-e$ is the electron charge, $+e$ is the charge on the nucleus.

Schrödinger's equation in this case has the form

$$\nabla^2\psi + \frac{2\mu}{\hbar^2}\left(E + \frac{e^2}{r}\right)\psi = 0. \tag{23}$$

We have to find those values of E, for which equation (23) has a solution, continuous over all space and satisfying the normalization condition

$$\int\int\int_{-\infty}^{\infty} |\psi(x, y, z)|^2 \, dx \, dy \, dz = 1. \tag{24}$$

We introduce a spherical system of coordinates with origin at the nucleus, which is assumed stationary

$$\frac{1}{r^2}\frac{\partial}{\partial r}\left(r^2 \frac{\partial \psi}{\partial r}\right) + \frac{1}{r^2}\nabla^2_{\theta,\varphi}\psi + \frac{2\mu}{\hbar^2}\left(E + \frac{e^2}{r}\right)\psi = 0, \tag{25}$$

and look for a solution in the form

$$\psi(r, \vartheta, \varphi) = \chi(r) Y_l^{(m)}(\vartheta, \psi). \tag{26}$$

Making use of the differential equation for the spherical functions $Y_l^{(m)}(\theta, \varphi)$:

$$\nabla^2_{\theta,\varphi} Y_l^{(m)}(\theta, \varphi) + l(l+1) Y_l^{(m)}(\theta, \varphi) = 0,$$

we obtain:

$$\frac{d^2\chi}{dr^2} + \frac{2}{r}\frac{d\chi}{dr} + \left[\frac{2\mu}{h^2}\left(E + \frac{e^2}{r}\right) - \frac{l(l+1)}{r^2}\right]\chi = 0. \tag{27}$$

Let us introduce as a unit of length the quantity

$$a = \frac{\hbar^2}{\mu e^2} 0.529 \times 10^{-8} \, \text{cm},$$

as a unit of energy the quantity

$$E_0 = \frac{\mu e^4}{\hbar} = \frac{e^2}{a}.$$

Assuming

$$\varrho = \frac{r}{a}, \qquad \varepsilon = \frac{E}{E_0} \, (\varepsilon < 0), \tag{28}$$

we rewrite equation (27) in the form

$$\frac{d^2\chi}{d\varrho^2} + \frac{2}{\varrho}\frac{d\chi}{d\varrho} + \left(2\varepsilon + \frac{2}{\varrho} - \frac{l(l+1)}{\varrho^2}\right)\chi = 0. \tag{29}$$

By means of the substitution

$$\chi = \frac{1}{\sqrt{\varrho}} \tag{30}$$

equation (28) reduces to the form

$$\frac{d^2y}{d\varrho^2} + \frac{1}{\varrho}\frac{dy}{d\varrho} + \left(2\varepsilon + \frac{2}{\varrho} - \frac{s^2}{4\varrho^2}\right)y = 0, \tag{31}$$

where

$$s = 2l + 1.$$

Introducing as an independent variable the quantity

$$x = \varrho \sqrt{(-8\varepsilon)}, \tag{32}$$

we obtain in place of (31) the equation

$$xy'' + y' - \left(\frac{x}{4} + \frac{s^2}{4x}\right) y + \lambda y = 0 \tag{33}$$

or

$$\frac{d}{dx}\left(x \frac{dy}{dx}\right) - \left(\frac{x}{4} + \frac{s^2}{4x}\right) y + \lambda y = 0, \tag{33'}$$

where

$$\lambda = \frac{1}{\sqrt{-2\varepsilon}}, \tag{34}$$

considered by us in § 2 (31).

The eigen-values found there were shown to be equal to

$$\lambda = n_r + \frac{s+1}{2},$$

and the eigen-functions were the generalized Chebyshev–Laguerre polynomials $Q_{n_r}^{*(s)}$:

$$y_{n_r}(x) = x^{\frac{s}{2}} e^{-\frac{x}{2}} Q_{n_r}^{*(s)}(x), \tag{35}$$

where $Q_{n_r}^{*(s)}(x)$ are given by formula (32), § 2.

Taking into account that

$$s = 2l + 1,$$

we obtain:

$$\lambda = n_r + l + 1 = n \qquad (n = 1, 2, \ldots). \tag{36}$$

The whole number n is called the principal quantum number, n_r the radial quantum number, l the azimuthal quantum number.

Replacing λ by its expression according to formulae (34) and (28), we obtain quantum values of the energy

$$E_n = -\frac{\mu e^4}{2\hbar^2 n^2}. \tag{37}$$

They depend only on the principal quantum number n.

Let us relate the energy E to a frequency by the relations $\hbar\,\omega = h\,\nu$, $E = -h\,\nu$, where $\nu = \omega/2\pi$ is the frequency. Then we have:

$$\nu = \frac{\mu e^4}{2\hbar^2 n^2 h} = \frac{R}{n^2}, \tag{38}$$

where $R = \mu e^4/2\hbar^2 h$ is the so-called Rydberg constant.

We can determine the frequencies of the spectral lines. The frequency ν_{nn_1} observable in a spectral line corresponds to a transition from a state with energy E_n to a state with energy E_{n_1}.

The frequency ν_{nn_1} of a quantum, emitted in such a quantum transition, equals

$$\nu_{nn1} = R\left(\frac{1}{n_1^2} - \frac{1}{n^2}\right). \tag{39}$$

Assuming $n_1 = 1$ and giving n the values $n = 2, 3, \ldots,$ we obtain a series of lines, comprising the so-called Lyman series

$$\nu_{nn_1} = R\left(1 - \frac{1}{n^2}\right).$$

Further, the values $n = 2$, $n = 3, 4, \ldots,$ give the Balmer series

$$\nu_{nn_1} = R\left(\frac{1}{2^2} - \frac{1}{n^2}\right),$$

the values $n = 3$, $n = 4, 5, \ldots$ the Paschen series

$$\nu_{nn_1} = R\left(\frac{1}{3^2} - \frac{1}{n^2}\right).$$

Let us write down the eigen-functions of the hydrogen atom-because of (26) it is sufficient to determine the radial functions $\chi(\varrho)$.

Making use of formulae (35), (32), (30), (34), (36), we may write:

$$\chi_{nl}(\varrho) = A_n\left(\frac{2\varrho}{n}\right)^l e^{-\frac{\varrho}{n}} Q^{*(2l+1)}_{n-l-1}\left(\frac{2\varrho}{n}\right), \tag{40}$$

where A_n is a normalizing factor, determinable from the condition

$$\int_0^\infty \varrho\chi^2_{nl}(\varrho)\,d\varrho = 1. \tag{41}$$

Evaluating Ar, we obtain the following expression for the normalized radial functions:

$$\chi_{nl}(\varrho) = \frac{2}{n}\left(\frac{2\varrho}{n}\right)^l e^{-\frac{\varrho}{n}} Q_{n-l-1}^{*(2l+1)}\left(\frac{2\varrho}{n}\right). \tag{42}$$

Using formulae (26) and (19) the normalized eigen-functions have the form

$$\varphi_{mnl} = \sqrt{\left[\frac{(2l+1)(l+m)!}{2\varepsilon_m \pi (l+m)!}\right]} Y_l^{(m)}(\vartheta, \varphi)\, \chi_{nl}(\varrho),$$

where $\chi_{nl}(\varrho)$ are given by formula (42).

The number m $(m = 0, \pm 1, \pm 2, \ldots, \pm l)$ is called the magnetic quantum number.

Since n_r is always positive $(n_r = 0, 1, 2, \ldots)$, for a given n

$$n = n_r + l + 1$$

the quantum number l cannot be greater than $n - 1$ $(l = 0, 1, 2, \ldots, n - 1)$. Therefore for a given value of the principal quantum number n the number l can take n values $l = 0, 1, \ldots, n - 1$, and to each value of l there correspond $2l + 1$ values of m. Hence it follows that to a given value of the energy E_n, i.e. to a given value n, there correspond

$$\sum_{l=0}^{n-1} (2l + 1) = 1 + 3 + 5 + \ldots + (2n - 1) = n^2$$

different eigen-functions. Thus, each energy level has a degeneracy n^2.

The discrete spectrum of the negative eigen-values of energy E_n consists of an infinite number of terms coming closer together at zero energy.

The second distinguishing feature of the problem under consideration for Schrödinger's equation is the presence of a continuous spectrum of positive eigenvalues [any positive number E is an eigen-value of equation (23)]. In this case the electron is no longer bound to the nucleus, but still moves in its field (the ionized hydrogen atom). We do not consider a proof of the existence of a continuous spectrum, referring the reader to the specific literature.*

* See, for instance, V. A. Fok, *Principles of Quantum Mechanics*, 1932; Courant and Hilbert, *Methods of Mathematical Physics*, vol. I chap. V, 1951.

TABLES OF THE ERROR INTEGRAL AND SOME CYLINDRICAL FUNCTIONS

In this section we give tables of some special functions, which we meet in solving boundary value problems of mathematical physics. The tables are accompanied by a summary of the simplest properties of the functions being considered.

The error integral

The error integral

$$\Phi(z) = \frac{2}{\sqrt{\pi}} \int_0^z e^{-\alpha^2} d\alpha .$$

Expansion in a series for small z:

$$\Phi(z) = \frac{2}{\sqrt{\pi}} \left(z - \frac{z^3}{1!\,3} + \frac{z^5}{2!\,5} - \dots \right).$$

Asymptotic formula for large z

Changing variables in the integral $1 - \Phi(z) = \frac{2}{\sqrt{\pi}} \int_z^\infty e^{-\alpha^2} d\alpha$,

$\alpha = z + \beta/2z$, we obtain:

$$1 - \Phi(z) = \frac{1}{\sqrt{\pi}} \frac{e^{-z^2}}{z} \int_0^\infty e^{-\beta - \frac{\beta^2}{4z^2}} d\beta .$$

Thus,

$$\lim_{z \to \infty} \frac{1 - \Phi(z)}{\frac{1}{\sqrt{\pi}} \frac{e^{-z^2}}{z}} = 1 , \text{ i.e. } 1 - \Phi(z) \sim \frac{1}{\sqrt{\pi}} \frac{e^{-z^2}}{z} .$$

If one expands $e^{-\frac{\beta^2}{4z^2}}$ in series

$$e^{-\frac{\beta^2}{4z^2}} = \sum_{n=0}^\infty \frac{(-1)^n}{(2z)^{2n}} \frac{\beta^{2n}}{n!} ,$$

and multiplies this series by $e^{-\beta}$ and integrates from zero to infinity, then one obtains a divergent series

$$\sum_{n=0}^{\infty} \frac{(-1)^n}{(2z)^{2n}} \frac{(2n)!}{n!} = \sum_{n=0}^{\infty} \frac{(-1)^n (n+1)(n+2)\dots 2n}{(2z)^{2n}}$$

This divergent series gives the asymptotic expansion of the function $1 - \Phi(z)$:

$$1 - \Phi(z) \simeq \frac{1}{\sqrt{\pi}} \frac{e^{-z^2}}{z} \left(1 - \frac{1}{2z^2} + \frac{3 \cdot 4}{(2z)^4} - \frac{4 \cdot 5 \cdot 6}{(2z)^6} + \dots\right).$$

Cylindrical functions

Series	Asymptotic formulae
1. *Bessel functions*	
$J_0(x) = 1 - \frac{\left(\frac{x}{2}\right)^2}{1!} + \frac{\left(\frac{x}{2}\right)^4}{(2!)^2} - \dots$	$J_0(x) = \sqrt{\left(\frac{2}{\pi x}\right)} \cos\left(x - \frac{\pi}{4}\right) + \dots$
$J_1(x) = \frac{x}{2}\left[1 - \frac{\left(\frac{x}{2}\right)^2}{1 \cdot 2} + \frac{\left(\frac{x}{2}\right)^4}{1 \cdot 2\ 2 \cdot 3} - \dots\right]$	$J_1(x) = \sqrt{\left(\frac{2}{\pi x}\right)} \sin\left(x - \frac{\pi}{4}\right) + \dots$
$J_\nu(x) = \left(\frac{x}{2}\right)^\nu \frac{1}{\Gamma(\nu+1)} - \dots$	$J_\nu(x) = \sqrt{\left(\frac{2}{\pi x}\right)} \cos\left(x - \frac{\pi}{2}\nu - \frac{\pi}{4}\right) + \dots$
2. *Neumann functions*	
$N_0(x) = \frac{2}{\pi} J_0(x)\left(\ln\frac{x}{2} + C\right) + \frac{2}{\pi}\left(\frac{x}{2}\right)^2 + \dots$	$N_0(x) = \sqrt{\left(\frac{2}{\pi x}\right)} \sin\left(x - \frac{\pi}{4}\right) + \dots$
$C = 0.577215\dots$ Euler's const.	
$N_1(x) = -\frac{2}{\pi x} + \frac{2}{\pi} J_1(x)\left(\ln\frac{x}{2} + C\right) + \dots$	$N_1(x) = -\sqrt{\left(\frac{2}{\pi x}\right)} \cos\left(x - \frac{\pi}{4}\right) + \dots$
$N_n(x) = -\frac{1}{\pi}\left(\frac{2}{x}\right)^n (n-1)! + \dots \quad (n > 1)$	$N_n(x) = \sqrt{\left(\frac{2}{\pi x}\right)} \times \sin\left(x - n\frac{\pi}{2} - \frac{\pi}{4}\right) + \dots$

3. *Hankel functions*

$$H_\nu^{(1)}(x) = J_\nu(x) + iN_\nu(x)$$

$$H_\nu^{(2)}(x) = J_\nu(x) - iN_\nu(x)$$

$$H_\nu^{(1)}(x) = \sqrt{\left(\frac{2}{\pi x}\right)} e^{i\left(x - \frac{\pi}{2}\nu - \frac{\pi}{4}\right)} + \ldots$$

$$H_\nu^{(2)}(x) = \sqrt{\left(\frac{2}{\pi x}\right)} e^{-i\left(x - \frac{\pi}{2}\nu - \frac{\pi}{4}\right)} + \ldots$$

4. *Functions of imaginary argument*

$$I_0(x) = J_0(ix) = = 1 + \frac{\left(\frac{x}{2}\right)^2}{1!} + \frac{\left(\frac{x}{2}\right)^4}{(2!)^2} + \ldots$$

$$I_0(x) = \sqrt{\left(\frac{1}{2\pi x}\right)} e^x + \ldots$$

$$I_1(x) = -iJ_1(ix) = = \frac{x}{2}\left[1 + \frac{\left(\frac{x}{2}\right)^2}{1\cdot 2} \frac{\left(\frac{x}{2}\right)^4}{1\cdot 2\cdot 2\cdot 3} + \ldots\right]$$

$$I_1(x) = \sqrt{\left(\frac{1}{2\pi x}\right)} e^x + \ldots$$

$$I_\nu(x) = (-i)^\nu J_\nu(ix) = = \left(\frac{x}{2}\right)^\nu \frac{1}{\Gamma(\nu+1)} + \ldots$$

$$I_\nu(x) = \sqrt{\left(\frac{1}{2\pi x}\right)} e^x + \ldots$$

$$K_0(x) = \frac{x}{2} iH_0^{(1)}(ix) = = -\left(\ln\frac{x}{2} + C\right) I_0(x) + \left(\frac{x}{2}\right)^2 + \ldots$$

$$K_0(x) = \sqrt{\left(\frac{\pi}{2x}\right)} e^{-x} + \ldots$$

$$K_1(x) = -\frac{\pi}{2} H_1^{(1)}(ix) = \frac{1}{x} + \ldots$$

$$K_1(x) = \sqrt{\left(\frac{\pi}{2x}\right)} e^{-x} + \ldots$$

$$K_n x = \frac{1}{2}\pi e^{\frac{1}{2}\pi n i} H_n^{(1)}(ix) = = \frac{(n-1)!}{2}\left(\frac{2}{x}\right)^n + \ldots$$

$$K_n(x) = \sqrt{\left(\frac{\pi}{2x}\right)} e^{-x} + \ldots$$

5. *Functions of semi-integral order*

$$J_{\frac{1}{2}}(x) = \sqrt{\left(\frac{2}{\pi x}\right)} \sin x; \qquad J_{-\frac{1}{2}}(x) = \sqrt{\left(\frac{2}{\pi x}\right)} \cos x;$$

$$J_{\frac{3}{2}}(x) = \sqrt{\left(\frac{2}{\pi x}\right)}\left(\frac{\sin x}{x} - \cos x\right); \qquad J_{-\frac{3}{2}}(x) = \sqrt{\left(\frac{2}{\pi x}\right)}\left(-\frac{\cos x}{x} - \sin x\right).$$

6. *Recurrence relations*

$$\frac{d}{dx}\left[x^{\nu} J_{\nu}(x)\right] = x^{\nu} J_{\nu-1}(x) \text{ and } \frac{d}{dx}\left[\frac{J_{\nu}(x)}{x^{\nu}}\right] = -\frac{J_{\nu+1}(x)}{x^{\nu}},$$

$$J_{\nu-1}(x) + J_{\nu+1}(x) = \frac{2\nu}{x} J_{\nu}(x),$$

$$J_0'(x) = -J_1(x); \qquad \int_0^x J_1(x)\,dx = 1 - J_0(x),$$

$$\frac{d}{dx}\left[x J_1(x)\right] = x J_0(x); \qquad \int_0^x x J_0(x)\,dx = x J_1(x).$$

Similar formulae exist for other cylindrical functions of a real argument.

For functions of an imaginary argument

$$I_{\nu-1}(x) - I_{\nu+1}(x) = \frac{2\nu}{x} I_{\nu}(x), \qquad K_{\nu-1}(x) - K_{\nu+1}(x) = -\frac{2\nu}{x} K_{\nu}(x),$$

$$I_0'(x) = I_1(x), \qquad K_0'(x) = -K_1(x).$$

7. *Wronskian determinant for cylindrical functions*

$$y_{\nu}(x) N_{\nu}'(x) - N_{\nu}(x) J_{\nu}'(x) = \frac{2}{\pi x'},$$

in particular $J_0(x) N_1(x) - N_0(x) J_1(x) = -2/\pi x$.

8. *Integral relations*

$$J_n(x) = \frac{1}{2\pi}\int_{-\pi}^{\pi} e^{-ix\sin\varphi + in\varphi}\,d\varphi =$$

$$= \frac{(-i)^n}{2\pi}\int_{-\pi}^{\pi} e^{ix\cos\varphi + in\varphi}\,d\varphi = \frac{(-i)^n}{\pi}\int_0^{\pi} e^{ix\cos\varphi}\cos n\varphi\,d\varphi,$$

$$K_n(x) = \frac{1}{2}\int_{-\infty}^{\infty} e^{-x\cosh\xi - in\xi}\,d\xi, \text{ in particular } K_0(x) = \int_0^{\infty} e^{-x\cosh\xi}\,d\xi.$$

9. *Integrals, containing Bessel functions*

$$\int_0^{\infty} e^{-\lambda z} J_0(\lambda\varrho)\,d\lambda = \frac{1}{\sqrt{(\varrho^2 + z^2)}} \quad (z > 0),$$

$$\int_0^{\infty} J_1(\lambda\varrho)\, e^{-t\lambda^2} \lambda^{\nu+1}\,d\lambda = \frac{1}{2t}\left(\frac{\varrho}{2t}\right)^{\nu} e^{-\frac{\varrho^2}{4t}},$$

$$\int^{\infty} J_0(\lambda\varrho)\,\frac{e^{-\sqrt{(\lambda^2 - k^2)}\,|z|}}{\sqrt{(\lambda^2 - k^2)}}\,\lambda\,d\lambda = \frac{e^{ik\sqrt{(\varrho^2+z^2)}}}{\sqrt{(\varrho^2 + z^2)}} = \frac{e^{ikr}}{r}.$$

10. *Differential equations, reducible to Bessel's equation*

$y'' + \frac{1}{x} y' - \left(1 + \frac{\nu^2}{x^2}\right) y = 0$;	$y = AI_\nu(x) + BK_\nu(x)$;
$y'' + xy = 0$;	$y = \sqrt{x} Z_{\frac{1}{3}}\left(\frac{2}{3}\sqrt{x^3}\right)$;
$y'' + x^m y = 0$;	$y = \sqrt{x} Z_{\frac{1}{m+2}}\left(\frac{2}{m+2}\sqrt{[x^{m+2}]}\right)$,

where Z is some solution of Bessel's equation.

Table 1. Error integral

$$\Phi(z) = \frac{2}{\sqrt{\pi}} \int_0^z e^{-\alpha^2}\, d\alpha \qquad 0 \leqslant z \leqslant 2.8$$

z	$\Phi(z)$	z	$\Phi(z)$	z	$\Phi(z)$	z	$\Phi(z)$
0.00	0.0000	0.40	0.4284	0.80	0.7421	1.20	0.9103
0.01	0.0113	0.41	0.4380	0.81	0.7480	1.21	0.9130
0.02	0.0226	0.42	0.4475	0.82	0.7538	1.22	0.9155
0.03	0.0338	0.43	0.4569	0.83	0.7595	1.23	0.9181
0.04	0.0451	0.44	0.4662	0.84	0.7651	1.24	0.9205
0.05	0.0564	0.45	0.4755	0.85	0.7707	1.25	0.9229
0.06	0.0676	0.46	0.4847	0.86	0.7761	1.26	0.9252
0.07	0.0789	0.47	0.4937	0.87	0.7814	1.27	0.9275
0.08	0.0901	0.48	0.5027	0.88	0.7867	1.28	0.9297
0.09	0.1013	0.49	0.5117	0.89	0.7918	1.29	0.9319
0.10	0.1125	0.50	0.5205	0.90	0.7969	1.30	0.9340
0.11	0.1236	0.51	0.5292	0.91	0.8019	1.31	0.9361
0.12	0.1348	0.52	0.5379	0.92	0.8068	1.32	0.9381
0.13	0.1459	0.53	0.5465	0.93	0.8116	1.33	0.9400
0.14	0.1569	0.54	0.5549	0.94	0.8163	1.34	0.9419
0.15	0.1680	0.55	0.5633	0.95	0.8209	1.35	0.9438
0.16	0.1790	0.56	0.5716	0.96	0.8254	1.36	0.9456
0.17	0.1900	0.57	0.5798	0.97	0.8299	1.37	0.9473
0.18	0.2009	0.58	0.5879	0.98	0.8342	1.38	0.9490
0.19	0.2118	0.59	0.5959	0.99	0.8385	1.39	0.9507
0.20	0.2227	0.60	0.6039	1.00	0.8427	1.40	0.9523
0.21	0.2335	0.61	0.6117	1.01	0.8468	1.41	0.9539
0.22	0.2443	0.62	0.6194	1.02	0.8508	1.42	0.9554
0.23	0.2550	0.63	0.6270	1.03	0.8548	1.43	0.9569
0.24	0.2657	0.64	0.6346	1.04	0.8586	1.44	0.9583
0.25	0.2763	0.65	0.6420	1.05	0.8624	1.45	0.9597
0.26	0.2869	0.66	0.6494	1.06	0.8661	1.46	0.9611
0.27	0.2974	0.67	0.6566	1.07	0.8698	1.47	0.9624
0.28	0.3079	0.68	0.6633	1.08	0.8733	1.48	0.9637
0.29	0.3183	0.69	0.6708	1.09	0.8768	1.49	0.9649
0.30	0.3286	0.70	0.6778	1.10	0.8802	1.50	0.9661
0.31	0.3389	0.71	0.6847	1.11	0.8835	1.51	0.9661
0.32	0.3491	0.72	0.6914	1.12	0.8868	1.6	0.9763
0.33	0.3593	0.73	0.6981	1.13	0.8900	1.7	0.9838
0.34	0.3694	0.74	0.7047	1.14	0.8931	1.8	0.9891
0.35	0.3794	0.75	0.7112	1.15	0.8961	1.9	0.9928
0.36	0.3893	0.76	0.7175	1.16	0.8991	2.0	0.9953
0.37	0.3992	0.77	0.7238	1.17	0.9020	2.1	0.9970
0.38	0.4090	0.78	0.7300	1.18	0.9048	2.2	0.9981
0.39	0.4187	0.79	0.7361	1.19	0.9076	2.3	0.9989
						2.4	0.9993
						2.5	0.9996
						2.6	0.9998
						2.7	0.9999
						2.8	0.9999

Table 2. Values of Bessel functions of zero and first order from $x = 0$ to $x = 12.00$

x	$J_0(x)$	$J_1(x)$	x	$J_0(x)$	$J_1(x)$	x	$J_0(x)$	$J_1(x)$
0.00	+1.000	+0.000	4.00	—0.397	—0.066	8.00	+0.172	+0.235
0.10	+0.997	+0.050	4.10	—0.389	—0.103	8.10	+0.148	+0.248
0.20	+0.990	+0.099	4.20	—0.377	—0.139	8.20	+0.122	+0.258
0.30	+0.977	+0.148	4.30	—0.361	—0.172	8.30	+0.096	+0.266
0.40	+0.960	+0.196	4.40	—0.342	—0.203	8.40	+0.069	+0.271
0.50	+0.938	+0.242	4.50	—0.321	—0.231	8.50	+0.042	+0.273
0.60	+0.912	+0.288	4.60	—0.296	—0.257	8.60	+0.015	+0.273
0.70	+0.881	+0.329	4.70	—0.269	—0.279	8.70	—0.013	+0.270
0.80	+0.846	+0.369	4.80	—0.240	—0.298	8.80	—0.039	+0.264
0.90	+0.808	+0.406	4.90	—0.210	—0.315	8.90	—0.065	+0.256
1.00	+0.765	+0.440	5.00	—0.178	—0.328	9.00	—0.090	+0.245
1.10	+0.720	+0.471	5.10	—0.144	—0.337	9.10	—0.114	+0.232
1.20	+0.671	+0.498	5.20	—0.110	—0.343	9.20	—0.137	+0.217
1.30	+0.620	+0.522	5.30	—0.076	—0.346	9.30	—0.158	+0.200
1.40	+0.567	+0.542	5.40	—0.041	—0.345	9.40	—0.177	+0.182
1.50	+0.512	+0.558	5.50	—0.007	—0.341	9.50	—0.194	+0.161
1.60	+0.455	+0.570	5.60	+0.027	—0.334	9.60	—0.209	+0.140
1.70	+0.398	+0.578	5.70	+0.060	—0.324	9.70	—0.222	+0.117
1.80	+0.340	+0.582	5.80	+0.092	—0.311	9.80	—0.232	+0.093
1.90	+0.282	+0.581	5.90	+0.122	—0.295	9.90	—0.240	+0.068
2.00	+0.224	+0.577	6.00	+0.151	—0.277	10.00	—0.246	+0.043
2.10	+0.167	+0.568	6.10	+0.177	—0.256	10.10	—0.249	+0.018
2.20	+0.110	+0.556	6.20	+0.202	—0.233	10.20	—0.250	—0.007
2.30	+0.056	+0.540	6.30	+0.224	—0.208	10.30	—0.248	—0.031
2.40	+0.002	+0.520	6.40	+0.243	—0.182	10.40	—0.243	—0.055
2.50	—0.048	+0.497	6.50	+0.260	—0.154	10.50	—0.237	—0.079
2.60	—0.097	+0.471	6.60	+0.274	—0.125	10.60	—0.228	—0.101
2.70	—0.142	+0.442	6.70	+0.285	—0.095	10.70	—0.216	—0.122
2.80	—0.185	+0.410	6.80	+0.293	—0.065	10.80	—0.203	—0.142
2.90	—0.224	+0.375	6.90	+0.298	—0.035	10.90	—0.188	—0.160
3.00	—0.260	+0.339	7.00	+0.300	—0.005	11.00	—0 171	—0 177
3.10	—0.292	+0.301	7.10	+0.299	+0.025	11.10	—0.153	—0.191
3.20	—0.320	+0.261	7.20	+0.295	+0.054	11.20	—0.133	—0.204
3.30	—0.344	+0.221	7.30	+0.288	+0.083	11.30	—0.112	—0.214
3.40	—0.364	+0.179	7.40	+0.279	+0.110	11.40	—0.090	—0.222
3.50	—0.380	+0.137	7.50	+0.266	+0.135	11.50	—0.068	—0.228
3.60	—0.392	+0.095	7.60	+0.252	+0.159	11.60	—0.045	—0.232
3.70	—0.399	+0.054	7.70	+0.235	+0.181	11.70	—0.021	—0.233
3.80	—0.403	+0.013	7.80	+0.215	+0.201	11.80	+0.002	—0.232
3.90	—0.402	—0.027	7.90	+0.194	+0.219	11.90	+0.025	—0.229
						12.00	+0.048	—0.223

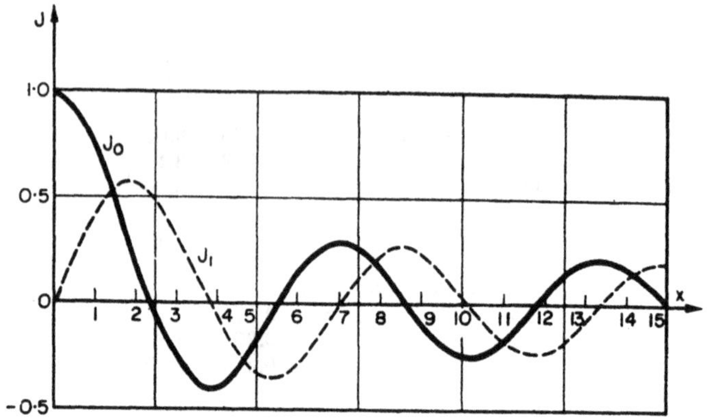

FIG. 1.

Graph of the functions $J_0(x)$ and $J_1(x)$ Bessel functions or cylindrical functions of the first kind.

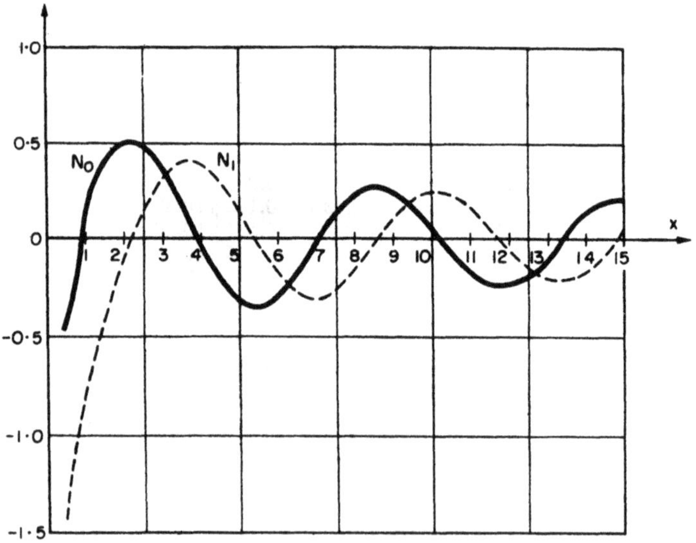

FIG. 2.

Graph of the functions $N_0(x)$ and $N_1(x)$ Neumann functions or cylindrical functions of the second kind.

TABLE 3. SUCCESSIVE ROOTS OF THE EQUATION $J_0(\mu_n) = 0$ AND THE CORRESPONDING VALUES $|J_1(\mu_n)|$

n	μ_n	$J_1(\mu_n)$	n	μ_n	$J_1(\mu_n)$
1	2.4048	0.5191	6	18.0711	0.1877
2	5.5201	0.3403	7	21.2116	0.1733
3	8.6537	0.2715	8	24.3525	0.1617
4	11.7915	0.2325	9	27.4935	0.1522
5	14.9309	0.2065	10	30.6346	0.1442

TABLE 4. VALUES OF THE FUNCTIONS $K_0(x)$ AND $K_1(x)$

x	$K_0(x)$	$K_1(x)$	x	$K_0(x)$	$K_1(x)$
0.1	2.4271	9.8538	3.1	0.0310	0.0356
0.2	1.7527	4.7760	3.2	0.0276	0.0316
0.3	1.3725	3.0560	3.3	0.0246	0.0281
0.4	1.1145	2.1844	3.4	0.0220	0.0250
0.5	0.9244	1.6564	3.5	0.0196	0.0222
0.6	0.7775	1.3028	3.6	0.0175	0.0198
0.7	0.6605	1.0503	3.7	0.0156	0.0176
0.8	0.5653	0.8618	3.8	0.0140	0.0157
0.9	0.4867	0.7165	3.9	0.0125	0.0140
1.0	0.4210	0.6019	4.0	0.0112	0.0125
1.1	0.3656	0.5098	4.1	0.0098	0.0111
1.2	0.3185	0.4346	4.2	0.0089	0.0099
1.3	0.2782	0.3725	4.3	0.0080	0.0089
1.4	0.2437	0.3208	4.4	0.0071	0.0079
1.5	0.2138	0.2774	4.5	0.0064	0.0071
1.6	0.1880	0.2406	4.6	0.0057	0.0063
1.7	0.1655	0.2094	4.7	0.0051	0.0056
1.8	0.1459	0.1826	4.8	0.0046	0.0051
1.9	0.1288	0.1597	4.9	0.0041	0.0045
2.0	0.1139	0.1399	5.0	0.0037	0.0040
2.1	0.1008	0.1227	5.1	0.0033	0.0036
2.2	0.0893	0.1079	5.2	0.0030	0.0032
2.3	0.0791	0.0950	5.3	0.0027	0.0029
2.4	0.0702	0.0837	5.4	0.0024	0.0026
2.5	0.0623	0.0739	5.5	0.0021	0.0023
2.6	0.0554	0.0653	5.6	0.0019	0.0021
2.7	0.0492	0.0577	5.7	0.0017	0.0019
2.8	0.0438	0.0511	5.8	0.0015	0.0017
2.9	0.0390	0.0453	5.9	0.0014	0.0015
3.0	0.0347	0.0402	6.0	0.0012	0.0013

INDEX

A CATALOG OF SELECTED

DOVER BOOKS

IN SCIENCE AND MATHEMATICS

Astronomy

CHARIOTS FOR APOLLO: The NASA History of Manned Lunar Spacecraft to 1969, Courtney G. Brooks, James M. Grimwood, and Loyd S. Swenson, Jr. This illustrated history by a trio of experts is the definitive reference on the Apollo spacecraft and lunar modules. It traces the vehicles' design, development, and operation in space. More than 100 photographs and illustrations. 576pp. 6 3/4 x 9 1/4. 0-486-46756-2

EXPLORING THE MOON THROUGH BINOCULARS AND SMALL TELESCOPES, Ernest H. Cherrington, Jr. Informative, profusely illustrated guide to locating and identifying craters, rills, seas, mountains, other lunar features. Newly revised and updated with special section of new photos. Over 100 photos and diagrams. 240pp. 8 1/4 x 11. 0-486-24491-1

WHERE NO MAN HAS GONE BEFORE: A History of NASA's Apollo Lunar Expeditions, William David Compton. Introduction by Paul Dickson. This official NASA history traces behind-the-scenes conflicts and cooperation between scientists and engineers. The first half concerns preparations for the Moon landings, and the second half documents the flights that followed Apollo 11. 1989 edition. 432pp. 7 x 10. 0-486-47888-2

APOLLO EXPEDITIONS TO THE MOON: The NASA History, Edited by Edgar M. Cortright. Official NASA publication marks the 40th anniversary of the first lunar landing and features essays by project participants recalling engineering and administrative challenges. Accessible, jargon-free accounts, highlighted by numerous illustrations. 336pp. 8 3/8 x 10 7/8. 0-486-47175-6

ON MARS: Exploration of the Red Planet, 1958-1978--The NASA History, Edward Clinton Ezell and Linda Neuman Ezell. NASA's official history chronicles the start of our explorations of our planetary neighbor. It recounts cooperation among government, industry, and academia, and it features dozens of photos from Viking cameras. 560pp. 6 3/4 x 9 1/4. 0-486-46757-0

ARISTARCHUS OF SAMOS: The Ancient Copernicus, Sir Thomas Heath. Heath's history of astronomy ranges from Homer and Hesiod to Aristarchus and includes quotes from numerous thinkers, compilers, and scholasticists from Thales and Anaximander through Pythagoras, Plato, Aristotle, and Heraclides. 34 figures. 448pp. 5 3/8 x 8 1/2. 0-486-43886-4

AN INTRODUCTION TO CELESTIAL MECHANICS, Forest Ray Moulton. Classic text still unsurpassed in presentation of fundamental principles. Covers rectilinear motion, central forces, problems of two and three bodies, much more. Includes over 200 problems, some with answers. 437pp. 5 3/8 x 8 1/2. 0-486-64687-4

BEYOND THE ATMOSPHERE: Early Years of Space Science, Homer E. Newell. This exciting survey is the work of a top NASA administrator who chronicles technological advances, the relationship of space science to general science, and the space program's social, political, and economic contexts. 528pp. 6 3/4 x 9 1/4. 0-486-47464-X

STAR LORE: Myths, Legends, and Facts, William Tyler Olcott. Captivating retellings of the origins and histories of ancient star groups include Pegasus, Ursa Major, Pleiades, signs of the zodiac, and other constellations. "Classic." – *Sky & Telescope*. 58 illustrations. 544pp. 5 3/8 x 8 1/2. 0-486-43581-4

A COMPLETE MANUAL OF AMATEUR ASTRONOMY: Tools and Techniques for Astronomical Observations, P. Clay Sherrod with Thomas L. Koed. Concise, highly readable book discusses the selection, set-up, and maintenance of a telescope; amateur studies of the sun; lunar topography and occultations; and more. 124 figures. 26 halftones. 37 tables. 335pp. 6 1/2 x 9 1/4. 0-486-42820-6

Chemistry

MOLECULAR COLLISION THEORY, M. S. Child. This high-level monograph offers an analytical treatment of classical scattering by a central force, quantum scattering by a central force, elastic scattering phase shifts, and semi-classical elastic scattering. 1974 edition. 310pp. 5 3/8 x 8 1/2. 0-486-69437-2

HANDBOOK OF COMPUTATIONAL QUANTUM CHEMISTRY, David B. Cook. This comprehensive text provides upper-level undergraduates and graduate students with an accessible introduction to the implementation of quantum ideas in molecular modeling, exploring practical applications alongside theoretical explanations. 1998 edition. 832pp. 5 3/8 x 8 1/2. 0-486-44307-8

RADIOACTIVE SUBSTANCES, Marie Curie. The celebrated scientist's thesis, which directly preceded her 1903 Nobel Prize, discusses establishing atomic character of radioactivity; extraction from pitchblende of polonium and radium; isolation of pure radium chloride; more. 96pp. 5 3/8 x 8 1/2. 0-486-42550-9

CHEMICAL MAGIC, Leonard A. Ford. Classic guide provides intriguing entertainment while elucidating sound scientific principles, with more than 100 unusual stunts: cold fire, dust explosions, a nylon rope trick, a disappearing beaker, much more. 128pp. 5 3/8 x 8 1/2. 0-486-67628-5

ALCHEMY, E. J. Holmyard. Classic study by noted authority covers 2,000 years of alchemical history: religious, mystical overtones; apparatus; signs, symbols, and secret terms; advent of scientific method, much more. Illustrated. 320pp. 5 3/8 x 8 1/2. 0-486-26298-7

CHEMICAL KINETICS AND REACTION DYNAMICS, Paul L. Houston. This text teaches the principles underlying modern chemical kinetics in a clear, direct fashion, using several examples to enhance basic understanding. Solutions to selected problems. 2001 edition. 352pp. 8 3/8 x 11. 0-486-45334-0

PROBLEMS AND SOLUTIONS IN QUANTUM CHEMISTRY AND PHYSICS, Charles S. Johnson and Lee G. Pedersen. Unusually varied problems, with detailed solutions, cover of quantum mechanics, wave mechanics, angular momentum, molecular spectroscopy, scattering theory, more. 280 problems, plus 139 supplementary exercises. 430pp. 6 1/2 x 9 1/4. 0-486-65236-X

ELEMENTS OF CHEMISTRY, Antoine Lavoisier. Monumental classic by the founder of modern chemistry features first explicit statement of law of conservation of matter in chemical change, and more. Facsimile reprint of original (1790) Kerr translation. 539pp. 5 3/8 x 8 1/2. 0-486-64624-6

MAGNETISM AND TRANSITION METAL COMPLEXES, F. E. Mabbs and D. J. Machin. A detailed view of the calculation methods involved in the magnetic properties of transition metal complexes, this volume offers sufficient background for original work in the field. 1973 edition. 240pp. 5 3/8 x 8 1/2. 0-486-46284-6

GENERAL CHEMISTRY, Linus Pauling. Revised third edition of classic first-year text by Nobel laureate. Atomic and molecular structure, quantum mechanics, statistical mechanics, thermodynamics correlated with descriptive chemistry. Problems. 992pp. 5 3/8 x 8 1/2. 0-486-65622-5

ELECTROLYTE SOLUTIONS: Second Revised Edition, R. A. Robinson and R. H. Stokes. Classic text deals primarily with measurement, interpretation of conductance, chemical potential, and diffusion in electrolyte solutions. Detailed theoretical interpretations, plus extensive tables of thermodynamic and transport properties. 1970 edition. 590pp. 5 3/8 x 8 1/2. 0-486-42225-9

Engineering

FUNDAMENTALS OF ASTRODYNAMICS, Roger R. Bate, Donald D. Mueller, and Jerry E. White. Teaching text developed by U.S. Air Force Academy develops the basic two-body and n-body equations of motion; orbit determination; classical orbital elements, coordinate transformations; differential correction; more. 1971 edition. 455pp. 5 3/8 x 8 1/2. 0-486-60061-0

INTRODUCTION TO CONTINUUM MECHANICS FOR ENGINEERS: Revised Edition, Ray M. Bowen. This self-contained text introduces classical continuum models within a modern framework. Its numerous exercises illustrate the governing principles, linearizations, and other approximations that constitute classical continuum models. 2007 edition. 320pp. 6 1/8 x 9 1/4. 0-486-47460-7

ENGINEERING MECHANICS FOR STRUCTURES, Louis L. Bucciarelli. This text explores the mechanics of solids and statics as well as the strength of materials and elasticity theory. Its many design exercises encourage creative initiative and systems thinking. 2009 edition. 320pp. 6 1/8 x 9 1/4. 0-486-46855-0

FEEDBACK CONTROL THEORY, John C. Doyle, Bruce A. Francis and Allen R. Tannenbaum. This excellent introduction to feedback control system design offers a theoretical approach that captures the essential issues and can be applied to a wide range of practical problems. 1992 edition. 224pp. 6 1/2 x 9 1/4. 0-486-46933-6

THE FORCES OF MATTER, Michael Faraday. These lectures by a famous inventor offer an easy-to-understand introduction to the interactions of the universe's physical forces. Six essays explore gravitation, cohesion, chemical affinity, heat, magnetism, and electricity. 1993 edition. 96pp. 5 3/8 x 8 1/2. 0-486-47482-8

DYNAMICS, Lawrence E. Goodman and William H. Warner. Beginning engineering text introduces calculus of vectors, particle motion, dynamics of particle systems and plane rigid bodies, technical applications in plane motions, and more. Exercises and answers in every chapter. 619pp. 5 3/8 x 8 1/2. 0-486-42006-X

ADAPTIVE FILTERING PREDICTION AND CONTROL, Graham C. Goodwin and Kwai Sang Sin. This unified survey focuses on linear discrete-time systems and explores natural extensions to nonlinear systems. It emphasizes discrete-time systems, summarizing theoretical and practical aspects of a large class of adaptive algorithms. 1984 edition. 560pp. 6 1/2 x 9 1/4. 0-486-46932-8

INDUCTANCE CALCULATIONS, Frederick W. Grover. This authoritative reference enables the design of virtually every type of inductor. It features a single simple formula for each type of inductor, together with tables containing essential numerical factors. 1946 edition. 304pp. 5 3/8 x 8 1/2. 0-486-47440-2

THERMODYNAMICS: Foundations and Applications, Elias P. Gyftopoulos and Gian Paolo Beretta. Designed by two MIT professors, this authoritative text discusses basic concepts and applications in detail, emphasizing generality, definitions, and logical consistency. More than 300 solved problems cover realistic energy systems and processes. 800pp. 6 1/8 x 9 1/4. 0-486-43932-1

THE FINITE ELEMENT METHOD: Linear Static and Dynamic Finite Element Analysis, Thomas J. R. Hughes. Text for students without in-depth mathematical training, this text includes a comprehensive presentation and analysis of algorithms of time-dependent phenomena plus beam, plate, and shell theories. Solution guide available upon request. 672pp. 6 1/2 x 9 1/4. 0-486-41181-8

Mathematics–Bestsellers

HANDBOOK OF MATHEMATICAL FUNCTIONS: with Formulas, Graphs, and Mathematical Tables, Edited by Milton Abramowitz and Irene A. Stegun. A classic resource for working with special functions, standard trig, and exponential logarithmic definitions and extensions, it features 29 sets of tables, some to as high as 20 places. 1046pp. 8 x 10 1/2. 0-486-61272-4

ABSTRACT AND CONCRETE CATEGORIES: The Joy of Cats, Jiri Adamek, Horst Herrlich, and George E. Strecker. This up-to-date introductory treatment employs category theory to explore the theory of structures. Its unique approach stresses concrete categories and presents a systematic view of factorization structures. Numerous examples. 1990 edition, updated 2004. 528pp. 6 1/8 x 9 1/4. 0-486-46934-4

MATHEMATICS: Its Content, Methods and Meaning, A. D. Aleksandrov, A. N. Kolmogorov, and M. A. Lavrent'ev. Major survey offers comprehensive, coherent discussions of analytic geometry, algebra, differential equations, calculus of variations, functions of a complex variable, prime numbers, linear and non-Euclidean geometry, topology, functional analysis, more. 1963 edition. 1120pp. 5 3/8 x 8 1/2. 0-486-40916-3

INTRODUCTION TO VECTORS AND TENSORS: Second Edition--Two Volumes Bound as One, Ray M. Bowen and C.-C. Wang. Convenient single-volume compilation of two texts offers both introduction and in-depth survey. Geared toward engineering and science students rather than mathematicians, it focuses on physics and engineering applications. 1976 edition. 560pp. 6 1/2 x 9 1/4. 0-486-46914-X

AN INTRODUCTION TO ORTHOGONAL POLYNOMIALS, Theodore S. Chihara. Concise introduction covers general elementary theory, including the representation theorem and distribution functions, continued fractions and chain sequences, the recurrence formula, special functions, and some specific systems. 1978 edition. 272pp. 5 3/8 x 8 1/2. 0-486-47929-3

ADVANCED MATHEMATICS FOR ENGINEERS AND SCIENTISTS, Paul DuChateau. This primary text and supplemental reference focuses on linear algebra, calculus, and ordinary differential equations. Additional topics include partial differential equations and approximation methods. Includes solved problems. 1992 edition. 400pp. 7 1/2 x 9 1/4. 0-486-47930-7

PARTIAL DIFFERENTIAL EQUATIONS FOR SCIENTISTS AND ENGINEERS, Stanley J. Farlow. Practical text shows how to formulate and solve partial differential equations. Coverage of diffusion-type problems, hyperbolic-type problems, elliptic-type problems, numerical and approximate methods. Solution guide available upon request. 1982 edition. 414pp. 6 1/8 x 9 1/4. 0-486-67620-X

VARIATIONAL PRINCIPLES AND FREE-BOUNDARY PROBLEMS, Avner Friedman. Advanced graduate-level text examines variational methods in partial differential equations and illustrates their applications to free-boundary problems. Features detailed statements of standard theory of elliptic and parabolic operators. 1982 edition. 720pp. 6 1/8 x 9 1/4. 0-486-47853-X

LINEAR ANALYSIS AND REPRESENTATION THEORY, Steven A. Gaal. Unified treatment covers topics from the theory of operators and operator algebras on Hilbert spaces; integration and representation theory for topological groups; and the theory of Lie algebras, Lie groups, and transform groups. 1973 edition. 704pp. 6 1/8 x 9 1/4. 0-486-47851-3

Mathematics–Logic and Problem Solving

PERPLEXING PUZZLES AND TANTALIZING TEASERS, Martin Gardner. Ninety-three riddles, mazes, illusions, tricky questions, word and picture puzzles, and other challenges offer hours of entertainment for youngsters. Filled with rib-tickling drawings. Solutions. 224pp. 5 3/8 x 8 1/2. 0-486-25637-5

MY BEST MATHEMATICAL AND LOGIC PUZZLES, Martin Gardner. The noted expert selects 70 of his favorite "short" puzzles. Includes The Returning Explorer, The Mutilated Chessboard, Scrambled Box Tops, and dozens more. Complete solutions included. 96pp. 5 3/8 x 8 1/2. 0-486-28152-3

THE LADY OR THE TIGER?: and Other Logic Puzzles, Raymond M. Smullyan. Created by a renowned puzzle master, these whimsically themed challenges involve paradoxes about probability, time, and change; metapuzzles; and self-referentiality. Nineteen chapters advance in difficulty from relatively simple to highly complex. 1982 edition. 240pp. 5 3/8 x 8 1/2. 0-486-47027-X

SATAN, CANTOR AND INFINITY: Mind-Boggling Puzzles, Raymond M. Smullyan. A renowned mathematician tells stories of knights and knaves in an entertaining look at the logical precepts behind infinity, probability, time, and change. Requires a strong background in mathematics. Complete solutions. 288pp. 5 3/8 x 8 1/2. 0-486-47036-9

THE RED BOOK OF MATHEMATICAL PROBLEMS, Kenneth S. Williams and Kenneth Hardy. Handy compilation of 100 practice problems, hints and solutions indispensable for students preparing for the William Lowell Putnam and other mathematical competitions. Preface to the First Edition. Sources. 1988 edition. 192pp. 5 3/8 x 8 1/2. 0-486-69415-1

KING ARTHUR IN SEARCH OF HIS DOG AND OTHER CURIOUS PUZZLES, Raymond M. Smullyan. This fanciful, original collection for readers of all ages features arithmetic puzzles, logic problems related to crime detection, and logic and arithmetic puzzles involving King Arthur and his Dogs of the Round Table. 160pp. 5 3/8 x 8 1/2. 0-486-47435-6

UNDECIDABLE THEORIES: Studies in Logic and the Foundation of Mathematics, Alfred Tarski in collaboration with Andrzej Mostowski and Raphael M. Robinson. This well-known book by the famed logician consists of three treatises: "A General Method in Proofs of Undecidability," "Undecidability and Essential Undecidability in Mathematics," and "Undecidability of the Elementary Theory of Groups." 1953 edition. 112pp. 5 3/8 x 8 1/2. 0-486-47703-7

LOGIC FOR MATHEMATICIANS, J. Barkley Rosser. Examination of essential topics and theorems assumes no background in logic. "Undoubtedly a major addition to the literature of mathematical logic." – *Bulletin of the American Mathematical Society.* 1978 edition. 592pp. 6 1/8 x 9 1/4. 0-486-46898-4

INTRODUCTION TO PROOF IN ABSTRACT MATHEMATICS, Andrew Wohlgemuth. This undergraduate text teaches students what constitutes an acceptable proof, and it develops their ability to do proofs of routine problems as well as those requiring creative insights. 1990 edition. 384pp. 6 1/2 x 9 1/4. 0-486-47854-8

FIRST COURSE IN MATHEMATICAL LOGIC, Patrick Suppes and Shirley Hill. Rigorous introduction is simple enough in presentation and context for wide range of students. Symbolizing sentences; logical inference; truth and validity; truth tables; terms, predicates, universal quantifiers; universal specification and laws of identity; more. 288pp. 5 3/8 x 8 1/2. 0-486-42259-3

Mathematics-Algebra and Calculus

VECTOR CALCULUS, Peter Baxandall and Hans Liebeck. This introductory text offers a rigorous, comprehensive treatment. Classical theorems of vector calculus are amply illustrated with figures, worked examples, physical applications, and exercises with hints and answers. 1986 edition. 560pp. 5 3/8 x 8 1/2. 0-486-46620-5

ADVANCED CALCULUS: An Introduction to Classical Analysis, Louis Brand. A course in analysis that focuses on the functions of a real variable, this text introduces the basic concepts in their simplest setting and illustrates its teachings with numerous examples, theorems, and proofs. 1955 edition. 592pp. 5 3/8 x 8 1/2. 0-486-44548-8

ADVANCED CALCULUS, Avner Friedman. Intended for students who have already completed a one-year course in elementary calculus, this two-part treatment advances from functions of one variable to those of several variables. Solutions. 1971 edition. 432pp. 5 3/8 x 8 1/2. 0-486-45795-8

METHODS OF MATHEMATICS APPLIED TO CALCULUS, PROBABILITY, AND STATISTICS, Richard W. Hamming. This 4-part treatment begins with algebra and analytic geometry and proceeds to an exploration of the calculus of algebraic functions and transcendental functions and applications. 1985 edition. Includes 310 figures and 18 tables. 880pp. 6 1/2 x 9 1/4. 0-486-43945-3

BASIC ALGEBRA I: Second Edition, Nathan Jacobson. A classic text and standard reference for a generation, this volume covers all undergraduate algebra topics, including groups, rings, modules, Galois theory, polynomials, linear algebra, and associative algebra. 1985 edition. 528pp. 6 1/8 x 9 1/4. 0-486-47189-6

BASIC ALGEBRA II: Second Edition, Nathan Jacobson. This classic text and standard reference comprises all subjects of a first-year graduate-level course, including in-depth coverage of groups and polynomials and extensive use of categories and functors. 1989 edition. 704pp. 6 1/8 x 9 1/4. 0-486-47187-X

CALCULUS: An Intuitive and Physical Approach (Second Edition), Morris Kline. Application-oriented introduction relates the subject as closely as possible to science with explorations of the derivative; differentiation and integration of the powers of x; theorems on differentiation, antidifferentiation; the chain rule; trigonometric functions; more. Examples. 1967 edition. 960pp. 6 1/2 x 9 1/4. 0-486-40453-6

ABSTRACT ALGEBRA AND SOLUTION BY RADICALS, John E. Maxfield and Margaret W. Maxfield. Accessible advanced undergraduate-level text starts with groups, rings, fields, and polynomials and advances to Galois theory, radicals and roots of unity, and solution by radicals. Numerous examples, illustrations, exercises, appendixes. 1971 edition. 224pp. 6 1/8 x 9 1/4. 0-486-47723-1

AN INTRODUCTION TO THE THEORY OF LINEAR SPACES, Georgi E. Shilov. Translated by Richard A. Silverman. Introductory treatment offers a clear exposition of algebra, geometry, and analysis as parts of an integrated whole rather than separate subjects. Numerous examples illustrate many different fields, and problems include hints or answers. 1961 edition. 320pp. 5 3/8 x 8 1/2. 0-486-63070-6

LINEAR ALGEBRA, Georgi E. Shilov. Covers determinants, linear spaces, systems of linear equations, linear functions of a vector argument, coordinate transformations, the canonical form of the matrix of a linear operator, bilinear and quadratic forms, and more. 387pp. 5 3/8 x 8 1/2. 0-486-63518-X

Mathematics–Probability and Statistics

BASIC PROBABILITY THEORY, Robert B. Ash. This text emphasizes the probabilistic way of thinking, rather than measure-theoretic concepts. Geared toward advanced undergraduates and graduate students, it features solutions to some of the problems. 1970 edition. 352pp. 5 3/8 x 8 1/2. 0-486-46628-0

PRINCIPLES OF STATISTICS, M. G. Bulmer. Concise description of classical statistics, from basic dice probabilities to modern regression analysis. Equal stress on theory and applications. Moderate difficulty; only basic calculus required. Includes problems with answers. 252pp. 5 5/8 x 8 1/4. 0-486-63760-3

OUTLINE OF BASIC STATISTICS: Dictionary and Formulas, John E. Freund and Frank J. Williams. Handy guide includes a 70-page outline of essential statistical formulas covering grouped and ungrouped data, finite populations, probability, and more, plus over 1,000 clear, concise definitions of statistical terms. 1966 edition. 208pp. 5 3/8 x 8 1/2. 0-486-47769-X

GOOD THINKING: The Foundations of Probability and Its Applications, Irving J. Good. This in-depth treatment of probability theory by a famous British statistician explores Keynesian principles and surveys such topics as Bayesian rationality, corroboration, hypothesis testing, and mathematical tools for induction and simplicity. 1983 edition. 352pp. 5 3/8 x 8 1/2. 0-486-47438-0

INTRODUCTION TO PROBABILITY THEORY WITH CONTEMPORARY APPLICATIONS, Lester L. Helms. Extensive discussions and clear examples, written in plain language, expose students to the rules and methods of probability. Exercises foster problem-solving skills, and all problems feature step-by-step solutions. 1997 edition. 368pp. 6 1/2 x 9 1/4. 0-486-47418-6

CHANCE, LUCK, AND STATISTICS, Horace C. Levinson. In simple, non-technical language, this volume explores the fundamentals governing chance and applies them to sports, government, and business. "Clear and lively ... remarkably accurate." – *Scientific Monthly.* 384pp. 5 3/8 x 8 1/2. 0-486-41997-5

FIFTY CHALLENGING PROBLEMS IN PROBABILITY WITH SOLUTIONS, Frederick Mosteller. Remarkable puzzlers, graded in difficulty, illustrate elementary and advanced aspects of probability. These problems were selected for originality, general interest, or because they demonstrate valuable techniques. Also includes detailed solutions. 88pp. 5 3/8 x 8 1/2. 0-486-65355-2

EXPERIMENTAL STATISTICS, Mary Gibbons Natrella. A handbook for those seeking engineering information and quantitative data for designing, developing, constructing, and testing equipment. Covers the planning of experiments, the analyzing of extreme-value data; and more. 1966 edition. Index. Includes 52 figures and 76 tables. 560pp. 8 3/8 x 11. 0-486-43937-2

STOCHASTIC MODELING: Analysis and Simulation, Barry L. Nelson. Coherent introduction to techniques also offers a guide to the mathematical, numerical, and simulation tools of systems analysis. Includes formulation of models, analysis, and interpretation of results. 1995 edition. 336pp. 6 1/8 x 9 1/4. 0-486-47770-3

INTRODUCTION TO BIOSTATISTICS: Second Edition, Robert R. Sokal and F. James Rohlf. Suitable for undergraduates with a minimal background in mathematics, this introduction ranges from descriptive statistics to fundamental distributions and the testing of hypotheses. Includes numerous worked-out problems and examples. 1987 edition. 384pp. 6 1/8 x 9 1/4. 0-486-46961-1

Mathematics–Geometry and Topology

PROBLEMS AND SOLUTIONS IN EUCLIDEAN GEOMETRY, M. N. Aref and William Wernick. Based on classical principles, this book is intended for a second course in Euclidean geometry and can be used as a refresher. More than 200 problems include hints and solutions. 1968 edition. 272pp. 5 3/8 x 8 1/2. 0-486-47720-7

TOPOLOGY OF 3-MANIFOLDS AND RELATED TOPICS, Edited by M. K. Fort, Jr. With a New Introduction by Daniel Silver. Summaries and full reports from a 1961 conference discuss decompositions and subsets of 3-space; n-manifolds; knot theory; the Poincaré conjecture; and periodic maps and isotopies. Familiarity with algebraic topology required. 1962 edition. 272pp. 6 1/8 x 9 1/4. 0-486-47753-3

POINT SET TOPOLOGY, Steven A. Gaal. Suitable for a complete course in topology, this text also functions as a self-contained treatment for independent study. Additional enrichment materials make it equally valuable as a reference. 1964 edition. 336pp. 5 3/8 x 8 1/2. 0-486-47222-1

INVITATION TO GEOMETRY, Z. A. Melzak. Intended for students of many different backgrounds with only a modest knowledge of mathematics, this text features self-contained chapters that can be adapted to several types of geometry courses. 1983 edition. 240pp. 5 3/8 x 8 1/2. 0-486-46626-4

TOPOLOGY AND GEOMETRY FOR PHYSICISTS, Charles Nash and Siddhartha Sen. Written by physicists for physics students, this text assumes no detailed background in topology or geometry. Topics include differential forms, homotopy, homology, cohomology, fiber bundles, connection and covariant derivatives, and Morse theory. 1983 edition. 320pp. 5 3/8 x 8 1/2. 0-486-47852-1

BEYOND GEOMETRY: Classic Papers from Riemann to Einstein, Edited with an Introduction and Notes by Peter Pesic. This is the only English-language collection of these 8 accessible essays. They trace seminal ideas about the foundations of geometry that led to Einstein's general theory of relativity. 224pp. 6 1/8 x 9 1/4. 0-486-45350-2

GEOMETRY FROM EUCLID TO KNOTS, Saul Stahl. This text provides a historical perspective on plane geometry and covers non-neutral Euclidean geometry, circles and regular polygons, projective geometry, symmetries, inversions, informal topology, and more. Includes 1,000 practice problems. Solutions available. 2003 edition. 480pp. 6 1/8 x 9 1/4. 0-486-47459-3

TOPOLOGICAL VECTOR SPACES, DISTRIBUTIONS AND KERNELS, François Trèves. Extending beyond the boundaries of Hilbert and Banach space theory, this text focuses on key aspects of functional analysis, particularly in regard to solving partial differential equations. 1967 edition. 592pp. 5 3/8 x 8 1/2. 0-486-45352-9

INTRODUCTION TO PROJECTIVE GEOMETRY, C. R. Wylie, Jr. This introductory volume offers strong reinforcement for its teachings, with detailed examples and numerous theorems, proofs, and exercises, plus complete answers to all odd-numbered end-of-chapter problems. 1970 edition. 576pp. 6 1/8 x 9 1/4. 0-486-46895-X

FOUNDATIONS OF GEOMETRY, C. R. Wylie, Jr. Geared toward students preparing to teach high school mathematics, this text explores the principles of Euclidean and non-Euclidean geometry and covers both generalities and specifics of the axiomatic method. 1964 edition. 352pp. 6 x 9. 0-486-47214-0

Mathematics–History

THE WORKS OF ARCHIMEDES, Archimedes. Translated by Sir Thomas Heath. Complete works of ancient geometer feature such topics as the famous problems of the ratio of the areas of a cylinder and an inscribed sphere; the properties of conoids, spheroids, and spirals; more. 326pp. 5 3/8 x 8 1/2. 0-486-42084-1

THE HISTORICAL ROOTS OF ELEMENTARY MATHEMATICS, Lucas N. H. Bunt, Phillip S. Jones, and Jack D. Bedient. Exciting, hands-on approach to understanding fundamental underpinnings of modern arithmetic, algebra, geometry and number systems examines their origins in early Egyptian, Babylonian, and Greek sources. 336pp. 5 3/8 x 8 1/2. 0-486-25563-8

THE THIRTEEN BOOKS OF EUCLID'S ELEMENTS, Euclid. Contains complete English text of all 13 books of the Elements plus critical apparatus analyzing each definition, postulate, and proposition in great detail. Covers textual and linguistic matters; mathematical analyses of Euclid's ideas; classical, medieval, Renaissance and modern commentators; refutations, supports, extrapolations, reinterpretations and historical notes. 995 figures. Total of 1,425pp. All books 5 3/8 x 8 1/2.
Vol. I: 443pp. 0-486-60088-2
Vol. II: 464pp. 0-486-60089-0
Vol. III: 546pp. 0-486-60090-4

A HISTORY OF GREEK MATHEMATICS, Sir Thomas Heath. This authoritative two-volume set that covers the essentials of mathematics and features every landmark innovation and every important figure, including Euclid, Apollonius, and others. 5 3/8 x 8 1/2. Vol. I: 461pp. 0-486-24073-8
Vol. II: 597pp. 0-486-24074-6

A MANUAL OF GREEK MATHEMATICS, Sir Thomas L. Heath. This concise but thorough history encompasses the enduring contributions of the ancient Greek mathematicians whose works form the basis of most modern mathematics. Discusses Pythagorean arithmetic, Plato, Euclid, more. 1931 edition. 576pp. 5 3/8 x 8 1/2.
0-486-43231-9

CHINESE MATHEMATICS IN THE THIRTEENTH CENTURY, Ulrich Libbrecht. An exploration of the 13th-century mathematician Ch'in, this fascinating book combines what is known of the mathematician's life with a history of his only extant work, the Shu-shu chiu-chang. 1973 edition. 592pp. 5 3/8 x 8 1/2.
0-486-44619-0

PHILOSOPHY OF MATHEMATICS AND DEDUCTIVE STRUCTURE IN EUCLID'S ELEMENTS, Ian Mueller. This text provides an understanding of the classical Greek conception of mathematics as expressed in Euclid's Elements. It focuses on philosophical, foundational, and logical questions and features helpful appendixes. 400pp. 6 1/2 x 9 1/4. 0-486-45300-6

BEYOND GEOMETRY: Classic Papers from Riemann to Einstein, Edited with an Introduction and Notes by Peter Pesic. This is the only English-language collection of these 8 accessible essays. They trace seminal ideas about the foundations of geometry that led to Einstein's general theory of relativity. 224pp. 6 1/8 x 9 1/4. 0-486-45350-2

HISTORY OF MATHEMATICS, David E. Smith. Two-volume history – from Egyptian papyri and medieval maps to modern graphs and diagrams. Non-technical chronological survey with thousands of biographical notes, critical evaluations, and contemporary opinions on over 1,100 mathematicians. 5 3/8 x 8 1/2.
Vol. I: 618pp. 0-486-20429-4
Vol. II: 736pp. 0-486-20430-8

Physics

THEORETICAL NUCLEAR PHYSICS, John M. Blatt and Victor F. Weisskopf. An uncommonly clear and cogent investigation and correlation of key aspects of theoretical nuclear physics by leading experts: the nucleus, nuclear forces, nuclear spectroscopy, two-, three- and four-body problems, nuclear reactions, beta-decay and nuclear shell structure. 896pp. 5 3/8 x 8 1/2. 0-486-66827-4

QUANTUM THEORY, David Bohm. This advanced undergraduate-level text presents the quantum theory in terms of qualitative and imaginative concepts, followed by specific applications worked out in mathematical detail. 655pp. 5 3/8 x 8 1/2. 0-486-65969-0

ATOMIC PHYSICS AND HUMAN KNOWLEDGE, Niels Bohr. Articles and speeches by the Nobel Prize–winning physicist, dating from 1934 to 1958, offer philosophical explorations of the relevance of atomic physics to many areas of human endeavor. 1961 edition. 112pp. 5 3/8 x 8 1/2. 0-486-47928-5

COSMOLOGY, Hermann Bondi. A co-developer of the steady-state theory explores his conception of the expanding universe. This historic book was among the first to present cosmology as a separate branch of physics. 1961 edition. 192pp. 5 3/8 x 8 1/2. 0-486-47483-6

LECTURES ON QUANTUM MECHANICS, Paul A. M. Dirac. Four concise, brilliant lectures on mathematical methods in quantum mechanics from Nobel Prize-winning quantum pioneer build on idea of visualizing quantum theory through the use of classical mechanics. 96pp. 5 3/8 x 8 1/2. 0-486-41713-1

THE PRINCIPLE OF RELATIVITY, Albert Einstein and Frances A. Davis. Eleven papers that forged the general and special theories of relativity include seven papers by Einstein, two by Lorentz, and one each by Minkowski and Weyl. 1923 edition. 240pp. 5 3/8 x 8 1/2. 0-486-60081-5

PHYSICS OF WAVES, William C. Elmore and Mark A. Heald. Ideal as a classroom text or for individual study, this unique one-volume overview of classical wave theory covers wave phenomena of acoustics, optics, electromagnetic radiations, and more. 477pp. 5 3/8 x 8 1/2. 0-486-64926-1

THERMODYNAMICS, Enrico Fermi. In this classic of modern science, the Nobel Laureate presents a clear treatment of systems, the First and Second Laws of Thermodynamics, entropy, thermodynamic potentials, and much more. Calculus required. 160pp. 5 3/8 x 8 1/2. 0-486-60361-X

QUANTUM THEORY OF MANY-PARTICLE SYSTEMS, Alexander L. Fetter and John Dirk Walecka. Self-contained treatment of nonrelativistic many-particle systems discusses both formalism and applications in terms of ground-state (zero-temperature) formalism, finite-temperature formalism, canonical transformations, and applications to physical systems. 1971 edition. 640pp. 5 3/8 x 8 1/2. 0-486-42827-3

QUANTUM MECHANICS AND PATH INTEGRALS: Emended Edition, Richard P. Feynman and Albert R. Hibbs. Emended by Daniel F. Styer. The Nobel Prize–winning physicist presents unique insights into his theory and its applications. Feynman starts with fundamentals and advances to the perturbation method, quantum electrodynamics, and statistical mechanics. 1965 edition, emended in 2005. 384pp. 6 1/8 x 9 1/4. 0-486-47722-3

Physics

INTRODUCTION TO MODERN OPTICS, Grant R. Fowles. A complete basic undergraduate course in modern optics for students in physics, technology, and engineering. The first half deals with classical physical optics; the second, quantum nature of light. Solutions. 336pp. 5 3/8 x 8 1/2. 0-486-65957-7

THE QUANTUM THEORY OF RADIATION: Third Edition, W. Heitler. The first comprehensive treatment of quantum physics in any language, this classic introduction to basic theory remains highly recommended and widely used, both as a text and as a reference. 1954 edition. 464pp. 5 3/8 x 8 1/2. 0-486-64558-4

QUANTUM FIELD THEORY, Claude Itzykson and Jean-Bernard Zuber. This comprehensive text begins with the standard quantization of electrodynamics and perturbative renormalization, advancing to functional methods, relativistic bound states, broken symmetries, nonabelian gauge fields, and asymptotic behavior. 1980 edition. 752pp. 6 1/2 x 9 1/4. 0-486-44568-2

FOUNDATIONS OF POTENTIAL THERY, Oliver D. Kellogg. Introduction to fundamentals of potential functions covers the force of gravity, fields of force, potentials, harmonic functions, electric images and Green's function, sequences of harmonic functions, fundamental existence theorems, and much more. 400pp. 5 3/8 x 8 1/2. 0-486-60144-7

FUNDAMENTALS OF MATHEMATICAL PHYSICS, Edgar A. Kraut. Indispensable for students of modern physics, this text provides the necessary background in mathematics to study the concepts of electromagnetic theory and quantum mechanics. 1967 edition. 480pp. 6 1/2 x 9 1/4. 0-486-45809-1

GEOMETRY AND LIGHT: The Science of Invisibility, Ulf Leonhardt and Thomas Philbin. Suitable for advanced undergraduate and graduate students of engineering, physics, and mathematics and scientific researchers of all types, this is the first authoritative text on invisibility and the science behind it. More than 100 full-color illustrations, plus exercises with solutions. 2010 edition. 288pp. 7 x 9 1/4. 0-486-47693-6

QUANTUM MECHANICS: New Approaches to Selected Topics, Harry J. Lipkin. Acclaimed as "excellent" (*Nature*) and "very original and refreshing" (*Physics Today*), these studies examine the Mössbauer effect, many-body quantum mechanics, scattering theory, Feynman diagrams, and relativistic quantum mechanics. 1973 edition. 480pp. 5 3/8 x 8 1/2. 0-486-45893-8

THEORY OF HEAT, James Clerk Maxwell. This classic sets forth the fundamentals of thermodynamics and kinetic theory simply enough to be understood by beginners, yet with enough subtlety to appeal to more advanced readers, too. 352pp. 5 3/8 x 8 1/2. 0-486-41735-2

QUANTUM MECHANICS, Albert Messiah. Subjects include formalism and its interpretation, analysis of simple systems, symmetries and invariance, methods of approximation, elements of relativistic quantum mechanics, much more. "Strongly recommended." – *American Journal of Physics*. 1152pp. 5 3/8 x 8 1/2. 0-486-40924-4

RELATIVISTIC QUANTUM FIELDS, Charles Nash. This graduate-level text contains techniques for performing calculations in quantum field theory. It focuses chiefly on the dimensional method and the renormalization group methods. Additional topics include functional integration and differentiation. 1978 edition. 240pp. 5 3/8 x 8 1/2. 0-486-47752-5

Physics

MATHEMATICAL TOOLS FOR PHYSICS, James Nearing. Encouraging students' development of intuition, this original work begins with a review of basic mathematics and advances to infinite series, complex algebra, differential equations, Fourier series, and more. 2010 edition. 496pp. 6 1/8 x 9 1/4. 0-486-48212-X

TREATISE ON THERMODYNAMICS, Max Planck. Great classic, still one of the best introductions to thermodynamics. Fundamentals, first and second principles of thermodynamics, applications to special states of equilibrium, more. Numerous worked examples. 1917 edition. 297pp. 5 3/8 x 8. 0-486-66371-X

AN INTRODUCTION TO RELATIVISTIC QUANTUM FIELD THEORY, Silvan S. Schweber. Complete, systematic, and self-contained, this text introduces modern quantum field theory. "Combines thorough knowledge with a high degree of didactic ability and a delightful style." – *Mathematical Reviews.* 1961 edition. 928pp. 5 3/8 x 8 1/2. 0-486-44228-4

THE ELECTROMAGNETIC FIELD, Albert Shadowitz. Comprehensive undergraduate text covers basics of electric and magnetic fields, building up to electromagnetic theory. Related topics include relativity theory. Over 900 problems, some with solutions. 1975 edition. 768pp. 5 5/8 x 8 1/4. 0-486-65660-8

THE PRINCIPLES OF STATISTICAL MECHANICS, Richard C. Tolman. Definitive treatise offers a concise exposition of classical statistical mechanics and a thorough elucidation of quantum statistical mechanics, plus applications of statistical mechanics to thermodynamic behavior. 1930 edition. 704pp. 5 5/8 x 8 1/4. 0-486-63896-0

INTRODUCTION TO THE PHYSICS OF FLUIDS AND SOLIDS, James S. Trefil. This interesting, informative survey by a well-known science author ranges from classical physics and geophysical topics, from the rings of Saturn and the rotation of the galaxy to underground nuclear tests. 1975 edition. 320pp. 5 3/8 x 8 1/2. 0-486-47437-2

STATISTICAL PHYSICS, Gregory H. Wannier. Classic text combines thermodynamics, statistical mechanics, and kinetic theory in one unified presentation. Topics include equilibrium statistics of special systems, kinetic theory, transport coefficients, and fluctuations. Problems with solutions. 1966 edition. 532pp. 5 3/8 x 8 1/2. 0-486-65401-X

SPACE, TIME, MATTER, Hermann Weyl. Excellent introduction probes deeply into Euclidean space, Riemann's space, Einstein's general relativity, gravitational waves and energy, and laws of conservation. "A classic of physics." – *British Journal for Philosophy and Science.* 330pp. 5 3/8 x 8 1/2. 0-486-60267-2

RANDOM VIBRATIONS: Theory and Practice, Paul H. Wirsching, Thomas L. Paez and Keith Ortiz. Comprehensive text and reference covers topics in probability, statistics, and random processes, plus methods for analyzing and controlling random vibrations. Suitable for graduate students and mechanical, structural, and aerospace engineers. 1995 edition. 464pp. 5 3/8 x 8 1/2. 0-486-45015-5

PHYSICS OF SHOCK WAVES AND HIGH-TEMPERATURE HYDRO DYNAMIC PHENOMENA, Ya B. Zel'dovich and Yu P. Raizer. Physical, chemical processes in gases at high temperatures are focus of outstanding text, which combines material from gas dynamics, shock-wave theory, thermodynamics and statistical physics, other fields. 284 illustrations. 1966–1967 edition. 944pp. 6 1/8 x 9 1/4. 0-486-42002-7